Strength and Conditioning

Strength and Conditioning

Biological Principles and Practical Applications

Edited by

Marco Cardinale, Rob Newton and Kazunori Nosaka

WILEY-BLACKWELL

A John Wiley & Sons, Ltd., Publication

Wiley-Blackwell is an imprint of John Wiley & Sons, formed by the merger of Wiley's global Scientific, Technical and Medical business with Blackwell Publishing.

Registered office: John Wiley & Sons Ltd, The Atrium, Southern Gate, Chichester, West Sussex, PO19 8SQ, UK

Editorial Offices: 9600 Garsington Road, Oxford, OX4 2DQ, UK
The Atrium, Southern Gate, Chichester, West Sussex, PO19 8SQ, UK
111 River Street, Hoboken, NJ 07030-5774, USA

For details of our global editorial offices, for customer services and for information about how to apply for permission to reuse the copyright material in this book please see our website at www.wiley.com/wiley-blackwell

Library of Congress Cataloguing-in-Publication Data

Strength and conditioning – biological principles and practical applications / edited by Marco Cardinale, Rob Newton, and Kazunori Nosaka.
p. ; cm.
Includes bibliographical references and index.
ISBN 978-0-470-01918-4 (cloth) – ISBN 978-0-470-01919-1 (pbk.)
1. Muscle strength. 2. Exercise–Physiological aspects. 3. Physical education and training. I. Cardinale, Marco. II. Newton, Rob (Robert U.) III. Nosaka, Kazunori.
[DNLM: 1. Physical Fitness. 2. Exercise. 3. Physical Endurance. QT 255]
QP321.S885 2011
613.7'11–dc22
2010029179

A catalogue record for this book is available from the British Library.

This book is published in the following electronic format: ePDF 9780470970003

Contents

Foreword by Sir Clive Woodward

High-level performance may seem easy to sport fans. Coaches and athletes know very well that it is very difficult to achieve. Winning does not happen in a straight line. It is the result of incredible efforts, dedication, passion, but also the result of the willingness to learn something new every day in every possible field which can help improve performance.

Strength and conditioning is fundamental for preparing athletes to perform at their best. It is also important to provide better quality of life in special populations and improve fitness. Scientific efforts in this field have improved enormously our understanding of various training methods and nutritional interventions to maximize performance. Furthermore, due to the advancement in technology, we are nowadays capable of measuring in real time many things able to inform the coach on the best way to improve the quality of the training programmes.

Winning the World Cup in 2003 was a great achievement. And in my view the quality of our strength and conditioning programme was second to none thanks to the ability of our staff to translate the latest scientific knowledge into innovative practical applications. The success of Team GB in Beijing has also been influenced by the continuous advancement of the strength and conditioning profession in the UK as well as the ability of our practitioners to continue their quest for knowledge in this fascinating field.

Now, even better and more information is available to coaches, athletes, sports scientists and sports medicine professionals willing to learn more about the field of strength and conditioning. *Strength and Conditioning: Biological Principles and Practical Applications* is a must read for everyone serious about this field. The editors, Dr Marco Cardinale, Dr Robert Newton and Dr Kazunori Nosaka are world leaders in this field and have collected an outstanding list of authors for all the chapters. Great scientists have produced an incredible resource to provide the reader with all the details needed to understand more about the biology of strength and conditioning and the guidance to pick the appropriate applications when designing training programmes.

When a group of sports scientific leaders from all over the world come together to produce a book about the human body and performance, the reader can be assured the material is at the leading edge of sports science. Success in sport is nowadays the result of a word class coach working with a world class athlete surrounded by the best experts possible in various fields. This does not mean the coach needs to delegate knowledge to others. A modern coach must be aware of the science to be able to challenge his/her support team and always stimulate the quest for best practice and innovation.

The book contains the latest information on many subjects including overtraining, muscle structure and function and its adaptive changes with training, hormonal regulation, nutrition, testing and training planning modalities, and most of all it provides guidance for the appropriate use of strength and conditioning with young athletes, paralympians and special populations.

I recommend that you read and use the information in this book to provide your athletes with the best chances of performing at their best.

Sir Clive Woodward
Director of Sport
British Olympic Association

Preface

In order to optimise athletic performance, athletes must be optimally trained. The science of training has evolved enormously in the last twenty years as a direct result of the large volume of research conducted on specific aspects and thanks to the development of innovative technology capable of measuring athletic performance in the lab and directly onto the field. Strength and conditioning is now a well recognised profession in many countries and professional organisations as well as academic institutions strive to educate in the best possible way strength and conditioning experts able to work not only with elite sports people but also with the general population. Strength and conditioning is now acknowledged as a critical component in the development and management of the elite athlete as well as broad application of the health and fitness regime of the general public, special populations and the elderly. Strength and conditioning it is in fact one of the main ways to improve health and human performance for everyone.

The challenge for everyone is to determine the appropriate type of training not only in terms of exercises and drills used, but most of all, to identify and understand the biological consequences of various training stimuli. Understanding how to apply the correct modality of exercise, the correct volume and intensity and the correct timing of various interventions is it in fact the "holy grail" of strength and conditioning. The only way to define it is therefore to understand more and more about the biological principles that govern human adaptations to such stimuli, as well as ways to measure and monitor specific adaptations to be able to prescribe the most effective and safe way of exercising. Modern advancements in molecular biology are now helping us understand with much greater insight how muscle adapts to various training and nutritional regimes. Technology solutions help us in measuring many aspects of human performance as it happens. We now have more advanced capabilities for prescribing "evidence-based" strength and conditioning programmes to everyone, from the general population to the elite athlete. Our aim with this book is to collect all the most relevant and up to date information on this topic as written by World leaders in this field and provide a comprehensive resource for everyone interested in understanding more about this fascinating field.

We have structured the book to provide a continuum of information to facilitate its use in educational settings as well as a key reference for strength and conditioning professionals and the interested lay public. In Section 1 we present the biological aspects of strength and conditioning. How muscles, bones and tendons work, how neural impulses allow muscle activity, how genetics and bioenergetics affect muscle function and how the cardiorespiratory system works. In section 2 we present the latest information on how various systems respond and adapt to various strength and conditioning paradigms to provide a better understanding of the implications of strength and conditioning programmes on human physiology. Section 3 provides a detailed description of various modalities to measure and monitor the effectiveness of strength and conditioning programmes as well as identifying ways for improving strength and conditioning programme prescriptions. Section 4 provides the latest information on practical applications and nutritional considerations to maximise the effectiveness of strength and conditioning programmes. Finally section 5 provides the latest guidelines to use strength and conditioning in various populations with the safest and most effective progressions.

The way the material has been presented varies among the chapters. We wanted our contributors to present their area of expertise for a varied audience ranging from sports scientists to coaches to postgraduate students. Authors from many countries have contributed to this book and whatever the style has been, we are confident the material can catch the attention of every reader.

ACKNOWLEDGEMENTS

The editors would like to thank Nicola Mc Girr, Izzy Canning, Fiona Woods, Gill Whitley the wonderful people in the John Wiley & Sons, Ltd. for the continuous support, guidance and help to make this book happen.

Marco Cardinale would like to thank the numerous colleagues contributing to this book as well as the coaches, athletes and sports scientists which over the years contributed to the development of strength and conditioning. Thank you, my dear wife Daniela and son Matteo for your love and support over the years and for putting up with my odd schedules and moods over the preparation of this project.

Robert Newton acknowledges the fantastic collaborators that he has had the honour to work with over the past three decades and the dedicated and hardworking students and postdocs that he has been privileged to mentor. Thank you, my dear wife Lisa and daughter Talani for all your support and patience through my obsession with research and teaching.

Ken Nosaka appreciates the opportunity to make this book with many authors, Marco, Rob, and the wonderful people in the John Wiley & Sons, Ltd. I have experienced how challenging it is to edit a book and to make an ideal book, but it was a good lesson. Because of many commitments, I have a limited time with my wife Kaoru and daughter Cocolo. I am so sorry, but your support and understanding are really appreciated, and would like to tell you that you make my life valuable.

List of Contributors

Per Aagaard
Institute of Sports Science and Clinical Biomechanics, University of Southern Denmark, Odense, Denmark

Chris R. Abbiss
School of Exercise, Biomedical and Health Sciences, Edith Cowan University, Joondalup, WA, Australia

Tim Ackland
School of Sport Science, Exercise and Health, University of Western Australia, WA, Australia

Jesper L. Andersen
Institute of Sports Medicine Copenhagen, Bispebjerg Hospital, Copenhagen, Denmark

David Bishop
Institute of Sport, Exercise and Active Living (ISEAL), School of Sport and Exercise Science,Victoria University, Melbourne, Australia

Anthony Blazevich
School of Exercise, Biomedical and Health Sciences, Edith Cowan University, Joondalup, WA, Australia

Marco Cardinale
British Olympic Medical Institute, Institute of Sport Exercise and Health, University College London, London, UK; School of Medical Sciences, University of Aberdeen, Aberdeen, Scotland, UK

Christian Cook
United Kingdom Sport Council, London, UK

Matthew Cook
English Institute of Sport, Manchester, UK

Stuart J. Cormack
Essendon Football Club, Melbourne, Australia

Prue Cormie
School of Exercise, Biomedical and Health Sciences, Edith Cowan University, Joondalup, WA, Australia

Blair Crewther
Imperial College, Institute of Biomedical Engineering, London UK

Paul Davies
English Institute of Sport, Bisham Abbey Performance Centre, Marlow, Bucks, UK

Vincenzo Denaro
Department of Orthopaedic and Trauma Surgery, Campus Bio-Medico University, Rome, Italy

Kevin De Pauw
Department of Human Physiology and Sports Medicine, Faculteit LK, Vrije Universiteit Brussel, Brussels, Belgium

Fred J. DiMenna
School of Sport and Health Sciences, University of Exeter, Exeter, UK

Avery D. Faigenbaum
Department of Health and Exercise Science, The College of New Jersey, Ewing, NJ, USA

Marco Gazzoni
LISiN, Dipartimento di Elettronica, Politecnico di Torino, Torino, Italy

Olivier Girard
ASPETAR – Qatar Orthopaedic and Sports Medicine Hospital, Doha, Qatar

Umile Giuseppe Longo
Department of Orthopaedic and Trauma Surgery, Campus Bio-Medico University, Rome, Italy

Albert Gollhofer
Institute of Sport and Sport Science, University of Freiburg, Germany

Urs Granacher
Institute of Sport Science, Friedrich Schiller University, Jena, Germany

Stuart Gray
School of Medical Sciences, College of Life Sciences and Medicine, University of Aberdeen, Aberdeen, UK

Markus Gruber
Department of Training and Movement Sciences, University of Potsdam, Germany

Mark Jarvis
English Institute of Sport, Birmingham, UK

Andrew M. Jones
School of Sport and Health Sciences, University of Exeter, Exeter, UK

Arimantas Lionikas
School of Medical Sciences, College of Life Sciences and Medicine, University of Aberdeen, Aberdeen, UK

Nicola A. Maffiuletti
Neuromuscular Research Laboratory, Schulthess Clinic, Zurich, Switzerland

Nicola Maffulli
Centre for Sports and Exercise Medicine, Queen Mary University of London, Barts and The London School of Medicine and Dentistry, Mile End Hospital, London, UK

Constantinos N. Maganaris
Institute for Biomedical Research into Human Movement and Health (IRM), Manchester Metropolitan University, Manchester, UK

Ron J. Maughan
School of Sport, Exercise and Health Sciences, Loughborough University, Loughborough, UK

Michael R. McGuigan
New Zealand Academy of Sport North Island, Auckland, New Zealand; Auckland University of Technology, New Zealand

Romain Meeusen
Department of Human Physiology and Sports Medicine, Faculteit LK, Vrije Universiteit Brussel, Brussels, Belgium

Thomas Muehlbauer
Institute of Exercise and Health Sciences, University of Basel, Switzerland

Marco V. Narici
Institute for Biomedical Research into Human Movement and Health (IRM), Manchester Metropolitan University, Manchester, UK

Robert U. Newton
School of Exercise, Biomedical and Health Sciences, Edith Cowan University, Joondalup, WA, Australia

Kazunori Nosaka
School of Exercise, Biomedical and Health Sciences, Edith Cowan University, Joondalup, WA, Australia

Jeremiah J. Peiffer
Murdoch University, School of Chiropractic and Sports Science, Murdoch, WA, Australia

Alberto Rainoldi
Motor Science Research Center, University School of Motor and Sport Sciences, Università degli Studi, di Torino, Torino, Italy

Aivaras Ratkevicius
School of Medical Sciences, College of Life Sciences and Medicine, University of Aberdeen, Aberdeen, UK

Jörn Rittweger
Institute for Biomedical Research into Human Movement and Health, Manchester Metropolitan University, Manchester, UK; Institute of Aerospace Medicine, German Aerospace Center (DLR), Cologne, Germany

William A. Sands
Monfort Family Human Performance Research Laboratory, Mesa State College – Kinesiology, Grand Junction, CO, USA

Andreas Schlumberger
EDEN Reha, Rehabilitation Clinic, Donaustauf, Germany

Christopher S. Shaw
School of Sport and Exercise Sciences, The University of Birmingham, UK

Jeremy Sheppard
Queensland Academy of Sport, Australia; School of Exercise, Biomedical and Health Sciences, Edith Cowan University, Joondalup, WA, Australia

Matt Spencer
Physiology Unit, Sport Science Department, ASPIRE, Academy for Sports Excellence, Doha, Qatar

Filippo Spiezia
Department of Orthopaedic and Trauma Surgery, Campus Bio-Medico University, Rome, Italy

Arthur Stewart
Centre for Obesity Research and Epidemiology, Robert Gordon University, Aberdeen, UK

Margaret E. Stone
Center of Excellence for Sport Science and Coach Education, East Tennessee State University, Johnson City, TN, USA

Michael H. Stone
Center of Excellence for Sport Science and Coach Education, East Tennessee State University, Johnson City, TN, USA; Department of Physical Education, Sport and Leisure Studies,University of Edinburgh, Edinburgh, UK; Edith Cowan University, Perth, Australia

Wolfgang Taube
Unit of Sports Science, University of Fribourg, Switzerland

Kevin D. Tipton
School of Sports Sciences, University of Stirling, Stirling, Scotland, UK

Valmor Tricoli
Department of Sport, School of Physical Education and Sport, University of São Paulo, São Paulo, Brazil

Henning Wackerhage
School of Medical Sciences, College of Life Sciences and Medicine, University of Aberdeen, Aberdeen, UK

Warren Young
School of Human Movement and Sport Sciences, University of Ballarat, Ballarat, Victoria. Australia

Section 1
Strength and Conditioning Biology

1.1 Skeletal Muscle Physiology

Valmor Tricoli, University of São Paulo, Department of Sport, School of Physical Education and Sport, São Paulo, Brazil

1.1.1 INTRODUCTION

The skeletal muscle is the human body's most abundant tissue. There are over 660 muscles in the body corresponding to approximately 40–45% of its total mass (Brooks, Fahey and Baldwin, 2005; McArdle, Katch and Katch, 2007). It is estimated that 75% of skeletal muscle mass is water, 20% is protein, and the remaining 5% is substances such as salts, enzymes, minerals, carbohydrates, fats, amino acids, and high-energy phosphates. Myosin, actin, troponin and tropomyosin are the most important proteins.

Skeletal muscles play a vital role in locomotion, heat production, support of soft tissues, and overall metabolism. They have a remarkable ability to adapt to a variety of environmental stimuli, including regular physical training (e.g. endurance or strength exercise) (Aagaard and Andersen, 1998; Andersen *et al.*, 2005; Holm *et al.*, 2008; Parcell *et al.*, 2005), substrate availability (Bohé *et al.*, 2003; Kraemer *et al.*, 2009; Phillips, 2009; Tipton *et al.*, 2009), and unloading conditions (Alkner and Tesch, 2004; Berg, Larsson and Tesch, 1997; Caiozzo *et al.*, 2009; Lemoine *et al.*, 2009; Trappe *et al.*, 2009).

This chapter will describe skeletal muscle's basic structure and function, contraction mechanism, fibre types and hypertrophy. Its integration with the neural system will be the focus of the next chapter.

1.1.2 SKELETAL MUSCLE MACROSTRUCTURE

Skeletal muscles are essentially composed of specialized contracting cells organized in a hierarchical fashion supported by a connective tissue framework. The entire muscle is surrounded by a layer of connective tissue called fascia. Underneath the fascia is a thinner layer of connective tissue called epimysium which encloses the whole muscle. Right below is the perimysium, which wraps a bundle of muscle fibres called fascicle (or fasciculus), thus; a muscle is formed by several fasciculi. Lastly, each muscle fibre is covered by a thin sheath of collagenous connective tissue called endomysium (Figure 1.1.1).

Directly beneath the endomysium lies the sarcolemma, an elastic membrane which contains a plasma membrane and a basement membrane (also called basal lamina). Sometimes, the term sarcolemma is used as a synonym for muscle-cell plasma membrane. Among other functions, the sarcolemma is responsible for conducting the action potential that leads to muscle contraction. Between the plasma membrane and basement membrane the satellite cells are located (Figure 1.1.2). Their regenerative function and possible role in muscle hypertrophy will be discussed later in this chapter.

All these layers of connective tissue maintain the skeletal muscle hierarchical structure and they combine together to form the tendons at each end of the muscle. The tendons attach muscles to the bones and transmit the force they generate to the skeletal system, ultimately producing movement.

1.1.3 SKELETAL MUSCLE MICROSTRUCTURE

Muscle fibres, also called muscle cells or myofibres, are long, cylindrical cells 1–100 μm in diameter and up to 30–40 cm in length. They are made primarily of smaller units called myofibrils which lie in parallel inside the muscle cells (Figure 1.1.3). Myofibrils are contractile structures made of myofilaments named actin and myosin. These two proteins are responsible for muscle contraction and are found organized within sarcomeres (Figure 1.1.3). During skeletal muscle hypertrophy, myofibrils increase in number, enlarging cell size.

A unique characteristic of muscle fibres is that they are multinucleated (i.e. have several nuclei). Unlike most body cells, which are mononucleated, a muscle fibre may have 250–300 myonuclei per millimetre (Brooks, Fahey and Baldwin, 2005; McArdle, Katch and Katch 2007). This is the result of the fusion of several individual mononucleated myoblasts (muscle's progenitor cells) during the human body's development. Together they form a myotube, which later differentiates into a myofibre. The plasma membrane of a muscle cell is often called sarcolemma and the sarcoplasm is equivalent to its cytoplasm. Some sarcoplasmic organelles such as the sarcoplasmic reticulum and the transverse tubules are specific to muscle cells.

Strength and Conditioning – Biological Principles and Practical Applications Marco Cardinale, Rob Newton, and Kazunori Nosaka.

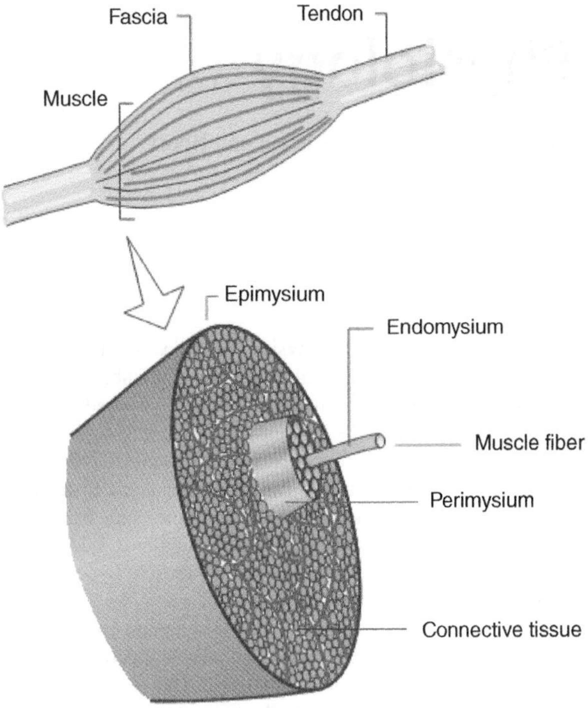

Figure 1.1.1 Skeletal muscle basic hierarchical structure. Adapted from Challis (2000)

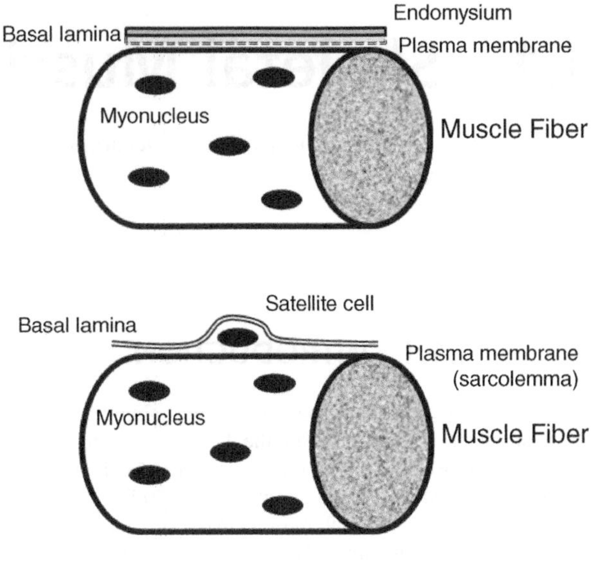

Figure 1.1.2 Illustration of a single multinucleated muscle fibre and satellite cell between the plasma membrane and the basal lamina

1.1.3.1 Sarcomere and myofilaments

A sarcomere is the smallest functional unit of a myofibre and is described as a structure limited by two Z disks or Z lines (Figure 1.1.3). Each sarcomere contains two types of myofilament: one thick filament called myosin and one thin filament called actin. The striated appearance of the skeletal muscle is the result of the positioning of myosin and actin filaments inside each sarcomere. A lighter area named the I band is alternated with a darker A band. The absence of filament overlap confers a lighter appearance to the I band, which contains only actin filaments, while overlap of myosin and actin gives a darker appearance to the A band. The Z line divides the I band in two equal halves; thus a sarcomere consists of one A band and two half-I bands, one at each side of the sarcomere. The middle of the A band contains the H zone, which is divided in two by the M line. As with the I band, there is no overlap between thick and thin filaments in the H zone. The M line comprises a number of structural proteins which are responsible for anchoring each myosin filament in the correct position at the centre of the sarcomere.

On average, a sarcomere that is between 2.0 and 2.25 µm long shows optimal overlap between myosin and actin filaments, providing ideal conditions for force production. At lengths shorter or larger than this optimum, force production is compromised (Figure 1.1.4).

The actin filament is formed by two intertwined strands (i.e. F-actin) of polymerized globular actin monomers (G-actin). This filament extends inward from each Z line toward the centre of the sarcomere. Attached to the actin filament are two regulatory proteins named troponin (Tn) and tropomyosin (Tm) which control the interaction between actin and myosin. Troponin is a protein complex positioned at regular intervals (every seven actin monomers) along the thin filament and plays a vital role in calcium ion (Ca^{++}) reception. This regulatory protein complex includes three components: troponin C (TnC), which binds Ca^{++}, troponin T (TnT), which binds tropomyosin, and troponin I (TnI), which is the inhibitory subunit. These subunits are in charge of moving tropomyosin away from the myosin binding site on the actin filament during the contraction process. Tropomyosin is distributed along the length of the actin filament in the groove formed by two F-actin strands and its main function is to inhibit the coupling between actin and myosin filaments from blocking active actin binding sites (Figure 1.1.5).

The thick filament is made mostly of myosin protein. A myosin molecule is composed of two heavy chains (MHCs) and two pairs of light chains (MLCs). It is the MHCs that determines muscle fibre phenotype. In mammalian muscles, the MHC component exists in four different isoforms (types I, IIA, IIB, and IIX), whereas the MLCs can be separated into two essential LCs and two regulatory LCs. Myosin heavy and light chains combine to fine tune the interaction between actin and myosin filaments during muscle contraction. A MHC is made of two distinct parts: a head (heavy meromyosin) and a tail (light meromyosin), in a form that may be compared to that of a golf club (Figure 1.1.6).

Approximately 300 myosin molecules are arranged tail-to-tail to constitute each thick filament (Brooks, Fahey and Baldwin, 2005). The myosin heads protrude from the filament;

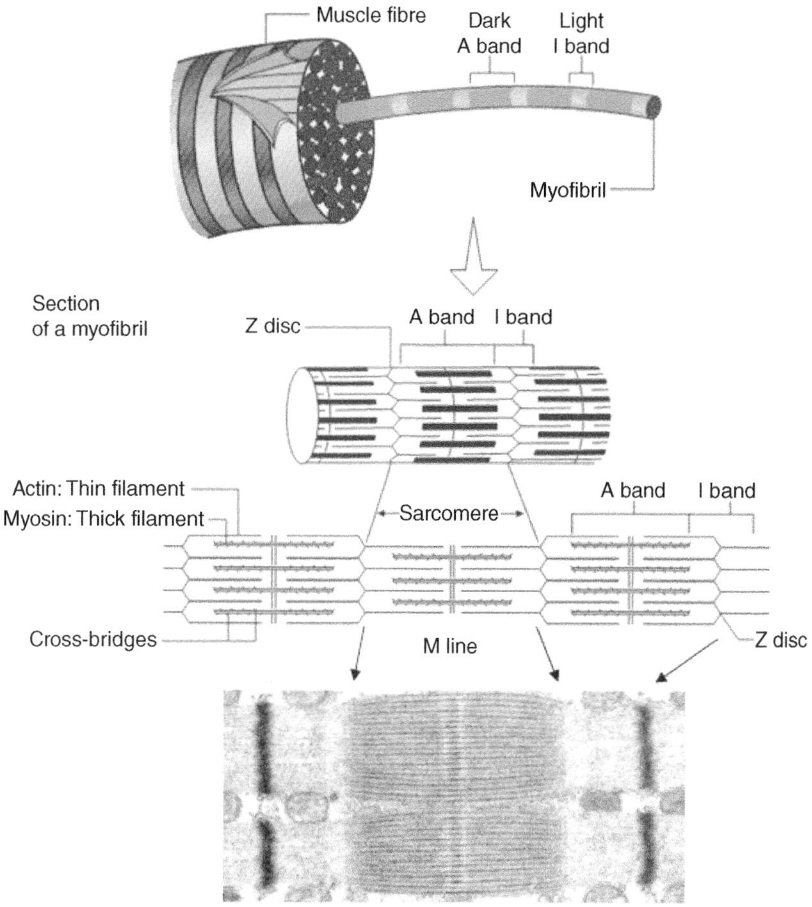

Figure 1.1.3 Representation of a muscle fibre and myofibril showing the structure of a sarcomere with actin and myosin filaments. Adapted from Challis (2000)

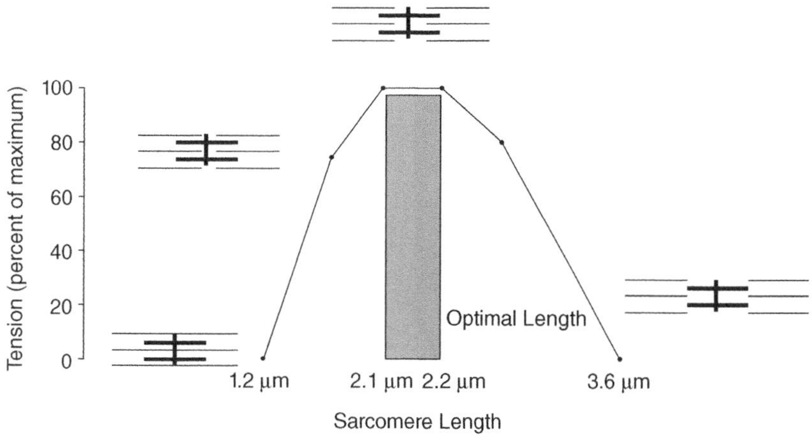

Figure 1.1.4 Skeletal muscle sarcomere length–tension relationship

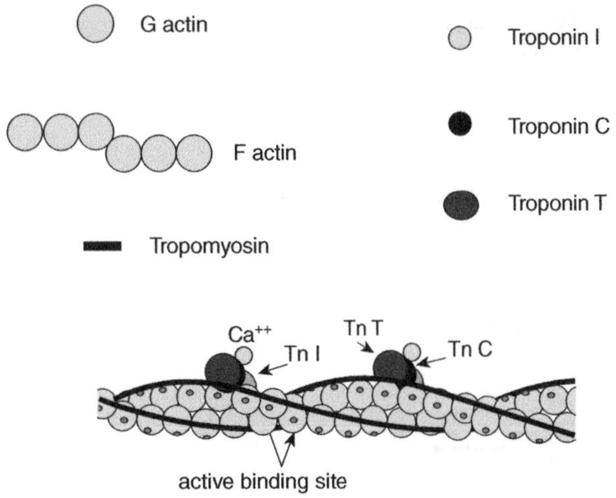

Figure 1.1.5 Schematic drawing of part of an actin filament showing the interaction with tropomyosin and troponin complex

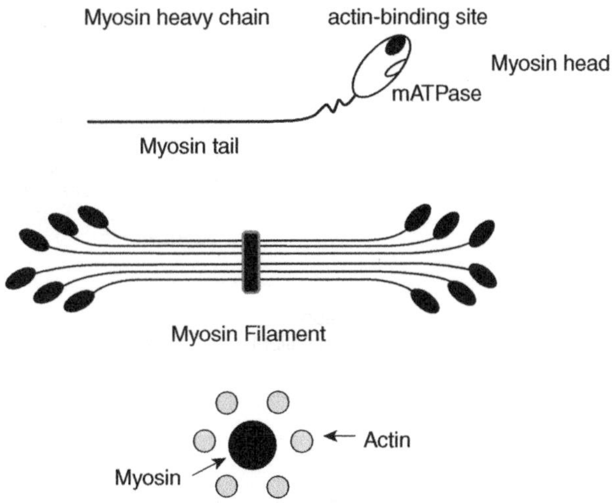

Figure 1.1.6 Schematic organization of a myosin filament, myosin heavy chain structure, and hexagonal lattice arrangement between myosin and actin filaments

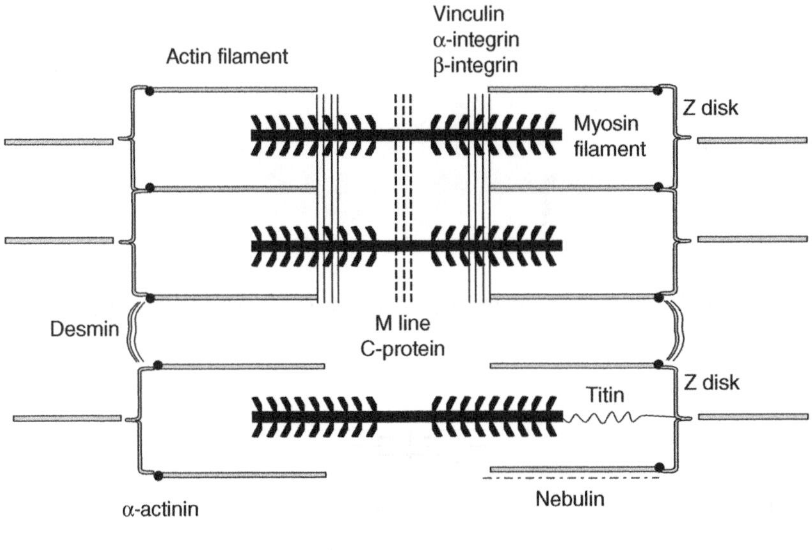

Figure 1.1.7 Illustration of the structural proteins in the sarcomere

each is arranged in a 60° rotation in relation to the preceding one, giving the myosin filament the appearance of a bottle brush. Each myosin head attaches directly to the actin filament and contains ATPase activity (see Section 1.1.5). Actin and myosin filaments are organized in a hexagonal lattice, which means that each myosin filament is surrounded by six actin filaments, providing a perfect match for interaction during contraction (Figure 1.1.6).

A large number of other proteins can be found in the interior of a muscle fibre and in sarcomeres; these constitute the so-called cytoskeleton. The cytoskeleton is an intracellular system composed of intermediate filaments; its main functions are to (1) provide structural integrity to the myofibre, (2) allow lateral force transmissions among adjacent sarcomeres, and (3) connect myofibrils to the cell's plasma membrane. The following proteins are involved in fulfilling these functions (Bloch and Gonzalez-Serratos, 2003; Hatze, 2002; Huijing, 1999; Monti *et al.*, 1999): vinculin, α- and β-integrin (responsible for the lateral force transmission within the fibre), dystrophin (links actin filaments with the dystroglycan and sarcoglycan complexes, which are associated with the sarcolemma and provide an extracellular connection to the basal lamina and the endomysium), α-actinin (attaches actin filaments to the Z disk), C protein (maintains the myosin

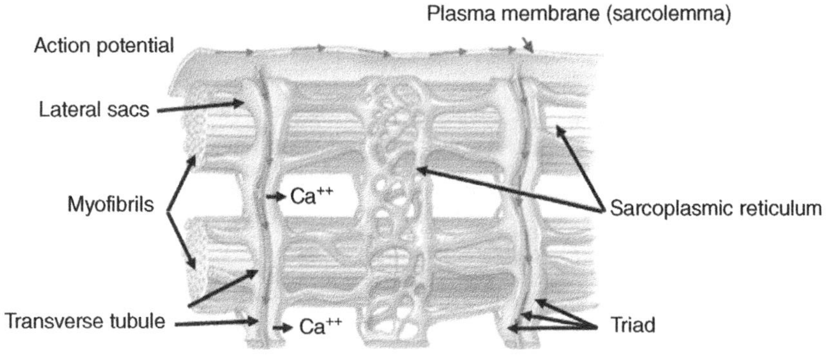

Figure 1.1.8 Representation of the transverse tubules (T tubules) and sarcoplasmic reticulum system

tails in a correct spatial arrangement), nebulin (works as a ruler, assisting the correct assembly of the actin filament), titin (contributes to secure the thick filaments to the Z lines), and desmin (links adjacent Z disks together). Some of these proteins are depicted in Figure 1.1.7.

1.1.3.2 Sarcoplasmic reticulum and transverse tubules

The sarcoplasmic reticulum (SR) of a myofibre is a specialized form of the endoplasmic reticulum found in most human body cells. Its main functions are (1) intracellular storage and (2) release and uptake of Ca^{++} associated with the regulation of muscle contraction. In fact, the release of Ca^{++} upon electrical stimulation of the muscle cell plays a major role in excitation–contraction coupling (E-C coupling; see Section 1.1.4). The SR can be divided into two parts: (1) the longitudinal SR and (2) the junctional SR (Rossi and Dirksen, 2006; Rossi *et al.*, 2008). The longitudinal SR, which is responsible for Ca^{++} storage and uptake, comprises numerous interconnected tubules, forming an intricate network of channels throughout the sarcoplasm and around the myofibrils. At specific cell regions, the ends of the longitudinal tubules fuse into single dilated sacs called terminal cisternae or lateral sacs. The junctional SR is found at the junction between the A and the I bands and represents the region responsible for Ca^{++} release (Figure 1.1.8).

The transverse tubules (T tubules) are organelles that carry the action potential (electrical stimulation) from the sarcolemma surface to the interior of the muscle fibre. Indeed, they are a continuation of the sarcolemma and run perpendicular to the fibre orientation axis, closely interacting with the SR. The association of a T tubule with two SR lateral sacs forms a structure called a triad which is located near each Z line of a sarcomere (Figure 1.1.8). This arrangement ensures synchronization between the depolarization of the sarcolemma and the release of Ca^{++} in the intracellular medium. It is the passage of an action potential down the T tubules that causes the release of intracellular Ca^{++} from the lateral sacs of the SR, causing muscle contraction; this is the mechanism referred to as E-C coupling (Rossi *et al.*, 2008).

1.1.4 CONTRACTION MECHANISM

1.1.4.1 Excitation–contraction (E-C) coupling

A muscle fibre has the ability to transform chemical energy into mechanical work through the cyclic interaction between actin and myosin filaments during contraction. The process starts with the arrival of an action potential and the release of Ca^{++} from the SR during the E-C coupling. An action potential is delivered to the muscle cell membrane by a motor neuron through the release of a neurotransmitter named acetylcholine (ACh). The release of ACh opens specific ion channels in the muscle plasma membrane, allowing sodium ions to diffuse from the extracellular medium into the cell, leading to the depolarization of the membrane. The depolarization wave spreads along the membrane and is carried to the cell's interior by the T tubules. As the action potential travels down a T tubule, calcium channels in the lateral sacs of the SR are opened, releasing Ca^{++} in the intracellular fluid. Calcium ions bind to troponin and, in the presence of ATP, start the process of skeletal muscle contraction (Figure 1.1.9).

1.1.4.2 Skeletal muscle contraction

Based on a three-state model for actin activation (McKillop and Geeves, 1993), the thin filament state is influenced by both Ca^{++} binding to troponin C (TnC) and myosin binding to actin. The position of tropomyosin (Tm) on actin generates a blocked, a closed, or an open state of actin filament activation. When intracellular Ca^{++} concentration is low, and thus Ca^{++} binding to TnC is low, the thin filament stays in a blocked state because Tm position does not allow actin–myosin interaction. However, the release of Ca^{++} by the SR and its binding to TnC changes the conformational shape of the troponin complex, shifting the actin filament from a blocked to a closed state. This transition allows a weak interaction of myosin heads with the actin filament, but during the closed state the weak interaction between myosin and actin does not produce force. Nevertheless, the weak interaction induces further movements in Tm, shifting the

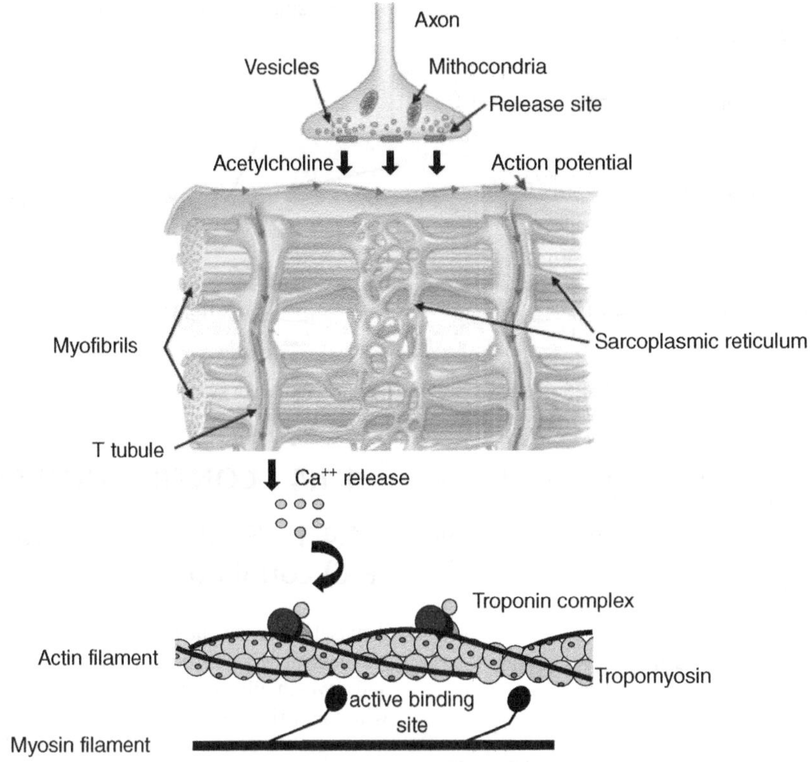

Figure 1.1.9 Excitation–contraction coupling and the sequence of events that will lead to skeletal muscle contraction

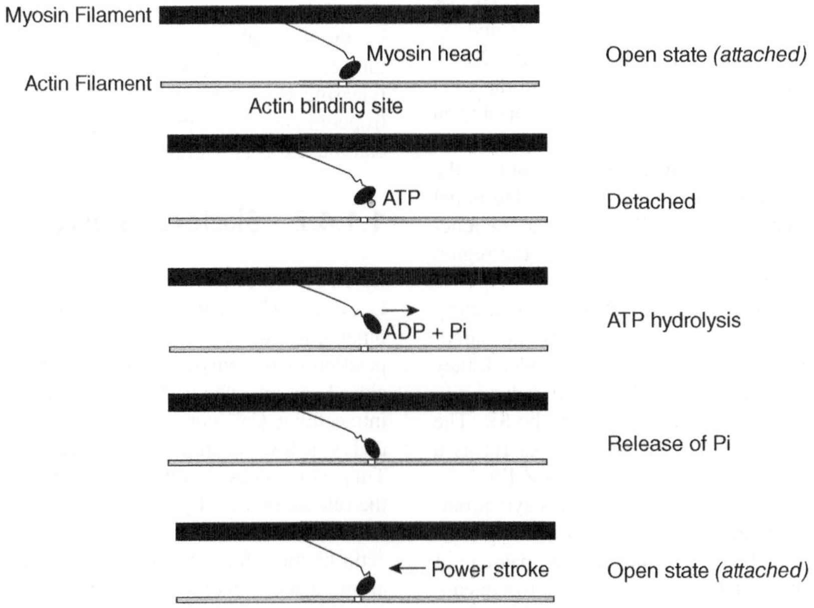

Figure 1.1.10 Cyclic process of muscle contraction

actin filament from the closed to the open state, resulting in a strong binding of myosin heads. In the open state, a high affinity between the myosin heads and the actin filament provides an opportunity to force production.

The formation of the acto-myosin complex permits the myosin head to hydrolyse a molecule of ATP and the free energy liberated is used to produce mechanical work (muscle contraction) during the cross-bridge power stroke (Vandenboom, 2004). Cross-bridges are projections from the myosin filament in the region of overlap with the actin filament. The release of ATP hydrolysis byproducts (Pi and ADP) induces structural changes in the acto-myosin complex that promote the cross-bridge cycle of attachment, force generation, and detachment. The force-generation phase is usually called the 'power stroke'. During this phase, actin and myosin filaments slide past each other and the thin filaments move toward the centre of the sarcomere, causing its shortening (Figure 1.1.10). Each cross-bridge power stroke produces a unitary force of 3–4 pico Newtons (1 pN = 10^{-12} N) over a working distance of 5–15 nanometres (1 nm = 10^{-9} m) against the actin filament (Finer, Simmons and Spudich, 1994). The resulting force and displacement per cross-bridge seem extremely small, but because billions of cross-bridges work at the same time, force output and muscle shortening can be large.

Interestingly, the myosin head retains high affinity for only one ligand (ATP or actin) at a time. Therefore, when an ATP molecule binds to the myosin head it reduces myosin affinity for actin and causes it to detach. After detachment, the myosin head rapidly hydrolyses the ATP and uses the free energy to reverse the structural changes that occurred during the power stroke, and then the system is ready to repeat the whole cycle (Figure 1.1.10). If ATP fails to rebind (or ADP release is inhibited), a 'rigour' cross-bridge is created, which eliminates the possibility of further force production. Thus, the ATP hydrolysis cycle is associated with the attachment and detachment of the myosin head from the actin filament (Gulick and Rayment, 1997). If the concentration of Ca^{++} returns to low levels, the muscle is relaxed by a reversal process that shifts the thin filament back toward the blocked state.

1.1.5 MUSCLE FIBRE TYPES

The skeletal muscle is composed of a heterogeneous mixture of different fibre types which have distinct molecular, metabolic, structural, and contractile characteristics, contributing to a range of functional properties. Skeletal muscle fibres also have an extraordinary ability to adapt and alter their phenotypic profile in response to a variety of environmental stimuli.

Muscle fibres defined as slow, type I, slow red or slow oxidative have been described as containing slow isoform contractile proteins, high volumes of mitochondria, high levels of myoglobin, high capillary density, and high oxidative enzyme capacity. Type IIA, fast red, fast IIA or fast oxidative fibres have been characterized as fast-contracting fibres with high oxidative capacity and moderate resistance to fatigue. Muscle fibres defined as type IIB, fast white, fast IIB (or IIx), or fast glycolytic present low mitochondrial density, high glycolytic enzyme activity, high myofibrilar ATPase (mATPase) activity, and low resistance to fatigue (Figure 1.1.11).

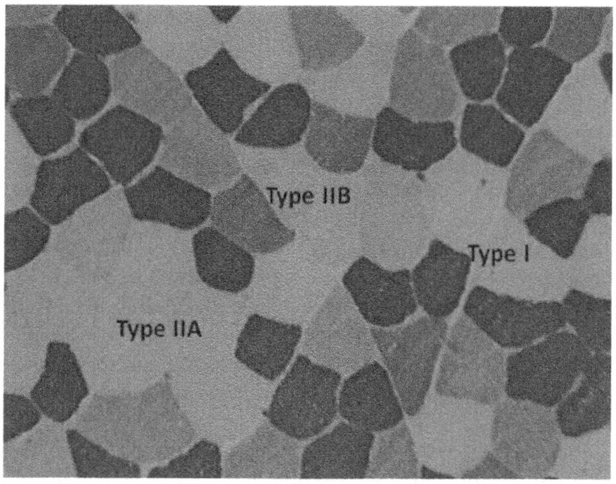

Figure 1.1.11 Photomicrograph showing three different muscle fibre types

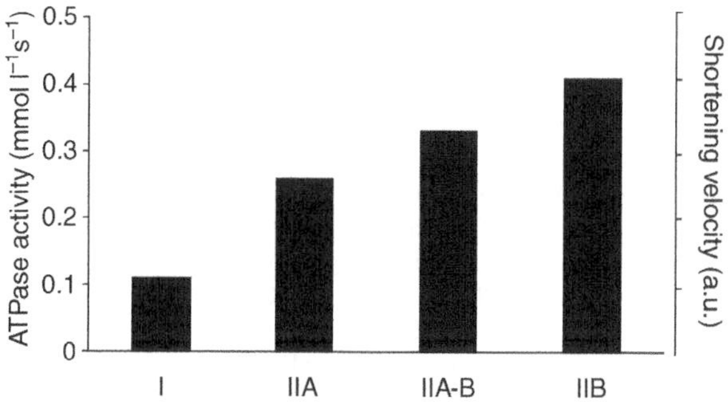

Figure 1.1.12 Representation of the relationship between skeletal muscle fibre type, mATPase activity, and muscle shortening velocity

Several methods, including mATPase histochemistry, immunohistochemistry, and electrophoretic analysis of myosin heavy chain (MHC) isoforms, have been used to identify myofibre types. The histochemical method is based on the differences in the pH sensitivity of mATPase activity (Brooke and Kaiser, 1970). Alkaline or acid assays are used to identify low or high mATPase activity. An ATPase is an enzyme that catalyses the hydrolysis (breakdown) of an ATP molecule. There is a strong correlation between mATPase activity and MHC isoform (Adams *et al.*, 1993; Fry, Allemeier and Staron, 1994; Staron, 1991; Staron *et al.*, 2000) because the ATPase is located at the heavy chain of the myosin molecule. This method has been used in combination with physiological measurements of contractile speed; in general the slowest fibres are classified as type I (lowest ATPase activity) and the fastest as type IIB (highest ATPase activity), with type IIA fibres expressing intermediate values (Figure 1.1.12). The explanation for these results is that the velocity of the interaction between myosin and actin depends on the speed with which the myosin heads are able to hydrolyse ATP molecules (see Section 1.1.4).

Immunohistochemistry explores the reaction of a specific antibody against a target protein isoform. In this method, muscle fibre types are determined on the basis of their immunoreactivity to antibodies specific to MHC isoforms. An electrophoretic analysis is a fibre typing method in which an electrical field is applied to a gel matrix which separates MHC isoforms according to their molecular weight and size (MHCIIb is the largest and MHCI is the smallest). Because of the difference in size, MHC isoforms migrate at different rates through the gel. Immunohistochemical and gel electrophoresis staining procedures are applied in order to visualize and quantify MHC isoforms.

Independent of the method applied, the results have revealed the existence of 'pure' and 'hybrid' muscle fibre types (Pette, 2001). A pure fibre type contains only one MHC isoform, whereas a hybrid fibre expresses two or more MHC isoforms. Thus, in mammalian skeletal muscles, four pure fibre types can be found: (1) slow type I (MHCIβ), (2) fast type IIA (MHCIIa), (3) fast type IID (MHCIId), and (4) fast type IIB (MHCIIb) (sometimes called fast type IIX fibres) (Pette and Staron, 2000). Actually, type IIb MHC isoform is more frequently found in rodent muscles, while in humans type IIx is more common (Smerdu *et al.*, 1994).

Combinations of these major MHC isoforms occur in hybrid fibres, which are classified according to their predominant isoform. Therefore, the following hybrid fibre types can be distinguished: type I/IIA, also termed IC (MHCIβ > MHCIIa); type IIA/I, also termed IIC (MHCIIa > MHCIβ); type IIAD (MHCIIa > MHCIId); type IIDA (MHCIId > MHCIIa); type IIDB (MHCIId > MHCIIb); and type IIBD (MHCIIb > MHCIId) (Pette, 2001). All these different combinations of MHCs contribute to generating the continuum of muscle fibre types represented below (Pette and Staron, 2000):

TypeI ↔ TypeIC ↔ TypeIIC ↔ **TypeIIA** ↔ TypeIIAD ↔ TypeIIDA ↔ **TypeIID** ↔ TypeIIDB ↔ TypeIIBD ↔ **TypeIIB**

Despite the fact that chronic physical exercise, involving either endurance or strength training, appears to induce transitions in the muscle fibre type continuum, MHC isoform changes are limited to the fast fibre subtypes, shifting from type IIB/D to type IIA fibres (Gillies *et al.*, 2006; Holm *et al.*, 2008; Putman *et al.*, 2004).The reverse transition, from type IIA to type IIB/D, occurs after a period of detraining (Andersen and Aagaard, 2000; Andersen *et al.*, 2005). However, it is still controversial whether training within physiological parameters can induce transitions between slow and fast myosin isoforms (Aagaard and Thorstensson, 2003). The amount of hybrid fibre increases in transforming muscles because the coexistence of different MHC isoforms in a myofibre leads it to adapt its phenotype in order to meet specific functional demands.

How is muscle fibre type defined? It appears that even in foetal muscle different types of myoblast exist and thus it is likely that myofibres have already begun their differentiation into different types by this stage (Kjaer *et al.*, 2003); innervation, mechanical loading or unloading, hormone profile, and ageing all play a major role in phenotype alteration.

The importance of innervation in the determination of specific myofibre types was demonstrated in a classic experiment of cross-reinnervation (for review see Pette and Vrbova, 1999): fast muscle phenotype became slow once reinnervated by a slow nerve, while a slow muscle became fast when reinnervated by a fast nerve. Nevertheless, motor nerves are not initially required for the differentiation of fast and slow muscle fibres, although innervation is later essential for muscle growth and survival. Motor innervation is also involved in fibre type differentiation during foetal development.

Curiously, unloaded muscles show a tendency to convert slow fibres to fast ones (Gallagher *et al.*, 2005; Trappe *et al.*, 2004, 2009). This shift in fibre type profile was observed in studies involving microgravity conditions (spaceflight or bedrest models). It appears that slow MHC isoforms are more sensitive to the lack of physical activity than fast isoforms (Caiozzo *et al.*, 1996, 2009; Edgerton *et al.*, 1995).

A hormonal effect on muscle fibre phenotypes is markedly observed with thyroid hormones. In skeletal muscles, thyroid hormones reduce MHCI gene transcription, while they stimulate transcription of MHCIIx and MHCIIb. The effects on the transcription of the MHCIIa gene are muscle-specific: transcription is activated in slow muscles such as soleus and repressed in fast muscles such as diaphragm (Baldwin and Haddad, 2001; DeNardi *et al.*, 1993). Thus, in general, reduced levels of thyroid hormone cause fast-to-slow shifts in MHC isoform expression, whereas high levels of thyroid hormone cause slow-to-fast shifts (Canepari *et al.*, 1998; Li *et al.*, 1996; Vadászová *et al.*, 2004).

In addition to muscle atrophy and a decrease in strength, ageing may cause fast-to-slow fibre type transitions (Canepari *et al.*, 2009; Korhonen *et al.*, 2006). Degenerative processes in the central nervous system (CNS) and/or peripheral nervous system (PNS), causing denervation (selective loss of fast α-motor neurons) and reinnervation (with slow α-motor neurons), physical inactivity, and altered thyroid hormone levels, may contribute to the observed atrophy and potential loss of fast muscle fibres in elderly people.

1.1.6 MUSCLE ARCHITECTURE

Skeletal muscle architecture is defined as the arrangement of myofibres within a muscle relative to its axis of force generation (Lieber and Fridén, 2000). Skeletal muscles present a variety of fasciculi arrangements but they can be described mostly as fusiforms or pennate muscles. A fusiform muscle has fibres organized in parallel to its force-generating axis and is described as having a parallel or longitudinal architecture. A pennate muscle has fibres arranged at an oblique angle to its force-generating axis (Figure 1.1.13).

Muscle fibre arrangement has a functional significance and the effect of muscle design on force production and contraction velocity is remarkable. In a fusiform muscle, the myofibres cause force production to occur directly at the tendon; the parallel arrangement allows fast muscle shortening. In a pennate

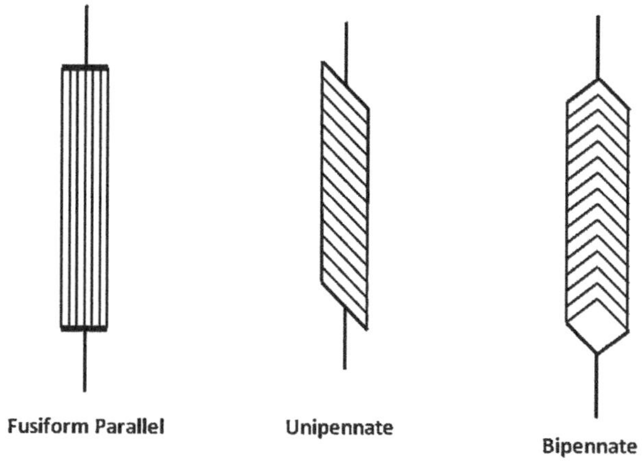

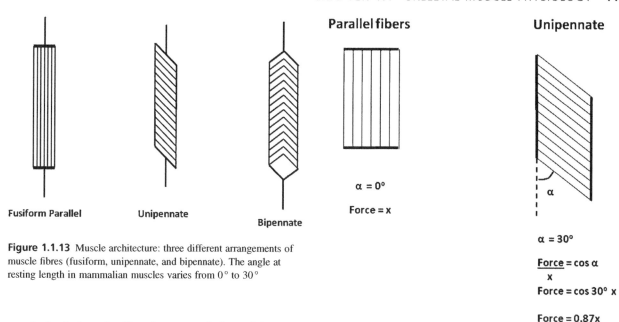

Figure 1.1.13 Muscle architecture: three different arrangements of muscle fibres (fusiform, unipennate, and bipennate). The angle at resting length in mammalian muscles varies from 0° to 30°

muscle, fascicule angle affects force transmission and decreases shortening velocity. Because fibres are orientated at an angle relative to the axis of force generation, not all fibre tensile force is transmitted to the tendons and only a component of fibre force production is actually transmitted along the muscle axis, in proportion to the pennation angle cosine (Figure 1.1.14). It seems that pennate arrangement is detrimental to muscle strength performance as it results in a loss of force compared with a muscle of the same mass and fibre length but zero pennation angle. However, a muscle with pennate fibres is able to generate great amounts of force. In fact, high-intensity strength training causes increases in pennation angle (Aagaard *et al.*, 2001; Kawakami, Abe and Fukunaga, 1993; Kawakami *et al.*, 1995; Reeves *et al.*, 2009; Seynnes, de Boer and Narici, 2007) which enhance the muscle's ability to pack more sarcomeres and myofibres, creating a large physiological cross-sectional area (PCSA). A PCSA is measured perpendicular to the fibre orientation, whereas an anatomical cross-sectional area (ACSA) is measured perpendicular to the muscle orientation. A large PCSA positively affects the force-generating capacity of the muscle, compensating for the loss in force transmission due to increases in pennation angle (Lieber and Fridén, 2000).

Two other important parameters in architectural analysis are muscle length and fibre length. Muscle velocity is proportional to muscle/fibre length. Muscles with similar PCSAs and pennation angles but different fibre lengths have different velocity outputs. The muscle with the longest fibres presents greater contraction velocity, while shorter muscles are more suitable for force production with low velocity. Thus, antigravity short extensor muscles are designed more for force production, while longer flexors are for long excursions with higher velocity. The soleus muscle, with its high PCSA and short fibre length, suitable for generating high force with small excursion, is a good example of an antigravity postural muscle, while the biceps femoris is an example of a long flexor muscle suitable for generating high velocity with long excursion.

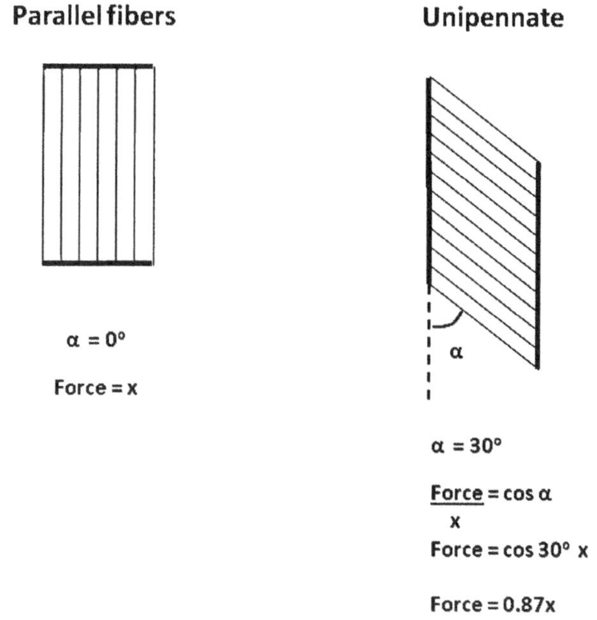

Figure 1.1.14 Representation of the effect of muscle fibre arrangement. Fibres orientated parallel to the axis of force generation transmit all of their force to the tendon. Fibres orientated at a 30° angle relative to the force-generation axis transmit part of their force to the tendon, proportional to the angle cosine. Adapted from Lieber (1992)

1.1.7 HYPERTROPHY AND HYPERPLASIA

A good example of skeletal muscle's ability to adapt to environmental stimuli is the hypertrophy that occurs after a period of strength training. Hypertrophy can be defined as the increase in muscle fibre size and/or muscle mass due to an accumulation of contractile and noncontractile proteins inside the cell (Figure 1.1.15). Increased rate of protein synthesis, decreased rate of protein breakdown, or a combination of both of these factors, is responsible for muscle hypertrophy (Rennie *et al.*, 2004). Most of us are in a state of equilibrium between muscle protein synthesis and protein degradation; thus muscle mass remains constant. Muscle size and muscle mass can also be enlarged by way of hyperplasia, which is an increase in the number of myofibres. However, the suggestion that new muscle fibres may form in human adults as a result of strength training is still highly controversial (Kadi, 2000). Hyperplasia has been demonstrated following strength training in some animal models (Antonio and Gonyea, 1993; 1994; Tamaki *et al.*, 1997) but the evidence in human subjects is unclear (Kelley, 1996; McCall *et al.*, 1996).

Hypertrophy-orientated strength training affects all muscle fibre types, however type II fast fibres have usually shown a more pronounced response than type I slow fibres (Aagaard

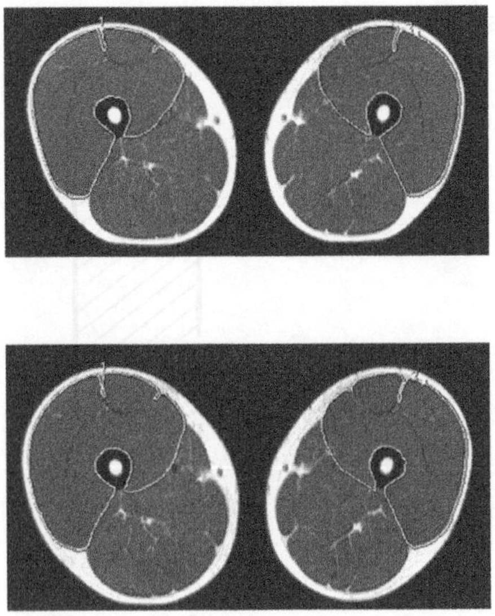

Figure 1.1.15 Magnetic resonance image of the quadriceps femoris before and after a period of strength training

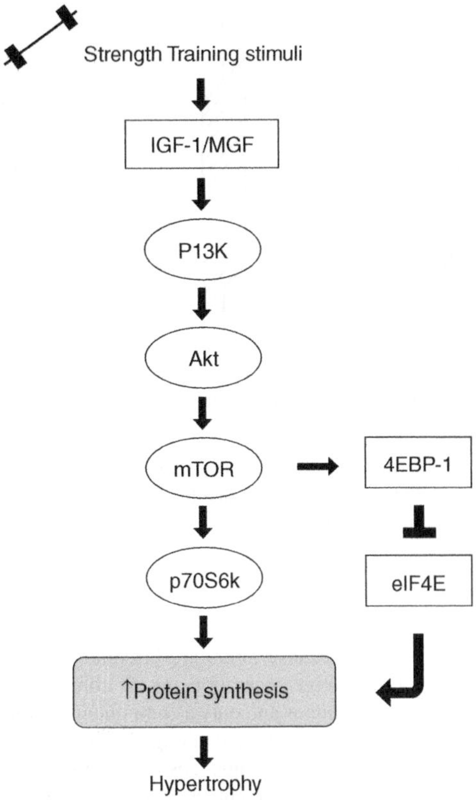

Figure 1.1.16 Schematic illustration of the effect of strength training stimuli on activation of the hypertrophy pathway

et al., 2001; Cribb and Hayes, 2006; Verdijk *et al.*, 2009). It appears that type II fibres possess a greater adaptive capacity for hypertrophy than type I fibres. This is interesting because it has been demonstrated that muscle protein synthesis does not differ between fibre types after a bout of strength training exercise (Mittendorfer *et al.*, 2005). Similarly, satellite cells, which are important contributors to muscle hypertrophy (see Section 1.1.8), are equally distributed in human vastus lateralis type I and type II fibres (Kadi, Charifi and Henriksson, 2006).

The question of how mechanical signals provided by strength training are translated into increased muscle protein synthesis and hypertrophy has not been fully answered. It is beyond the scope of this chapter to deeply explain the possible mechanism and intracellular pathways involved in muscle hypertrophy (for more details see Spangenburg, 2009; Zanchi and Lancha, 2008). In brief, high tension applied over skeletal muscles during strength training stimulates the release of growth factors such as insulin-like growth factor 1 (IGF-1). It has been shown that this hormone is involved in muscle hypertrophy through autocrine and/or paracrine mechanisms (Barton-Davis, Shoturma and Sweeney, 1999). IGF-1 is a potent activator of the protein kinase B (Akt)/mammalian target of rapamycin (mTOR) signalling pathway, which is a key regulator in protein synthesis. Activation of Akt is mediated by the IGF-1/phosphatidylinositol-3 kinase (PI3K) pathway. Once activated, Akt activates mTOR, which in turn activates 70 kDa ribosomal S6 protein kinase (p70S6k), a positive regulator of protein translation (Baar and Esser, 1999). The degree of activation of p70S6k is closely associated with the subsequent protein synthesis and muscle growth. mTOR also inhibits the activity of 4E binding protein 1 (4E-BP1), a negative regulator of the protein-initiation factor eIF-4E, which is involved in translation initiation and like p70S6k contributes to enhanced protein synthesis (Koopman *et al.*, 2006). Any alteration in mTOR activation will result in changes in p70s6k activation and 4EBP-1 inactivation, ultimately affecting the initiation of protein synthesis and causing increases in muscle size and mass (Figure 1.1.16). It appears that muscle tension per se may also stimulate skeletal muscle growth through a mechanism called mechano-transduction; the regulation of the Akt/mTOR signalling pathway in response to mechanical stimuli is still not well understood however (Rennie *et al.*, 2004).

1.1.8 SATELLITE CELLS

Adult skeletal muscle contains a cell population with stem cell-like properties called satellite cells (SCs) (Hawke, 2005). These are located at the periphery of myofibres, between the basal lamina and the sarcolemma (Figure 1.1.2) (Vierck *et al.*, 2000; Zammit, 2008). Skeletal muscle SCs are undifferentiated quiescent myogenic precursors that display self- renewal properties (Huard, Cao and Qu-Petersen, 2003), which means that they can generate daughter cells which can become new SCs (Kadi *et al.*, 2004; Zammit and Beauchamp, 2001). They serve as reserve cells and are recruited when myofibre growth and/or regeneration after injury is needed. In response to signals associated with muscle damage, mechanical loading, and exercise, SCs leave the quiescent state, become activated, and reenter the cell cycle (Hawke and Garry, 2001; Tidball, 2005). After activation, these cells proliferate and migrate to the site of injury to repair or replace damaged myofibres by fusing together to create a myotube or by fusing to existing myofibres (Hawke, 2005; Machida and Booth, 2004).

It appears that the fusion of SCs to existing myofibres is also critical for increases in fibre cross-sectional area. This physiological event takes into consideration the concept of the myonuclear domain or DNA unit, which suggests that each myonucleus manages the production of mRNA and protein synthesis for a specific volume of sarcoplasm (Adams, 2002; Kadi and Thornell, 2000; Petrella *et al.*, 2008). It has been proposed that there may be a limit on the amount of expansion a myonuclear domain can undergo during hypertrophy (i.e. a domain ceiling size) (Kadi *et al.*, 2004). A limit indicates that an expansion of the myonuclear domain toward a threshold >2000 μm^2/nucleus (Petrella *et al.*, 2006), or between 17 and 25% of the fibre's initial size (Kadi *et al.*, 2004), may put each myonucleus under greater strain, thus increasing the demand for new myonuclei to make continued growth possible (Petrella *et al.*, 2008).

This implies that increases in muscle fibre size (i.e. hypertrophy) must be associated with a proportional increase in myonucleus number. However, muscle fibres are permanently differentiated, and therefore incapable of producing additional myonuclei through mitosis (Hawke and Garry, 2001; Machida and Booth, 2004). Nevertheless, it has been shown that myonucleus number increases during skeletal muscle hypertrophy, in order to maintain the cytoplasm-to-myonucleus ratio (Allen *et al.*, 1995; Petrella *et al.*, 2006), and the only alternative source of additional myonuclei is the pool of SCs (Kadi, 2000; Rosenblatt and Parry, 1992). In adult skeletal muscles, it has been observed that satellite cell-derived myonuclei are incorporated into muscle fibres during hypertrophy (Kadi and Thornell, 2000). Following the initial increase in muscle fibre size that occurs via increased mRNA activity and protein accretion, the incorporation of additional myonuclei in muscle fibres represents an important mechanism for sustaining muscle fibre enlargement (Petrella *et al.*, 2006). Incorporated satellite cell-derived myonuclei are no longer capable of dividing, but they can produce muscle-specific proteins that increase myofibre size (Allen, Roy and Edgerton, 1999), resulting in hypertrophy.

The mechanisms by which skeletal muscle SCs are activated and incorporated into growing myofibres are still unclear but it has been suggested that they are modulated by cytokines and autocrine/paracrine growth factors (Hawke and Garry, 2001; Petrella *et al.*, 2008). For example, strength exercises induce damage to the sarcolemma, connective tissue, and muscle fibre structural and contractile proteins, initiating an immune response which then attracts macrophages to the damaged area. Macrophages secrete a number of cytokines, which regulate the SCs pool. In order to regenerate the damaged myofibres, SCs differentiate and fuse together to generate a myotube, which then fuses to the damaged muscle fibre, repairing the injury.

Despite the differences between skeletal muscle regeneration and hypertrophy, both processes share similarities regarding SCs activation, proliferation, and differentiation. It is believed that muscle-loading conditions (i.e. strength training) stimulate the release of growth factors (IGF-1 and MGF), which affects SC activation, proliferation and differentiation (Adams and Haddad, 1996; Bamman *et al.*, 2007; Vierck *et al.*, 2000). Insulin-like growth factor-1 (IGF-1) and its isoform mechano-growth factor (MGF) are potent endocrine and skeletal muscle autocrine/paracrine growth factors. IGF-1 is upregulated in response to hypertrophic signals in skeletal muscle and promotes activation, proliferation and fusion of the SCs. During skeletal muscle hypertrophy, SCs fuse to the existing myofibres, essentially donating their nuclei, whereas in regeneration from more extensive muscle damage SCs fuse together to generate a new myotube, which repairs damaged fibres (Hawke, 2005).

The ageing process decreases the number of SCs and negatively affects their proliferative capacity; these two factors may be related to the atrophy and poor muscle regeneration described in some elderly individuals (Crameri *et al.*, 2004; Kadi *et al.*, 2004).

References

Aagaard P. and Andersen J.L. (1998) Correlation between contractile strength and myosin heavy chain isoform composition in human skeletal muscle. *Med Sci Sports Exerc*, **30** (8), 1217–1222.

Aagaard P., Andersen J.L., Dyhre-Poulsen P. *et al.* (2001) A mechanism for increased contractile strength of human pennate muscle in response to strength training: changes in muscle architecture. *J Physiol*, **15** (534 Pt 2), 613–623.

Aagaard P. and Thorstensson A. (2003) Neuromuscular aspects of exercise: adaptive responses evoked by strength training. In: Kjaer M. (eds), *Textbook of Sports Medicine*, Blackwell, London; 70–106.

Adams G.R., Hather B.M., Baldwin K.M. and Dudley G.A. (1993) Skeletal muscle myosin heavy chain composition and resistance training. *J Appl Physiol*, **74** (2), 911–915.

Adams G.R. and Haddad F. (1996) The relationships among IGF-1,DNA content, and protein accumulation during skeletal muscle hypertrophy. *J Appl Physiol*, **81**, 2509–2516.

Adams G.R. (2002) Invited review: autocrine/paracrine IGF-I and skeletal muscle adaptation. *J Appl Physiol*, **93**, 1159–1167.

Alkner B.A. and Tesch P.A. (2004) Knee extensor and plantar flexor muscle size and function following 90 days of bed rest with or without resistance exercise. *Eur J Appl Physiol*, **93** (3), 294–305.

Allen D.L., Monke S.R., Talmadge R.J. *et al.* (1995) Plasticity of myonuclear number in hypertrophied and atrophied mammalian skeletal muscle fibers. *J Appl Physiol*, **78** (5), 1969–1976.

Allen D.L., Roy R.R. and Edgerton V.R. (1999) Myonuclear domains in muscle adaptation and disease. *Muscle Nerve*, **22**, 1350–1360.

Andersen J.L. and Aagaard P. (2000) Myosin heavy chain IIX overshoot in human skeletal muscle. *Muscle Nerve*, **23**, 1095–1104.

Andersen L.L., Andersen J.L., Magnusson P. *et al.* (2005) Changes in the human muscle force-velocity relationship in response to resistance training and subsequent detraining. *J Appl Physiol*, **99**, 87–94.

Antonio J. and Gonyea W.J. (1993) Role of muscle fiber hypertrophy and hyperplasia in intermittently stretched avian muscle. *J Appl Physiol*, **74**, 1893–1898.

Antonio J. and Gonyea W.J. (1994) Muscle fiber splitting in stretch-enlarged avian muscle. *Med Sci Sports Exerc*, **26**, 973–977.

Baar K. and Esser K. (1999) Phosphorylation of p70(S6k) correlates with increased skeletal muscle mass following resistance exercise. *Am J Physiol*, **276**, C120–C127.

Baldwin K.M. and Haddad F. (2001) Invited review: effects of different activity and inactivity paradigms on myosin heavy chain gene expression in striated muscle. *J Appl Physiol*, **90**, 345–357.

Bamman M.M., Petrella J.K., Kim J.S. *et al.* (2007) Cluster analysis tests the importance of myogenic gene expression during myofiber hypertrophy in humans. *J Appl Physiol*, **102**, 2232–2239.

Barton-Davis E.R., Shoturma D.I. and Sweeney H.L. (1999) Contribution of satellite cells to IGF-I induced hypertrophy of skeletal muscle. *Acta Physiol Scand*, **167** (4), 301–315.

Berg H.E., Larsson L. and Tesch P.A. (1997) Lower limb skeletal muscle function after 6 wk of bed rest. *J Appl Physiol*, **82** (1), 182–188.

Bloch R.J. and Gonzalez-Serratos H. (2003) Lateral force transmission across costameres in skeletal muscle. *Exerc Sport Sci Rev*, **31** (2), 73–78.

Bohé J., Low A., Wolfe R.R. and Rennie M.J. (2003) Human muscle protein synthesis is modulated by extracellular, not intramuscular amino acid availability: a dose–response study. *J Physiol*, **552** (1), 315–324.

Brooke M.H. and Kaiser K.K. (1970) Three myosin adenosine triphosphatase systems: the nature of their pH lability and sulfhydryl dependence. *J Histochem Cytochem*, **18**, 670–672.

Brooks G.A., Fahey T.D. and Baldwin K.M. (2005) *Exercise Physiology: Human Bioenergetics and Its Applications*, McGraw-Hill, Boston.

Caiozzo V.J., Haddad F., Baker M.J. *et al.* (1996) Microgravity-induced transformations of myosin isoforms and contractile properties of skeletal muscle. *J Appl Physiol*, **81** (1), 123–132.

Caiozzo V.J., Haddad F., Lee S. *et al.* (2009) Artificial gravity as a countermeasure to microgravity: a pilot study examining the effects on knee extensor and plantar flexor muscle groups. *J Appl Physiol*, **107** (1), 39–46.

Canepari M., Cappelli V., Pellegrino M.A. *et al.* (1998) Thyroid hormone regulation of MHC isoform composition and myofibrillar ATPase activity in rat skeletal muscles. *Arch Physiol Biochem*, **106** (4), 308–315.

Canepari M., Pellegrino M.A., D'Antona G. and Bottinelli R. (2009) Single muscle fiber properties in aging and disuse. *Scand J Med Sci Sports*, **20** (1), 10–19.

Challis J.H. (2000) Muscle-tendon architecture and athletic performance, in *Biomechanics in Sport: Performance Enhancement and Injury Prevention* (ed. V.M. Zatsiorsky), Blackwell Science, Oxford., pp. 33–55.

Crameri R.M., Langberg H., Magnusson P. *et al.* (2004) Changes in satellite cells in human skeletal muscle after a single bout of high intensity exercise. *J Physiol*, **558**, 333–340.

Cribb P.J. and Hayes A. (2006) Effects of supplement timing and resistance exercise on skeletal muscle hypertrophy. *Med Sci Sports Exerc*, **38** (11) 1918–1925.

DeNardi C., Ausoni S., Moretti P. *et al.* (1993) Type 2X-myosin heavy chain is coded by a muscle fiber type-specific and developmentally regulated gene. *J Cell Biol*, **123** (4), 823–835.

Edgerton V.R., Zhou M.Y., Ohira Y. *et al.* (1995) Human fiber size and enzymatic properties after 5 and 11 days of spaceflight. *J Appl Physiol*, **78** (5), 1733–1739.

Finer J.T., Simmons R.M. and Spudich J.A. (1994) Single myosin molecule mechanics: piconewton forces and nanometre steps. *Nature*, **10** (368 6467):113–119.

Fry A.C., Allemeier C.A. and Staron R.A. (1994) Correlation between percentage fiber type area and myosin heavy chain content in human skeletal muscle. *Eur J Appl Physiol*, **68**, 246–251.

Gallagher P., Trappe S., Harber M. *et al.* (2005) Effects of 84-days of bedrest and resistance training on single muscle fibre myosin heavy chain distribution in human vastus lateralis and soleus muscles. *Acta Physiol Scand*, **185** (1), 61–69.

Gillies E.M., Putman C.T. and Bell G.J. (2006) The effect of varying the time of concentric and eccentric muscle actions during resistance training on skeletal muscle adaptations in women. *Eur J Appl Physiol*, **97** (4), 443–453.

Gulick A.M. and Rayment I. (1997) Structural studies on myosin II: communication between distant protein domains. *Bioessays*, **19** (7), 561–569.

Hatze H. (2002) Fundamental issues, recent advances, and future directions in myodynamics. *J Electromyogr Kinesiol*, **12** (6), 447–454.

Hawke T.J. (2005) Muscle stem cells and exercise training. *Exerc Sport Sci Rev*, **33** (2), 63–68.

Hawke T.J. and Garry D.J. (2001) Myogenic satellite cells: physiology to molecular biology. *J Appl Physiol*, **91**, 534–551.

Holm L., Reitelseder S., Pedersen T.G. *et al.* (2008) Changes in muscle size and MHC composition in response to resistance exercise with heavy and light loading intensity. *J Appl Physiol*, **105** (5), 1454–1461.

Huard J., Cao B. and Qu-Petersen Z. (2003) Muscle-derived stem cells: potential for muscle regeneration. *Birth Defects Res Part C*, **69**, 230–237.

Huijing P.A. (1999) Muscle as a collagen fiber reinforced composite: a review of force transmission in muscle and whole limb. *J Biomech*, **32**, 329–345.

Kadi F. (2000) Adaptation of human skeletal muscle to training and anabolic steroids. *Acta Physiol Scand*, **168** (Suppl.), 1–52.

Kadi F., Charifi N., Denis C. and Lexell J. (2004) Satellite cells and myonuclei in young and elderly women and men. *Muscle Nerve*, **29**, 120–127.

Kadi F., Charifi N. and Henriksson J. (2006) The number of satellite cells in slow and fast fibres from human vastus lateralis muscle. *Histochem Cell Biol*, **126**, 83–87.

Kadi F. and Thornell L.E. (2000) Concomitant increases in myonuclear and satellite cell content in female trapezius muscle following strength training. *Histochem Cell Biol*, **113**, 99–103.

Kawakami Y., Abe T. and Fukunaga T. (1993) Muscle-fiber pennation angles are greater in hypertrophied than normal muscles. *J Appl Physiol*, **74**, 2740–2744.

Kawakami Y., Abe T., Kuno S. and Fukunaga T. (1995) Training induced changes in muscle architecture and tension. *Eur J Appl Physiol*, **72**, 37–43.

Kelley G. (1996) Mechanical overload and skeletal muscle fiber hyperplasia: a meta-analysis. *J Appl Physiol*, **81** (4), 1584–1588.

Kjær M., Kalimo H. and Saltin B. (2003) Skeletal muscle: physiology, training and repair after injury, in *Textbook of Sports Medicine* (eds M. Kjær, M. Krogsgaard, P. Magnusson *et al.*), Blackwell Science Ltd, pp. 49–69.

Koopman R., Zorenc A.H., Gransier R.J. *et al.* (2006) Increase in S6K1 phosphorylation in human skeletal muscle following resistance exercise occurs mainly in type II muscle fibers. *Am J Physiol Endocrinol Metab*, **290** (6), E1245–E1252.

Korhonen M.T., Cristea A., Alén M. *et al.* (2006) Aging, muscle fiber type, and contractile function in sprint-trained athletes. *J Appl Physiol*, **101** (3), 906–917.

Kraemer W.J., Hatfield D.L., Volek J.S. *et al.* (2009) Effects of amino acids supplement on physiological adaptations to resistance training. *Med Sci Sports Exerc*, **41** (5), 1111–1121.

Lemoine J.K., Haus J.M., Trappe S.W. and Trappe T.A. (2009) Muscle proteins during 60-day bedrest in women: impact of exercise or nutrition. *Muscle Nerve*, **39** (4), 463–471.

Li X., Hughes S.M., Salviati G. *et al.* (1996) Thyroid hormone effects on contractility and myosin composition of soleus muscle and single fibres from young and old rats. *J Physiol*, **15** (494 Pt 2):555–567.

Lieber R.L. (1992) *Skeletal Muscle Structure and Function: Implications for Rehabilitation and Sports Medicine*, Williams and Wilkins, Baltimore.

Lieber R.L. and Friden J. (2000) Functional and clinical significance of skeletal muscle architecture. *Muscle Nerve*, **23**, 1647–1666.

Machida S. and Booth F.W. (2004) Insulin-like growth factor 1 and muscle growth: implication for satellite cell proliferation. *Proc Nutr Soc*, **63**, 337–340.

McArdle W.D., Katch F.I. and Katch V.L. (2007) *Exercise Physiology: Energy, Nutrition and Human Performance*, Lippincott Williams & Wilkins, Baltimore.

McCall G.E., Byrnes W.C., Dickinson A. *et al.* (1996) Muscle fiber hypertrophy, hyperplasia, and capillary density in college men after resistance training. *J Appl Physiol*, **81** (5), 2004–2012.

McKillop D.F. and Geeves M.A. (1993) Regulation of the interaction between actin and myosin subfragment 1: evidence for three states of the thin filament. *Biophys J*, **65** (2), 693–701.

Mittendorfer B., Andersen J.L., Plomgaard P. *et al.* (2005) Protein synthesis rates in human muscles: neither anatomical location nor fibre-type composition are major determinants. *J Physiol*, **15** (563 Pt 1): 203–211.

Monti R.J., Roy R.R., Hodgson J.A. and Edgerton V.R. (1999) Transmission of forces within mammalian skeletal muscles. *J Biomech*, **32**, 371–380.

Parcell A.C., Sawyer R.D., Drummond M.J. *et al.* (2005) Single-fiber MHC polymorphic expression is unaffected by sprint cycle training. *Med Sci Sports Exerc*, **37** (7), 1133–1137.

Petrella J.K., Kim J., Cross J.M. *et al.* (2006) Efficacy of myonuclear addition may explain differential myofiber growth among resistancetrained young and older men and women. *Am J Physiol Endocrinol Metab*, **291**, E937–E946.

Petrella J.K., Kim J., Mayhew D.L. *et al.* (2008) Potent myofiber hypertrophy during resistance training in humans is associated with satellite cell-mediated myonuclear addition: a cluster analysis. *J Appl Physiol*, **104**, 1736–1742.

Pette D. (2001) Historical perspectives: plasticity of mammalian skeletal muscle. *J Appl Physiol*, **90**, 1119–1124.

Pette D. and Staron R.S. (2000) Myosin isoforms, muscle fiber types, and transitions. *Microsc Res Tech*, **50**, 500–509.

Pette D. and Vrbova M.D.G. (1999) What does chronic electrical stimulation teach us about muscle plasticity? *Muscle Nerve*, **22**, 666–677.

Phillips S.M. (2009) Physiologic and molecular bases of muscle hypertrophy and atrophy: impact of resistance exercise on human skeletal muscle (protein and exercise dose effects). *Appl Physiol Nutr Metab*, **34** (3), 403–410.

Putman C.T., Xu X., Gillies E. *et al.* (2004) Effects of strength, endurance and combined training on myosin heavy chain content and fibre-type distribution in humans. *Eur J Appl Physiol*, **92** (4–5), 376–384.

Reeves N.D., Maganaris C.N., Longo S. and Narici M.V. (2009) Differential adaptations to eccentric versus conventional resistance training in older humans. *Exp Physiol*, **94** (7), 825–833.

Rennie M.J., Wackerhage H., Spangenburg E.E. and Booth F.W. (2004) Control of the size of human muscle mass. *Annu Rev Physiol*, **66**, 799–828.

Rosenblatt J.D. and Parry D.J. (1992) Gamma irradiation prevents compensatory hypertrophy of overloaded mouse extensor digitorum longus muscle. *J Appl Physiol*, **73** (6), 2538–2543.

Rossi A.E. and Dirksen R.T. (2006) Sarcoplasmic reticulum: the dynamic calcium governor of muscle. *Muscle Nerve*, **33**, 715–731.

Rossi D., Barone V., Giacomello E. *et al.* (2008) The sarcoplasmic reticulum: an organized patchwork of specialized domains. *Traffic*, **9**, 1044–1049.

Seynnes O.R., de Boer M. and Narici M.V. (2007) Early skeletal muscle hypertrophy and architectural changes in response to high-intensity resistance training. *J Appl Physiol*, **102**, 368–373.

Smerdu V., Karsch-Mizarachi I., Campione M. *et al.* (1994) Type IIx myosin heavy chain transcripts are expressed in type IIb fibers of human skeletal muscle. *Am J Physiol*, **267**, C1723–C1728.

Spangenburg E.E. (2009) Changes in muscle mass with mechanical load: possible cellular mechanisms. *Appl Physiol Nutr Metab*, **34**, 328–335.

Staron R.S. (1991) Correlation between myofibrilar ATPase activity and myosin heavy chain composition in single human muscle fibers. *Histochemistry*, **96**, 21–24.

Staron R.S., Hagerman F.C., Hikida R.S. *et al.* (2000) Fiber type composition of the vastus lateralis muscle of young men and women. *J Histochem Cytochem*, **48** (5), 623–629.

Tamaki T., Akatsuka A., Tokunaga M. *et al.* (1997) Morphological and biochemical evidence of muscle hyperplasia following weight-lifting exercise in rats. *Am J Physiol*, **273**, C246–C256.

Tidball J.G. (2005) Mechanical signal transduction in skeletal muscle growth and adaptation. *J Appl Physiol*, **98** (5), 1900–1908.

Tipton K.D., Elliott T.A., Ferrando A.A. *et al.* (2009) Stimulation of muscle anabolism by resistance exercise and ingestion of leucine plus protein. *Appl Physiol Nutr Metab*, **34** (2), 151–161.

Trappe S., Costill D., Gallagher P. *et al.* (2009) Exercise in space: human skeletal muscle after 6 months aboard the International Space Station. *J Appl Physiol*, **106**, 1159–1168.

Trappe S., Trappe T., Gallagher P. *et al.* (2004) Human single muscle fibre function with 84 day bed-rest and resistance exercise. *J Physiol*, **557**, (Pt 2):501–513.

Vadászová A., Zacharová G., Machácová K. *et al.* (2004) Influence of Thyroid Status on the Differentiation of Slow and Fast Muscle Phenotypes. *Physiol Res*, **53** (Suppl. 1), S57–S61.

Vandenboom R. (2004) The myofibrillar complex and fatigue: a review. *Can J Appl Physiol*, **29** (3), 330–356.

Verdijk L.B., Gleeson B.G., Jonkers R.A. *et al.* (2009) Skeletal muscle hypertrophy following resistance training is accompanied by a fiber type-specific increase in satellite cell content in elderly men. *J Gerontol A Biol Sci*, **64** (3), 332–339.

Vierck J., O'Reilly B., Hossner K. *et al.* (2000) Satellite cell regulation following myotrauma caused by resistance exercise. *Cell Biol Int*, **24** (5), 263–272.

Zammit P.S. (2008) All muscle satellite cells are equal, but are some more equal than others? *J Cell Sci*, **121**, 2975–2982.

Zammit P.S. and Beauchamp J.R. (2001) The skeletal muscle satellite cell: stem cell or son of stem cell? *Differentiation*, **68**, 193–204.

Zanchi N.E. and Lancha A.H. Jr. (2008) Mechanical stimuli of skeletal muscle: implications on mTOR/p70s6k and protein synthesis. *Eur J Appl Physiol*, **102** (3), 253–263.

1.2 Neuromuscular Physiology

Alberto Rainoldi[1] and Marco Gazzoni[2], [1]University School of Motor and Sport Sciences, Università degli Studi di Torino, Motor Science Research Center, Torino, Italy, [2]LISiN, Politecnico di Torino, Dipartimento di Elettronica, Torino, Italy

1.2.1 THE NEUROMUSCULAR SYSTEM

Movement involves the interaction between the nervous system and the muscles. The nervous system generates the signals which tell the muscles to activate and the muscles provide the proper force.

1.2.1.1 Motor units

The skeletal muscles consist of up to a million muscle fibres. To coordinate the contraction of all these fibres, they are subdivided into functional units: the motor units. A motor unit consists of one motoneuron (located in the spinal cord and brainstem) and the fibres which it innervates (Figure 1.2.1).

The fibres belonging to a single motor unit are localized within a region of the muscle cross-section (up to 15%) and are intermingled with fibres belonging to other motor units.

Moreover, some evidence exists for the muscle organization in neuromuscular compartments; that is, regions of muscle supplied by a primary branch of the muscle nerve. A single muscle can consist of several distinct regions, each with a different physiological function (English, 1984; Fleckenstein *et al.*, 1992).

Motor unit types

Many criteria and terminologies have been used to classify the types of fibre or motor unit (see Chapter 1.1). Histochemical, biochemical, and molecular properties are used to measure the mechanisms responsible for physiological properties (e.g. contraction speed, magnitude of force, fatigue resistance). Direct physiological measurements of contraction speed, magnitude of force, and fatigue resistance have also been used to distinguish motor unit types.

Although the measurement of motor unit properties in human muscle results in a less discrete classification scheme than that for muscle fibres, motor unit activity is often discussed in terms of the following scheme. The smaller slow-twitch motor units have been called 'tonic units'. Histochemically, they are the smaller units (type I), and metabolically they have fibres rich in mitochondria, are highly capillarized, and therefore have a high capacity for aerobic metabolism. Mechanically, they produce twitches with a low peak tension and a long time to peak (60–120 ms). The larger fast-twitch motor units are called 'phasic units' (type II). They have less mitochondria, are poorly capillarized, and therefore rely on anaerobic metabolism. The twitches have larger peak tensions in a shorter time (10–50 ms).

Although commonly adopted, the motor unit scheme described above does not appear to be appropriate for human muscles on the basis of the following considerations: (1) investigators have been unable to clearly distinguish motor unit types in human muscle (Fuglevand, Macefield and Bigland-Ritchie, 1999); (2) the first motor units activated in a voluntary contraction can be either slow-twitch or fast-twitch (Van Cutsem *et al.*, 1997); and (3) the cross-sectional area of human muscle fibres often does not increase from type I to type II (Harridge *et al.*, 1996; Miller *et al.*, 1993).

1.2.1.2 Muscle receptors

A number of different peripheral receptors are located within muscle, tendon, and fascia. Their role is to provide afferent information from the periphery to the CNS following the typical loop of any omeostatic system. They in fact continuously provide feedback to allow the CNS to properly maintain human body 'stability'.

In general, afferent axons are classified with respect to their cross-section in four groups, from the larger, with faster conduction velocity (group I), to the smaller, with slower conduction velocity (group IV).

Muscle spindles

Spindles are fusiform organs which lie in parallel with skeletal muscle fibres. There are more than 25 000 spindles throughout the human body; their role is to provide feedback information to changes in muscle length (Proske, 1997). Each spindle is innervated by gamma motonenurons; such neural input comes from the spinal cord. Group I afferents end in a spiral shape on

Strength and Conditioning – Biological Principles and Practical Applications Marco Cardinale, Rob Newton, and Kazunori Nosaka.

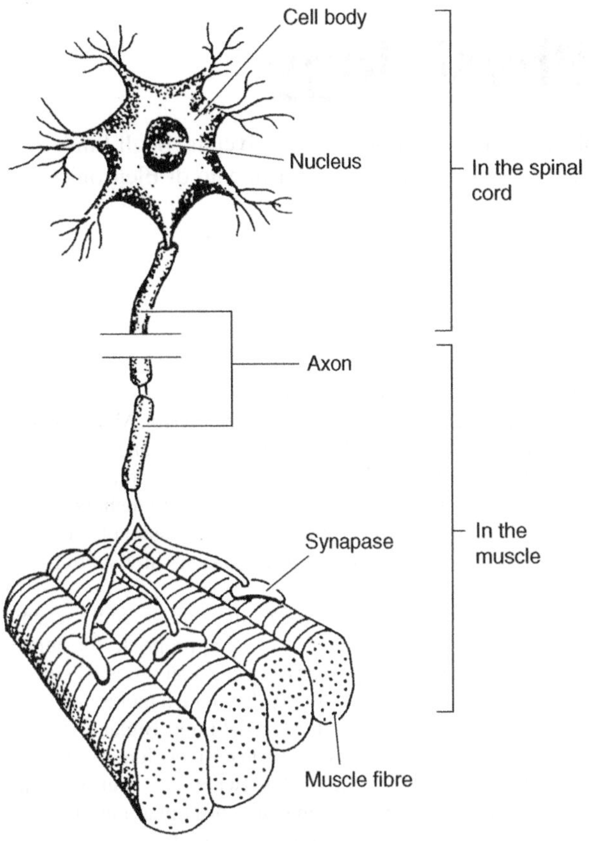

Figure 1.2.1 A motor unit consists of its motor nerve, which branches out to form connections to many muscle fibres through synapses, called motor end plates

the muscle spindle; when the fibre is stretched, the spiral is modified and an action potential is transmitted to the CNS. In the same way, when the whole muscle is passively stretched, the spindle spirals 'feel' the deformation and forward the information to the CNS.

Tendon organs

These are placed within the tendon and, due to their position, are often described as 'organs in series with skeletal muscle fibres', whereas muscle spindles are in parallel with them, as described above. These organs are activated as a consequence of (active or passive) stretching of the muscle, thus of the connective tissue attachments, and of the group I afferents which are branched in the myotendinous junction. For these reasons, these organs are considered as muscle force sensors; their threshold of excitation depends on the mode of activation (active or passive).

Joint receptors

Unlike muscle spindles and tendon organs, joint receptors vary in relation to their location and function. They are usually served by neurons with group II, III, and IV afferents. For instance, Ruffini-ending mechanoreceptors are able to monitor joint position and displacement, angular velocity, and intra-articular pressure (Johannsson, Sjölander and Sojka, 1991). Pacinian corpuscles seem to be able to detect acceleration of the joint (Bell, Bolanowsky and Holmes, 1994). Golgi endings are corpuscles with a behaviour similar to that of tendon organs. They play a protective role monitoring tension in ligaments, especially at the extremes of the range of motion. In general, joint receptors seem to act on a single joint indirectly, modulating the activity of the muscle spindle and thus influencing the α-motor neuron output.

1.2.1.3 Nervous and muscular conduction velocity

Conduction velocity is the velocity at which action potentials propagate along the fibre. Conduction velocities of both axonal and sarcolemmal action potentials vary with the diameter of the axon. The larger the diameter, the greater the conduction velocity.

Nervous and muscular propagations are sustained by two different mechanisms: the saltatory conduction in myelinated fibres from one node of Ranvier to the next, and the depolarization front induced by sodium and potassium currents, respectively. As a consequence, two different average conduction velocities are available, the former in the range 90–105 m/s and the latter in the range 3–5 m/s. These two ranges correspond to two extremely different tasks: (1) to provide information to the CNS at a velocity fast enough to ensure proper feedback (within safe reaction time and throughout the body); (2) to allow fast contraction of muscle fibres, which can have a maximum emi-length of 15 cm (as in the case of the longest muscle, the sartorius muscle, for instance). Since muscle fibres can change their cross-section area (hypertrophy) and, in some conditions, their number (hyperplasia), the conduction velocity and thus the MVC can be increased even if the fibre type remains unchanged.

1.2.1.4 Mechanical output production

The force that a muscle exerts during a contraction depends on the excitation provided by the nervous system (the number of motor neurons that are activated and the rates at which they discharge), the mechanical properties of the muscle, and the muscle architecture.

As described in Chapter 1.1, for a given level of excitation, the muscle force depends on muscle length (force–length relation) and on the rate of change in length (force–velocity relation).

The basic contractile properties of muscle, as characterized by the force–length and force–velocity relations of the single fibre, are influenced by the way in which the fibres are organized to form a muscle (Lieber and Friden, 2000; Russell, Motlagh and Ashley, 2000). Muscle fibres can be in series, in parallel, or at an angle to the line of pull of the muscle (angle of pennation). In-series arrangement maximizes the maximum shortening velocity and the range of motion of the muscle, whereas in-parallel arrangement maximizes the force the muscle can exert. In pennated muscles, the contribution of the fibres to the muscle force varies with the cosine of the angle of pennation.

Motor units and muscle twitch

The contractile property of a motor unit is described by the force–time response (twitch) to a single excitatory input (Figure 1.2.2a). All motor units have the same characteristic twitch shape, described by: (1) the time for the tension to reach the maximum (contraction time), (2) the magnitude of the peak force, and (3) the time the force takes to decline to one-half of its peak value (half-relaxation time).

Force–frequency relation

Motor units are rarely activated to produce individual twitches. They usually receive as input several action potentials, resulting in overlapping twitch responses, producing a force 1.5–10.0 times greater than the twitch force. The degree to which the twitches summate depends on the rate at which the action potentials are discharged, producing the force–frequency relation (Figure 1.2.3b). With an increase in the frequency of the action potentials, the force profile (tetanus) changes from an irregular profile (unfused) (Figure 1.2.2c) to a smooth plateau (fused tetanus) (Figure 1.2.2d). The capability of the motor unit to exert force is assessed from the peak force of a single fused tetanus, not from the single twitch.

Force–length relation

The cross-bridge theory of muscle contraction suggests that muscle force is generated by cross-bridges extending from the thick filaments to the thin filaments. As the length of the muscle changes, the number of actin binding sites available for the cross-bridges changes, and this influence the force a muscle can exert. The relationship between maximum tetanic force and sarcomere length is illustrated in Figure 1.2.4.

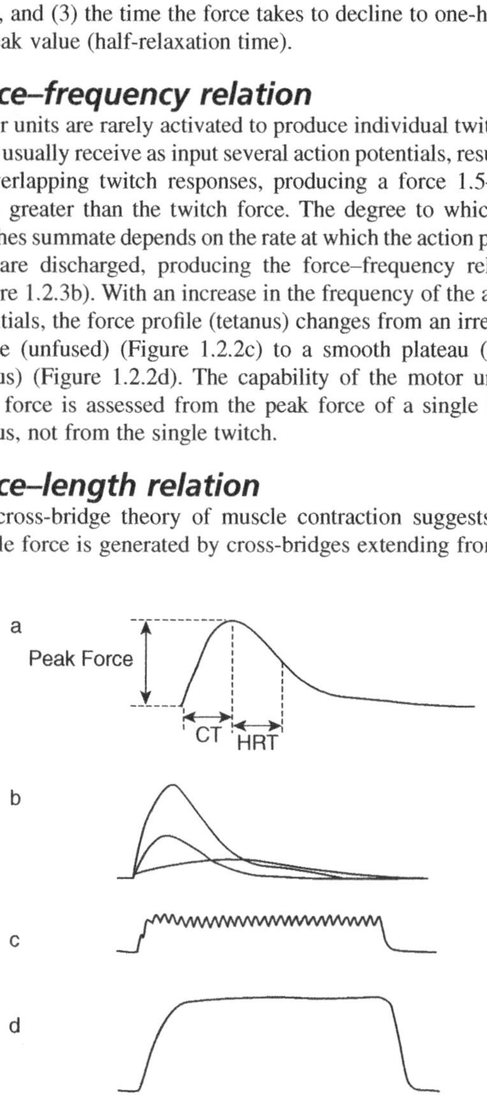

Figure 1.2.2 Twitch and tetanic responses of motor units in a cat hindlimb muscle. (a) Motor unit twitch (contraction time (CT) = 24 ms; half relaxation time (HRT) = 21 ms; peak force = 0.03 N). (b) Twitch responses for fast- and slow-twitch motor units. (c) Unfused tetanus. (d) Fused tetanus

Force–velocity relation

Hill (1938) characterized the force–velocity relationship and emphasized the importance of this parameter in the study of muscle function.

The force–velocity curve relation is reported in Figure 1.2.5. It shows a decrease of force with an increase in the speed at which the muscle shortens and an increase of force with an increase of muscle-lengthening velocity. Isometric contractions are along the zero-velocity axis of this graph and can be considered as a special case within the range of possible velocities. It should be noted that this curve represents the characteristics at a certain muscle length.

It is important to underline that the force–length and force–velocity relations describe the force exerted by a muscle at a given length or at a given constant rate of change in length. During movement these conditions are rarely satisfied and the force exerted by muscle often deviates from that expected on the basis of the described relations.

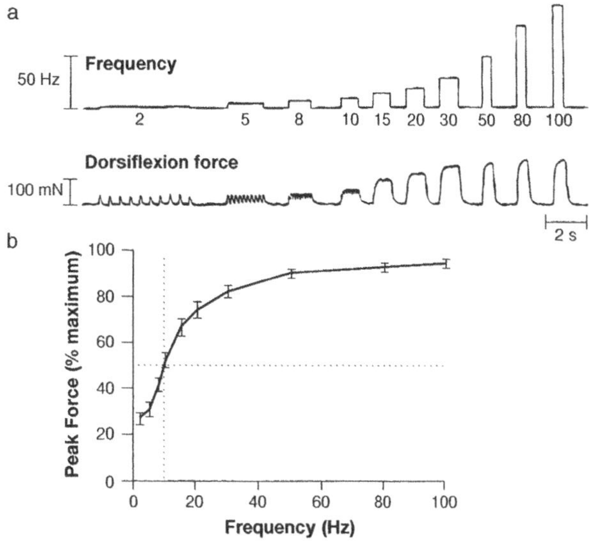

Figure 1.2.3 Force–frequency relation for motor units in human toe extensor muscles. (a) Motor units activated by intraneural stimulation (upper trace) evoke a force in the dorsiflexor muscles (lower trace). (b) Force was normalized to the maximum value for each motor unit to produce the force–frequency relation for 13 motor units. The increase in force when action potential rate goes from 5 to 10 Hz is not the same as that due to increasing the rate from 20 to 25 Hz, even though there is a difference of 5 Hz in each case. The greatest increase in force (steepest slope) occurs at intermediate discharge rates (9–12 Hz). Reproduced from Macefield, Fuglevand & Bigland-Ritchie. J Neurophysiol. 1996 Jun;75(6):2509–19.

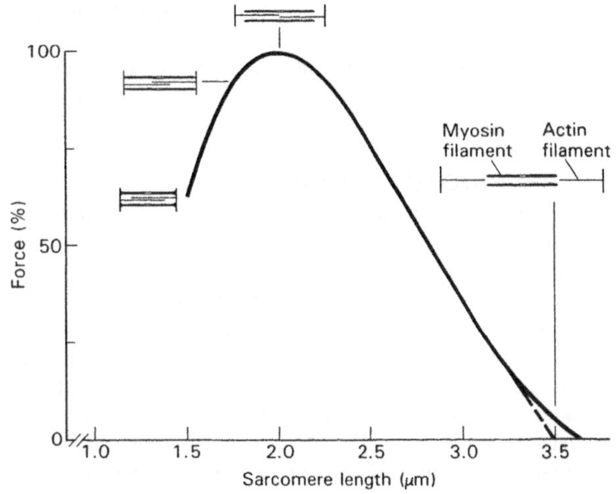

Figure 1.2.4 Variation of maximum tetanic force with sarcomer length. Insets show degree of filament overlap of four different sarcomer lengths. At resting length, about 2.5 μm, there is a maximum number of cross-bridges between the filaments, and therefore a maximum tension is possible. As the muscle lengthens, the filaments are pulled apart, the number of cross-bridges reduces, and tension decreases. At full length, about 4.0 μm, there are no cross-bridges and the tension reduces to zero. As the muscle shortens to less than resting length, there is an overlapping of the cross-bridges that results in a reduction of tension, which continues until a full overlap occurs, at about 1.5 μm. From Edman and Reggiani (1987)

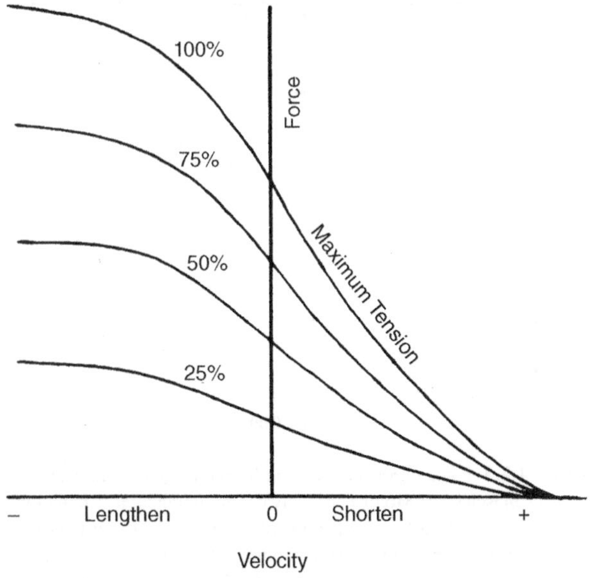

Figure 1.2.5 Force–velocity characteristics of skeletal muscle, showing a decrease of tension as muscle shortens and an increase as it lengthens. All such characteristics must be taken as the muscle shortens or lengthens at a given length, and the length must be reported. A family of curves results if different levels of muscle activation are plotted. Figure shows curves at 25, 50, 75, and 100% levels of activation

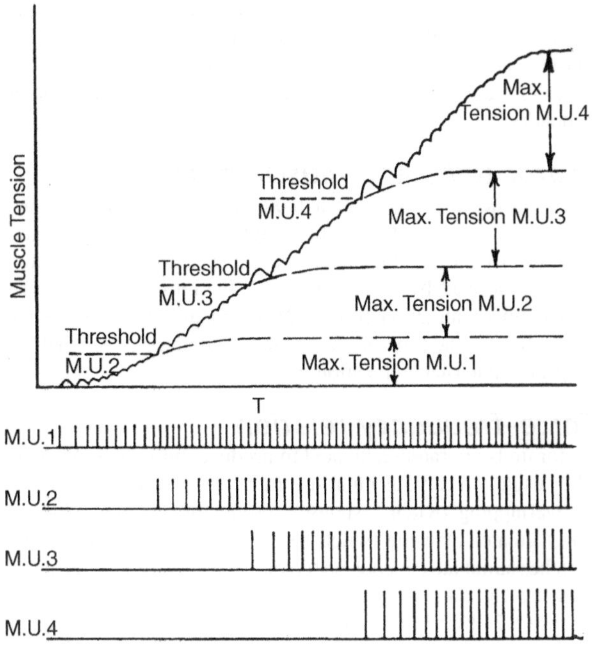

Figure 1.2.6 Example of a force curve resulting from the recruitment of motor units. The smallest motor unit (MU 1) is recruited first, usually at an initial frequency ranging from about 5 to 13 Hz. Tension increases as MU 1 fires more rapidly, until a certain tension is reached, at which MU 2 is recruited. Here MU 2 starts firing at its initial low rate and further tension is achieved by the increased firing of both MU 1 and 2. At a certain tension MU 1 will reach its maximum firing rate (15–60 Hz) and will therefore be generating its maximum tension

1.2.1.5 Muscle force modulation

Each muscle has a finite number of motor units, each of which is controlled by a separate nerve ending. Excitation of each unit is an all-or-nothing event.

As previously stated, the force that a muscle can exert is varied through: (1) the number of motor units that are active (motor unit recruitment), (2) the stimulation rate of motor units (discharge rate modulation).

Motor unit recruitment

In 1938, Denny-Brown and Pennybacker reported that a movement always appears to be accomplished by the activation of motor units in a relatively fixed sequence (orderly recruitment). To increase the force exerted by a muscle, additional motor units are activated (recruited); once a motor unit is recruited, it remains active until the force declines. To reduce the exerted force, motor units are sequentially inactivated (derecruited) in reverse order; that is, the last motor unit recruited is the first derecruited (Figure 1.2.6). The recruitment of motor units follows the size principle (Henneman *et al.*, 1974), which states that the order in which the motor neurons are activated is

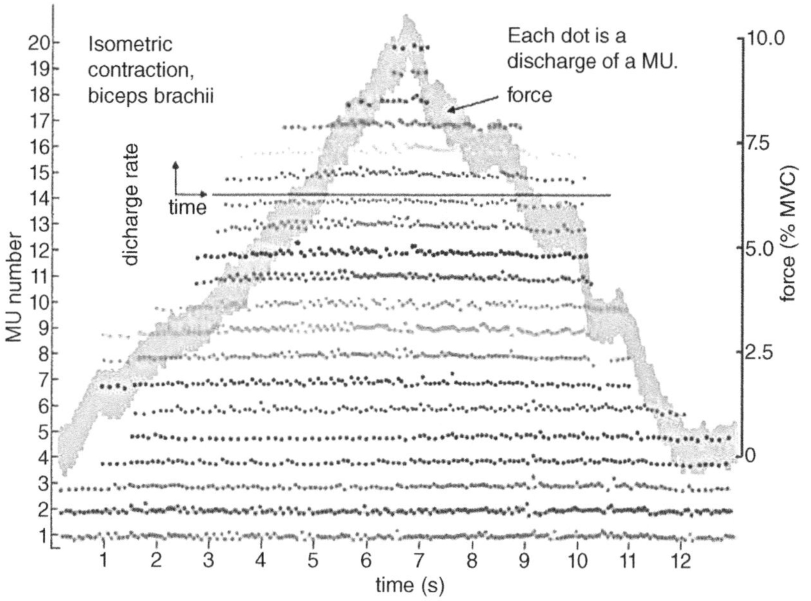

Figure 1.2.7 Recruitment and discharge pattern during an abductor digiti minimi muscle contraction in which force increases to 10% of maximum. Each dot represents the instantaneous discharge frequency of a motor unit (inverse of the interspike interval) and each horizontal line represents the discharge rate of a motor unit over time. The order of motor unit recruitment and derecruitment can be seen, associated with the produced muscle force

dictated by the motor neuron size, proceeding from smallest to largest. Figure 1.2.7 depicts an example of a force curve resulting from the recruitment of motor units. The process of increasing tension, reaching new thresholds, and recruiting another, larger motor unit continues until maximum voluntary contraction is reached. Force is reduced by the reverse process: successive reduction of firing rates and dropping out of the larger units first.

Discharge rate

When a motor unit is recruited and the force exerted by the muscle continues to increase, the rate at which the motor neuron discharges usually increases (Figure 1.2.8).

The minimum rate at which motor neurons discharge action potentials repetitively during voluntary contractions is about 5–7 Hz. Maximum discharge rates vary across muscles (35–40 Hz for first dorsal interosseus, 25–35 Hz for adductor digiti minimi and adductor pollicis, 20–25 Hz for biceps brachii and extensor digitorum communis, 11 Hz for soleus).

The contribution of motor unit recruitment to muscle force varies between muscles. In some muscles, all motor units are recruited when the force reaches about 50% of maximum, in other muscles recruitment continues up to 85% of the maximum force (De Luca *et al.*, 1982; Kukulka and Clamann, 1981; Van Cutsem *et al.*, 1997). The increase in muscle force beyond the upper limit of motor unit recruitment is accomplished entirely through variation in the discharge rate of action potentials.

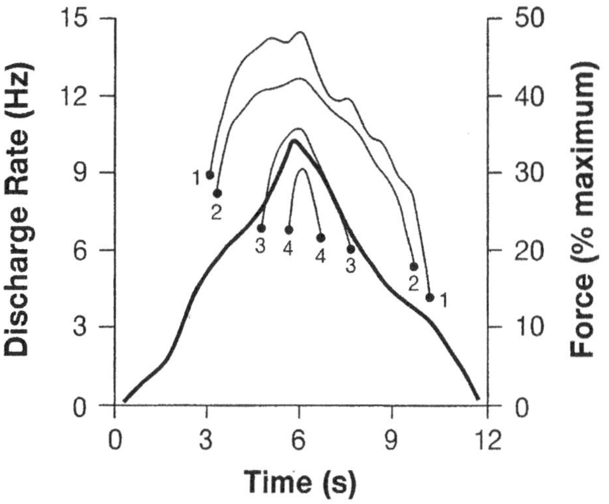

Figure 1.2.8 Modulation of discharge rate by four motor units during a gradual increase and then decrease in the force (thick line) exerted by the knee extensor muscles. Each thin line represents the activity of a single motor unit, with recruitment occurring at the leftmost dot on each thin line. MU 1, for example, was recruited at a force of about 18% of maximum with an initial discharge rate of 9 Hz, which increased to 15 Hz at the peak force (Person and Kudina, 1972)

Experimental examples of abnormal motor unit recruitment

Although the orderly recruitment of motor units has been verified in many studies, it can vary under some conditions.

At the motor unit level, changes in recruitment order have been observed as a consequence of manipulation of sensory feedback (Garnett and Stephens, 1981; Kanda, Burke and Walmsley, 1977) and during performance of lengthening contractions (Nardone, Romanò and Schieppati, 1989). One way to demonstrate this effect is to determine the recruitment order of pairs of motor units in the absence and presence of a perceptible cutaneous sensation elicited by electrical stimulation of the skin. In the absence of this stimulus, one motor unit is recruited at a lower force than the other. In the presence of the cutaneous sensation, however, the unit with the higher recruitment threshold is activated first. This reversal of recruitment order is useful if the cutaneous sensation requires a rapid response.

At the whole-muscle level one example of an alteration in recruitment order is the paw-shake response. This behaviour is elicited in animals when a piece of tape sticks to a paw; the animal's response is to vigorously shake the paw in an attempt to remove the tape. In normal activities, such as standing, walking, and running, the force exerted by soleus (slow-twitch muscle) remains relatively constant across tasks, while that exerted by medial gastrocnemius (fast-twitch muscle) increases with the power demand of the task. In the paw-shake response fast-twitch muscles are mainly activated. The recruitment change in a group of synergist muscles allows the animal to move its limb rapidly.

Westad, Westgaard and De Luca (2003) observed abnormal motor unit recruitment in human trapezius; low-threshold motor units recruited at the start of the contraction were observed to stop firing while motor units of higher recruitment threshold stayed active. Casale *et al.* (2009) observed that recruitment during voluntary contractions was altered in fibromyalgic patients. Experimental findings suggest that fibromyalgic patients use compensatory motor strategies to compensate chronic fatigue, recruiting motor units in a different manner than that predicted by Henneman's principle.

Discharge patterns

In addition to motor unit recruitment and discharge rate, the force exerted by a muscle is influenced by the discharge pattern of motor units. The discharge patterns known as 'common drive' and 'motor unit synchronization' refer to the temporal relation of the action potentials firing among different motor units. Common drive describes the correlated variation in the average discharge rate of concurrently active motor. Motor unit synchronization quantifies the amount of correlation between the timing of the individual action potentials discharged by active motor units.

1.2.1.6 Electrically elicited contractions

Muscle contraction can be inducted by selective electrical stimulation (NMES) of a nerve branch or of a motor point of a muscle; this allows 'disconnection' of the investigated portion of muscle from the CNS and control of motor unit activation frequency. By changing the intensity of the electrical stimulation, it is possible to change the number of activated motor units. The effectiveness of NMES was proved to be greater in unhealthy subjects (for instance, during and post immobilization) than in normal people, for whom voluntary exercise was found more favourable (Bax, States and Verhagen, 2005; Delitto and Snyder-Mackler, 1990).

In contrast to the order in which motor units are activated during voluntary contractions, the classic view is that large-diameter axons are more easily excited by imposed electric fields, which would reverse the activation order of motor units in electrically evoked contractions compared with voluntary contractions. However, some studies demonstrate that neuromuscular electrical stimulation tends to activate motor units in a similar order to that which occurs during voluntary contractions, with more variability with respect to voluntary contractions.

Recent works (Collins, 2007) demonstrate that the electrically elicited contractions do not only act on the peripheral system (that is, behind the motor junction), but that the simultaneous depolarization of sensory axons can also contribute to the contractions by the synaptic recruitment of spinal motoneurons.

1.2.2 MUSCLE FATIGUE

Fatigue is a daily experience corresponding to lack of maintenance of a physical activity. This is actually the definition of 'mechanical fatigue', which refers intuitively to output (force, torque, motor task) reduction. This visible impairment is induced by a physiological process, which can imply different sites and contributions to the fatigue; they can be schematically listed as follows:

- Central fatigue (of the primary motor cortex and of central drive to motor neurons).
- Fatigue of the neuromuscular junction (in the activation of muscle and motor units and in neuromuscular propagation).
- Peripheral fatigue (of the excitation–contraction coupling, of the availability of metabolic substrates, of muscle blood flow).

1.2.2.1 Central and peripheral fatigue

The mechanical output production, as described above, can decrease during exertion for a number of reasons. Twenty years ago a technique was devised to highlight the amount of central drive reduction (Hales and Gandevia, 1988); the issue of spinal and supraspinal factors in human muscle fatigue was clearly described in a recent extended review (Gandevia, 2001). The technique consisted in the so-called 'twitch interpolation': supramaximal electric shocks are applied to the nerve during an MVC task. If such an electrical stimulation increases the force output, the central drive during the contraction is not maximal and the subject is showing central fatigue. In such a way it is also possible to directly assess the ability of a subject to provide true MVC. Moreover, it is possible to monitor the time course of other physical variables such as force or torque, power, angular velocity of a joint, motor unit firing rate, and motor unit synchronization.

Prolonged activity can also impair neuromuscular propagation, decreasing force output due to a failure of axonal action potential, a failure of the excitation–secretion coupling, a depletion of the neurotransmitter, or a decrease of the sensitivity of postsynaptic receptors (Spira, Yarom and Parnas, 1976).

Peripheral fatigue is defined as the changes occurring within the muscle, behind the neuromuscular junction. Thus it is possible to study the electromyographic fatigue; that is, all the changes occurring in the EMG signal during a sustained voluntary contraction or an electrically elicited contraction (M wave). Starting from such signals, it is possible to monitor the variation in time of the conduction velocity of muscle fibre action potentials (or of the M wave), which represents the most physiological parameter actually available from the EMG signal (Farina *et al.*, 2004). Innovative techniques (see below) allow tracking of single motor units, enabling us to obtain the conduction velocity distribution of the pool of active and recordable motor units and to visualize bidimensional maps of electrical activity over the skin.

Further fatigue within the muscle is also related to variation in metabolic substrates. This means a number of changes that can occur in active muscle fibres: depletion of glycogen (Hermansen, Hultman and Saltin, 1967), increase of H^+ and P_i concentrations (Vøllestad, 1995), reduction of intracellular pH (Brody *et al.*, 1991), and reduction of oxygen supply due to the increase of intramuscular pressure (Sjøgaard, Savard and Juel, 1988).

1.2.2.2 The role of oxygen availability in fatigue development

As described above, the amount of oxygen supply is one of the pivotal factor affecting contraction and causing fatigue. Casale *et al.* (2004) concluded that acute exposure to hypobaric hypoxia did not significantly affect muscle fibre membrane properties (no peripheral effect) but that it did impact on motor unit control properties (central control strategies for adaptation). Hence the lack of oxygen induced by high altitude was able to induce central effects aimed to recruiting more fast motor units than the equivalent effects at sea level, allowing maintenance of the requested force task. Thus it is possible to assess whether variation in EMG manifestations of fatigue are induced by a central or a peripheral adaptation by means of two different contraction modalities (voluntary or electrically induced).

A specific protocol was recently designed to highlight the effect of oxygen availability in endurance- and power-trained athletes. As depicted in Figure 1.2.9, fatigue index increased

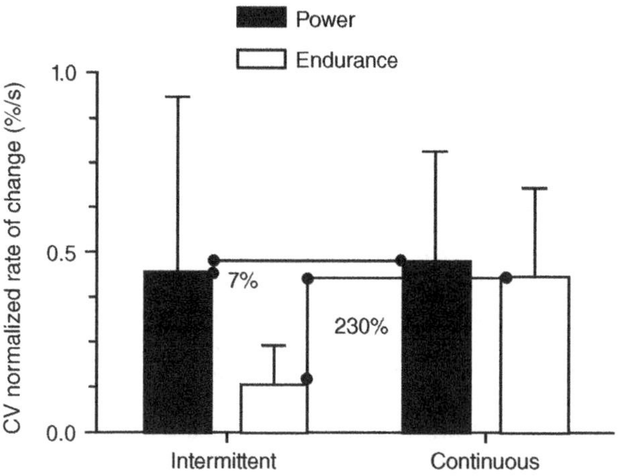

Figure 1.2.9 Normalized (with respect to initial values) rates of change of CV in continuous and intermittent contractions for both groups of athletes. The increment of fatigue in the continuous with respect to the intermittent contraction was greater in the endurance than in the power athletes. Mean ± SD absolute values are reported

by 200% when endurance athletes were asked to move from intermittent (characterized by 1 s of rest in between) to continuous contraction (of the same workload), whereas power-trained athletes did not modify their fatigue rate (Rainoldi *et al.*, 2008). That approach is now proposed as a noninvasive technique to distinguish between two extremely different phenotypes.

1.2.3 MUSCLE FUNCTION ASSESSMENT

1.2.3.1 Imaging techniques

The advent of modern imaging techniques offers new approaches for monitoring musculoskeletal function and structure. The most widely used techniques are T1- and T2-weighted magnetic resonance (MR) imaging, cine phase-contrast MR imaging, MR elastography, and ultrasonography (Segal, 2007).

T1- and T2-weighted MR imaging is useful for determining muscle cross-sectional area noninvasively and *in vivo*, allowing monitoring adaptations in muscle cross-sectional area consequent to disease, immobilization, rehabilitation, or exercise.

Muscle MR imaging has been used to analyse which muscles or regions of muscles are active during functional tasks, but low-level muscle activity is detected by EMG before it can be detected by MR imaging (Cheng *et al.*, 1995).

T2 time mapping showed the capability, under some circumstances, to evaluate the spatial localization of activity within portions of muscles, allowing the study of neuromuscular compartments (Akima *et al.*, 2000; Segal and Song, 2005).

One potentially important use for MR imaging is the monitoring of the adaptation of muscle activation in response to fatigue, CNS injuries, diseases, or therapies (Akima *et al.*, 2002; Pearson, Fouad and Misiaszek, 1999).

Cine phase-contrast MR imaging has been proposed to visualize the differential displacement of muscle fibres within a muscle during movement (Pappas *et al.*, 2002; Pelc *et al.*, 1994) and to provide muscle architectural information (Finni *et al.*, 2003) (Figures 1.2.10 and 1.2.11).

MR elastography is a method that allows the estimation of the mechanical properties of tissues during both passive and active movements (Basford *et al.*, 2002; Heers *et al.*, 2003). A high correlation between MR elastography wavelengths and electromyographic activity has been found (Heers *et al.*, 2003), suggesting that MR elastography is a valid approach for the noninvasive assessment of muscle properties in the context of muscle activity.

Ultrasonography has been used to examine musculotendinous structure and movement (Fukashiro *et al.*, 1995; Kawakami, Ichinose and Fukunaga, 1998). This approach does not require a special room, as MR imaging does; it requires only an ultrasonography transducer, sophisticated analysis software, and, most importantly, skilful application. The technique is completely noninvasive, is carried out *in vivo*, and is essentially real-time.

1.2.3.2 Surface EMG as a muscle imaging tool

Surface EMG (sEMG) has been considered the gold standard for the noninvasive measurement of muscle activity in human subjects and provides a standard method for monitoring temporal information about muscle activity. In the last 10 years, research activity has been moved toward sEMG recording with two-dimensional array systems, the so-called high-density sEMG. This technique, by using multiple closely spaced electrodes overlying an area of the skin, provides not only temporal information, but also spatial distribution of the electrical activity over the muscle. It thus opens new possibilities for studying muscle characteristics, transforming the amount and quality of information actually available. High-density sEMG allows the

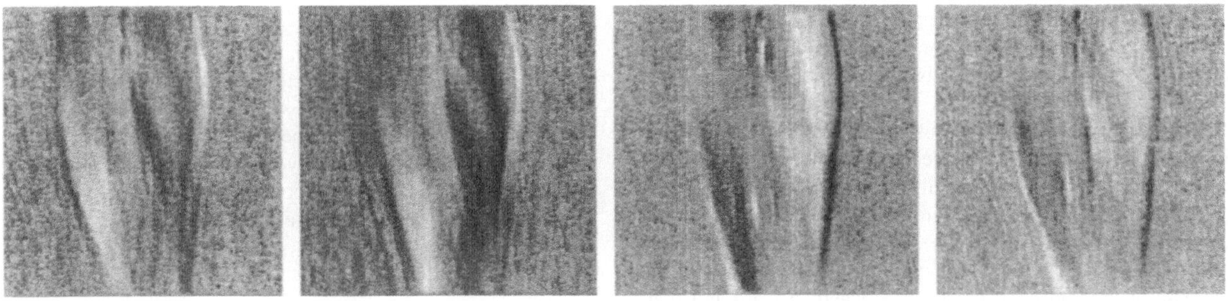

Figure 1.2.10 Examples of the cine phase contrast images obtained using the velocity-encoded MRI pulse sequence. The optical density of each pixel is proportional to the velocity of the volume of tissue represented by the pixel. Each image is the same sagittal section of the leg of a normal subject during an isometric plantarflexion action. The left side of the image is anterior and the right side is posterior. The velocities and direction of the movement of a given point are reflected in the optical densities, with the lower scale of optical densities representing movement in one direction and the upper scale of optical densities representing movements in the opposite direction. Note the difference in velocities within and between the muscles of the anterior and posterior compartments (Lai, Sinha, Hodgson and Edgerrton, unpublished observations)

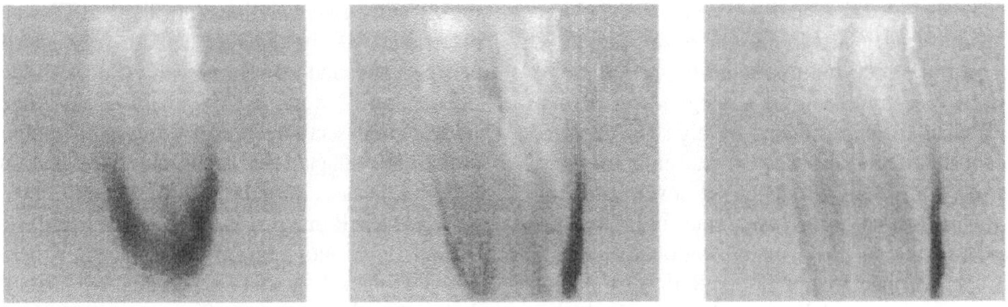

Figure 1.2.11 Examples of the cine phase contrast images using the velocity-encoded MRI pulse sequence at different sagittal planes of the leg of a normal subject during an isometric action. Left side of the image is anterior, right side is posterior (Lai, Sinha, Hodgson and Edgerrton, unpublished observations)

estimation of muscle anatomical characteristics such as muscle fibre length, muscle fibre orientation, and localization of the innervation zones (Farina and Merletti, 2004; Lapatki *et al.*, 2006). Moreover, it allows the estimation of motor unit location (Drost *et al.*, 2007; Roeleveld and Stegeman, 2002), decomposition of the sEMG interference signal into the constituent trains of motor unit action potentials (MUAPs), and analysis of single motor unit properties (De Luca *et al.*, 2006; Holobar and Zazula, 2004; Kleine *et al.*, 2007). The information that can be extracted from high-density sEMG signals has been shown to be relevant for both physiological studies and clinical applications (Figures 1.2.12 and 1.2.13).

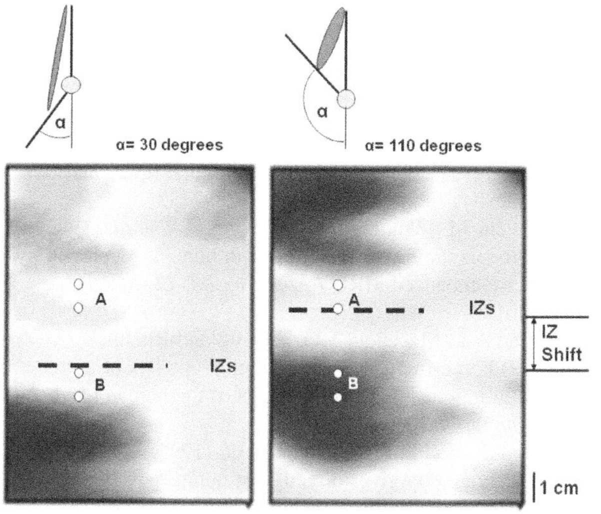

Figure 1.2.12 Change of activity distribution (interpolated RMS map) over the biceps brachii during a slow flexion of the elbow. As the elbow flexes by 80° the innervation zone (IZ) moves in the proximal direction by 2–3 cm. It is evident that the single electrode pair at A would indicate a decrease of muscle activity while an electrode pair placed in B would indicate an increase. This demonstrates the need for 2D electrode arrays

1.2.3.3 Surface EMG for noninvasive neuromuscular assessment

A huge effort was made over the last 20 years to assess the repeatability and affordability of the sEMG technique based on the differential montage (with a couple or an array of electrodes). A number of clear findings are now available in different fields. For an in-depth description we refer the reader to Merletti and Parker (2004) and to the following reviews: Drost *et al.* (2006), Farina *et al.* (2004), Merletti, Rainoldi and Farina (2001), Merletti, Farina and Gazzoni (2003), Mesin, Merletti and Rainoldi (2009), Roeleveld and Stegeman (2002), Staudenmann *et al.* (2010), Stegeman *et al.* (2000), Zwarts and Stegeman (2003), Zwarts, Drost and Stegeman (2000) and Zwarts *et al.* (2004), among others, which cover different applications and topics.

It is possible to conclude that if the maximum care were applied in order to reduce the number of confounding factors then *any* change in the EMG signal would contribute to a wider definition of electromyographic fatigue. Hence we conclude that the time is right to accept the use of such an approach in recording the complexity of the human system. To study human movement means studying the neuromuscular system from a number of different points of view. The electric signal generated during a contraction is now ready to be disentangled.

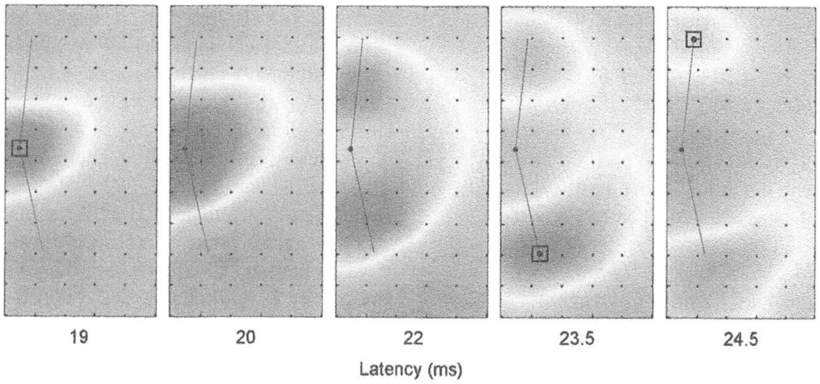

Figure 1.2.13 Example of the use of high-density SEMG for the determination of endplate position and muscle-fibre orientation. A sequence of interpolated monopolar amplitude maps illustrates topographically the initiation of the potential (latencies 19 and 20) and its conduction in the upper and lower part of the DAO muscle. Endplate position and muscle-fibre orientation were determined by localizing the grid areas of maximal amplitude (area size was 1/2 IED in both dimensions) in the interpolated monopolar amplitude maps at the latencies of MUAP initiation or termination. Reproduced from Lapatki *et al.*, J Neurophysiol. 2006 Jan;95(1):342–54.

References

Akima H., Foley J.M., Prior B.M. *et al.* (2002) Vastus lateralis fatigue alters recruitment of musculus quadriceps femoris in humans. *J Appl Physiol*, **92**, 679–684.

Akima H., Ito M., Yoshikawa H. and Fukunaga T. (2000) Recruitment plasticity of neuromuscular compartments in exercised tibialis anterior using echo-planar magnetic resonance imaging in humans. *Neurosci Lett*, **296**, 133–136.

Basford J.R., Jenkyn T.R., An K.N. *et al.* (2002) Evaluation of healthy and diseased muscle with magnetic resonance elastography. *Arch Phys Med Rehabil*, **83**, 1530–1536.

Bax L., States F. and Verhagen A. (2005) Does neuromuscular electrical stimulation strengthen the quadriceps femoris? A systematic review of randomized control trials. *Sport Med*, **35**, 191–212.

Bell J., Bolanowsky S. and Holmes M.H. (1994) The structure and function of Pacinian corpuscles: a review. *Prog Neurobiol*, **42**, 79–128.

Brody L.R., Pollock M.T., Roy S.H. *et al.* (1991) pH-induced effects on median frequency and conduction velocity of the myoelectric signal. *J Appl Physiol*, **71** (5), 1878–1885.

Casale R., Farina D., Merletti R. and Rainoldi A. (2004) Myoelectric manifestations of fatigue during exposure to hypobaric hypoxia for 12 days. *Muscle Nerve*, **30** (5), 618–625.

Casale R., Sarzi-Puttini P., Atzeni F. *et al.* (2009) Central motor control failure in fibromyalgia: a surface electromyography study. *BMC Musculoskelet Disord*, **1**, 10–78.

Cheng H.A., Robergs R.A., Letellier J.P. *et al.* (1995) Changes in muscle proton transverse relaxation times and acidosis during exercise and recovery. *J Appl Physiol*, **79**, 1370–1378.

Collins D.F. (2007) Central contributions to contractions evoked by tetanic neuromuscular electrical stimulation. *Exerc Sport Sci Rev*, **35** (3), 102–109.

De Luca C.J., Adam A., Wotiz R. *et al.* (2006) Decomposition of surface EMG signals. *J Neurophysiol*, **96** (3), 1646–1657.

De Luca C.J., LeFever R.S., McCue M.P. and Xenakis A.P. (1982) Control scheme governing concurrently active human motor units during voluntary contractions. *J Physiol*, **329**, 129–142.

Delitto A. and Snyder-Mackler L. (1990) Two theories of muscle strength augmentation using percutaneous electrical stimulation. *Phys Ther*, **70** (3), 158–164.

Drost G., Kleine B.U., Stegeman D.F. *et al.* (2007) Fasciculation potentials in high-density surface EMG. *J Clin Neurophysiol*, **24** (3), 301–307.

Drost G., Stegeman D.F., van Engelen B.G. and Zwarts M.J. (2006) Clinical applications of high-density surface EMG: a systematic review. *J Electromyogr Kinesiol*, **16** (6), 586–602. Edman K.A. and Reggiani C. (1987) The sarcomere length–tension relation determined in short segments of intact muscle fibres of the frog. *J Physiol*, **385**, 709–732.

English A.W. (1984) An electromyographic analysis of compartments in cat lateral gastrocnemius muscle during unrestrained locomotion. *J Neurophysiol*, **52** (1), 114–125.

Farina D. and Merletti R. (2004) Methods for estimating muscle fibre conduction velocity from surface electromyographic signals. *Med Biol Eng Comput*, **42** (4), 432–445.

Farina D., Merletti R. and Enoka R.M. (2004) The extraction of neural strategies from the surface EMG. *J Appl Physiol*, **96** (4), 1486–1495.

Finni T., Hodgson J.A., Lai A.M. *et al.* (2003) Mapping of movement in the isometrically contracting human soleus muscle reveals details of its structural and functional complexity. *J Appl Physiol*, **95**, 2128–2133.

Fleckenstein J.L., Watumull D., Bertocci L.A. *et al.* (1992) Finger-specific flexor recruitment in humans: depiction by exercise-enhanced MRI. *J Appl Physiol*, **72** (5), 1974–1977.

Fuglevand A.J., Macefield V.G. and Bigland-Ritchie B. (1999) Force-frequency and fatigue properties of motor units in muscles that control digits of the human hand. *J Neurophysiol*, **81** (4), 1718–1729.

Fukashiro S., Itoh M., Ichinose Y. *et al.* (1995) Ultrasonography gives directly but noninvasively elastic characteristic of human tendon in vivo. *Eur J Appl Physiol Occup Physiol*, **71**, 555–557.

Gandevia S.C. (2001) Spinal and supraspinal factors in human muscle fatigue. *Physiol Rev*, **81** (4), 1725–1789.

Garnett R. and Stephens J.A. (1981) Changes in the recruitment threshold of motor units produced by cutaneous stimulation in man. *J Physiol*, **311**, 463–473.

Hales J.P. and Gandevia S.C. (1988) Assessment of maximal voluntary contraction with twitch interpolation: an instrument to measure twitch responses. *J Neurosci Methods*, **25** (2), 97–102.

Harridge S.D., Bottinelli R., Canepari M. *et al.* (1996) Whole-muscle and single-fibre contractile properties and myosin heavy chain isoforms in humans. *Pflugers Arch*, **432** (5), 913–920.

Heers G., Jenkyn T., Dresner M.A. *et al.* (2003) Measurement of muscle activity with magnetic resonance elastography. *Clin Biomech*, **18**, 537–542.

Henneman E., Clamann P.H., Gillies J.D. and Skinner R.D. (1974) Rank order of motoneurons within a pool: law of combination. *J Neurophysiol*, **37**, 1338–1349.

Hermansen L., Hultman E. and Saltin B. (1967) Muscle glycogen during prolonged severe exercise. *Acta Physiol Scand*, **71** (2), 129–139.

Hill A.V. (1938) The heat of shortening and the dynamic constants of muscle. *Proc R Soc Lond B Biol Sci*, **126**, 136–195.

Holobar A. and Zazula D. (2004) Correlation-based decomposition of surface electromyograms at low contraction forces. *Med Biol Eng Comput*, **42** (4), 487–495.

Johannsson H., Sjölander P. and Sojka P. (1991) Receptors in the knee joint ligaments and their role in the biomechanics of the joint. *CRC Crit Rev Biomed Eng*, **18**, 341–368.

Kanda K., Burke R.E. and Walmsley B. (1977) Differential control of fast and slow twitch motor units in the decerebrate cat. *Exp Brain Res*, **29** (1), 57–74.

Kawakami Y., Ichinose Y. and Fukunaga T. (1998) Architectural and functional features of human triceps surae muscles during contraction. *J Appl Physiol*, **85**, 398–404.

Kleine B.U., van Dijk J.P., Lapatki B.G. *et al.* (2007) Using two-dimensional spatial information in decomposition of surface EMG signals. *J Electromyogr Kinesiol*, **17** (5), 535–548.

Kukulka C.G. and Clamann H.P. (1981) Comparison of the recruitment and discharge properties of motor units in human brachial biceps and adductor pollicis during isometric contractions. *Brain Res*, **24** (219, 1):45–55.

Lapatki B.G., Oostenveld R., Van Dijk J.P. *et al.* (2006) Topographical characteristics of motor units of the lower facial musculature revealed by means of high-density surface EMG. *J Neurophysiol*, **95** (1), 342–354.

Lieber R.L. and Fridén J. (2000) Functional and clinical significance of skeletal muscle architecture. *Muscle Nerve*, **23** (11), 1647–1666.

Merletti R., Farina D. and Gazzoni M. (2003) The linear electrode array: a useful tool with many applications. *J Electromyogr Kinesiol*, **13** (1), 37–47.

Merletti R. and Parker P. (2004) *Electromyography: Physiology, Engineering and Noninvasive Applications*, John Wiley and Sons/IEEE Press Publication, USA.

Merletti R., Rainoldi A. and Farina D. (2001) Surface electromyography for noninvasive characterization of muscle. *Exerc Sport Sci Rev*, **29** (1), 20–25.

Mesin L., Merletti R. and Rainoldi A. (2009) Surface EMG: the issue of electrode location. *J Electromyogr Kinesiol*, **19** (5), 719–726.

Miller A.E., MacDougall J.D., Tarnopolsky M.A. and Sale D.G. (1993) Gender differences in strength and muscle fiber characteristics. *Eur J Appl Physiol Occup Physiol*, **66** (3), 254–262.

Nardone A., Romanò C. and Schieppati M. (1989) Selective recruitment of high-threshold human motor units during voluntary isotonic lengthening of active muscles. *J Physiol*, **409**, 451–471.

Pappas G.P., Asakawa D.S., Delp S.L. *et al.* (2002) Nonuniform shortening in the biceps brachii during elbow flexion. *J Appl Physiol*, **92**, 2381–2389.

Pearson K.G., Fouad K. and Misiaszek J.E. (1999) Adaptive changes in motor activity associated with functional recovery following muscle denervation in walking cats. *J Neurophysiol*, **82**, 370–381.

Pelc N.J., Sommer F.G., Li K.C. *et al.* (1994) Quantitative magnetic resonance flow imaging. *Magn Reson Q*, **10**, 125–147.

Person R.S. and Kudina L.P. (1972) Discharge frequency and discharge pattern of human motor units during voluntary contraction of muscle. *Electroencephalogr Clin Neurophysiol*, **32** (5), 471–483.

Proske U. (1997) The mammalian muscle spindle. *News Physiol Sci*, **12**, 37–42.

Rainoldi A., Gazzoni M., Gollin M. and Minetto M.A. (2008) Neuromuscular responses to continuous and intermittent voluntary contractions. XVII ISEK Congress, Niagara Falls, Ontario, Canada, June 18–21.

Roeleveld K. and Stegeman D.F. (2002) What do we learn from motor unit action potentials in surface electromyography? *Muscle Nerve*, **25** (Suppl. 11), S92–S97.

Russell B., Motlagh D. and Ashley W.W. (2000) Form follows function: how muscle shape is regulated by work. *J Appl Physiol*, **88** (3), 1127–1132.

Segal R.L. (2007) Use of imaging to assess normal and adaptive muscle function. *Phys Ther*, **87** (6), 704–718.

Segal R.L. and Song A.W. (2005) Nonuniform activity of human calf muscles during an exercise task. *Arch Phys Med Rehabil*, **86**, 2013–2017.

Sjøgaard G., Savard G. and Juel C. (1988) Muscle blood flow during isometric activity and its relation to muscle fatigue. *Eur J Appl Physiol Occup Physiol*, **57** (3), 327–335.

Spira M.E., Yarom Y. and Parnas I. (1976) Modulation of spike frequency by regions of special axonal geometry and by synaptic inputs. *J Neurophysiol*, **39** (4), 882–899.

Staudenmann D., Roeleveld K., Stegeman D.F. and van Dieën J.H. (2010) Methodological aspects of SEMG recordings for force estimation: a tutorial and review. *J Electromyogr Kinesiol*, **20** (3), 375–387.

Stegeman D.F., Zwarts M.J., Anders C. and Hashimoto T. (2000) Multi-channel surface EMG in clinical neurophysiology. *Suppl Clin Neurophysiol*, **53**, 155–162.

Van Cutsem M., Feiereisen P., Duchateau J. and Hainaut K. (1997) Mechanical properties and behaviour of motor units in the tibialis anterior during voluntary contractions. *Can J Appl Physiol*, **22** (6), 585–597.

Vøllestad N.K. (1995) Metabolic correlates of fatigue from different types of exercise in man. *Adv Exp Med Biol*, **384**, 185–194.

Westad C., Westgaard R.H. and De Luca C.J. (2003) Motor unit recruitment and derecruitment induced by brief increase in contraction amplitude of the human trapezius muscle. *J Physiol*, **552**, 645–656.

Zwarts M.J., Drost G. and Stegeman D.F. (2000) Recent progress in the diagnostic use of surface EMG for neurological diseases. *J Electromyogr Kinesiol*, **10** (5), 287–291.

Zwarts M.J., Lapatki B.G., Kleine B.U. and Stegeman D.F. (2004) Surface EMG: how far can you go? *Suppl Clin Neurophysiol*, **57**, 111–119.

Zwarts M.J. and Stegeman D.F. (2003) Multichannel surface EMG: basic aspects and clinical utility. *Muscle Nerve*, **28** (1), 1–17.

1.3 Bone Physiology

Jörn Rittweger, Institute for Biomedical Research into Human Movement and Health, Manchester Metropolitan University, Manchester, UK and Institute of Aerospace Medicine, German Aerospace Center (DLR), Cologne, Germany

1.3.1 INTRODUCTION

Bones are living organs. Yet they have traditionally been regarded as a matter of anatomical study rather than being of interest to the physiologist. Access to the cells residing in their hard tissue is inherently difficult to obtain and recent research suggests that their cells originate from stem cells located far from their final location. Therefore, despite the fact that the outlines of our bones are conserved long after death, bone physiology remains an elusive process, and its study requires interdisciplinary approaches, plus a good degree of 'indirect' thinking.

This chapter begins with an overview of bone anatomy (=playground), going on to discuss the different bone cells (=players), then a brief introduction to bone mechanics is given (=scope of the game) and finally current concepts of bones adaption are discussed (=tactics of the game).

1.3.2 BONE ANATOMY

The principle role of bones is to provide mechanical rigidity. The skeleton provides the muscles with a framework to act upon. Bones also protect soft tissue organs (e.g. brain, heart). In itself, the term 'bone' is ambiguous, as it describes an organ (e.g. the femur or the mandible), a tissue (e.g. compact or trabecular bone), and a material.

1.3.2.1 Bones as organs

Bones first evolved as cartilaginous organs in fish. The conquest of land stipulated larger musculoskeletal forces and thus the evolution of bones as more rigid instruments. Accordingly, the cartilaginous fish bones have been replaced by 'proper' bone tissue in land-living vertebrates. Interestingly, ontogeny reenacts the phylogenetic replacement of cartilaginous tissue, as bones developmentally emerge from mesenchyme or cartilage by intramembranous and endochondral ossification, respectively. Such transformations of other tissues into bone do not normally take place later in life, but there are some exceptions to this rule. In *myositis ossificans*, for example, bone tissue emerges within skeletal muscle, often after trauma (Geschickter and Maseritz, 1938). In *fibrodysplasia ossificans progressiva*, a rare autosomal dominant disorder with impaired BMP-4 signalling, virtually all soft tissues are turned into bone (Kaplan *et al.*, 2006; Shore *et al.*, 2006). Finally, and much more commonly, osteophytes (or bone spurs) in osteoarthritis develop from fibrocartilage (Gilbertson, 1975). These examples may demonstrate that the capability of ossification is retained throughout life, but that the signals to induce this process are normally lacking.

Most bone surfaces are covered with periosteum, an epithelium containing blood vessels and nerve fibres. Among the nerve fibres, there are many nociceptors, which mediate pain perception in fractures and contusions.

There are two different kinds of mechanical interface to bones. First, joint surfaces are covered by hyaline cartilage with low frictional coefficient, typically around 0.001 (Özkaya and Nordin, 1998). Joints can thus transmit only compressive forces. Second, entheses constitute contact zones with tendons, ligaments or skeletal muscle and can be either fibrous or fibrocartilaginous. They convey tensile and shear forces. Their projections into the bone are known as Sharpey's fibres.

All bones in land-living vertebrates have a comparatively smooth outer surface. This layer is called the cortex (Latin for 'bark'). Bone cortices can be wafer-thin in some places, but examples up to 30 cm thick have been found in some extinct sauropods. With the exception of air-filled wing bones in some birds, the interior of all bones is filled up with marrow. This can be either fatty or red (= haematopoietic). Marrow fat is probably of some importance, as it is spared during starvation. However, its precise physiological role is unknown (Currey, 2003).

Bones have different shapes. They can be long (e.g. femur), short (e.g. calcaneus), flat bones (e.g. sternum), irregular (e.g. vertebra) or sesamoid[1] (e.g. patella). Within the long bones,

[1]A sesmoid bone is one that is engulfed by a tendon and is primarily loaded in tension. It serves to reduce shear strains where the line of action of a force is diverted.

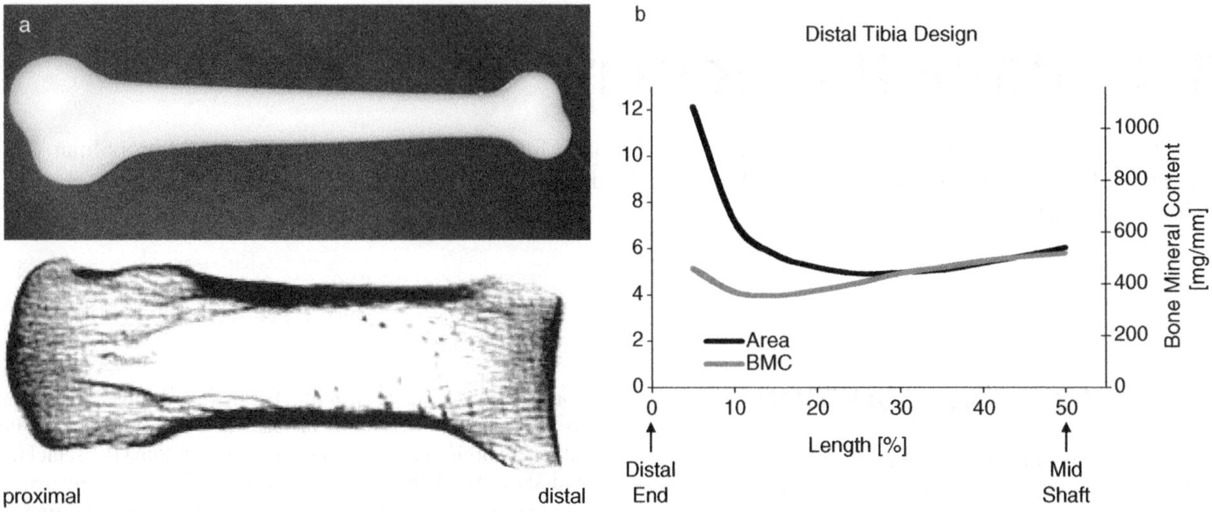

Figure 1.3.1 Principle design of long bones. (a) Replica of a 'typical' bone, utilized for nonscientific purposes in the street carnival of Cologne, Germany, in 2009. Anatomically, no such bone exists. Nevertheless, this mock-up is recognized as a bone even under conditions of impaired brain function (carnival). Therefore, it seems to be exemplary for the principle design of long bones. (b) Total bone area and bone mineral content (BMC), as assessed in the distal tibia of 10 young healthy males by peripheral quantitative computed tomography (pQCT). As can be seen, there is a disproportionate increase in total bone area towards the distal end of the tibia, without an increase in bone mineral content. This is because of the trabacularization towards the end, which helps to spread the forces over a larger surface. Unpublished data by Rittweger *et al.* (c) pQCT scan of a human metatarsal. Fine struts of trabecular bone branch of the cortex in order to support the joint surface. The presence of trabeculae explains how total bone area in Figure 1.3.1b can be increased without increases in bone area

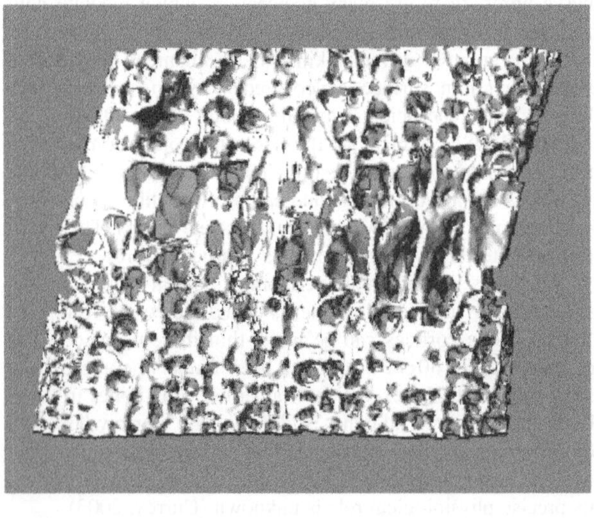

Figure 1.3.2 Trabecular bone of a vertebral body in a rat, assessed by micro-CT. Whilst rods predominate close to the endplates (top and bottom), mainly plates are found in the central part. Image courtesy of Jürg Gasser

there are three different zones. The zone beyond the growth plate (=physis) is called the epiphysis, the zone next to it the metaphysis, and the central part of the shaft is the diaphysis.

The principle design of long bones is so typical that it is recognized by children. To understand this design, we have to consider the fact that joint cartilage is a substantially weaker material than bone (Yamada and Evans, 1970). When physiologically loaded, the same forces pass along the axis of the long bones and across the joints.[2] In order to support the forces, therefore, joint surfaces must be larger than bone cross-sectional areas (see Figure 1.3.1a,b). On the other hand, bone material is approximately twice as dense[3] as the rest of our body. In order to reduce the bone's mass, therefore, the sub-cortical bone of our joints is 'spongy' and light, rather than compact and heavy (see Figure 1.3.1c), and the shaft is slender.

1.3.2.2 Bone tissue

Bone tissue is either compact or trabecular. Trabecular bone, also called spongy bone or cancellous bone, is made up of interconnected rods and plates and can be found in the sub-cortical zones of the joints and in the interior parts of flat bones and irregular bones. Its bone volume fraction is typically around 25%, although this varies widely and can reach values above 50% in horse's leg bones (Figure 1.3.2).

Compact[4] bone is a tissue in which the bone material is densely packed. It can be found in the shafts of the long bones. The tissue organization of compact bone implies a low surface-to-volume ratio, which reduces accessibility of compact bone

[2]This is because the forces that load our bones arise from muscle contractions, and these muscles pass one or more joints.

[3]The term 'dense' refers to Archimedes' density (=mass per volume).

[4]The terms 'compact' and 'cortical' are often used interchangeably, which strictly speaking is not correct.

tissue and may be one of the reasons for the low turnover rate observed in the deeper layers of compact bone (Parfitt, 2002).

Both cancellous bone and compact bone tissue are made up of stacked layers known as lamellae. In lamellar compact bone, these lamellae are stacked in parallel upon each other (see Figure 1.3.4) such that the 'grain' is rotated by a constant amount between layers. In Haversian bone, lamellae are arranged in a pattern of concentric rings (see Figure 1.3.3). In woven bone this regular pattern is lost and the lamellae are without any clear organization.[5] Woven bone formation occurs only under extreme overloading and may perhaps be understood as some kind of 'panic' reaction to this.

Since bone material is solid, specific pores within the compact bone tissue are required to permit blood and nerve supply. They are called Volkmann canals in lamellar bone and Haversian canals in Haversian bone.

1.3.2.3 The material level: organic and inorganic constituents

Bone, as a material, consists of an organic and an inorganic phase. Each of these contributes in a different way to the overall material properties, and two simple experiments can inform us about the respective role of each phase. First, demineralization[6] of a bone dissolves the mineral (or inorganic) phase (= portion) to leave the organic phase behind. This is a fluffy fabric, very similar to a tendon, which is rigid (and strong) in tension, but not in compression or shear loading. Conversely, burning the organic phase[7] and keeping the mineral phase in place alters a bone's colour slightly, but leaves its shape intact. It continues to be considerably rigid in compression, but a gentle knock can atomize it – the material has become brittle, similar to glass. Assigning single properties to the bone material's constituent is a useful first didactic step. However, it should be realized that a material's overall properties ultimately arise from the interplay of all its constituents (Currey, 2002).

The organic phase constitutes approximately two thirds of the bone material's volume, but only one third of its dry mass. Its main constituent is type I collagen. The remarkable tensile strength of that protein is mainly down to its aminic bond. Every third amino acid is glycine, and among the rest there is a large fraction of proline or hydroxyproline. Unlike many other proteins, type I collagen does not form an α-helix, but rather combines three strands to form a triple helix. The formation starts with procollagen synthesis in the osteoblasts. After exocytosis, a polypeptide is cleaved off the C-terminus

[5]The so-called callus, which bridges the gap in fracture healing, is, anatomically speaking, made up of woven bone, as is the microcallus sometimes seen in trabecular bone.

[6]Bone can be demineralized *in vitro* by acid or chelating agents. It is important to realize that genuine demineralization does not occur *in vivo*. Rather, bone resorption encompasses simultaneous demineralization and degradation of the organic phase (see Section 1.3.3.1). It is therefore inappropriate to apostrophize bone resorption, or even bone loss, as 'demineralization', as is sometimes done in the literature.

[7]This can be done by 'ashing' it at 500 °C.

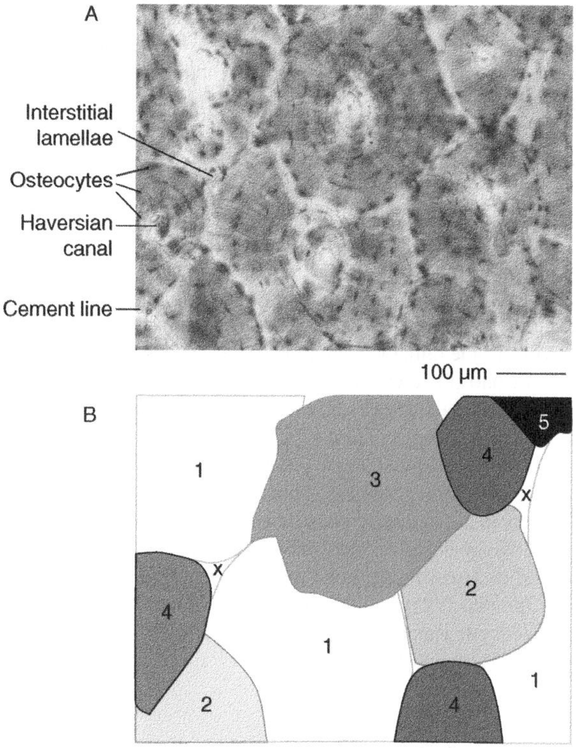

Figure 1.3.3 Compact bone tissue. (a) Undecalcified section of human compact bone after fuchsin staining. Transmitted light microscopy with 100x magnification. Haversian systems, also called secondary osteons, can be recognized by the surrounding cement line (bright). Importantly, there are only a few osteocytic connections across the cement line. The Haversian canal in their centre constitutes a conduit for arterial, venous and nerve supply. The dense network of canaliculi between the osteocytes can be appreciated, as well as the concentric organization of their cell bodies around the Haversian canal. (b) Schematic of the same slide as (a), with an analysis of the overlapping of different Haversian systems in order to reveal the tissue's history. In this slide, osteons delineated with smaller numbers are older than those delineated with larger numbers. When deciding which osteon is older and which younger, the continuity of aligned osteocytes is a key criterion. Sometimes, intracortical remodelling leaves 'chunks' of bone behind that have no connection to a Haversian or a Volkmann canal. These bits are referred to as 'interstitial lamellae' (marked by 'x'). Apparently, the osteocytes contained in them are derived from the dense canalicular network of the Haversian system. There seems to be an accumulation of linear microcracks within interstitial bone (Diab and Vashishth, 2007), which pinpoints the crucial role that osteocytic communication may play in the targeting of remodelling. Moreover, linear microcracks are more detrimental to the risk of fracture than the diffuse microdamage that is prevailing in other parts of bone and in the young (Diab *et al.*, 2006). It may therefore be that lack of remodelling of interstitial bone is a key factor in osteoporotic fractures

and from the N-terminus to generate tropocollagen. This is then enzymatically incorporated into the existing tropocollagen network by so-called cross-links. In addition to enzymatic cross-linking, which is a physiological process, there is also non-enzymatic cross-linking by advanced glycation end-products (AGEs). AGE cross-linking reduces a material's toughness (Garnero *et al.*, 2006; Wang *et al.*, 2002). It becomes increasingly prevalent with age (Saito *et al.*, 1997) and also in bone with osteoporotic fractures (Saito *et al.*, 2006; Shiraki *et al.*, 2008). It therefore seems possible that reduced toughness by AGEs is a contributing mechanism to osteoporotic fractures.

Most of the mineral (= inorganic) phase consists of a tertiary calcium phosphate crystal that is stoichiometrically similar to hydroxyl apatite. However, in about 5% of the crystals the phosphate group is replaced by carbonate, and the calcium can also be replaced by sodium, potassium and other cations. We should therefore speak of 'bone apatite'. The high-order elements in the inorganic phase cause considerable X-ray attenuation. X-ray-based methods have therefore become a standard method of assessing the bone mineral.[8] However, not all of the bone mineral is crystalline, and there is a narrow seam along the osteocytic canals that can readily exchange fluids and ions (Arnold, Frost and Buss, 1971).

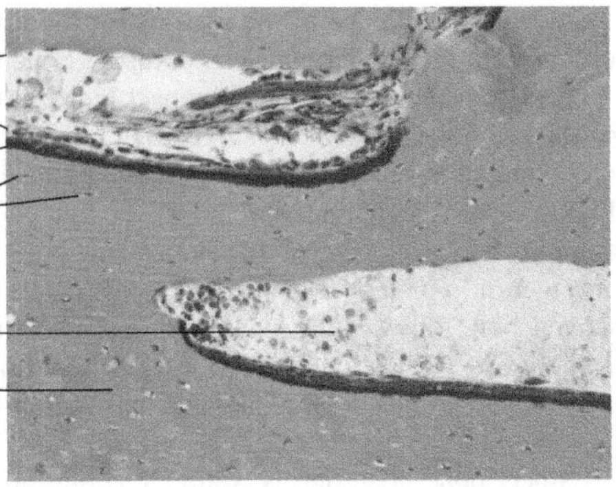

Figure 1.3.4 Trabecular bone tissue. Undecalcified human trabecular bone from the iliac crest in an eight-year-old boy. Goldner trichrome staining and transmitted light microscopy (100×). The reader is kindly asked to first look at the figure in a systematic way; that is, from greater distance. You will then see that there are two darker surfaces located on 'top' of a trabeculum, and two scalloped surfaces at its 'bottom'. The dark surfaces consist of osteoid, which has recently been laid down by the osteoblast arrays. The scalloped surfaces, on the other hand, have undergone resorption. Together, resorption at the bottom and formation at the top engender a drift of the bone structure towards the top. This is typical of modelling. In this case the 'top' is the external aspect of the *os ileum*, and the modelling drift will mediate enlargement of the pelvis during growth. Note that bone formation takes place on a smooth surface (that is, one that has not yet undergone resorption), hence we can be sure that we are dealing with modelling and not remodelling. Note also that the bone lamellae (visible by the bright streaks) have variable orientation throughout the section. Slide by courtesy of Rose Travers and Frank Rauch

1.3.3 BONE BIOLOGY

Bone, at the material, tissue and organ levels, is generated and maintained by four different kinds of cell. Osteoblasts can generate bone material and osteoclasts destroy it. Osteoblasts derive either from bone marrow stromal cells (Aubin and Liu, 1996) or through transdifferentiation from the lining cells that cover most bone surfaces. When forming bone, some of the osteoblasts end-differentiate into osteocytes, which reside deep within the bone (see Figures 1.3.3 and 1.3.4). When becoming quiescent, osteoblasts reduce their height and differentiate back into lining cells to stop bone formation altogether.

1.3.3.1 Osteoclasts

The term 'osteoclast' derives from the Greek words for 'bone' and 'broken'. In fact, these cells degrade, dissolve and resorb bone material. They do so in a tightly sealed extracellular space that isolates the 'osteolytic cocktail' (Teitelbaum, 2000). This cocktail contains H^+ and proteolytic enzymes (e.g. collagenases, cathepsins), which together erode the surface. Transcytosis takes the solutes from the sealed zone to the basolateral domain into a venole adjacent to the osteoclast. The void space, resulting from an osteoclast's 'bite', is called a Howship lacuna, and surface erosion by osteoclasts generates a scalloped surface that can be thought of as a concatenation of such lacunae (see Figure 1.3.4).

The more active osteoclasts are, the more nuclei they have. The average lifespan of an osteoclast is 12 days (Parfitt *et al.*, 1996), after which the cell undergoes apoptosis. Osteoclasts have receptors and respond to parathyroid hormone, calcitonin and interleukin 6. *In vitro* studies suggest that osteoclasts are activated by receptor activator of nuclear factor κ B (RANK), released by osteoblasts, which interacts with RANK-ligand (RANKL) expressed on the osteoclasts' membrane. Osteoprotegerin (OPG) is a decoy RANK receptor expressed by osteoblasts which inhibits osteoclast differentiation and activity.

1.3.3.2 Osteoblasts

The term 'osteoblast' derives from the Greek for 'bone' and 'germ'. Osteoblasts bear some similarity with fibroblasts in that they generate collagen to enhance the extracellular matrix. However, it is the privilege of the osteoblasts to induce mineralization of that collagenous tissue to form bone material. Naturally, this can take place only on existing bone surfaces. For some reason, osteoblasts never work in isolation but rather in conjunction with many other cells (see Figure 1.3.4). As

[8] Hence the terms 'bone mineral content' and 'bone mineral density'.

stated above, osteoblasts can differentiate back and forth into lining cells. They can also transdifferentiate into osteocytes, but not back again.

The process of bone formation involves two steps. First, a mix of different proteins, the so-called osteoid, is laid down. This osteoid consists of collagen (>90%), elastine and glycosaminoglykanes; proteins with mechanical competence. In addition to these, other proteins such as TGF-ß, osteocalcin and osteopontin, which may have a regulatory function, are also involved. The second step is the mineralization of the osteoid. This seems to take around 10 days to begin (so-called mineralization lag time). Osteoblasts are involved in this process too, providing the enzymes required (e.g. alkaline phosphatase and osteocalcin). After this initiation, bone mineralization continues more or less automatically. Importantly, crystallized apatite can only be resolved by resorption. Hence, older bone material has a greater concentration of minerals than younger material.

1.3.3.3 Osteocytes

Osteocytes reside deep within the bone tissue. They must rely upon diffusion for exchange of nutrients and gases. To this end, each osteocyte entertains 60 dendritic connections (Boyde, 1972) running through the so-called 'canaliculi' (see Figure 1.3.3). The volume required for this network and also for the lacunae housing the osteocytes' cell bodies amounts to approximately 2% of the total bone volume (Frost, 1960).

It is thought that osteocytes help to control the trafficking of noncrystalline bone mineral (Talmage, 2004; Talmage *et al.*, 2003). Osteocytes possess receptors specific to parthyroid hormone (PTH) (Divieti *et al.*, 2001), suggesting that these cells may respond to endocrine signals and thus mediate serum calcium homoeostasis.

Osteocytes, lining cells and osteoblasts are all interconnected via connexons (gap junctions), forming a functional syncytium that even extends into the stromal cells in the marrow (Palazzini *et al.*, 1998; Palumbo *et al.*, 1990, 2001). It is very likely that this osteocyte network constitutes a means of communication and information processing (see Section 1.3.5.4).

1.3.4 MECHANICAL FUNCTIONS OF BONE

1.3.4.1 Material properties

Bone material is superior to all other materials of the human body in terms of elasticity, strength and toughness.[9] Figure 1.3.5 illustrates what these terms mean. As shown in the figure, increases in strain and in stress are related. Strain is a measure of the material's deformation, and stress is the tension (force per unit area) generated by the strain. There are two different regions to be discerned in the diagram. Within the so-called

[9]With the exception of teeth.

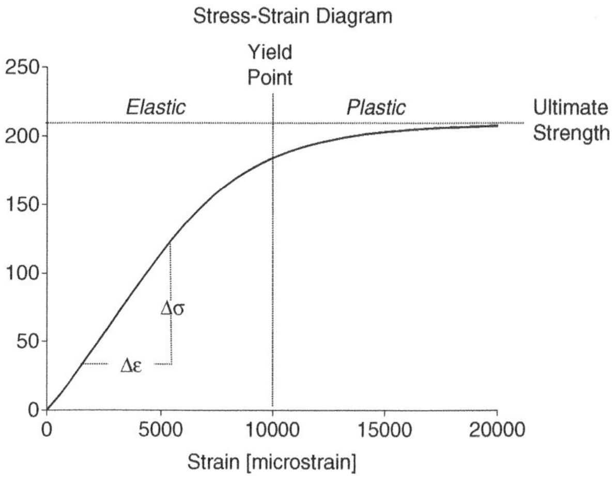

Figure 1.3.5 Stress–strain diagram. Strain (that is, deformation of a solid material) is quantified by the dimensionless number ε. One microstrain is deformation by 10^{-6} of the original length. Within the material, the strain causes a stress (σ). This is given as force per unit area. Two different portions can be discerned in the curve, one in which strains are elastic (i.e. the energy is stored and returned) and one in which they are plastic (i.e. the strain energy is not returned, but rather transformed into material damage and heat). Young's modulus is given as $E = \Delta\sigma/\Delta\varepsilon$ in the elastic region of the stress–strain curve. The stress at which the material fails is also referred to as the material's ultimate strength. Values in this diagram denote typical values for compact bone in line with its typical loading direction

elastic region (for strains below the yield point) energy is elastically stored. In the plastic region the material absorbs some of the deformation energy. This is often in the form of material microdamage.

The elastic modulus (also called Young's modulus or material stiffness) is defined as the slope of the stress–strain curve in the elastic region. Resilience is the material's capacity to elastically store energy; it is given by the area under the curve within the elastic region. The ultimate strength of a material is the stress at which it fails. The toughness of a material is the total amount of energy it can store and/or absorb before failure; it is given by the area under the curve of both the elastic and the plastic regions. A brittle material is one with small toughness.

When considering these relationships (shown in Figure 1.3.5), four points should be borne in mind. First, the elastic modulus of bone material increases with strain rate (i.e. the speed of deformation). Second, the elastic modulus increases with the degree of mineralization (Currey, 1984). Third, bone's material properties are different when loaded in compression, tension and shear. Bone's ultimate strength, for example, is 180 MPa in compression, but only 130 MPa in tensile loading. Finally, and most importantly, bone material is anisotropic. For example, elastic modulus and ultimate strength are up to four times greater when loaded along the lamellae than perpendicular to them (Liu, Weiner and Wagner, 1999; Liu,

Wagner and Weiner, 2000). Normally, the orientation of lamellae in cortical bone is in parallel with the bone's main axis and the anisotropic behaviour is therefore aligned to maximize the whole bone's strength. By contrast, in trabecular bone the lamella can be at virtually any angle in relation to the trabeculum's axis (see Figure 1.3.4). It is therefore difficult to predict the mechanical behaviour of bone without a full account of its anisotropy. Accordingly, the validity of finite-element models based on microCT data, as used by some scientists, is questionable.

Another important material property is that of resistance to fatigue. In bone, for example, failure stress decreases from 180 MPa when loaded once to 140 MPa when loaded a thousand times, and to below 100 MPa when loaded a million times (Gray, 1974). This is explained by the accumulation of microdamage (Cotton *et al.*, 2005). Functionally, this leads to reduced stiffness, strength and toughness. In a homogeneous material, such as steel, cracks tend to propagate by themselves once they exceed a length typical for that material (=Griffith length). This is usually not the case for composite materials, such as bone, where crack propagation is inhibited by several mechanisms (Peterlik *et al.*, 2006). Bone therefore has comparatively greater toughness and can accumulate more plastic strains before failure.

Remodelling (see Section 1.3.5.2) is the physiological process that helps to repair bone microdamage. Unfortunately, this does not always work. Fatigue fractures[10] occur in response to repetitive loading, typically in military recruits and athletes. It is thought that a chronic change in habitual loading patterns leads to the generation of microdamage, which promotes remodelling activity. The excavation of the damaged tissue necessarily raises the stresses experienced nearby, which in turn potentiates the generation of new microdamage, finally giving rise to a positive feedback loop or a vicious circle. Material fatigue may similarly be involved in osteoporosis (Burr *et al.*, 1997), as the increased risk of fracture in older age can only partly be explained by reductions in bone mass (Johnell *et al.*, 2005; Kanis *et al.*, 2007). Furthermore, microdamage accumulation with increasing age is well documented (Diab *et al.*, 2006; Li *et al.*, 2005) and it is therefore logical to expect some contribution towards fracture propensity in osteoporotic patients.

The pertinent question now is how far bone's material properties vary. As outlined above, bone becomes stiffer with increasing mineralization. One might therefore expect increased material stiffness in older age. However, this does not seem to occur (Zioupos and Currey, 1998); rather, old people's bones are characterized by greater variation in their material properties and it is possible that this contributes to reduced toughness (Zioupos and Currey, 1998). With the exception of genetic disorders such as osteogenesis imperfecta (see Section 1.3.5.3), and disregarding evolutionary specialization of hard tissues such as antler, tooth and tusk, the material properties of humans have to be considered as fairly constant within individuals. As a consequence, adaptation of bones (as organs) to environmental stimuli has to occur through structural adjustment.

[10] Fatigue fractures are sometimes also called stress fractures. This term is misleading as all fractures involve some kind of mechanical stress.

1.3.4.2 Structural properties

There are three different ways in which force can act upon solid materials, namely compression, tension and shear (see Figure 1.3.6a). In bending, there is a combination of compression (on the concave side; see Figure 1.3.6c) and tension (on the convex side). Moreover, due to Poisson's effect (see Figure 1.3.6b), there is tension and shear even in mere compressive (=uniaxial) loading. Hence, in reality any force application will elicit all three kinds of strain within a solid material.

Adaptation to compression and tensile loading occurs through alteration of the cross-sectional area perpendicular to the line of action. More precisely, structural strength is given by the product of cross-sectional area and the material's ultimate strength. Structural adaptation to bending is a bit more complicated. As depicted in Figure 1.3.6c, a neutral plane exists in the centre of a beam the length of which does not change during bending. Accordingly, as there is no strain in this plane, it does not generate any stress and therefore does not contribute to the beam's resistance to bending. Conversely, the strains

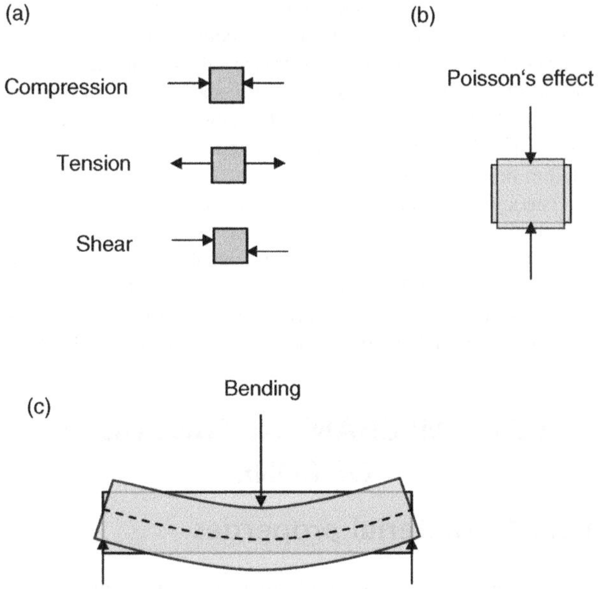

Figure 1.3.6 Illustration of the different ways in which forces can act upon a material. Compression and tension are the uniaxial loading patterns (a). However, even simple compression will cause a cross-expansion, known as Poisson's effect (b). Hence there are shear strains even in uniaxial loading. Bending causes a complex strain pattern (c). Note that the upper edge of the beam is shortened, whereas the lower edge becomes longer. The neutral plane (dashed line) does not change in length

Table 1.3.1 Example of the structural properties in four different beams and in the human tibia. All have identical cross-sectional areas, and thus also identical strength in compression and tension. Their strength in bending, which is given by the moment of resistance (W), varies among the different structures. W_X is the moment of resistance for bending around the horizontal axis, and W_Y is for bending around the vertical axis. Because of its rotational symmetry, $W_X = W_Y$ for the cylinders. This is not the case for the rectangular cylinder. Moreover, W is greater when the cylinder is hollow, just because the available material is distributed more 'eccentrically'. Accordingly, a flagpole is well designed to resist bending moments with wind from all directions. In this sense, the tibia can be regarded as a combination of a hollow cylinder (distributing its material far from the central fibre) and a rectangular beam with some degree of rotational asymmetry. This asymmetry accounts for bending moments introduced by the eccentric attachment of the calf musculature, which imposes considerable forces upon the tibia. Adapted from Rittweger *et al.* (2000)

	Full cylinder	Hollow cylinder	Square beam	Rectangular beam	Human tibia
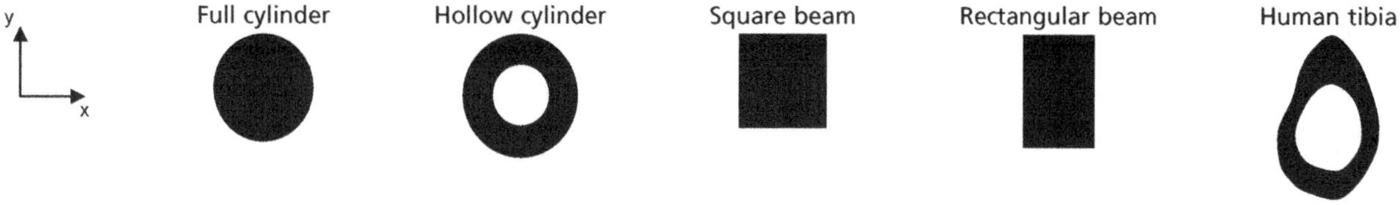					
Size	R = 1.13	r = 0.65 R = 1.30	h = b = 2	h = 2.42 b = 1.65	
A (cm^2)	4	4	4	4	4
W_Y (cm^3)	1.13	1.62	1.33	1.1	1.39
W_X (cm^3)	1.13	1.62	1.33	1.62	1.62

become increasingly larger towards the edges of the cross-section. More precisely, the contribution of any material particle to the beam's bending stiffness increases with the square of its distance from the neutral plane. The mathematical construct derived from this principle to quantify the rigidity in bending is the axial moment of inertia (I) and the structural strength in bending is given by the axial moment of resistance[11](W).

In practice, structures are strong in bending when W is large; that is, when there is much material located far away from the neutral plane. This is illustrated in Table 1.3.1, in which the principle capability of the tibia to resist bending moments is documented.

Architects have developed the concept of 'tensintegrity' through imitation of nature. This approach reduces bending loads and attempts to replace them with compressive and tensile forces. However, this is not always entirely possible and bending does occur to some extent in the musculoskeletal system (Biewener *et al.*, 1983; Hartman *et al.*, 1984). Accordingly, our bones are adapted to it (Rittweger *et al.*, 2000).

1.3.5 ADAPTIVE PROCESSES IN BONE

The notion that bones are capable of adapting to varying environmental conditions is relatively new. Much of our current understanding is owed to Harold M. Frost. Whilst previous researchers had focussed mainly on the activity of specific cells, that is osteoblasts and osteoclasts, Frost considered the interplay between these cells to account for adaptive processes in bone. Two kinds of such interplay can be discerned, namely modelling and remodelling. Whilst modelling prevails before puberty, bone turnover is mainly through remodelling during the later stages of life. Yet modelling still occurs in old age (Erben, 1996).

[11] Also known as the section modulus.

1.3.5.1 Modelling

Modelling can be compared to the work of a sculptor. It is characterized by drifts (Frost, 1990). As exemplified in Figure 1.3.4, these drifts occur such that envelopes undergoing formation oppose other envelopes undergoing formation. As a result, solid bone structures can change their shape gradually. For example, longitudinal growth is associated with enlargement of the outer (=periosteal) and inner (=endocortical) diameters in the diaphysis, which involves formation on the periosteal and resorption on the endocortical envelopes, together enlarging the shaft diameter.

Typically modelling leads to accrual of bone material, and it thus enforces a bone's structure. However, modelling is not a mere synonym for formation, as it employs both formation *and* resorption. Modelling is also thought to optimize the structure of trabecular networks (Huiskes *et al.*, 2000) and it can in addition help to neutralize bone flexure after fractures (Frost, 2004).

1.3.5.2 Remodelling

Remodelling replaces old bone material with new material. It thus prevents microdamage accumulation (Frost, 1960; Mori and Burr, 1993). The so-called ARF sequence of remodelling consists of the *A*ctivation of osteoclasts, the *R*esorption of old bone, and the *F*ormation of new bone (Takahashi, Epker and Frost, 1964). Bone remodelling is effectuated by so-called basic multicellular units (BMU). These consist of a few osteoclasts, which resorb the old bone material at the front of the BMU. An artery and two veins are linked to them for blood supply.[12] At the back of the BMU, the resorption cavity is filled up with new bone material by a cluster of osteoblasts. The entire ARF sequence is thought to take 90–120 days to complete.

Remodelling can either be stochastic or targeted. Stochastic remodelling is thought to serve calcium homoeostasis and is spatially unspecific. Targeted remodelling, by contrast, is specifically induced by bone microdamage (Burr *et al.*, 1985). BMUs tunnel their way in line with the principal stress (Hert, Fiala and Petrtyl, 1994), possibly attracted to their target by osteocyte apoptosis (Martin, 2007) occurring ahead of the BMUs (Burger, Klein-Nulend and Smit, 2003) but also within the vicinity of recent microcracks (Follet *et al.*, 2007; Noble *et al.*, 1997, 2003).

Once an osteoclast is activated, osteoblasts are recruited, either from lining cells or from osteoblast progenitors, to travel behind the osteoclast. It is thought by many authors that osteoblastic and osteoclastic activities within a BMU are coupled to each other. This belief is based on *in vitro* studies showing that bone residues emerging from osteoclastic resorption, such as TGF-β (Pfeilschifter *et al.*, 1990) and osteocalcin (Mundy *et al.*, 1982), constitute a stimulus of chemotaxis and of differentiation to osteoblasts. Osteoclasts, conversely, are thought to be controlled by osteoblasts through RANK and OPG (see Section 1.3.3.1). However, as pointed out by Gasser (2006), it is questionable whether such osteoblast/osteoclast coupling does occur *in vivo*, as bone resorption and formation within the BMU are separated in space and time, so that it is difficult to consider how any coupling might work *in vivo*. A much simpler hypothesis proposes that bone formation within the existing BMU is primarily controlled by mechanical strain (Huiskes *et al.*, 2000; Smit and Burger, 2000), for example via Sost/sclerostin expression (see Section 1.3.5.4).

With some imagination, the history of the remodelling sequence can be deciphered under the microscope (see Figure 1.3.3). In compact bone, it leads to the generation of secondary osteons, also called the Haversian systems (Havers, 1691). The BMU's bone balance is normally slightly negative. More precisely, of the 0.5 mm^3 bone turned over by each BMU, 0.003 mm^3 or 0.6% remains void (Frost, 2004). However, that loss is increased under disuse conditions and after menopause, when it can be almost complete in extreme cases.

[12]It should be noted that there is also a nerve fibre alongside the vessels in the BMU. Evidence suggests that sympathetic nervous fibres modulate the effects of mechanical stimuli upon the remodelling process, although it is currently unknown what exactly these modulatory effects are (Marenzana and Chenu, 2008).

1.3.5.3 Theories of bone adaptation

Within the realm of zoology, bone strains seem to be limited to about 2000 microstrains, with remarkable similarity across different species (Biewener, 1990; Biewener and Taylor, 1986; Rubin and Lanyon, 1984). It is important to bear in mind that bone material properties are very similar across species, too. There seems to be an evolutionary design principle, therefore, to adjust whole-bone stiffness to the mechanical forces exerted upon it; in other words, form follows function (Thompson, 1917; Wolff, 1899).

However, bones are not only genetically adapted to fulfil their mechanical role; physiological processes also allow adaptation of bone stiffness and strength to environmental changes. In a classic experiment Rubin and Lanyon (1987) demonstrated that application of strains above a certain level induces accrual of bone tissue, whilst lack of such strains leads to bone loss. This has been conceptualized in a number of feedback control theories (Beaupre, Orr and Carter, 1990; Fyhrie and Schaffler, 1995; Huiskes *et al.*, 2000), of which the mechanostat theory (Frost, 1987b) is probably the most well-known. According to that theory, bone accrual by modelling is induced when strains are in excess of the minimal effective strain for modelling (MESm; see Figure 1.3.7a). When strains remain below the minimal effective strain for remodelling (MESr), conversely, BMU-based remodelling is associated with increasingly negative bone balance, leading to removal of unnecessary bone. This can be understood as a negative feedback control system, helping to adapt bone stiffness and thus strength to variable forces (see Figure 1.3.7b).

Whilst the mechanostat theory is intuitive as well as formal, it leaves a couple of important aspects open, and it contradicts some empirical findings. For example, it makes no assumption about the nature or the number of strain cycles. Evidence, however, suggests that strain rate affects bone adaptive processes independently of strain magnitude (Mosley and Lanyon, 1998). Moreover, the number of strain cycles seems to be important (Qin, Rubin and McLeod, 1998) as a large number of strain cycles with small magnitude may be as effective as a small number of large-magnitude cycles (Rubin *et al.*, 2001). Another observation not explained by the mechanostat theory is that bone losses phase out after three to eight years of complete immobilization (Eser *et al.*, 2004; Zehnder *et al.*, 2004). To accommodate for this phenomenon, a theory of cellular accommodation has been proposed (Schriefer *et al.*, 2005), which proclaims that both the rapidity and the magnitude of a change are meaningful.

Notwithstanding whatever theory can best explain bone adaptive processes, there are two intriguing clinical examples to be discussed in this context. As mentioned above, osteogenesis imperfecta (OI) is a heritable disorder, characterized by fragile bones and caused by various different mutations in the gene coding for type-1 collagen. OI is associated with enhanced mineralization of bone (Boyde *et al.*, 1999), which leads to greater elastic modulus (Weber *et al.*, 2006). Therefore the same stresses will cause smaller strains. Interestingly, patients with

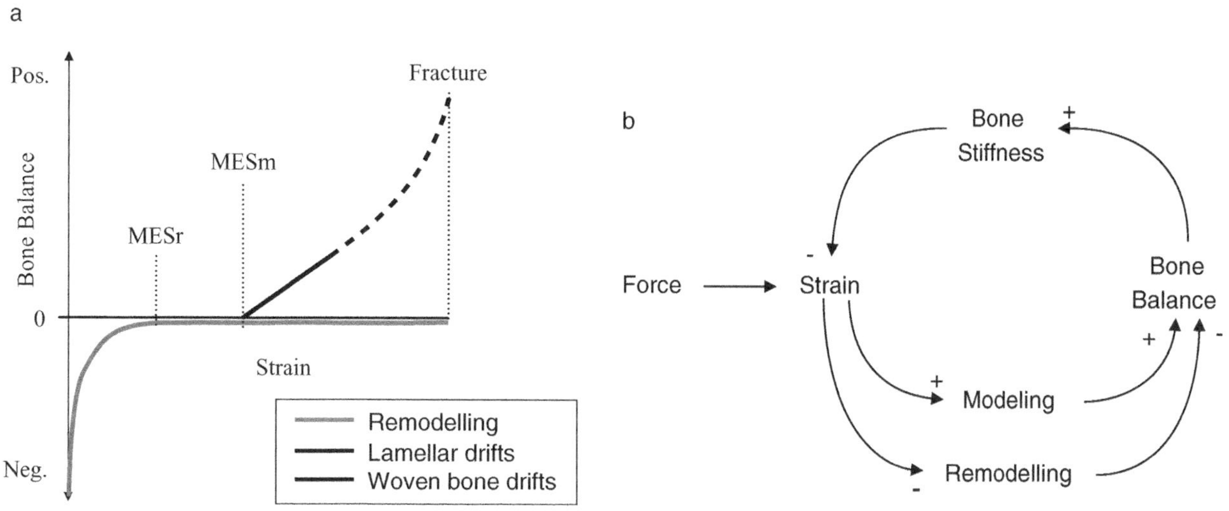

Figure 1.3.7 Mechanostat theory. (a) Schematic according to the original proposal (Frost, 1987a, 2003). According to the theory, habitual strains control modelling and remodelling. With strains above the MESr threshold, remodelling-related bone losses are minimized. Below MESr, however, remodelling is associated with an increasingly negative bone balance. When strains exceed the MESm threshold, modelling is turned on, leading to gains in lamellar bone (most typical in humans) or woven bone formation. (b) Representation as a negative feedback control system. For example, an increase in habitual force will lead to positive bone balance, and thus enhanced bone stiffness. This in turn takes the bone strains back to the set value. As a result, the bone adapts to the increased force. It is important to recognize that the strain is regulated to be constant in this model. By convention, arrows ending in '+' denote concordant effects, whilst '–' denotes discordant effects

OI have reduced bone mass (Rauch and Glorieux, 2004). X-linked hypophosphataemic rickets (XLHR) can be regarded as the 'contrary' condition, with bone hypomineralization due to excessive renal phosphorus excretion. Bone strains in this condition will therefore be larger than normal. In line with expectations, XLHR patients seem to have enhanced bone mineral density and mass in their forearm and lumbar spine (Oliveri *et al.*, 1991; Shore, Langman and Poznanski, 2000). Taken together, these observations do indeed suggest that strain-related signals govern whole-bone stiffness and strength (Rauch, 2006).

1.3.5.4 Mechanotransduction

Given the fundamental importance of strain-related signals for bone, the question arises as to how these signals are converted into biological information. This process is referred to as mechanotransduction. It comprises the 'recognition' of strains (or a related signal) within the bone, their processing, and communication with the surface on which either formation or resorption is to take place.

Earlier studies have focussed on the direct effects of mechanical strains on osteocytes. Stretch-related calcium influx (Ziambaras *et al.*, 1998) and its propagation through gap junctions in the osteocyte network is one possible mechanism. Likewise, insterstitial fluid flow (IFF) in the canalicular system could constitute an elegant mechanism to amplify mechanical signals. Indeed, *in vitro* studies with cultured osteoblasts suggest that the effect of IFF could be more important than the strain signal itself (Owan *et al.*, 1997). In addition, it has to be considered that IFF carries ions that give rise to piezoelectric and electrokinetic potentials (Guzelsu and Regimbal, 1990; Walsh and Guzelsu, 1991). IFF effects are thought to be further mediated by nitric oxide (Bacabac *et al.*, 2004) and prostaglandin deliberation (Jiang and Cheng, 2001) to stimulate osteoblastic bone formation. Within the osteocyte, there is accumulating evidence for an involvement of Sost/sclerostin expression (Robling *et al.*, 2008), which in turn impinges on osteoblastic bone formation, probably via LRP5/Wnt signalling pathways.

In conclusion, science has started to unravel the events involved in bone mechanotransduction. The question therefore is how the different mechanisms found to be responsive to strains interact. As highlighted by Harold Frost (1993), the phenomenon of 'flexure neutralization' cannot be explained by a single mechanotransductive mechanism. To solve this problem, a 'three-way rule' considers local strains *and* the whole-bone deformation. In support of such an explanation, the marrow cavity seems to be an ideal candidate for discerning whole-bone compression from whole-bone elongation. Indeed, intramedullary hydraulic pressure oscillations (60 mmHg) have been shown to prompt an increase in bone cross-section Qin *et al.* 2003.

1.3.6 ENDOCRINE INVOLVEMENT OF BONE

1.3.6.1 Calcium homoeostasis

Of the 1–2 kg of calcium in the human body, only 1 g resides outside the bones. The skeleton constitutes a large reservoir with the theoretical potential to overwhelm the body with calcium. Proper regulation of calcium fluxes into and out of bone is therefore paramount to the *milieu intérieur*.

In mammals, the key players in the regulation of calcium homoeostasis are vitamin D and its derivatives, as well as parathyroid hormone (PTH).[13] PTH is secreted into the blood from the parathyroid glands. PTH induces tubular reabsorption of calcium and excretion of phosphorus, as well as D-hormone synthesis in the kidney. It has no direct effect upon intestinal absorption of calcium. However, PTH does promote renal production of D-hormone, which in turn enhances intestinal calcium absorption. Chronically elevated PTH levels favour bone resorption, but bone formation can outweigh bone resorption when PTH levels peak intermittently. This engenders *de novo* bone formation (Reeve *et al.*, 1980), most likely through modelling (Gasser, 2006). Consequently, recombinant PTH is given to enhance bone mass and strength in patients with osteoporosis (Neer *et al.*, 2001; Reeve, 1996). Calcitonin acts antagonistically to PTH, but seems to be less important for the control of calcium homoeostasis in mammals.

Vitamin D (cholecalciferol) is derived from nutritional intake (Holick, 2005), and more importantly[14] from UV light-induced production in the skin from 7-dehydroxycholesterin. It is transformed in two enzymatic steps in the liver and in the kidney into the D-hormone (1,25-hydroxy-cholecalciferol). D-hormone is an antagonist to PTH in some sense, as it inhibits formation of PTH in the parathyroid glands. On the other hand, D-hormone stimulates the uptake of calcium in the intestine and kidney, and it also fosters bone mineralization. Besides its effects upon calcium-controlling organs, vitamin D and its derivatives have effects upon many other organs and cells in the human body, including skeletal muscle and the breast glands.

It is important to realize that bone formation and resorption are relatively slow instruments for the government of calcium homoeostasis. For example, it takes 10 for osteoclast recruitment takes 8 days (Eriksen, Axelrod and Melsen, 1994). It therefore seems that more rapid mechanisms are required for an accurate control of calcium homoeostasis (Borgens, 1984; Parfitt, 2003).

Within the mineral phase, insolulubility of the apatite crystals causes a concentration gradient for ions between the extracellular fluid (ECF) and the crystalline phase. This drives calcium out of the ECF and into the bone (Talmage *et al.*, 2003). Bone material therefore has to be considered primarily as a sink rather than as a source for calcium. In order to counteract this continuous ion efflux from the ECF (Rubinacci *et al.*, 2000), calcium is actively transported by the osteocyte-lining cell continuum from the bone material into the ECF (Marenzana *et al.*, 2005).

[13] Interestingly, fish do not have PTH and rely solely upon calcitonin for regulation of calcium homeostasis.

[14] The relative contribution of nutritional intake and endogenous vitamin D generation depends upon latitude, skin exposure and nutrition, including 'fortified' processed foods, all of which vary largely around the globe.

1.3.6.2 Phosphorus homoeostasis

The human body contains approximately 600 g of phosphorus, 85% of which is located within the bones. Phosphorus is involved in many biochemical reactions and biological processes and its concentration within the cell is relatively high (1–2 mM). Nutrition is normally rich in phosphorus and intestinal absorption is usually around 80%. Both intestinal absorption and renal reabsorption are controlled by D-hormone. Serum levels of phosphorus can vary considerably, causing fluctuations of 1 mM in physiological stimuli such as ingestion of carbohydrates. Although there are some endocrine disorders that affect phosphorus homoeostasis, it is usually not as crucial as calcium homoeostasis.

1.3.6.3 Oestrogens

The public is increasingly aware that menopause, and thus oestrogen withdrawal, causes bone loss and subsequently osteoporotic fractures. Yet there is only an incomplete understanding of oestrogen's effects upon bone, particularly in relation to exercise. Oestrogens are a family of steroid hormones, the most active compound of which is 17-β oestradiol (E2). They rapidly diffuse through the cell membrane to bind to tissue-specific intracellular receptors and to unfold their genomic effects. There are two different oestrogen receptors, ER-α and ER-β, which are thought to both be present in bone cells but to transmit different effects.

Ovariectomy in a rat is frequently used as a model to mimic the effects of oestrogen withdrawal upon bone. It leads to the depletion of the trabecluar bone compartment (Kalu *et al.*, 1991) and to endocortical resorption (Kalu *et al.*, 1989), but also to periosteal bone formation (Turner, Vandersteenhoven and Bell, 1987). Consequently, postmenopausal women have wider bones with reduced cortical area, but more-or-less preserved bending and torsional stiffness, as compared to premenopausal women (Ferretti *et al.*, 1998). It is clear, therefore, on an observational level, that women in their fertile period have more bone mineral than is required for mechanical reasons, and it seems likely that evolution has foreseen this as a necessary calcium reservoir for pregnancy and breast-feeding (Schiessl, Frost and Jee, 1998).

There is currently no consensus regarding the involvement of oestrogens and oestrogen receptors in the mechanical adaptation of bone. Stringent evidence has been put forward to suggest that oestrogen suppresses any bone response to exercise (Jarvinen *et al.*, 2003). In stark contrast, other authors have shown that ER-α receptors are essential mediators of the skeleton's response to mechanical strains (Lee *et al.*, 2003). In further contrast, Hertrampf *et al.* (2007) propose that ER-α mediates those effects of oestrogen upon bone that are independent of exercise, whilst the exercise effects are mediated via ER-β.

However, there is a growing understanding that oestrogens suppress osteocyte apoptosis, most likely by antioxidative effects (Mann *et al.*, 2007), and thus probably also the rate of bone remodelling. This could also explain the generally greater cortical bone mineral density observed in women (Wilks *et al.*, 2008). Finally, involvement of oestrogens and androgens in the skeletal sexual dimorphism is more or less well understood, and so is the role of E2 in growth-plate closure.

References

Arnold J.S., Frost H.M. and Buss R.O. (1971) The osteocyte as a bone pump. *Clin Orthop*, **78**, 47.

Aubin J.E. and Liu F. (1996) The osteoblast lineage, in *Principles of Bone Biology* (eds J.P. Bilezikian, L.G. Raisz and G.A. Rodan), San Diego, pp. 51–67.

Bacabac R.G., Smit T.H., Mullender M.G. *et al.* (2004) Nitric oxide production by bone cells is fluid shear stress rate dependent. *Biochem Biophys Res Commun*, **315**, 823.

Beaupre G.S., Orr T.E. and Carter D.R. (1990) An approach for time-dependent bone modeling and remodeling: theoretical development. *J Orthop Res*, **8**, 651–661.

Biewener A.A. (1990) Biomechanics of mammalian terrestrial locomotion. *Science*, **250**, 1097.

Biewener A.A. and Taylor C.R. (1986) Bone strain: a determinant of gait and speed? *J Exp Biol*, **123**, 383.

Biewener A.A., Thomason J., Goodship A. and Lanyon L.E. (1983) Bone stress in the horse forelimb during locomotion at different gaits: a comparison of two experimental methods. *J Biomech*, **16**, 565.

Borgens R.B. (1984) Endogenous ionic currents traverse intact and damaged bone. *Science*, **225**, 478–482.

Boyde A. (1972) Scanning electron microscope studies of bone, in *The Biochemistry and Physiology of Bone* (ed. G.H. Bourne), Academic Press, New York, pp. 259–310.

Boyde A., Travers R., Glorieux F.H. and Jones S.J. (1999) The mineralization density of iliac crest bone from children with osteogenesis imperfecta. *Calcif Tissue Int*, **64**, 185–190.

Burger E.H., Klein-Nulend J. and Smit T.H. (2003) Strain-derived canalicular fluid flow regulates osteoclast activity in a remodelling osteon: a proposal. *J Biomech*, **36**, 1453–1459.

Burr D.B., Forwood M.R., Fyhrie D.P. *et al.* (1997) Bone microdamage and skeletal fragility in osteoporotic and stress fractures. *J Bone Miner Res*, **12**, 6.

Burr D.B., Martin R.B., Schaffler M.B. and Radin E.L. (1985) Bone remodeling in response to in vivo fatigue microdamage. *J Biomech*, **18**, 189–200.

Cotton J.R., Winwood K., Zioupos P. and Taylor M. (2005) Damage rate is a predictor of fatigue life and creep strain rate in tensile fatigue of human cortical bone samples. *J Biomech Eng*, **127**, 213.

Currey J.D. (1984) Effects of differences in mineralization on the mechanical properties of bone. *Philos Trans R Soc Lond B Biol Sci*, **304**, 509.

Currey J.D. (2002) *Bones: Structure and Mechanics*, Princeton University Press, Princeton.

Currey J.D. (2003) The many adaptations of bone. *J Biomech*, **36**, 1487.

Diab T., Condon K.W., Burr D.B. and Vashishth D. (2006) Age-related change in the damage morphology of human cortical bone and its role in bone fragility. *Bone*, **38**, 427.

Diab T. and Vashishth D. (2007) Morphology, localization and accumulation of in vivo microdamage in human cortical bone. *Bone*, **40**, 612–618.

Divieti P., Inomata N., Chapin K. *et al.* (2001) Receptors for the carboxyl-terminal region of pth(1-84) are highly expressed in osteocytic cells. *Endocrinology*, **142**, 916–925.

Erben R.G. (1996) Trabecular and endocortical bone surfaces in the rat: modeling or remodeling? *Anat Rec*, **246**, 39.

Eriksen E.F., Axelrod D.W. and Melsen F. (1994) Bone histology and bone histomorphometry, in *Bone Histomorphometry* (eds D.W. Eriksen and F. Melsen), Raven Press, New York, pp. 13–20.

Eser P., Frotzler A., Zehnder Y. *et al.* (2004) Relationship between the duration of paralysis and bone structure: a pQCT study of spinal cord injured individuals. *Bone*, **34**, 869–880.

Ferretti J.L., Capozza R.F., Cointry G.R. *et al.* (1998) Bone mass is higher in women than in men per unit of muscle mass but bone mechanostat would compensate for the difference in the species. *Bone*, **23**, S471.

Follet H., Li J., Phipps R.J. *et al.* (2007) Risedronate and alendronate suppress osteocyte apoptosis following cyclic fatigue loading. *Bone*, **40**, 1172–1177.

Frost H.M. (1960) Presence of microscopic cracks 'in vivo' in bone. *Henry Ford Hosp Med Bull*, **8**, 25.

Frost H.M. (1987a) Bone 'mass' and the 'mechanostat': a proposal. *Anat Rec*, **219**, 1–9.

Frost H.M. (1987b) The mechanostat: a proposed pathogenic mechanism of osteoporoses and the bone mass effects of mechanical and nonmechanical agents. *Bone Miner*, **2**, 73.

Frost H.M. (1990) Skeletal structural adaptations to mechanical usage (SATMU): 1. Redefining Wolff's law: the bone modeling problem. *Anat Rec*, **226**, 403.

Frost H.M. (1993) Wolff's law: an 'MGS' derivation of Gamma in the Three-Way Rule for mechanically controlled lamellar bone modeling drifts. *Bone Miner*, **22**, 117.

Frost H.M. (2003) Bone's mechanostat: a 2003 update. *Anat Record Part A*, **275A**, 1081.

Frost H.M. (2004) *The Utah Paradigm of Skeletal Physiology*, ISMNI, Athens.

Fyhrie D.P. and Schaffler M.B. (1995) The adaptation of bone apparent density to applied load. *J Biomech*, **28**, 135–146.

Garnero P., Borel O., Gineyts E. *et al.* (2006) Extracellular post-translational modifications of collagen are major determinants of biomechanical properties of fetal bovine cortical bone. *Bone*, **38**, 300–309.

Gasser J.A. (2006) Coupled or uncoupled remodeling, is that the question? *J Musculoskelet Neuronal Interact*, **6**, 128–133.

Geschickter C.F. and Maseritz I.H. (1938) Myositis ossificans. *J Bone Joint Surg Am*, **20**, 661–674.

Gilbertson E.M. (1975) Development of periarticular osteophytes in experimentally induced osteoarthritis in the dog. A study using microradiographic, microangiographic, and fluorescent bone-labelling techniques. *Ann Rheum Dis*, **34**, 12–25.

Gray R.J.K.G.K. (1974) Compressive fatigue behaviour of bovine compact bone. *J Biomech*, **7**, 292.

Guzelsu N. and Regimbal R.L. (1990) The origin of electrokinetic potentials in bone tissue: the organic phase. *J Biomech*, **23**, 661–672.

Hartman W., Schamhardt H.C., Lammertink J.L. and Badoux D.M. (1984) Bone strain in the equine tibia: an in vivo strain gauge analysis. *Am J Vet Res*, **45**, 880–884.

Havers C. (1691) *Osteologia Nova*, Samuel Smith, London.

Hert J., Fiala P. and Petrtyl M. (1994) Osteon orientation of the diaphysis of the long bones in man. *Bone*, **15**, 269–277.

Hertrampf T., Gruca M.J., Seibel J. *et al.* (2007) The bone-protective effect of the phytoestrogen genistein is mediated via ER alpha-dependent mechanisms and strongly enhanced by physical activity. *Bone*, **40**, 1529–1535.

Holick M.F. (2005) The vitamin D epidemic and its health consequences. *J Nutr*, **135**, S2739–S2748.

Huiskes R., Ruimerman R., van Lenthe G.H. and Janssen J.D. (2000) Effects of mechanical forces on maintenance and adaptation of form in trabecular bone. *Nature*, **405**, 704.

Jarvinen T.L.N., Kannus P., Pajamaki I. *et al.* (2003) Estrogen deposits extra mineral into bones of female rats in publerty, but simultaneously seems to suppress the responsiveness of female skeleton to mechanical loading. *Bone*, **32**, 642.

Jiang J.X. and Cheng B. (2001) Mechanical stimulation of gap junctions in bone osteocytes is mediated by prostaglandin E2. *Cell Commun Adhes*, **8**, 283–288.

Johnell O., Kanis J.A., Oden A. *et al.* (2005) Predictive value of BMD for hip and other fractures. *J Bone Miner Res*, **20**, 1185–1194.

Kalu D.N., Liu C.C., Hardin R.R. and Hollis B.W. (1989) The aged rat model of ovarian hormone deficiency bone loss. *Endocrinology*, **124**, 7–16.

Kalu D.N., Salerno E., Liu C.C. *et al.* (1991) A comparative study of the actions of tamoxifen, estrogen and progesterone in the ovariectomized rat. *Bone Miner*, **15**, 109–123.

Kanis J.A., Oden A., Johnell O. *et al.* (2007) The use of clinical risk factors enhances the performance of BMD in the prediction of hip and osteoporotic fractures in men and women. *Osteoporos Int*, **18**, 1033–1046.

Kaplan F.S., Fiori J., Ahn S.D.L.P. *et al.* (2006) Dysregulation of the BMP-4 signaling pathway in fibrodysplasia ossificans progressiva. *Ann N Y Acad Sci*, **1068**, 54–65.

E.M., K., Jessop, H., Suswillo, R., Zaman, G. and Lanyon and L. (2003) Endocrinology: bone adaptation requires oestrogen receptor-alpha. *Nature*, **424**, 389.

Li J., Miller M.A., Hutchins G.D. and Burr D.B. (2005) Imaging bone microdamage in vivo with positron emission tomography. *Bone*, **37**, 819.

Liu D., Wagner H.D. and Weiner S. (2000) Bending and fracture of compact circumferential and osteonal lamellar bone of the baboon tibia. *J Mater Sci Mater Med*, **11**, 49–60.

Liu D., Weiner S. and Wagner H.D. (1999) Anisotropic mechanical properties of lamellar bone using miniature cantilever bending specimens. *J Biomech*, **32**, 647–654.

Mann V., Huber C., Kogianni G. *et al.* (2007) The antioxidant effect of estrogen and Selective Estrogen Receptor Modulators in the inhibition of osteocyte apoptosis in vitro. *Bone*, **40**, 674–684.

Marenzana M. and Chenu C. (2008) Sympathetic nervous system and bone adaptive response to its mechanical environment. *J Musculoskelet Neuronal Interact*, **8**, 111–120.

Marenzana M., Shipley A.M., Squitiero P. *et al.* (2005) Bone as an ion exchange organ: evidence for instantaneous cell-dependent calcium efflux from bone not due to resorption. *Bone*, **37**, 545.

Martin R.B. (2007) Targeted bone remodeling involves BMU steering as well as activation. *Bone*, **40**, 1574–1580.

Mori S. and Burr D.B. (1993) Increased intracortical remodeling following fatigue damage. *Bone*, **14**, 103–109.

Mosley J.R. and Lanyon L.E. (1998) Strain rate as a controlling influence on adaptive modeling in response to dynamic loading of the ulna in growing male rats. *Bone*, **23**, 313.

Mundy G.R., Rodan S.B., Majeska R.J. *et al.* (1982) Unidirectional migration of osteosarcoma cells with osteoblast characteristics in response to products of bone resorption. *Calcif Tissue Int*, **34**, 542–546.

Neer R.M., Arnaud C.D., Zanchetta J.R. *et al.* (2001) Effect of parathyroid hormone (1-34) on fractures and bone mineral density in postmenopausal women with osteoporosis. *N Engl J Med*, **344**, 1434.

Noble B.S., Peet N., Stevens H.Y. *et al.* (2003) Mechanical loading: biphasic osteocyte survival and targeting of osteoclasts for bone destruction in rat cortical bone. *Am J Physiol Cell Physiol*, **284**, C934–C943.

Noble B.S., Stevens H., Loveridge N. and Reeve J. (1997) Identification of apoptotic changes in osteocytes in normal and pathological human bone. *Bone*, **20**, 273–282.

Özkaya, N. and Nordin M. (1998) *Fundamentals of Biomechanics*, Springer, New York.

Oliveri M.B., Cassinelli H., Bergada C. and Mautalen C.A. (1991) Bone mineral density of the spine and radius shaft in children with X-linked hypophosphatemic rickets (XLH). *Bone Miner*, **12**, 91–100.

Owan I., Burr D.B., Turner C.H. *et al.* (1997) Mechanotransduction in bone: osteoblasts are more responsive to fluid forces than mechanical strain. *Am J Physiol*, **273**, C810.

Palazzini S., Palumbo C., Ferretti M. and Marotti G. (1998) Stromal cell structure and relationships in perimedullary spaces of chick embryo shaft bones. *Anat Embryol (Berl)*, **197**, 349–357.

Palumbo C., Ferretti M., Ardizzoni A. *et al.* (2001) Osteocyte-osteoclast morphological relationships and the putative role of osteocytes in bone remodeling. *J Musculoskelet Neuronal Interact*, **1**, 327–332.

Palumbo C., Palazzini S. and Marotti G. (1990) Morphological study of intercellular junctions during osteocyte differentiation. *Bone*, **11**, 401–406.

Parfitt A.M. (2002) Misconceptions (2): turnover is always higher in cancellous than in cortical bone. *Bone*, **30**, 807–809.

Parfitt A.M. (2003) Misconceptions (3): calcium leaves bone only by resorption and enters only by formation. *Bone*, **33**, 259.

Parfitt A.M., Mundy G.R., Roodman G.D. *et al.* (1996) A new model for the regulation of bone resorption, with particular reference to the effects of bisphosphonates. *J Bone Miner Res*, **11**, 150.

Peterlik H., Roschger P., Klaushofer K. and Fratzl P. (2006) From brittle to ductile fracture of bone. *Nat Mater*, **5**, 52–55.

Pfeilschifter J., Erdmann J., Schmidt W. *et al.* (1990) Differential regulation of plasminogen activator and plasminogen activator inhibitor by osteotropic factors in primary cultures of mature osteoblasts and osteoblast precursors. *Endocrinology*, **126**, 703–711.

Qin Y.X., Rubin C.T. and McLeod K.J. (1998) Nonlinear dependence of loading intensity and cycle number in the maintenance of bone mass and morphology. *J Orthop Res*, **16**, 482–489.

Qin Y.X., Kaplan T., Saldanha A. and Rubin C. (2003) Fluid pressure gradients, arising from oscillations in intramedullary pressure, is correlated with the formation of bone and inhibition of intracortical porosity. *J Biomech*, **36** (10), 1427–1437.

Rauch F. (2006) Material matters: a mechanostat-based perspective on bone development in osteogenesis imperfecta and hypophosphatemic rickets. *J Musculoskelet Neuronal Interact*, **6**, 142–146.

Rauch F. and Glorieux F.H. (2004) Osteogenesis imperfecta. *Lancet*, **363**, 1377–1385.

Reeve J. (1996) PTH: a future role in the management of osteoporosis? *J Bone Miner Res*, **11**, 440–445.

Reeve J., Meunier P.J., Parsons J.A. *et al.* (1980) Anabolic effect of human parathyroid hormone fragment on trabecular bone in involutional osteoporosis: a multicentre trial. *Br Med J*, **280**, 1340–1344.

Rittweger J., Beller G., Ehrig J. *et al.* (2000) Bone-muscle strength indices for the human lower leg. *Bone*, **27**, 319–326.

Robling A.G., Niziolek P.J., Baldridge L.A. *et al.* (2008) Mechanical stimulation of bone in vivo reduces osteocyte expression of Sost/sclerostin. *J Biol Chem*, **283**, 5866–5875.

Rubin C.T. and Lanyon L.E. (1984) Dynamic strain similarity in vertebrates: an alternative to allometric limb bone scaling. *J Theor Biol*, **107**, 321.

Rubin C.T. and Lanyon L.E. (1987) Kappa Delta Award paper. Osteoregulatory nature of mechanical stimuli: function as a determinant for adaptive remodeling in bone. *J Orthop Res*, **5**, 300–310.

Rubin C.T., Sommerfeldt D.W., Judex S. and Qin Y. (2001) Inhibition of osteopenia by low magnitude, high-frequency mechanical stimuli. *Drug Discov Today*, **6**, 848–858.

Rubinacci A., Benelli F.D., Borgo E. and Villa I. (2000) Bone as an ion exchange system: evidence for a pump-leak mechanism devoted to the maintenance of high bone K(+). *Am J Physiol Endocrinol Metab*, **278**, E15–E24.

Saito M., Fujii K., Mori Y. and Marumo K. (2006) Role of collagen enzymatic and glycation induced cross-links as a determinant of bone quality in spontaneously diabetic WBN/Kob rats. *Osteoporos Int*, **17**, 1514–1523.

Saito M., Marumo K., Fujii K. and Ishioka N. (1997) Single-column high-performance liquid chromatographic-fluorescence detection of immature, mature, and senescent cross-links of collagen. *Anal Biochem*, **253**, 26–32.

Schiessl H., Frost H.M. and Jee W.S. (1998) Estrogen and bone-muscle strength and mass relationships. *Bone*, **22**, 1–6.

Schriefer J.L., Warden S.J., Saxon L.K. *et al.* (2005) Cellular accommodation and the response of bone to mechanical loading. *J Biomech*, **38**, 1838–1845.

Shiraki M., Kuroda T., Tanaka S. *et al.* (2008) Nonenzymatic collagen cross-links induced by glycoxidation (pentosidine) predicts vertebral fractures. *J Bone Miner Metab*, **26**, 93–100.

Shore R.M., Langman C.B. and Poznanski A.K. (2000) Lumbar and radial bone mineral density in children and adolescents with X-linked hypophosphatemia: evaluation with dual X-ray absorptiometry. *Skeletal Radiol*, **29**, 90–93.

Shore E.M., Xu M., Feldman G.J. *et al.* (2006) A recurrent mutation in the BMP type I receptor ACVR1 causes inherited and sporadic fibrodysplasia ossificans progressiva. *Nat Genet*, **38**, 525–527.

Smit T.H. and Burger E.H. (2000) Is BMU-coupling a strain-regulated phenomenon? A finite element analysis. *J Bone Miner Res*, **15**, 301.

Takahashi H., Epker B. and Frost H.M. (1964) Resorption precedes formative activity. *Surg Forum*, **15**, 437.

Talmage R.V. (2004) Perspectives on calcium homeostasis. *Bone*, **35**, 577.

Talmage R.V., Matthews J.L., Mobley H.T. and Lester G.E. (2003) Calcium homeostasis and bone surface proteins, a postulated vital process for plasma calcium control. *J Musculoskelet Neuronal Interact*, **3**, 194–200.

Teitelbaum S.L. (2000) Bone resorption by osteoclasts. *Science*, **289**, 1504–1508.

Thompson D.A. (1917) *On Growth and Form*, Cambridge University Press, Cambridge.

Turner R.T., Vandersteenhoven J.J. and Bell N.H. (1987) The effects of ovariectomy and 17 beta-estradiol on cortical bone histomorphometry in growing rats. *J Bone Miner Res*, **2**, 115–122.

Walsh W.R. and Guzelsu N. (1991) Electrokinetic behavior of intact wet bone: compartmental model. *J Orthop Res*, **9**, 683–692.

Wang X., Shen X., Li X. and Agrawal C.M. (2002) Age-related changes in the collagen network and toughness of bone. *Bone*, **31**, 1–7.

Weber M., Roschger P., Fratzl-Zelman N. *et al.* (2006) Pamidronate does not adversely affect bone intrinsic material properties in children with osteogenesis imperfecta. *Bone*, **39**, 616–622.

Wilks D., Winwood K., Gilliver S.F. *et al.* (2009) Bone mass and geometry of the tibia and the radius of Master sprinters, middle and long distance runners, race walkers, and sedentary control participants: a pQCT study. *Bone*, **45** (1), 91–97. PMID 19332164.

Wolff J. (1899) Die Lehre von der functionellen Knochengestalt. *Arch Pathol Anat Physiol*, **155**, 256.

Yamada H. and Evans F.G. (1970) *Strength of Biological Materials*, The Williams & Wilkins Company, Baltimore.

Zehnder Y., Luthi M., Michel D. *et al.* (2004) Long-term changes in bone metabolism, bone mineral density, quantitative ultrasound parameters, and fracture incidence after spinal cord injury: a cross-sectional observational study in 100 paraplegic men. *Osteoporos Int*, **15**, 180–189.

Ziambaras K., Lecanda F., Steinberg T.H. and Civitelli R. (1998) Cyclic stretch enhances gap junctional communication between osteoblastic cells. *J Bone Miner Res*, **13**, 218.

Zioupos P. and Currey J.D. (1998) Changes in the stiffness, strength, and toughness of human cortical bone with age. *Bone*, **22**, 57–66.

1.4 Tendon Physiology

Nicola Maffulli[1], Umile Giuseppe Longo[2], Filippo Spiezia[2] and Vincenzo Denaro[2], [1]Queen Mary University of London, Centre for Sports and Exercise Medicine, Barts, Mile End Hospital, The London School of Medicine and Dentistry, London, UK, [2]Campus Bio-Medico University, Department of Orthopaedic and Trauma Surgery, Rome, Italy

1.4.1 TENDONS

Tendons are part of a musculotendinous unit; they transmit forces from muscle to rigid bone levers, producing joint motion and enhancing joint stability (Kvist, 1994), and are made of connective tissue.

Tendons are subjected to high forces, their material proprieties are higher than muscles, and given their proprioceptive properties, they help to maintain posture (Benjamin, Qin and Ralphs, 1995). Loads make tendons more rigid, thus they absorb less energy but are more effective at moving heavy loads (Fyfe and Stanish, 1992). At low rates of loading, the tendons are more viscous, absorbing more energy, and being less effective at moving loads (Fyfe and Stanish, 1992). Tendons concentrate the pull of muscle on a small area. Hence the muscle can change the direction of pull and act from a distance. Tendons also produce an optimal distance between the muscle belly and the joint without requiring an extended length of muscle between the origin and insertion. Tendons have a peculiar anatomy on which their physical properties depend (Kastelic, Galeski and Baer, 1978).

The number, size and orientation of the collagen fibres, as well as their thickness and internal fibrillar organization determine the strength of a tendon (Oxlund, 1986).

Collagen, glycosaminoglycans, noncollagenous proteins, cells, and water are abundant in tendons (Longo, Ronga and Maffulli, 2009a, 2009b). About 90–95% of the cellular elements of tendons are tenoblasts and tenocytes. These are aligned in rows between collagen fibre bundles (Longo *et al.*, 2008b). Tenoblasts transform into mature tenocytes, and are highly metabolically immature, spindle-shaped cells, with numerous cytoplasmic organelles (Maffulli *et al.*, 2008). Tenocytes produce extracellular matrix proteins (Longo *et al.*, 2007, 2008a, 2009).

Sports or work activity may modify the alignment of the fibres of the tendon.

The majority of the tendon fibres run in the direction of stress, with a spiral component. Some fibres may run perpendicular to the line of stress (Jozsa *et al.*, 1991). Fibres with relatively small diameter can run the full length of a long tendon (Kirkendall and Garrett, 1997); those with a diameter greater than 1500 A may not extend the full length (O'Brien, 2005). Tendons may be flattened or rounded. They may be found at the origin, insertion or form tendinous intersections within a muscle. An aponeurosis is a flattened tendon, consisting of several layers of densely arranged collagen fibres. The fascicles are parallel in one layer but run in different directions in adjacent layers. The interior of the tendon consists mainly of longitudinal fibrils, with some transverse and horizontal collagen fibrils (Chansky and Iannotti, 1991). Transmission and scanning electron microscopy demonstrate that collagen fibrils are orientated longitudinally, transversely and horizontally. Crossing each other, the longitudinal fibrils form spirals and plaits (Chansky and Iannotti, 1991; Jozsa *et al.*, 1991).

Tendons may give rise to fleshy muscles: for example, the semimembranous tendon has several expansions that form ligaments, including the oblique popliteal ligament of the knee and the fascia covering the popliteus muscle.

Segmental muscles that develop from myotomes often have tendinous intersections. In certain areas, each segment has its own blood and nerve supply, as is the case for the rectus abdominis, the hamstrings and the sterno-clido-mastoid.

Tendons which cross articular surfaces or bone may have sesamoid bones, already present as cartilaginous nodules in the foetus. There is occasionally a sesamoid in the lateral head of the gastrocnemius (fabella), in the tibialis anterior, opposite the distal aspect of the medial cuneiform, and in the tibialis posterior below the plantar calcaneo navicular ligament, the 'spring ligament'. The long heads of the biceps brachii, the FHL (flexor hallucis longus) and the popliteus are perfect examples of how the tendons may be intracapsular. In these cases, tendons are surrounded by the synovial membrane of the joint, and this membrane extends for a variable distance beyond the joint itself.

When tendons pass over bony prominences or lie in grooves, they may be covered by fibrous sheaths or retinacula to prevent them from bowstringing when the muscle contracts. Otherwise

Strength and Conditioning – Biological Principles and Practical Applications Marco Cardinale, Rob Newton, and Kazunori Nosaka.

reflection pulleys may hold tendons as they pass over a curved area.

Tendons may be enclosed in synovial membrane when they run in fibro-osseous tunnels or pass under retinacula, the fascial slings. A film of fluid separates two continuous, concentric layers of the membrane. The visceral layer surrounds the tendon and the parietal layer is attached to the adjacent connective tissues. A mesotendon is present as a tendon invaginates into the sheath.

In the fibro-osseous sheaths of the phalanges of the hand and foot, the synovial folds are called the vincula longa and vincula brevia. Vincula contain the blood vessels, which supply the flexor tendons inside the sheaths. The lining of the sheath secretes synovial fluid and is responsible for inflammatory reactions through cellular proliferation and the formation of more fluid, which may result in adhesions and restriction of movement between the two layers.

The plantaris tendon in the gastrosoleus complex is a classic example of supernumerary tendon.

1.4.2 THE MUSCULOTENDINOUS JUNCTION

In the mesenchyme, tendons develop independently; their connection with the muscle appears later. The junctional area between the muscle and the tendon is called the myotendinous junction. During transmission of muscular contractile force to the tendon, this area is subjected to great mechanical stress. The collagen fibres of a tendon extend into the body of the muscle, increasing the anchoring surface area. Tendinous fibres are disposed to direct the force produced by the muscular contraction to the point of insertion.

The tendon elongates at the musculotendinous junction, and the growth plate of a muscle is considered to be the musculotendinous junction. It contains cells that deposit collagen and can elongate rapidly, and also contains nerve receptors and organs of Golgi. Terminal expansions of muscle fibres may be present here, as shown by electron microscopy. These ends have a highly indented sarcolemma, with a dense internal layer of cytoplasm into which the actin filaments of the adjacent sarcomeres are inserted. Muscle tears tend to occur at the musculotendinous junction (Garrett, 1990). The sites of muscle tears can be explained by variations in the extent of the tendon into the muscle at the origin and insertion. The adductor longus tendon may present variations in the shape and extent. In the hamstrings, tendinous intersections denoting the original myotomes are found.

1.4.3 THE OSEOTENDINOUS JUNCTION

The osteotendinous junction (OTJ) presents a gradual transition from tendon to fibrocartilage to lamellar bone. This area can be divided into four zones: pure fibrous tissue, unmineralized fibrocartilage, mineralized fibrocartilage, and bone. The outer limit of the mineralized fibrocartilage, the tidemark, consists of basophilic lines (cement or blue lines). These lines are usually smoother than at the osteo-chondral junction. On the tendon side of the tidemark there are chondrocytes. Tendon fibres can extend as far as the osteo-chondral junction. Rare blood vessels may cross from bone to tendon. If the attachment is very close to the articular cartilage, the zone of fibrocartilage is continuous with the articular cartilage. The chemical composition of fibrocartilage is age-dependant, both in the OTJ and in other fibrocartilagenous zones of the tendon.

Smooth mechanical transition is allowed by osteogenesis at the tendon–bone junction. The periosteum has osteogenic potential, except where tendons are inserted. Dense collagen fibres connect the periosteum to the underlying bone. During bone growth, collagen fibres from the tendon are anchored deeper into the deposited bone. In stress fractures of the tibia, variations in hot spots on bone scans may be explained by variations in the attachments of the tendon to bone (Ekenman *et al.*, 1995).

The occupation and sports activity of an individual may influence the structure of the attachment zone of a tendon. For example, the insertion of the biceps of a window cleaner, who works with his forearm pronated, will differ from that of an individual who works with the forearm supinated.

1.4.4 NERVE SUPPLY

Sensory nerves from the overlying superficial nerves and from nearby deep nerves supply the tendons. The nerve supply is mostly, but not all, afferent. Afferent receptors can be found near the musculotendinous junction (Stilwell, 1957), either on the surface or in the tendon. Longitudinal plexus enter via the septa of the endotendon or the mesotendon if there is a synovial sheath. Branches may reach the surface or the interior of a tendon, passing from the paratenon via the epitenon.

There are four types of receptor. Type I receptors (pressure receptors or Ruffini corpuscles) are very sensitive to stretch and adapt slowly (Stilwell, 1957). Type II receptors (Vater–Pacini corpuscles), are activated by any movement. Type III receptors (Golgi tendon organs) are mechanoreceptors. They consist of unmyelinated nerve endings encapsulated by endoneural tissue. They lie in series with the extra fusal fibres and monitor increases in muscle tension rather than length. Muscle contraction produces pressure; the lamellated corpuscles respond to this stimulus, with the amount of pressure depending on the force of contraction. They may provide a more finely tuned feedback. Type IV receptors are the free nerve endings that act as pain receptors.

1.4.5 BLOOD SUPPLY

The blood supply of tendons is divided into three regions: (1) the musculotendinous junction; (2) the length of the tendon; and (3) the tendon–bone junction. The blood vessels originate from vessels in the perimysium, periosteum and via the paratenon and mesotendon.

Superficial vessels in the surrounding tissues allow the supply of blood to the musculotendinous junction. Small arteries branch and supply both muscles and tendons without anastamosis between the capillaries.

The paratenon is the main blood source to the middle portion of the tendon. The main blood supply in tendons that are exposed to friction and are enclosed in a synovial sheath is via the vincula. Blood vessels in the paratenon run transversely towards the tendon and branch several times before running parallel to the long axis of the tendon. The vessels enter the tendon along the endotendon; the arterioles run longitudinally, flanked by two venules. Capillaries loop from the arterioles to the venules, but they do not penetrate the collagen bundles.

Vessels supplying the bone–tendon junction supply the lower third of the tendon. There is no direct communication between the vessels because of the fibrocartilaginous layer between the tendon and bone, but there is some indirect anastamosis between the vessels.

At sites of friction, torsion, or compression, the blood supply of tendons is compromised; this happens in the tibialis posterior, supraspinatus, and Achilles tendons (Frey, Shereff and Greenidge, 1990; Ling, Chen and Wan, 1990). In the critical zone of the supraspinatus there can be an area of hypervascularity secondary to low-grade inflammation with neovascularization due to mechanical irritation (Ling, Chen and Wan, 1990).

There is an avascular region at the metacarpo-phalangeal joint and the proximal interphalangeal joint, possibly resulting from the mechanical forces exerted at these zones (Zhang *et al.*, 1990).

The Achilles tendon is the thickest and strongest tendon. It is approximately 15 cm long and on its anterior surface it receives the muscular fibres from the soleus for several centimetres. As the tendon descends, it twists; this produces an area of stress in the tendon, which is most marked 2–6 cm above the insertion, the area of poor vascularity, a common site of tendon ailments (Barfred, 1971).

The blood supply of the Achilles tendon consists mainly of longitudinal arteries that course the length of the tendon. This tendon is supplied at its musculotendinous junction, along the length of the tendon, and at its junction with bone.

1.4.6 COMPOSITION

Tendons consist of 30% collagen and 2% elastin embedded in an extracellular matrix containing 68% water and tenocytes. Elastin contributes to the flexibility of the tendon. The collagen protein, tropocollagen, forms 65–80% of the mass of dry-weight tendons and ligament.

Tendons are relatively avascular and appear white. A tendon is a roughly uniaxial composite mainly comprising type I collagen in an extracellular matrix composed largely of mucopolysaccharides and a proteoglycan gel (Kastelic, Galeski and Baer, 1978).

Ligaments and tendons consist mainly of type I collagen. Ligaments have 9–12% type III collagen and are more cellular than tendons (Khan *et al.*, 1999). Type II collagen is found abundantly in the fibrocartilage at the attachment zone of the tendon (the OTJ) and is also present in tendons that wrap around bony pulleys. Collagen consists of clearly defined, parallel, and wavy bundles and has a characteristic reflective appearance under polarized light between the collagen bundles.

1.4.7 COLLAGEN FORMATION

The structural unit of collagen is the tropocollagen. It is a long, thin protein, 280 nm long and 1.5 nm wide, consisting mainly of type I collagen. Tropocollagen is formed as procollagen in fibroblast cells and is secreted and cleaved extracellularly. Successively, it becomes collagen.

The α-chain comprises 100 amino acids. There are three α-chains, which are surrounded by a thin layer of proteoglycans and glycosaminoglycans Two of the α-chains are identical (α1); one differs slightly (α2). The three-polypeptide chains each form a left-handed helix. Hydrogen bonds connect the chains together to form a rope-like right-handed superhelix. A collagen molecule has a rod-like shape. Almost two thirds of the collagen molecule consists of three amino acids, glycine (33%), proline (15%), and hydroxyproline (15%). Each α-chain consists of a repeating triplet of glycine and two other amino acids. Glycine is found at every third residue, while proline (15%) and hydroxyproline (15%) occur frequently at the other two positions. Glycine enhances stability by forming hydrogen bonds between the three chains. Collagen also contains two amino acids, hydroxyproline and hydrozylysine (1.3%), which are not often found in other proteins. Hydroxyproline and hydroxylysine increase the strength of collagen. Hydroxyproline is also essential in the hydrogen bonding between the polypeptide chains, while hydroxylysine is essential in the covalent cross-linking of tropocollagen into bundles of various sizes. The domains are non-helical peptides placed at both ends of procollagen. Peptides enzymatically cleave the domains to form tropocollagen.

1.4.8 CROSS-LINKS

Electrostatic cross-linking chemical bonds stabilize tropocollagen molecules and hold them together. Hydroxyproline is important in hydrogen bonding (intramolecularly) between the polypeptide chains. Hydroxylysine is also involved in covalent (intermolecular) cross-linking between adjacent tropocollagen molecules. Lysyl-oxidase is the key enzyme; it is a rate-limiting step of collagen cross-linking. Cross-links of hydroxylysin are the most prevalent intermolecular cross-links in native insoluble collagen, and have a paramount role in creating the tensile strength of collagen, allowing energy absorption, and increasing collagen's resistance to proteases.

Shortly after the synthesis of collagen, fibres acquire all the cross-links they will have. Cross-links are at their maximum in early postnatal life and reach their minimum at physical maturity. Newly synthesized collagen molecules are stabilized by reducible cross-links, but their numbers decrease during maturation. Non-reducible cross-links are found in mature collagen, which is stiffer, stronger, and more stable. Reduction of cross-links makes collagen fibres weak and friable. Hence, cross-linking of collagen is one of the best biomarkers of ageing.

Cross-linking substances are removed by metabolic processes in early life but accumulate in old age.

1.4.9 ELASTIN

Elastin does not contain much hydroxyproline or lysine, but is rich in glycine and proline. It does not form helices and is hydrophobic. It has a large content of valine and contains desmosine and isodesmonine, which form cross-links between the polypeptides, but no hydroxylysine. Elastin contributes to the flexibility of a tendon.

1.4.10 CELLS

Tenocytes and tenoblasts or fibroblasts are tendon cells. Tenocytes are flat, tapered cells, spindle-shaped longitudinally and stellate in cross-section. Tenocytes may be found sparingly in rows between collagen fibrils. They posses cell processes that form a three-dimensional network extending through the extracellular matrix. The communication is via cell processes, and these cells may be motile (Kraushaar and Nirschl, 1999).

Tenoblasts are spindle-shaped or stellate cells with long tapering eosinophillic flat nuclei. Tenoblasts are motile and highly proliferative. They have well-developed rough endoplasmic reticulum, on which the precursor polypeptides of collagen, elastin, proteoglycans, and glycoproteins, are synthesized. Tendon fibroblasts (tenoblasts) in the same tendon may have different functions. The epitenocyte functions as a modified fibroblast with a well-developed capacity of repair.

1.4.11 GROUND SUBSTANCE

Ground substance is composed of proteoglycans and glycoproteins which surround the collagen fibres. Gliding and cross-tissue interactions are allowed by the high viscosity of ground substance, which provides the structural support, lubrication, and spacing of the fibres. Nutrients and gases diffuse through ground substance. It regulates the extracellular assembly of procollagen into mature collagen. Water makes up 60–80% of its total weight. Less than 1% of the total dry weight of tendon is composed of proteoglycans and glycoproteins. These proteins are involved with intermolecular and cellular interactions and maintain the water within the tissues. Proteoglycans and glycoproteins also play an important role in the formation of fibrils and fibres. The covalent cross-links between the tropocollagen molecules reinforce the fibrillar structure.

The water-binding capacity of these macromolecules is important. Most proteoglycans are orientated at 90° to collagen, and each molecule of proteoglycan can interact with four collagen molecules. Others are randomly arranged to lie parallel to the fibres, but they only interact with one fibre (Scott, 1988). The matrix is constantly being turned over and remodelled by the fibroblasts and by degrading enzymes (collagenases, proteoglycanase, glycoaminoglycanase, and other proteases).

The proteogylcans and glycoproteins consist of two components, glycosaminogylcans (GAGs) and structural glycoproteins. The main proteogylcans in tendons associated with glycosaminoglycans are dermatan sulfate, hyaluronate, chondroitin 4 sulfates, and chondroitin 6 sulfates. Other proteoglycans found in tendons include biglycan, decorin, and aggrecan. Aggrecan is a chondroitin sulfate bearing large proteoglycan in the tension regions of tendons (Vogel *et al.*, 1994). The glycoproteins consist mainly of protein, such as fibronectin, to which carbohydrates are attached.

Fibronectins are high-molecular weight noncollagenous extracellular glycocoproteins. Fibronectin plays a role in cellular adhesion (cell–cell and cell–substrate) and cell migration. It may be essential for the organization of collagen I and III fibrils into bundles, and may act as a template for collagen fibre formation during the remodelling phase.

Hyaluronate is a high-molecular weight matrix glycosaminoglycan, which interacts with fibronectin to produce a scaffold for cell migration. It later replaces fibronectin.

Integrins are extracellular matrix-binding proteins with specific cell-surface receptors. Large amounts of aggrecan and biglycan develop at points where tendons wrap around bone and are subjected to compressive and tensional loads. TGF-β may be involved in differentiation of regions of tendon subjected to compression as compressed tendon contains both decorin and biglycan, whereas tensional tendons contain primarily decorin (Vogel and Hernandez, 1992). The synthesis of proteoglycans begins in the rough endoplasmic reticulum, where the protein portion is synthesized. Glycosylation starts in the rough endoplasmic reticulum and is completed in the Golgi complex, where sulfation takes place. The turnover of proteoglycans is rapid, from 2 to 10 days. Lysosomal enzymes degrade the proteoglycans, and lack of specific hydrolases in the lysososmes results in their accumulation.

When newly formed, the ground matrix appears vacuolated. The formation of tropocollagen and of extracellular matrix are closely inter-related. The proteoglycans in the ground substance seem to regulate fibril formation as the content of proteoglycans decreases in tendons when the tropocollagen has reached its ultimate size. An adequate amount of ground substance is necessary for the aggregation of collagenous proteins into the shape of fibrils.

1.4.12 CRIMP

Crimp represents a regular sinusoidal pattern in the matrix. Collagen fibrils in the rested, nonstrained state are not straight, but wavy or crimped. Both tendons and ligaments have the main feature of crimp. The periodicity and amplitude of crimp is structure-specific (Viidik, 1973). It is visualized under polarized light. Crimp provides a buffer in which slight longitudinal elongation can occur without fibrous damage, and acts as a shock absorber along the length of the tissue. Different patterns of crimping exist: straight or undulated in a planar wave pattern.

The fibre bundles are interwoven without regular orientation and the tissues are irregularly arranged. Fibres are regularly arranged and have an orderly parallel arrangement if tension is only in one direction. The parallel orientation of collagen fibres is lost in tendinopathy. A decrease in collagen fibre diameter and in the overall density of collagen may be found. Collagen microtears may be surrounded by erythrocytes, fibrin, and fibronectin deposits. In a normal tendon, collagen fibres in tendons are tightly bundled in a parallel fashion. In tendinopathic samples there is unequal and irregular crimping, loosening, and increased waviness of collagen fibres, with an increase in type III (reparative) collagen.

Many factors can affect collagen production: heredity, diet, nerve supply, inborn errors, and hormones. Corticosteroids also inhibit the production of new collagen. Insulin, oestrogen, and testosterone may increase collagen production.

Osteogenesis imperfecta, Ehlers–Danos syndrome, scurvy, and progressive systemic sclerosis are collagen disorders. Muscles and tendons atrophy and the collagen content decreases when the nerve supply to the tendon is interrupted. Inactivity also results in increased collagen degradation, decreased tensile strength, and decreased concentration of metabolic enzymes. Given the reduction of enzymes essential for the formation of collagen with age, repair of soft tissue is delayed in the older age groups. Exercise increases collagen synthesis, the number and size of the fibrils, and the concentration of metabolic enzymes. Physical training increases the tensile and maximum static strength of tendons.

Cell and tissue organization, cell–cell and cell–matrix communication, cell proliferation and apoptosis, matrix remodelling, cell migration, and many other cell behaviours in both physiological and pathophysiological conditions in all the musculoskeletal tissues (bone, cartilage, ligament, and tendon) are modulated by dynamic stresses and strains.

Changes in tissue structure and function following application of mechanical forces, including gravity, tension, compression, hydrostatic pressure, and fluid shear stress are the main subjects of mechanobiology (Wang, 2006). The ability of tendons to alter their structure in response to mechanical loading is called tissue mechanical adaptation, and it is effected by cells in tissues at all times. It is not yet clear how cells perceive dynamic stresses and strains and convert them into biochemical signals which lead to tissue adaptive physiological or pathological changes.

Tendons show viscoelastic behaviour and adapt their functional and mechanical features to dynamic stresses and strains. The mechanical behaviour of tendons is unique to their peculiar structural characteristics. Viscoelasticity is the property of materials that exhibit both viscous and elastic characteristics when undergoing deformation (Butler *et al.*, 1978). Collagen, water, and interactions between collagenous proteins and noncollagenous proteins (i.e. proteoglycans) determine the viscoelastic features of tendons. Tendons are therefore more deformable at low strain rates, but they are less effective in load transfer. Tendons are less deformable at high strain rates, with high degree of stiffness, and they become more effective in moving high loads (Jozsa and Kannus, 1997).

Structural changes in tendons are caused by dynamic stresses and strains imposed on them. Sports determine an increase in the cross-sectional area and tensile strength of tendons, and tenocytes increase the production of type I collagen (Langberg, Rosendal and Kjaer, 2001; Michna and Hartmann, 1989; Suominen, Kiiskinen and Heikkinen, 1980; Tipton *et al.*, 1975). Inappropriate physical exercise or occupation leads to tendon overuse injuries and tendinopathy (Khan and Maffulli, 1998; Maffulli, Khan and Puddu, 1998). There is greater matrix–collagen turnover in growing chickens, resulting in reduced maturation of tendon collagen during high-intensity exercise (Curwin, Vailas and Wood, 1988). In an animal model following high-intensity training there is an increased IGF-I immunoreactivity (Hansson *et al.*, 1988). IGF-I acts as a potent stimulator of mitogenesis and protein synthesis (Simmons *et al.*, 2002) and therefore may be used as a protein marker for remodelling activities of the tendon.

Physical training results in an increased turnover of type I collagen in local connective tissue of the peritendinous Achilles tendon region (Langberg, Rosendal and Kjaer, 2001). Immobilization decreases the total weight, tensile strength, and stiffness of tendons (Maffulli and King, 1992).

Tendon overuse injuries (Amiel *et al.*, 1982) are common in occupational and athletic settings (Almekinders and Temple, 1998). An important causative factor for tendinopathy is excessive mechanical loading (Lippi, Longo and Maffulli, 2009). Repetitive strains below the failure threshold of tendons may cause microinjuries to the tendon, possibly with episodes of tendon inflammation (Khan and Maffulli, 1998). Human tendons following repetitive mechanical loading present increased levels of PGE2 (Langberg, Rosendal and Kjaer, 2001). Studies show that *in vitro* repetitive mechanical loading of human tendon fibroblasts increases the production of PGE2 (Almekinders, Banes and Ballenger, 1993) and LTB4 (Li *et al.*, 2004). Prolonged PGE1 administration produces peri- and intra-tendinous degeneration (Sullo *et al.*, 2001). Degenerative changes within the rabbit patellar tendon have been reported after repeated exposure of the tendon to PGE2 (Khan, Li and Wang, 2005). Clinical studies have reported no significant differences in the mean concentrations of PGE2 between tendons with tendinopathy and normal tendons. However, there were higher concentrations of the excitatory neurotransmitter glutamate in tendinopathic Achilles tendons (Alfredson, Thorsen and Lorentzon, 1999). Repetitive submaximal strains below failure threshold are probably responsible for tendon microinjuries and episodes of PG-mediated tendon inflammation. It is therefore possible that when microinjuries become clinically evident, the PGEs are no longer present in the tendon.

Overuse of tendons can lead to detrimental changes in tendon tissue structure and result in tendinopathic changes (Longo *et al.*, 2007, 2008a, 2009). Despite the clear clinical relevance of mechanotransduction signalling pathways in tenocytes, the mechanisms by which these cells perceive and respond to mechanical stimuli are poorly understood. Calcium ions play an important role in mechanotransduction and act as one of the primary second messengers utilized by cells to convert mechanical signals to biochemical signals (Bootman *et al.*, 2001). Preliminary data have shown upregulation of calcium signalling pathways in human tenocytes exposed to fluid flow-induced shear stress (el Haj *et al.*, 1999), and it has been proposed that mechanosensitive and voltage-gated ion channels may play a key role in the initial responses of human tenocytes to mechanical load (Banes *et al.*, 1995). Both mechanosensitive and voltage-gated ion channels may have key roles in mechanotransduction signalling pathways in their connective tissues, including bone (el Haj *et al.*, 1999), smooth muscle (Holm *et al.*, 2000), and heart cells (Niu and Sachs, 2003). Voltage-operated calcium channels (VOCCs) permit the influx of extracellular calcium in response to changes in membrane potential and form the basis of electrical signalling in excitable cells (Catterall, 2000). VOCCs also are potentially mechanosensitive (Lyford *et al.*, 2002). Mechanosensitive members of the tandem pore domain potassium channel family (Patel and Honore, 2001), including TREK-1 and TREK-2, may be involved in mechanotransduction signalling pathways in smooth-muscle cells (Koh *et al.*, 2001), heart cells (Aimond *et al.*, 2000), and bone cells (Hughes *et al.*, 2006). TREK-1 channels produce a spontaneously active background leak potassium conductance to hyperpolarize the cell membrane potential and regulate electrical excitability (Heurteaux *et al.*, 2004). TREK-1 is sensitive to membrane stretch, lysophosphatidylcholines, lysophosphatidic acids, polyunsaturated fatty acids, intracellular pH, temperature, and a range of clinically relevant compounds including general and local anaesthetics (Gruss *et al.*, 2004).

Human tenocytes express a diverse array of ion channels, including L-type VOCCs and TREK-1 (Magra *et al.*, 2007). VOCCs are likely to be a key mediator of calcium signalling events in human tenocytes, whereas TREK-1 could potentially perform a number of roles in tenocytes, ranging from osmoregulation and cell volume control, to control of resting membrane potentials levels of electrical excitability, to the direct detection of mechanical stimuli. TREK-1 and VOCC channels could be potential targets for pharmacological management of chronic tendinopathies (Magra *et al.*, 2007).

References

Aimond F., Rauzier J.M., Bony C. and Vassort G. (2000) Simultaneous activation of p38 MAPK and p42/44 MAPK by ATP stimulates the K+ current ITREK in cardiomyocytes. *J Biol Chem*, **275** (50), 39, 110–116.

Alfredson H., Thorsen K. and Lorentzon R. (1999) In situ microdialysis in tendon tissue: high levels of glutamate, but not prostaglandin E2 in chronic Achilles tendon pain. *Knee Surg Sports Traumatol Arthrosc*, **7** (6), 378–381.

Almekinders L.C. and Temple J.D. (1998) Etiology, diagnosis, and treatment of tendonitis: an analysis of the literature. *Med Sci Sports Exerc*, **30** (8), 1183–1190.

Almekinders L.C., Banes A.J. and Ballenger C.A. (1993) Effects of repetitive motion on human fibroblasts. *Med Sci Sports Exerc*, **25** (5), 603–607.

Amiel D., Woo S.L., Harwood F.L. and Akeson W.H. (1982) The effect of immobilization on collagen turnover in connective tissue: a biochemical-biomechanical correlation. *Acta Orthop Scand*, **53** (3), 325–332.

Banes A.J., Tsuzaki M., Yamamoto J. *et al.* (1995) Mechanoreception at the cellular level: the detection, interpretation, and diversity of responses to mechanical signals. *Biochem Cell Biol*, **73** (7–8), 349–365.

Barfred T. (1971) Experimental rupture of the Achilles tendon. Comparison of various types of experimental rupture in rats. *Acta Orthop Scand*, **42** (6), 528–543.

Benjamin M., Qin S. and Ralphs J.R. (1995) Fibrocartilage associated with human tendons and their pulleys. *J Anat*, **187** (Pt 3), 625–633.

Bootman M.D., Collins T.J., Peppiatt C.M. *et al.* (2001) Calcium signalling: an overview. *Semin Cell Dev Biol*, **12** (1), 3–10.

Butler D.L., Grood E.S., Noyes F.R. and Zernicke R.F. (1978) Biomechanics of ligaments and tendons. *Exerc Sport Sci Rev*, **6**, 125–181.

Catterall W.A. (2000) Structure and regulation of voltage-gated Ca2+ channels. *Annu Rev Cell Dev Biol*, **16**, 521–555.

Chansky H.A. and Iannotti J.P. (1991) The vascularity of the rotator cuff. *Clin Sports Med*, **10** (4), 807–822.

Curwin S.L., Vailas A.C. and Wood J. (1988) Immature tendon adaptation to strenuous exercise. *J Appl Physiol*, **65** (5), 2297–2301.

Ekenman I., Tsai-Fellander L., Johansson C. and O'Brien M. (1995) The plantar flexor muscle attachments on the tibia. A cadaver study. *Scand J Med Sci Sports*, **5** (3), 160–164.

el Haj A.J., Walker L.M., Preston M.R. and Publicover S.J. (1999) Mechanotransduction pathways in bone: calcium fluxes and the role of voltage-operated calcium channels. *Med Biol Eng Comput*, **37** (3), 403–409.

Frey C., Shereff M. and Greenidge N. (1990) Vascularity of the posterior tibial tendon. *J Bone Joint Surg Am*, **72** (6), 884–888.

Fyfe I. and Stanish W.D. (1992) The use of eccentric training and stretching in the treatment and prevention of tendon injuries. *Clin Sports Med*, **11** (3), 601–624.

Garrett W.E. Jr. (1990) Muscle strain injuries: clinical and basic aspects. *Med Sci Sports Exerc*, **22** (4), 436–443.

Gruss M., Bushell T.J., Bright D.P. *et al.* (2004) Two-pore-domain K+ channels are a novel target for the anesthetic gases xenon, nitrous oxide, and cyclopropane. *Mol Pharmacol*, **65** (2), 443–452.

Hansson H.A., Engstrom A.M., Holm S. and Rosenqvist A.L. (1988) Somatomedin C immunoreactivity in the Achilles tendon varies in a dynamic manner with the mechanical load. *Acta Physiol Scand*, **134** (2), 199–208.

Heurteaux C., Guy N., Laigle C. *et al.* (2004) TREK-1, a K+ channel involved in neuroprotection and general anesthesia. *EMBO J*, **23** (13), 2684–2695.

Holm A.N., Rich A., Sarr M.G. and Farrugia G. (2000) Whole cell current and membrane potential regulation by a human smooth muscle mechanosensitive calcium channel. *Am J Physiol Gastrointest Liver Physiol*, **279** (6), G1155–G1161.

Hughes S., Magnay J., Foreman M. *et al.* (2006) Expression of the mechanosensitive 2PK+ channel TREK-1 in human osteoblasts. *J Cell Physiol*, **206** (3), 738–748.

Jozsa L. and Kannus P. (1997) Human tendons: anatomy, physiology, and pathology. *Human Kinetics*, **164**.

Jozsa L., Kannus P., Balint J.B. and Reffy A. (1991) Three-dimensional ultrastructure of human tendons. *Acta Anat (Basel)*, **142** (4), 306–312.

Kastelic J., Galeski A. and Baer E. (1978) The multicomposite structure of tendon. *Connect Tissue Res*, **6** (1), 11–23.

Khan K.M. and Maffulli N. (1998) Tendinopathy: an Achilles' heel for athletes and clinicians. *Clin J Sport Med*, **8** (3), 151–154.

Khan K.M., Cook J.L., Bonar F. *et al.* (1999) Histopathology of common tendinopathies. Update and implications for clinical management. *Sports Med*, **27** (6), 393–408.

Khan M.H., Li Z. and Wang J.H. (2005) Repeated exposure of tendon to prostaglandin-E2 leads to localized tendon degeneration. *Clin J Sport Med*, **15** (1), 27–33.

Kirkendall D.T. and Garrett W.E. (1997) Function and biomechanics of tendons. *Scand J Med Sci Sports*, **7** (2), 62–66.

Koh S.D., Monaghan K., Sergeant G.P. *et al.* (2001) TREK-1 regulation by nitric oxide and cGMP-dependent protein kinase. An essential role in smooth muscle inhibitory neurotransmission. *J Biol Chem*, **276** (47), 44–338.

Kraushaar B.S. and Nirschl R.P. (1999) Tendinosis of the elbow (tennis elbow). Clinical features and findings of histological, immunohistochemical, and electron microscopy studies. *J Bone Joint Surg Am*, **81** (2), 259–278.

Kvist M. (1994) Achilles tendon injuries in athletes. *Sports Med*, **18** (3), 173–201.

Langberg H., Rosendal L. and Kjaer M. (2001) Training-induced changes in peritendinous type I collagen turnover determined by microdialysis in humans. *J Physiol*, **534** (Pt 1): 297–302.

Li Z., Yang G., Khan M. *et al.* (2004) Inflammatory response of human tendon fibroblasts to cyclic mechanical stretching. *Am J Sports Med*, **32** (2), 435–440.

Ling S.C., Chen C.F. and Wan R.X. (1990) A study on the vascular supply of the supraspinatus tendon. *Surg Radiol Anat*, **12** (3), 161–165.

Lippi G., Longo U.G. and Maffulli N. (2009) Genetics and sports. *Br Med Bull*, **93**, 27–47.

Longo U.G., Franceschi F., Ruzzini L. *et al.* (2007) Light microscopic histology of supraspinatus tendon ruptures. *Knee Surg Sports Traumatol Arthrosc*, **15** (11), 1390–1394.

Longo U.G., Franceschi F., Ruzzini L. *et al.* (2008a) Histopathology of the supraspinatus tendon in rotator cuff tears. *Am J Sports Med*, **36** (3), 533–538.

Longo U.G., Oliva F., Denaro V. and Maffulli N. (2008b) Oxygen species and overuse tendinopathy in athletes. *Disabil Rehabil*, **30** (20–22), 1563–1571.

Longo U.G., Ronga M. and Maffulli N. (2009a) Achilles tendinopathy. *Sports Med Arthrosc*, **17** (2), 112–126.

Longo U.G., Ronga M. and Maffulli N. (2009b) Acute ruptures of the achilles tendon. *Sports Med Arthrosc*, **17** (2), 127–138.

Longo U.G., Franceschi F., Ruzzini L. *et al.* (2009) Characteristics at haematoxylin and eosin staining of ruptures of the long head of the biceps tendon. *Br J Sports Med*, **43** (8), 603–607.

Lyford G.L., Strege P.R., Shepard A. *et al.* (2002) alpha(1C) (Ca(V)1.2) L-type calcium channel mediates mechanosensitive calcium regulation. *Am J Physiol Cell Physiol*, **283** (3), C1001–C1008.

Maffulli N. and King J.B. (1992) Effects of physical activity on some components of the skeletal system. *Sports Med*, **13** (6), 393–407.

Maffulli N., Khan K.M. and Puddu G. (1998) Overuse tendon conditions: time to change a confusing terminology. *Arthroscopy*, **14** (8), 840–843.

Maffulli N., Longo U.G., Franceschi F. *et al.* (2008) Movin and Bonar scores assess the same characteristics of tendon histology. *Clin Orthop Relat Res*, **466** (7), 1605–1611.

Magra M., Hughes S., Haj A.J. and Maffulli N. (2007) VOCCs and TREK-1 ion channel expression in human tenocytes. *Am J Physiol Cell Physiol*, **292** (3), C1053–C1060.

Michna H. and Hartmann G. (1989) Adaptation of tendon collagen to exercise. *Int Orthop*, **13** (3), 161–165.

Niu W. and Sachs F. (2003) Dynamic properties of stretch-activated K+ channels in adult rat atrial myocytes. *Prog Biophys Mol Biol*, **82** (1–3), 121–135.

O'Brien M. (2005) Anatomy of tendons, in *Tendon Injuries: Basic Science and Clinical Medicine* (eds N. Maffulli, P. Renström and W.B. Leadbetter), Springer-Verlag, London, pp. 3–13, Chapter 1.

Oxlund H. (1986) Relationships between the biomechanical properties, composition and molecular structure of connective tissues. *Connect Tissue Res*, **15** (1–2), 65–72.

Patel A.J. and Honore E. (2001) Properties and modulation of mammalian 2P domain K+ channels. *Trends Neurosci*, **24** (6), 339–346.

Scott J.E. (1988) Proteoglycan-fibrillar collagen interactions. *Biochem J*, **252** (2), 313–323.

Simmons J.G., Pucilowska J.B., Keku T.O. and Lund P.K. (2002) IGF-I and TGF-beta1 have distinct effects on phenotype and proliferation of intestinal fibroblasts. *Am J Physiol Gastrointest Liver Physiol*, **283** (3), G809–G818.

Stilwell D.L. Jr. (1957) The innervation of tendons and aponeuroses. *Am J Anat*, **100** (3), 289–317.

Sullo A., Maffulli N., Capasso G. and Testa V. (2001) The effects of prolonged peritendinous administration of PGE1 to the rat Achilles tendon: a possible animal model of chronic Achilles tendinopathy. *J Orthop Sci*, **6** (4), 349–357.

Suominen H., Kiiskinen A. and Heikkinen E. (1980) Effects of physical training on metabolism of connective tissues in young mice. *Acta Physiol Scand*, **108** (1), 17–22.

Tipton C.M., Matthes R.D., Maynard J.A. and Carey R.A. (1975) The influence of physical activity on ligaments and tendons. *Med Sci Sports*, **7** (3), 165–175.

Viidik A. (1973) Functional properties of collagenous tissues. *Int Rev Connect Tissue Res*, **6**, 127–215.

Vogel K.G. and Hernandez D.J. (1992) The effects of transforming growth factor-beta and serum on proteoglycan synthesis by tendon fibrocartilage. *Eur J Cell Biol*, **59** (2), 304–313.

Vogel K.G., Sandy J.D., Pogany G. and Robbins J.R. (1994) Aggrecan in bovine tendon. *Matrix Biol*, **14** (2), 171–179.

Wang J.H. (2006) Mechanobiology of tendon. *J Biomech*, **39** (9), 1563–1582.

Zhang Z.Z., Zhong S.Z., Sun B. and Ho G.T. (1990) Blood supply of the flexor digital tendon in the hand and its clinical significance. *Surg Radiol Anat*, **12** (2), 113–117.

1.5 Bioenergetics of Exercise

Ron J. Maughan, School of Sport, Exercise and Health Sciences, Loughborough University, Loughborough, UK

1.5.1 INTRODUCTION

All life requires the continuous expenditure of energy. Even at rest, the average human body consumes about 250 ml of oxygen each minute; this oxygen is used in the chemical reactions that provide the energy necessary to maintain physiological function. During exercise, the muscles require additional energy to generate force or to do work, the heart has to work harder to increase blood supply, the respiratory muscles face an increased demand for moving air in and out of the lungs, and the metabolic rate must increase accordingly. In sustained exercise, an increased rate of energy turnover must be maintained for the whole duration of the exercise and for some time afterwards; depending on the task and the fitness level of the individual, this may be from 5 to 20 times the resting metabolic rate. In very high-intensity activity, the demand for energy may be more than 100 times the resting level, though such intense efforts can be sustained for only very short periods of time. These observations immediately raise several questions: what is this energy used for, where does it come from, what happens when the energy demand exceeds the energy supply, and more.

Though technically it belongs in the field of exercise biochemistry, an understanding of fuel consumption and the processes involved in energy supply is fundamental to exercise physiology and sports nutrition.

1.5.2 EXERCISE, ENERGY, WORK, AND POWER

As mentioned above, the energy demand of the resting metabolism of the average human is met by a rate of oxygen consumption of about 250 ml/min. At rest, the body is in a relative steady state, at least with regard to oxygen content, and all of the energy is supplied by oxidative metabolism, though there will be energy supply that does not involve oxygen consumption in some tissues. The oxygen consumption and the energy expenditure therefore tally very closely; they are in effect different ways of expressing the same thing. In some tissues, however, energy is generated without oxygen being consumed. Red blood cells do not have any mitochondria; these are the subcellular structures that house the metabolic machinery which allows energy to be derived from metabolic fuels in the presence of oxygen. Red blood cells meet their energy needs by breaking down glucose to lactate in a series of reactions that do not involve oxygen. Other tissues, however, remove that lactate from the bloodstream, either by using it as a fuel in the presence of oxygen or by converting it to other valuable metabolic compounds.

In some forms of exercise, not all of the energy demand is met by oxidative metabolism. The rate of oxygen consumption increases more or less linearly as the power output increases, but there is a finite limit to the amount of oxygen that an individual can use: this is identified as the maximum oxygen uptake, or VO_{2max}, and is discussed further below. An individual's VO_{2max} is often referred to as their aerobic capacity and is one of the defining physiological characteristics. Human skeletal muscle, however, can perform work in the absence of an adequate supply of oxygen as a consequence of its ability to generate energy without any oxygen consmuption. In these situations, other ways of expressing the metabolic rate must be used. An oxygen consumption of 250 ml/min is equivalent to an energy expenditure of about 5 kJ/min (or about 1.3 kcal/min for those who use the calorie, as many nutritionists do). A rate of energy expenditure is an expression of power, and the SI unit of power is the Watt (W), which is equal to 1 J/s; 5 kJ/min is therefore equivalent to about 83 W (Winter and Fowler, 2009). Oxidation of carbohydrate generates about 5 kcal for each litre of oxygen consumed; the corresponding figure when fat is the fuel is about 4.7 kcal. A 70 kg runner moving at a speed of 15 km/h will require about 3.5 l of oxygen per minute, or about 1.17 kW. It is important to recognize, however, that not all muscular activity, and therefore not all energy expenditure, relates to the performance of external work that can be measured, though it does require an input of chemical energy.

It is often convenient to think of rates of energy expenditure in terms of the amount of oxygen that would have to be consumed if all of the energy demand were to be met by energy generated by oxidative metabolism. It is important, too, to

Strength and Conditioning – Biological Principles and Practical Applications Marco Cardinale, Rob Newton, and Kazunori Nosaka.

recognize that many of the physiological responses to exercise are proportional not to the rate of oxygen uptake (or the rate of power output), but rather to the energy demand as a fraction of the individual's aerobic capacity. A survey of a large and diverse sample population will show little correspondence between heart rate and power output during a cycle ergometer test, for example. The relationship will be just the same if oxygen uptake is substituted for power output. Recalculating the data, however, will show a relatively good relationship between heart rate and the fraction of aerobic capacity that is employed. The energy demand will be very similar for all individuals at the same power output, but the physiological responses will be very different.

When energy is required by the body, it is made available by the hydrolysis of the high-energy phosphate bonds in the adenosine triphosphate (ATP) molecule. Storage of ATP in substantial amounts is not possible–the amount of chemical energy stored in each molecule of ATP is rather small and it would be inefficient to store more because of the mass that would have to be carried. All of the body's energy-supplying mechanisms are geared towards the resynthesis of ATP; although the ATP concentration within the cell remains almost constant, this is only because the rates of breakdown and resynthesis are balanced. The greater the energy demand, the faster the turnover rate. The amount of energy released by hydrolysis ATP is about 31 kJ per mole of ATP. Given that the molecular weight of ATP is about 507, we can calculate that a resting individual with an energy turnover of 83W (which is 83 J/s) must be breaking down about 1.4 g of ATP every second. As the ATP concentration in the cells remains constant, enough energy must be fed back into the system to regenerate ATP at precisely the same rate. Our runner in the example above must break down about half a kilogram of ATP every minute to maintain her pace. Given that the total ATP content of the body is about 50 g, this means that each molecule of ATP in the body turns over on average about once every six seconds.

Ensuring that this energy is made available at exactly the rate it is required is one of the major challenges to the body during exercise.

1.5.3 SOURCES OF ENERGY

All of the energy used by the human body is ultimately derived from the chemical energy in the foods that we eat. This chemical energy may be used immediately, or it may be stored (primarily as fat, protein, or carbohydrate) for later use, as discussed below.

The hydrolysis of ATP to release energy is catalysed by a number of different ATPase enzymes, each of which has a specific function and a specific location in the cell. At rest, when no contractile activity is taking place, the energy demand is relatively low and most of the energy is used for the maintenance of electrical and ionic balances across the muscle membrane and across various internal membranes. The intracellular concentration of potassium is high and the intracellular sodium concentration is low; the concentrations are reversed outside the cell. Maintaining these gradients requires energy. Biosynthetic reactions such as protein synthesis and glycogen storage also require an input of chemical energy.

In muscle, energy from the hydrolysis of ATP by myosin ATPase activates specific sites on the force-developing elements, which try to increase the amount of overlap between the actin and myosin filaments that make up the main structural elements of the muscle. If the load on the muscle is less than the force that can be generated, the muscle will shorten; if not, force will be generated without any shortening, and so no external work will be done. Energy is still required whether or not shortening of the muscle takes place. On activation of the muscle, calcium is released into the cytoplasm from the storage sites in the sarcoplasmic reticulum, and this calcium must be removed in order for the muscle to relax and return to a resting state. Active reuptake of calcium ions by the sarcoplasmic reticulum requires ATP, as these ions must be pumped uphill against a concentration gradient.

The mechanisms involved in the breakdown and resynthesis of ATP can be divided into four main components:

1. ATP is broken down under the influence of a specific ATPase to adenosine diphosphate (ADP) and inorganic phosphate (P_i) in order to yield energy for muscle activity or to power other reactions. This is the immediate source of energy and all other energy-producing reactions must channel their output through this mechanism.
2. Phosphocreatine (PCr) is broken down to creatine and phosphate; the phosphate group is not liberated as P_i, but is transferred directly to an ADP molecule to re-form ATP. This reaction is catalysed by the enzyme creatine kinase, which is present in skeletal muscle at very high activities, allowing the reaction to occur rapidly.
3. Glucose-6-phosphate, derived from muscle glycogen or from glucose taken up from the bloodstream, is converted to lactate and produces ATP by substrate-level phosphorylation reactions. These reactions do not require oxygen, and so are commonly referred to as 'anaerobic'.
4. The products of carbohydrate, lipid, protein, and alcohol metabolism can enter the tricarboxylic acid (TCA) cycle (also known as the Krebs cycle, after Sir Hans Krebs, who first described it) in the mitochondria and be oxidized to carbon dioxide and water. This process is known as oxidative phosphorylation and yields energy for the synthesis of ATP.

The purpose of the last three mechanisms is to regenerate ATP at sufficient rates to prevent a significant fall in the intramuscular ATP concentration. If the ATP concentration falls, the concentrations of ADP and adenosine monophosphate (AMP) will rise. The concentration ratio of ATP to ADP and AMP is a marker for the energy status of the cell. If the ratio is high, the cell is in effect 'fully charged'. This energy charge is monitored in every cell: a fall in the ATP concentration or a rise in the concentration of ADP or AMP will activate the metabolic pathways necessary to increase ATP production. This is achieved by activation or inhibition of key regulatory enzymes through changes in the concentration of the adenine nucleotides.

The first three mechanisms identified above are all anaerobic even when they occur in an environment where there is a plentiful supply of oxygen. Each uses only one specific substrate for energy production (i.e. ATP, PCr and glucose-6-phosphate), though the glucose-6-phosphate may be derived from either glucose or glycogen. Glycerol, which is released into the circulation when triglycerides are broken down to make their component fatty acids available, can enter the glycolytic pathway in liver and some other tissues. It does not appear to be an energy source for skeletal muscle during exercise, probably because of the lack of the enzyme glycerol kinase in muscle.

The aerobic (oxygen-using) processes in the mitochondria metabolize a variety of substrates. The sarcoplasm contains a variety of enzymes which can convert carbohydrates, lipids, and the carbon skeletons of amino acids derived from proteins into substrate, primarily a 2-carbon acetyl group linked to coenzyme A (acetyl CoA); this can be completely oxidized in the mitochondria, resulting in production of ATP.

1.5.3.1 Phosphagen metabolism

If a muscle is poisoned with cyanide, which binds irreversibly to the iron atoms in the enzyme cytochrome oxidase (as well as to the iron atoms in haemoglobin and myoglobin, which blocks oxygen transport), no energy can be provided by oxidative metabolism. If iodoacetic acid, which is an inhibitor of glyceraldehyde 3-phosphate dehydrogenase (one of the enzymes involved in the glycolytic pathway) and thus prevents ATP formation by glycolysis, is also present, the muscle will still appear to function normally for a short period of time before fatigue occurs. This tells us that the muscle has another source of energy which allows it to continue to generate force in this situation; because fatigue occurs rapidly, this also tells us that the capacity of this alternative energy source is limited. This source is the intramuscular store of ATP and PCr (also known as creatine phosphate); together these are referred to as the phosphagens.

The most important property of the phosphagens is that the energy store they represent is available to the muscle almost immediately. With only a single reaction involved and an enzyme with a very high activity, the PCr in muscle can be used to resynthesise ATP at a very high rate. This high rate of energy transfer corresponds to the ability to produce rapid forceful actions by the muscles. The major disadvantage of this system is its limited capacity: the total amount of energy available is small. The major limitation again is the amount of creatine phosphate that can be stored in the muscle: increasing stores increases the osmolality inside the cells, which means that more water moves into the muscles to maintain osmotic equilibrium. The use of creatine supplements, however, does allow a relatively small increase in the amount of creatine phosphate stored in the muscle, and this is associated with improvements in strength and power. The use of creatine supplements is discussed in more detail elsewhere.

It may be important to note that the creatine kinase reaction is actually slightly more complex than its usual representation. Most textbooks show a simplified version of the equation as follows:

$$PCr + ADP \rightarrow ATP + Cr$$

At physiological pH and in the ionic environment of the cell cytoplasm, however, PCr, ADP, and ATP are all dissociated to some degree and carry a positive or negative charge, so a more realistic representation of the reaction is:

$$PCr^{2+} + ADP^{3-} + H^{+} \rightarrow ATP^{4-} + Cr$$

From this, it is apparent that a hydrogen ion is consumed when a phosphate group is transferred from PCr to ADP; that is, the environment becomes more alkaline. This was actually observed in early investigations of muscle metabolism, when the muscle was stimulated to contract after being poisoned with iodoacetic acid and cyanide as described above. This may be more than just an interesting aside, as the muscle will normally be generating large amounts of hydrogen ions as a result of the breakdown of glycogen to lactate at the same time as the demand for energy from the phosphagen store is at its peak. These hydrogen ions result in acidification of the cytoplasm, with negative consequences for some of the key enzymes of glycolysis as well as for the control of actin and myosin interactions. Buffering some of these hydrogen ions by breaking down PCr will allow more energy to be produced by glycolysis before the pH in the muscle becomes dangerously low.

If no energy source other than the phosphagens is available to the muscle, fatigue will occur rapidly. During short sprints lasting only a few seconds (perhaps up to about 4–5), no slowing down occurs over the last few metres – full power can be maintained all the way – and the energy requirements are met by breakdown of the phosphagen stores. Over longer distances, running speed begins to fall off, as these stores become depleted and power output declines; the other energy sources cannot regenerate ATP fast enough for maximum power to be maintained. However, the rate of recovery from a short sprint is quite rapid, and a second burst can be completed at the same speed after only 2–3 minutes of recovery, provided that an adequate supply of oxygen is available to allow for regeneration of the PCr by production of ATP via oxidative metabolism. For longer sprints, much longer recovery periods are needed before the ability to produce a maximum performance is restored. These observations tell us something about the rate of restoration of the pre-exercise chemical state of the muscle.

During sustained exercise, creatine and creatine phosphate play a further important role in the cell, though this is often neglected. Most of the ATP used by muscle cells is generated by oxidative phosphorylation inside the mitochondria, but the highest demand for ATP utilization during muscle contraction occurs in the cytoplasm. The mitochondrial membrane is relatively impermeable to ATP and to ADP, so there must be an

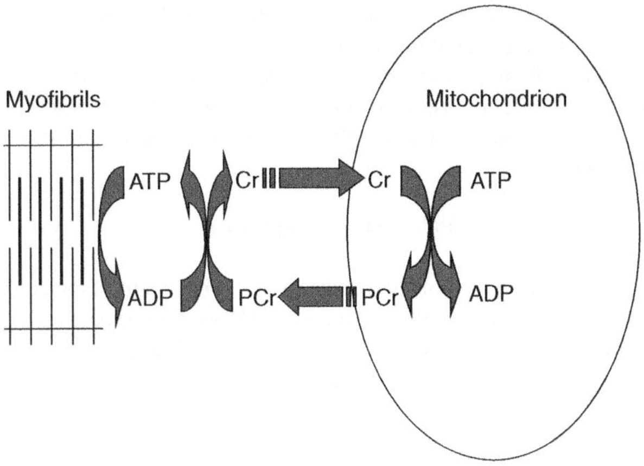

Figure 1.5.1 In endurance exercise, most of the ATP hydrolysis takes place in the cytoplasm of muscle cells but oxidative generation of ATP takes place within the mitochondria. Because the mitochondrial membrane is relatively impermeable to ATP, translocation of energy equivalents is achieved by a creatine-phosphocreatine shuttle.

alternative way of shuttling the energy produced by oxidative metabolism to the sites where it is required. The primary way in which this is achieved is by the transfer of ATP equivalents across the mitochondrial membrane in the form of PCr (Figure 1.5.1).

1.5.3.2 The glycolytic system

Human skeletal muscle can exert force without the use of oxygen as a consequence of its ability to generate energy anaerobically. Two separate systems are available to the muscle to permit this; these are the phosphagen or high-energy phosphate system described above and the glycolytic system. Because the glycolytic system depends on the production of lactate while the phosphagen system involves no lactate formation, they are sometimes referred to as the lactic and the alactic systems, respectively. The terms 'lactic acid' and 'lactate' are often used interchangeably, but although lactic acid is perhaps a more descriptive term, clearly indicating the acidic nature of the molecule, lactate is more accurate and will be used here. At physiological pH lactic acid is essentially entirely dissociated to form lactate and hydrogen ions.

Under normal conditions, an isolated muscle incubated in an organ bath and made to contract by application of an electrical stimulus clearly does not fatigue after only a few seconds of effort, so a source of energy other than the phosphagens must be available. This is derived from glycolysis, which is the name given to the pathway involving the breakdown of glucose (or glycogen). The end product of this series of chemical reactions, which starts with a glucose molecule ($C_6H_{12}O_6$) containing six carbon atoms, is two 3-carbon molecules of pyruvate ($C_3H_3O_3^-$ or CH_3COCOO^-). This process does not use oxygen, but does result in a small amount of energy in the form of ATP being available to the muscle from reactions involving substrate-level phosphorylation. For the reactions to proceed, the pyruvate must be removed; when the rate at which energy is required can be met aerobically, pyruvate is converted to carbon dioxide and water by oxidative metabolism in the mitochondria. In some situations the pyruvate is removed by conversion to lactate, anaerobically, leading to the system being referred to as the lactate anaerobic system.

Table 1.5.1 Capacity (the amount of work that can be done) and power (the rate at which work can be done) of the energy supplying metabolic processes available to human skeletal muscle. These values are expressed per kg of muscle. They are approximations only and will be greatly influenced by training status and other factors.

	Capacity (J/kg)	Power (W/kg)
APT/CP Hydrolysis	400	800
Lactate formation	1000	325
Oxidative metabolism		200

Activation of the glycolytic system occurs almost instantaneously at the onset of exercise, despite the number of reactions involved. All of the enzymes are present in the cytoplasm of skeletal muscle at relatively high activities, and the primary substrate (glycogen) is normally readily available at high concentrations.

The total capacity of the glycolytic system for producing energy in the form of ATP is large in comparison with the phosphagen system (Table 1.5.1). In high-intensity exercise the muscle glycogen stores are broken down rapidly, with a correspondingly high rate of lactate formation; some of the lactate diffuses out of the muscle fibres where it is produced and appears in the blood. A large part, but not all, of the muscle glycogen store can be used for anaerobic energy production during high-intensity exercise, and supplies the major part of the energy requirement for maximum-intensity efforts lasting from 20 seconds to 5 minutes. For shorter durations, the phosphagens are the major energy source, while oxidative metabolism becomes progressively more important as the duration (or distance) increases.

Although the total capacity of the glycolytic system is greater than that of the phosphagen system, the rate at which it can produce energy (ATP) is lower (Table 1.5.1). The power output that can be sustained by this system is therefore correspondingly lower, and it is for this reason that maximum speeds cannot be sustained for more than a few seconds; once the phosphagens are depleted, the intensity of exercise must necessarily fall.

The rate of lactate formation is dependant primarily on the intensity of the exercise, but more on the relative exercise intensity (%VO_{2max}) than the absolute intensity. It is set by the metabolic characteristics of the skeletal muscle and the recruitment patterns of the various muscle fibre types within the skeletal muscle.

The factors that integrate the metabolic response to exercise so that energy supply can meet energy demand as closely as possible are outlined below.

1.5.3.3 Aerobic metabolism: oxidation of carbohydrate, lipid, and protein

Other pathways of regenerating ATP rely mainly on the provision of carbohydrate or lipid substrates from intramuscular stores or from the circulation, and their subsequent breakdown (catabolism) to yield energy through the reactions of the electron transport chain and oxidative phosphorylation. To regenerate ATP from lipid (fat) catabolism, oxygen is required, as there is no mechanism that allows the energy content of the fat molecules to be made available in the absence of oxygen. This is in contrast to carbohydrate catabolism, which can occur with or without the use of oxygen via the glycolytic pathway described above.

The catabolism of glucose begins with anaerobic glycolysis, which yields two molecules of pyruvate (or lactate) and two molecules of ATP for each molecule of glucose that enters the glycolytic pathway (although three molecules of ATP are derived for each glucose moiety if the initial substrate is muscle glycogen). In aerobic metabolism, mostly pyruvate (rather than lactate) is formed and the 3-carbon pyruvate molecule is first converted to a 2-carbon acetyl group which is bonded to an activator group, coenzyme A (CoA), to form acetyl CoA. This reaction is catalysed by the enzyme pyruvate dehydrogenase (PDH). PDH is a complex enzyme that exists in both active and inactive forms and is a key regulator in the integration of fat and carbohydrate use through control of the rate of entry of CHO fuel into the oxidative metabolic pathways. The catabolism of fatty acids also results in the formation of 2-carbon acetyl units in the form of acetyl CoA. Acetyl CoA is the major entry point for metabolic fuels into the TCA cycle.

In comparison to the catabolism of carbohydrate and lipid, the breakdown of protein is usually a relatively minor source of energy for exercise. Over the course of a day, the body remains in protein balance, so the rate of breakdown and oxidation of the amino acids contained in protein must match the amount of these amino acids present in the diet. Protein typically accounts for about 12–15% of total energy intake, so about 12–15% of energy must come from protein breakdown over the day. In most situations, protein catabolism contributes less than about 5% of the energy provision during physical activity. Protein catabolism can provide both ketogenic and glycogenic amino acids, which may eventually be oxidized either by deamination and conversion into one of the intermediate substrates in the TCA cycle, or by conversion to pyruvate or acetoacetate and eventual transformation to acetyl CoA. However, protein catabolism can become an increasingly important source of energy for exercise during starvation and in the later stages of very prolonged exercise, when the availability of CHO is limited.

1.5.4 THE TRICARBOXYLIC ACID (TCA) CYCLE

The main function of the TCA cycle is to degrade the acetyl CoA substrate to carbon dioxide and hydrogen atoms; the metabolic machinery that allows this to take place is located within the mitochondria. The hydrogen atoms are then oxidized via the electron transport (respiratory) chain, allowing oxidative phosphorylation and the subsequent regeneration of ATP from ADP. The reactions of the TCA cycle will not be considered in detail here and the interested reader is referred elsewhere (e.g. Maughan and Gleeson, 2010).

The TCA cycle begins with the 2-carbon acetyl CoA, which reacts with a 4-carbon molecule of oxaloacetate to form citrate. Citrate then undergoes a series of reactions, during which it loses two of these carbon atoms as CO_2, resulting in regeneration of oxaloacetate, which is then free to react with another acetyl CoA molecule and begin the cycle all over again. Some of the energy released in these reactions is captured as ATP, and one ATP molecule is formed during each turn of the cycle. More of the available energy is captured by adenine nucleotides (FAD and NAD). The overall reaction involving each molecule of acetyl CoA is as follows:

$$\text{acetyl CoA} + \text{ADP} + 3\text{NAD}^+ + \text{FAD} \rightarrow 2\text{CO}_2 + \text{ATP} + 3\text{NADH} + 3\text{H}^+ + \text{FADH}_2$$

Note that oxygen itself does not participate directly in the reactions of the TCA cycle. In essence, the most important function of the TCA cycle is to generate hydrogen atoms for their subsequent passage to the electron transport chain by means of NAD^+ and FAD. The aerobic process of electron transport–oxidative phosphorylation regenerates ATP from ADP, thus conserving some of the chemical energy contained within the original substrates in the form of high-energy phosphates. As long as there is an adequate supply of O_2, and substrate is available, NAD^+ and FAD are continuously regenerated and TCA metabolism proceeds.

The electron transport chain is a series of linked carrier molecules which remove electrons from hydrogen and eventually pass them to oxygen; oxygen also accepts hydrogen to form water. Much of the energy generated in the transfer of electrons from hydrogen to oxygen is trapped or conserved as chemical potential energy in the form of high-energy phosphates, the remainder is lost as heat.

In terms of the energy conservation of glucose metabolism, the overall reaction starting with glucose as the fuel can be summarized as follows:

$$\text{glucose} + 6\,\text{O}_2 + 38\,\text{ADP} + 38\,\text{P}_\text{i} \rightarrow 6\,\text{CO}_2 + 38\,\text{ATP}$$

The corresponding equation for oxidation of a typical fatty acid can be expressed as:

$$\text{palmitate} + 23\,\text{O}_2 + 130\,\text{ADP} + 130\,\text{P}_\text{i} \rightarrow 6\,\text{CO}_2 + 146\,\text{H}_2\text{O} + 130\,\text{ATP}$$

It may seem from this that fat is the best fuel to use, especially as it is present in almost unlimited amounts, in contrast to the small body stores of carbohydrate, but there are several considerations to bear in mind. First, carbohydrate is a much more versatile fuel, as it can be used to generate energy in the absence of oxygen. Second, the maximum rate of oxidative energy supply using carbohydrate as a fuel is typically about twice as high as that achieved when fat is used. Third, carbohydrate is a more 'efficient' fuel in that the oxygen cost is less than that when fat is used. Even though this difference is relatively small (about 5.01 kcal/l for CHO as opposed to about 4.7 kcal/l for fat), it may be crucial when the oxygen supply is limited. A greater oxygen demand by the working muscles means that they need a greater blood flow, and this may result in diversion of blood away from other tissues.

1.5.5 OXYGEN DELIVERY

As the rate of energy demand increases in exercise, so the rate of oxygen consumption also increases, at least up until a finite point is reached beyond which there is no further increase. At low- to moderate-intensity exercise, there is a good linear relationship between power output and oxygen consumption. At high-power outputs, however, the curve becomes exponential. In well-motivated and reasonably fit individuals, a point is reached at which there is no further increase in oxygen consumption even when the power output is increased. This point was defined by Hill and Lupton in 1923 as the individual's maximum oxygen uptake (VO_{2max}). Clearly, the energy demand continues to increase as the power output is increased, and the shortfall in energy provision is met by anaerobic metabolism. Where no plateau is reached, the highest oxygen uptake achieved is referred to as the peak oxygen uptake (VO_{2peak}).

There has been, and remains, much debate as to what limits the maximum oxygen uptake. In some experimental models, it may be limited by the ability of the lungs to oxygenate the arterial blood, by the ability of the muscles to use oxygen, or perhaps by other factors. The evidence, however, is overwhelmingly in support of the limitation lying in the ability of the heart to supply blood to the working muscles (Ekblom and Hermansen, 1968). Forty years of further research and much debate have not altered this viewpoint (Levine, 2008).

There is some inertia in the oxygen transport/oxygen utilization system, and whereas the demand for energy increases more or less immediately at the start of exercise, it takes some time for oxygen uptake to reach its peak level. The kinetics of oxygen uptake at the onset of exercise has been investigated extensively and it is clear that many factors will influence this response (Jones and Poole, 2005). In moderate exercise at constant power output there is a rapid increase in the rate of oxygen consumption at the onset of exercise, with a relatively steady state being achieved after about 2–3 minutes. Over time, however, there will be a slight upward drift in the oxygen consumption, accompanied by an upward drift in heart rate and other physiological parameters. Many factors contribute to this, including a progressive increase in the contribution of fat oxidation to oxidative energy supply relative to the contribution of CHO. For each litre of oxygen used, CHO oxidation will provide 5.01 kcal (21.0 kJ) of energy, while fat oxidation provides about 4.69 kcal (19.6 kJ), which is about 6% less (Jéquier *et al.*, 1987).

The shortfall in energy supply in the first few minutes of exercise is met by anaerobic metabolism, as outlined above. Without this, the rate of acceleration at the onset of exercise would be set by the rate of oxidative metabolism. Rapid acceleration would be impossible, and full speed would not be reached until after 2–3 minutes. In evolutionary terms this would not be a good survival strategy. Humans are well equipped for both rapid accelerations (relying on anaerobic metabolism) and sustained exercise (relying on aerobic metabolism). Both of these assets are important where success in escape from predators and the pursuit of prey will determine survival of the individual and of the species.

One of the key adaptations to endurance training is an increase in the capacity for energy supply by oxidative metabolism, and a high VO_{2max} has long been recognized as one of the distinguishing characteristics of the elite endurance athlete. This is accompanied by changes in the cardio-respiratory system and within the trained muscles themselves. Cross-sectional comparisons of individuals of differing training status, and longitudinal studies of the adaptations to a short-term endurance programme, both show that:

- Endurance training is accompanied by an increased maximum cardiac output.
- Endurance training is accompanied by increases in the capacity of the muscles to generate energy by oxidative metabolism.

Both of these adaptations tend to occur in parallel, accounting at least in part for the debate as to whether oxygen supply or oxygen use is the limitation to VO_{2max}.

1.5.6 ENERGY STORES

Energy intake comes from the food we eat, specifically from the energy-containing macronutrients, carbohydrate, fat, protein, and alcohol. Alcohol cannot be stored and must be metabolized immediately, but the other fuels can either be used immediately as energy sources or stored for later use. Protein is not stored, in the sense that all proteins are functionally important molecules (e.g. structural proteins, enzymes, ion channels, receptors, contractile proteins, etc.), and the concentration of most free amino acids in intracellular and extracellular body fluids is too low for these to be of any significance. Loss of structural and functional proteins inevitably involves some loss of functional capacity, but this may be tolerable when no alternative energy source is available, as in periods of prolonged starvation. Though it is usually accepted that there is no protein store in the human body, it must be remembered that the rate of protein turnover in gut tissues is extremely high: gut mucosal protein synthesis occurs over 30 times faster than that of skeletal muscle (Charlton, Ahlman and Nair, 2000). In the early stages of fasting, there will be some loss in intestinal tissue mass, making the amino acids released from breakdown of gut tissue available for protein synthesis in other tissues. This probably occurs after only a few hours without food, and it may therefore be viewed as a store of protein.

Carbohydrates are stored in the body in only very limited and rather variable amounts. The size of the carbohydrate stores is very much influenced by the diet and exercise activity in the preceding hours and days. The availability of endogenous carbohydrate is not a problem for the organism when regular intake of carbohydrate foods is possible, or when the demand for carbohydrate is low, but regular hard exercise can pose a major challenge. Glycogen is the storage form of carbohydrate in animals, analogous to the storage of starch in plants. The glycogen molecule is a branched polymer consisting of many (typically 50000–100000) molecules of glucose joined together by α1,4-bonds; branch points are introduced into the chains of glucose molecules when an additional α1,6-bond is formed. The advantage of these polymers is that relatively large amounts of carbohydrate can be stored without the dramatic increase in osmolality that would occur if the same amount of carbohydrate were present as glucose monomers.

Skeletal muscle contains a significant store of glycogen, which can be seen under the electron microscope as small granules in the sarcoplasm. The glycogen content of skeletal muscle at rest in well-fed humans is approximately 14–18 g per kg wet mass (about 80–100 mmol glucosyl units/kg). The liver also contains glycogen; about 80–120 g is stored in the liver of an adult human in the post-absorptive state, which can be released into the circulation in order to maintain the blood glucose concentration at about 5 mM (0.9 g per litre). The liver glycogen store falls rapidly during fasting and during prolonged exercise, as glucose is released into the circulation to maintain the blood glucose concentration (Hultman and Greenhaff, 2000). Considering that glucose can be oxidized at rates of 2–4 g/min during prolonged exercise, it is apparent that the blood glucose cannot be considered as any form of carbohydrate 'store' but is rather only the transit form that allows glucose to be moved between tissues.

Early studies of fuel use during exercise relied on measures of oxygen uptake and the respiratory exchange ratio. While these allowed determination of the rates of fat and CHO oxidation, they did not enable the sources of these fuels to be determined. Application of the muscle biopsy technique to the study of human metabolism began in Scandinavia the early1960s. In a classic study (Bergstrom and Hultman, 1966), a one-legged cycling model was used to show that prolonged exercise resulted in depletion of the glycogen store in the exercise muscle but not in the resting muscle. Although there were only two subjects (the authors themselves), they were also able to show that subsequent re-feeding of a high-carbohydrate diet for three days resulted in abnormally high levels of glycogen storage in the exercise muscles but had no effect on the resting muscles. Soon afterwards, it was shown that exercise capacity was closely related to the size of the pre-exercise muscle glycogen store (Ahlborg *et al.*, 1967; Bergstrom *et al.*, 1967). These observations and others published at around the same time were to set the tone for much of the sports nutrition advice given to athletes up until the present day. More recently, other methodologies have helped to elucidate the contribution of fuels to energy metabolism: these include isotopic tracer methods and, more recently, the use of non-invasive methods such as magnetic resonance spectroscopy.

The major storage form of energy in the human body is lipid, which is stored as triglyceride (triacylglycerol), mainly in adipose tissue. The total storage capacity for lipid is extremely large, and for most practical purposes the amount of energy stored in the form of fat is far in excess of that required for any exercise task (Table 1.5.2). Even in the leanest individual, at least 2–3 kg of triglyceride is stored. Such low levels would seldom be encountered in healthy individuals, but they are not uncommon in elite male marathon runners (Pollock *et al.*, 1977). Women generally store much more fat than men, but elite female athletes also have a very low body fat content (Wilmore, Brown and Davis, 1977). It is not unusual to see triglyceride accounting for 30–40% of total body mass in sedentary individuals. These adipose tissue depots are located mainly under the skin (subcutaneous depots) or within the abdominal cavity.

Triglyceride mobilized from the adipose tissue stores must first be broken down by a lipase enzyme to release free fatty acids (FFAs) into the circulation for uptake by working muscle. Skeletal muscle also contains some triacylglycerol stored within the muscle cells, as well as some adipose cells within the muscle. The intramuscular triglyceride (IMTG) is present as small lipid droplets within the cell and many of these are in close proximity to the mitochondria, where they will be oxidized when required. These IMTG stores can account for about 20% of total energy supply, or about 30–50% of the total amount of fat oxidized during prolonged moderate-intensity exercise (Stellingwerff *et al.*, 2007). This source of fuel may become relatively more important after exercise training, when the capacity for fat oxidation is greatly increased. Study of the role of the triglyceride stored within the muscle cells has been hindered by the difficulties in making accurate measurements of these relatively small stores of fat, but their significance is now being appreciated. This is in part due to improved methods for measuring IMTG; these methods include measures on muscle biopsy samples (biochemical analysis, electron microscopy, and histochemistry), as well as newer, non-invasive alternatives (computed tomography, magnetic resonance spectroscopy, magnetic resonance imaging). The contribution of

Table 1.5.2 Energy stores in a typical 70 kg man in a fed and rested state. These can vary greatly between individuals and can also vary substantially over time within an individual.

Carbohydrate stores	
Blood glucose	3–5 g
Liver glycogen	80–100 g
Muscle glycogen	300–400 g
Fat stores	
Adipose tissue	3–20 kg
Muscle triglyceride	500 g

intramuscular lipid to total energy turnover during a period of prolonged exercise will be influenced by many factors, including training status, pre-exercise diet, and gender (Roepstorff, Vistisen and Kiens, 2005). While the adipose triglyceride may amount to 10–50% of body mass, the intramuscular store is much smaller, at about 300 g. The IMTG content of highly oxidative muscle fibres is much higher than that of the fast-twitch, glycolytic fibres (Schrauwen-Hinderling *et al.*, 2006). As with muscle glycogen stores, the IMTG store is influenced by diet and can be increased by eating a high-fat diet (Zehnder *et al.*, 2006). This in turn leads to an increased reliance on this fuel source during exercise. Increased levels of intramuscular fat are associated with obesity, diabetes, and insulin resistance, but stores are also higher in well-trained endurance athletes, who normally have low levels of adipose tissue triglyceride (Tarnopolsky *et al.*, 2006).

Lipid stores in the body are far larger than those of carbohydrate, and lipid is a more efficient storage form of energy, releasing 37 kJ for each gram oxidized, compared with 16 kJ/g for carbohydrate. Each gram of carbohydrate stored also retains about 1–3 g of water, further decreasing the efficiency of carbohydrate as an energy source. The energy cost of running a marathon for a 70 kg runner is about 12 000 kJ; if this could be achieved by the oxidation of fat alone, the total amount of lipid required would be about 320 g, whereas 750 g of carbohydrate would be required if carbohydrate oxidation were the sole source of energy. Apart from considerations of the weight to be carried, this amount of carbohydrate exceeds the total amount normally stored in the liver and the muscles combined. By comparison, we can calculate the amount of ATP required to run a marathon if the only energy source available to the muscle were ATP. The energy cost of running can be expressed as the oxygen demand, which is equivalent to about 600 l of oxygen consumption for the 70 kg runner quoted above.

For simplicity, we might assume that all the energy comes from oxidation of carbohydrate. The equation for ATP generation from glucose oxidation is:

$$C_6H_{12}O_6 + 6\,O_2 \rightarrow 6\,CO_2 + 6\,H_2O + 36\text{ ATP}$$

Substituting the molecular weight of the reactants, we get:

$$180\text{ g glucose} + 192\text{ g } O_2 \rightarrow 264\text{ g } CO_2 + 108\text{ g } H_2O + 18.25\text{ kg ATP}$$

One mole of a gas occupies 22.4 l at standard temperature and pressure, so we can recalculate this equation as:

$$180\text{ g glucose} + 134.4\text{ l } O_2 \rightarrow 134.4\text{ l } CO_2 + 108\text{ g } H_2O + 18.25\text{ kg ATP}$$

From this, we can calculate that 0.136 kg of ATP is resynthesized for each litre of O_2 consumed. Because 600 l of O_2 was used for the marathon, this means that the total mass of ATP required to complete the distance was 82 kg, more than the runner's total body mass. Compared to this, the amounts of fat and carbohydrate consumed are rather small. It is not surprising then that in most situations carbohydrate and lipids supply most of the energy required to regenerate the ATP needed to fuel exercise.

1.5.7 CONCLUSION

The rate of energy provision to cells must equal the rate of energy consumption to prevent potentially damaging disturbances to cell homeostasis. Muscle cells are uniquely equipped to cope with sudden changes in the rate of energy supply and with high rates of energy supply. Cells use the energy released by the hydrolysis of the terminal phosphate bond of the ATP molecule, but the cellular ATPO content cannot fall much below the resting level without irreversible cell damage. Three main mechanisms contribute to ensuring that ATP is resynthesised as fast as it is hydrolysed. These are: 1. the transfer of a phosphate group from creatine phosphate; 2. substrate level phosphorylation involving degradation of glycogen or glucose to pyruvate; 3. oxidative phosphoyrlation. Various combinations of these three metabolic pathways allow for the very high energy demands of a 100 m spring and for the sustained demands of a marathon run. From a biochemical perspective, fatigue is simply an inability to meet the body's energy demand.

References

Ahlborg B., Bergstrom J., Ekelund L.-G. and Hultman E. (1967) Muscle glycogen and muscle electrolytes during prolonged physical exercise. *Acta Physiol Scand*, **70**, 129–142.

Bergstrom J. and Hultman E. (1966) Muscle glycogen synthesis after exercise: an enhancing factor localized to the muscle cells in man. *Nature*, **210**, 309–310.

Bergstrom J., Hermansen L., Hultman E. and Saltin B. (1967) Diet, muscle glycogen and physical performance. *Acta Physiol Scand*, **71**, 140–150.

Charlton M., Ahlman B. and Nair K.S. (2000) The effect of insulin on human small intestinal mucosal protein synthesis. *Gastroenterology*, **118**, 299–306.

Ekblom B. and Hermansen L. (1968) Cardiac output in athletes. *J Appl Physiol*, **25**, 619–625.

Hill A.V. and Lupton H. (1923) Muscular exercise, lactic acid, and the supply and utilization of oxygen. *Q J Med*, **16**, 135–171.

Hultman E. and Greenhaff P.L. (2000) Carbohydrate metabolism in exercise, in *Nutrition in Sport* (ed. R.J. Maughan), Blackwell, Oxford, pp. 85–96.

Jéquier E., Acheson K. and Schutz Y. (1987) Assessment of energy expenditure and fuel utilization in man. *Ann Rev Nutr*, **7**, 187–208.

Jones A.M. and Poole D.C. (2005) *Oxygen Uptake Kinetics in Sport, Exercise and Medicine*, Routledge, London.

Levine B.D. (2008) VO2max: what do we know and what do we still need to know? *J Physiol*, **586**, 25–34.

Maughan R.J. and Gleeson M. (2010) *The Biochemical Basis of Sports Performance*, 2nd edn. Oxford University Press, Oxford.

Pollock M.L., Gettman L.R., Jackson A. *et al.* (1977) Body composition of elite class distance runners. *Ann NY Acad Sci*, **301**, 361–370.

Roepstorff C., Vistisen B. and Kiens B. (2005) Intramuscular triacylglycerol in energy metabolism during exercise in humans. *Exerc Sport Sci Rev*, **33**, 182–188.

Schrauwen-Hinderling V.B., Hesselink M.K., Schrauwen P. and Kooi M.E. (2006) Intramyocellular lipid content in human skeletal muscle. *Obesity*, **14**, 357–367.

Stellingwerff T., Boon H., Jonkers R.A. *et al.* (2007) Significant intramyocellular lipid use during prolonged cycling in endurance-trained males as assessed by three different methodologies. *Am J Physiol Endocrinol Metab*, **292**, E1715–E1723.

Tarnopolsky M.A., Rennie C.D., Robertshaw H.A. *et al.* (2006) Influence of endurance exercise training and sex on intramyocellular lipid and mitochondrial ultrastructure, substrate use, and mitochondrial enzyme activity. *Am J Physiol Regul Integr Comp Physiol*, **292**, R1271–R1278.

Wilmore J.H., Brown C.H. and Davis J.A. (1977) Body physique and composition of the female distance runner. *Ann NY Acad Sci*, **301**, 764–776.

Winter E.M. and Fowler N. (2009) Exercise defined and quantified according to the Systeme International d'Unites. *J Sports Sci*, **27**, 447–460.

Zehnder M., Christ E.R., Ith M. *et al.* (2006) Intramyocellular lipid stores increase markedly in athletes after 1.5 days lipid supplementation and are utilized during exercise in proportion to their content. *Eur J Appl Physiol*, **98**, 341–354.

1.6 Respiratory and Cardiovascular Physiology

Jeremiah J. Peiffer[1] and Chris R. Abbiss[2,3,4], [1]Murdoch University, School of Chiropractic and Sports Science, Murdoch, WA, Australia, [2]Edith Cowan University, School of Exercise, Biomedical and Health Sciences, Joondalup, WA, Australia, [3]Australian Institute of Sport, Department of Physiology, Belconnen, ACT, Australia, [4]Commonwealth Scientific and Industrial Research Organisation, Division of Materials Science and Engineering, Belmont, VIC, Australia

1.6.1 THE RESPIRATORY SYSTEM

1.6.1.1 Introduction

Often overlooked, the respiratory system is an important component in overall health as well as in athletic performance. Its key role is to facilitate the exchange of oxygen (O_2) from the ambient air into the blood stream, in addition to removing metabolic carbon dioxide (CO_2) from the body. The respiratory system may appear simplistic, but in actuality respiratory function is complex, involving a combination of mechanical actions directed by neural drive. The following pages present a basic overview of this system. After reading this section you should be able to explain: (1) the basic anatomy of the respiratory system, (2) gas exchange in the lungs and the role of diffusion, (3) how ventilation occurs at rest and during exercise, and (4) the mechanical and neural controls of ventilation.

1.6.1.2 Anatomy

The respiratory system comprises two sections, the conducting zone and the respiratory zone (Figure 1.6.1). During ventilation, ambient air will enter the nose or mouth and travel through the conducting zone, passing first through the trachea and then proceeding into the primary bronchus. From the primary bronchus, air will descend into the left and right bronchi and further into the smaller, yet more abundant, bronchioles. The conducting zone provides an area for air filtration, humidification, and warming; nonetheless, no gas exchange occurs there, and thus it is commonly referred to as dead space.

After travelling through the conducting zone, ventilated air will proceed into the respiratory zone. From the bronchioles, the oxygen-rich air passes through the respiratory bronchioles, into the alveolar ducts, and finally into the alveolar sac. Thousands of alveoli, positioned like grapes on a grape vine, make up each alveolar sac, and it is possible for the lungs to contain more than 600+ million individual alveoli. A thin membranous tissue surrounds each alveoli, which when filled with air will stretch, further reducing the alveoli wall thickness (like a balloon being filled with air). Helping alveolar expansion, the internal walls of each alveoli are covered in a substance called surfactant which acts to decrease surface tension. On the outer surface of the alveoli is a complex network of capillaries. Due to the thickness of the alveolar membrane, and the close proximity to the capillaries, the alveoli are the primary point of gas exchange in the lung.

1.6.1.3 Gas exchange

The exchange of O_2 and CO_2 between the lungs and the blood is essential to human life. As highlighted in the previous section, alveolar structure and the close proximity of the surrounding capillaries provide an optimal environment for gas exchange. The movement of O_2 and CO_2 between the alveoli and blood occurs through the diffusion of these molecules through the alveolar wall, across the interstitial fluid, and through the capillary wall (Figure 1.6.2). The control of diffusion is determined by a combination of factors, including: (1) diffusion distance, (2) surface area, and (3) pressure gradients.

Diffusion distance and surface area

The thickness of the alveolar wall (approximately 0.1 micron) and the small volume of interstitial fluid between the alveoli and the capillaries result in very low resistance for the diffusion of O_2 and CO_2 (McArdle, Katch and Katch, 2004). Additionally,

Strength and Conditioning – Biological Principles and Practical Applications Marco Cardinale, Rob Newton, and Kazunori Nosaka.

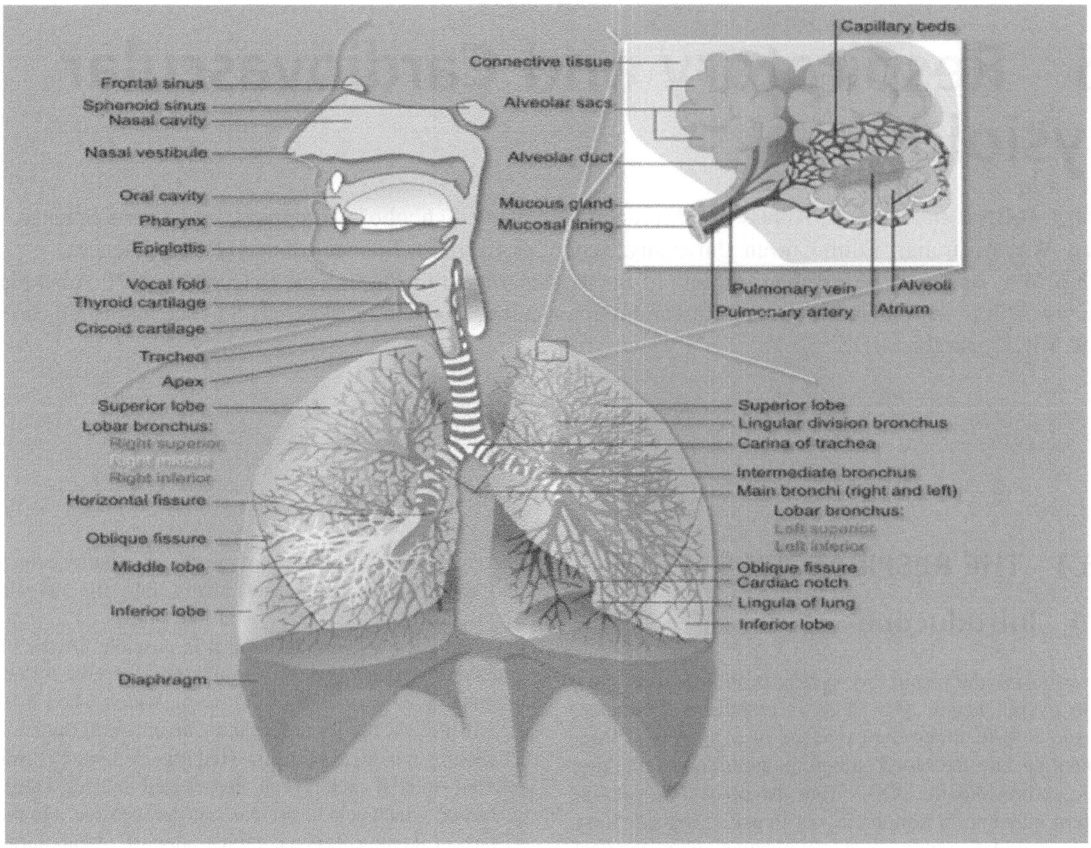

Figure 1.6.1 The anatomy of the human respiratory system

the sheer number of alveoli (600+ million) comprising the alveolar sacs provides a massive surface area on which gas exchange can occur; if all the alveoli in the lung were removed from the body and laid flat on the ground, their surface area would be great enough to cover half a standard tennis court. It is important to note, however, that disease can affect the diffusion distance and/or the surface area available for diffusion, resulting in a decrease in gas exchange (Comroe *et al.*, 1964).

Pressure gradients

The rate of diffusion for O_2 and CO_2 is influenced by the pressure gradients for each in the lungs and the blood. When the pressure gradient between the alveoli and the blood is high, O_2 will rapidly cross the alveolar membrane (Figure 1.6.2a). Nevertheless, this process is not without end. Diffusion is governed by Henry's law, which states that the diffusion of a molecule across a membrane is influenced by not only the pressure difference between the two mediums but also the ability of the molecule to dissolve in the accepting medium. The movement of O_2 and CO_2 from areas of high to low pressure will alter the original pressure of each medium (Figure 1.6.2b), lowering the pressure gradient and decreasing the movement of molecules. At a terminal point, the partial pressures between the two areas will be in equilibrium (at similar pressures) and the movement of molecules will cease (Figure 1.6.2c).

In the human body, the pressure gradients that control the movement of O_2 are a product of the environment. For instance, ambient air primarily comprises three types of gas: nitrogen (N_2), oxygen (O_2), and carbon dioxide (CO_2). As a percentage, N_2 makes up the majority (79%), followed by O_2 (21%), and then CO_2 (0.03%). At sea level, the atmospheric pressure of air is approximately 760 mmHg. When combined with the percentage composition of air, the partial pressure of each gas can be calculated. For example, the partial pressure of O_2 (P_{O2}) at sea level is: 760 mmHg × 0.2095 = 160 mmHg. Using the same technique, P_{CO2} (0.2 mmHg) and P_{N2} (600 mmHg) can be calculated.

While the ambient air P_{O2} can easily be calculated as in the previous example, upon entering the body the P_{O2} is immediately reduced. In the conducting zone of the respiratory system, ambient air becomes 100% saturated with water vapour. In the presence of water vapour, all other partial gas pressures are diluted, and therefore the total pressure of the air entering the body is reduced by the P_{H2O} (47 mmHg) at body temperature (37 °C). Thus, ambient air entering the respiratory zone has a P_{O2} of 150 mmHg, or 10 mmHg lower than in sea-level ambient

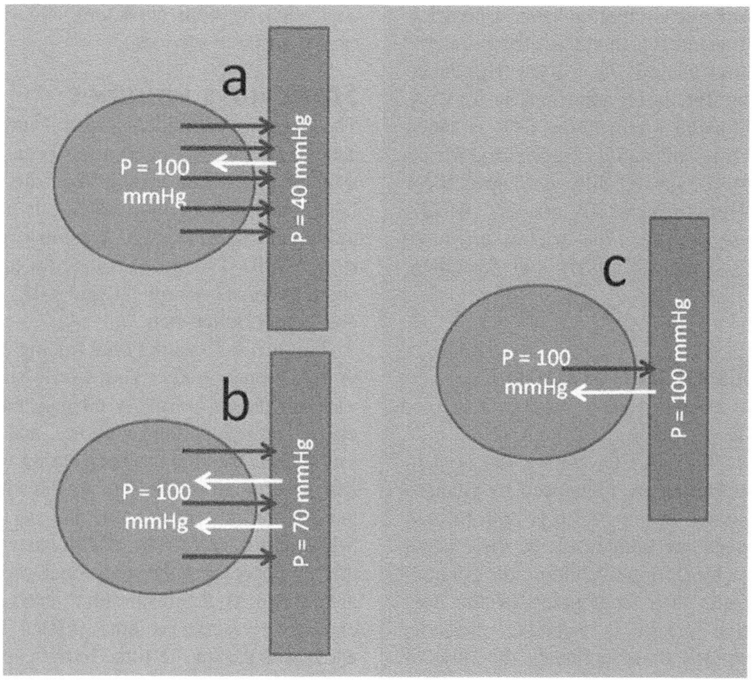

Figure 1.6.2 Schematic of gas exchange between the alveoli and venous blood. As alveolar pressure is high (a), O_2 is transported rapidly from the alveoli to the blood. However, as O_2 dissolves in the blood the pressure gradients begin to equilibrate (b), until the pressure gradient is in equilibrium (c). Circles = alveoli and boxes = capillaries

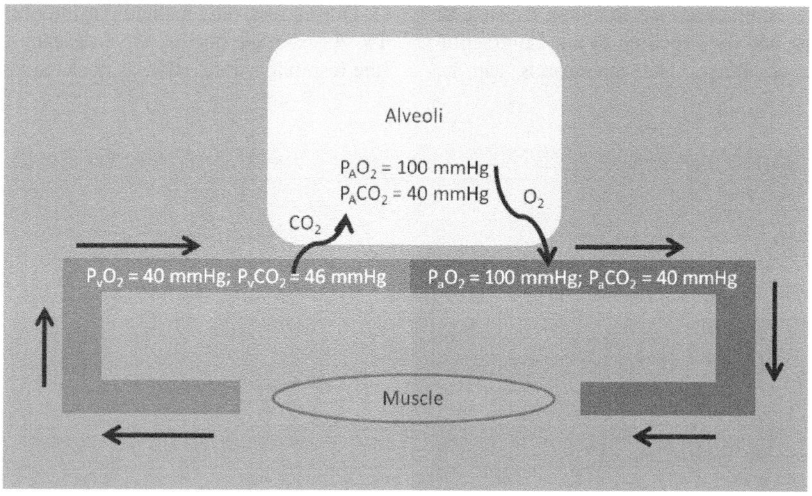

Figure 1.6.3 Schematic representation of the transport of O_2 and CO_2 between the alveoli and blood. P_a = partial pressure in the arteries, P_v = partial pressures in the vein, P_A = partial pressure in the alveoli

air. At the alveoli (primary site of diffusion), P_{O2} will again have been reduced due to some diffusion of O_2 outside the alveoli and the increase in diffused CO_2 from the blood. Indeed, at the time of diffusion in the alveoli, the final alveolar partial pressure (P_A) of O_2 is approximately 105 mmHg, or 55 mmHg less than in sea-level ambient air (Brooks *et al.*, 2000; West, 1962).

Even with a 55 mmHg loss in pressure, the pressure gradient for O_2 is still large enough to facilitate the transport of O_2 from the alveoli to the blood (West, 1962). As shown in Figure 1.6.3, blood returning from the systemic circulation has a venous P_{O2} of 40 mmHg and a P_{CO2} of 46 mmHg; therefore, the pressures in the alveoli (P_{AO2} = 100 mmHg and P_{ACO2} = 40 mmHg) create a diffusion gradient of 60 mmHg for O_2 and 6 mmHg for CO_2. Surprisingly, while the diffusion gradient for CO_2 is much lower than that for O_2, CO_2 is still rapidly transported across the alveolar membrane due to a greater diffusion capacity for CO_2 compared with O_2 (approximately 20x greater) (Brooks *et al.*, 2000). Upon exiting the lung, the arterial pressures of O_2 (P_{aO2}) and CO_2 (P_{aCO2}) are 100 mmHg and 40 mmHg, respectively.

1.6.1.4 Mechanics of ventilation

Negative pressure

Similar to the exchange of O_2 and CO_2 within the alveoli, the mechanisms behind ventilation are influenced by changes in pressure gradients. As shown in Figure 1.6.1, both the left and right lobes of the lungs are connected, at their distal ends, to the diaphragm muscle. During inhalation, the contraction of the diaphragm results in a lengthening of the lung (pulling downward), which creates a negative pressure. Opening the mouth during this time facilitates the flow of higher-pressure room air into the respiratory system. Imagine you have covered the top of a syringe with your finger; pulling down on the plunger creates suction by increasing the volume in the chamber and creating negative pressure, much like the diaphragm and lung. During expiration, the diaphragm relaxes and the small muscles between the ribs (the intercostal muscles) contract around the lung, causing an increase in lung pressure and the expelling of air. Contraction and relaxation of the diaphragm and intercostals can be controlled by both conscious and subconscious means (discussed in later sections).

Static lung volumes

The size of an individual's lung is influenced predominately by genetic factors. Although it has been suggested that endurance training can increase lung size, this statement is completely false. Instead, total lung volume is primarily a function of an individual's gender, age, and most importantly height (Cotes *et al.*, 1979). The normal range for total lung volume in males and females is between 3.0 and 6.0 L, with the greatest volumes observed in taller men.

Figure 1.6.4 is a fictional tracing of the static lung volumes in the human body. The word 'static' suggests that these volumes do not normally change, but certain pulmonary disorders can reduce lung volumes, and recently specific respiratory muscle training has been shown to increase some volumes. During normal ventilation, we breathe in a rhythmic fashion; the air that is ventilated in and out of the lung is known as tidal volume (TV). The additional volumes above and below the TV represents the additional capacity for ventilation and are known as the inspiratory reserve volume (IRV) and the expiratory reserve volume (ERV). Together, the IRV, ERV, and TV make up the forced vital capacity (FVC). You should notice that a small volume of air remains below the FVC. This space is known as the residual volume (RV) and is a nonventilating volume of air; it is essential as it helps to maintain lung structural integrity. The total lung capacity (TLC) therefore comprises both ventilating (FVC) and nonventilating (RV) areas of the lung.

During rest, only a small portion of the FVC is used by the TV. As exercise begins, the frequency of breathing increases (the respiratory rate, RR), as does the volume of air ventilated

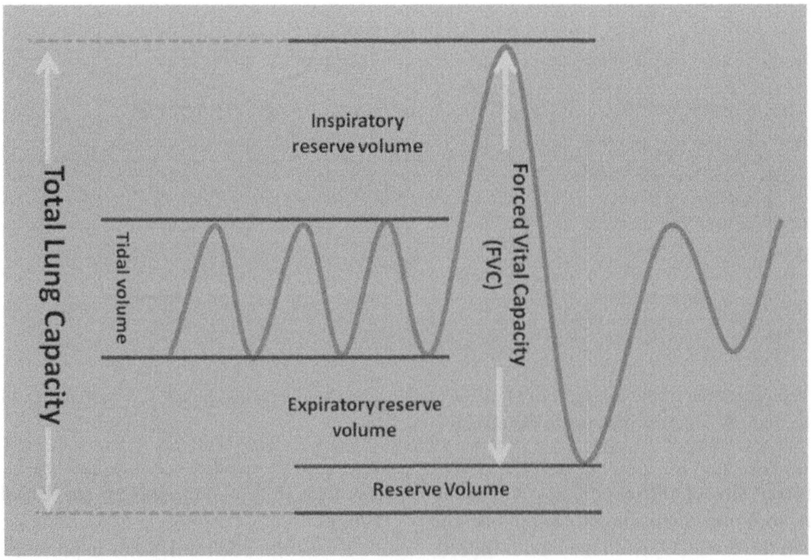

Figure 1.6.4 Diagram of lung volumes measured from helium dilution spirometery

per breath. Together, these increase the minute ventilation (V_E) (explained in detail later). The increase in TV encroaches into both the IRV and the ERV. Nevertheless, more IRV is utilized during the increase in TV than ERV. It should be noted, however, that even during intense exercise TV rarely exceeds 60% of the IRV (Guenette *et al.*, 2007).

Dynamic lung volumes

During a given exercise bout, static lung volumes do not change. For this reason, increasing pulmonary ventilation to meet the demands of exercise requires rapidly increasing the rate of ventilation. For example, during maximal exercise in highly trained individuals, V_E can increase to approximately 37 times the resting rate (rest = 6 l/min; maximal = 220 l/min). Such high ventilatory rates are achievable due to changes in dynamic lung volumes. Compared to static volumes, dynamic lung volumes have a greater capacity to change with training status (Doherty and Dimitriou, 1997). Measuring dynamic lung volume can indicate an individual's capacity to reach high ventilatory rates during exercise; common measurements include forced expiratory volume over one second (FEV_1) and maximal voluntary ventilation (MVV). During measures of FEV_1, individuals are instructed to inhale maximally and then exhale as fast and forcefully as they can for six seconds. The FEV_1 is the volume of air expired during the first second of expiration. This value is compared with the FVC; in normal healthy individuals the FEV_1 : FVC ratio should be greater than 80% (Figure 1.6.5). Ratios recorded below this value may be indicative of pulmonary obstructive disease (Comroe *et al.*, 1964). The measurement of MVV can indicate an individual's maximal ability to ventilate. During this test, individuals are required to breathe as deeply and rapidly as possible for 15 seconds. The total amount of ventilation (in L) is then extrapolated to represent a one-minute value (l/min), which is compared to age- and gender-normative values. In individuals with pulmonary disorders MVV is usually less than 50% of the age and gender norms (Kenney *et al.*, 1995).

As previously stated, with the appropriate training both static and dynamic lung volumes can be altered. With the use of specifically designed respiratory training devices, respiratory muscle training can increase both FEV_1 and FVC (Wells *et al.*, 2005). In addition, respiratory muscle training has been shown to increase the force of inspired and expired ventilation, and MVV (Wells *et al.*, 2005). Further, respiratory muscle training can increase endurance performance and recovery from high-intensity exercise (Romer, McConnell and Jones, 2002a, 2002b). While the increase in performance is not directly linked to changes in lung volumes, other proposed mechanisms have been suggested for the change in performance, including: decreases in respiratory blood flow, decreases in perceived exertion, and lower respiratory muscle fatigue.

1.6.1.5 Minute ventilation

At rest, there is little demand on the pulmonary system to increase the availability of O_2 and the removal of CO_2. For this reason, in a resting state pulmonary V_E is quite low (~6 l/min); it is a product of a low TV (~0.5 L per breath) and RR (12 breaths per minute) (Table 1.6.1). During exercise, there is an increase in the demand for O_2 and in the need to remove accumulated CO_2. For this reason, V_E is increased by increasing both

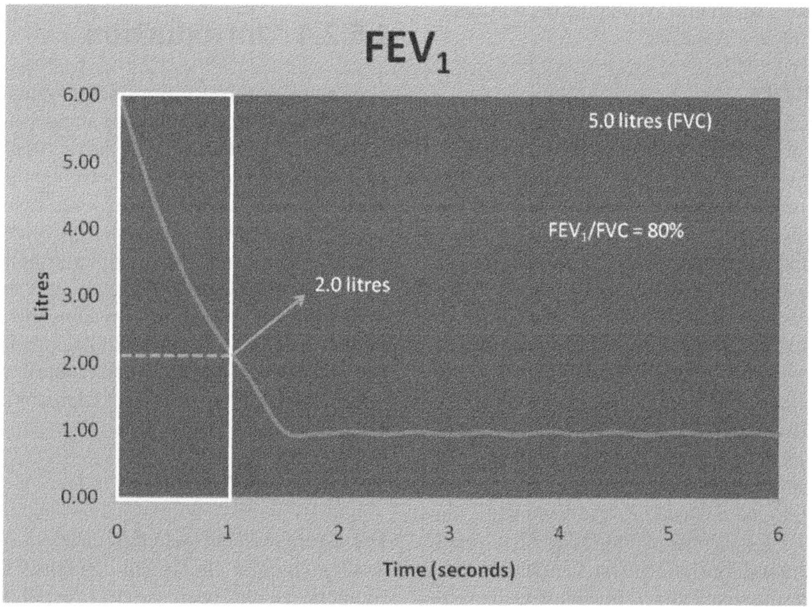

Figure 1.6.5 Theoretical tracing from a pulmonary function test

Table 1.6.1 Typical V_E and TV measurements in a male at rest and during moderate and maximal cycling exercise

	Resting	Moderate exercise	Maximal exercise
V_E (l/min)	6	77	207
TV (l/breath)	0.5	2.6	5.0
RR (breaths/minute)	12	28	41

the TV and the RR. Although V_E can be increased by increases in either of these values alone, the ability to obtained high minute ventilatory rates, as seen during maximal exercise, requires a combination of both.

1.6.1.6 Control of ventilation

Ventilatory control is achieved on both a conscious and an subconscious level. In most circumstances, breathing is controlled primarily by subconscious mechanisms, mediated by complex feedback signals from chemical receptors located in the periphery and brain, as well as by mechanical receptors in the lungs, both of which continuously send signals to the respiratory centre of the brain (medulla), otherwise known as the central controller (for review see Corne and Bshouty, 2005). However, humans are capable of overriding normal subconscious control of their ventilation; during instances of prolonged holding of breath or purposeful hyperventilation, humans can consciously override central and peripheral feedback signalling. Nevertheless, at some point the subconscious drive will overcome the conscious drive and control of ventilation will once again return to a level of homoeostasis.

Chemical receptors

Ventilatory chemical receptors are located both in the periphery and centrally within the brain. The primary function of chemical receptors is to monitor the levels of P_{aO2}, P_{aCO2}, and blood pH. In the periphery, chemical receptors are located in both the aortic arch and the right and left carotid arteries. The close proximity of the aortic and carotid chemical receptors to the left ventricle allows for the rapid assessment of the blood being pumped to the periphery. The response of peripheral receptors to P_{aO2} is hyperbolic in nature; therefore, small changes in P_{aO2}, within normal ranges (>75 mmHg), have little effect on peripheral receptors and thus little influence on ventilation. Nevertheless, if P_{aO2} falls below 75 mmHg the action of the peripheral receptors is increased in an exponential manner. In contrast to P_{aO2}, chemical receptors have a significantly greater sensitivity to P_{aCO2} and pH. At a normal P_{aO2} (100 mmHg), the response to changes in P_{aCO2} is linear, indicating that small changes in P_{aCO2} can have large effects on ventilation (i.e. increasing P_{aCO2} = increasing ventilation). In addition, the production of metabolic CO_2 is often accompanied by an increase in H^+ and thus a decrease in pH. Low pH levels are associated with an increase in peripheral chemical receptor sensitivity to P_{aCO2}. Therefore, during exercise where metabolic CO_2 production is high, the level of P_{aCO2} is a major driving force for peripheral chemical receptor control of ventilation.

While the peripheral chemical receptors are responsive to changes in blood P_{aO2}, P_{aCO2}, and pH, central chemical receptors respond primarily to increases and decreases in P_{aCO2} and the resultant changes in cerebral spinal fluid pH. Central chemical receptors are located in the medulla and several places in the brain stem. Due to their heightened sensitivity to P_{aCO2} and pH, the effect of central chemical receptors on ventilation is large. Indeed, central chemical receptors account for over 70% of the chemical receptor control of ventilation.

Mechanical receptors

Although a large majority of ventilatory control is dictated by central and peripheral chemical receptors, additional mechanical receptors found in the lungs can also mediate respiratory action. Within the lung, two receptors have been observed which produce afferent feedback to the central controller. Located within the smooth muscle of the lung are peripheral stretch receptors, myelinated receptors sensitive to changes in lung volumes. In circumstances where lung volumes are high (e.g. exercise), the peripheral stretch receptors send signals to the central controller resulting in a shortened inspiration and a lengthened expiration. Conversely, located in the upper airways are flow-sensitive receptors called rapidly adapting receptors; these are also myelinated, and are primarily sensitive to low flow rates. During instances of low flow ventilation, these receptors are stimulated to signal the central controller to upregulate the rate of inspiration.

1.6.2 THE CARDIOVASCULAR SYSTEM

1.6.2.1 Introduction

The key role of the cardiovascular system is to provide all cells of the body with a continuous stream of oxygen and nutrients in order to maintain cellular energy transfer. Just as important a function is the removal of metabolic byproducts from the site of energy release, which is achieved by cardiovascular circulation. The cardiovascular system involves gas exchange from lungs to blood and the distribution of this blood around the body. The following pages present a basic overview of the cardiovascular system. After reading this section you should be able to explain: (1) the basic anatomy of the cardiovascular system, (2) the function and control of the cardiovascular system, and (3) the process of capillary exchange.

1.6.2.2 Anatomy

The cardiovascular system is made up of a complex and continuous vascular circuit and comprises the following five key components: the heart, arteries, capillaries, veins, and blood (Figure 1.6.6). As the name suggests, the heart (*cardio*) and the various blood vessels (*vascular*) are central to the cardiovascular system. The heart provides circulatory blood flow,

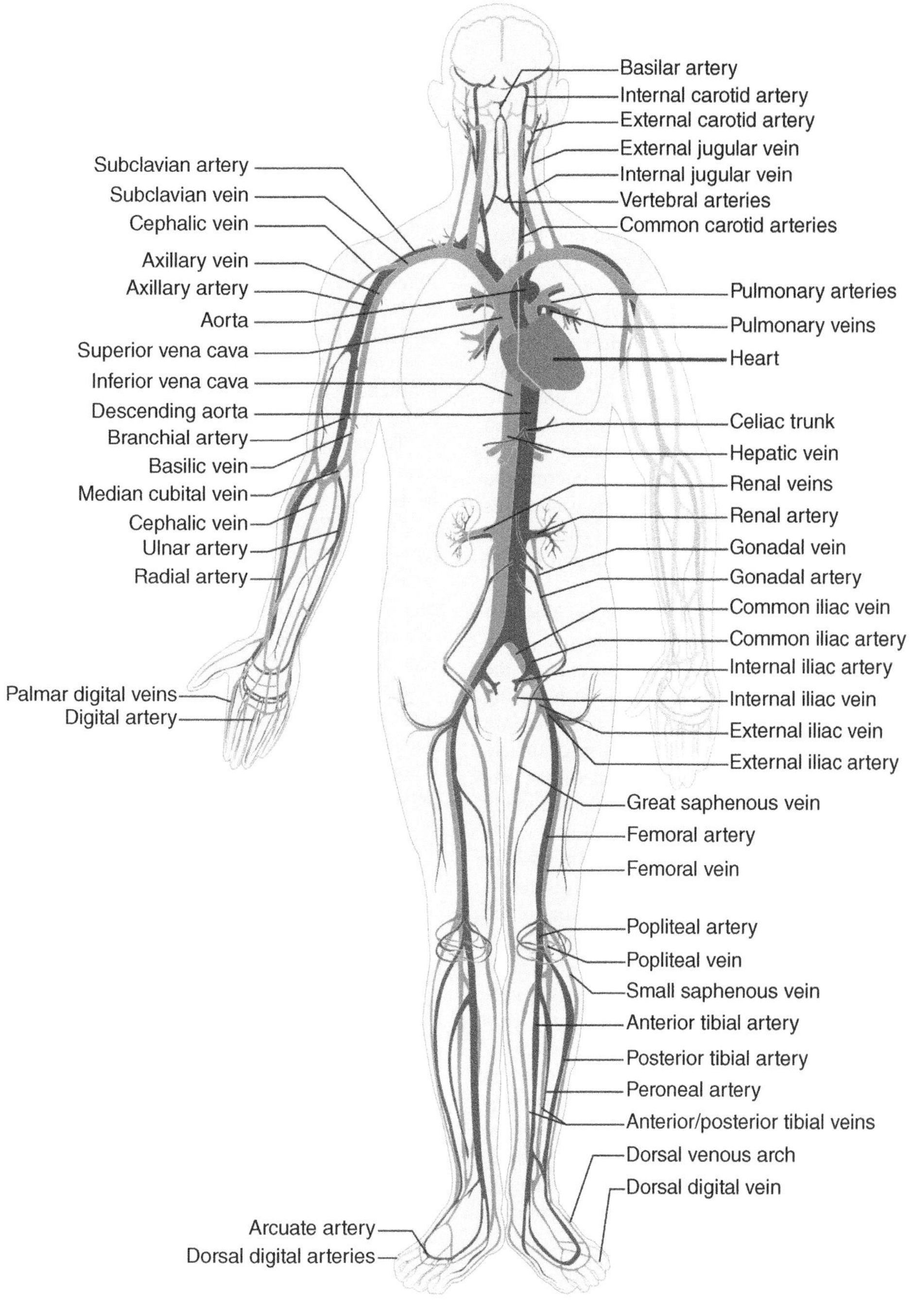

Figure 1.6.6 Schematic of the human cardiovascular system, consisting of the heart and systemic vascular circuit. Dark grey shading represents oxygen-rich blood, while light grey shading denotes deoxygenated blood

while the vascular provides a network for blood transport. The arteries make up the high-pressure distribution circuit that delivers the oxygen-rich blood quickly to the areas in need, the capillaries are the gas-exchange vessels allowing the diffusion of blood gases and nutrients into and out of the cells, and the veins are the low-pressure collection and return vessels. An often overlooked but very important part of the cardiovascular system is the blood, which acts as a medium for the transport of nutrients, wastes, gases, vitamins, and hormones around the body.

1.6.2.3 The heart

The heart is a hollow and efficient muscular organ which acts as a pump, providing the thrust required to transport blood around the body. This pump is situated in the mid-centre of the chest cavity, about two-thirds of the way to the left of midline. This four-chamber organ weighs approximately 300–350 g and is typically about the size of a closed fist. Despite its modest size and weight, the heart has incredible strength and endurance. At rest, the heart beats approximately 100 000 times per day, pumping some 4000–5500 L of blood, while the working heart of an endurance athlete has been calculated to be able to pump up to 35–40 L in a minute (Powers and Howley, 2007). The majority of the heart's mass is composed of muscle; this muscle, the myocardium, is a form of striated muscle similar to that of skeletal muscle. In contrast, however, the multinucleated individual cells or fibres of the heart are shorter than skeletal muscle and interconnect in latticework fashion via intercalated discs. These intercellular connections allow electrical impulses to be transmitted throughout the myocardium, enabling the heart to function as a single unit (McArdle, Katch and Katch, 2004).

The heart contains four separate chambers or cavities, but functionally it acts as two separate pumps: the right side of the heart contains only deoxygenated blood returning from the body, while the left side contains oxygenated blood returning from the lungs. The two superior chambers of the heart are referred to as the right and left atria, while the lower chambers are the right and left ventricles. During a single cardiac cycle each chamber in the heart experiences periods of contraction and relaxation. The relaxation phase where the chamber fills with blood is called diastole, while the contraction phase where blood is pushed into an adjacent chamber or arterial trunk is called systole. Throughout the cardiac cycle circulating blood from the body enters the heart through the superior and inferior vena cava and proceeds into the right atrium. During each heart beat, blood is pumped from the right atrium into the right ventricle. Blood is contained within the right ventricle by a one-way flow valve termed the tricuspid valve, which separates the right atrium and right ventricle. From the right ventricle, blood is pumped through the pulmonary artery to the lungs, where gas exchange occurs. Re-oxygenated blood returning from the lungs then enters the left atrium through the pulmonary veins. This blood is pumped into the left ventricle through the bicuspid valve, which is the term for the atrioventricular valve separating the left atrium and left ventricle.

Cardiac function and control

The transfer of blood throughout the heart results from changes in pressure caused by contracting myocardium. Blood will flow from one chamber to another only if the pressure gradients differ. This is important since atria and ventricular pressure gradients are dependent upon carefully timed muscular contractions. The heart is innervated by the autonomic nervous system and autonomic neurons may control the rate of cardiac cycle, though they do not initiate the contraction. Instead, the conducting system of the heart, which comprises specialized muscle tissue, initiates and distributes the electrical impulses responsible for the contraction of cardiac muscle fibres. The sinoatrial node (SA node) is located on the posterior wall of the right atrium and is responsible for the spontaneous electrical activity resulting in myocardial contraction. Due to its role, the SA node is often termed the natural pacemaker or cardiac pacemaker. Depolarization of the SA node spreads across the artia, resulting in atrial contraction. This electrical impulse also depolarizes the atrioventricular (AV) node, which is located at the inferior floor of the right atrium. From the AV node, a tract of conducting fibres called the atrioventricular bundle or bundle of His travels to the ventricular septum and enters the right and left bundle branches. These branches diverge into small conduction myofibres called Purkinje fibres, which spread the electrical impulse throughout the ventricles, resulting in ventricular contraction (Tortora and Anagnostakos, 1987).

The study of these electrical signals may provide physicians with valuable information on the function of an individual's heart. Such electrical activity can be recorded on an electrocardiogram (ECG) and used to monitor changes or potential problems in heart activity (Figure 1.6.7). Throughout each cardiac cycle a number of important deflection points may be observed on a typical ECG trace. The first deflection point is the P wave and represents depolarization of the atria. The QRS complex represents the depolarization of the ventricles and the following T wave represents ventricular repolarisation. Atrial repolarisation is often not observed because it occurs at the same time as ventricular depolarization and is therefore masked by the QRS complex. Examination of the magnitude of waves and the time between deflection points may provide significant information on the general function of the heart (Tortora and Anagnostakos, 1987).

Stroke volume (SV) refers to the amount of blood ejected from the left ventricle per contraction. This volume of blood is

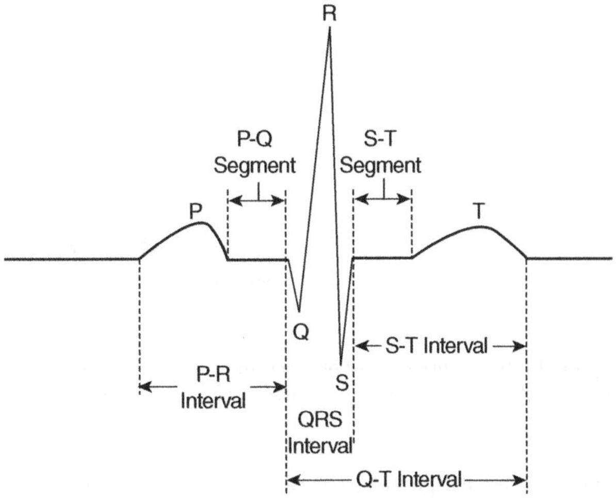

Figure 1.6.7 Example of a resting electrocardiogram detailing the various phases of the cardiac cycle

simply the difference between the amount of blood present in the ventricle after filling (end diastolic volume, EDV) and the volume of blood remaining after ventricular contraction (end systolic volume, ESV). The proportion of blood ejected with each beat relative to the filled ventricle (i.e. EDV) is termed the ejection fraction. The total volume of blood pumped from the left ventricle each minute is referred to as the cardiac output (Q). Cardiac output is therefore the product of heart rate (HR) and stroke volume (SV). Calculations for EF, SV, and Q are shown below:

$$EF = (SV/EDV) \times 100\%$$

$$SV = EDV - ESV$$

$$Q = HR \times SV$$

Based on the above equation, cardiac output may be altered by changes in either heart rate or stroke volume. Changing heart rate is the body's principal mechanism for short-term changes in blood supply. Heart rate and stroke volume may be regulated by many factors, including temperature, emotional state, chemicals, gender, and age. However, the most important control of heart rate and strength of contraction is the autonomic nervous system.

Autonomic control

Within the medulla of the brain is a group of neurons called the cardioacceleratory centre, from which sympathetic fibres travel down the spinal cord to the heart via the cardiac nerves. When stimulated, nerve impulses travel from the cardioacceleratory centre along the sympathetic fibres, causing the release of norepinephrine and resulting in an increase in the rate and strength of myocardial contraction. Apposing sympathetic tone is parasympathetic stimulus, which also arises from the medulla (cardioinhibitory centre). When this centre is stimulated, electrical impulses travel along the parasympathetic fibres to the heart via the vagus nerve, causing the release of acetylcholine, which decreases heart rate. The control of cardiac output is therefore the result of apposing sympathetic and parasympathetic tone (Rowell, 1997). In order to regulate heart rate and strength of contraction, cardioacceleratory and cardioinhibitory centres are stimulated in response to changes in blood pressure, which are detected by baroreceptors located in the carotid artery (carotid sinus reflex), arch of the aorta (aortic reflex), and the superior and inferior vena cava (atrial reflex) (Delp and O'Leary, 2004; Tortora and Anagnostakos, 1987).

1.6.2.4 The vascular system

The vascular system comprises a series of blood vessels which form a network to transport blood from the heart to the various tissues of the body, then back to the heart. These vessels differ in physiological characteristics and function and can be separated into arteries, arterioles, capillaries, venules, and veins. Arteries are the large, high-pressure vessels that transport blood from the heart. Artery walls are constructed of layers of connective tissue and smooth muscle. These vessels are highly elastic and expand to accommodate the rapid ejection of blood from the ventricles. When the ventricles relax the elastic arterial walls recoil, moving blood onwards. Further from the heart, distributing arteries contain more smooth muscle than elastic fibres, allowing greater vasoconstriction or vasodilation to regulate blood flow to the periphery. Arteries branch into smaller vessels called arterioles that deliver blood to the capillaries.

Capillaries are microscopic vessels that permit the exchange of nutrients and wastes between blood and all cells in the body. Capillaries are therefore found near every cell in the body, although the density differs depending on the activity of the tissue. On average, skeletal muscle contains approximately 2000–3000 capillaries per square millimetre of tissue. A ring of smooth muscle at the origin of the capillary (pre-capillary sphincter) controls the regulation of blood flow through a process termed autoregulation (Delp and O'Leary, 2004; Wilmore and Costill, 1994).

Dexoygenated blood exiting the capillaries flows into the venous system through small veins called venules. Venules transport blood to the veins, from where it is returned to the heart. Once leaving the capillaries, blood pressure is drastically reduced. For this reason many veins contain valves that permit blood to flow in only one direction. Despite this, the low pressure of the venous system results in a considerable volume of the total blood being contained within the veins (60–65% of total blood). As such the veins have been referred to as blood reservoirs. The impact of blood pooling in the veins has received increasing attention in recent years, with the thought that facilitating venous return will improve the removal of waste products, improving performance and promoting faster recovery from exercise.

Capillary exchange

The velocity of blood throughout the vascular system is dictated by the pressure of various vessels. Blood pressure is greatest in the aorta (100 mmHg) and gradually decreases as it passes through the arteries (100–40 mmHg), arterioles (40–25 mmHg), capillaries (25–12 mmHg), venules (12–8 mmHg), veins (10–5 mmHg), and back to the right atrium (0 mmHg) (McArdle, Katch and Katch, 2004; Powers and Howley, 2007). Although pressure gradually decreases, the velocity of blood is lowest at the capillaries and increases as it enters the venous system. This increase in velocity is the result of lower peripheral resistance in venous system. The low velocity of blood through the capillaries is important in allowing the movement of nutrients and wastes between blood and tissue. The exchange of water and dissolved substances through capillary walls is dependent on a number of opposing forces: hydrostatic and osmotic pressure. Blood hydrostatic pressure (weight of water in blood) tends to move fluid from capillaries to the interstitial fluid and averages approximately 30 mmHg at the arterial and 15 mmHg at the venous end. This flow of fluid is opposed by interstitial fluid hydrostatic pressure (0 mmHg), tending to move fluid in the opposite direction. Blood osmotic pressure (28 mmHg) is related to the presence of non-diffusible proteins in the blood, which tends to move fluid from the interstitial fluid to the blood. Opposing this, interstitial fluid osmotic pressure (6 mmHg) tends to move fluid from capillaries to the interstitial fluid. The balance between osmotic and hydrostatic pressure dictates the movement of fluid so that fluid moves from the capillaries to the interstitial fluid at the arterial end and vice versa at the venous end (Tortora and Anagnostakos, 1987).

1.6.2.5 Blood and haemodynamics

The blood is a vital part of the cardiovascular system and acts as a medium for many important functions, including transportation, temperature regulation, and acid–base (pH) balance. The total volume of blood within an individual's body typically ranges from 5 to 6 L in males and 4 to 5 L in females (Wilmore and Costill, 1994). This blood comprises approximately 55–60% plasma, with the remainder constituting formed elements. Plasma is primarily made up of water (~90%), with the remainder being plasma proteins (~7%) and cellular nutrients, antibodies, electrolytes, hormones, enzymes, and wastes (~3%) (Wilmore and Costill, 1994). The formed elements are the red blood cells (erythrocytes, 99%), white blood cells (leukocytes), and platelets (thrombocytes). The ratio between formed elements (i.e. red blood cells) and plasma is referred to as the haematocrit. Due to reductions in plasma volume, haemoconcentrations may change, resulting in an increase in haematocrit with little or no change in the total number of red blood cells. The red blood cells, which make up 99% of the formed elements, have a life span of approximately four weeks and are responsible for oxygen transportation. Approximately 99% of the oxygen transported in the blood is chemically bound to haemoglobin, which is a protein found in red blood cells. Each red blood cell contains approximately 250 million haemoglobin molecules, which may each bind to four oxygen molecules. Consequently, the oxygen-carrying capacity of blood is extremely high, with the average healthy male being able to transport approximately 200 ml of oxygen per litre of blood when haemoglobin is 100% saturated with oxygen.

The relative saturation of haemoglobin depends upon the haemoglobin's affinity for oxygen, which is influenced by a number of factors, including the partial pressure of oxygen to which the haemoglobin is exposed (P_{O2}), blood temperature, pH, and carbon dioxide, carbon monoxide, and 2,3 diphosphoglyceric acid (2–3 DPG) concentrations. The relationship between oxygen saturation and the P_{O2} in the blood may be explained by the oxyhaemoglobin dissociation curve, which is presented in Figure 1.6.8. This shows that the percentage of haemoglobin saturation (% HbO_2) increases dramatically to a partial pressure of 40 mmHg, after which it slows to a plateau of approximately 90–100 mmHg. This relationship is important, and is responsible for the transportation of O_2 from the lungs to the blood ('loading') and from the haemoglobin to the tissue ('unloading'). High P_{O2} in the alveoli of the lungs results in an increase in arterial O_2 pressure and thus the formation of oxyhaemoglobin. In contrast, low P_{O2} in the tissues deceases capillary P_{O2}, resulting in unloading of O_2 and the formation of deoxyhaemoglobin. At rest, tissue O_2 demands are low, resulting in relatively high venous P_{O2} (40 mmHg), and consequently only about 25% of the O_2 bound to haemoglobin is unloaded to the tissues. However, with intense exercise P_{O2} may dramatically drop (18–20 mmHg), resulting in 90% of oxygen being unloaded (Powers and Howley, 2007). The affinity of oxygen to haemoglobin may be influenced not only by P_{O2} but also by blood pH (Figure 1.6.8). An increase in blood acidity (i.e. reduction in pH) reduces oxygen's affinity to haemoglobin by weakening the bond between them. This influence is referred to as the Bohr effect and causes an increased unloading of O_2

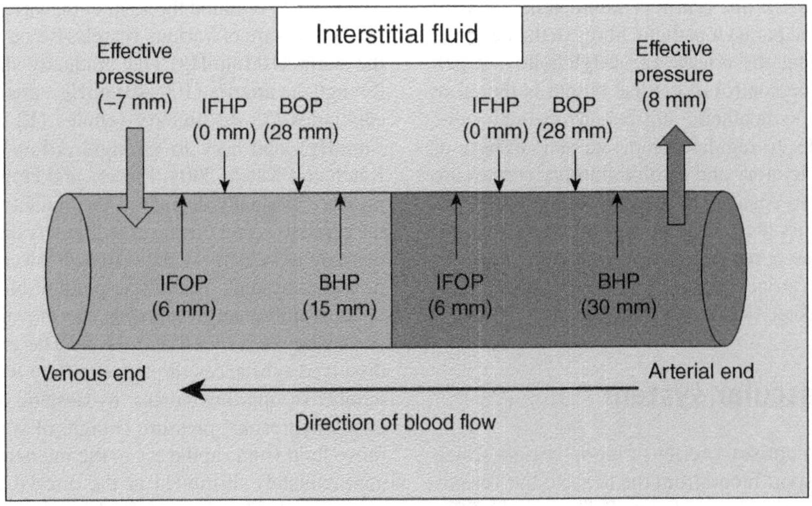

Figure 1.6.8 Hydrostatic and osmotic pressures involved in the movement of fluids into the interstitial fluid (out of capillaries; filtration) and into capillaries (out of interstitial fluid; reabsorption). IFHP = interstitial fluid hydrostatic pressure, IFOP = interstitial fluid osmotic pressure, BOP = blood osmotic pressure, BHP = blood hydrostatic pressure

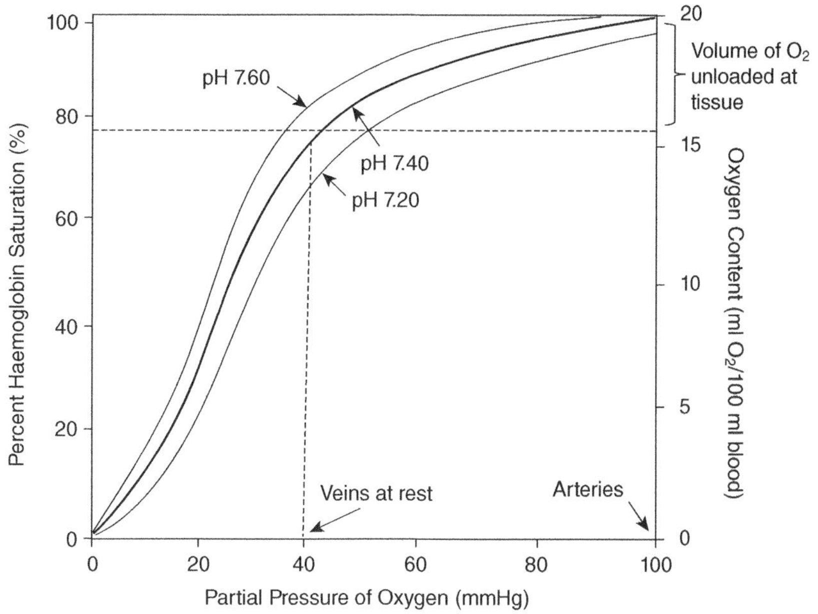

Figure 1.6.9 The relationship between percentage oxyhaemoglobin saturation and partial pressure of oxygen in the blood, known as the oxyhaemoglobin dissociation curve. Also shown is the influence of pH on haemoglobin affinity for oxygen. Increasing blood temperature has a similar effect to decreasing pH, resulting in a rightward shift in the curve

to the tissues. Likewise, an increase in blood temperature weakens the bond between O_2 and haemoglobin, resulting in the deoxyhaemoglobin dissociation curve shifting to the right. Since active tissue produces greater heat and high acid levels (reduced pH), these shifts facilitate the unloading of O_2 to the active tissue. 2–3 DPG is created by red blood cells during glycolysis and may bind to haemoglobin, reducing its affinity for oxygen and enabling unloading of O_2 in tissue capillaries (Powers and Howley, 2007). Concentrations of 2–3 DPG increase at altitude and in anaemic populations; this is likely to be an important adaptive mechanism.

At rest the oxygen content in the blood ranges from approximately 20 ml per 100 ml of blood in the arteries to 14 ml per 100 ml of mixed blood in the veins. The difference between arterial and venous oxygen content is referred to as the arterial venous oxygen difference (a–v O_2 difference) and reflects the oxygen taken up by active tissue. At rest the a–v O_2 difference is approximately 6 ml (20: 14 ml), although this may increase up to 16–18 ml per 100 ml of blood during intense exercise (Powers and Howley, 2007). This increase in the a–v O_2 difference reflects a greater oxygen extraction by the working tissue (i.e. a decrease in O_2 content of venous blood), as seen in Figure 1.6.9. Throughout exercise the arterial oxygen content remains relatively stable, although it has been reported to reduce slightly during high-intensity maximal exercise in elite athletes, possibly due to blood flow rate exceeding the rate of oxygen diffusion from alveoli to haemoglobin (red blood cell transit time) (Chapman, Emery and Stager, 1999; Wilmore and Costill, 1994).

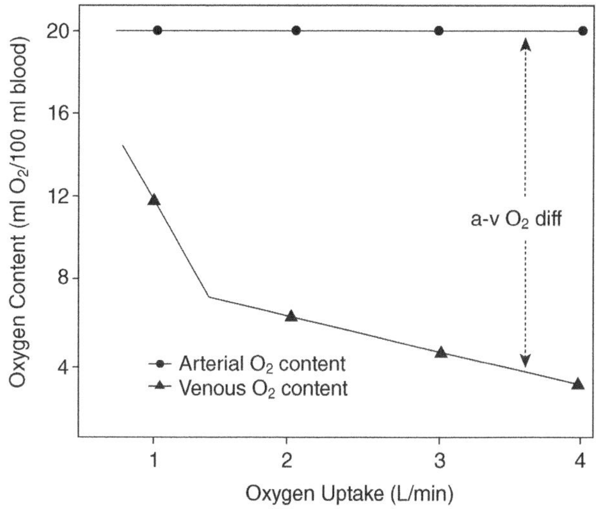

Figure 1.6.10 Changes in arterial and venous oxygen content during low to high exercise intensities

The relationship between the a–v O_2 difference and the total volume of blood being pumped to active (i.e. cardiac) tissue gives an indication of the total oxygen being consumed by the active tissue (Figure 1.6.10). This oxygen uptake may be explained by the Fick equation:

$$VO_2 = Q \times (a - v\ O_2\ \text{difference})$$

1.6.3 CONCLUSION

This chapter reviewed the anatomy, function, and control of the cardiovascular respiratory system. In particular, it outlined the mechanisms for gas exchange in the lungs, the transportation of blood in the body, and the exchange of gas and nutrients at the cellular level.

References

Brooks G.A., Fahey T.D., White T.P. and Baldwin K.M. (2000) *Exercise Physiology: Human Bioenergetics and Its Applications*, Mayfield Publishing Co, USA.

Chapman R.F., Emery M. and Stager J.M. (1999) Degree of arterial desaturation in normoxia influences VO2max decline in mild hypoxia. *Med Sci Sports Exerc*, **31**, 658–663.

Comroe J.H., Forster R.E., Dubois A.B. *et al.* (1964) *The Lung: Clinical Physiology and Pulmonary Function Tests*, Year Book Medical Publishers, Chicago.

Corne S. and Bshouty Z. (2005) Basic principles of control of breathing. *Respir Care Clin N Am*, **11**, 147–172.

Cotes J.E., Dabbs J.M., Hall A.M. *et al.* (1979) Sitting height, fat-free mass and body fat as reference variables for lung function in healthy British children: comparison with stature. *Annu Hum Biol*, **6**, 307–314.

Delp M.D. and O'Leary D.S. (2004) Integrative control of the skeletal muscle microcirculation in the maintenance of arterial pressure during exercise. *J Appl Physiol*, **97**, 1112–1118.

Doherty M. and Dimitriou L. (1997) Comparison of lung volume in Greek swimmers, land based athletes, and sedentary controls using allometric scaling. *Br J Sports Med*, **31**, 337–341.

Guenette J.A., Witt J.D., McKenzie D.C. *et al.* (2007) Respiratory mechanics during exercise in endurance-trained men and women. *J Physiol*, **581**, 1309–1322.

Kenney W.L., Mahler D.A., Humphrey R.H. and Bryant C.X. (1995) *Acsm's Guidelines for Exercise Testing and Prescription: American College of Sports Medicine*, Lippincott Williams & Wilkins.

McArdle W.D., Katch F.I. and Katch V.L. (2004) *Exercise Physiology: Energy, Nutrition and Human Performance*, Lippincott Williams & Wilkins.

Powers S.K. and Howley E.T. (2007) *Exercise Physiology: Theory and Application to Fitness and Performance*, McGraw-Hill, New York.

Romer L.M., McConnell A.K. and Jones D.A. (2002a) Effects of inspiratory muscle training on time-trial performance in trained cyclists. *J Sports Sci*, **20**, 547–562.

Romer L.M., McConnell A.K. and Jones D.A. (2002b) Effects of inspiratory muscle training upon recovery time during high intensity, repetitive sprint activity. *Int J Sports Med*, **23**, 353–360.

Rowell L.B. (1997) Neural control of muscle blood flow: importance during dynamic exercise. *Clin Exp Pharmacol Physiol*, **24**, 117–125.

Tortora G.J. and Anagnostakos N.P. (1987) *Principles of Anatomy and Physiology*, Harper & Row, New York.

Wells G.D., Plyley M., Thomas S. *et al.* (2005) Effects of concurrent inspiratory and expiratory muscle training on respiratory and exercise performance in competitive swimmers. *Eur J Appl Physiol*, **94**, 527–540.

West J.B. (1962) Regional differences in gas exchange in the lung of erect man. *J Appl Physiol*, **17**, 893–898.

Wilmore J.H. and Costill D.L. (1994) *Physiology of Sport and Exercise*, Human Kinetics, Champaign.

1.7 Genetic and Signal Transduction Aspects of Strength Training

Henning Wackerhage, Arimantas Lionikas, Stuart Gray and Aivaras Ratkevicius, School of Medical Sciences, College of Life Sciences and Medicine, University of Aberdeen, Aberdeen, UK

1.7.1 GENETICS OF STRENGTH AND TRAINABILITY

1.7.1.1 Introduction to sport and exercise genetics

The aim of this chapter is to explain how genetic variation can influence muscle mass, strength, and trainability, and how signal transduction pathways mediate the adaptation to strength or resistance training.

Genetics is the science of heredity; it can be applied to study exercise-related traits such as strength, maximal oxygen uptake, and motor skills. Traits such as grip strength depend on both genetic variation and various environmental factors, or in other words, nature and nurture. Classical sport and exercise geneticists aim to quantify the genetic contribution to a trait. Why is this important? There are many answers to this question. For example, the heritability of a trait informs us whether it is more dependent on genetically endowed talent or on environmental factors such as training and nutrition. Also, if a trait's heritability estimate (heritability can vary between 0 or 0% and 1 or 100%) is high then it makes sense to search for the genetic variations that explain it.

Heritability has traditionally been estimated by performing twin or family studies. In twin studies a trait such as grip strength is measured in monozygotic twin pairs (with identical DNA) and dizygotic pairs (with 50% shared DNA). The heritability, h^2, of hand grip strength can be derived based on the intraclass correlations (that is, the correlation coefficient, r, between twin 1 and twin 2 of each pair in the monozygotic (MZ) and dizygotic (DZ) populations) in the following manner: $h^2 = (r_{MZ} - r_{DZ}) / (1 - r_{DZ})$ (Newman, Freeman and Holzinger, 1937).

Searching for the genes underlying heritability is a difficult task because there are over 3 billion base pairs and over 20 000 genes in the human genome. One of the strategies for achieving this is the candidate gene approach, which is often used when an exceptional phenotype is observed. An example where this strategy has successfully been employed is in a study where researchers hypothesized that a mutation in the myostatin gene was responsible for the unusually high muscle mass observed in a child (Schuelke *et al.*, 2004). The researchers based their hypothesis on the knowledge that myostatin knockout mice display hyperplasia and hypertrophy (McPherron, Lawler and Lee, 1997) and that myostatin mutations occur 'naturally' in some cattle breeds (McPherron and Lee, 1997). After analysing the myostatin gene in the affected child, the researchers identified a mutation in the myostatin gene which was likely to cause the phenotype and was not present in the normal population. Disadvantages of this method are that it only works well with exceptional, monogenic phenotypes and where a candidate gene hypothesis can be proposed, for example from a transgenic mouse model showing a similar phenotype.

Another strategy is to perform so-called genome-wide association studies. In brief, these studies are based on the concept that variations in relevant genetic markers are associated with phenotypic variation. Currently a dense set of genomic markers such as single nucleotide polymorphisms (SNPs) is available. By studying the effects of each SNP on a phenotype such as handgrip strength or trainability one can discover genomic regions that can affect the phenotype.

The advantage of this approach is that, unlike the candidate gene approach, no prior information on gene function is needed. This method has the potential to discover novel genes and the biological pathways that influence a phenotype. The limitation of this approach is that usually thousands of subjects are needed for such studies.

1.7.1.2 Heritability of muscle mass, strength, and strength trainability

In this section we will review the genetic determinants of muscle mass, strength, and trainability. Muscle mass and strength depend on environmental factors, such as physical

Strength and Conditioning – Biological Principles and Practical Applications Marco Cardinale, Rob Newton, and Kazunori Nosaka.

activity, health, and nutrition, and on genetic factors. 'Genetic factors' means that people inherit different alleles or variants of the genes that affect a trait such as muscle strength. Several studies indicate that approximately half of the variation in human muscle strength is inherited (Arden and Spector, 1997; Carmelli and Reed, 2000; Frederiksen *et al.*, 2002; Perusse *et al.*, 1987; Reed *et al.*, 1991; Silventoinen *et al.*, 2008; Thomis *et al.*, 1997).

The physiological mechanism underlying the variation in strength may be related to the proportion of different fibre types in skeletal muscle. Slow-twitch (type I) and fast-twitch (type II) fibres differ in force, power, velocity, rate of relaxation, and fatigability. There is a wide degree of variation in the proportion of fibre types in specific skeletal muscles. For example, the proportion of type I fibres ranges from 16% to 97% in human vastus lateralis muscle (Simoneau and Bouchard, 1989; Staron *et al.*, 2000). Due to limitations with human studies, a wide range of heritability of fibre type proportions has been reported (Komi *et al.*, 1977; Simoneau and Bouchard, 1995). However, rodent studies provide convincing evidence of the importance of genetic variation on the proportion of fibre types (Nimmo, Wilson and Snow, 1985; Suwa, Nakamura and Katsuta, 1999; van der Laarse, Diegenbach and Maslam, 1984; Vaughan *et al.*, 1974). For example, 37% of fibres in soleus muscle of the C57BL/6 strain are type I versus 58% in the CPB-K strain (van der Laarse 1984).

Another physiological mechanism that affects muscle strength and size is the number of fibres within a given muscle. For example, Lexell, Taylor and Sjostrom (1988) counted 393 000–903 000 fibres in the vastus lateralis in a cohort of males with a mean age of 19 years. The design of this study does not allow quantification of how much of this variation is due to genetic factors and how much is due to environmental factors. Animal studies, however, provide evidence that genetic variation can significantly influence muscle fibre numbers (Nimmo, Wilson and Snow, 1985; Suwa, Nakamura and Katsuta, 1999; van der Laarse, Diegenbach and Maslam, 1984; Vaughan *et al.*, 1974). For example, in population of mice derived from the C3H/He and SWR strains, heritability of fibres counted in the soleus muscle was around 50% (Nimmo, Wilson and Snow, 1985).

Strength and resistance training is the main method utilized to increase both muscle mass and strength. Individuals vary in their ability to adapt to strength training, or in other words their strength trainability differs. A study carried out on more than 500 young, untrained, healthy individuals confirmed that the same strength training programme results in greatly varying strength and muscle size adaptations (Hubal *et al.*, 2005). After 12 weeks of progressive resistance training the elbow flexor strength increased on average by 54%, and muscle size by 19%. However, the extent of change varied from 0% to +250% for strength and from −2% to +59% for muscle size (Hubal *et al.*, 2005).

Why does strength trainability vary that much? The answer is unknown, but a potential explanation is again the variation in the proportion of fast- and slow-twitch fibres. Several studies report that type II fibres hypertrophy to a greater extent than type I fibres in response to strength training (Hather *et al.*, 1991; Houston *et al.*, 1983; MacDougall *et al.*, 1980; Staron *et al.*, 1990). Thus individuals with more type II fibres may experience a greater muscle hypertrophy after a strength training programme than individuals with more type I fibres. Another possible explanation is that allelic variation of the genes relevant for the adaptive response to training *per se* plays an important role. A twin study examining the role of the genetic factors on strength gain following resistance training showed that ~20% of variation in strength gain after training is explained by the genetic variation in trainability and is not related to variation at the baseline (Thomis *et al.*, 1998).

To summarize, muscle strength in a population varies due to genetic and environmental factors affecting the development and maintenance of skeletal muscle and also because of genetic factors that affect the adaptation to external stimuli such as strength training. Many questions remain, including 'What and how many genetic variants have a significant effect on strength?', 'What is the effect size of genetic variants?', and 'What is the frequency of these genetic variants in different populations?' Answers to these questions are needed in order to allow us to develop meaningful applications such as genetic tests to predict strength or strength trainability.

Mutations in a single gene can disrupt the function of the coded protein or preclude its translation affecting the phenotype in a profound way and even causing disease. Duchenne muscular dystrophy is an example of such a disease. Understanding the heredity of such conditions enabled accurate predictions of whether the mutant allele, if inherited, would result in disease. This led to expectations that variation in a phenotype such as strength might be affected in a similar manner, where allelic variant of some major gene determines whether strength-generating ability is high or low. Such a scenario, however, is rare. For example, allelic variants of the ACTN3 (Walsh *et al.*, 2008) and CNTF (Roth *et al.*, 2001) genes, implicated in variation in muscle strength, exerted rather small effects, accounting for ~1% of phenotypic variation.

In the polygenic model of genetic effects, many genes together influence the phenotype through a small effect of each. Studies on skeletal muscle in mice support the polygenic model in strength as a number of the genomic regions known as quantitative trait loci (QTL), which affect muscle mass, were identified on different chromosomes of the mouse genome (Brockmann *et al.*, 1998; Lionikas *et al.*, 2003; Masinde *et al.*, 2002). Identification of the genes underlying these QTL will offer researchers new candidate genes and biological pathways for examination of their effects on human muscles; this will eventually lead to identification of the genes underlying high heritability estimates and understanding of the constellations of alleles resulting in high or low muscle strength. Even while the underlying genes remain elusive, acknowledgement of the polygenic nature of the genetic effects is important because it means that predictions of strength performance for an individual based on his or her alleles of one gene are of a very limited value.

Genetic testing is currently accessible to the general public (Genetic Technologies Ltd, 2007). It is often assumed that such tests can predict athletic ability. Perhaps the most popular gene offered for testing is ACTN3. It has been discovered that mutation in this gene resulting in a premature stop codon is common in Caucasians (North *et al.*, 1999). It was also noted that the frequency of the mutant X allele, which encodes the stop codon, among elite power athletes is lower and among endurance athletes higher than in the general population (Yang *et al.*, 2003). While a genetic test can determine the alleles of ACTN3 carried by an individual, the predictive value of such information is limited because the gene accounts only for ~1% of strength variation (Walsh *et al.*, 2008). Testing with a genetic test for a single gene effect will not provide much information. One can however envisage that in the future useful information may be obtained by testing for multiple genes which all affect, for example, muscle power. At present screening for 50 m sprinting ability, vertical jumps, or 1 RM strength tests is a more informative assessment of athletic potential. Ethical and legal implications also have to be taken into account when deciding whether or not to perform genetic tests on athletes. This has been reviewed as part of the BASES position stand on 'Genetic Research and Testing in Sport and Exercise Science' (Williams *et al.*, 2009).

1.7.2 SIGNAL TRANSDUCTION PATHWAYS THAT MEDIATE THE ADAPTATION TO STRENGTH TRAINING

1.7.2.1 Introduction to adaptation to exercise: signal transduction pathway regulation

In the last 20 years exercise physiologists have used molecular biology techniques to address the question 'What are the mechanisms that make skeletal muscle and other organs adapt to exercise?' The answer is: signal transduction pathways. Such pathways:

1. sense signals associated with exercise via sensor proteins;
2. integrate this information, often by kinase-mediated phosphorylation and phosphatase-mediated dephosphorylation, as well as other forms of protein–protein interaction;
3. regulate the transcription and translation (protein synthesis) of genes, or other functions such as protein breakdown, cell division, and cell death (Figure 1.7.1).

We will now explain these three steps. Sensor proteins are comparable to sensor organs such as the eye, ear, and nose.

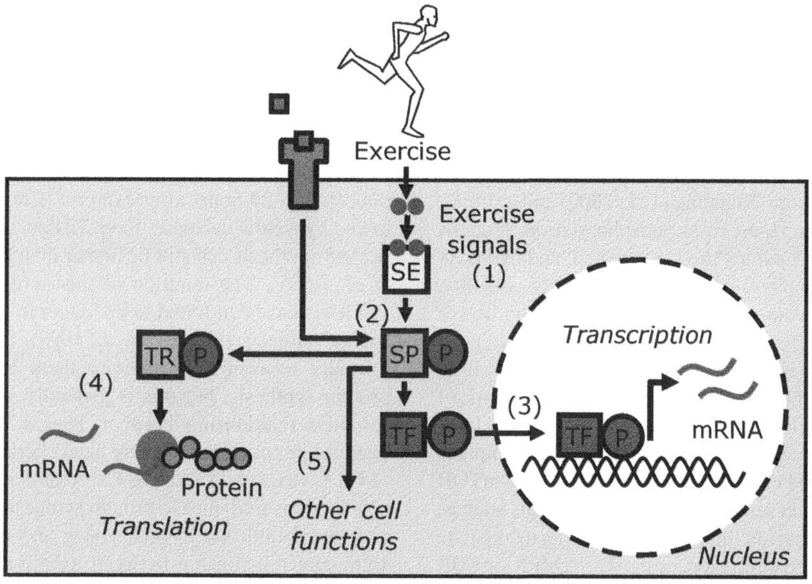

Figure 1.7.1 Model detailing adaptation to exercise. (1) Exercise changes many intra- and extracellular signals, which are sensed by sensor proteins (SE). (2) This information is then conveyed and integrated by a network of signalling proteins (SPs). (3) These signalling proteins regulate functions such as transcription, where a gene is copied into mRNA, (4) translation, which is the synthesis of new protein from mRNA, which takes place in ribosomes, and (5) other cell functions such as cell division. Transcription is regulated by DNA-binding transcription factors (TF) whereas translation is regulated by translational regulators (TR)

Table 1.7.1 Five sensor proteins

Sensor protein	Binding sites for	Mediates adaption to
AMPK	AMP, glycogen	increased energy turnover
HIF prolyl-hydroxylases	oxygen	hypoxia
calmodulin	calcium	muscle contraction
adrenergic receptors	adrenaline	systemic exercise changes
titin kinase	force/stretch[a]	passive muscle stretch

[a]Stretch increases ATP binding to titin kinase, which primes the enzyme for subsequent autophosphorylation (Puchner *et al.*, 2008).

During strength training many molecules bind to the sensor proteins and variables such as force, temperature, and length change are sensed intracellularly by them. These small molecules will change the conformation and/or activity of the sensor proteins and thus trigger a signalling cascade. Table 1.7.1 gives five examples of sensor proteins.

One can imagine how 'rest' signals (low AMP, high glycogen, low calcium, low adrenaline, low force/stretch) differ from 'exercise' signals (high AMP, low glycogen, high calcium, high adrenaline, high force/stretch) and how consequently a unique set of signalling proteins is activated by exercise.

Signalling proteins, like the nervous system, convey the information gathered by the sensor proteins, compute it, and regulate outputs. The main class of signalling proteins are proteins that can use a phosphate group from an ATP to phosphorylate or dephosphorylate a serine, threonine, or tyrosine on one or more downstream proteins. The enzymes that phosphorylate proteins are termed 'kinases' while those which dephosphorylate proteins are termed 'phosphatases'. There are 518 kinase genes in the human genome (Manning *et al.*, 2002) and a similar number of phosphatases. Apart from phosphorylation, proteins can also be ubiquitinated, sumoylated, acetylated, and hydroxylated (this list is incomplete) affecting the activity, stability, localization, and/or protein–protein interactions of the downstream protein.

The resultant system is a network of thousands of signalling and auxiliary proteins. This network, similar to the network of neurons in the brain, not only conveys information linearly from sensor proteins to the downstream endpoints, but also computes such information. A good example of this is the AMPK–mTOR signalling network, which senses and computes signals such as AMP, glycogen, amino acids, and muscle loading to regulate protein synthesis and ribosome biogenesis. The AMPK–mTOR system will be discussed in more detail below.

At the bottom of the cellular signalling network are proteins that regulate nearly all cellular functions such as transcription factors and translational regulators. Transcription factors bind to DNA binding sites and recruit RNA polymerase to genes whose transcription is up- or downregulated by exercise. Translational regulators assemble ribosomes, which then translate the mRNAs into proteins and make a cell grow. Other proteins regulate protein breakdown, the activity and location of transporter proteins such as glucose transporters, the synthesis of new DNA, or the aggregation of DNA to chromosomes before and during cell division and even cellular death.

1.7.2.2 Human protein synthesis and breakdown after exercise

After resistance exercise, signal transduction pathways regulate changes in protein synthesis and breakdown. Before discussing the signalling pathways involved we will first focus on these downstream effects. A positive protein balance (net protein accretion) is essential for muscle growth:

$$\text{net protein accretion} = \text{protein synthesis} - \text{protein breakdown}$$

A positive protein balance can occur through an increase in protein synthesis, a decrease in protein breakdown, or any combination of the two that results in net protein accretion. Throughout daily life there is a continuous turnover of protein within skeletal muscle, with synthesis and breakdown both occurring at varying levels depending on age, nutritional status, and contractile activity. During exercise, both resistance and endurance, the majority of ATP is used in the development of force and power within the muscle. This means that many other 'non-essential cellular processes', including protein synthesis, are reduced during the initial recovery period after exercise. The reduction of protein synthesis during acute muscle contraction was first observed in a perfused rat hindlimb (Bylund-Fellenius *et al.*, 1984).

After resistance exercise, once contraction-related ATP requirements are back to basal levels, there is a marked increase in protein synthesis. Depending on the intensity at which exercise is performed, protein synthesis will be approximately doubled 12–24 hours after exercise (Chesley *et al.*, 1992), followed by a slow decrease. Even 72 hours after exercise, muscle protein synthesis can still be higher than before exercise (Miller *et al.*, 2005). The duration of this period of elevated protein synthesis has been found to be longer in untrained compared to trained subjects (Phillips *et al.*, 1999). In response to a lower intensity of contractile activity there is a slight increase in protein synthesis, but this is generally short-lived. This dose-response relationship between muscle protein synthesis and exercise intensity has recently been characterized in both young (24 ± 6 years) and old (70 ± 5 years) subjects. Kumar *et al.* (2009) measured muscle protein synthesis 1–2 hours after exercise and found that in both young and old subjects there is a sigmoidal relationship between exercise intensity and muscle protein synthesis (Figure 1.7.2). The protein synthesis response was consistently smaller in the older population. One major finding of this study was that exercise intensities above 60% 1 RM were required for protein synthesis to be maximized. This indirectly supports strength training practice (ACSM, 2009). These data also make sense in light of a meta-analysis that showed that resistance training with 60% of 1 RM in untrained

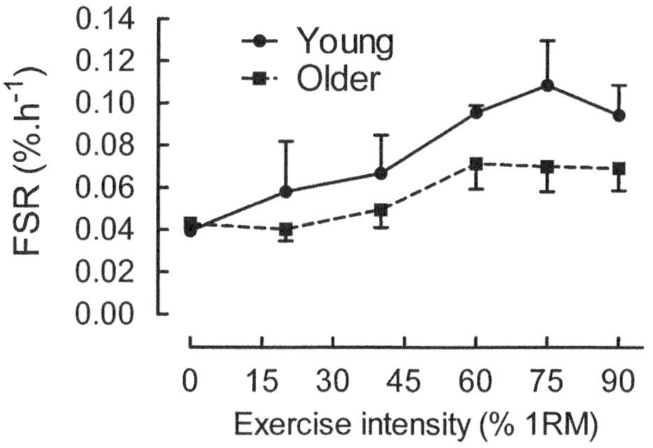

Figure 1.7.2 Dose–response relationship of myofibrillar protein synthesis (FSR, fractional synthetic rate, % h^{-1}), measured at 1–2 h post-exercise for five young men (24 ± 6 years) and five older men (70 ± 5 years) at each intensity. Reproduced from Kumar *et al.* (2009) with permission from Blackwell Publishing

subjects and 80% of 1 RM in trained subjects is most effective at increasing muscle strength (Rhea *et al.*, 2003).

We mentioned previously that protein breakdown must also be taken into account when looking at protein accretion within the muscles. Intuitively one might think that resistance training decreases protein breakdown to stimulate additional growth, but in reality this is probably not the case. It appears that protein breakdown is not altered during exercise (Tipton *et al.*, 1999) but, like protein synthesis, is elevated after resistance exercise in fasted subjects (Biolo *et al.*, 1995). The magnitude of this increase is approximately 25–50%, peaking approximately 3 hours after exercise, leaving the muscles in a slightly improved (compared to rest) but still negative protein balance. The duration of the increase in muscle protein breakdown is believed to be shorter than that observed for protein synthesis (Phillips *et al.*, 1997). In order to achieve a positive protein balance a meal needs to be ingested, as this will elevate protein synthesis above breakdown (Tipton *et al.*, 1999).

1.7.2.3 The AMPK–mTOR system and the regulation of protein synthesis

The AMPK–mTOR signalling system is now well established as the major regulator of protein synthesis in response to resistance exercise. Several practical implications arise from our understanding of this system. The AMPK–mTOR system and myostatin signalling are shown in Figure 1.7.3.

The mTOR pathway is a complicated pathway which regulates the assembly of the ribosome and the subsequent synthesis of proteins, which is also known as translation. mTOR forms two complexes with other proteins, termed mTORC1 and mTORC2 (Proud, 2007). The mTOR pathway not only stimulates many steps that lead to an upregulation of protein synthesis (Proud, 2007) but also regulates the capacity for protein synthesis by producing more ribosomes, which are the cellular machines that synthesize proteins (Mayer and Grummt, 2006). We will not discuss this pathway in detail, but will focus on those elements and functions that are important in regulating the adaptation of skeletal muscle to resistance exercise.

The first evidence of the involvement of the mTOR pathway in mediating the growth adaptation to resistance exercise was obtained by studying high-frequency electrically stimulated rat muscles. Using this model it was demonstrated that p70 S6k, which is downstream of mTOR, was increasingly phosphorylated after high-intensity stimulation and that the activity of p70 S6k correlated with the magnitude of muscle growth (Baar and Esser, 1999). It was later shown that mTOR was necessary for muscle hypertrophy (Bodine *et al.*, 2001a) and that the PKB/Akt-mTOR signalling cascade was activated in response to IGF-1 (a known muscle growth factor) signalling (Rommel *et al.*, 2001). A considerable amount of evidence has been accumulated to support the hypothesis that IGF-1 splice variants play a key role in stimulating muscle hypertrophy after resistance exercise by activating the mTOR pathway (Goldspink, 2005). However, recent mechanistic studies have cast doubts on the importance of IGF-1 and suggest that mTOR activation after resistance exercise is instead dependent on phospholipase D (PLD) (O'Neil *et al.*, 2009). The exact signal that activates mTOR after resistance exercise in a PLD-dependent or possibly independent manner is currently unknown. One candidate sensor is titin kinase (Lange *et al.*, 2005), where a stretch/force-sensing mechanism has been identified (Puchner *et al.*, 2008). The mTOR pathway is also activated by nutrients, which has important implications for whether and when nutrients should be taken by athletes engaging in resistance exercise. The mediators of the nutrient response are insulin and essential amino acids, especially leucine. Like IGF-1, insulin activates mTOR via receptors and PKB/Akt, but the mechanism by which amino acids activate mTOR is still not entirely known. It probably involves small GTPase proteins (Avruch *et al.*, 2009).

AMPK is a kinase that is activated by elevated AMP which increases during exercise and triggers the upstream kinases to phosphorylate AMPK at Thr172 and inhibited by glycogen (Wojtaszewski *et al.*, 2002). The activation of AMPK by high AMP and low glycogen is dependent on AMP and glycogen binding sites on AMPK subunits (Hudson *et al.*, 2003). AMPK not only regulates many acute and chronic adaptations to endurance exercise but also inhibits translation initiation and elongation via TSC2 (Inoki, Zhu and Guan, 2003) and eEF2 (Horman *et al.*, 2002). In other words, AMPK activity reduces protein synthesis and muscle growth, and this mechanism can explain the reduction of protein synthesis during acute exercise (Bylund-Fellenius *et al.*, 1984). Many studies suggest that AMPK does indeed inhibit translational signalling, protein synthesis, and growth in skeletal muscle (Bolster *et al.*, 2002; Dreyer *et al.*, 2006; Thomson, Fick and Gordon, 2008), but a recent study suggests that it may not always be involved (Rose *et al.*, 2009). One can imagine a tug of war where protein synthesis is reduced via AMPK and increased via mTOR.

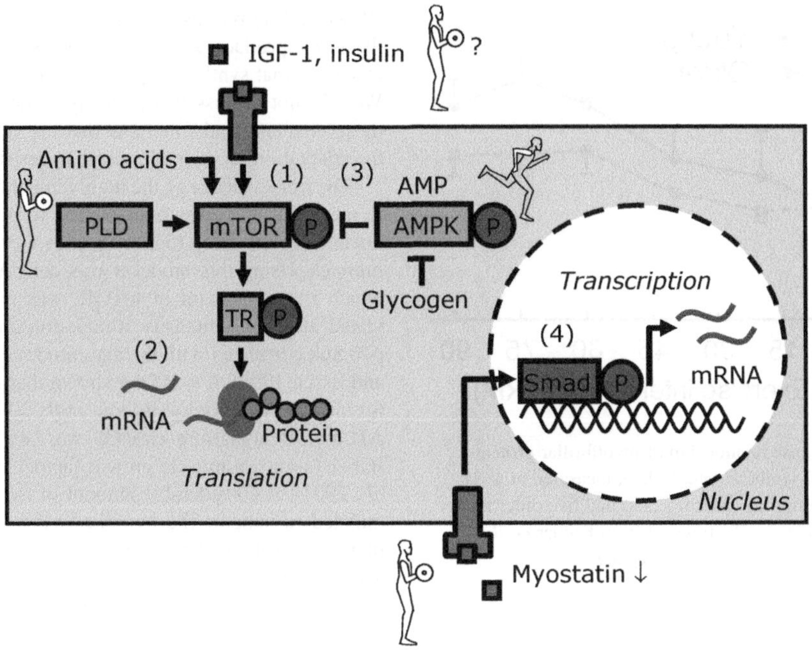

Figure 1.7.3 Model detailing adaptation to exercise. (1) mTOR is activated by resistance exercise via phospholipase D (PLD), amino acids, and endocrine factors such as IGF-1 (although changes in IGF-1 are probably not required for the adaptation to resistance exercise) and insulin. (2) mTOR then activates translation factors, resulting in increased translation initiation and elongation of protein synthesis. (3) mTOR is inhibited by AMPK, which is activated by an increase in the concentration of AMP and by a low glycogen concentration. Thus endurance exercise and low glycogen due to exercise or diet will inhibit mTOR and protein synthesis. (4) Myostatin expression usually decreases after resistance exercise. This results in decreased Smad2/3 phosphorylation, translocation to the nucleus, and decreased muscle-growth inhibitory gene expression. Details and references are given in the text

1.7.2.4 Potential practical implications

To summarize, mTOR and protein synthesis are more active when AMP is low, glycogen is high, after resistance exercise, and when nutrient uptake leads to increased concentrations of amino acids and insulin. What are the practical implications?

Concurrent endurance and strength training where muscle hypertrophy is the goal

Hickson reported that combined strength and endurance training led to less strength increase over 10 weeks than strength training alone (Baar, 2006; Hickson, 1980). Assuming that the decreased strength increase in the strength and endurance training group was due to less hypertrophy, this can be explained by the inhibition of protein synthesis by AMPK and/or other mechanisms. So what can be done for sports such as rowing or rugby where muscle mass, strength, and endurance need to be developed in parallel? Judging purely from the signalling data, strength training for muscle hypertrophy should be avoided when muscle glycogen is low. Thus, especially after high-intensity endurance exercise where glycogen is nearly depleted after 20 minutes or after >1 hour of medium intensity exercise (Gollnick, Piehl and Saltin, 1974), it will probably be important to increase the glycogen content before engaging in resistance exercise for hypertrophy. This is because of the likely inhibition of protein synthesis by low glycogen. Also, prolonged cooling-down exercise should be avoided for the muscles that have been strength trained due to the AMPK activation.

Concurrent endurance and strength training where muscle hypertrophy is not a goal

In other sports, such as boxing or Alpine climbing, endurance and strength have to be increased but muscle mass and body weight gains need to be minimized. A potential strategy here is to increase AMPK activity during or after resistance exercise to minimize the growth response but to maintain the neuromuscular adaptation to exercise. Potential strategies to reduce the growth effect of resistance exercise are to avoid protein or amino acid intake before, during, and directly after resistance exercise, as these would increase the growth response (Esmarck *et al.*, 2001; Tipton *et al.*, 2001). Also, performing endurance exercise and especially glycogen-depleting types of exercise (Gollnick, Piehl and Saltin, 1974) before or shortly after resistance exercise will activate AMPK and should thus reduce the growth response more than exercise that is performed well after resistance exercise.

1.7.2.5 Myostatin–Smad signalling

The counterpart of the mTOR pathway is the myostatin–Smad system. Its key component is a muscle-secreted protein or myokine that has been termed 'myostatin'. Myostatin is a member of the transforming growth factor-β (TGF-β) family (Lee, 2004; McPherron, Lawler and Lee, 1997) and inhibits muscle growth. Myostatin knockout mice have a 2–3 times higher muscle mass than the wildtype due to both hyperplasia and hypertrophy of muscle fibres. Various natural mutations of the myostatin gene have been reported. These include muscular cattle breeds such as the Belgian Blue or Piedmontese (McPherron and Lee, 1997) and a human mutation (Schuelke *et al.*, 2004). In contrast, overexpression of myostatin causes cachexia in mice (Zimmers *et al.*, 2002).

Expression of myostatin mRNA can be detected pre- and postnatally. It is regulated by environmental stimuli: most (Roth *et al.*, 2003; Walker *et al.*, 2004; Zambon *et al.*, 2003) but not all (Willoughby, 2004) studies show that resistance and endurance exercise reduce myostatin mRNA and/or protein. In contrast, myostatin expression increases in response to muscle immobilization (Wehling, Cai and Tidball, 2000). Glucocorticoids also promote the transcription of myostatin (Ma *et al.*, 2001).

Myostatin is expressed as a larger precursor protein that is cut to yield functionally active ~24kDa myostatin dimers (Lee, 2004). After secretion, myostatin can bind and be neutralized by serum proteins such as growth and differentiation factor-associated serum protein-1 (gasp1), follistatin, and follistatin-related gene (FLRG) (Lee, 2004). Thus, myostatin signalling depends not only on the concentration of myostatin protein but also on its processing and the availability of inhibitory binding partners. Active myostatin binds to activin IIA and IIB receptors (Lee and McPherron, 2001), which then recruit and phosporylate activin type I receptors, increasing their kinase activity. The type I receptor phosphorylates the receptor-regulated Smads (R-Smads), Smad2 and Smad3 (Bogdanovich *et al.*, 2002). This activated complex of R-Smads dissociates from the receptor and binds to the common Smad, termed Smad4. The Smad complexes translocate to the nucleus and bind to short DNA stretches termed 'Smad-binding elements' and regulate the expression of hundreds of genes (Massague, Seoane and Wotton, 2005). Myostatin inhibits the proliferation and differentiation of mononucleated cells, such as satellite cells, which are important for muscle fibre formation and regeneration (Thomas *et al.*, 2000). It is unclear how a decrease in myostatin signalling promotes muscle growth in adult muscle. Hypertrophy requires a positive protein balance and there is little information on the effect of myostatin on protein synthesis or breakdown. To our knowledge only one *in vitro* study has demonstrated an inhibition of protein synthesis by myostatin and this was in a muscle cell culture model (Taylor *et al.*, 2001).

Myostatin inhibitors are potential treatments for muscle atrophy in disease and ageing but may equally be misused as doping agents. Anti-myostatin antibodies have been used to successfully improve muscle function in a mouse model of muscular dystrophy (Bogdanovich *et al.*, 2002). The myostatin Smad system has advantages for drug developers as it can be targeted extracellularly and as is a muscle-specific signalling protein. A study where a monoclonal anti-myostatin MYO-029 antibody was used in muscular dystrophy patients has been completed but no significant muscle size or other treatment effects have been reported (Wagner *et al.*, 2008). In another study, single muscle fibre contractile properties were improved in patients who received MYO-029, but the patient numbers (n = 5) and controls (n = 1) were small (Krivickas, Walsh and Amato, 2009)

To summarize, myostatin is the opponent of mTOR signalling when it comes to controlling of muscle growth. Myostatin expression is regulated by environmental sitmuli such as exercise but it is largely unclear how reduced myostatin signalling promotes hypertrophy.

1.7.2.6 Signalling associated with muscle protein breakdown

Skeletal muscle protein breakdown occurs through a number of different proteolytic pathways. Proteolysis can be simple defined as the process in which proteins, in this case myofibrillar proteins, are broken down into peptides or amino acids by cellular enzymes called proteases. In skeletal muscle these pathways include lysosomes, ubiquitin-proteasome-dependant systems, calcium-activated proteases, caspases, and metalloproteinases. The precise role of each mechanism in the skeletal muscle adaptations associated with exercise remains to be fully elucidated. Indeed only a handful of studies have investigated changes in these systems during exercise.

The muscle-specific ubiquitin ligases muscle atrophy F box (MAFbx) and muscle-specific really-interesting novel gene finger protein 1 (MuRF-1) have been implicated in the control of protein breakdown, with both sharing forkhead box 3A (FOXO3A) as a transcription factor. Indeed, in cell culture overexpression of MAFbx results in fibres of a smaller diameter and in mice with both MAFbx and MuRF-1 genes knockout demonstrate atrophy sparing when the sciatic nerve is cut (Bodine *et al.*, 2001b). Similarly, when FOXO3A is constitutively active in myotubes it activates MAFbx transcription, resulting in a reduction in fibre diameter (Sandri *et al.*, 2004). In response to resistance exercise, MAFbx and MuRF-1 have been found to be rapidly elevated, with a suppression of MAFbx and FOXO3A 4–24 hours after exercise (Louis *et al.*, 2007; Raue *et al.*, 2007). Further studies have also found that MAFbx mRNA was decreased for up to 24 hours after a bout of resistance exercise (Kostek *et al.*, 2007), indicating that it may not be involved in the upregulation of muscle protein breakdown in response to exercise. Further support for such an assertion was found by Greenhaff *et al.* (2008), who discovered that MAFbx was not associated with changes in protein breakdown. Clearly the involvement of these ubiquitin ligases in the control of protein breakdown during exercise remains to be elucidated.

Further studies have demonstrated that increasing intracellular Ca^{2+} concentration in skeletal muscles, *ex vivo*, will lead to an increase in protein breakdown (Baracos, Greenberg and Goldberg, 1986; Goodman, 1987; Kameyama and Etlinger, 1979; Zeman *et al.*, 1985). From these findings a logical hypothesis would be that stimulation of protein breakdown via Ca^{2+} would occur through the Ca^{2+}-activated proteases. However, in testing this hypothesis recent work has found that calpains, Ca^{2+}-activated proteases, are not activated in skeletal muscle during exercise (Murphy, Snow and Lamb, 2006). A final mechanism was suggested in 1984 when it was demonstrated by Bylund-Fellenius *et al.* (1984) that in muscles in which protein breakdown was inhibited there was a degradation of high-energy phosphates and an accumulation of lactate. This led the authors to conclude that protein degradation might be linked to the energy status of the muscle and the level of metabolite accumulation.

It appears unlikely, therefore, that one single pathway controls the rapid remodelling of myofibrillar proteins after a bout of exercise, but rather that multiple pathways probably contribute to various degrees.

1.7.2.7 Satellite-cell regulation during strength training

Skeletal muscle fibres are multinucleated cells with many hundreds of nuclei in each myofibre, each of which has a volume of cytoplasm, or the so-called nuclear domain. It has been hypothesized that the ratio between a nucleus and the volume of muscle fibre is relatively constant. Indeed, there is a parallel increase in both muscle fibre volume and nuclear number in response to functional overload in rats (Roy *et al.*, 1999), meaning that the nuclear domain will remain relatively constant. It has been suggested that in humans there is a ceiling size to the myonuclear domain of around $2000\,\mu m^2$, after which no growth of the myofibre is possible (Kadi *et al.*, 2004). In that case it is only possible for the skeletal muscle to grow with the addition of further nuclei, which poses a problem as skeletal muscle cells are post-mitotic (do not divide anymore) and thus external cells must provide these additional nuclei.

Satellite cells are skeletal muscle-specific stem cells and are found between the basal lamina and the cell membrane of each myofibre (Mauro, 1961). Satellite cells play a major role in the regulation of muscle development, postnatal growth, and injury repair (Kuang, Gillespie and Rudnicki, 2008), and potentially in the hypertrophy response to exercise (Rosenblatt and Parry, 1993), although this is not universally accepted (O'Connor and Pavlath, 2007). Evidence supporting the role of satellite cells in hypertrophy includes the findings that satellite cells proliferate, and re-enter the cell cycle, in response to various hypertrophic stimuli, including synergistic ablation, stretch overload, testosterone, and importantly resistance exercise. For example, markers of satellite-cell activation and cell-cycle activity, such as cyclin D1, and of differentiation, such as p21 and MyoD, are increased after a single bout of resistance exercise (Bickel *et al.*, 2005) and more prolonged strength training (Kadi *et al.*, 2004). The time course of satellite-cell proliferation lasts from approximately 1–2 days after resistance exercise up to a week, which is before any significant increase in myofibre size will occur. In order to investigate the physiological significance of the increase in satellite cell numbers, various studies have used local gamma irradiation to inhibit satellite-cell function, with the classical study being performed by Rosenblatt and Parry (1993). After such treatment hypertrophy is either completely prevented or drastically reduced in response to a hypertrophic signal, suggesting a role for satellite cells in hypertrophy, depending upon the stimuli applied. Furthermore, in response to disuse during spaceflight, immobilization, hind-limb suspension, denervation, and ageing, the number of myonuclei in a single muscle fibre decreases, due to death by apoptosis, highlighting a likely role for satellite cells in atrophy as well as hypertrophy. On the other hand, as mentioned above, several studies have indicated that hypertrophy is possible without the activation of satellite cells. These studies have used pharmacological β-adrenergic agonists, which have an anabolic effect in skeletal muscle, finding that an increase in muscle size/mass or protein content occurs without any changes in DNA content (a surrogate for muscle nuclei). In order to fully resolve the inconsistencies in these findings and reveal the true role of satellite cells in skeletal muscle hypertrophy, more advanced methods will need to be developed to determine whether satellite-cell fusion is necessary for skeletal muscle hypertrophy.

1.7.2.8 What we have not covered

Due to a lack of space we have had to leave out important molecular mechanisms that link skeletal muscle and other organs to strength and the adaptation to strength training. Here we briefly mention these mechanisms to give an idea of the larger picture.

Besides skeletal muscle, the major strength and strength-adaptation organ is the nervous system. It controls the number of contracting muscle fibres by activating α-motor neurones and thus motor units, and the intensity of fibre contraction by regulating the discharge rate of motor units in response to training (Duchateau, Semmler and Enoka, 2006). The cause for such neural adaptations is unclear but it seems likely that it will involve genes that are regulated by neuronal activity and which consequently change synapse or neuron behaviour (Flavell and Greenberg, 2008). The double challenge for researchers in this field is to identify the mechanisms responsible for such behaviour in the complex nervous system and to explore the complex signalling within the cells therein.

Finally, bones and tendons adapt to resistance exercise in a manner similar to skeletal muscle. In bone the challenge is to identify the signal transduction pathways that link mechanical signals associated with bone loading with increases in collagen synthesis and tissue mineralization (Scott *et al.*, 2008). Tendons adapt to exercise via an increase in collagen synthesis and modifications of the proteins at the extracellular matrix (Kjaer *et al.*, 2005). Again, signal transduction pathways link exercise to these processes, and these remain to be fully elucidated.

References

ACSM (2009) American College of Sports Medicine position stand. Progression models in resistance training for healthy adults. *Med Sci Sports Exerc*, **41**, 687–608.

Arden N.K. and Spector T.D. (1997) Genetic influences on muscle strength, lean body mass, and bone mineral density: a twin study. *J Bone Miner Res*, **12**, 2076–2081.

Avruch J., Long X., Ortiz-Vega S. *et al.* (2009) Amino acid regulation of TOR complex 1. *Am J Physiol Endocrinol Metab*, **296**, E592–E602.

Baar K. (2006) Training for endurance and strength: lessons from cell signaling. *Med Sci Sports Exerc*, **38**, 1939–1944.

Baar K. and Esser K. (1999) Phosphorylation of p70(S6k) correlates with increased skeletal muscle mass following resistance exercise. *Am J Physiol*, **276**, C120–C127.

Baracos V., Greenberg R.E. and Goldberg A.L. (1986) Influence of calcium and other divalent cations on protein turnover in rat skeletal muscle. *Am J Physiol Endocrinol Metab*, **250**, E702–E710.

Bickel C.S., Slade J., Mahoney E. *et al.* (2005) Time course of molecular responses of human skeletal muscle to acute bouts of resistance exercise. *J Appl Physiol*, **98**, 482–488.

Biolo G., Maggi S.P., Williams B.D. *et al.* (1995) Increased rates of muscle protein turnover and amino acid transport after resistance exercise in humans. *Am J Physiol*, **268**, E514–E520.

Bodine S.C., Stitt T.N., Gonzalez M. *et al.* (2001a) Akt/mTOR pathway is a crucial regulator of skeletal muscle hypertrophy and can prevent muscle atrophy in vivo. *Nat Cell Biol*, **3**, 1014–1019.

Bodine S.C., Latres E., Baumhueter S. *et al.* (2001b) Identification of ubiquitin ligases required for skeletal muscle atrophy. *Science*, **294**, 1704–1708.

Bogdanovich S., Krag T.O., Barton E.R. *et al.* (2002) Functional improvement of dystrophic muscle by myostatin blockade. *Nature*, **420**, 418–421.

Bolster D.R., Crozier S.J., Kimball S.R. and Jefferson L.S. (2002) AMP-activated protein kinase suppresses protein synthesis in rat skeletal muscle through down-regulated mammalian target of rapamycin (mTOR) signaling. *J Biol Chem*, **277** (23), 977–980.

Brockmann G.A., Haley C.S., Renne U. *et al.* (1998) Quantitative trait loci affecting body weight and fatness from a mouse line selected for extreme high growth. *Genetics*, **150**, 369–381.

Bylund-Fellenius A.C., Ojamaa K.M., Flaim K.E. *et al.* (1984) Protein synthesis versus energy state in contracting muscles of perfused rat hindlimb. *Am J Physiol Endocrinol Metab*, **246**, E297–E305.

Carmelli D. and Reed T. (2000) Stability and change in genetic and environmental influences on hand-grip strength in older male twins. *J Appl Physiol*, **89**, 1879–1883.

Chesley A., MacDougall J.D., Tarnopolsky M.A. *et al.* (1992) Changes in human muscle protein synthesis after resistance exercise. *J Appl Physiol*, **73**, 1383–1388.

Dreyer H.C., Fujita S., Cadenas J.G. *et al.* (2006) Resistance exercise increases AMPK activity and reduces 4E-BP1 phosphorylation and protein synthesis in human skeletal muscle. *J Physiol*, **576**, 613–624.

Duchateau J., Semmler J.G. and Enoka R.M. (2006) Training adaptations in the behavior of human motor units. *J Appl Physiol*, **101**, 1766–1775.

Esmarck B., Andersen J.L., Olsen S. *et al.* (2001) Timing of postexercise protein intake is important for muscle hypertrophy with resistance training in elderly humans. *J Physiol*, **535**, 301–311.

Flavell S.W. and Greenberg M.E. (2008) Signaling mechanisms linking neuronal activity to gene expression and plasticity of the nervous system. *Annu Rev Neurosci*, **31**, 563–590.

Frederiksen H., Gaist D., Petersen H.C. *et al.* (2002) Hand grip strength: a phenotype suitable for identifying genetic variants affecting mid- and late-life physical functioning. *Genet Epidemiol*, **23**, 110–122.

Genetic Technologies Ltd. (2007) ACTN3 sport performance testTM, http://www.gtpersonal.com.au/sports_performance.php/ (Accessed 17th June 2010).

Goldspink G. (2005) Mechanical signals, IGF-I gene splicing, and muscle adaptation. *Physiology (Bethesda)*, **20**, 232–238.

Gollnick P.D., Piehl K. and Saltin B. (1974) Selective glycogen depletion pattern in human muscle fibres after exercise of varying intensity and at varying pedalling rates. *J Physiol*, **241**, 45–57.

Goodman M.N. (1987) Differential effects of acute changes in cell Ca^{2+} concentration on myofibrillar and non-myofibrillar protein breakdown in the rat extensor digitorum longus muscle in vitro. Assessment by production of tyrosine and N tau-methylhistidine. *Biochem J*, **241**, 121–127.

Greenhaff P.L., Karagounis L.G., Peirce N. *et al.* (2008) Disassociation between the effects of amino acids and insulin on signaling, ubiquitin ligases, and protein turnover in human muscle. *Am J Physiol Endocrinol Metab*, **295**, E595–E604.

Hather B.M., Tesch P.A., Buchanan P. and Dudley G.A. (1991) Influence of eccentric actions on skeletal muscle adaptations to resistance training. *Acta Physiol Scand*, **143**, 177–185.

Hickson R.C. (1980) Interference of strength development by simultaneously training for strength and endurance. *Eur J Appl Physiol Occup Physiol*, **45**, 255–263.

Horman S., Browne G., Krause U. *et al.* (2002) Activation of AMP-activated protein kinase leads to the phosphorylation of elongation factor 2 and an inhibition of protein synthesis. *Curr Biol*, **12**, 1419–1423.

Houston M.E., Froese E.A., Valeriote S.P. *et al.* (1983) Muscle performance, morphology and metabolic capacity during strength training and detraining: a one leg model. *Eur J Appl Physiol Occup Physiol*, **51**, 25–35.

Hubal M.J., Gordish-Dressman H., Thompson P.D. *et al.* (2005) Variability in muscle size and strength gain after unilateral resistance training. *Med Sci Sports Exerc*, **37**, 964–972.

Hudson E.R., Pan D.A., James J. *et al.* (2003) A novel domain in AMP-activated protein kinase causes glycogen storage bodies similar to those seen in hereditary cardiac arrhythmias. *Curr Biol*, **13**, 861–866.

Inoki K., Zhu T. and Guan K.L. (2003) TSC2 mediates cellular energy response to control cell growth and survival. *Cell*, **115**, 577–590.

Kadi F., Schjerling P., Andersen L.L. *et al.* (2004) The effects of heavy resistance training and detraining on satellite cells in human skeletal muscles. *J Physiol*, **558**, 1005–1012.

Kameyama T. and Etlinger J.D. (1979) Calcium-dependent regulation of protein synthesis and degradation in muscle. *Nature*, **279**, 344–346.

Kjaer M., Langberg H., Miller B.F. *et al.* (2005) Metabolic activity and collagen turnover in human tendon in response to physical activity. *J Musculoskelet Neuronal Interact*, **5**, 41–52.

Komi P.V., Viitasalo J.H., Havu M. *et al.* (1977) Skeletal muscle fibres and muscle enzyme activities in monozygous and dizygous twins of both sexes. *Acta Physiol Scand*, **100**, 385–392.

Kostek M.C., Chen Y.W., Cuthbertson D.J. *et al.* (2007) Gene expression responses over 24 h to lengthening and shortening contractions in human muscle: major changes in CSRP3, MUSTN1, SIX1, and FBXO32. *Physiol Genomics*, **31**, 42–52.

Krivickas L.S., Walsh R. and Amato A.A. (2009) Single muscle fiber contractile properties in adults with muscular dystrophy treated with MYO-029. *Muscle Nerve*, **39**, 3–9.

Kuang S., Gillespie M.A. and Rudnicki M.A. (2008) Niche regulation of muscle satellite cell self-renewal and differentiation. *Cell Stem Cell*, **2**, 22–31.

Kumar V., Selby A., Rankin D. *et al.* (2009) Age-related differences in the dose-response relationship of muscle protein synthesis to resistance exercise in young and old men. *J Physiol*, **587**, 211–217.

Lange S., Xiang F., Yakovenko A. *et al.* (2005) The kinase domain of titin controls muscle gene expression and protein turnover. *Science*, **308**, 1599–1603.

Lee S.J. (2004) Regulation of muscle mass by myostatin. *Annu Rev Cell Dev Biol*, **20**, 61–86.

Lee S.J. and McPherron A.C. (2001) Regulation of myostatin activity and muscle growth. *Proc Natl Acad Sci U S A*, **98**, 9306–9311.

Lexell J., Taylor C.C. and Sjostrom M. (1988) What is the cause of the ageing atrophy? Total number, size and proportion of different fiber types studied in whole vastus lateralis muscle from 15- to 83-year-old men. *J Neurol Sci*, **84**, 275–294.

Lionikas A., Blizard D.A., Vandenbergh D.J. *et al.* (2003) Genetic architecture of fast- and slow-twitch skeletal muscle weight in 200-day-old mice of the C57BL/6J and DBA/2J lineage. *Physiol Genomics*, **16**, 141–152.

Louis E., Raue U., Yang Y. *et al.* (2007) Time course of proteolytic, cytokine, and myostatin gene expression after acute exercise in human skeletal muscle. *J Appl Physiol*, **103**, 1744–1751.

Ma K., Mallidis C., Artaza J. *et al.* (2001) Characterization of 5'-regulatory region of human myostatin gene: regulation by dexamethasone in vitro. *Am J Physiol Endocrinol Metab*, **281**, E1128–E1136.

MacDougall J.D., Elder G.C., Sale D.G. *et al.* (1980) Effects of strength training and immobilization on human muscle fibres. *Eur J Appl Physiol Occup Physiol*, **43**, 25–34.

Manning G., Whyte D.B., Martinez R. *et al.* (2002) The protein kinase complement of the human genome. *Science*, **298**, 1912–1934.

Masinde G.L., Li X., Gu W. *et al.* (2002) Quantitative trait loci that harbor genes regulating muscle size in (MRL/MPJ x SJL/J) F(2) mice. *Funct Integr Genomics*, **2**, 120–125.

Massague J., Seoane J. and Wotton D. (2005) Smad transcription factors. *Genes Dev*, **19**, 2783–2810.

Mauro A. (1961) Satellite cell of skeletal muscle fibers. *J Biophys Biochem Cytol*, **9**, 493–495.

Mayer C. and Grummt I. (2006) Ribosome biogenesis and cell growth: mTOR coordinates transcription by all three classes of nuclear RNA polymerases. *Oncogene*, **25**, 6384–6391.

McPherron A.C. and Lee S.J. (1997) Double muscling in cattle due to mutations in the myostatin gene. *Proc Natl Acad Sci U S A*, **94**, 12–457.

McPherron A.C., Lawler A.M. and Lee S.J. (1997) Regulation of skeletal muscle mass in mice by a new TGF-beta superfamily member. *Nature*, **387**, 83–90.

Miller B.F., Olesen J.L., Hansen M. *et al.* (2005) Coordinated collagen and muscle protein synthesis in human patella tendon and quadriceps muscle after exercise. *J Physiol*, **567**, 1021–1033.

Murphy R.M., Snow R.J. and Lamb G.D. (2006) {micro}-Calpain and calpain-3 are not autolyzed with exhaustive exercise in humans. *Am J Physiol Cell Physiol*, **290**, C116–C122.

Newman H., Freeman F. and Holzinger J. (1937) *Twins: A Study of Heredity and Environment*, University of Chicago Press, Chicago.

Nimmo M.A., Wilson R.H. and Snow D.H. (1985) The inheritance of skeletal muscle fibre composition in mice. *Comp Biochem Physiol A Comp Physiol*, **81**, 109–115.

North K.N., Yang N., Wattanasirichaigoon D. *et al.* (1999) A common nonsense mutation results in alpha-actinin-3 deficiency in the general population. *Nat Genet*, **21**, 353–354.

O'Connor R.S. and Pavlath G.K. (2007) Point:counterpoint: satellite cell addition is/is not obligatory for skeletal muscle hypertrophy. *J Appl Physiol*, **103**, 1099–1100.

O'Neil T.K., Duffy L.R., Frey J.W. and Hornberger T.A. (2009) The role of phosphoinositide 3-kinase and phosphatidic acid in the regulation of mTOR following eccentric contractions. *J Physiol*, **587** (14), 3691–3701.

Perusse L., Lortie G., Leblanc C. *et al.* (1987) Genetic and environmental sources of variation in physical fitness. *Ann Hum Biol*, **14**, 425–434.

Phillips S.M., Tipton K.D., Aarsland A. *et al.* (1997) Mixed muscle protein synthesis and breakdown after resistance exercise in humans. *Am J Physiol*, **273**, E99–107.

Phillips S.M., Tipton K.D., Ferrando A.A. and Wolfe R.R. (1999) Resistance training reduces the acute exercise-induced increase in muscle protein turnover. *Am J Physiol*, **276**, E118–E124.

Proud C.G. (2007) Signalling to translation: how signal transduction pathways control the protein synthetic machinery. *Biochem J*, **403**, 217–234.

Puchner E.M., Alexandrovich A., Kho A.L. *et al.* (2008) Mechanoenzymatics of titin kinase. *Proc Natl Acad Sci U S A*, **105** (13), 385–390.

Raue U., Slivka D., Jemiolo B. *et al.* (2007) Proteolytic gene expression differs at rest and after resistance exercise between young and old women. *J Gerontol Ser A Biol Sci Med Sci*, **62**, 1407–1412.

Reed T., Fabsitz R.R., Selby J.V. and Carmelli D. (1991) Genetic influences and grip strength norms in the NHLBI twin study males aged 59–69. *Ann Hum Biol*, **18**, 425–432.

Rhea M.R., Alvar B.A., Burkett L.N. and Ball S.D. (2003) A meta-analysis to determine the dose response for strength development. *Med Sci Sports Exerc*, **35**, 456–464.

Rommel C., Bodine S.C., Clarke B.A. *et al.* (2001) Mediation of IGF-1-induced skeletal myotube hypertrophy by PI(3) K/Akt/mTOR and PI(3)K/Akt/GSK3 pathways. *Nat Cell Biol*, **3**, 1009–1013.

Rose A.J., Alsted T.J., Jensen T.E. *et al.* (2009) A Ca(2+)-calmodulin-eEF2K-eEF2 signalling cascade, but not AMPK, contributes to the suppression of skeletal muscle protein synthesis during contractions. *J Physiol*, **587**, 1547–1563.

Rosenblatt J.D. and Parry D.J. (1993) Adaptation of rat extensor digitorum longus muscle to gamma irradiation and overload. *Pflugers Arch*, **423**, 255–264.

Roth S.M., Schrager M.A., Ferrell R.E. *et al.* (2001) CNTF genotype is associated with muscular strength and quality in humans across the adult age span. *J Appl Physiol*, **90**, 1205–1210.

Roth S.M., Martel G.F., Ferrell R.E. *et al.* (2003) Myostatin gene expression is reduced in humans with heavy-resistance strength training: a brief communication. *Exp Biol Med (Maywood)*, **228**, 706–709.

Roy R.R., Monke S.R., Allen D.L. and Edgerton V.R. (1999) Modulation of myonuclear number in functionally overloaded and exercised rat plantaris fibers. *J Appl Physiol*, **87**, 634–642.

Sandri M., Sandri C., Gilbert A. *et al.* (2004) Foxo transcription factors induce the atrophy-related ubiquitin ligase atrogin-1 and cause skeletal muscle atrophy. *Cell*, **117**, 399–412.

Schuelke M., Wagner K.R., Stolz L.E. *et al.* (2004) Myostatin mutation associated with gross muscle hypertrophy in a child. *N Engl J Med*, **350**, 2682–2688.

Scott A., Khan K.M., Duronio V. and Hart D.A. (2008) Mechanotransduction in human bone: in vitro cellular physiology that underpins bone changes with exercise. *Sports Med*, **38**, 139–160.

Silventoinen K., Magnusson P.K., Tynelius P. *et al.* (2008) Heritability of body size and muscle strength in young adulthood: a study of one million Swedish men. *Genet Epidemiol*, **32**, 341–349.

Simoneau J.A. and Bouchard C. (1989) Human variation in skeletal muscle fiber-type proportion and enzyme activities. *Am J Physiol*, **257**, E567–E572.

Simoneau J.A. and Bouchard C. (1995) Genetic determinism of fiber type proportion in human skeletal muscle. *FASEB J*, **9**, 1091–1095.

Staron R.S., Malicky E.S., Leonardi M.J. *et al.* (1990) Muscle hypertrophy and fast fiber type conversions in heavy resistance-trained women. *Eur J Appl Physiol Occup Physiol*, **60**, 71–79.

Staron R.S., Hagerman F.C., Hikida R.S. *et al.* (2000) Fiber type composition of the vastus lateralis muscle of young men and women. *J Histochem Cytochem*, **48**, 623–629.

Suwa M., Nakamura T. and Katsuta S. (1999) Muscle fibre number is a possible determinant of muscle fibre composition in rats. *Acta Physiol Scand*, **167**, 267–272.

Taylor W.E., Bhasin S., Artaza J. *et al.* (2001) Myostatin inhibits cell proliferation and protein synthesis in C2C12 muscle cells. *Am J Physiol Endocrinol Metab*, **280**, E221–E228.

Thomas M., Langley B., Berry C. *et al.* (2000) Myostatin, a negative regulator of muscle growth, functions by inhibiting myoblast proliferation. *J Biol Chem*, **275** (40), 235–243.

Thomis M.A., Van L.M., Maes H.H. *et al.* (1997) Multivariate genetic analysis of maximal isometric muscle force at different elbow angles. *J Appl Physiol*, **82**, 959–967.

Thomis M.A., Beunen G.P., Maes H.H. *et al.* (1998) Strength training: importance of genetic factors. *Med Sci Sports Exerc*, **30**, 724–731.

Thomson D.M., Fick C.A. and Gordon S.E. (2008) AMPK activation attenuates S6K1, 4E-BP1, and eEF2 signaling responses to high-frequency electrically stimulated skeletal muscle contractions. *J Appl Physiol*, **104**, 625–632.

Tipton K.D., Ferrando A.A., Phillips S.M. *et al.* (1999) Postexercise net protein synthesis in human muscle from orally administered amino acids. *Am J Physiol*, **276**, E628–E634.

Tipton K.D., Rasmussen B.B., Miller S.L. *et al.* (2001) Timing of amino acid-carbohydrate ingestion alters anabolic response of muscle to resistance exercise. *Am J Physiol Endocrinol Metab*, **281**, E197–E206.

van der Laarse W.J., Crusio W.E., Maslam S. and van Abeelen J.H.F. (1984) Genetic architecture of numbers of fast and slow muscle fibres in the mouse soleus muscle. *Heredity*, **53**, 643–647.

Vaughan H.S., Aziz U., Goldspink G. and Nowell N.W. (1974) Sex and stock differences in the histochemical myofibrillar adenosine triphosphatase reaction in the soleus muscle of the mouse. *J Histochem Cytochem*, **22**, 155–159.

Wagner K.R., Fleckenstein J.L., Amato A.A. *et al.* (2008) A phase I/IItrial of MYO-029 in adult subjects with muscular dystrophy. *Ann Neurol*, **63**, 561–571.

Walker K.S., Kambadur R., Sharma M. and Smith H.K. (2004) Resistance training alters plasma myostatin but not IGF-1 in healthy men. *Med Sci Sports Exerc*, **36**, 787–793.

Walsh S., Liu D., Metter E.J. *et al.* (2008) ACTN3 genotype is associated with muscle phenotypes in women across the adult age span. *J Appl Physiol*, **105**, 1486–1491.

Wehling M., Cai B. and Tidball J.G. (2000) Modulation of myostatin expression during modified muscle use. *FASEB J*, **14**, 103–110.

Williams A.G., Wackerhage H., Miah A. *et al.* (2009) Genetic Research and Testing in Sport and Exercise Science, British Association of Sport and Exercise Sciences Position Stand, http://www.bases.org.uk/ Molecular-Exercise-Physiology (accessed 7 August 2009).

Willoughby D.S. (2004) Effects of heavy resistance training on myostatin mRNA and protein expression. *Med Sci Sports Exerc*, **36**, 574–582.

Wojtaszewski J.F., Jorgensen S.B., Hellsten Y. *et al.* (2002) Glycogen-dependent effects of 5-aminoimidazole-4-carboxamide (AICA)-riboside on AMP-activated protein kinase and glycogen synthase activities in rat skeletal muscle. *Diabetes*, **51**, 284–292.

Yang N., MacArthur D.G., Gulbin J.P. *et al.* (2003) ACTN3 genotype is associated with human elite athletic performance. *Am J Hum Genet*, **73**, 627–631.

Zambon A.C., McDearmon E.L., Salomonis N. *et al.* (2003) Time- and exercise-dependent gene regulation in human skeletal muscle. *Genome Biol*, **4**, R61.

Zeman R.J., Kameyama T., Matsumoto K. *et al.* (1985) Regulation of protein degradation in muscle by calcium. Evidence for enhanced nonlysosomal proteolysis associated with elevated cytosolic calcium. *J Biol Chem*, **260** (13), 619–624.

Zimmers T.A., Davies M.V., Koniaris L.G. *et al.* (2002) Induction of cachexia in mice by systemically administered myostatin. *Science*, **296**, 1486–1488.

1.8 Strength and Conditioning Biomechanics

Robert U. Newton, School of Exercise, Biomedical and Health Sciences, Edith Cowan University, Joondalup, WA, Australia

1.8.1 INTRODUCTION

Biomechanics is the science of applying mechanical principles to biological systems such as the human body. As such biomechanics has much to offer the field of strength and conditioning. A basic knowledge of biomechanics is essential for the specialist involved in testing and training people because mechanical principles dictate the production of movement and the outcomes in terms of performance. Strength and conditioning methods often involve both simple and quite complex equipment and apparatus which the athlete will manipulate and interact with. These interactions and subsequent demands placed on the athlete's neuromuscular system are determined by biomechanical principles. As we will see, the human body is constructed of a series of biological machines that allow us to produce an incredible range of movement and the resulting sport, training, dance, and everyday activities. For the sake of simplicity, some of the biomechanical concepts will be explained with slight variations from the strict mechanical definitions. In the strength and conditioning field it is more important to convey meaning and for the concepts to be understood than to be pedantic about the terminology.

The purpose of this chapter is to convey some key biomechanical concepts essential for understanding how we produce movement, how various exercise machines function and interact with the performer, and how tests and equipment are used to assess performance.

1.8.1.1 Biomechanics and sport

All sporting movement results from the application of forces generated by the athlete's muscles working the bony levers and other machines of the skeletal system. Human movement is incredibly complex in all the myriad activities in which we participate, and this is perhaps most eloquently expressed in the extraordinary exploits of athletes. For example, jumping vertically into the air is a common movement that most of us perform without thought, but the underlying mechanical phenomena which contribute to jumping are numerous and complex. Understanding and application of biomechanics is integral to the strength and conditioning specialist and will make for more effective and safer exercise programme designs. This process occurs at several levels:

1. Biomechanics knowledge increases understanding of how a particular movement is performed and the key factors that limit performance.
2. A major aspect is technique analysis because greater movement understanding brings better ability to identify inefficient and/or dangerous techniques.
3. The range of equipment available for strength and conditioning is considerable and expanding. Biomechanical analysis will differentiate useful and well-designed equipment from that which is potentially dangerous or undesirable. Understanding of biomechanics also permits the design of new equipment and exercises based in science.
4. From a biomechanical perspective the athlete can be examined as a machine with neural control and feedback, mechanical actuators, biological structures with certain properties, and interaction with the environment and equipment. This brings excellent understanding of factors limiting performance and mechanisms of injury. The result is better training programme design.

1.8.2 BIOMECHANICAL CONCEPTS FOR STRENGTH AND CONDITIONING

It is important to develop a basic understanding of some key mechanical variables and relationships so that we can explore in more detail biomechanical concepts of strength and conditioning exercise, performance testing, and test interpretation.

Strength and Conditioning – Biological Principles and Practical Applications Marco Cardinale, Rob Newton, and Kazunori Nosaka.

1.8.2.1 Time, distance, velocity, and acceleration

Time

We all experience the passing of time in the fact that we remember what we have just experienced and know that events will occur in the future. Time is measured in various units, but principally seconds, minutes, hours, days, months, and years are used in strength and conditioning to represent different quantities of time. For very short events time is usually measured in milliseconds.

Distance

Distance is a slightly different mechanical quantity to displacement but the terms are usually used interchangeably. In strength and conditioning we usually refer to the length of a path over which a body or an object moves, and this is correctly termed 'distance'. Distance is a measure of a change in position or location in space and has two forms, linear and angular. Linear distance is usually measured in metres (m) and is the length of the path travelled. Angular distance refers to how far an object has rotated and is normally measured in degrees. You may also see angular distance measured in radians. A radian is equal to approximately 57°. Displacement is not always the length of the path taken but rather is the length of a straight line between the initial and final position. We can illustrate the distinction between the two terms using the bench press. When lowering the barbell to the chest and returning the arms to the extended position the initial and final position are the same, so the displacement of the barbell is zero. Distance is the length of path down and up, so will be say 1.6 m.

Velocity

Velocity (often termed speed) is the rate of change of distance over time. It is calculated as the distance moved divided by the time taken to cover that distance and is expressed in units of metres per second (m/s).

Acceleration

Acceleration is the rate of change of velocity with time. It refers not only to changes in rate of movement but also to changes in direction. The change of running direction when performing a cutting manoeuvre requires an acceleration other than zero, even though the speed of movement might be maintained. Acceleration is calculated as the change in velocity divided by the time over which this occurred. The measurement unit is usually metres per second per second (m/s/s).

It should be noted that both velocity and acceleration refer not just to linear motion but also to angular motion or rotation.

The units of measurement of these physics quantities are defined according to established standards called the International System of Units, abbreviated as 'SI units'. All of the countries of the world except three (Burma, Liberia, and the United States) have adopted SI as their system of measurement. Units of measurement are often the result of equations used to calculate the quantity. For example, velocity is displacement divided by time and so has units of m/s, or more commonly $m.s^{-1}$. The dot between the m and the s means product or multiplication, while the −1 superscript to the seconds means inverse or 1 over seconds. They are the same units, just expressed differently mathematically.

1.8.2.2 Mass, force, gravity, momentum, work, and power

Mass

The quantity of matter that makes up a given object. This sounds a quite abstract concept but fortunately it is easy to measure because we often equate mass with weight. The two concepts are not the same, because weight is actually the force generated by gravity (defined below) acting on the mass. Mass is measured in units of kilograms (kg).

Inertia

The resistance of an object to changes in its state of motion. In other words, if the object is stationary it will resist being moved; if it is already moving it will resist changes in the direction or speed of movement. Inertia in linear terms is measured in kilograms and is functionally the same as mass. For example, a sprinting football player who weighs 120 kg is much harder to stop than a gymnast who weighs 50 kg.

Force

Force can be most easily visualized as a push or pull for linear force or a twist for rotating force (termed torque). Force is measured in units called Newtons (N), or Newton metres (N.m) in the case of torque. To produce a change in motion (acceleration) there must be application of a force or torque. The amount of change that results depends on the inertia of the object. To calculate the total force of resistance at a given time point the following formula is used:

$$F = m(a + g)$$

where F = force, m = mass of barbell or dumbbell, a = instantaneous acceleration, g = acceleration due to gravity (9.81 m/s/s).

Torque is calculated as the force applied multiplied by the length of the lever arm:

$$T = F \times d$$

As an example, a 10 kg weight placed on the ankles 0.5 m from the centre of the knee joint and with the lower leg (shank) parallel to the floor creates a $10 \times 9.81 \times 0.5 = 49.05$ N.m of torque about the knee.

Gravity

A specific force produced because of the enormous size of the planet Earth. Simply because of the mass of the Earth there is an attractive force on all objects near to it, and that is why when you throw a ball into the air it falls back to the ground. The most common form of resistance training uses this principle: weight training is lifting and lowering objects against the force of gravity.

Momentum

The quantity of motion that an object possesses, momentum is calculated as the velocity multiplied by the mass. The units are kilogram metres per second (kg.m/s). This concept has relevance to strength and conditioning across a range of aspects, including injury mechanisms and assessment of performance. For example, it is often useful to express sprinting ability not just in terms of time or velocity but in terms of momentum. This is because an athlete with mass 100 kg running at 10 m/s has considerably more momentum than an 80 kg athlete running at the same velocity. In collision sports such as rugby or football it is momentum that usually determines the outcome. Momentum is also important to consider during resistance training because the load that is applied to the athlete is not simply the mass but also how fast the weight is moving.

Impulse

The product of the force applied and the time over which this force is acting. The units of measurement are Newton seconds (N.s). This is a very important parameter for strength and conditioning because of what has been termed the impulse–momentum relationship. For most athletic movements, achieving a high velocity of release (in jumping, sprinting, throwing, kicking, striking) is a critical performance outcome. This velocity is determined entirely by the impulse imparted to the body, ball, or implement. As an example, the velocity of takeoff in the vertical jump and thus the resulting jump height is the result of the impulse that the athlete can apply to the ground.

Work

Calculated as the force applied multiplied by the distance moved. It is a useful concept in strength and conditioning. For example, in coaching, the volume of a weight-training workout is most accurately calculated as the amount of work completed. Work is measured in units called joules (J).

Power

The rate of doing work, power can be calculated as work divided by the time over which it is completed. Power can also be calculated as the force applied multiplied by the velocity. Power is measured in units called Watts (W), although it is also frequently expressed in horsepower (hp), whereby 1 hp = 745.7 W.

1.8.2.3 Friction

Friction is the force that resists two surfaces sliding over each other. The concept is important for strength and conditioning because we use friction in exercise equipment (e.g. weighted sled towing), and it is also a factor in weight training (e.g. chalk applied to hands). Friction is a force and so is measured in Newtons. It is related to the nature of the two surfaces and the force pressing them together; it should be noted that it is the interaction between *both* surfaces. For example, a smooth-soled basketball shoe has excellent grip on the polished wooden floor of the court but little grip when used on a grassed soccer field. An example which illustrates how increasing the force pushing two surfaces together affects friction is the Monarch cycle ergometer, which is commonly used for exercise testing and training. When you adjust the tension on the belt around the wheel, the belt is pushed harder on to the wheel rim, increasing friction and thus the exertion required to push the pedals.

1.8.3 THE FORCE–VELOCITY–POWER RELATIONSHIP

Most sports require the application of maximal power output rather than force. Understanding the mechanical quantity of power and factors contributing to its generation is invaluable to the strength and conditioning specialist. Due to the structure of muscle and how it lengthens and shortens while developing tension, the amount of force that can be produced is related to the velocity at which the muscle is changing length (Figure 1.8.1). The following discussion assumes that the muscle is being maximally activated. Several important points for strength and conditioning arise from this phenomenon:

1. The lowest tension can be developed during concentric (shortening) contractions.
2. When the velocity is zero, this is termed an isometric contraction, whereby the muscle is generating tension but no movement is occurring. Greater tension is developed during isometric contractions than during concentric contractions.

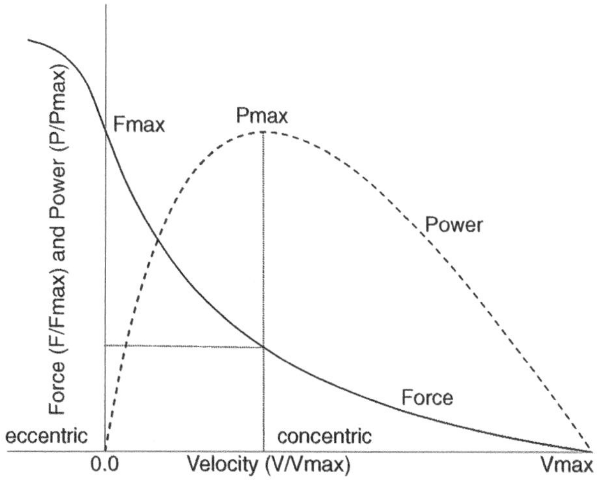

Figure 1.8.1 Force–velocity–power relationship for muscle

3. Negative velocities indicate that the muscle is lengthening while developing tension; this is termed eccentric. Muscle can generate greatest tension while it is lengthening. Note that at faster velocities of lengthening the force no longer increases and may even decrease slightly, most likely due to reflex inhibition.

4. Positive power can only be produced during concentric contractions and the faster the velocity of muscle shortening, the less force the muscle can develop. When the muscle is contracting against a high external load, force is high but velocity will be low. If the load is light then high contraction velocity can be achieved but force is low. Because of this interplay of force and velocity, the product of the two is maximized at a point in between. In fact, power output is greatest when the muscle is shortening at about 30% of the maximum shortening velocity. This by definition corresponds to a force of approximately 30% of that produced during an isometric contraction.

We observe this phenomenon regularly in the weight room. For example, when squatting heavier loads, the speed of movement decreases. For maximal lifts the velocity may be very low, in fact it might approach zero in certain phases of the range of motion as the lifter tries to maximize the force applied. Also, when performing training aimed at increasing power output, lighter loads tend to be used, but moved at much higher velocities, say 30–50% of isometric maximum.

1.8.4 MUSCULOSKELETAL MACHINES

The muscles develop tension and this is applied to the bones to produce or resist movement. Essentially these systems are a form of machine, taking the tension developed by the muscle and converting it to motion of a bone around a joint. There are many examples of such machines in the human body and when accurately controlled by the computer which is our nervous system the result is the bewildering array of human movements that we observe each day. In this case the musculoskeletal machines function to change high force production and small range of muscle length change into low force but high velocity and range of motion at the end of the bone opposite that to which the muscle attaches.

1.8.4.1 Lever systems

The most common musculoskeletal machine in the body is the **lever system**. The muscle pulls on the bone, which may rotate about the joint. A lever is a common machine in everyday use (using a screwdriver to open a can of paint, a tyre lever to dislodge a hubcap). In these examples, the length of the force arm is much greater than that of the resistance arm and so pushing on the lever with say 100 N results in 1000 N applied on the paint lid to push it open (Figure 1.8.2). But note that the handle end of the screwdriver must move through 10× the

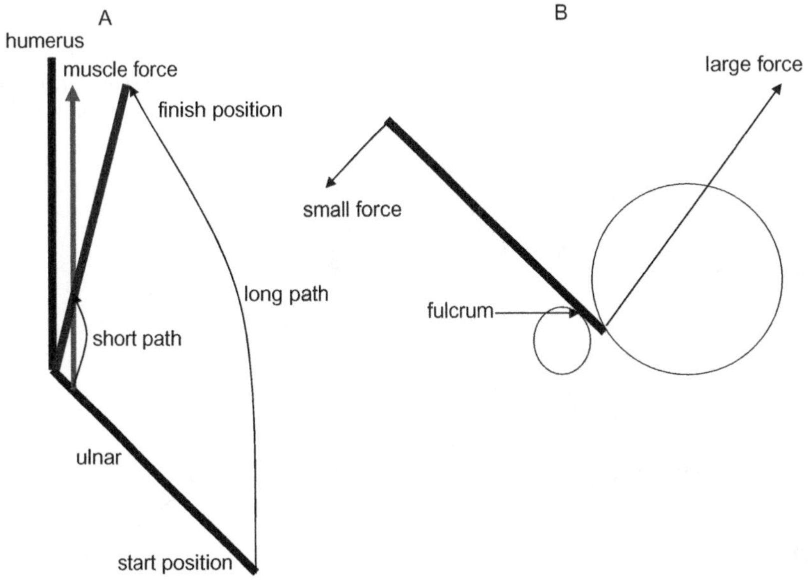

Figure 1.8.2 Examples of basic lever systems designed for high range and speed of movement (a) or high force (b). For the elbow flexion (a), note that a small shortening of the muscle results in a large displacement at the hand, however much higher force is required by the muscle to produce force at the hand. In (b), a long lever is used to move a heavy object. A large range of movement and less force at the long end results in high force but limited range of movement applied to the object being moved

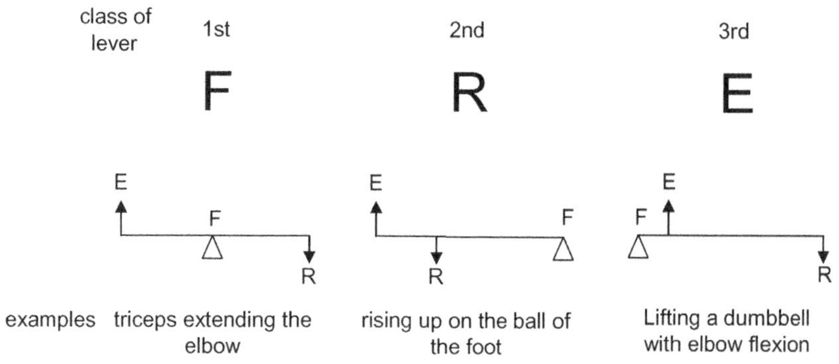

Figure 1.8.3 Arrangement of fulcrum, effort, and resistance for the three different classes of lever

distance of the blade end under the lid. This tradeoff of distance and force is called mechanical advantage. Interestingly, the musculoskeletal machines of the human body usually work at a mechanical disadvantage in terms of force but advantage in terms of distance moved and thus velocity capability: we are designed for speed rather than high force production (Figure 1.8.2).

There are three possible types of lever and all of these can be observed in the human body. Class of lever is based on the arrangement of the fulcrum (F; pivot point or joint), resistance (R; point of application of force against the resistance), and effort (E; muscle attachment and thus point of application of the muscle force). A good way to remember the three classes of lever is 'FRE' (Figure 1.8.3). In a first-class lever, the fulcrum (F) is in the middle. In a second-class lever the resistance (R) is in the middle. In a third-class lever the effort (E) is in the middle. In the human body, the third-class lever is the most common because it is the best design for large range and speed of motion. First-class is the next most common because depending on whether the fulcrum is closer to the effort or the resistance, first-class levers can favour either force production or range of motion. Second-class levers are the least common in the human because such an arrangement always results in mechanical advantage and thus reduced speed and range of motion. Remember, we are designed for speed and range of motion – not force production.

1.8.4.2 Wheel-axle systems

Wheel-axle machines only occur at ball and socket joints; the hip and shoulder are particularly important examples and are commonly involved in strength and conditioning. When either of these joints is moving in internal or external rotation the muscles and bones are working as a wheel-axle system. When the tendon of the agonist muscle wraps partially around the end of the bone, muscle tension will tend to rotate the bone about its long axis, similar to the axle in an automobile.

Let's examine the example of the shoulder joint being internally rotated. The tendon of the pectoralis major muscle inserts on the humerus, and when the shoulder is externally rotated the tendon is partially wrapped around the bone. Muscle shortening internally rotates the humerus, forearm, and hand. If the elbow is flexed, the hand will travel through a much greater distance than the upper arm, and hence the system acts as a machine, trading force production at the proximal end of the humerus for greater distance and velocity of movement at the hand.

1.8.5 BIOMECHANICS OF MUSCLE FUNCTION

A number of key aspects of how muscle functions are important for understanding the biomechanics of strength and conditioning. Force or power applied is determined by a complex range of neural and mechanical interactions within the muscle, between muscle and tendon, and between muscle and the machines of the skeleton.

1.8.5.1 Length–tension effect

Muscle generates tension by the interaction of myosin and actin filaments; this process results in the muscle being able to generate different quantities of tension at varying lengths. This is termed the length–tension effect, which is an inverted 'U' relationship such that greatest tension is produced when the muscle is at or near its resting length, and substantially less contractile tension can be developed when the muscle is stretched or shortened above or below this length (Figure 1.8.4). The actual force output of the muscle is slightly different to this because as the muscle stretches beyond the resting length another force increases in contribution: the elastic recoil of the stretched muscle and tendon (Figure 1.8.4).

1.8.5.2 Muscle angle of pull

Muscles pull on the bones to produce movement, resist movement, or maintain a certain body position. This force acting on the bony levers generates torque. The muscle force is most efficiently converted to torque when the angle at which the force is applied is at 90° to the long axis of the bone. At angles other than perpendicular, only a portion of the muscle force is directed to producing joint rotation, with the remainder tending to either pull the joint together (stabilizing force) or apart (dislocating force) (Figure 1.8.5). Muscle angle of pull has a very large impact on the effectiveness of the developed muscle tension. For example, when the angle of pull is 30° to the long axis of the bony lever, the contractile force is only 50% efficient in terms of transfer into torque about the joint.

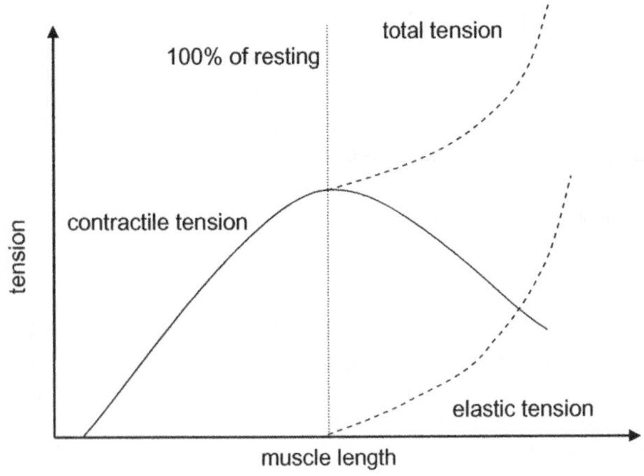

Figure 1.8.4 Length–tension effect for skeletal muscle, showing both contractile tension and elastic components

1.8.5.3 Strength curve

Combination of the length–tension effect and angle of pull results in the strength curve for the particular movement. It is important to note that the strength curve is the net combination of all the muscles, bones, and joints involved in the movement: for a single joint movement, such as elbow extension, only the triceps brachii length and angle of pull combine to produce the strength curve; for a more complex, multijoint movement, such as bench press, all the muscles involved in shoulder horizontal adduction and elbow extension combine to produce the strength curve. The result is a range of different-shaped strength curves for each movement. The most common is 'ascending', which describes movements such as squat and bench press. In both cases, from the bottom to the top of the lift the force-generating capacity of the musculo-skeletal system increases.

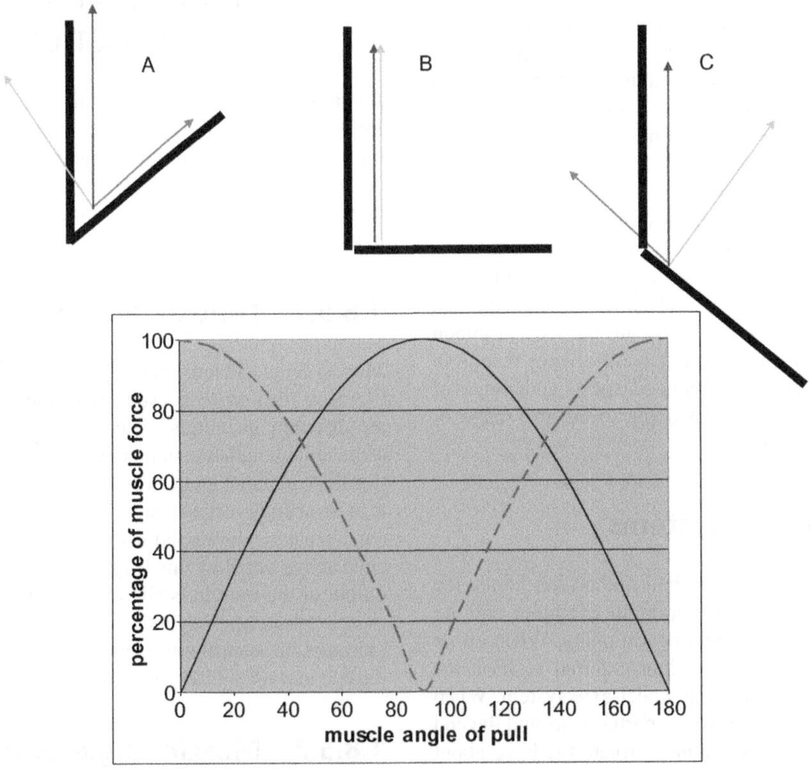

Figure 1.8.5 Muscle angle of pull determines percentage of muscle force contributing to joint rotation (solid line) or to dislocating and stabilizing forces (dashed line). (a) Angle of pull is less than 90° and dislocating force is evident. (b) Angle of pull is 90° and 100% of muscle force is contributing to joint rotation, with no dislocating or stabilizing force. (c) Angle of pull is greater than 90° and a stabilizing force is evident

Figure 1.8.6 Line of resistance for the arm curl. The white arrow represents the resistance vector and the red arrow represents the component producing torque about the joint. The length of the red arrow indicates the amount of resistance force applied. Note that the resistance torque is maximized when the forearm is horizontal and is considerably less at higher or lower angles

1.8.5.4 Line and magnitude of resistance

The direction of the force that must be overcome during resistance training is termed the line of resistance. When training with free weights this is always vertically down because the resistance is the gravitational weight force acting on the barbell or dumbbell. Using combinations of levers and pulleys, the line of resistance for a given exercise can be in any direction. For example, for a pull-down machine, a pulley is used to reverse the direction of the downwards gravitational force of the weight stack to provide a vertically upward force. Similarly, pulleys are used in a seated row machine to direct the line of resistance horizontally (Figure 1.8.6).

For essentially linear movements such as squat, shoulder and bench press, the resistance of a free weight stays constant except for changes in velocity. This is because, as we have already seen, force is equal to mass times (a + g). The result is that when the velocity of movement is constant, acceleration is zero, and so the resistance is simply the weight force of the barbell. However, when increasing velocity at the start of the lift (positive acceleration) the resistance to overcome is higher than the weight force alone. At the top of the lift, when velocity is decreasing (negative acceleration), the resistance to overcome is actually less than the weight force of the barbell.

The changes in magnitude of resistance become even more complex for rotational movements such as arm curls or knee extensions with free weights. The resistance torque about the joint will change in a sine wave fashion so that it is greatest when the limb is horizontal and zero when the limb is vertical.

1.8.5.5 Sticking region

Combining the strength curve with the line of resistance for a given movement results in varying degree of difficulty through the range of motion. So even though the external resistance might be constant, some parts of the lift are more difficult than others. This is termed the 'sticking region' and is the point in the movement at which the lift is most likely to fail. For example, in a standing elbow flexion (arm curl) exercise using a dumbell, the sticking region is around 90–110° joint angle. Interestingly, this is a region of high torque capability in the strength curve, but it is also where the resistive torque generated by the dumbbell is at its peak. For the bench press, the sticking region is 5–10 cm off the chest. This is a linear lift and the line and magnitude of resistance do not change appreciably. However, at this point the strength curve is at a low point due to inefficient angle of pull for the pectoralis major. Also, at this point the initial high force generated at the changeover from eccentric to concentric phases has dissipated (see Section 1.8.8) and mechanical advantage has not yet started to increase appreciably.

1.8.5.6 Muscle architecture, strength, and power

Muscles in the human body have a range of designs or architectures specific to the functional demands required. The two main divisions are fusiform and pennate, and each has several subtypes. Fusiform muscles (e.g. biceps brachialis) have muscle fibres and fascicles running in parallel to a tendon at origin and insertion, longer fibre length, and are able to generate good range of shortening and in particular higher velocity of shortening. The architecture of pennate muscles involves muscle fibres running obliquely into the tendon. Such a structure allows more muscle fibres to be packed into the muscle and this favours higher force output.

The angle at which the muscle fibres are aligned relative to the tendon is termed the pennation angle and this also has impact on the relative ability of the muscle in terms of force versus shortening velocity. A higher pennation angle favours force production, while a lower angle allows the muscle to produce greater range of motion and velocity of contraction.

It is important for the strength and conditioning specialist to understand the concepts of muscle architecture and pennation angle due to their implications for function and thus performance. Of particular interest is the fact that pennation angle can be altered through training. For example, Blazevich *et al.* (2003) have demonstrated changes in muscle pennation angle in as little as 5 weeks of resistance training. Heavy resistance training leads to an increase in pennation angle and thus increased strength capacity. Higher-velocity ballistic training produces adaptations of decreased pennation angle and thus increased velocity and power output. The fact that such changes occur in as little as 5 weeks of training has considerable importance for the periodization of training programme design.

1.8.5.7 Multiarticulate muscles, active and passive insufficiency

Many muscles of the body are multiarticulate: they cross more than one joint and therefore can produce movement at each of the joints they cross. This arrangement has great benefits in terms of efficiency and effectiveness of muscle contraction. However, there are two considerations for strength and conditioning. A muscle can only shorten by a certain amount, typically up to 50% of its resting length. The result is that if the muscle is already shortened about one joint then it cannot contract very forcefully to produce movement over the other joint that it crosses. This is termed 'active insufficiency' and has significance for resistance training. In selecting certain exercises, one can change the emphasis on a given muscle group. For example, when training the calf muscles, ankle plantar flexion can be performed with the knee extended (standing calf raise) or flexed (seated calf raise), switching the training emphasis from the gastrocnemius to the soleus. In a standing calf raise the gastrocnemius is lengthened over the knee joint and so is the primary muscle producing the ankle plantar flexion. However, in a seated calf raise the gastrocnemius is already shortened about the knee joint and cannot contribute much force about the ankle. The soleus becomes the prime mover in this case.

Similarly, a muscle can only be stretched to a certain extent. Multiarticulate muscles are lengthened over all the joints they cross. If a given muscle is already in a lengthened state to allow movement to the end of the range for one joint then it may not be possible to lengthen it further to permit full range of motion at the other joint it crosses. This is termed 'passive insufficiency'. For an application of this principle to exercise we will again use the gastrocnemius and soleus. To adequately stretch the muscles of the calf one has to perform two stretches: one with the knee extended, which tends to place the gastrocnemius on stretch, and the other with the knee bent to stretch the soleus. As soon as you bend the knee, the gastrocnemius is no longer fully lengthened about the knee and so can allow more dorsiflexion of the ankle. The soleus then becomes the limiting muscle and so is effectively stretched.

1.8.6 BODY SIZE, SHAPE, AND POWER-TO-WEIGHT RATIO

Consideration of body size and shape is important to understanding individual differences in strength and power. For example, long trunk and short arms and legs favours powerlifting and weightlifting performance because the shorter lever lengths permit greater force production if all else is equal. However, fast speed of movement (e.g. throwing or kicking) is better achieved by an athlete who has longer relative limb lengths. In general, the larger the athlete's body size and mass, the greater their strength capability because absolute strength is very much dependent on total muscle mass and cross-sectional area. In many sports, relative strength and power are more important than absolute measures. These measures are the strength-to-weight and power-to-weight ratio, respectively, and are simple measures to calculate. For example, relative strength for a given movement is simply the 1RM divided by body mass. If we assume a 1RM squat of 120 kg at a body mass of 80 kg the relative strength for this movement is 1.5. In any sport in which projection of the body is important (e.g. vertical jumping, sprinting) the power-to-weight ratio of the athlete is a critical measure because it determines the amount of acceleration and thus peak velocity that can be achieved. For example, an athlete who produced 6000 W in a vertical jump at a body mass of 100 kg has a power-to-weight ratio of 60 W/kg.

The centre of mass (COM) is that point about which all of the matter of the body is evenly distributed in all directions. When standing upright with the arms at the sides in the anatomical position this point is in the middle of the abdomen, just below the navel. The human body can move and take all manner of positions and so the COM also moves and can even lie outside the body. The concept of COM becomes important when discussing balance and stability.

1.8.7 BALANCE AND STABILITY

It is useful to have an understanding of the biomechanics of balance and stability when analysing strength and conditioning movements. Balance is the process of controlling the body position and movements in either static or dynamic equilibrium for a given purpose. Stability is the ease or difficulty with which this equilibrium can be disturbed. Because all exercises require balance, manipulating stability can impact the degree of difficulty and injury risk, or can be used to introduce a different training stimulus, for example balance training.

1.8.7.1 Factors contributing to stability

Stability is determined by:

1. Base of support: the physical dimensions of the area defined by the points of support. For example, when performing the military press exercise the base of support is the area

defined by the rectangle enclosing both feet. If the feet are placed further apart then the base of support is larger and stability is higher. If the split position is adopted, the area increases even further, particularly in the anterior–posterior plane (Figure 1.8.7). When seated on a bench with the feet wide apart the area and thus stability is even greater.

2. Mass of the object or body: heavier objects are inherently more stable because the inertia (resistance to change in state of motion) is higher, as is the gravitational weight force. For example, placing weight plates on a Smith machine or power rack makes the device more stable.
3. Height of the COM: as noted above, for a human standing with their arms at their sides, the COM is in the centre of the body, roughly 5 cm below the navel. Raising their arms over their head lifts the COM; bending at their hips and knees lowers it. The higher the COM, the less stable the object.
4. Location of the COM relative to the edges of the base of support: if a line dropped down from the COM is close to an edge of the base of support the body will have low stability in that direction.

With this knowledge the strength and conditioning specialist can alter the body position and the equipment used to increase or decrease stability. For example, if an older person is having difficulty maintaining balance while performing arm curls then seating them on a bench will make the exercise easier. This is because the base of support has been increased and the COM lowered. If an athlete is unstable when performing the triceps press down exercise, spreading the feet and splitting them one forward and one back, bending slightly at the knees, will give much better stability. There are also instances where less stability is desirable to increase the balance demands of the task. This can be achieved for the dumbbell shoulder press, for example, simply by standing on one leg.

1.8.7.2 Initiating movement or change of motion

When an athlete attempts to accelerate quickly or change direction, they will decrease their stability in the direction of movement. This means any given force they exert in that direction will result in rapid acceleration. For example, in starting the sprint, the athlete leans as far forward over the hands so that the line of gravity (straight line down from the COM) falls very close to the front edge of their base of support.

1.8.8 THE STRETCH–SHORTENING CYCLE

Almost all human movement involves a preparatory or 'windup' action in the opposing direction, followed by the intended movement. This sequence involves a lengthening and stretching

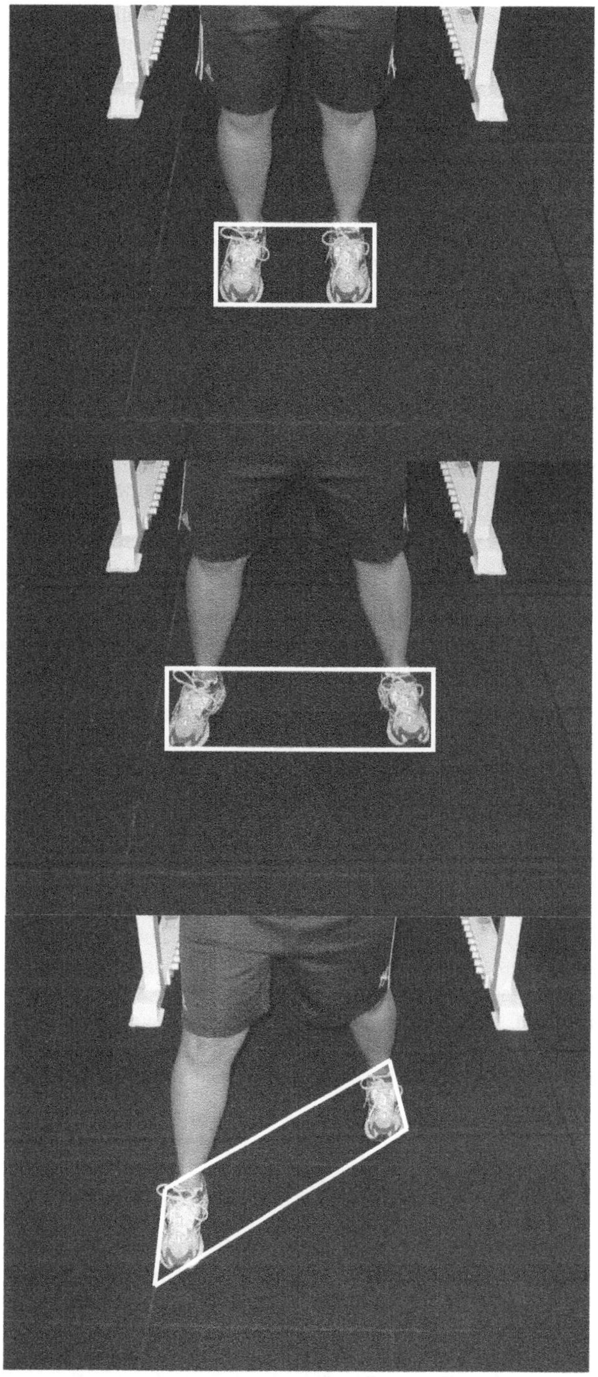

Figure 1.8.7 Base of support for standing feet spread (narrow and wide stance) versus split position. Shifting from narrow to wide stance greatly increases stability in the side-to-side directions. Moving to the split position increases stability in the forward and backward directions

of the muscles used to produce the movement, followed by a shortening, and so has been termed the stretch–shortening cycle (SSC). This phenomenon is important in strength and conditioning because such an action is performed to some extent in all training exercises. In fact, some exercises such as plyometrics are designed specifically to develop SSC ability. The preparatory movement involves eccentric muscle contraction and this phase potentiates the subsequent concentric movement by around 15–20%. That is, movements performed without a pre-stretching produce 15–20% less impulse than true SSC movements. There are around six mechanisms proposed (Walshe *et al.* 1998) to explain this phenomenon, but the predominant factor is that a higher level of force (termed 'pre-load') can be generated at the start of the concentric movement if pre-stretching is performed. Although the stretch reflex and storage and recovery of elastic strain energy have also frequently been touted, the role of these two mechanisms is probably less important, particularly in short-duration ballistic movements such as squat, vertical jump, and throwing.

1.8.9 BIOMECHANICS OF RESISTANCE MACHINES

The variety of equipment developed for strength and conditioning is quite remarkable. This equipment is all designed to change the direction of resistance and in some cases vary the magnitude of resistance through the range of movement. Most use the gravitational weight force but there is also a plethora of machines that use elastic, hydraulic, aerodynamic drag, or pneumatic resistance. How these machines interact with the human body is determined by biomechanical principles. While some have been well designed, others have not incorporated good biomechanics and the result has been poor effectiveness or even injury risk. Understanding the mechanics underlying a piece of resistance training or conditioning equipment will assist in initial purchase decisions as well as exercise selection. Following is a discussion of the predominant resistance machine designs.

1.8.9.1 Free weights

Working against the inertia and gravitational weight force of a freely moving mass such as a dumbell or barbell represents the most simple but perhaps most frequently used form of resistance training. The biomechanics of such equipment are similarly straightforward, with the resistance acting vertically downward at all times. The force can be calculated as $F = mg$, where m = the mass and g = acceleration due to gravity (9.81 m/s/s). When the weight is also being accelerated there is an additional resistance due to the inertia of the object and this extra force can be calculated as $F = ma$. For rapid movements this component of the resistance to movement can become quite large, for example in the clean and jerk or snatch. As already described, when the weight is decreasing in velocity of movement the acceleration is negative and the resistance to overcome is actually reduced. This contributes to why the top part of a squat or bench press requires less effort than earlier in the concentric phase.

1.8.9.2 Gravity-based machines

A disadvantage of free weights is that the line of resistance is always vertically downwards. A second problem is that the resistance (that is, the weight force of the dumbell or barbell) is constant. This does not match the strength curves for various movements and as such the amount of weight that can be lifted is limited to that which can be successfully moved through the sticking region previously discussed. For this reason several machines have been developed, all using the gravitational weight force but modifying the resistance curve to more closely match the strength curve for the movement. The biomechanics of these techniques will be discussed shortly.

Pulleys

The most basic modification was to build machines which allowed the weight force to be redirected. Early versions used standard weight plates but then incorporated cables and pulleys to provide vertically upward or horizontally directed resistance.

Pin-loaded

To overcome the problems of shifting weight plates on and off these early machines, a weight stack was incorporated, with a pin used to select the resistance. This improved ease of use, helped to keep the weight room tidier, and reduced injury risk from lifting weight plates on and off the machines.

Levers and cams

The next problem to overcome was the mismatch of the resistance profile to the strength curve for the movement. For example, in the bench press using free weights or constant resistance machines the load lifted is limited to what can be moved through the bottom part of the range. Once the load is off the chest, the effort required decreases as the ascending strength curve exceeds the constant resistance load by an increasing proportion. In this scenario, the neuromuscular system is not stressed as much in the upper part of the range of motion and the intensity is limited by the sticking region. To overcome this problem a range of variable resistance devices were developed. The initial designs involved sliding levers; an example was the universal variable resistance. The principle was simple: as bench- or leg-press movement proceeded from the bottom to the top position, the lever arm would slide out, changing the point of application of the load on the weight stack, increasing the moment arm and thus the amount of resistance to be overcome by the lifter.

Later examples used variable radius cams, for example Nautilus, to achieve the same result. In this case a cable or chain wrapping around a cam with a changing radius altered the length of the resistance arm and thus the amount of resistance that the lifter must overcome. Using biomechanical principles the cams could be designed to match the strength curve for various training exercises.

One of the most recent implementations of the variable resistance concept in resistance training is the Hammerstrength range of equipment. The biomechanics of this equipment is quite simple and effective. The weight plates are placed on horns located at the end of a lever. In the bottom position the lever is rotated out of the vertical position, and as in the concepts we discussed earlier regarding line of resistance, not all of the weight force of the plates is directed against the lifter. As the lift is executed, the lever arm moves toward the horizontal and so the vertically downwards weight force moves closer to 90° to the lever arm and the amount of load transferred to the lifter increases, matching more closely to the strength curve for the movement.

1.8.9.3 Hydraulic resistance

Hydraulic resistance devices use a hydraulic ram and the resistance of oil (hydraulic fluid) being forced though a small aperture. Two different technologies are used. One involves flow control; the size of the aperture is adjusted, thus changing the velocity at which the fluid can flow and therefore the speed of movement. As greater force is exerted against the machine the resistance increases to match as the fluid is incompressible and velocity of flow is somewhat independent of the pressure. The second type uses a pressure-release valve configuration. The valve is initially held closed by a tensioned spring but when a force greater than the valve resistance is applied the valve opens and the fluid is permitted to flow. This technology provides a more constant resistance, similar to free weights, compared to the quasi-constant velocity of flow-control systems.

It is important to note that both systems are passive and thus provide a purely concentric exercise mode. This equipment has found a niche in circuit training programme designs and with special populations because there is reduced muscle soreness (no eccentric phase), it is easy to use, and no momentum is built up, thus reducing the risk of injury.

1.8.9.4 Pneumatic resistance

As the name suggests, this equipment uses air pressure (pneumatic) to provide resistance to both concentric and eccentric exercise. A pneumatic ram similar to an enormous syringe is

Table 1.8.1 Biomechanical comparison of machines versus free weights

Free Weights	Machines
Lower stability means greater balance control required; this may have added training effect	Higher stability makes exercise easier for novices and certain populations where demands of balance control may be risky or compromise strength gains
Line of resistance is constant and vertically downward	Line of resistance can be altered to any plane and direction
Resistance force is constant and proportional to the mass, and does not necessarily match the strength curve for a specific movement	Resistance force can be varied through the use of levers and cams in an attempt to match strength curve
Resistance is more specific to the free masses that must usually be manipulated in sport performance and tasks of everyday living	Controlling the plane of movement, varying the resistance, and changing the line of resistance reduces specificity of training
While free weights do not match strength curves, their *free*, three-dimensional movement does allow for wide variation in weight, height, lever length, and gender	Even the best application of biomechanics can only approximate the *average* person, and deviations in height, weight, gender, and lever lengths result in a considerable mismatch of machine and body mechanics
Lifting free weights, particularly in a standing position, requires considerable activation of muscles acting as fixators and stabilizers, which increases training efficiency and strengthens muscles in these important roles	While less activation of muscles other than the agonists is required when using machines, this allows concentration on the agonist muscles for greater activation
Both eccentric and concentric phases of free-weight lifts can generate considerable momentum, which must be controlled at the end of range by the lifter to avoid injury	Machines, and in particular hydraulic and pneumatic machines, reduce the amount of momentum generated; training may be safer in this aspect
Require more effort loading the bar, lifting plates on and off, lifting barbells and dumbbells from the rack; all may be a source of injury if not performed with good ergonomics	Load is easily selected and changed
Not a biomechanical issue, but administratively free weights require more effort to keep orderly and tidy	Easy to keep tidy and safe exercise environment
If the lifter cannot complete a lift, some exercises such as bench and squat are difficult to escape from under the bar	Weight stack just drops back and stops if lifter cannot fails the attempt

preloaded to a certain pressure using compressed air. The higher the pressure, the greater the resistance provided. When the movement is performed, the ram compresses the air even further, increasing pressure and thus resistance. This provides an ascending resistance curve, which matches reasonably well the ascending strength curve of most human movements. On the return movement the athlete is working to control the speed of the expanding gas and so these machines allow eccentric as well as concentric phases.

1.8.9.5 Elastic resistance

Examples range from equipment as simple as the theraband to devices such as the Vertimax. All use the resistance that is developed when an elastic material such as rubber or bungee cord is stretched. Such devices also provide an ascending resistance curve because elastic tension is greater the more the material is deformed, and they allow for eccentric exercise as the muscles act to control the speed of elastic recoil.

1.8.10 MACHINES VS FREE WEIGHTS

It is an interesting biomechanical discussion to compare the advantages and disadvantages of machines versus free weights for resistance training. Clearly each has benefits and it is really a matter of understanding the different biomechanical principles and then selecting the equipment and exercises appropriate to the individual. An evaluation is included in Table 1.8.1.

1.8.11 CONCLUSION

Biomechanics has considerable application in understanding many aspects of strength and conditioning. From an appreciation of the determinants of friction and how an athlete can develop force against the ground, to the design principles of the wide array of resistance equipment available, biomechanics knowledge will enable the strength and conditioning professional to increase their effectiveness and safety.

References

Blazevich A.J., Gill N.D., Bronks R. and Newton R.U. (2003) Training-specific muscle architecture adaptation after 5-wk training in athletes. *Medicine & Science in Sports & Exercise*, **35** (12), 2013–2022.

Walshe A.D., Wilson G.J. and Ettema G.J. (1998) Stretch-shorten cycle compared with isometric preload: contributions to enhanced muscular performance. *Journal of Applied Physiology*, **84**, 97–106.

Section 2
Physiological adaptations to strength and conditioning

2.1 Neural Adaptations to Resistance Exercise

Per Aagaard, University of Southern Denmark, Institute of Sports Science and Clinical Biomechanics, Odense, Denmark

2.1.1 INTRODUCTION

Resistance training induces adaptive changes in the morphology and architecture of human skeletal muscle while also leading to adaptive changes in nervous system function. In turn, all these changes appear to contribute to the marked increase in maximal contractile muscle force and power that can be seen with resistance (strength) training in both young and ageing persons, including even very frail and old (>80 years) individuals. The adaptive change in neural function has been evaluated by use of muscle electromyography (EMG) measurements, which recently have included single motor unit recording and measurements of evoked spinal reflex responses (Hoffman reflex, V-wave) and transcranial brain stimulation (TMS, TES). Also, interpolated muscle twitch recording superimposed on to maximal muscle contraction has been used to assess the magnitude of central activation of the muscle fibres, while peripheral nerve stimulation has recently been used to examine aspects of intermuscular synergist inhibition.

As discussed in this chapter, different lines of evidence exist to demonstrate that resistance training can induce substantial changes in nervous system function. Notably, the adaptive alteration in neuromuscular function elicited by resistance training seems to occur both in untrained individuals, including frail elderly individuals and patients, as well as in highly trained athletes. In all of these wildly differing individuals the training-induced enhancement in neuromuscular capacity leads to improved mechanical muscle function, which in turn results in an improved functional performance in various activities of daily living.

2.1.2 EFFECTS OF STRENGTH TRAINING ON MECHANICAL MUSCLE FUNCTION

2.1.2.1 Maximal concentric and eccentric muscle strength

The influence of strength training on the maximal contraction strength of human muscle *in vivo* has been extensively investigated for the concentric part of the moment–velocity curve (Aagaard *et al.*, 1996; Caiozzo, Perrine and Edgerton, 1981; Colliander and Tesch, 1990; Costill *et al.*, 1979; Coyle *et al.*, 1981; Lesmes *et al.*, 1978; Moffroid and Whipple, 1970; Seger, Arvidson and Thorstensson, 1998). Also, data exist for the training-induced change in maximal eccentric muscle strength (Aagaard *et al.*, 1996; Colliander and Tesch, 1990; Higbie *et al.*, 1996; Hortobagyi *et al.*, 1996a, 1996b; Komi and Buskirk, 1972; Narici *et al.*, 1989; Seger, Arvidson and Thorstensson, 1998; Spurway *et al.*, 2000).

Following strength training using concentric muscle contractions alone, maximal muscle strength and power were reported to increase at the specific velocity employed during training (Aagaard *et al.*, 1994; Caiozzo, Perrine and Edgerton, 1981; Kanehisa and Miyashita, 1983; Kaneko *et al.*, 1983; Moffroid and Whipple, 1970; Tabata *et al.*, 1990). These and similar observations have been taken to indicate a specificity of training velocity and training load. However, it is questionable whether a generalized concept of training specificity should exist, since maximal muscle strength and power may also increase at velocities lower than the actual velocity of training

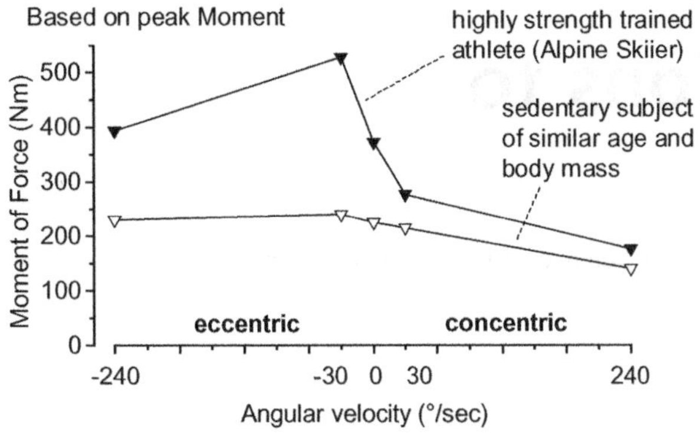

Figure 2.1.1 Maximal concentric, eccentric, and isometric muscle strength obtained by use of isokinetic dynamometry (KinCom) for the quadriceps muscle in a highly trained strength athlete (male National Team Alpine skier) and in a young untrained individual matched for age and body mass. Notice the marked difference in maximal eccentric muscle strength between the trained and the untrained subject

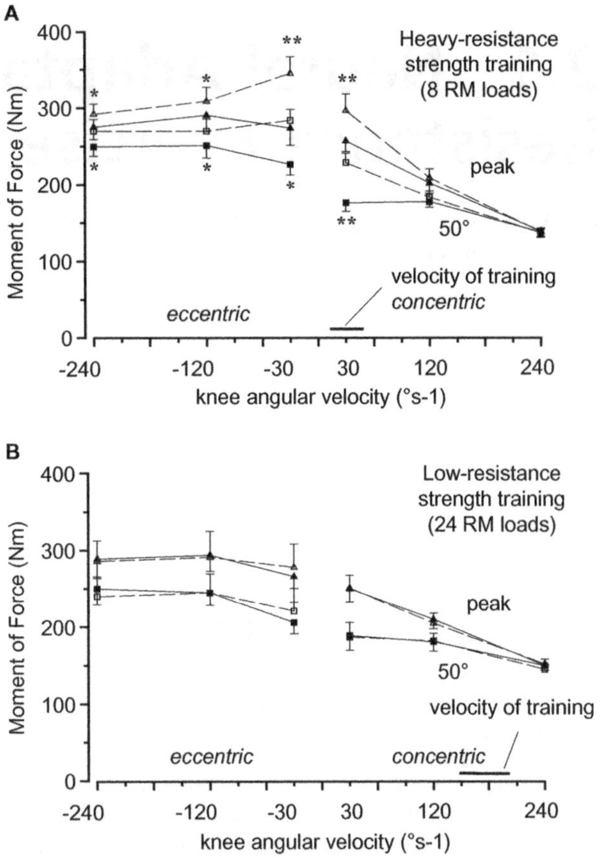

Figure 2.1.2 Maximal concentric and eccentric muscle strength (quadriceps femoris muscle) measured by isokinetic dynamometry (KinCom) before (full lines, closed symbols) and after (broken lines, open symbols) 12 weeks of heavy-resistance (a) or low-resistance (b) strength training in a group of elite soccer players. Peak moment (triangles) and isoangular moment (at 50°; 0° = full knee extension) were obtained, and the velocity used during training is also depicted. All training was performed using a hydraulic leg extension device (Hydra fitness). Post > pre: * $P < 0.05$, ** $P < 0.01$. Data from Aagaard *et al.* (1996)

(Costill *et al.*, 1979; Coyle *et al.*, 1981; Housh and Housh, 1993; Lesmes *et al.*, 1978) as well as at higher velocity (Aagaard *et al.*, 1994; Colliander and Tesch, 1990; Housh and Housh, 1993). These conflicting findings are likely caused by a variable involvement of learning effects (Rutherford and Jones, 1986). Thus, using different training and/or data-recording devices tends to reduce the influence of learning and hence may provide a more valid measure of the adaptation in neuromuscular function elicited by strength training.

Untrained individuals typically demonstrate a levelling-off (plateauing) in maximal muscle strength during slow concentric muscle contraction, whereas strength-trained individuals do not (Figure 2.1.1). Notably, the plateauing in maximal muscle strength can be removed by heavy-resistance strength training (Aagaard *et al.*, 1996; Caiozzo, Perrine and Edgerton, 1981) (Figure 2.1.2). Strength training with low external loads and high contraction speeds appears to have no effect on the plateauing phenomenon (Figure 2.1.2), suggesting that heavy muscle loadings (>80% 1RM) should be used to diminish or fully remove the influence of this force-inhibiting mechanism.

Heavy-resistance strength training (loads >85% 1RM) appears to evoke significant gains in maximal eccentric muscle strength (Aagaard *et al.*, 1996, 2000a; Andersen *et al.*, 2005; Colliander and Tesch, 1990; Duncan *et al.*, 1989; Higbie *et al.*, 1996; Hortobagyi *et al.*, 1996a, 1996b; Komi and Buskirk, 1972; Narici *et al.*, 1989; Seger, Arvidson and Thorstensson, 1998; Spurway *et al.*, 2000,). Moreover, strength training using maximal eccentric muscle contractions or coupled concentric–eccentric contractions leads to greater gains in maximal eccentric muscle strength than concentric training alone (Colliander and Tesch, 1990; Duncan *et al.*, 1989; Higbie *et al.*, 1996 Hortobagyi *et al.*, 1996a, 1996b; Norrbrand *et al.*, 2008). On the other hand, maximal eccentric muscle strength seems to remain unaffected by low-resistance strength training (Aagaard *et al.*, 1996; Duncan *et al.*, 1989; Holm *et al.*, 2008; Takarada *et al.*, 2000) (Figure 2.1.2b), suggesting that the involvement of very high muscle forces during training is probably a key stimulus of adaptive changes in maximal eccentric muscle strength.

2.1.2.2 Muscle power

Maximal muscle strength is typically increased by 20–40% in response to heavy-resistance strength training of medium to moderate duration (8–16 weeks) (Aagaard *et al.*, 1996; Colliander and Tesch, 1990; Häkkinen *et al.*, 1998b). Slightly greater gains in maximal contractile muscle power of 20–50% have been observed when heavy-resistance strength training was performed in young untrained persons, elite soccer players,

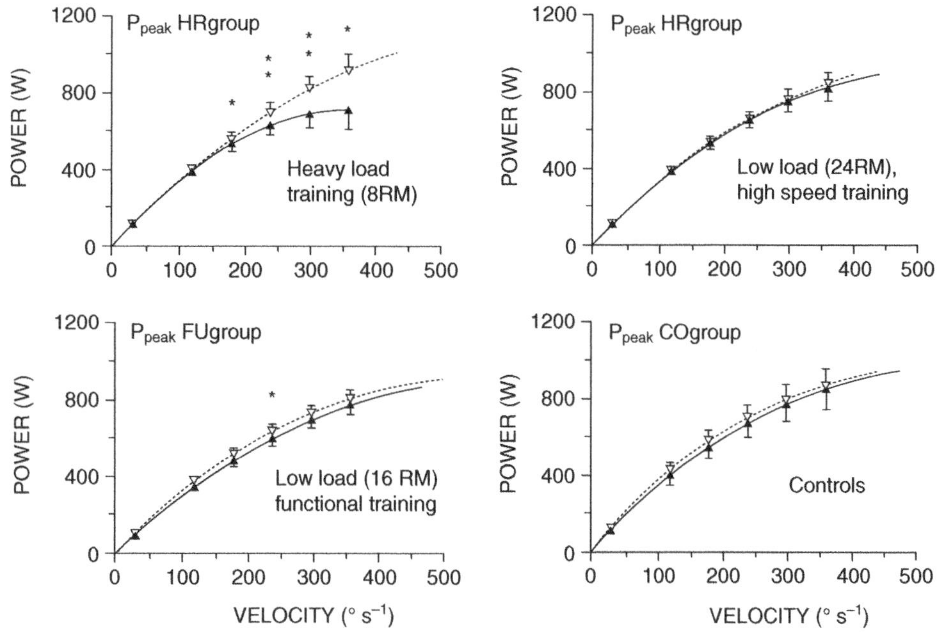

Figure 2.1.3 Peak muscle power (quadriceps femoris muscle) measured by non-isokinetic dynamometry (Hill's flywheel) before (full lines, closed symbols) and after (broken lines, open symbols) 12 weeks of heavy-resistance (HR group), low-resistance (LR group), or functional-resistance (FU group; loaded kicking movements) strength training in elite soccer players. A group of non-training controls was also tested (CO group). Post > pre: * $P < 0.05$, ** $P < 0.01$. Data adapted from Aagaard *et al.* (1994)

and ageing individuals (Aagaard *et al.*, 1994; Caserotti *et al.*, 2008; De Vos *et al.*, 2005; Duchateau and Hainaut, 1984; Kaneko *et al.*, 1983; Toji *et al.*, 1997) (Figure 2.1.3). Some of these studies employed explosive-type training that involved maximal intentional acceleration of the external load, which may constitute a key element for achieving large gains in maximum muscle power output. Notably, strength training using heavy-resistance muscle loadings (>80% 1RM) appears to evoke similar or greater increases in maximum muscle power compared to low-resistance strength training (Aagaard *et al.*, 1994; De Vos *et al.*, 2005; Duchateau and Hainaut, 1984) (Figure 2.1.3).

2.1.2.3 Contractile rate of force development

Rapid force capacity ('explosive muscle strength') can be measured as the maximal contractile rate of force development (RFD) during maximal voluntary muscle contraction (Aagaard *et al.*, 2002b; Schmidtbleicher and Buehrle, 1987) (Figure 2.1.4). RFD reflects the ability of the neuromuscular system to generate very steep increases in muscle force within fractions of a second at the onset of contraction, which has important functional significance for the force and power generated during rapid, forceful movements (Aagaard *et al.*, 2002b; Suetta *et al.*, 2004). A high RFD is vital not only to the trained athlete but also to the elderly individual who needs to counteract sudden perturbations in postural balance. As discussed in more detail below, resistance training leads to parallel increases in RFD and neuromuscular activity (EMG amplitude, rate of EMG rise) in the initial phase (0–200 ms) of muscle contraction (Aagaard *et al.*, 2002b; Del Balso and Cafarelli, 2007; Schmidtbleicher and Buehrle, 1987; Van Cutsem, Duchateau and Hainaut, 1998).

2.1.3 EFFECTS OF STRENGTH TRAINING ON NEURAL FUNCTION

The adaptive plasticity of the neuromuscular system in response to strength training can be ascribed to changes in a large number of neural and muscular factors. Thus, it is well documented that strength training can evoke marked changes in muscle morphology and nervous system function (Aagaard, 2003; Duchateau, Semmler and Enoka, 2006; Folland and Williams, 2007; Sale, 1992). This section will focus on changes in neural function induced by strength training, while the adaptation in various muscular factors is addressed in Chapter 2.2.

In brief, the neural adaptation mechanisms involved with strength training include changes in spinal motor neuron recruitment and rate coding (firing frequency), motor neuron excitability, corticospinal excitability, and coactivation of

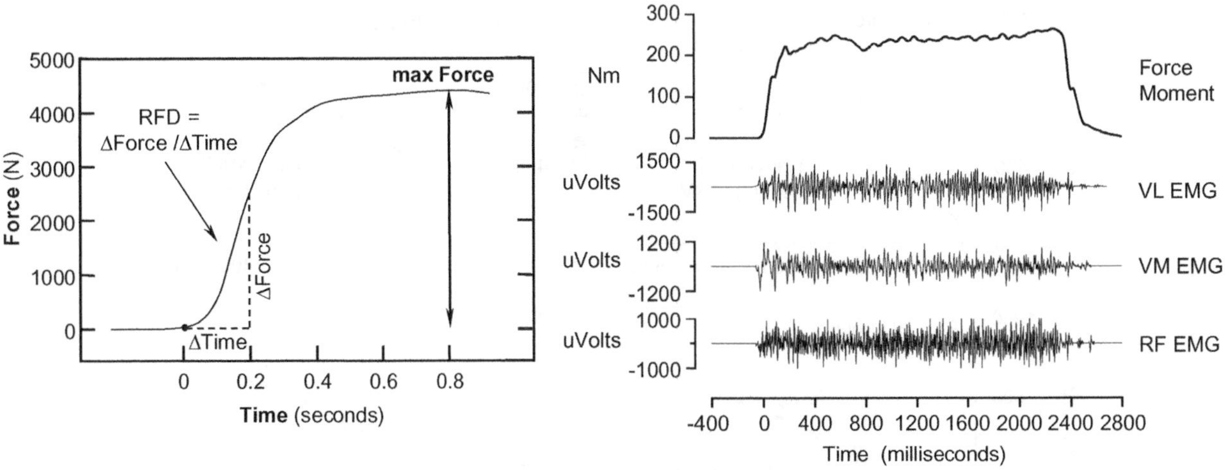

Figure 2.1.4 Rapid muscle strength is measured as the contractile rate of force development (RFD); that is, calculated as the slope of the force–time curve (left panel) or the moment–time curve (right panel, top graph) during the phase of rising muscle force (0–200 ms). RFD is strongly influenced by the magnitude of neuromuscular activity in the agonist muscles, typically recorded as surface EMG signals (right panel; quadriceps muscle; vastus lateralis = VL, vastus medialis = VM, rectus femoris = RF). Data adapted from Aagaard *et al.* (2002b)

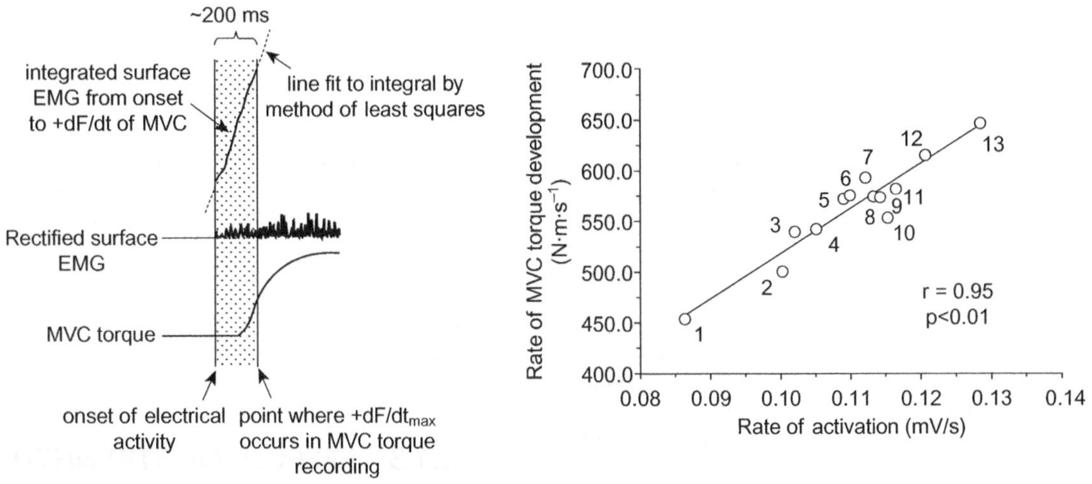

Figure 2.1.5 The magnitude of contractile RFD is strongly influenced by the level of neuromuscular activity in the agonist muscles. When the slope of the integrated EMG (iEMG) curve is determined (left panel, top graph), a strong linear relationship is found to exist between the rate of accumulated EMG activity (ΔiEMG/Δt) and the rate of rise in contractile force production (RFD = rate of torque development) (right panel). Reproduced from Del Balso & Cafarelli. 2007. J. Appl. Physiol. © American Physiological Society

antagonist muscles (Aagaard, 2003; Duchateau, Semmler and Enoka, 2006; Sale, 1992). A line of experimental evidence has been provided with the use of EMG recording during *in vivo* muscle contraction, although inherent methodological limitations may exist with the recording of surface EMG. Measurements of evoked spinal responses (H-reflex, V-wave) can be used to address the adaptability in spinal circuitry function with strength training. In addition, training-induced changes in transmission efficacy in descending corticospinal pathways have recently been examined.

2.1.3.1 Maximal EMG amplitude

A widely used assessment tool in the evaluation of neural function with training is the recording of EMG signals during maximal voluntary contraction (Figure 2.1.4) The EMG interference signal obtained by surface electrode recording constitutes a complex outcome of motor unit recruitment and motor neuron firing frequency (rate coding). In addition, the net EMG signal amplitude is highly influenced by the specific summation pattern of the individual motor unit action potentials (MUAPs) (Day and Hulliger, 2001; Farina, Merletti and Enoka, 2004), which among other things is affected by the degree of motor unit synchronization (Keenan *et al.*, 2005).

Increased neuromuscular activity evidenced by elevated EMG amplitudes has been observed following weeks to months of strength training, which may reflect an increased efferent neural drive to the muscle fibres (Aagaard *et al.*, 2000a, 2002a, 2002b; Andersen *et al.*, 2005; Barry, Warman and Carson, 2005; Del Balso and Cafarelli, 2007; Häkkinen, Alén and Komi, 1985a; Häkkinen, Komi and Alén, 1985b; Häkkinen *et al.*, 1987, 1998a, 1998b, 2000, 2001; Häkkinen and Komi, 1983, 1986; Hortobagyi *et al.*, 1996b, 2000; Moritani and DeVries, 1979; Narici *et al.*, 1989; Petrella *et al.*, 2007; Reeves, Narici and Maganaris, 2004; Schmidtbleicher and Buehrle, 1987; Suetta *et al.*, 2004; Van Cutsem, Duchateau and Hainaut, 1998). However, inherent methodological constraints are involved with the recording and interpretation of muscle-surface EMG, since the interference summation pattern of the single MUAPs is not likely to directly reflect the neuronal output of the active motor neurons (Day and Hulliger, 2001; Farina, Merletti and Enoka, 2004; Keenan *et al.*, 2005; Yao, Fuglevand and Enoka, 2000). Consequently, a number of studies have failed to demonstrate increased maximal EMG amplitude during MVC following resistance training (Blazevich *et al.*, 2008; Cannon and Cafarelli, 1987; Narici *et al.*, 1996; Thorstensson *et al.*, 1976), which could at least in part also be due to altered skin and muscle tissue conduction properties (i.e. changes in subcutaneous fat thickness, altered muscle fibre pennation angles). To some extent, however, it is possible to surpass some of the inherent methodological limitations associated with the recording of muscle-surface EMG by employing measurements of evoked spinal motor neuron responses (H-reflex, V-wave), peripheral nerve stimulation, transcranial magnetic or electrical stimulation (TMS, TES), or by using intramuscular EMG recordings.

2.1.3.2 Contractile RFD: changes in neural factors with strength training

A strong influence of efferent neuromuscular activity on contractile RFD has been demonstrated in a number of studies (Figure 2.1.5) and parallel gains in contractile RFD and neuromuscular activity evaluated by surface EMG have been observed following strength training (Aagaard *et al.*, 2002b; Barry, Warman and Carson, 2005; Del Balso and Cafarelli, 2007; Häkkinen, Alén and Komi, 1985a; Häkkinen, Komi and Alén, 1985b; Häkkinen *et al.*, 1998a, 2001; Häkkinen and Komi, 1986; Schmidtbleicher and Buehrle, 1987; Suetta *et al.*, 2004; Van Cutsem, Duchateau and Hainaut, 1998) (Figure 2.1.6). However, not just heavy-resistance strength training but also maximal ballistic training using lower loads can lead to increased RFD (Duchateau and Hainaut, 1984; Van Cutsem, Duchateau and Hainaut, 1998). In their classical study Behm and Sale (1993) compared dynamic and isometric ballistic training performed with maximal intentional acceleration of the training load, and observed that RFD increased similarly with both types of training. This finding suggests that the use of an intended ballistic effort may be more important for inducing increases in RFD than the actual type of muscle contraction performed.

Muscle fibre innervation frequency has a strong positive influence on RFD, which can be observed in single muscle fibres (Metzger and Moss, 1990a, 1990b) as well as in human muscles *in vivo* (De Haan, 1998; Grimby, Hannerz and Hedman, 1981; Nelson, 1996). Notably, RFD continues to increase at

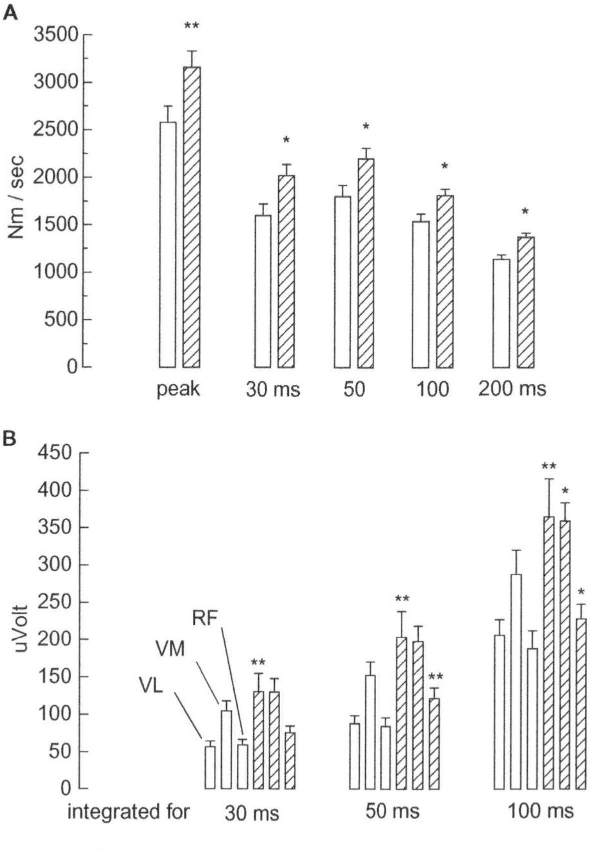

Figure 2.1.6 Contractile RFD (a) and neuromuscular activity (EMG amplitude) (b) obtained before (open bars) and after (hatched bars) 14 weeks of periodized heavy-resistance strength training. Open bars = before training, hatched bars = after training. Post > pre: * $P < 0.05$, ** $P < 0.01$. Note the gain in RFD following training, which was accompanied by marked increases in neuromuscular activity. Data adapted from Aagaard *et al.* (2002b)

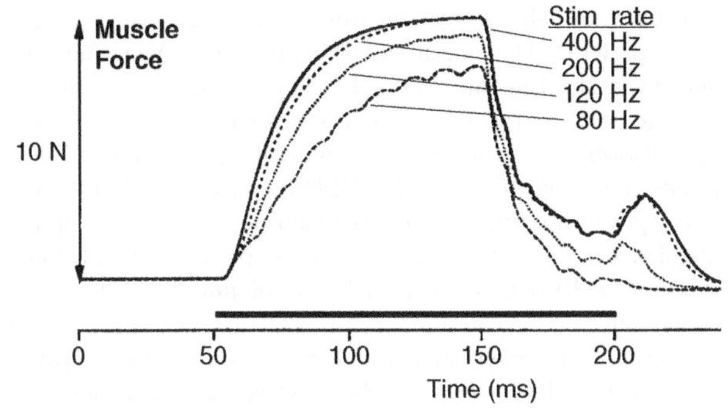

Figure 2.1.7 Influence of muscle fibre innervation frequency on contractile RFD. The graph shows the development in isometric muscle force when using electrical stimulation of the nerve innervating the rat medial gastrocnemius muscle at various stimulation frequencies (80, 120, 200, and 400 Hz). The 200 Hz innervation frequency was sufficient to elicit full tetanic fusion and maximal muscle force production, whereas a greater RFD was achieved in the phase of rising muscle force by using 400 Hz innervation frequency. Data from De Haan (1998)

innervation rates that are higher than full tetanic fusion rate (De Haan, 1998; Grimby, Hannerz and Hedman, 1981; Nelson, 1996) (Figure 2.1.7). Consequently, the occurrence of very high (i.e. supramaximal) motor neuron firing frequency at the onset of rapid muscle force production (Van Cutsem, Duchateau and Hainaut, 1998) likely plays a functional role to increase maximal RFD rather than increasing maximal contraction force per se, in order to increase the level of contractile force exertion in the initial phase of contraction (0–200 ms).

Strength training can lead to elevated motor neuron firing frequency during maximal voluntary contraction. For instance, Van Cutsem, Duchateau and Hainaut (1998) observed a marked rise in the maximal firing rate of motor neurons innervating the tibialis muscle at the onset of maximal, forceful contraction following 12 weeks of ballistic-type resistance training (Figure 2.1.8). Importantly, maximal motor neuron firing frequency has been reported to increase in both young and elderly individuals following resistance training (Kamen and Knight, 2004; Van Cutsem, Duchateau and Hainaut, 1998). Untrained elderly demonstrate lower maximal motor neuron firing rates than young subjects (Kamen and Knight, 2004; Klass, Baudry and Duchateau, 2008; Patten, Kamen and Rowland, 2001), but no age-related differences could be detected following a period of strength training (Kamen and Knight, 2004; Patten, Kamen and Rowland, 2001). In other words, the training-induced increase in maximal motor neuron firing frequency is effective in overruling the age-related decline in maximal motor unit discharge rate.

A sixfold elevated incidence of discharge doublet firing was noticed in the firing pattern of single motor units following ballistic-type resistance training (pre: 5.2%, post: 32.7% of all motor units) (Van Cutsem, Duchateau and Hainaut, 1998) (Figure 2.1.9). This adaptive change in motor neuron firing pattern takes advantage of the catch-like property of skeletal muscle. Specifically, the presence of discharge doublet firing with inter-spike intervals <5–10 ms (firing frequency > 100–200 Hz) is known to result in a marked increase in contractile force and RFD at the onset of contraction (Binder-MacLeod and Kesar, 2005; Burke, Rundomin and Zajack, 1976; Moritani and Yoshitake, 1998) (Figure 2.1.9).

Recurrent Renshaw inhibition of spinal motor neurons has been considered as a limiting factor for the maximal discharge rate, and is thought to have a regulating influence on the reciprocal Ia-inhibitory pathway (Hultborn and Pierrot-Deseilligny, 1979). Animal experiments show that Renshaw cells receive several types of supraspinal synaptic input that can enhance as well as depress the recurrent pathway (Hultborn, Lindström and Wigström, 1979; Pierrot-Deseilligny and Morin, 1980). Compared with tonic steady-force contractions, Renshaw cell activity is more inhibited during maximal phasic muscle contractions, resulting in reduced recurrent inhibition in the latter condition (Pierrot-Deseilligny and Morin, 1980). Consequently, the use of explosive-type resistance training (i.e. contractions involving high RFD) may be optimal for evoking changes in the maximum firing rate of spinal motor neurons. This effect may be restricted to large proximal muscle groups, since recurrent inhibition appears to be absent in the smaller distal muscles of the hands and feet.

Resistance training can lead to tendon hypertrophy and increased muscle tendon stiffness (Arampatzis, Karamanidis and Albracht, 2007; Kongsgaard *et al.*, 2007). The training-induced increase in tendon stiffness is likely to also contribute to the observed rise in RFD, since contractile RFD is positively affected by the stiffness of the series-elastic force-transmitting structures (Bojsen-Møller *et al.*, 2005; Wilkie, 1950).

There are several important functional consequences of increasing RFD by means of strength training; the training-induced rise in RFD allows for an enhanced acceleration of movement, elevated limb speed during short-lasting movements, and increased muscle force and power during fast movements. Notably, the training-induced rise in contractile RFD is not only important to the athlete but also for elderly individuals when walking at high horizontal speed (crossing a busy street, for example) (Suetta *et al.*, 2004) and to ensure an optimal postural balance (Izquierdo *et al.*, 1999b).

2.1.3.3 Maximal eccentric muscle contraction: changes in neural factors with strength training

Eccentric muscle contractions involve exertion of muscle fibre force during simultaneous muscle lengthening. For untrained individuals, the shape of the contractile force–velocity relationship *in vivo* deviates markedly during maximal eccentric contraction from that obtained in isolated muscle (Figure 2.1.10). In contrast, strength-trained individuals appear to be able to produce very high levels of maximal eccentric muscle force compared to untrained subjects (see Figure 2.1.1). High levels

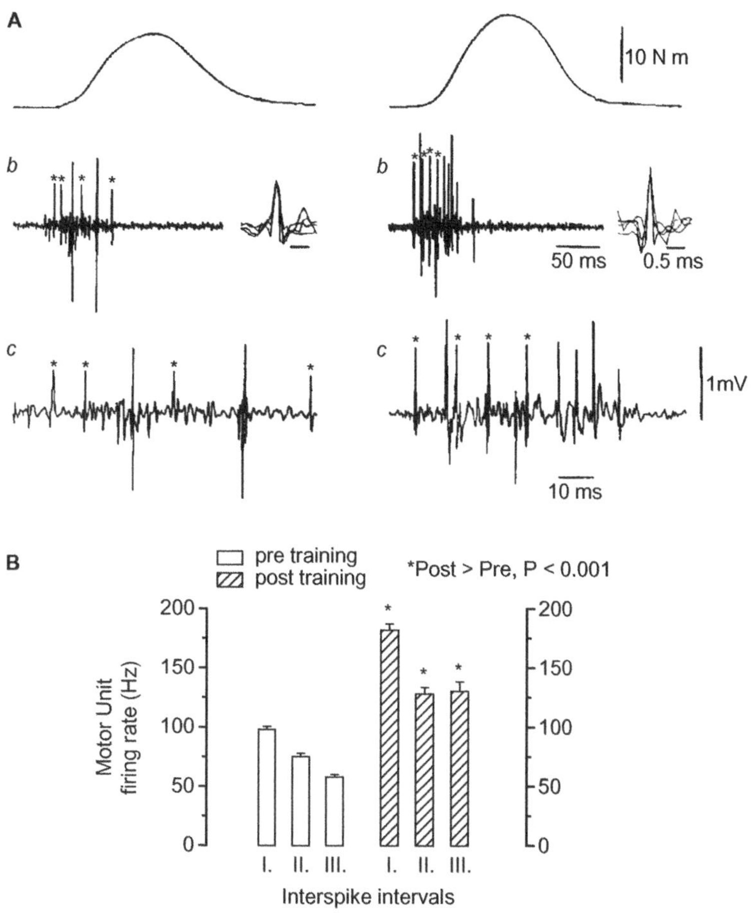

Figure 2.1.8 (a) Motor unit firing patterns recorded in the ankle dorsiflexor muscles (TA) during maximal ballistic dynamic contractions (40% MVC load) before (left-side graphs) and after (right-side graphs) 12 weeks of ballistic-type strength training. Note that shorter inter-spike intervals and a higher sustained firing rate were observed following training. From Van Cutsem, Duchateau and Hainaut (1998). (b) Motor unit firing rates measured at the onset of contraction before (open bars) and after (hatched bars) ballistic-type strength training. Note the marked increase (~100%) in motor unit firing frequency. Data reported by Van Cutsem, Duchateau and Hainaut (1998)

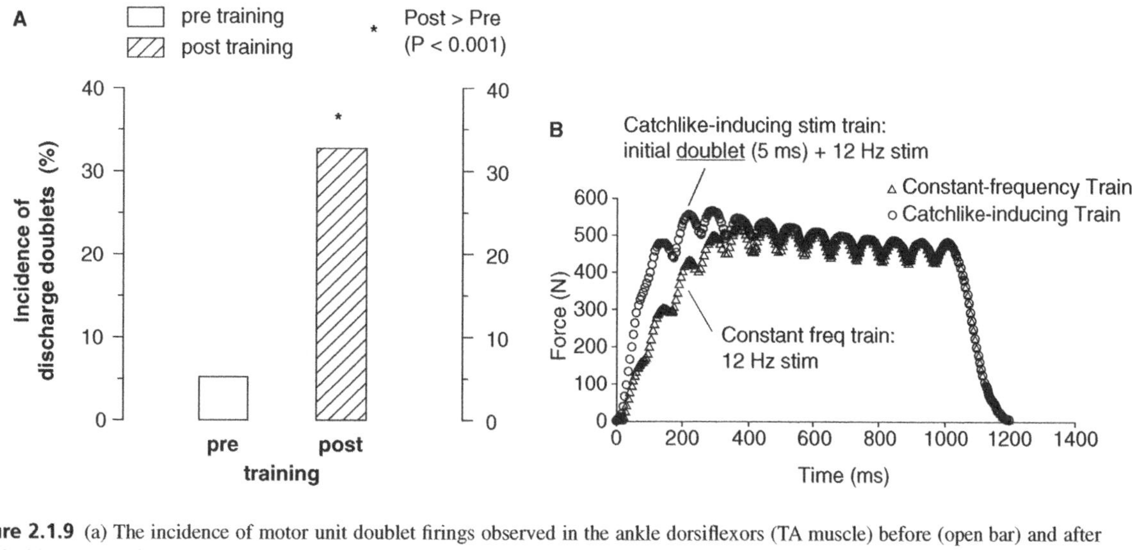

Figure 2.1.9 (a) The incidence of motor unit doublet firings observed in the ankle dorsiflexors (TA muscle) before (open bar) and after (hatched bar) 12 weeks of ballistic-type strength training. Note that doublet firing was sixfold elevated after the period of training. Data reported by Van Cutsem, Duchateau and Hainaut (1998). (b) Effect of doublet firing on contractile RFD (quadriceps femoris muscle). Addition of an initial doublet (5 ms inter-spike interval) to an innervation pattern of constant frequency stimulation (12 Hz) results in a highly elevated RFD during the initial phase of rising muscle force. A similar albeit less marked trend can be observed during tetanic stimulation. Reproduced from Binder-Macleod & Kesar. 2005

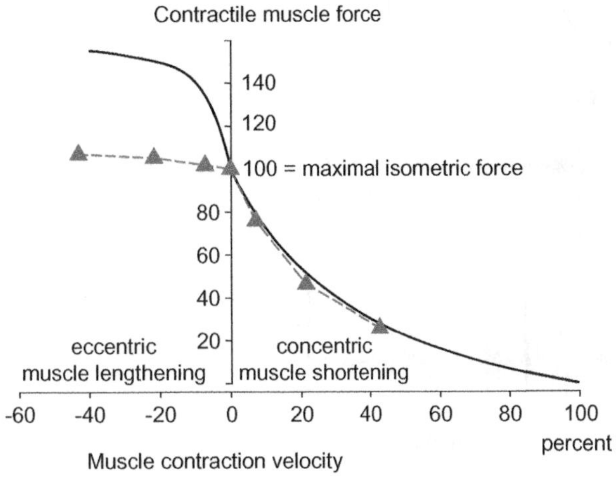

Figure 2.1.10 The force–velocity relationship obtained for isolated muscle when activated by electrical stimulation (full line; Edman, 1988; Hill, 1938; Katz, 1939) and when measured *in vivo* during maximal voluntary contraction efforts in physically fit human subjects using isokinetic dynamometry (triangles, broken line; Westing, Seger and Thorstensson, 1990). Positive velocities denote muscle shortening (concentric contraction), negative velocities denote muscle lengthening (eccentric contractions), and zero velocity denotes isometric muscle contraction. All force values are expressed relative to maximal isometric force. Note that during eccentric muscle contraction there is a marked deviation between *in vivo* and *in vitro* muscle force capacity

of eccentric muscle strength are required in many types of sports and exercise, as this provides an enhanced capacity to decelerate movements in very short time and thereby perform fast stretch–shortening cycle (SSC) actions (e.g. rapid jumping), while also allowing rapid shifts in movement direction (e.g. fast side-cutting movements). Further, high eccentric strength in antagonist muscles provides enhanced capacity to decelerate and halt movements at the end-ROM to protect ligaments (e.g. ACL) and joint capsule structures (Aagaard *et al.*, 1998). High levels of eccentric antagonist muscle strength also allow a more rapid limb deceleration during fast ballistic movements, which yields more time for limb acceleration and hence allows a higher movement speed to be reached (Jaric *et al.*, 1995).

It has been suggested that a neural regulatory mechanism that limits the recruitment and/or discharge rate of motor units exists during maximal voluntary eccentric muscle contraction (Aagaard *et al.*, 2000a). In support of this notion, the neuromuscular activity (EMG amplitude) recorded during maximal eccentric contractions appears to be markedly reduced compared to concentric contractions (Aagaard *et al.*, 2000a; Amiridis *et al.*, 1996; Andersen *et al.*, 2005; Duclay and Martin, 2005; Duclay *et al.*, 2008; Kellis and Baltzopoulos, 1998; Komi *et al.*, 2000; Westing, Cresswell and Thorstensson, 1991) (Figure 2.1.11). Superimposed muscle-twitch recordings have indicated that central activation is substantially reduced during maximal eccentric muscle contraction in untrained individuals (Babault *et al.*, 2001; Webber and Kriellaars, 1997). In further support of an inhibitory mechanism, reduced evoked spinal motor neuron responses (H-reflex amplitude; see Section 2.1.3.4) have been observed during maximal eccentric muscle

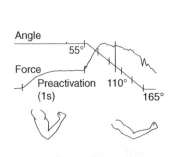

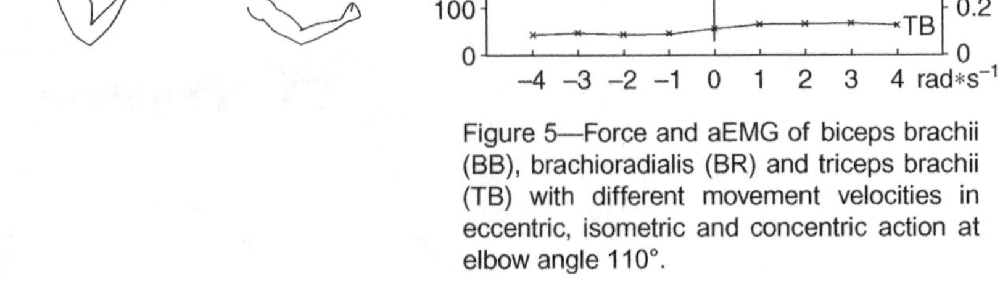

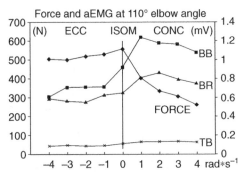

Figure 2.1.11 Right panel: maximal voluntary muscle strength (diamond symbols, left vertical axis) and neuromuscular activity (EMG amplitude, right vertical axis) for the elbow flexors (BB = biceps brachii, BR = brachioradialis) and the antagonist elbow extensors (TB = triceps brachii) during concentric (positive velocities), isometric (zero velocity), and eccentric (negative velocities) contraction performed in an isokinetic dynamometer. Left panel: example of data recording during eccentric muscle contraction. Data from Komi *et al.* (2000)

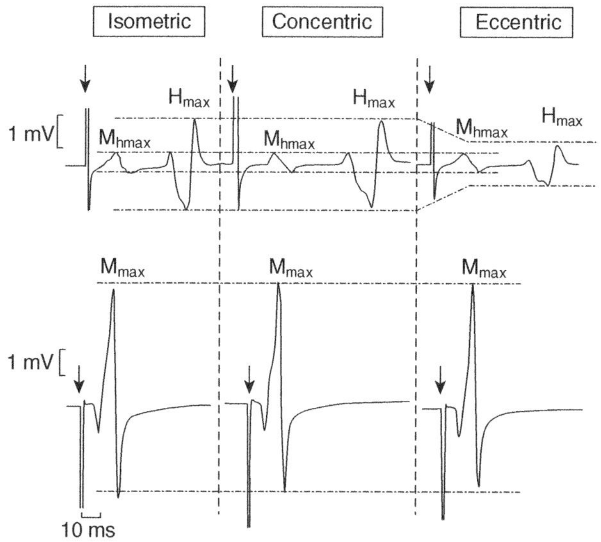

Figure 2.1.12 H-reflex responses recorded in the soleus muscle during isometric, concentric, and eccentric plantarflexor contractions of maximal voluntary effort (top panel). Note the depression in H-reflex amplitude during maximal eccentric contraction, suggesting reduced spinal motor neuron excitability and/or increased presynaptic or postsynaptic inhibition. The finding that maximal M-wave (M_{max}) remained unchanged across contraction modes (bottom panel) verifies that the depressed H-reflex response during eccentric contraction was not a recording artefact. Data from Duclay *et al.* (2008)

contraction in untrained individuals (Duclay *et al.*, 2008) (Figure 2.1.12).

Importantly, the apparent suppression in motor neuron activation during maximal eccentric contraction can be downregulated or removed by use of heavy-resistance strength training. Thus, the observed suppression in EMG amplitude during maximal eccentric contraction is partially abolished in parallel with a gain in maximal eccentric muscle strength after periods of heavy-resistance strength training, in turn resulting in a marked increase in maximal eccentric muscle strength (Aagaard *et al.*, 2000a; Andersen *et al.*, 2005) (Figure 2.1.13a). The gain in maximal eccentric muscle strength induced by heavy-resistance strength appears to be strongly associated with the corresponding increase in neuromuscular activity, indicating the pivotal importance of neural adaptation in allowing this change to occur (Figure 2.1.13b).

The specific neural pathways responsible for the suppression in neuromuscular activity during maximal eccentric contraction remain unidentified. During maximal voluntary muscle contraction, efferent motor-neuronal output is influenced by central descending pathways, afferent inflow from group Ib Golgi organ afferents, group Ia and II muscle spindle afferents, group III muscle afferents, and by recurrent Renshaw inhibition. All of these pathways may exhibit adaptive plasticity with training.

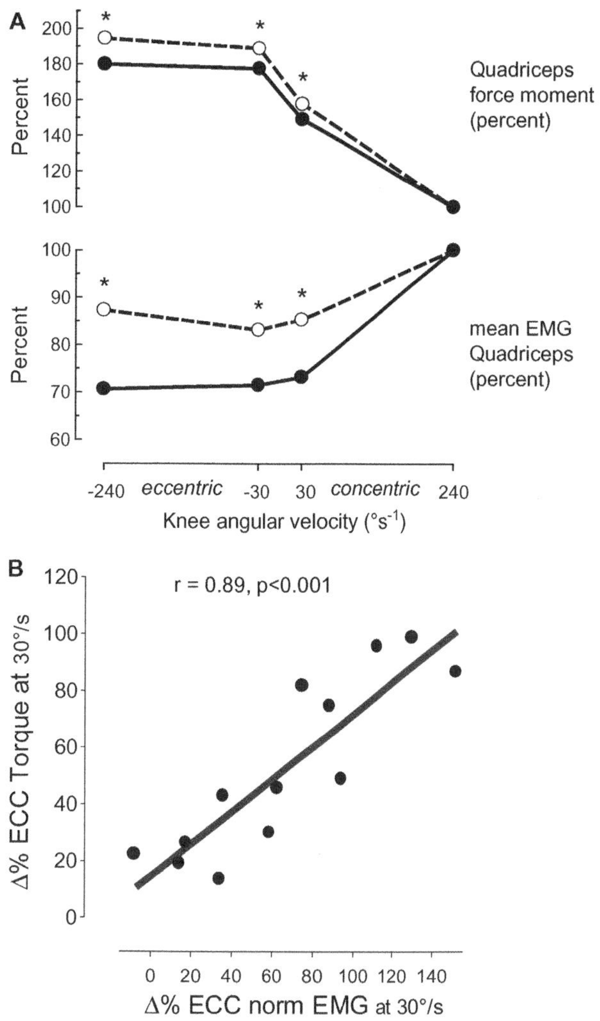

Figure 2.1.13 (a) Maximal contraction strength and neuromuscular activity during maximal eccentric (negative velocities) and concentric (positive velocities) muscle contractions before (solid lines) and after (broken lines) 14 weeks of strength training. All values are normalized relative to fast concentric contraction. Notice that the suppression in neuromuscular activity during eccentric and slow concentric contraction prior to training was reduced following training. Data from Aagaard *et al.* (2000a). (b) The training-induced gain in maximal eccentric muscle strength is strongly related to the corresponding rise in neuromuscular activity. Data from Andersen *et al.* (2005)

Interestingly, recent experiments showed that evoked spinal V-wave responses (see Section 2.1.3.4) increased during maximal eccentric plantarflexor contraction following heavy-resistance (maximal eccentric) strength training (Duclay *et al.*, 2008), indicating that supraspinal and/or spinal neuronal pathways were altered with resistance training. One likely mechanism for the marked increase in eccentric muscle strength with resistance training could be a downregulation in spinal inhibitory interneuron activity mediated via Golgi organ Ib afferents. Furthermore, the H-reflex response appears to be markedly suppressed during both active and passive muscle lengthening compared to shortening (Duclay *et al.*, 2008; Pinniger *et al.*, 2000) (see Figure 2.1.12), suggesting the presence of presynaptic inhibition of Ia afferents during eccentric muscle contraction. A training-induced reduction in presynaptic inhibition of Ia afferents would therefore result in an elevated excitatory inflow to the spinal motor neuron pool during maximal eccentric muscle contraction, which would increase maximal eccentric muscle strength.

Important functional consequences emerge when maximal eccentric muscle strength is increased by means of resistance training. Thus, elevated eccentric muscle strength enables more rapid performance of deceleration and SSC movements, side-cutting, and jumping actions. Furthermore, training-induced gains in eccentric antagonist muscle are important to ligament (ACL) and joint protection during sports and exercise. Elevated maximal eccentric muscle strength also allows elderly individuals to perform daily events such as stair descent in a safer manner.

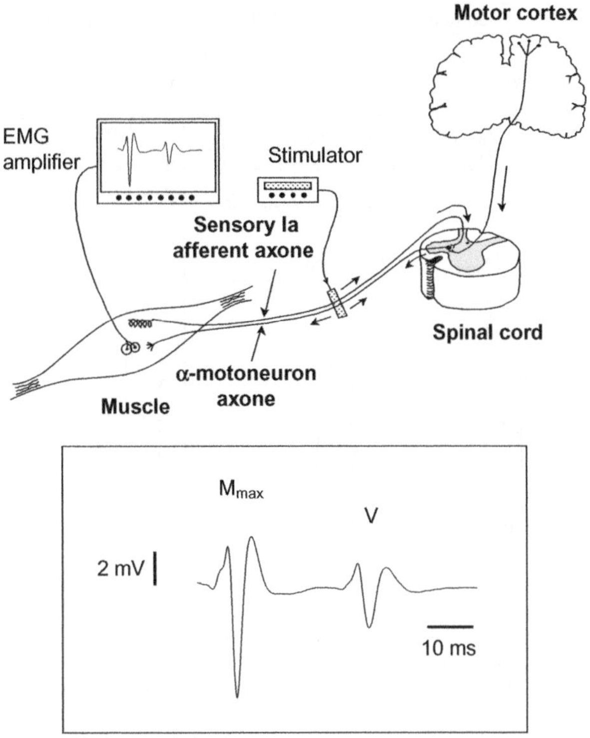

Figure 2.1.14 V-wave responses recorded in the soleus muscle by electrical stimulation of Ia afferent axons in the peripheral nerve (n. tibialis) during maximal voluntary muscle contraction. The peak-to-peak amplitude of the V-wave reflects the magnitude of central descending motor drive as well as the excitability state of spinal motor neurones, including presynaptic inhibition of Ia afferent synapses and postsynaptic inhibition. Adapted from Aagaard, 2002a

2.1.3.4 Evoked spinal motor neuron responses

The Hoffman (H) reflex can be used to examine training-induced changes in spinal circuitry function at rest and during active contraction, as the size of the evoked H-reflex amplitude reflects the level of spinal motor neuron excitability and the magnitude of presynaptic inhibition of muscle-spindle Ia afferents (Nielsen and Kagamihara, 1992; Schieppati, 1987). The V-wave is a variant of the H-reflex that can be elicited when supramaximal stimulation of the peripheral nerve is superimposed on to voluntary muscle contraction (Aagaard *et al.*, 2002a; Hultborn and Pierrot-Deseilligny, 1979; Upton, McComas and Sica, 1971) (Figure 2.1.14). When obtained during maximal muscle contraction, the alteration in H-reflex and V-wave amplitude may be used to quantify the training-induced rise in efferent output from spinal motor neurons (V-wave) and motor neuron excitability and/or presynaptic inhibition (H-reflex, V-wave) (Aagaard *et al.*, 2002a; Del Balso and Cafarelli, 2007; Sale *et al.*, 1983). Importantly, the evoked motor neuron responses (peak-to-peak amplitude or area) are normalized to the maximal M-wave amplitude in order to eliminate or diminish the measuring bias normally associated with repeated surface EMG recording.

Elevated V-wave and H-reflex amplitudes have been observed during maximal muscle contraction following months of resistance training (Aagaard *et al.*, 2002a; Duclay *et al.*, 2008; Sale *et al.*, 1983) (Figure 2.1.15). A rise in V-wave amplitude also was found after short-term (3–4 weeks) strength training (Del Balso and Cafarelli, 2007; Fimland *et al.*, 2009a, 2009b, 2010), as well as following eccentric strength training (Duclay *et al.*, 2008). While the peak-to-peak amplitude of the H-reflex recorded during MVC increased following 14 weeks of dynamic strength training (Aagaard *et al.*, 2002a), elevated H-reflex responses were recorded at submaximal force levels after shorter periods (3–5 weeks) of strength training (Holtermann *et al.*, 2007; Lagerquist, Zehr and Docherty, 2006). Altogether these findings suggest that the neural adaptation to strength training can be elicited within a relatively short time frame of training, and additionally that neural adaptation may take place throughout the continued time course of training (see Figure 2.1.16).

When post-training V-waves were recorded at muscle force levels corresponding to pre-level MVC, no change in the evoked V-wave response could be detected, whereas a rise in V-wave amplitude was found at post-training MVC (Del Balso and Cafarelli, 2007) (Figure 2.1.16). This finding clearly illustrates and underlines that a close cause–response relationship exists between the change in efferent motor neuron output and the gain in maximal muscle force induced by resistance training.

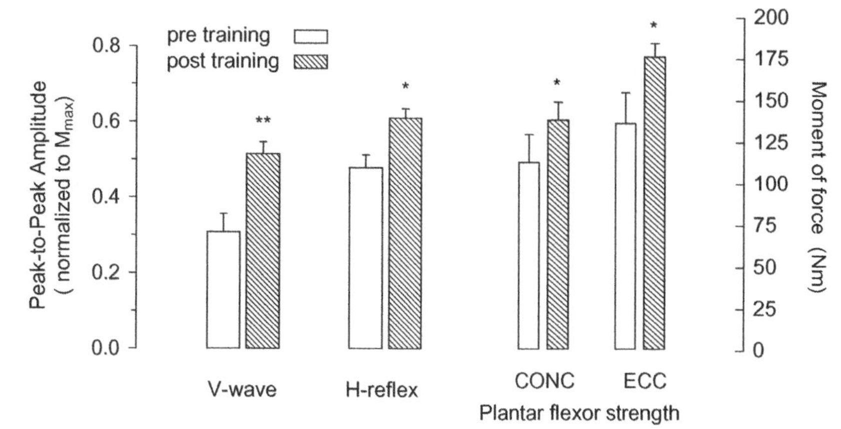

Figure 2.1.15 Changes in spinal evoked H-reflex and W-wave responses, and and in maximal muscle strength elicited by 14 weeks of strength training. Enhanced H-reflex and V-wave responses were observed following the period of training, indicating that neural adaptive changes occurred at both spinal and supraspinal levels. Pre > post: * $P < 0.05$, ** $P < 0.01$. Adapted from Aagaard, 2002a

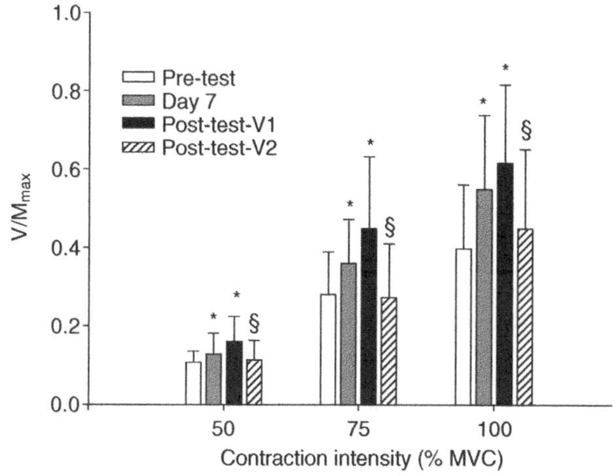

Figure 2.1.16 Evoked spinal V-wave responses recorded in the soleus muscle during isometric plantarflexions at 50, 75, and 100% MVC. * Greater than pre-test ($P < 0.05$). Post-test V2: V-wave response recorded post-training at pre-MVC. Reproduced from Del Balso & Cafarelli. 2007. J. Appl. Physiol. © American Physiological Society

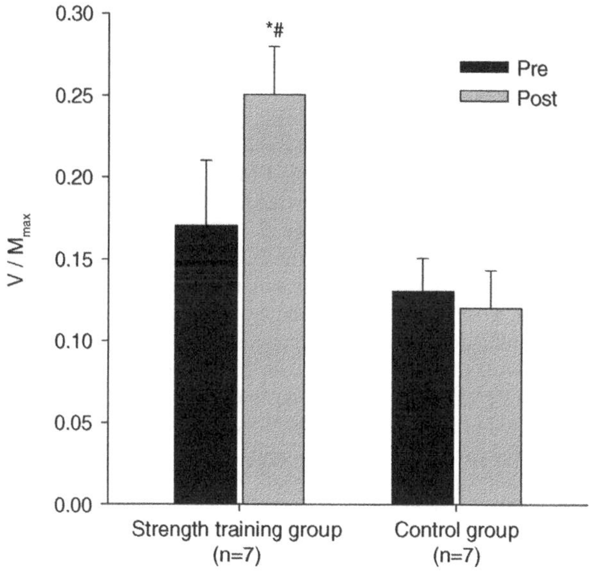

Figure 2.1.17 Evoked spinal V-wave responses recorded in the soleus muscle during maximal isometric plantarflexion in multiple sclerosis patients before (black bars) and after (grey bars) four weeks of heavy-resistance strength training. Pre > post: * $P < 0.05$, # different from controls $P < 0.05$. Reproduced from Fimland. 2010, Eur. J. App. Physiol. © American Physiological Society

Recent data have shown that strength training can lead to enhanced spinal circuitry function in patients with neuropathy such as multiple sclerosis (MS) (Fimland *et al.*, 2010). MS is a neuro-degenerative disease affecting the nervous system, leading to loss of myelinization and destruction of peripheral axons, and consequently MS patients demonstrate 30–70% reduced muscle strength compared to healthy control subjects (Ng *et al.*, 2004; Rice *et al.*, 1992). Elevated V-wave responses and increased MVC were recently observed in middle-aged MS patients following 15 sessions of heavy-resistance strength training, suggesting that strength training may provide an effective tool for preventing or retarding the gradual decline in motor function typically seen with this condition (Figure 2.1.17).

The training-induced rise in V-wave and H-reflex amplitude indicates the presence of enhanced neural drive in descending corticospinal pathways, elevated motor neuron excitability, reduced presynaptic inhibition of Ia afferents, and/or reduced postsynaptic motor neuron inhibition (Aagaard, 2003). When obtained in resting conditions, however, the H-reflex response appear to remain unaltered by resistance training (Aagaard *et al.*, 2002a; Del Balso and Cafarelli, 2007; Duclay *et al.*, 2008; Scaglioni *et al.*, 2002), suggesting that the training induced change in spinal circuitry function does not involve chronic changes in neuroanatomy (e.g. increased number of Ia afferent terminals, etc.), since these would be expected to cause changes in resting H-reflex amplitude.

2.1.3.5 Excitability in descending corticospinal pathways

In recent years TMS and TES have been increasingly used to assess training-induced changes in the transmission efficacy in descending motor pathways from the cerebral cortex to the muscle fibres, including the spinal cord. Via analysis of the evoked EMG responses recorded in the target muscle (so-called motor-evoked potentials, MEPs) (Figure 2.1.18), information can be obtained about the level of corticospinal excitability, cortical recruitment threshold, recruitment gain, as well as intra-cortical inhibition and facilitation.

Somewhat equivocal results have been obtained with these newly-developed techniques. MEP_{max} and the input–output slope obtained by TMS in the biceps brachii muscle during low-level contraction (5% of max EMG) remained unchanged following short-term (4 weeks) strength training (for definition of TMS parameters see Figure 2.1.18b), whereas marked increases were found in isometric and dynamic MVC (Jensen, Marstrand and Nielsen, 2005). In a more recent study, however, maximal MEP amplitude evoked by TMS during low-force tonic contraction (10% MVC) increased by 32% in response to four weeks of isometric strength training of the ankle dorsiflexors (TA muscle), suggesting that the period of strength training led to an increased excitability in descending corticospinal pathways (Griffin and Cafarelli, 2007). In addition, task- and training-specific enhancements in MEP size and MEP recruitment were recently reported following four weeks of ballistic-type strength training for the ankle dorsiflexors and plantarflexors (Beck *et al.*, 2007). MEP recruitment evaluates the input–output properties of the motor system, including excitatory synaptic transmission efficacy in the motor cortex and density of corticospinal projections to the contracting muscles. The authors concluded that supraspinal sites were involved in the adaptation to ballistic ('explosive-type') strength training, and that such training is capable of altering corticospinal facilitation, possibly by changing recruitment gain (Beck *et al.*, 2007).

In contrast, strength training has also been reported to decrease MEP_{max} and the slope of the input–output relation at rest (Lundbye Jensen, Marstrand and Nielsen, 2005), and to result in a reduced ratio between evoked MEP size and muscle torque production or level of background EMG during tonic contraction of moderate intensity (40–60% of MVC), with no changes observed at lower contraction intensities (Carroll, Riek and Carson, 2002). Similar findings emerged for TES-induced MEPs (Carroll, Riek and Carson, 2002). TES (transcranial electrical stimulation) recruits corticospinal neurones directly by electrical depolarization of their axones at deeper sites in the brain, whereas TMS (transcranial magnetic stimulation) recruits corticospinal cells via lateral transsynaptic excitation in the cortical layer. The TES-induced MEP response is therefore much less influenced by the excitability state of the motor cortex than that produced using TMS, and the above findings of similar changes in TMS- and TES-evoked MEP properties consequently led to the suggestion that subcortical rather than cortical sites might be responsible for the adaptive plasticity with strength training (Carroll, Riek and Carson, 2002).

These contradictory findings, manifested by enhanced versus reduced MEPs respectively, on the influence of strength training on cerebral motor cortex function may be related to the different muscles examined (i.e. hand versus lower leg), differences in the motor tasks used during training and testing including differences in contraction intensity, and possibly the specific TMS/TES-stimulation paradigm used.

2.1.3.6 Antagonist muscle coactivation

Coactivation of antagonist muscles is an inherent element in normal human movement (Aagaard *et al.*, 2000b; DeLuca and Mambrito, 1987; Smith, 1981). The presence of antagonist-muscle coactivation protects ligaments from excessive strain at the end-range of joint motion (Draganich and Vahey, 1990; More *et al.*, 1993), ensures a homogenous distribution of compression forces over the articular surfaces of the joint (Baratta *et al.*, 1988), and results in increased joint and limb stiffness (Milner and Cloutier, 1993). In addition, an adequately high level of antagonist muscle strength is important for the execution of fast, ballistic limb movements. Thus, high eccentric antagonist strength allows for a shortened phase of limb deceleration at the end of movement, thereby increasing the time available for limb acceleration, resulting in an increased maximal movement velocity (Jaric *et al.*, 1995). Elevated agonist–antagonist muscle coactivation is typically observed during unstable motor tasks or during the anticipation of compensatory muscle forces (DeLuca and Mambrito, 1987). Increased antagonist muscle coactivation can be observed in elderly individuals during stair walking (Hortobagyi and DeVita, 2000; Larsen *et al.*, 2008), and elevated levels of antagonist muscle coactivation have been implicated as a risk factor for exercise-induced rib stress fracture in elite rowers (Vinther *et al.*, 2006). Conversely, a rise in hamstring antagonist coactivation during active knee extension results in reduced strain and stress forces in the anterior cruciate ligament (ACL), potentially protecting against ACL injury (Aagaard *et al.*, 2000b; Draganich and Vahey, 1990; More *et al.*, 1993). Increasing attention has been directed to the aspect of medial versus lateral antagonist muscle coactivation, since female elite soccer players with reduced medial hamstring muscle coactivation during rapid, forceful side-cutting manoeuvres were more

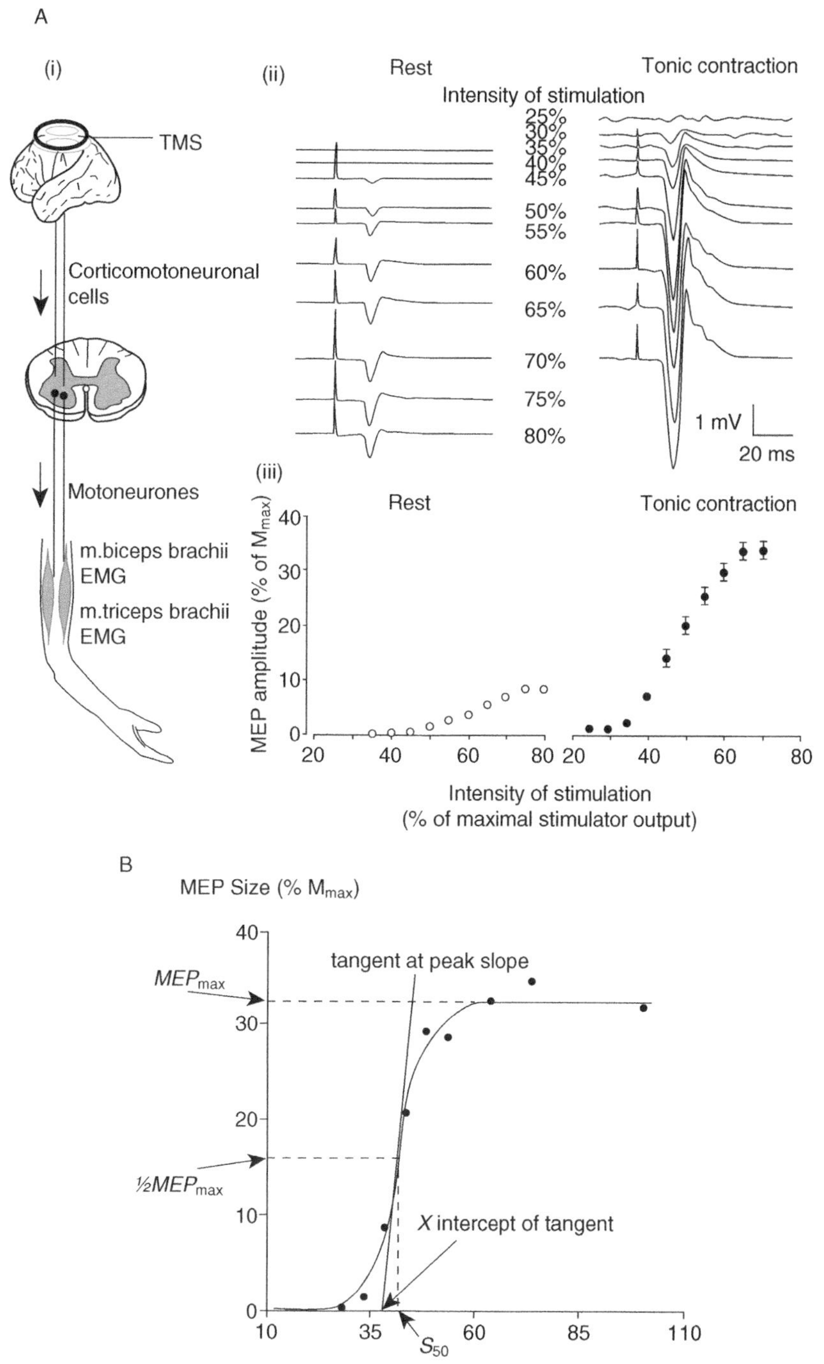

Figure 2.1.18 (a) Transcranical magnetic stimulation (TMS) (i) gives rise to evoked motor potential (MEP) in the muscle (ii), where steeper input–output relationships are observed during active muscle contraction compared to rest (iii). From Jensen, Marstrand and Nielsen (2005). (b) Analysed features of the MEP input–output relationship. Reproduced from Carroll, 2002. J. Physiol. © John Wiley & Sons, Ltd.

likely to sustain serious ACL injury compared to players demonstrating a more uniform medial-to-lateral coactivation pattern (Zebis *et al.*, 2009).

It remains to be settled whether strength training per se leads to altered patterns of antagonist coactivation. Decreased antagonist coactivation may seem desirable since it leads to increased net joint moment (agonist muscle moment minus antagonist muscle moment). On the other hand, a decrease in antagonist muscle coactivation might lead to a reduced stabilization of the joint, as discussed above. Following strength training, the magnitude of antagonist muscle coactivation has been reported to decrease (Carolan and Cafarelli, 1992; Häkkinen *et al.*, 1998a, 2000, 2001), increase (Baratta *et al.*, 1988), or remain unchanged (Aagaard *et al.*, 2000b, 2001; Colson *et al.*, 1999; Häkkinen *et al.*, 1998a, 2000, 2001; Hortobagyi *et al.*, 1996b; Reeves, Maganaris and Narici, 2005; Valkeinen *et al.*, 2000). During maximal static knee extension (isometric quadriceps contraction), older individuals appear to demonstrate elevated coactivation of the antagonist hamstring muscles than younger subjects (Häkkinen *et al.*, 1998a, 2000; Izquierdo *et al.*, 1999a; Macaluso *et al.*, 2002). In elderly individuals, antagonist hamstring coactivation was found to decrease in response to six months of heavy-resistance strength training, reaching a level similar to that observed in middle-aged subjects (20–25% of maximal agonist activity), which remained unaltered with training (Häkkinen *et al.*, 1998a, 2000). Antagonist muscle coactivation is markedly elevated when uncertainty exists in the motor task (DeLuca and Mambrito, 1987). Consequently, elderly subjects, and to a lesser extent young individuals, may occasionally demonstrate very high levels of coactivation (30–45% of maximal agonist activity) during the initial pre-training assessment of MVC. This may, at least in part, explain why some studies have

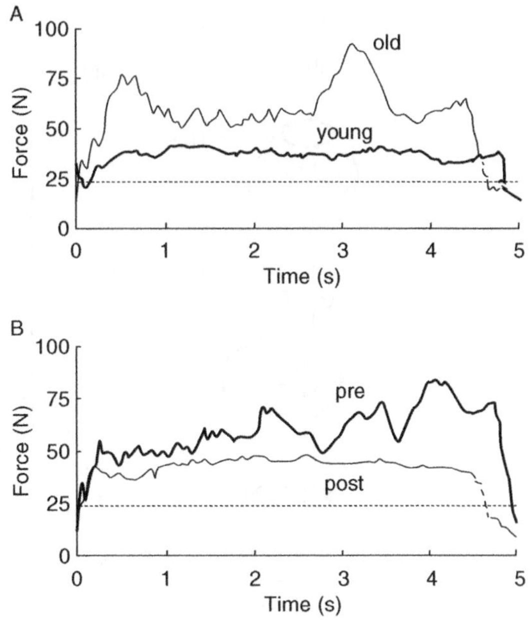

Figure 2.1.19 Force steadiness and force accuracy. Tracking of 25-N target force during 5 sec slow (15 °/s) eccentric quadriceps contraction (post 10 familiarization trials). (a) Typical force tracings obtained in old and young subjects demonstrating increased SD(force) and reduced force accuracy in old versus young individuals. (b) Reduced SD(force) and improved force accuracy in old subject after a period of strength training. Reproduced from Hortobagyi *et al.*, 2001

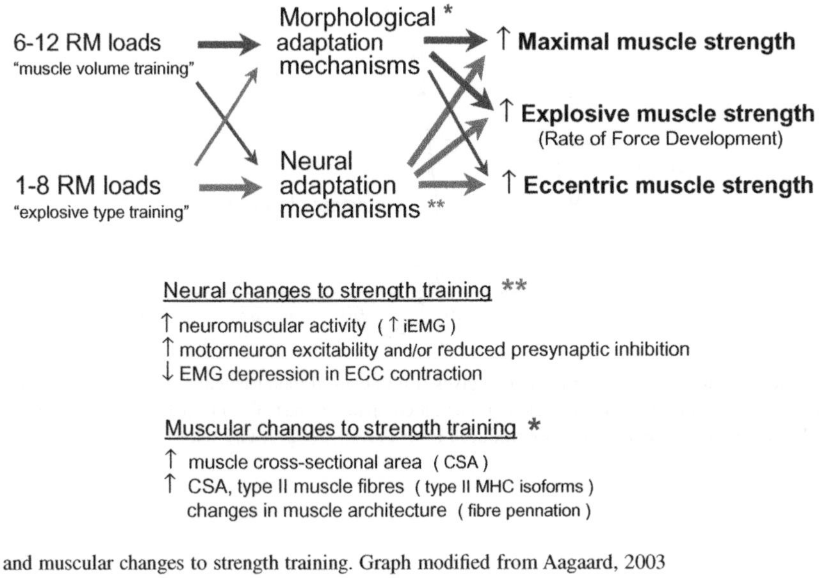

Figure 2.1.20 Neural and muscular changes to strength training. Graph modified from Aagaard, 2003

consistently observed a decrease in antagonist muscle coactivation following strength training when the majority have not been able to confirm this adaptation.

2.1.3.7 Force steadiness, fine motor control

The ability for fine motor control can be evaluated by analysis of the steadiness of constant-force muscle contraction. Steadiness can be measured as the variance (SD) in the fluctuation in muscle force while a subject tries to sustain a certain target force (i.e. 10% of max force).

Notably, greater force error and less steady muscle force production (elevated SD) can be seen during submaximal constant-force contractions in elderly compared to young subjects (Hortobagyi *et al.*, 2001; Tracy and Enoka, 2002) (Figure 2.1.19). Substantial reductions in force error and force variability (i.e. diminished SD(force)) have been observed following strength training in ageing individuals (Hortobagyi *et al.*, 2001; Tracy, Byrnes and Enoka, 2004; Tracy and Enoka, 2006) (Figure 2.1.19). These findings demonstrate that strength training can lead to improved force steadiness and fine motor control in the elderly.

2.1.4 CONCLUSION

Different lines of evidence have been presented in this chapter to demonstrate that strength training elicits a substantial variety of beneficial alterations in nervous system function. Specifically, a huge range of training-induced plasticity appears to exist at spinal, supraspinal, and cortical levels. The neural adaptation to strength training comprises gains in maximal muscle strength, rapid force capacity (RFD), and maximal eccentric muscle strength (Figure 2.1.20). Most importantly, the adaptive improvement in neuronal function elicited by resistance training can occur both in untrained individuals, including frail elderly individuals and patients, and in highly trained athletes. In all individuals the training-induced enhancement in neuromuscular capacity leads to improved mechanical muscle function, which in turn results in an improved functional performance in various activities of daily living, including sports and physical exercise.

References

Aagaard P. (2003) Training-induced changes in neural function. *Exerc Sports Sci Rev*, **31** (2), 61–67.

Aagaard P., Simonsen E.B., Trolle M. *et al.* (1994) Effects of different strength training regimes on moment and power generation during dynamic knee extension. *Eur J Appl Physiol*, **69**, 382–386.

Aagaard P., Simonsen E.B., Trolle M. *et al.* (1996) Specificity of training velocity and training load on gains in isokinetic knee joint strength. *Acta Physiol Scand*, **156**, 123–129.

Aagaard P., Simonsen E.B., Magnusson P. *et al.* (1998) A new concept for isokinetic hamstring/quadriceps strength ratio. *Am J Sports Med*, **26**, 231–237.

Aagaard P., Simonsen E.B., Andersen J.L. *et al.* (2000a) Neural inhibition during maximal eccentric and concentric quadriceps contraction: effects of resistance training. *J Appl Physiol*, **89**, 2249–2257.

Aagaard P., Simonsen E.B., Andersen J.L. *et al.* (2000b) Antagonist muscle coactivation during isokinetic knee extension. *Scand J Med Sci Sports*, **10**, 58–67.

Aagaard P., Simonsen E.B., Andersen J.L. *et al.* (2001) Changes in antagonist muscle coactivation pattern during maximal knee extension following heavy-resistance strength training. *J Physiol*, **531P**, 62P (abstract).

Aagaard P., Simonsen E.B., Andersen J.L. *et al.* (2002a) Neural adaptation to resistance training: changes in evoked V-wave and H-reflex responses. *J Appl Physiol*, **92**, 2309–2318.

Aagaard P., Simonsen E.B., Andersen J.L. *et al.* (2002b) Increased rate of force development and neural drive of human skeletal muscle following resistance training. *J Appl Physiol*, **93**, 1318–1326.

Amiridis I.G., Martin A., Morlon B. *et al.* (1996) Co-activation and tension-regulating phenomena during isokinetic knee extension in sedentary and highly skilled humans. *Eur J Appl Physiol*, **73**, 149–156.

Andersen L.L., Andersen J.L., Magnusson S.P. and Aagaard P. (2005) Neuromuscular adaptations to detraining following resistance training in previously untrained subjects. *Eur J Appl Physiol*, **93**, 511–518.

Arampatzis A., Karamanidis K. and Albracht K. (2007) Adaptational responses of the human Achilles tendon by modulation of the applied cyclic strain magnitude. *J Exp Biol*, **210**, 2743–2753.

Babault N., Pousson M., Ballay Y. and Van Hoecke J. (2001) Activation of human quadriceps femoris during isometric, concentric, and eccentric contractions. *J Appl Physiol*, **91**, 2628–2634.

Baratta R., Solomonow M., Zhou B.H. *et al.* (1988) Muscular coactivation: the role of the antagonist musculature in maintaining knee stability. *Am J Sports Med*, **16**, 83–87.

Barry B.K., Warman G.E. and Carson R.G. (2005) Age-related differences in rapid muscle activation after rate of force development training of the elbow flexors. *Exp Brain Res*, **162**, 122–132.

Beck S., Taube W., Gruber M. *et al.* (2007) Task-specific changes in motor evoked potentials of lower limb muscles after different training interventions. *Brain Res*, **1179**, 51–60.

Behm D.G. and Sale D.G. (1993) Intended rather than actual movement velocity determines velocity-specific training response. *J Appl Physiol*, **74**, 359–368.

Binder-Macleod S. and Kesar T. (2005) Catchlike property of skeletal muscle: recent findings and clinical implications. *Muscle Nerve*, **31**, 681–693.

Blazevich A.J., Horne S., Cannavan D. *et al.* (2008) Effect of contraction mode of slow-speed resistance training on the maximum rate of force development in the human quadriceps. *Muscle Nerve*, **38**, 1133–1146.

Bojsen-Møller J., Magnusson S.P., Rasmussen L. *et al.* (2005) Muscle performance during maximal isometric and dynamic contractions is influenced by the stiffness of the tendinous structures. *J Appl Physiol*, **99**, 986–994.

Burke R.E., Rundomin P. and Zajack F.E. (1976) The effect of activation history on tension production by individual muscle units. *Brain Res*, **18**, 515–529.

Caiozzo V.J., Perrine J.J. and Edgerton V.R. (1981) Training induced alterations of the in vivo force-velocity relationship in human muscle. *J Appl Physiol*, **51**, 750–754.

Cannon G. and Cafarelli E. (1987) Neuromuscular adaptations to strength training. *J Appl Physiol*, **63**, 2396–2402.

Carolan B. and Cafarelli E. (1992) Adaptations in co-activation in response to isometric resistance training. *J Appl Physiol*, **73**, 911–917.

Carroll T.J., Riek S. and Carson R.G. (2002) The sites of neural adaptation induced by resistance training in humans. *J Physiol*, **544** (2), 641–652.

Caserotti P., Aagaard P., Larsen J.B. and Puggaard P. (2008) Explosive heavy-resistance training in old and very old adults:changes in rapid muscle force, strength and power. *Scand J Med Sci Sports*, **18**, 773–782.

Colliander E.B. and Tesch P.A. (1990) Effects of eccentric and concentric muscle actions in resistance training. *Acta Physiol Scand*, **140**, 31–39.

Colson S., Pousson M., Martin A. and Van Hoecke J. (1999) Isokinetic elbow flexion and coactivation following eccentric training. *J Electromyogr Kinesiol*, **9**, 13–20.

Costill D.L., Coyle E.F., Fink W.F. *et al.* (1979) Adaptations in skeletal muscle following strength training. *J Appl Physiol*, **46**, 96–99.

Coyle E.F., Feiring D.C., Rotkis T.C. *et al.* (1981) Specificity of power improvements through slow and fast isokinetic training. *J Appl Physiol*, **51**, 1437–1442.

Day S.J. and Hulliger M. (2001) Experimental simulation of cat electromyogram: evidence for algebraic summation of motor-unit action-potential trains. *J Neurophysiol*, **86**, 2144–2158.

De Haan A. (1998) The influence of stimulation frequency on force-velocity characteristics of in situ rat medial gastrocnemius muscle. *Exp Physiol*, **83**, 77–84.

De Vos N.J., Singh N.A., Ross D.A. *et al.* (2005) Optimal load for increasing muscle power during explosive resistance training in older adults. *J Gerontol A Biol Sci Med Sci*, **60**, 638–647.

Del Balso C. and Cafarelli E. (2007) Adaptations in the activation of human skeletal muscle induced by short-term isometric resistance training. *J Appl Physiol*, **103**, 402–411.

DeLuca C.J. and Mambrito B. (1987) Voluntary control of motor units in human antagonist muscles: coactivation and reciprocal activation. *J Neurophysiol*, **58**, 525–542.

Draganich L.F. and Vahey J.W. (1990) An in vitro study of anterior cruciate ligament strain induced by quadriceps and hamstring forces. *J Orthop Res*, **8**, 57–63.

Duchateau J. and Hainaut K. (1984) Isometric or dynamic training: differential effects on mechanical properties of a human muscle. *J Appl Physiol*, **56**, 296–301.

Duchateau J., Semmler J.G. and Enoka R.M. (2006) Training adaptations in the behavior of human motor units. *J Appl Physiol*, **101**, 1766–1775.

Duclay J. and Martin A. (2005) Evoked H-reflex and V-wave responses during maximal isometric, concentric, and eccentric muscle contraction. *J Neurophysiol*, **94**, 3555–3562.

Duclay J., Martin A., Robbe A. and Pousson M. (2008) Spinal reflex plasticity during maximal dynamic contractions after eccentric training. *Med Sci Sports Exerc*, **40**, 722–734.

Duncan P.W., Chandler J.M., Cavanaugh D.K. *et al.* (1989) Mode and speed specificity of eccentric and concentric exercise training. *J Orthop Sports Phys Ther*, **11**, 70–75.

Edman K.A.P. (1988) Double-hyperbolic force-velocity relation in frog muscle fibres. *J Physiol*, **404**, 301–321.

Farina D., Merletti R. and Enoka R.M. (2004) The extraction of neural strategies from the surface EMG. *J Appl Physiol*, **96**, 1486–1495.

Fimland M.S., Helgerud J., Gruber M. *et al.* (2009a) Functional maximal strength training induces neural transfer to single-joint tasks. *Eur J Appl Physiol*, **107**, 21–29.

Fimland M.S., Helgerud J., Solstad G.M. *et al.* (2009b) Neural adaptations underlying cross-education after unilateral strength training. *Eur J Appl Physiol*, **107**, 723–730.

Fimland M.S., Helgerud J., Gruber M. *et al.* (2010) Enhanced neural drive after maximal strength training in multiple sclerosis patients. *Eur J Appl Physiol*, in press.

Folland J.P. and Williams A.G. (2007) The adaptations to strength training: morphological and neurological contributions to increased strength. *Sports Med*, **37**, 145–168.

Griffin L. and Cafarelli E. (2007) Transcranial magnetic stimulation during resistance training of the tibialis anterior muscle. *J Electromyogr Kinesiol*, **17**, 446–452.

Grimby L., Hannerz J. and Hedman B. (1981) The fatigue and voluntary discharge properties of single motor units in man. *J Physiol*, **316**, 545–554.

Häkkinen K. and Komi P.V. (1983) Electromyographic changes during strength training and detraining. *Med Sci Sports Exerc*, **15**, 455–460.

Häkkinen K. and Komi P.V. (1986) Training induced changes in neuromuscular performance under voluntary and reflex conditions. *Eur J Appl Physiol*, **55**, 147–155.

Häkkinen K., Alén M. and Komi P.V. (1985a) Changes in isometric force and relaxation time, EMG and muscle fibre characteristics of human skeletal muscle during training and detraining. *Acta Physiol Scand*, **125**, 573–585.

Häkkinen K., Komi P.V. and Alén M. (1985b) Effect of explosive type strength training on ismetric force- and relaxation-time, electromyographic and muscle fibre characteristics of leg extensor muscles. *Acta Physiol Scand*, **125**, 587–600.

Häkkinen K., Komi O.V., Alén M. and Kauhanen H. (1987) EMG, muscle fibre and force production characteristics during a 1 year training period in elite lifters. *Eur J Appl Physiol*, **56**, 419–427.

Häkkinen K., Kallinen M., Izquierdo M. *et al.* (1998a) Changes in agonist-antagonist EMG, muscle CSA, and force during strength training in middle-aged and older people. *J Appl Physiol*, **84**, 1341–1349.

Häkkinen K., Newton R.U., Gordon S.E. *et al.* (1998b) Changes in muscle morphology, electromyographic activity, and force production characteristics during progressive strenght training in young and older men. *J Gerontol A Biol Sci Med Sci*, **53**, B415–B423.

Häkkinen K., Alén M., Kallinen M. *et al.* (2000) Neuromuscular adaptation during prolonged strength training, detraining and re-strength-training in middle-aged and elderly people. *Eur J Appl Physiol*, **83**, 51–62.

Häkkinen K., Kraemer W.J., Newton R.U. and Alen M. (2001) Changes in electromyographic activity, muscle fibre and force production characteristics during heavy resistance/power strength training in middle-aged and older men and women. *Acta Physiol Scand*, **171**, 51–62.

Higbie E.J., Cureton K.J., Warren G.L. and Prior B.M. (1996) Effects of concentric and eccentric training on muscle strength, cross-sectional area and neural activation. *J Appl Physiol*, **81**, 2173–2181.

Hill A.V. (1938) The heat of shortening and the dynamic constants of muscle. *Proc R Soc Lond Ser B*, **126**, 136–195.

Holm L., Reitelseder S., Pedersen T.G. *et al.* (2008) Changes in muscle size and MHC composition in response to resistance exercise with heavy and light loading intensity. *J Appl Physiol*, **105**, 1454–1461.

Holtermann A., Roeleveld K., Engstrøm M. and Sand T. (2007) Enhanced H-reflex with resistance training is related to increased rate of force development. *Eur J Appl Physiol*, **101**, 301–312.

Hortobagyi T. and DeVita P. (2000) Muscle pre- and coactivity during downward stepping are associated with leg stiffness in aging. *J Electromyogr Kinesiol*, **10**, 117–126.

Hortobagyi T., Barrier J., Beard B. *et al.* (1996a) Greater initial adaptations to submaximal muscle lengthening than maximal shortening. *J Appl Physiol*, **81**, 1677–1682.

Hortobagyi T., Hill J.P., Houmard J.A. *et al.* (1996b) Adaptive responses to muscle lengthening and shortening in humans. *J Appl Physiol*, **80**, 765–772.

Hortobagyi T., Dempsey L., Fraser D. *et al.* (2000) Changes in muscle strength, muscle fibre size and myofibrillar gene expression after immobilization and retraining in humans. *J Physiol*, **524** (1), 293–304.

Hortobagyi T., Tunnel D., Moody J. *et al.* (2001) Low- or high-intensity strength training partially restores impaired quadriceps force accuracy and steadiness in aged adults. *J Gerontol Biol Sci*, **56A**, B38–B47.

Housh D.J. and Housh T.J. (1993) The effects of unilateral velocity-specific concentric strength training. *J Orthop Sports Phys Ther*, **17**, 252–256.

Hultborn H. and Pierrot-Deseilligny E. (1979) Changes in recurrent inhibition during voluntary soleus contractions in man studies by an H-reflex technique. *J Physiol*, **297**, 229–251.

Hultborn H., Lindström S. and Wigström H. (1979) On the function of recurrent inhibition in the spinal cord. *Exp Brain Res*, **37**, 399–403.

Izquierdo M., Ibañez J., Gorostiaga E. *et al.* (1999a) Maximal strength and power characteristics in isometric and dynamic actions of the upper and lower extremities in middle-aged and older men. *Acta Physiol Scand*, **167**, 57–68.

Izquierdo M., Aguado X., Gonzalez R. *et al.* (1999b) Maximal and explosive force production capacity and balance performance in men of different ages. *Eur J Appl Physiol*, **9**, 260–267.

Jaric S., Ropret R., Kukolj M. and Ilic D.B. (1995) Role of antagonist and antagonist muscle strength in performance of rapid movements. *Eur J Appl Physiol*, **71**, 464–468.

Jensen J.L., Marstrand P.C.D. and Nielsen J.B. (2005) Motor skill training and strength training are associated with different plastic changes in the central nervous system. *J Appl Physiol*, **99**, 1558–1568.

Kamen G. and Knight C.A. (2004) Training-related adaptations in motor unit discharge rate in young and older adults. *J Gerontol Med Sci*, **59A**, 1334–1338.

Kanehisa H. and Miyashita M. (1983) Effect of isometric and isokinetic muscle training on static strength and dynamic power. *Eur J Appl Physiol Occup Physiol*, **50**, 365–371.

Kaneko M., Fuchimoto T., Toji H. and Suei K. (1983) Training effect of different loads on the force-velocity relationship and mechanical power output in human muscle. *Scand J Sports Sci*, **5** (2), 50–55.

Katz B. (1939) The relation between force and speed in muscular contraction. *J Physiol*, **96**, 45–64.

Keenan K.G., Dario Farina K.S., Maluf R.M. and Enoka R.M. (2005) Influence of amplitude cancellation on the simulated surface electromyogram. *J Appl Physiol*, **98**, 120–131.

Kellis E. and Baltzopoulos V. (1998) Muscle activation differences between eccentric and concentric isokinetic exercise. *Med Sci Sports Exerc*, **30**, 1616–1623.

Klass M., Baudry S. and Duchateau J. (2008) Age-related decline in rate of torque development is accompanied by lower maximal motor unit discharge frequency during fast contractions. *J Appl Physiol*, **104**, 739–746.

Komi P.V. and Buskirk E.R. (1972) Effect of eccentric and concentric muscle conditioning on tension and electrical activity of human muscle. *Ergonomics*, **15**, 417–434.

Komi P.V., Linnamo V., Silventoinen P. and Sillanpää M. (2000) Force and EMG power spectrum during eccentric and concentric actions. *Med Sci Sports Exerc*, **32**, 1757–1762.

Kongsgaard M., Reitelseder S., Pedersen T.G. *et al.* (2007) Region specific patellar tendon hypertrophy in humans following resistance training. *Acta Physiol Scand*, **191**, 111–121.

Lagerquist O., Zehr E.P. and Docherty D. (2006) Increased spinal reflex excitability is not associated with neural plasticity underlying the cross-education effect. *J Appl Physiol*, **100**, 83–90.

Larsen A.H., Puggaard L., Hämäläinen U. and Aagaard P. (2008) Comparison of ground reaction forces and antagonist muscle coactivation during stair walking with ageing. *J Electromyogr Kinesiol*, **18**, 568–580.

Lesmes G.R., Costill D.L., Coyle E.F. and Fink W.J. (1978) Muscle strength and power changes during maximal isokinetic training. *Med Sci Sports*, **10**, 256–269.

Macaluso A., Nimmo M.A., Foster J.E. *et al.* (2002) Contractile muscle volume and agonist-antagonist coactivation account for differences in torque between young and older women. *Muscle Nerve*, **25**, 858–863.

Metzger J.M. and Moss R.L. (1990a) Calcium-sensitive cross-bridge transitions in mammalian fast and slow skeletal muscle fibres. *Science*, **247**, 1088–1090.

Metzger J.M. and Moss R.L. (1990b) pH modulation of the kinetics of a Ca2+-sensitive cross-bridge state transition in mammalian single skeletal muscle fibres. *J Physiol*, **428**, 751–764.

Milner T.E. and Cloutier C. (1993) Compensation for mechanically unstable loading in voluntary wrist movement. *Exp Brain Res*, **94**, 522–532.

Moffroid M. and Whipple R. (1970) Specificity of speed of exercise. *Phys Ther*, **50**, 692–1700.

More R.C., Karras B.T., Neiman R. *et al.* (1993) Hamstrings – an anterior cruciate ligament protagonist. *Am J Sports Med*, **21**, 231–237.

Moritani T. and DeVries H.A. (1979) Neural factors versus hypertrophy in the time course of muscle strength gain. *Am J Phys Med Rehabil*, **58**, 115–130.

Moritani T. and Yoshitake Y. (1998) The use of electromyography in applied physiology. *J Electromyogr Kinesiol*, **8**, 363–381.

Narici M.V., Roig S., Landomi L. *et al.* (1989) Changes in force, cross-sectional area and neural activation during strength training and detraining of the human quadriceps. *Eur J Appl Physiol*, **59**, 310–319.

Narici M.V., Hoppeler H., Kayser B. *et al.* (1996) Human quadriceps cross-sectional area, torque and neural activation during 6 months strength training. *Acta Physiol Scand*, **157**, 175–186.

Nelson A.G. (1996) Supramaximal activation increases motor unit velocity of unloaded shortening. *J Appl Biomech*, **12**, 285–291.

Ng A.V., Miller R.G., Gelinas D. *et al.* (2004) Functional relationships of central and peripheral muscle alterations in multiple sclerosis. *Muscle Nerve*, **29**, 843–852.

Nielsen J. and Kagamihara Y. (1992) The regulation of disynaptic reciprocal Ia inhibition during co-contraction of antagonist muscles in man. *J Physiol*, **456**, 373–391.

Norrbrand L., Fluckey J.D., Pozzo M. and Tesch P.A. (2008) Resistance training using eccentric overload induces early adaptations in skeletal muscle size. *Eur J Appl Physiol*, **102**, 271–281.

Patten C., Kamen G. and Rowland D.M. (2001) Adaptations in maximal motor unit discharge rate to strength training in young and older adults. *Muscle Nerve*, **24**, 542–550.

Petrella J.K., Kim J.S., Tuggle S.C. and Bamman M.M. (2007) Contributions of force and velocity to improved power with progressive resistance training in young and older adults. *Eur J Appl Physiol*, **99**, 343–351.

Pierrot-Deseilligny E. and Morin C. (1980) Evidence for supraspinal influences on Renshaw inhibition during motor activity in man, in *Spinal and Supraspinal Mechanisms of Voluntary Motor Control and Locomotion*, vol. 8 (ed. J.E. Desmedt), Karger, Basel, pp. 142–169.

Pinniger G.J., Steele J.R., Thorstensson A. and Cresswell A.G. (2000) Tension regulation during lengthening and shortening actions of the human soleus muscle. *Eur J Appl Physiol*, **81**, 375–383.

Reeves N.D., Narici M.V. and Maganaris C.N. (2004) In vivo human muscle structure and function: adaptations to resistance training in old age. *Exp Physiol*, **89**, 675–689.

Reeves N.D., Maganaris C.N. and Narici M.V. (2005) Plasticity of dynamic muscle performance with strength training in elderly humans. *Muscle Nerve*, **31**, 355–364.

Rice C.L., Vollmer T.L. and Bigland-Ritchie B. (1992) Neuromuscular responses of patients with multiple sclerosis. *Muscle Nerve*, **15**, 1123–1132.

Rutherford O.M. and Jones D.A. (1986) The role of learning and coordination in strength training. *Eur J Appl Physiol*, **55**, 100–105.

Sale D.G. (1992) Neural adaption to strength training, in *Strength and Power in Sports, IOC Medical Commision* (ed. P.V. Komi), Blackwell Scientific Publications, Oxford, pp. 249–265.

Sale D.G., MacDougall J.D., Upton A. and McComas A. (1983) Effect of strength training upon motoneuron excitability in man. *Med Sci Sports Exerc*, **15**, 57–62.

Scaglioni G., Ferri A., Minetti A.E. *et al.* (2002) Plantar flexor activation capacity and H reflex in older adults: adaptations to strength training. *J Appl Physiol*, **92**, 2292–2302.

Schieppati M. (1987) The Hoffmann reflex: a means for assessing spinal reflex excitability and its descending control in man. *Prog Neurobiol*, **28**, 345–376.

Schmidtbleicher D. and Buehrle M. (1987) Neuronal adaptation and increase of cross-sectional area studying different strength training methods, in *Biomech* (ed. B. Johnson), X-B, Human Kinetics Publishers, Champaign, IL, pp. 615–620.

Seger J.Y., Arvidson B. and Thorstensson A. (1998) Specific effects of eccentric and concentric training on muscle strength and morphology in humans. *Eur J Appl Physiol*, **79**, 49–57.

Smith A.M. (1981) The coactivation of antagonist muscles. *Can J Physiol Pharmacol*, **59**, 733–747.

Spurway N.C., Watson H., McMillan K. and Connolly G. (2000) The effect of strength training on the apparent inhibition of eccentric force production in voluntary activated human quadriceps. *Eur J Appl Physiol*, **82**, 374–380.

Suetta C., Aagaard P., Rosted A. *et al.* (2004) Training-induced changes in muscle CSA, muscle strength, EMG, and rate of force development in elderly subjects after long-term unilateral disuse. *J Appl Physiol*, **97**, 1954–1961.

Tabata I., Atomi Y., Kanehisa H. and Miyashita M. (1990) Effect of high-intensity endurance training on isokinetic muscle power. *Eur J Appl Physiol Occup Physiol*, **60**, 254–258.

Takarada Y., Takazawa H., Sato Y. *et al.* (2000) Effects of resistance exercise combined with moderate vascular occlusion on muscular function in humans. *J Appl Physiol*, **88**, 2097–2106.

Thorstensson A., Karlsson J., Viitasalo J.H.T. *et al.* (1976) Effect of strength training on EMG of human skeletal muscle. *Acta Physiol Scand*, **98**, 232–236.

Toji H., Suei K. and Kaneko M. (1997) Effects of combined training loads on relations among force, velocity, and power development. *Can J Appl Physiol*, **22**, 328–336.

Tracy B.L. and Enoka R.M. (2002) Older adults are less steady during submaximal isometric contractions with the knee extensor muscles. *J Appl Physiol*, **92**, 1004–1012.

Tracy B.L. and Enoka R.M. (2006) Steadiness training with light loads in the knee extensors of elderly adults. *Med Sci Sports Exerc*, **38**, 735–745.

Tracy B.L., Byrnes W.C. and Enoka R.M. (2004) Strength training reduces force fluctuations during anisometric contractions of the quadriceps femoris muscles in old adults. *J Appl Physiol*, **96**, 1530–1540.

Upton A.R.M., McComas A.J. and Sica R.E.P. (1971) Potentation of 'late' responses evoked in muscles during effort. *J Neurol Neurosurg Psychiat*, **34**, 699–711.

Valkeinen H., Alen M., Häkkinen A. *et al.* (2000) Changes in unilateral knee extension and flexion force and agonist-antagonist EMG during strength training in fibromyalgia women and healthy women. Proc. 5th Ann. Congr. Eur. Coll. Sports Sci. II (abstract), p. 764.

Van Cutsem M., Duchateau J. and Hainaut K. (1998) Changes in single motor unit behavior contribute to the increase in contraction speed after dynamic training in humans. *J Physiol*, **513** (1), 295–305.

Vinther A., Kanstrup I.L., Christiansen E. *et al.* (2006) Exercise-induced rib stress fractures: potential risk factors related to thoracic muscle co-contraction and movement pattern. *Scand J Med Sci Sports*, **16**, 188–196.

Webber S. and Kriellaars D. (1997) Neuromuscular factors contributing to in vivo eccentric moment generation. *J Appl Physiol*, **83**, 40–45.

Westing S.H., Seger J.Y. and Thorstensson A. (1990) Effects of electrical stimulation on eccentric and concentric torque-velocity relationships during knee extension in man. *Acta Physiol Scand*, **140**, 17–22.

Westing S.H., Cresswell A.G. and Thorstensson A. (1991) Muscle activation during maximal voluntary eccentric and concentric knee extension. *Eur J Appl Physiol*, **62**, 104–108.

Wilkie D.R. (1950) The relation between force and velocity in human muscle. *J Physiol*, **110**, 249–280.

Yao W., Fuglevand A.J. and Enoka R.M. (2000) Motor-unit synchronization increases EMG amplitude and decreases force steadiness of simulated contractions. *J Neurophysiol*, **83**, 441–452.

Zebis M.K., Andersen L.L., Bencke J. *et al.* (2009) Identification of athletes at future risk of Anterior Cruciate Ligament ruptures by neuromuscular screening. *Am J Sports Med*, **37**, 1967–1973.

2.2 Structural and Molecular Adaptations to Training

Jesper L. Andersen, Institute of Sports Medicine Copenhagen, Bispebjerg Hospital, Copenhagen, Denmark

2.2.1 INTRODUCTION

There are many reasons for conducting resistance exercise strength training: improvement of athletic performance, general health, rehabilitation, counteraction of ageing-induced muscle atrophy, or the simple pleasure of working the muscles after a long day at the office (Fry, 2004).

Ultimately strength training is conducted with the purpose of making muscles stronger. The path to reaching this goal is bifurcated and can in principle be reached by changes in neuromuscular recruitment pattern or by increasing muscle size. In practice, it is usually a combination of the two. This chapter deals with adaptations of the *muscle* when submitted to strength training.

Why do muscles grow when subjected to resistance training? Why is growth more pronounced with certain types and intensities of resistance training? Why does ingestion of relatively small amounts of protein after resistance training facilitate muscle growth? Does human skeletal muscle change fibre type composition with strength training? What is the role of satellite cells in muscle hypertrophy? How does concurrent training influence the adaptations in the muscle? These are all areas of human skeletal muscle adaptation to strength training in which development has occurred in recent years. It is emphasized that it is not the ambition of this chapter to delve deeply into specific areas but rather to put spotlight on selected issues of current interest. A number of recommendations are provided, which are principally aimed at athletes, but can with simple adjustments also be useful for elderly people or in a wide range of rehabilitation programs.

2.2.2 PROTEIN SYNTHESIS AND DEGRADATION IN HUMAN SKELETAL MUSCLE

Passing any gym in which strength training is taking place, the sight of people moving their protein-shakes along with them as they switch from one machine to another is more the rule than the exception. This phenomenon was not observed 10–15 years ago. The massage of protein supplementation has certainly found a foothold in the sporting and training community. But does it have any significant effect on the development of muscle mass? To try to answer that question, we have to start out by looking at the more fundamental mechanisms related to growth and reduction of skeletal muscle.

The skeletal muscle mass is maintained when there is an equilibrium between muscle protein synthesis and muscle protein breakdown. A disturbance of the balance in favour of one side over the other will lead to either muscle hypertrophy or muscle atrophy. Since the purpose of strength training is to increase muscle strength and in most cases to increase muscle volume, it is obvious that it is unfavourable for muscle protein degradation to exceed muscle protein synthesis, which eventually leads to muscle atrophy. Nevertheless, this scenario can occur for instance in an overtraining situation, and will always occur during detraining or tapering periods. While the first of these two scenarios is of course very undesirable, the latter is controlled and calculated, and in most cases the atrophy signals will not dominate, but will be dampened and insufficiently large to maintain the hypertrophied state of the well-trained athlete's muscles.

To achieve growth of the muscle fibres a positive net protein balance has to occur; *during the post-exercise phase muscle protein synthesis has to exceed muscle protein breakdown.* It is technically difficult to exactly measure protein breakdown during resistance training, but the few attempts made suggest that no major differences from the nontraining situation occur (Durham *et al.*, 2004; Kumar *et al.*, 2009a; Tipton *et al.*, 2001). Thus it hasn't convincingly been established what happens with muscle protein synthesis during resistance training, but extracting data from the few human studies conducted and extrapolating from various animal studies it seems that most kinds of muscle activity, and especially resistance-related muscle activity, will decrease protein synthesis during the actual training phase (Bylund-Fellenius *et al.*, 1984; Dreyer *et al.*, 2006; Fujita *et al.*, 2009). Therefore a negative net protein balance during the actual resistance training session

Strength and Conditioning – Biological Principles and Practical Applications Marco Cardinale, Rob Newton, and Kazunori Nosaka.

seems probable. Nevertheless, the change in protein synthesis and the breakdown that happens during exercise are relatively small and occur within a limited timeframe, and are therefore of less importance than the changes that are manifested in the hours and days after the training session.

In the initial post-exercise phase muscle protein synthesis is increased, but so is muscle protein breakdown (Dreyer *et al.*, 2006; Fujita *et al.*, 2009; Kumar *et al.*, 2009a; Phillips *et al.*, 1997). The actual curves signifying protein synthesis and protein breakdown follow specific patterns; one common feature is that after strenuous resistance training exercise the protein synthesis is elevated for approximately 48 hours (Kumar *et al.*, 2009a), and sometimes as much as 72 hours (Miller *et al.*, 2005), whereas the protein breakdown is back to pre-training levels within 24 hours after exercise (Phillips *et al.*, 1997). As described later, the post-training protein synthesis in the first hours after training is greatly influenced by the fed state of the subject during and immediately after the training bout, whereas the protein breakdown is much less affected by the feeding status (Rasmussen and Phillips, 2003). A crude general rule is that if the subject is in a fed stage during or immediately after the training session then the protein synthesis in the post-exercise phase will at all times exceed the protein breakdown, while the exact opposite is the case if the subject is in a fasted state (Kumar *et al.*, 2009a; Phillips *et al.*, 1997).

Providing protein to the system in conjunction with resistance training will at least double the increase in protein synthesis as compared to no protein provided (Kumar *et al.*, 2009a; Rennie *et al.*, 1982). There is great scientific, as well as commercial, interest in the quality (proportions of different amino acids and digestibility) of the supplementary protein to be ingested in connection with resistance exercise, which mean that much time, effort, and money has been invested in trying to design the optimum amino acid compound for use in resistance training (Tang and Phillips, 2009). Nevertheless, whether the protein comes from a regular meal or a protein supplementation product may not have a major influence. The composition of the amino acids (e.g. if the protein ingested originates from whey, casein, or soy) will definitely have some modulating importance (Cuthbertson *et al.*, 2005; Tang and Phillips, 2009; Tang *et al.*, 2009), but it seem as if the most important factor is that protein is ingested. Furthermore, there seems to be a dose–response relation between the amount of protein provided and the increase in protein synthesis, but only up to a certain amount of ingested protein. Thus, studies indicate that there is a relatively linier correlation between amount of dietary protein obtained in connection with resistance exercise and increase in protein synthesis, though it is also evident that this correlation plateaus out at around 20 g of ingested protein (for an 85 kg human male) and that any increase in protein intake above this 20 g limit will have no significant influence on protein synthesis rate (Moore *et al.*, 2009; Tang and Phillips, 2009). Furthermore, there is evidence that protein in conjunction with resistance training will increase the length of time in which the increase in protein synthesis is elevated, compared to a situation in which resistance training is carried out with no protein supplementation, or in which protein is provided but no resistance exercise is conducted (Bohé *et al.*, 2001; Kumar *et al.*, 2009a, 2009b; Miller *et al.*, 2005).

Consensus on the timing of ingestion of protein in connection with resistance training is not evident at present. Several studies have claimed that ingestion right before and immediately after exercise is somewhat superior to the alternative in enhancing protein synthesis. It is likely that the difference between the two models is relatively small. The key point is still to have the supply of protein in close connection with the exercise training (reviewed in Kumar *et al.*, 2009a). On the other hand, most available evidence suggests that a delay in the ingestion of protein will diminish the protein synthesis. A delay of at least two hours has proved significantly less efficient than ingestion immediately after training in long-term manifestation of muscle fibre hypertrophy (Esmarck *et al.*, 2001).

Very recent data elucidate the correlation between resistance training intensity and the magnitude of increase in protein synthesis. In one resistance training study the same total amount of work was preformed with both relatively low resistance (20–40% of 1RM) and very high resistance (~90% of 1RM). Both types of training evoked an increase in protein synthesis, but although the same total amount of work was performed, the 20–40% of 1 RM training gave a significantly smaller increase in protein synthesis than training in the 60–90% of 1 RM area (Kumar *et al.*, 2009b). Interestingly, it seems as if there is a plateau of the increased protein synthesis in the 60–90% of 1 RM area, which in practical terms mean that if one is training with the prime aim of increasing muscle mass, from a protein synthesis viewpoint there is no major difference between training at 60% of 1RM and at 90% of 1RM. In practical terms it is already an established 'truth' among coaches and athletes that the optimal training intensity for muscle hypertrophy is somewhere in the above-mentioned area. The novel information is that we can now confirm this and furthermore establish that the intensity as such probably doesn't play a major role so long as it is between ~60 and 90% of 1 RM.

Another recent study on the response of muscle protein synthesis after resistance training seems to add to our knowledge of the differences between the trained and the untrained state. In this study a group of young male subjects conducted resistance training exclusively for one leg, while the other leg served as an untrained control (Tang *et al.*, 2008). After eight weeks the trained leg showed significant hypertrophy as well as increased strength. Interestingly, a measurement of protein synthesis in the two legs showed significant differences. The untrained leg had a lower initial increase in protein synthesis than the trained leg when submitted to resistance exercise, but the protein synthesis of the untrained leg was still significantly elevated 28 hours after exercise, in contrast to the trained leg, in which the protein synthesis at this point in time was no longer elevated from pre-exercise values. The conclusion is that the protein synthesis in the trained muscle reacts promptly and strongly, but fades faster than that in the muscles of the untrained leg, given equal relative amounts of resistance training (Tang *et al.*, 2008). Thus, this reaction pattern is in agreement with the general concept of reaction to 'stress', in the sense that a muscle that is accustomed to stress (exercise) will react promptly but also deal with the stimulus faster. Similar results have been obtained when comparing muscle protein turnover in subjects accustomed to resistance training versus untrained subjects (Phillips *et al.*, 1999). A general rule seems to be that, although high in the early phase, the *overall* protein synthesis in a trained muscle after resistance training is less than in an untrained muscle, but protein degradation is less after resistance training in trained muscle compared to untrained muscle.

A study design aiming to examine the differences between resistance and endurance training in the trained and untrained state has provided additional information on muscle protein synthesis patterns after various types of training (Wilkinson *et al.*, 2008). In this study Wilkinson *et al.* found that protein synthesis of both the myofibrillar and the mitochondrial protein fractions increased after resistance training in the untrained state, whereas in the trained state only the synthesis of the myofibrillar proteins increased. After endurance training only the synthesis of the mitochondrial proteins increased, in both the trained and the untrained state. Several different interpretations of these differences in reaction pattern in protein synthesis and degradation are possible, but one might be that the trained muscle needs, and can tolerate, more frequent stimulus than the untrained muscle in order to respond in a favourable manner. Furthermore, the anabolic response of resistance training is reduced with increased training. These results also imply that the frequency and intensity of resistance training need to increase progressively as the muscle becomes better trained, to keep up the well-trained homoeostasis of the muscle. Finally, as concluded by Wilkinson *et al.* (2008) and later expanded upon by Kumar *et al.* (2009a), chronic resistance exercise can modify the protein synthesis response of the protein fractions towards an exercise phenotypic response, or as Wilkinson *et al.* put it, 'These findings provide evidence that human skeletal muscle protein synthetic response to different modes of exercise is specific for proteins needed for structural and metabolic adaptations to the particular exercise stimulus, and that this specificity of response is altered with training to be more specific' (Wilkinson *et al.*, 2008).

2.2.3 MUSCLE HYPERTROPHY AND ATROPHY

Often the entire increase in strength in the early phase of a resistance training programme initiated in untrained subjects is contributed to adaptations of the neural drive (Aagaard *et al.*, 2002; Gabriel, Kamen and Frost, 2006). There is no doubt that significant modifications in the nervous system in the first weeks of heavy resistance training occur, initiating strength gains. The question is whether these changes alone are responsible for the increased strength or whether muscle hypertrophy kicks in almost immediately and contributes to the strength gain after days or a few weeks of training. Most experiments can't show a significant increase in muscle hypertrophy earlier than four weeks into a training programme, and often six to eight weeks are need before the hypertrophy becomes significant (Staron *et al.*, 1994). No doubt the main contribution to strength gain in the early phase arises from the neuromuscular adaptations, but it is somewhat strange that no muscle hypertrophy should happen before four to eight weeks of intensive training when it can be shown through acute studies that protein synthesis is significantly upregulated and a positive net protein balance occurs hours after a single exercise bout (Kumar *et al.*, 2009a). Likewise, signalling factors that vouch for muscle hypertrophy are increased within the same time frame (Bickel *et al.*, 2005; Bodine, 2006; Coffey and Hawley, 2007; Favier, Benoit and Freyssenet, 2008).

Some newer studies may cast additional light on the problem. In a study by Seynnes, de Boer and Narici (2007), a small group of subjects were followed closely at the very beginning of a strength training programme. In this study hypertrophy measured via MRI was registered after just 20 days of training, and it could be calculated that the rate of muscle hypertrophy in the 20-day period was as much as 0.2%/day. In the same period the overall gain in MVC was in the magnitude of ~1%/day (Seynnes, de Boer and Narici, 2007). Unfortunately, no muscle biopsies were obtained and thus no measurements of hypertrophy in the individual muscle fibres are available. Another study involving resistance training over an eight-week period showed hypertrophy of the type II fibres, but not of the type I fibres, after four weeks of training (Woolstenhulme *et al.*, 2006), but overall the studies reporting significant hypertrophy in fibre level after less than four weeks are scarce. Even though fibre hypertrophy can't be documented significantly in the first weeks of training, this doesn't mean that it doesn't occur. One could get the idea that the problem lies not so much in the fact that there is no hypertrophy in the very early phase of a strength training programme, but that we are not able to measure a potential hypertrophy in a sensitive and sufficiently accurate manner. Calculations based on the magnitude of fibre hypertrophy registered at week 8–12 into a resistance training programme imply that the contribution of hypertrophy to increase in muscle strength in the initial phase could be as much as ~20% (Seynnes, de Boer and Narici, 2007).

Hypertrophy may not always be what is required when strength training is conducted, but quite often it is. Strength training in all its forms, performed with a reasonable amount of load, will in the untrained or moderately trained muscle inevitably lead to some muscle hypertrophy, but it is equally clear that different types of training regimens don't lead to the same amount of hypertrophy. It seems as if the relative loading of the muscle is very important to the outcome. In a review, Fry (2004) estimates that 18–35% of the variance of the hypertrophic outcome after strength training is determined by the relative intensity of loading during exercise, estimated from studies using a loading between 40 and 95% of 1 RM. The variance is greater in the type II fibres (35%) than in the type I fibres (18%). Nevertheless, overall ~25% of the hypertrophic response in a mixed human skeletal muscle is determined by the loading. This underscores the fact that loading is one of the most (probably the most) important factors when designing strength training programmes, both if the prime goal is increased muscle mass and if it is increased muscle strength with as little increase in muscle mass as possible.

If the prime goal *were* hypertrophy, what would be the optimal relative loading? Data suggest that maximal hypertrophy occurs with loads from 80 to 95% of 1 RM (Fry, 2004). As mentioned earlier, the maximal protein synthesis seems to occur over a broad range from 60 to 90% of 1 RM. At first glance it could seem that there is a minor divergence between the range of maximal hypertrophy and highest protein synthesis, especially in the 'low' relative loading area. Can we explain this? Working at 60% of 1 RM would typically result in three to six times (variable with different exercises) more repetitions for the muscle per exercise than working at 90% of 1 RM. Thus, the total number of contractions per training session is much higher for the 'low' loads compared to the high loads. If we assume that the two different training sessions increase protein synthesis equally (which most data from the literature seem to imply), and the release of intrinsic growth factors within the muscle is in the same magnitude (which may be more questionable, but is poorly described in the literature), then the difference between the outcomes of the two different training regimens could arise from differences in protein degradation. Maybe the higher number of contractions conducted in the area around 60% of 1 RM provides more wear and tear on the contractile proteins, leading to a greater protein breakdown than the fewer contractions in the 80–95% of 1 RM area, resulting in slightly less *net* protein synthesis. So far this is only speculation and to this author's knowledge the two scenarios have not been tried out in the same study (involving measurement of both protein synthesis and muscle fibre hypertrophy) in a manner that could substantiate these speculations.

Relative loading is a very important factor in strength training. Variables such as type of exercise, order of exercises, total volume of exercises, and rest between sets and training sessions can all be regulated in a training regimen (Fleck and Kraemer, 2004). Likewise, variables such as speed of contraction, contraction time per rep, working to failure or not, the choice between exercising in machines or with free weights, and overall periodization principles will also influence the end result (Fry, 2004). To work all these factors into a general equation is very difficult, if not impossible, although interesting attempts have been made to illustrate how apparently similar training regimes can be very different when some of the above-mentioned parameters are modulated (Toigo and Boutellier, 2006). Another important factor is the number of repetitions conducted per exercise or training session; by linking this with loading some simple rules of adaptation of the muscle can be made. High load/low reps will provide the greatest increase in strength the fastest, compared to medium load/medium reps and low load/high reps. Hypertrophy might be a little higher with high load/low reps compared to medium load/medium reps, though not necessarily much higher, but certainly higher than in low load/high reps (Campos *et al.*, 2002; Fry, 2004).

As described above, strength training will in most situations lead to muscle hypertrophy, but as is also evident, that the amount of hypertrophy is dependent on how the training is conducted. It has become evident that strength training does not affect the different muscle fibre types equally; the relative loading will determine how much the different fibre types are influenced. Heavy resistance training over an extended period (e.g. three months) will initiate hypertrophy in the range of 5–15% in slow type I fibres and 15–25% in fast type II fibres (Andersen and Aagaard, 2000). Data on the magnitude of relative hypertrophy found in the literature vary, depending on the amount and type of strength training conducted, as well as on the starting point of the subjects examined. Nevertheless, a rough estimate is that the fast type II fibres will hypertrophy twice as much as the slow type I fibres. Since fast type II fibres cover a larger relative cross-sectional area of the muscle, the end result is not only a stronger muscle but also a faster muscle. In this scenario it is assumed that no major change in fibre-type distribution occurs, a matter that we shall return to. But there is a twist to the story: when studies using different amounts of relative loading are compared (Fry, 2004), it becomes evident that hypertrophy increases as the relative loading increases, apparently following a linear regression line, though this regression line is different for the two major fibre types. The increase in hypertrophy in the slow type I fibres observed as the loading increases appears as a gentle line, whereas the increase in hypertrophy seen in fast type II fibres follows a steeper line. Thus, the difference in hypertrophic response between the two major fibre types increases as the loading increases (Fry, 2004). Undoubtedly the same amount of training will evoke less hypertrophy in already well (strength) trained subjects. Thus, the background of the individual who undertakes the strength training is important.

When planning strength training for an athlete or a patient it is important to know and take into account their training background: a certain amount/volume of training might introduce significant muscle hypertrophy in an athlete or a patient with no prior strength training experience, whereas an athlete who has undertaken large amounts of resistance training might experience regular atrophy of their muscles if they conduct the same amount and type of resistance training prescribed for a less experienced person, simply because the stimulus to their muscles and nervous system will be less intense than the muscle–CNS signalling that they normally receive. One of the key points of a hypertrophied muscle is that it is not in equilibrium and will strive towards a less hypertrophied status if the stimulus is lowered or removed. Thus, the 'problem' that many athletes face is centred around questions like: How do I counteract atrophy form a very hypertrophic state during the competition period? How fast does the hypertrophied muscle decrease (and it does) if I remove the resistance training? How much training is enough the keep up the gained strength and hypertrophy, or at least to minimize the loss? These are questions that should be addressed more closely in the coming years.

2.2.3.1 Changes in fibre type composition with strength training

We know that a muscle's ability to conduct a fast and forceful contraction contributes positively to performance in certain athletic events. Within muscle physiology it has been know for many years that the maximum speed at which a muscle can contract is to a large extent given by the its composition of fast and slow muscle fibres (Bottinelli and Reggiani, 2000; Harridge, 1996). Likewise, the maximum force and power produced by the single muscle fibre is strongly positively related to its content of fast myosin (Bottinelli *et al.*, 1999), which can also be observed during *in vivo* muscle contraction in the intact human (Aagaard and Andersen, 1998).

Human skeletal muscle fibres contain different proteins facilitating contraction. Some proteins are structural, having the task of maintaining the physical structure of the fibre as force is produced, whereas others are involved in the actual contractile process (Schiaffino and Reggiani, 1996). The two main proteins involved in muscle contraction are myosin (the thick filament) and actin (the thin filament). In the human skeletal muscle actin only exists in a singular form. Myosin (or more precise the heavy chain of the myosin molecule, MHC), on the other hand, exists in three different forms (known as isoforms; essentially different versions of the same protein taking care of the same task) (Schiaffino and Reggiani, 1994). Each of these MHC isoforms, when present in the fibre, provides it with specific functional characteristics, the primary one being contraction velocity. A number of other proteins contribute or modulate the outcome but the absolute most important determinant of the fibre's contraction velocity is the MHC isoform within it. The three MHC isoforms present in physiologically relevant amounts in human skeletal limp muscles are MHC I, MHC IIA, and MHC IIX (the latter is often referred to in older literature as 'IIB'; Smerdu *et al.*, 1994) (Schiaffino and Reggiani, 1996). Thus, muscle fibres can be separated into different fibre types with specific contraction characteristics via identification of the MHC isoform(s) present in the individual fibres. The three different MHC isoforms should in principle result in three different muscle fibre types, but in the human skeletal muscle two MHC isoforms are often present alongside each other in the same fibre. Under normal circumstances only MHC I/MHC IIA and MHC IIA/MHC IIX are expressed together in the same muscle fibre. This results in five different fibre types: fibres containing only MHC I, only MHC IIA, and only MHC IIX, which constitute the 'pure' fibre types, and 'hybrid fibres' co-expressing MHC I and MHC IIA as well as MHC IIA and MHC IIX. Additionally the hybrid fibres can contain various relative amounts of the two isoforms, and thus a continuum of fibre types from the pure MHC I to the pure MHC IIX fibre can be found in the human skeletal muscle (Andersen, Klitgaard and Saltin, 1994).

Maximum contraction velocity of single human skeletal muscle fibres can be determined experimentally, and shows that fibres containing MHC I are the slowest and fibres containing MHC IIX are the fastest. Thus, contraction velocity for the different fibre types is: MHC I < MHC I/IIA hybrids < MHC IIA < MHC IIA/IIX hybrids < MHC IIX (Bottinelli, 2001; Harridge *et al.*, 1996). The difference in maximum shortening velocity (when determined in single fibres) between fibres containing only one of the three MHC isoforms is in the order of magnitude of 1 : 3 : 8 (MHC I : MHC IIA : MHC IIX), where co-expression hybrid fibres are placed in between (Fitts and Widrick, 1996; Harridge, 2007). These data are the result of experiments conducted on isolated single fibres at relatively low temperature (15–18 °C). Recent data conducted at more physiological relevant temperature (35 °C) seem to indicate that the true difference is less, and in the magnitude of 1 : 2 between MHC I and MHC II fibres (Lionikas, Li and Larsson, 2006).

A number of studies have shown a strong correlation between fibre type composition of an intact muscle and the velocity properties of the muscle, both in different muscles with different fibre type compositions in the same individual (Harridge, 1996; Harridge *et al.*, 1996) and in the same muscle between different individuals with different fibre type compositions (Aagaard and Andersen, 1998; Tihanyi, Apor and Fekete, 1982; Yates and Kamon, 1983). The relationship between fibre type composition and muscle contractile velocity does not emerge at slow contraction velocities, because slow fibres in this situation have ample time to build force up to more or less the same level as the fast fibres (Aagaard and Andersen, 1998). Consequently, the close relationship between maximal concentric muscle strength and the percentage of MHC II in intact human skeletal muscle first becomes apparent at high contraction velocities (Aagaard and Andersen, 1998). In practical terms this mean that a person with a relatively large proportion of fast fibres will be able to achieve higher muscle force and power output during fast movements, including the early acceleration phase, than a person with a low relative proportion of fast fibres. Likewise, muscles characterized by a high relative proportion of MHC II content are substantially more 'explosive' (i.e. demonstrate a greater rate of force development, RFD) than muscles with low relative MHC II content, demonstrating an enhanced capacity for rapid force production (Harridge *et al.*, 1996).

With the above in mind we know that a person with a high relative amount of MHC II, all other things being equal, will be more suited for sports in which fast, explosive-type movements performed over shorter periods of time are crucial. It becomes interesting to know whether it is possible through training to change the MHC isoform composition in human skeletal muscle.

From a huge collection of animal studies it is known that manipulation of the MHC isoform content of muscle fibres, as well as of the intact muscles, is possible (Pette and Staron, 2000). In humans, a number of critical conditions can introduce large changes in MHC compositions in skeletal muscle. A spinal cord injury leading to paralysis will after a while lead to an almost complete abolishment of the slow MHC isoform in the affected muscles, leaving the muscle to exclusively express the two fast MHC isoforms (Andersen *et al.*, 1996), which tells us that significant switches between expression of fast and slow MHC isoforms are possible in most skeletal muscles.

Nevertheless, the above-described scenario of a complete change in expression from slow to fast MHC isoforms after a spinal cord injury, along with the often drastic manipulations used in animal studies, does not explain what happens within the frame of 'physical training'. What is the range of changes in MHC isoform composition in the muscles when we do strength training? Numerous studies have shown that heavy-resistance exercise training will decrease the expression of MHC IIX in human skeletal muscle and simultaneously increase the expression of MHC IIA, whereas the expression of MHC I is less affected (Adams *et al.*, 1993; Andersen and Aagaard, 2000; Hather *et al.*, 1991). This is a solid observation and a general consensus on this point has emerged (Fry, 2004; Folland and Williams, 2007). Likewise, cessation of resistance training will induce, or re-induce, MHC IIX at the expense of MHC IIA (Andersen and Aagaard, 2000; Andersen *et al.*, 2005). Whether or not the number of fibres expressing MHC I is increased or decreased after strength training is debateable, but most likely there is no or only very subtle change in the *number* of fibres expressing MHC I (Andersen and Aagaard, 2000; Fry, 2004). Thus, the general rule of MHC isoform plasticity in human skeletal muscle appears to be that introduction of, or increase in, the amount of resistance training leads to a decrease in MHC IIX and an increase in MHC IIA, while a withdrawal or decrease in resistance training leads to an increase in MHC IIX and a decrease in MHC IIA. MHC I seems to be relatively unaffected by resistance training (Andersen and Aagaard, 2000; Fry, 2004).

The disappearance of MHC IIX with strength training might seem unfavourable since this MHC isoform has the fastest contraction velocity and highest power production. Removal of MHC IIX from the muscle should lead to a slowing and reduced power output of the muscle. Theoretically this is the case when looking at the isolated individual fibre, but in terms of the capacity of the intact muscle this apparent slowing is, in most athletic settings, more than outweighed by the increase in contractile strength, power, and RFD of the trained muscle (Aagaard, 2004). In consequence, maximal unloaded limb movement speed is observed to increase (Aagaard *et al.*, 2003; Schmidtbleicher and Haralambie, 1981) or remain unaltered (Andersen *et al.*, 2005) following 3–4 months of heavy-resistance strength training. The increase in muscle force, power, and RFD following heavy-resistance strength training is to a large extent caused by the fast fibres demonstrating a twofold greater hypertrophy than the slow fibres in response to heavy-resistance training (Aagaard *et al.*, 2001; Andersen and Aagaard, 2000; Kosek *et al.*, 2006). Moreover, a differentiated hypertrophy of the fast and slow fibres with heavy resistance training, in favour of the fast fibres, will eventually give rise to not only a bigger muscle but also a muscle in which a relatively larger proportion of the cross-sectional area is occupied by fast fibres and there is thus also a higher relative amount of MHC II (Aagaard, 2004; Andersen and Aagaard, 2000).

Data from our lab indicate that heavy-resistance training followed by de-training can evoke a boost in proportions of the MHC IIX isoform. In a strength training study examining a group of young healthy males, we saw that the percentage of MHC IIX in the vastus lateralis muscle of the subjects decreased from 9% to only 2% during the three-month training period. Interestingly, the percentage of MHC IIX subsequently increased to 17% after an additional period of three months of de-training (Andersen and Aagaard, 2000). Thus, the relative level of MHC IIX after three months of de-training was significantly higher than both the level after training and the level before the training period (Andersen and Aagaard, 2000). In a similar study, we found that the MHC IIX boost after de-training was accompanied by a parallel increase in RFD in the trained muscles (Andersen *et al.*, 2005). However, de-training also resulted in a loss of muscle mass to levels comparable with those observed prior to the training period. This apparent boosting of the MHC IIX isoform with de-training (tapering) is potentially interesting if the goal of a long-term training programme is to increase the relative amount of MHC IIX in the muscle of a specific athlete, typically an athlete competing in an athletic event in which no endurance work is necessary, and contractile speed, power, and/or explosiveness (RFD) is dominantly favoured. At this point in time we do not know how the muscle will react beyond the experimental period of three months, but it can be expected that the level of MHC IIX will eventually return to the original pre-training value (Staron *et al.*, 1991).

Is a high relative amount of MHC IIX in the major skeletal muscles desirable in situations other than an athlete's participation in relatively specialized compositions? It has been established that muscle fibres containing predominantly the MHC IIX isoform rely on a metabolism that enables them to produce very high amounts of energy in a short time (i.e. exerting very high power), but only over a very limited period (seconds) (Harridge, 1996; Harridge *et al.*, 1996). In consequence, the MHC IIX-containing fibres need to rest to avoid exhaustion. They will not get sufficient rest in the major ball sports (with the possible exception of e.g. American football), or other sports in which continuous work over longer periods are in demand. Therefore, fibres containing MHC IIA may be preferable for athletes who compete in events in which a relatively fast but also somewhat enduring muscle is desirable; that is, most ball games, 400–1500 m running, rowing, kayaking, cycling events like sprint and team pursuit, and so on. Training to meet these conditions is much 'easier' to plan than training to provoke fibres to express exclusively MHC IIX. However, if the aim is to produce a very fast 100 m or 200 m sprinter (i.e. targeting the latter training regime) then training involving hours of continuous work at a moderate aerobic level should be avoided, as this type of exercise can lead to an increased number of fibres expressing MHC I (Schaub *et al.*, 1989) and/or fibres co-expressing MHC I and MHC IIA. Furthermore, aerobic exercise of the above nature may partially blunt the hypertrophic muscle response from concurrent resistance training (Baar, 2006; Coffey *et al.*, 2009; Glowacki *et al.*, 2004; Nader, 2006). Planning of the training schedule will also have to include some type of tapering leading up to competition.

In those cases in which some type of short-term muscle endurance is also needed, training exercises should comprise high-intensity intermittent work along with substantial amounts of resistance exercise; the former gives rise to improved short-term endurance of the MHC IIA fibres, while the latter produces a preferential hypertrophy in the MHC II muscle fibres in general. Ultimately, the result is a muscle which is optimized towards the highest possible relative amount of MHC IIA at the expense of both MHC I and MHC IIX. Of course, this scenario favours athletes who have a relatively high amount of type II fibres to begin with. Whether or not these type II fibres contain

MHC IIA or MHC IIX at the point of origin is of less importance, since the inherent transformation MHC IIX → MHC IIA will be introduced through resistance training.

2.2.4 WHAT IS THE SIGNIFICANCE OF SATELLITE CELLS IN HUMAN SKELETAL MUSCLE?

Unlike many other cell types in the human body, muscle fibres are multinucleated. The very elongated nature of the muscle fibres supports an organization in which each of the hundreds or thousands of nuclei along the length of the individual fibres governs a finite area or volume (Kadi *et al.*, 2004, 2005; Petrella *et al.*, 2008). This volume is termed the myonuclear domain. Thus, each nucleus regulates gene transcription and ultimately protein synthesis of its specific domain. If a fibre grows, and no new myonuclei are added, each existing myonucleus has to support a myonuclear domain that has increased in size. All other things being equal, this could potentially bring down efficiency.

Myonuclei in the adult skeletal muscle cannot divide. Thus, if there is need for supplementation of myonuclei due to damage repair or in situations in which substantial hypertrophy is ongoing, the individual muscle fibre needs a source of new myonuclei. This is achieved by the satellite cells: quiescent undifferentiated progenitor cells located outside the sarcolemma but under the basal membrane of the individual skeletal muscle fibres (Hawke, 2005; Kadi *et al.*, 2005). Unlike myonuclei, the satellite cells have the potential to go into mitosis and divide. If activated, the satellite cells will proliferate and replace damaged myonuclei or supplement the existing pool of myonuclei. There is substantial evidence to suggest that satellite cells have a capacity for asymmetric cell division to allow for self-renewal in order to maintain the precursor pool (Conboy and Rando, 2002). In practical terms this means that when an activated satellite cell divides, one of the two new cells will return to a quiescent state, ready for a new division, while the other one will proceed to become a valid myonucleus, taking part in the machinery of the muscle fibre to which it belongs. Furthermore, in some situations satellite cells have the potential to deliver nuclei for the formation of new muscle fibres.

This process has become evident in animal studies, in which damage of adult muscle fibres evokes a neoformation of muscle fibres (Hawke, 2005). Whether such a neoformation occurs after extensive resistance training is still unclear and a matter of speculation. Thus, whether or not a potential neoformation has any physiological significance in the general muscle hypertrophy of resistance-trained human skeletal muscle is debateable, but most likely satellite cells only facilitate formation of new segments in already existing damaged fibres, as well as delivering additional myonuclei to already existing muscle fibres, without adding truly new fibres.

Today it is possible, in human skeletal muscle biopsies, to stain and count the number of myonuclei as well as the number of quiescent and activated satellite cells, and with a relatively high certainty to separate the three (Mackey *et al.*, 2009). Using different markers unique to either myonuclei or quiescent and activated satellite cells in combination with the physical location of the nucleus, it has become evident that resistance training is a potent stimulus for the activation of skeletal muscle satellite cells (Kadi *et al.*, 2004, 2005; Mackey *et al.*, 2009; Petrella *et al.*, 2008).

We have conducted a study in which we biopsied the vastus lateralis muscle of a group of young men during 90 days of heavy-resistance exercise training followed by 90 days of de-training (Kadi *et al.*, 2005). We observed muscle fibre hypertrophy of ~6% after 30 days and ~17% after 90 days of training. Likewise, an increased number of satellite cells could be observed in the training period, followed by a return to pre-training values in the de-training period. The interesting part is that even though it was possible to show an activation of the satellite cells, it was not possible to measure an increase in the number of myonuclei during either the training or the de-training period. Thus, it might be that the satellite cells were activated, proliferated, but not differentiated. At least the new satellite cells did not end up as a measurable increase in myonuclei (Kadi *et al.*, 2005). Thus, it seems as if each myonucleus was able to support a larger cytoplasmic area; consequently the size of the myonuclear domains was increased (Kadi *et al.*, 2004). These data, along with other similar findings (Petrella *et al.*, 2006) seem to imply that a certain amount of hypertrophy of the muscle fibres can be managed by the existing myonuclei simply by expanding their myonuclear domains. The question is, what happens if the hypertrophic stimulus continues and hypertrophy of the existing fibres expands beyond the ~20% observed in our study?

Data from several research groups suggest that a limit of hypertrophy in the magnitude of ~25% seems to constitute some kind of ceiling: a ceiling that has to be broken before a regular and significant increase in the number of myonuclei sets in (Kadi *et al.*, 2004, 2005). Petrella *et al.* (2006) have suggested this ceiling could be around 2000 μm^2 per nucleus.

A very recent study has shed new light on this myonuclear domain ceiling. In this study a relatively large group of subjects performed 16 weeks of hypertrophy-inducing resistance training (Petrella *et al.*, 2008). Afterwards the subjects could be subdivided into three groups, having muscle fibre hypertrophy of ~50%, ~25% and ~0%. Of course, the group with no or very little hypertrophy did not reach the myonuclear domain ceiling. The ~25% group expanded the myonuclear domain up to 2000 μm^2 per nucleus, while the ~50% group showed a myonuclear ceiling domain of 2250 μm^2 per nucleus. These data indicate that any potential myonuclear ceiling domain is likely to vary from person to person (Petrella *et al.*, 2008). Moreover, the study showed that the availability of satellite cells in the untrained muscle is important to the hypertrophic potential, since the ~50% group had significantly more satellite cells in their vastus lateralis muscle prior to training; in addition they were able to expand the pool of satellite cells by a greater amount than the two other groups.

Together, these studies may provide some explanation as to why muscles in different individuals hypertrophy at very different rates even when subjected to exactly the same training stimulus.

On the speculative level, the whole concept of a myonuclear domain ceiling can lead to a muscle fibre hypertrophic scenario that follows a gearing pathway. First gear is the initial phase, in which the muscle fibre increases the size without any major increase in new myonuclei; this phase involves relatively fast hypertrophy, indicating that an increase in the net protein synthesis is managed by the already existing machinery. Second gear sets in when the myonuclear ceiling domain is reached; in this gear the continuing hypertrophy develops more slowly since new myonuclei have to be incorporated in order to further expand growth. In the late stage of the hypertrophic process the muscle fibres drive in both gears simultaneously. It should be noted that the proliferation for the later differentiation of the satellite cells appears to start early in the initiation phase of the resistance training programme, preparing the muscle fibre for the situation that might arise in the future (Petrella *et al.*, 2008). On an even more speculative note, it could be imagined that the period in which the gearshift happens could involve a plateau in the hypertrophy.

Of interest in the study by Petrella and co-workers (Petrella *et al.*, 2008) is the observation that *the proliferation of satellite cells far surpassed the myonuclear addition* in the ~50% group. It is suggested that this 'overproduction' of satellite cells could lead to a stockpile of precursor satellite cells. Such a stockpile, if kept available and not discarded, could act as a potential booster in a situation where resistance training is resumed after a period with no or very limited resistance training, explaining at least in part the phenomenon known as 'muscle memory'. Muscle memory is the observation among bodybuilders and strength athletes that if one has already been through a period of intense hypertrophic-inducing training, it is 'easier' to get this hypertrophic effect again, after a period of loss of muscle mass, than it is when starting from scratch. On the other hand, our own studies seem to indicate that the number of satellite cells is not increased above baseline in the de-training phase (Kadi *et al.*, 2005), which speaks against the idea of the stockpile of satellite cells. Thus it is more likely, but difficult to verify, that the existing satellite cells store some knowledge of earlier damage or stress (for lack of a better way of putting it) that makes them react more rapidly the second time they are subjected to the resistance training stimulus.

2.2.5 CONCURRENT STRENGTH AND ENDURANCE TRAINING: CONSEQUENCES FOR MUSCLE ADAPTATIONS

Very few sports are solely strength-related, for example almost all ball games require skills related to muscle strength but also demand high anaerobic and aerobic capacity. Thus, physical training for these sports involves a combination of strength and endurance qualities. Of course, focus can be placed on one or the other, but the total training effort will always reflect the need for both. Although there is overlap in the molecular response of the skeletal muscle to endurance and resistance training, it has become evident that skeletal muscle is capable of distinguishing between the intensity, duration, and nature of contractions (Kumar *et al.*, 2009a).

A substantial number of studies have tried to examine how concurrent strength and endurance training might influence the adaptation of either quality compared to conducting the two types of training separately. Given the topic of this chapter, it is important to try to elucidate how concurrent endurance training will affect the strength and hypertrophy response that is initiated by strength training alone. Although not new, the question is still difficult to answer. Nevertheless the arrival of molecular biology has given rise to some possible explanations for a few of the adaptive responses that can be seen in long-term training studies (Bell *et al.*, 2000; Glowacki *et al.*, 2004; Häkkinen *et al.*, 2003; Izquierdo *et al.*, 2005; Kraemer *et al.*, 1995). Around 1980 Robert C. Hickson published a number articles opening up the area scientifically. He suggested that conducting concurrent strength and endurance training didn't affect the adaptation of cardiovascular qualities (VO_{2max} etc.) (Hickson, 1980), and later experiments have shown that endurance performance is likely to benefit from concurrent strength training if it is conducted in the right manner (Hoff, Helgerud and Wisløff, 1999; Østerås, Helgerud and Hoff, 2002; Rønnestad, Hansen and Raastad, 2009; Støren *et al.*, 2008). Hickson also showed that concurrent strength and endurance training was likely to at least dampen some of adaptations in muscle strength and hypertrophy when compared to strength training alone (Hickson, 1980). Long-term studies conducted since have to a certain extent reached conclusions along the same lines (Häkkinen *et al.*, 2003).

What has been more difficult to explain is how signalling pathways in the muscle following endurance training might inhibit the signalling pathway initiated by resistance training. Probably the most robust explanation revolves around a complex of regulation pathways with offspring in the activation of the AMPK pathway through endurance training; this seems to block the Akt/mTOR pathway, which is partly responsible for the increase in protein synthesis and eventual hypertrophy after resistance training (Bodine, 2006). This idea was first developed in animal models (Baar and Esser, 1999; Bodine *et al.*, 2001; Haddad and Adams, 2002; Pallafacchina *et al.*, 2002) and later in human training models (Baar, 2006; Bodine, 2006; Coffey and Hawley, 2007; Nader, 2006; Rivas, Lessard and Coffey, 2009). We now know that concurrent endurance training to some extent compromises the benefits of resistance training. Nevertheless, it is obvious that many athletes have to conduct endurance training along with strength training in order to fulfil the demands of the sport in which they are competing.

The next question is, how do we plan concurrent training so that we get the most out of the invested effort? For example, if both qualities are trained on the same days, what would be the preferred order if muscle hypertrophy were to be given high priority? A very recent study has looked at concurrent endurance and strength training and how the order in which the training is conducted affects stimuli at the mRNA level. In this study a group of reasonably well-trained male subjects was, on different days, subjected to either endurance training followed by resistance training or vice versa, in a cross-over design. Muscle biopsies were obtained prior to exercise, after the first exercise bout, after the second excise bout, and again three hours post-exercise (Coffey *et al.*, 2009). The experiment shows that the phosphorylation of mTOR is downregulated in the post-training phase if endurance training is undertaken after strength training, potentially downregulating protein synthesis. If endurance training is undertaken first, this is not the case. Furthermore, phosphorylation of Akt1 is upregulated after resistance training (positive for protein synthesis), but what is interesting is that a post-endurance training bout does not prevent a later increase in Akt1 phosphorylation initiated by the subsequent resistance training bout. Furthermore, the experiment shows that mRNA for the two ubiqutin ligases, Atrogin and MuRF, which target protein for later degradation and thus create an environment of increased protein degradation, increases when endurance exercise is undertaken after resistance exercise; the reverse is true when the exercise bouts are carried out the other way around. Together, these data seem to indicate that resistance training should be conducted after endurance training if the purpose is to provide better circumstances for an overall positive ratio in the protein synthesis/degradation balance.

Although there is still a great deal of work to be conducted in the area of concurrent endurance and resistance training, a picture is starting to form. (1) If endurance training is undertaken prior to resistance training then there is a possibility that the hypertrophic response seen with resistance training will only be somewhat diminished, but the overall negative effect might be relatively small and thus only relevant when the the focus is on muscle hypertrophy. (2) If resistance training is conducted prior to endurance training then there is an effect on the hypertrophic stimulus provided through the resistance training; furthermore, studies seems to indicate that the break on the hypertrophic response, most likely through accelerated protein degradation, is more pronounced than when endurance training is conducted prior to resistance training.

As a preliminary conclusion our advice to athletes is: end your concurrent training session with the type of training that has the highest priority. For example, athletes who train with the purpose of optimizing muscle strength and muscle hypertrophy, but still need to conduct concurrent endurance training, should carry out their resistance training last.

References

Aagaard P. (2004) Making muscles stronger: exercise, nutrition, drugs. *J Musculoskelet Neuronal Interact*, **4**, 165–174.

Aagaard P. and Andersen J.L. (1998) Correlation between contractile strength and myosin heavy chain isoform composition in human skeletal muscle. *Med Sci Sports Exerc*, **30**, 1217–1222.

Aagaard P., Andersen J.L., Dyhre-Poulsen P. *et al.* (2001) A mechanism for increased contractile strength of human pennate muscle in response to strength training: changes in muscle architecture. *J Physiol*, **534**, 613–623.

Aagaard P., Simonsen E.B., Andersen J.L. *et al.* (2002) Increased rate of force development and neural drive of human skeletal muscle following resistance training. *J Appl Physiol*, **93**, 1318–1326.

Aagaard P., Simonsen E.B., Andersen J.L. *et al.* (2003) Changes in maximal unloaded knee extension velocity induced by resistance training. *Med Sci Sports Exerc*, **35**, S369 (abstract).

Adams G.R., Hather B.M., Baldwin K.M. and Dudley G.A. (1993) Skeletal muscle myosin heavy chain composition and resistance training. *J Appl Physiol*, **74**, 911–915.

Andersen J.L. and Aagaard P. (2000) Myosin heavy chain IIX overshooting in human skeletal muscle. *Muscle Nerve*, **23**, 1095–1104.

Andersen J.L., Klitgaard H. and Saltin B. (1994) Myosin heavy chain isoforms in single fibres from m. vastus lateralis of sprinters: influence of training. *Acta Physiol Scand*, **151**, 135–142.

Andersen J.L., Mohr T., Biering-Sørensen F. *et al.* (1996) Myosin heavy chain isoform transformation in single fibres from m. vastus lateralis in spinal cord injured individuals: effects of long-term functional electrical stimulation (FES). *Pflugers Arch*, **431**, 513–518.

Andersen L.L., Andersen J.L., Magnusson S.P. *et al.* (2005) Changes in the human force-velocity relationship in response to resistance training and subsequent detraining. *J Appl Physiol*, **99**, 87–94.

Baar K. (2006) Training for endurance and strength: lessons from cell signaling. *Med Sci Sports Exerc*, **38**, 1939–1944.

Baar K. and Esser K. (1999) Phosphorylation of p70(S6k) correlates with increased skeletal muscle mass following resistance exercise. *Am J Physiol*, **276**, C120–C127.

Bell G.J., Syrotuik D., Martin T.P. *et al.* (2000) Effect of concurrent strength and endurance training on skeletal muscle properties and hormone concentrations in humans. *Eur J Appl Physiol*, **81**, 418–427.

Bickel C.S., Slade J., Mahoney E. *et al.* (2005) Time course of molecular responses of human skeletal muscle to acute bouts of resistance exercise. *J Appl Physiol*, **98**, 482–488.

Bodine S.C. (2006) mTOR signaling and the molecular adaptation to resistance exercise. *Med Sci Sports Exerc*, **38**, 1950–1957.

Bodine S.C., Stitt T.N., Gonzalez M. *et al.* (2001) Akt/mTOR pathway is a crucial regulator of skeletal muscle hypertrophy and can prevent muscle atrophy in vivo. *Nat Cell Biol*, **3**, 1014–1019.

Bohé J., Low J.F., Wolfe R.R. and Rennie M.J. (2001) Latency and duration of stimulation of human muscle protein synthesis during continuous infusion of amino acids. *J Physiol*, **532**, 575–579.

Bottinelli R. (2001) Functional heterogeneity of mammalian single muscle fibres: do myosin isoforms tell the whole story? *Pflugers Arch*, **443**, 6–17.

Bottinelli R. and Reggiani C. (2000) Human skeletal muscle fibres: molecular and functional diversity. *Prog Biophys Mol Biol*, **73**, 195–262.

Bottinelli R., Pellegrino M.A., Canepari R. *et al.* (1999) Specific contributions of various muscle fibre types to human muscle performance: an in vitro study. *J Electromyogr Kinesiol*, **9**, 87–95.

Bylund-Fellenius A.C., Ojamaa K.M., Flaim K.E. *et al.* (1984) Protein synthesis versus energy state in contracting muscles of perfused rat hindlimb. *Am J Physiol*, **246**, E297–E305.

Campos G.E., Luecke T.J., Wendeln H.K. *et al.* (2002) Muscular adaptations in response to three different resistance-training regimens: specificity of repetition maximum training zones. *Eur J Appl Physiol*, **88**, 50–60.

Coffey V.G. and Hawley J.A. (2007) The molecular bases of training adaptation. *Sports Med*, **37**, 737–763.

Coffey V.G., Pilegaard H., Garnham A.P. *et al.* (2009) Consecutive bouts of diverse contractile activity alter acute responses in human skeletal muscle. *J Appl Physiol*, **106**, 1187–1197.

Conboy I.M. and Rando T.A. (2002) The regulation of Notch signaling controls satellite cell activation and cell fate determination in postnatal myogenesis. *Dev Cell*, **3**, 397–409.

Cuthbertson D., Smith K., Babraj J. *et al.* (2005) Anabolic signaling deficits underlie amino acid resistance of wasting, aging muscle. *FASEB J*, **19**, 422–424.

Dreyer H.C., Fujita S., Cadenas J.G. *et al.* (2006) Resistance exercise increases AMPK activity and reduces 4E-BP1 phosphorylation and protein synthesis in human skeletal muscle. *J Physiol*, **576**, 613–624.

Durham W.J., Miller S.L., Yeckel C.W. *et al.* (2004) Leg glucose and protein metabolism during an acute bout of resistance exercise in humans. *J Appl Physiol*, **97**, 1379–1386.

Esmarck B., Andersen J.L., Olsen S. *et al.* (2001) Timing of postexercise protein intake is important for muscle hypertrophy with resistance training in elderly humans. *J Physiol*, **535**, 301–311.

Favier F.B., Benoit H. and Freyssenet D. (2008) Cellular and molecular events controlling skeletal muscle mass in response to altered use. *Pflugers Arch*, **456**, 587–600.

Fitts R.H. and Widrick J.J. (1996) Muscle mechanics: adaptations with exercise-training. *Exerc Sport Sci Rev*, **24**, 427–473.

Fleck S.J. and Kraemer W.J. (2004) *Designing Resistance Training Programs*, 3rd edn, Human Kinetics, Champaign, IL.

Folland J.P. and Williams A.G. (2007) The adaptations to strength training: morphological and neurological contributions to increased strength. *Sports Med*, **37**, 145–168.

Fry A.C. (2004) The role of resistance exercise intensity on muscle fibre adaptations. *Sports Med*, **34**, 663–679.

Fujita S., Dreyer H.C., Drummond M.J. *et al.* (2009) Essential amino acid and carbohydrate ingestion before resistance exercise does not enhance postexercise muscle protein synthesis. *J Appl Physiol*, **106**, 1730–1739.

Gabriel D.A., Kamen G. and Frost G. (2006) Neural adaptations to resistive exercise: mechanisms and recommendations for training practices. *Sports Med*, **36**, 133–149.

Glowacki S.P., Martin S.E., Maurer A. *et al.* (2004) Effects of resistance, endurance, and concurrent exercise on training outcomes in men. *Med Sci Sports Exerc*, **36**, 2119–2127.

Haddad F. and Adams G.R. (2002) Selected contribution: acute cellular and molecular responses to resistance exercise. *J Appl Physiol*, **93**, 394–403.

Häkkinen K., Alen M., Kraemer W.J. *et al.* (2003) Neuromuscular adaptations during concurrent strength and endurance training versus strength training. *Eur J Appl Physiol*, **89**, 42–52.

Harridge S.D. (1996) The muscle contractile system and its adaptation to training, in *Human Muscular Function During Dynamic Exercise* (eds P. Marconnet, B. Saltin, P. Komi *et al.*), Medicine and Sports Science, Karger, Basel, pp. 82–94.

Harridge S.D. (2007) Plasticity of human skeletal muscle: gene expression to in vivo function. *Exp Physiol*, **92**, 783–797.

Harridge S.D.R., Bottinelli R., Canepari M. *et al.* (1996) Whole-muscle and single-fibre contractile properties and myosin heavy chain isoforms in humans. *Pflugers Arch*, **432**, 913–920.

Hather B.M., Tesch P.A., Buchanan P. and Dudley G.A. (1991) Influence of eccentric actions on skeletal muscle adaptations to resistance training. *Acta Physiol Scand*, **143**, 177–185.

Hawke T.J. (2005) Muscle stem cells and exercise training. *Exerc Sport Sci Rev*, **33**, 63–68.

Hickson R.C. (1980) Interference of strength development by simultaneously training for strength and endurance. *Eur J Appl Physiol Occup Physiol*, **45**, 255–263.

Hoff J., Helgerud J. and Wisløff U. (1999) Maximal strength training improves work economy in trained female cross-country skiers. *Med Sci Sports Exerc*, **31**, 870–877.

Izquierdo M., Häkkinen K., Ibáñez J. *et al.* (2005) Effects of combined resistance and cardiovascular training on strength, power, muscle cross-sectional area, and endurance markers in middle-aged men. *Eur J Appl Physiol*, **94**, 70–75.

Kadi F., Schjerling P., Andersen L.L. *et al.* (2004) The effects of heavy resistance training and detraining on satellite cells in human skeletal muscles. *J Physiol*, **558**, 1005–1012.

Kadi F., Charifi N., Denis C. *et al.* (2005) The behaviour of satellite cells in response to exercise: what have we learned from human studies? *Pflugers Arch*, **451**, 319–327.

Kosek D.J., Kim J.S., Petrella J.K. *et al.* (2006) Efficacy of 3 days/wk resistance training on myofiber hypertrophy and myogenic mechanisms in young vs. older adults. *J Appl Physiol*, **101**, 531–544.

Kraemer W.J., Patton J.F., Gordon S.E. *et al.* (1995) Compatibility of high-intensity strength and endurance training on hormonal and skeletal muscle adaptations. *J Appl Physiol*, **78**, 976–989.

Kumar V., Atherton P., Smith K. and Rennie M.J. (2009a) Human muscle protein synthesis and breakdown during and after exercise. *J Appl Physiol*, **106**, 2026–2039.

Kumar V., Selby A., Rankin D. *et al.* (2009b) Age-related differences in the dose-response relationship of muscle protein synthesis to resistance exercise in young and old men. *J Physiol*, **15**, 211–217.

Lionikas A., Li M. and Larsson L. (2006) Human skeletal muscle myosin function at physiological and non-physiological temperatures. *Acta Physiol*, **186**, 151–158.

Mackey A.L., Kjaer M., Charifi N. *et al.* (2009) Assessment of satellite cell number and activity status in human skeletal muscle biopsies. *Muscle Nerve*, **40**, 455–465.

Miller B.F., Olesen J.L., Hansen M. *et al.* (2005) Coordinated collagen and muscle protein synthesis in human patella tendon and quadriceps muscle after exercise. *J Physiol*, **567**, 1021–1033.

Moore D.R., Robinson M.J., Fry J.L. *et al.* (2009) Ingested protein dose response of muscle and albumin protein synthesis after resistance exercise in young men. *Am J Clin Nutr*, **89**, 161–168.

Nader G.A. (2006) Concurrent strength and endurance training: from molecules to man. *Med Sci Sports Exerc*, **38**, 1965–1970.

Østerås H., Helgerud J. and Hoff J. (2002) Maximal strength-training effects on force-velocity and force-power relationships explain increases in aerobic performance in humans. *Eur J Appl Physiol*, **88**, 255–263.

Pallafacchina G., Calabria E., Serrano A.L. *et al.* (2002) A protein kinase B-dependent and rapamycin-sensitive pathway controls skeletal muscle growth but not fiber type specification. *Proc Natl Acad Sci U S A*, **99**, 9213–9218.

Petrella J.K., Kim J.S., Cross J.M. *et al.* (2006) Efficacy of myonuclear addition may explain differential myofiber growth among resistance-trained young and older men and women. *Am J Physiol Endocrinol Metab*, **291**, E937–E946.

Petrella J.K., Kim J.S., Mayhew D.L. *et al.* (2008) Potent myofiber hypertrophy during resistance training in humans is associated with satellite cell-mediated myonuclear addition: a cluster analysis. *J Appl Physiol*, **104**, 1736–1742.

Pette D. and Staron R.S. (2000) Myosin isoforms, muscle fiber types, and transitions. *Microsc Res Tech*, **50**, 500–509.

Phillips S.M., Tipton K.D., Aarsland A. *et al.* (1997) Mixed muscle protein synthesis and breakdown after resistance exercise in humans. *Am J Physiol*, **273**, E99–107.

Phillips S.M., Tipton K.D., Ferrando A.A. and Wolfe R.R. (1999) Resistance training reduces the acute exercise-induced increase in muscle protein turnover. *Am J Physiol*, **276**, E118–E124.

Rasmussen B.B. and Phillips S.M. (2003) Contractile and nutritional regulation of human muscle growth. *Exerc Sport Sci Rev*, **31**, 127–131.

Rennie M.J., Edwards R.H., Halliday D. *et al.* (1982) Muscle protein synthesis measured by stable isotope techniques in man: the effects of feeding and fasting. *Clin Sci*, **63**, 519–523.

Rivas D.A., Lessard S.J. and Coffey V.G. (2009) mTOR function in skeletal muscle: a focal point for overnutrition and exercise. *Appl Physiol Nutr Metab*, **34**, 807–816.

Rønnestad B.R., Hansen E.A. and Raastad T. (2009) Strength training improves 5-min all-out performance following 185 min of cycling. *Scand J Med Sci Sports*, Epub ahead of print.

Schaub M.C., Brunner U.T., Von Schulthess C. *et al.* (1989) Adaptation of contractile proteins in human heart and skeletal muscles. *Biomed Biochim Acta*, **48**, S306–S312.

Schiaffino S. and Reggiani C. (1994) Myosin isoforms in mammalian skeletal muscle. *J Appl Physiol*, **77**, 493–501.

Schiaffino S. and Reggiani C. (1996) Molecular diversity of myofibrillar proteins: gene regulation and functional significance. *Physiol Rev*, **76**, 371–423.

Schmidtbleicher D. and Haralambie G. (1981) Changes in contractile properties of muscle after strength training in man. *Eur J Appl Physiol*, **46**, 221–228.

Seynnes O.R., de Boer M. and Narici M.V. (2007) Early skeletal muscle hypertrophy and architectural changes in response to high-intensity resistance training. *J Appl Physiol*, **102**, 368–373.

Smerdu V., Karsch-Mizrachi I., Campione M. *et al.* (1994) Type IIx myosin heavy chain transcripts are expressed in type IIb fibers of human skeletal muscle. *Am J Physiol*, **267**, C1723–C1728.

Staron R.S., Leonardi M.J., Karapondo D.L. *et al.* (1991) Strength and skeletal muscle adaptations in heavy-resistance-trained women after detraining and retraining. *J Appl Physiol*, **70**, 631–640.

Staron R.S., Karapondo D.L., Kraemer W.J. *et al.* (1994) Skeletal muscle adaptations during early phase of heavy-resistance training in men and women. *J Appl Physiol*, **76**, 1247–1255.

Støren O., Helgerud J., Støa E.M. and Hoff J. (2008) Maximal strength training improves running economy in distance runners. *Med Sci Sports Exerc*, **40**, 1087–1092.

Tang J.E. and Phillips S.M. (2009) Maximizing muscle protein anabolism: the role of protein quality. *Curr Opin Clin Nutr Metab Care*, **12**, 66–71.

Tang J.E., Perco J.G., Moore D.R. *et al.* (2008) Resistance training alters the response of fed state mixed muscle protein synthesis in young men. *Am J Physiol Regul Integr Comp Physiol*, **294**, R172–R178.

Tang J.E., Moore D.R., Kujbida G.W. *et al.* (2009) Ingestion of whey hydrolysate, casein, or soy protein isolate: effects on mixed muscle protein synthesis at rest and following resistance exercise in young men. *J Appl Physiol*, **107**, 987–992.

Tihanyi J., Apor P. and Fekete G. (1982) Force-velocity-power characteristics and fiber composition in human knee extensor muscles. *Eur J Appl Physiol Occup Physiol*, **48**, 331–343.

Tipton K.D., Rasmussen B.B., Miller S.L. *et al.* (2001) Timing of amino acid-carbohydrate ingestion alters anabolic response of muscle to resistance exercise. *Am J Physiol Endocrinol Metab*, **281**, E197–E206.

Toigo M. and Boutellier U. (2006) New fundamental resistance exercise determinants of molecular and cellular muscle adaptations. *Eur J Appl Physiol*, **97**, 643–663.

Wilkinson S.B., Phillips S.M., Atherton P.J. *et al.* (2008) Differential effects of resistance and endurance exercise in the fed state on signalling molecule phosphorylation and protein synthesis in human muscle. *J Physiol*, **586**, 3701–3717.

Woolstenhulme M.T., Conlee R.K., Drummond M.J. *et al.* (2006) Temporal response of desmin and dystrophin proteins to progressive resistance exercise in human skeletal muscle. *J Appl Physiol*, **100**, 1876–1882.

Yates J.W. and Kamon E. (1983) A comparison of peak and constant angle torque-velocity curves in fast and slow-twitch populations. *Eur J Appl Physiol Occup Physiol*, **51**, 67–74.

2.3 Adaptive Processes in Human Bone and Tendon

Constantinos N. Maganaris, Jörn Rittweger and Marco V. Narici, Manchester Metropolitan University, Institute for Biomedical Research into Human Movement and Health (IRM), Manchester, UK

2.3.1 INTRODUCTION

Bones and tendons have certain structural and functional similarities. Both materials are composites and are subjected to forces generated during joint movement and locomotion, and both exhibit mechanical behaviour adaptable to the conditions of functional loading they experience.

The plasticity of bones is a characteristic relevant to the management of fractures, a major concern in old age worldwide (Baron *et al.*, 1996; Gullberg, Johnell and Kanis, 1997). The fracture risk in older people is, at least partly, due to the reduction in bone material (Kanis *et al.*, 2001), and also to deterioration of the bone material's intrinsic properties (Diab *et al.*, 2006; Nalla *et al.*, 2004; Shiraki *et al.*, 2008). The term 'osteoporosis' summarizes these bone-related alterations that predispose to fractures. As discussed in detail in Chapter 1.3, bones can adapt to mechanical stimuli by adding material where it is needed and removing it where it is dispensable.[1] Hence, one might expect that 'increasing' the mechanical stimulation of bone should lead to 'bigger' and stronger bones. Accordingly, exercise should constitute a straightforward means to prevent osteoporosis and fractures (Kohrt *et al.*, 2004; Rittweger, 2006).

The first part of this chapter will discuss the extent to which this is achievable and address the issue of how exactly bones behave in response to mechanical loading in the form of exercise and physical activity, or the lack of it. The second part of the chapter is dedicated to tendons, which also adapt to the presence or absence of mechanical loading and change their properties accordingly. This part will discuss in detail the functional role of tendons and review recent knowledge on the application of *in vivo* techniques for quantifying human tendon mechanical properties and their adaptations to mechanical loading in the form of strength training and chronic physical activity, or chronic unloading caused by pathology and experimental manipulation in healthy humans.

[1]Bone material properties, by contrast, seem to be fairly constant within species; see Currey (2002), Ruffoni *et al.* (2007).

2.3.2 BONE

2.3.2.1 Origin of musculoskeletal forces

Mechanical engineers design structures so that they can safely tolerate the peak loads applied to them. Biomechanical analyses suggest that the peak forces in our musculoskeletal system are generated by muscle contractions, rather than by gravitational loading. This is illustrated for the human ankle joint in Figure 2.3.1.

In hopping, walking, and running, loads are typically applied through the ball of the foot. As illustrated in Figure 2.3.1, the tensile force generated by the calf muscles and transmitted through the Achilles tendon is three times larger than the ground reaction force.[2] As a consequence, the gravitational forces of the body weight are small in comparison to the forces generated by muscle contractions. This is due to the fact that our muscles work against short levers, ranging between 1 : 2 to 1 : 10 (Martin, Burr and Sharkey, 1998; Özkaya and Nordin, 1998). The situation is even more aggravated in the upper extremity, where body weight does not usually contribute at all to the loading of bones. In this context, the concept of 'weight bearing' seems dispensable, or even misleading. Rather, we have to acknowledge that it is the muscular forces that the skeleton has to adapt to. In the lower extremity, those

[2]We are considering only the net moments and forces here. In reality, one has to assume that the antagonistic muscle groups will be co-contracting at the same time (see Baratta *et al.*, 1988), and that the calf muscle and tendon forces are consequently larger than those net forces.

Strength and Conditioning – Biological Principles and Practical Applications Marco Cardinale, Rob Newton, and Kazunori Nosaka.

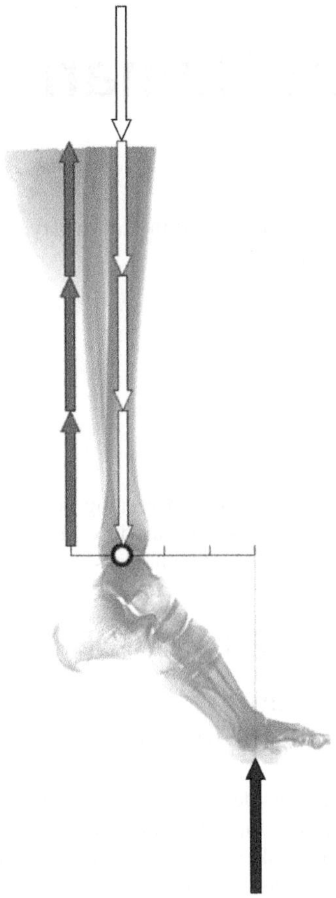

Figure 2.3.1 Biomechanical analysis of the loading patterns around the ankle joint. The force introduced via the ball of the foot operates a lever that is approximately three times longer than the Achilles tendon. Therefore, the Achilles tendon force will be three times larger than the ground reaction force (indicated by the black arrows). If we neglect the mass of the foot and the lower shank, another force component will be transmitted through the tibia to support the body mass (indicated by the fourth white arrow). Peak ground reaction forces during one-legged hopping range around 2.5 kN in young men with 70 kg body mass (Runge *et al.*, 2004). Accordingly, the calf muscles generate a force of 7.5 kN, and the tibia is loaded with 10 kN (equivalent to the weight of one tonne). Weight, on the other hand, is defined as mass × gravitational acceleration. Hence, the force contribution to the loading of the tibia, as by the body weight, is 70 kg × 9.81 m/s/s, or 0.7 kN, or only 7% of the total loading

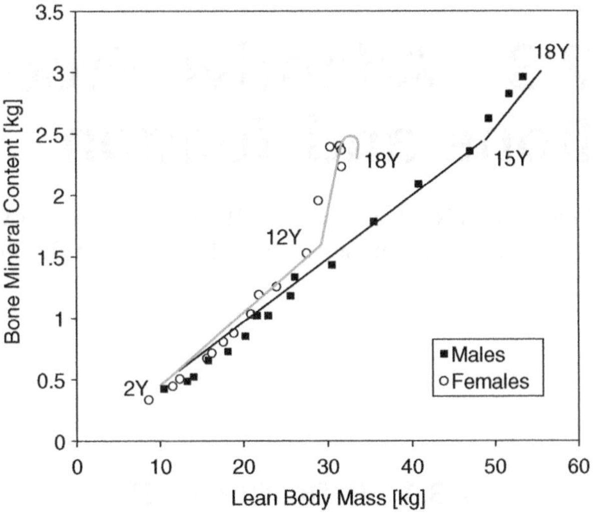

Figure 2.3.2 Muscle–bone relationship, as assessed by whole-body dual x-ray absorptiometry. Bone mineral content (BMC) and lean body mass (most of which is muscle mass) are directly proportional to each other in boys between 2 and 15 years of age, and in girls between 2 and 12 years of age. After the onset of puberty, this relationship is lost in girls, who accrue more BMC than is explicable by lean body mass. Data from Zanchetta, Plotkin and Alvarez-Filgueira (1995) and adopted according to Schiessl, Frost and Jee (1998)

muscle forces are obviously generated with body mass as an inertial resistor.

Evidence suggests that the tibia strength is indeed adapted to calf muscle cross-section as a surrogate measure of muscle strength (Rittweger *et al.*, 2000). More generally, there seems to be a stoichiometric relationship between musculature and skeleton in the human body, as demonstrated in Figure 2.3.2. It can also be seen from this figure that during puberty girls start to accrue more bone mass than explicable by muscle mass. It seems reasonable to assume that this surplus constitutes a calcium reservoir to be utilized during lactation (Kalkwarf and Specker, 1995; Kalkwarf *et al.*, 1996; Schiessl, Frost and Jee, 1998).

The question arises how many, and what kind of, deformation cycles are required to optimally build stronger bones. In growing rats (Umemura *et al.*, 1997), but also in the turkey ulna (Rubin and Lanyon, 1987), four to five cycles per day seem to elicit a maximal or near-maximal response. On the other hand, it has been argued that many cycles of comparatively small deformation magnitude can have the same effect upon bone as a large number of relatively small deformations (Rubin *et al.*, 2001). Moreover, strain rate is known to affect bone adaptive processes independent of strain magnitude (Mosley and Lanyon, 1998). It is of interest in this context that muscles themselves generate higher-frequency components (Huang, Rubin and McLeod, 1999), and that the rate of force development by a muscle varies across individuals and contraction types, as for example in different kinds of sports.

2.3.2.2 Effects of immobilization on bone

Immobilization leads to rapid and profound bone losses. This is best demonstrated in individuals with spinal-cord injury

(SCI), where muscles are paralysed due to nervous traffic disruption. Biochemical markers of bone resorption are strongly increased immediately after the injury (Maimoun *et al.*, 2005), returning to baseline values within a few years (Reiter *et al.*, 2007; Zehnder *et al.*, 2004). Importantly, the bone losses occur only in the paralysed limbs (Biering-Sorensen, Bohr and Schaadt, 1990; Frey-Rindova *et al.*, 2000; Tsuzuku, Ikegami and Yabe, 1999), and a new steady state is reached after approximately five years (Eser *et al.*, 2004). The losses are more pronounced in the epiphyses[3] than in the diaphyses[4] (Eser *et al.*, 2004). It should be realized that some muscle contractions still take place in the paralysed limbs, and it has been proposed that the residual state reflects bone adaptation to the musculature's force-generating potential (Rittweger *et al.*, 2006b). In that sense, one can interpret the bone losses after SCI as an adaptive process.

Similarly, bone losses are observed in many other clinical disorders, such as stroke (Jorgensen *et al.*, 2000) and anterior cruciate ligament (Leppala *et al.*, 1999), as well as other soft-tissue knee injuries (Sievanen, Heinonen and Kannus, 1996). Bone loss is also a major concern for long-term space missions. It probably owes to the inability to elicit forceful muscle contractions under microgravity conditions that bone mass is readily lost during space flight (Mack *et al.*, 1967; Oganov *et al.*, 1992; Rambaut *et al.*, 1975); experimental bed rest is a ground-based model to simulate this (LeBlanc *et al.*, 1990; Rittweger *et al.*, 2005a; Vico *et al.*, 1987). The past decade has seen vigorous efforts to develop countermeasures against such immobilization-induced bone losses (Pavy-Le Traon *et al.*, 2007). Research has yielded unfavourable results for endurance exercise (Grigoriev *et al.*, 1992), and for resistive exercise when it is only performed for two to three days per week (Rittweger *et al.*, 2005a; Watanabe *et al.*, 2004). However, when performed on a daily basis, resistive exercise with (Rittweger *et al.*, 2010) or without (Shackelford *et al.*, 2004) whole-body vibration (WBV) has proved to be highly effective.

Bone losses incurred during bed rest are recovered within one or two years (Rittweger and Felsenberg, 2009). This recovery is remarkable in three ways. First, the accrual is initially extremely rapid (see Figure 2.3.3a), comparable to the pubertal growth spurt in quantitative terms. Second, the recovery is stimulated by merely habitual activity and does not require specific measures. The ease with which bone is accrued during rehabilitation is therefore in stark contrast to the hardships of preventing bed rest-induced bone losses by regular exercise (Rittweger *et al.*, 2006a), and also to the difficulty of increasing bone mass by regular training (see Section 2.3.2.3). Third, the bone mineral gained during the recovery period matches exactly the losses incurred during bed rest (see Figure 2.3.3b). This is the more surprising as bed rest-induced bone losses show great inter-individual variability. It therefore seems that bone adaptation is both highly efficient and tightly controlled.

[3]Close to the joints.
[4]The shafts of long bones.

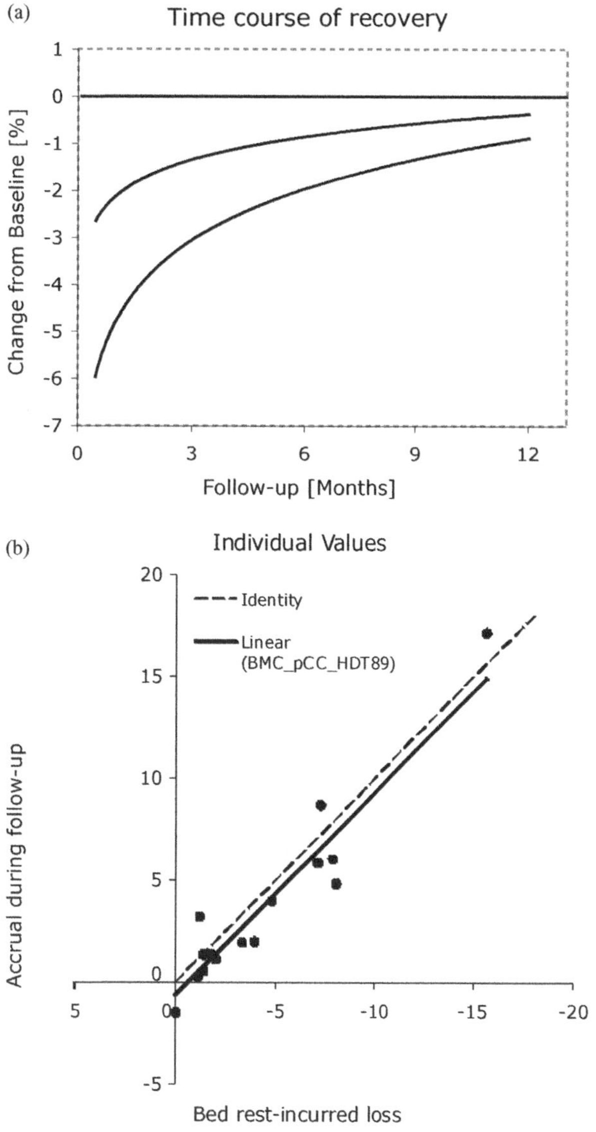

Figure 2.3.3 Recovery of losses in bone mineral content that occurred in the distal tibia epiphysis in response to 90 days of strict bed rest. Diagrams have been adopted from Rittweger and Felsenberg (2009). (a) Values are given as percentage changes from baseline for the control group (Ctrl), which did bed rest only, without any exercise, and for the group that did resistive exercise on a gravity-independent dynamometer two to three times per week (Ex). A curvilinear time course of recovery was perceivable in all individuals. It has been fitted here by an exponential function, compatible with a first-order regulatory system. (b) Comparison of intra-individual losses at the end of bed rest and the accrual of bone mass during 12 months' follow-up after the bed rest. As can be seen, the two were closely correlated ($R^2 = 0.87$), indicating that re-accrual individually matched previous losses. Upper curved line = Ex; lower curved line = Ctrl

2.3.2.3 Effects of exercise on bone

It is widely held that exercise is beneficial for bone health (Kohrt *et al.*, 2004). This view is based on a great number of interventional studies that have demonstrated increases in bone strength through physical exercise (Bassey *et al.*, 1998; Friedlander *et al.*, 1995; Heinonen *et al.*, 1996; Lohman *et al.*, 1995; Maddalozzo and Snow, 2000; Snow-Harter *et al.*, 1992; Vincent and Braith, 2002). Importantly, the exercise effects are site-specific (Adami *et al.*, 1999; Bass *et al.*, 2002; Heinonen *et al.*, 1996; Kannus *et al.*, 1994; Kontulainen *et al.*, 1999; Johannsen *et al.*, 2003; Winters-Stone and Snow, 2006); that is, they occur in the loaded bones only.

Bone adaptation consists in structural modification, rather than alteration of bone material properties (Haapasalo *et al.*, 2000). As such, there is an increase in area and cross-sectional moment of inertia (see Chapter 1.3) of cortical bone, and also an enhancement of the density of the trabecular (Haapasalo *et al.*, 2000; Nikander *et al.*, 2005, 2006; Wilks *et al.*, 2009a). Quantitatively, exercise effects seem to be larger in the diaphyses than in the epiphyses. Epiphyses are rich in trabecular bone, which might indicate that trabecular bone is in stronger relation to endocrine and metabolic processes than cortical bone (Hagino *et al.*, 1993; Rittweger *et al.*, 2009). However, it must be considered that the mechanical competence of trabecular bone increases by the power of 2–3 of its apparent density (Ebbesen, Thomsen and Mosekilde, 1997), meaning that relatively small increases in trabecular bone mineral density elicit large increases in stiffness and strength. In fact, when considering this nonlinear relationship, adaptation to exercise seems to be proportional in the epiphyses and diaphyses (Wilks *et al.*, 2009a), in keeping with the view that forces and force-related signals induced by exercise are proportionate within the different portions of bones.

Of course, it is important to consider the type of exercise. Generally exercises involving large muscular forces are found to be more effective than exercises with low force levels (Bassey and Ramsdale, 1994; Dickerman, Pertusi and Smith, 2000; Maddalozzo and Snow, 2000; Tsuzuku, Ikegami and Yabe, 1998; Vincent and Braith, 2002). Accordingly, sprinters have stronger bones than endurance runners (Bennell *et al.*, 1997; Wilks *et al.*, 2009a), and weightlifters have particularly strong bones (Dickerman, Pertusi and Smith, 2000; Karlsson, Johnell and Obrant, 1993). Moreover, exercises involving impacts, in particular with unconventional patterns, seem to have stronger effects upon bone strength than low-impact exercises (Nikander *et al.*, 2005, 2006). This is in line with observations from animal studies of strain rate having an osteogenic effect per se (Mosley and Lanyon, 1998). Accordingly, alpine skiers, gymnasts, and ball sports players have enhanced bone strength. Similarly, vibratory stimuli, as applied through WBV platforms (Cardinale and Rittweger, 2006), seem to induce bone accrual (Gilsanz *et al.*, 2006; Gusi, Raimundo and Leal, 2006; Rubin *et al.*, 2004; Verschueren *et al.*, 2004) independently of muscular effects (Verschueren *et al.*, 2004).

By contrast, exercises without impact loading seem to be of little help for bone. Cycling, for example, has traditionally been associated with poor bone health (Medelli *et al.*, 2009; Nichols, Palmer and Levy, 2003). To support that notion, it is observed that bone losses occur during the competitive season in cyclists (Barry and Kohrt, 2008), whilst gymnasts gain bone mass during the competitive season (Snow *et al.*, 2001). However, recent studies by Wilks *et al.* (2009a) and Wilks, Gilliver and Rittweger (2009b) have demonstrated that the leg bones of track cyclists are as strong as those of track runners, which may put the relative importance of impact loading for bone adaptation into some perspective.

As to the time course of bone adaptive processes, it should be considered that bone is lost more rapidly than it is accrued. For example, bone loss starts on the second day of bed rest (Baecker *et al.*, 2003), but recovery from 2- or 3-month bed rest occurs only after 12–24 months (Rittweger and Felsenberg, 2009; Rittweger *et al.*, 2010). Accordingly, exercise-induced gains are readily lost after exercise discontinuation (Winters and Snow, 2000), although it seems that some long-term benefits in non-exercisers may persist (Kontulainen *et al.*, 1999, 2004).

Although they are not adaptive processes in a strict sense, we will also discuss here the potential adverse effects that exercise can have upon tendons and bone.

Fatigue fractures[5] occur occasionally in military recruits as well as in runners and other athletes. It has to be assumed that repetitive strains promote microdamage generation (see Chapter 1.3), which necessitates bone remodelling.[6] Fatigue fractures occur when the repair process cannot keep up with the rate of microdamage generation. Often, but not always, pain alerts against imminent fatigue fractures.

The female athlete triad is a condition characterized by poor bone status, amenorrhoea and caloric restriction (Otis *et al.*, 1997). It is frequent in sports where success relies upon leanness, for example in distance running and ballet dancing. Reduced bone strength is usually caused by oestrogen depletion, which cannot be balanced by the positive osteogenic exercise effects (Bass *et al.*, 1994). Although little is known regarding the risk of fracture in affected women, the condition is regarded as potentially dangerous and should receive medical attention.

2.3.2.4 Bone adaptation across the life span

We are ignorant as to whether the effects of immobilization upon bone are modulated across the life span. However, there is ample evidence to suggest that the relative efficacy of exercise in enhancing bone strength changes as a function of age. The principles outlined in the last section are based on studies in adults, and we will now briefly discuss how far these apply

[5]Also sometimes misleadingly called 'stress' fractures.

[6]This is probably the reason for the reduced cortical bone mineral density in athletes (by 2% or so, see Haapasalo *et al.*, 2000), which is more pronounced in endurance runners than in sprinters (Wilks *et al.*, 2009a).

during childhood and puberty, during menopause, and in old age.

Ample evidence suggests that exercise during childhood and adolescence increases bone mass and strength (Hughes *et al.*, 2007; Specker and Binkley, 2003; Witzke and Snow, 2000); most of these studies have focussed on jumping exercises. Although very little is known about other exercises, one study suggests that resistance training may be more beneficial to bone than hopping during childhood (Morris *et al.*, 1997). Importantly, the bone's response to jumping exercises seems to be enhanced by supplementary calcium intake (Bass *et al.*, 2007; Specker and Binkley, 2003), underlining the importance of nutrition. Although little is known about the effects of puberty in boys, there is clear evidence in girls that exercise effects are most pronounced during early pubertal stages (Mackelvie *et al.*, 2001; Matthews *et al.*, 2006; Petit *et al.*, 2002), and that the efficacy diminishes after menarche (Bass *et al.*, 1998; Heinonen *et al.*, 2000). Thus, puberty constitutes a window of opportunity in which exercise is most beneficial toward building strong bones for a life time (Hind and Burrows, 2007).

The time after menopause is normally associated with bone losses in the order of 10% (Recker *et al.*, 2000), which are caused by oestrogen withdrawal (Kalu *et al.*, 1991). Evidence suggests that exercises may become less effective after menopause (Bassey *et al.*, 1998). However, a number of exercise intervention studies were able to mitigate (Heinonen *et al.*, 1998), to prevent (Maddalozzo and Snow, 2000; Milliken *et al.*, 2003; Rhodes *et al.*, 2000), or even to reverse (Verschueren *et al.*, 2004) post-menopausal bone losses, showing that the female skeleton in principle retains responsiveness to physical stimuli even after menopause.

Old age is often associated with increasing bone losses (Heaney *et al.*, 2000). However, these losses do not seem to occur uniformly in the skeleton, but are rather pronounced in the axial skeleton and the upper extremity, whilst the lower leg is usually spared (Riggs *et al.*, 2004). Numerous training studies suggest that the skeleton of older people is susceptible to exercise stimuli, and that positive bone responses can be obtained (Verschueren *et al.*, 2004; Vincent and Braith, 2002). However, a recent study in master runners and race-walkers suggests that osteogenic exercise benefits seem to diminish with increasing age (Wilks *et al.*, 2009c). Future studies will have to elucidate whether this is due to decreased responsiveness of bone, or whether the signals associated with running and other exercises themselves diminish.

2.3.3 TENDON

2.3.3.1 Functional and mechanical properties

The primary role of tendons is to transmit contractile forces to the skeleton in order to generate joint movement. In doing so, tendons do not behave as rigid bodies, but as springs, and this has several important functional implications.

First, having a muscle attached to a compliant tendon makes it more difficult to control the joint position (Rack and Ross, 1984). Consider, for example, an external oscillating force applied to a joint at a certain angle: maintaining the joint would require generation of a constant contractile force in the muscle. If the tendon of the muscle were very compliant, its length would change by the external oscillating load, even if the muscle were held at a constant length. This would result in failure to maintain the joint at the angle desired. The association between tendon compliance and joint position control is also relevant to upright posture. Traditionally, balance of the body during stance has been considered to be modulated by stretch reflexes in the calf muscles, such that during forward sway, the calf muscles are stretched, and the reflex-mediated contraction plantar flexes the ankle and tilts the body backwards. However, this requires a very stiff Achilles tendon that can faithfully convert ankle dorsal flexion to calf-muscle lengthening, and recent experiments have shown that this does not occur. Instead, it has been shown that that calf-muscle length change is usually poorly or negatively correlated with bodily sway, because the Achilles tendon is compliant relative to the bodily load (for a review see Loram, Maganaris and Lakie, 2009). This means that calf-muscle spindles are unable to register bodily sway while standing, and knowledge of body position is required to appropriately modulate muscle activity and maintain balance.

Second, the elongation of a tendon during a static muscle contraction is accompanied by an equivalent shortening in the muscle. For a given contractile force, a more extensible tendon will allow the muscle to shorten more. This extra shortening will cause a shortening in the sarcomeres of the muscle. If the average muscle sarcomere operates in the ascending limb of the force–length relation, having a more extensible tendon will result in less contractile force. In contrast, if the average sarcomere operates in the descending limb of the force–length relation, having a more extensible tendon will result in more contractile force (Zajac, 1989).

Third, stretching a tendon results in elastic energy storage, and most of this energy is returned once the tensile load is removed. This passive mechanism of energy provision operates in tendons in the feet of legged mammals during terrestrial locomotion, thus saving metabolic energy that would otherwise be needed to displace the body ahead (Alexander, 1988). However, in tendons that stretch and recoil repeatedly under physiological conditions (e.g. the Achilles tendon), the heat lost may result in cumulative tendon thermal damage and injury, predisposing the tendon to ultimately rupture. Indeed, *in vivo* measurements and modelling-based calculations during exercise indicate that highly stressed, spring-like tendons may develop temperature levels above the 42.5 °C threshold for fibroblast viability (Wilson and Goodship, 1994). These findings are in line with the degenerative lesions often observed in the core of tendons acting as elastic energy stores, indicating that hyperthermia may be involved in the pathophysiology of exercise-induced tendon trauma.

Most of our knowledge of tendon mechanical properties is based on tests using isolated, non-human material. Two methods have traditionally been used in biomechanics investigations: (1)

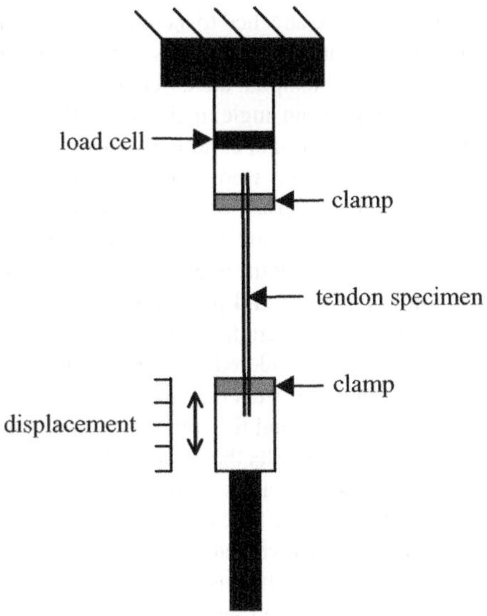

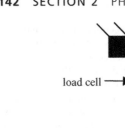

Figure 2.3.4 Schematic illustration of tensile testing apparatus

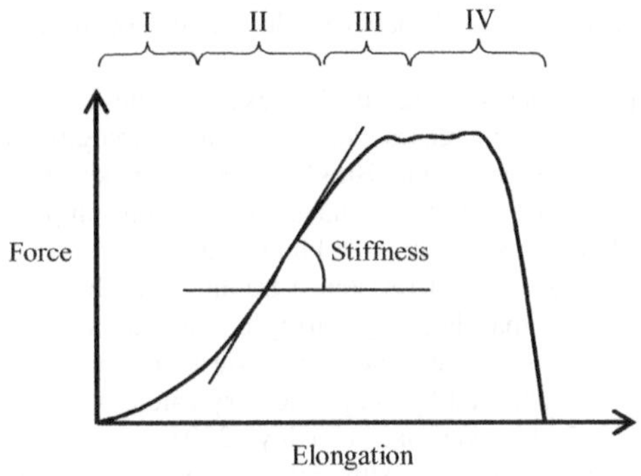

Figure 2.3.5 Typical force–elongation curve of a tendon pulled by a load exceeding the tendon elastic limit. I = toe region, II = linear region, III and IV = failure regions. Stiffness is the slope of the curve in the linear region

the free-vibration method, which is based on quantifying the decay in oscillation amplitude that takes place after a transient load is applied to a specimen (Ettema, Goh and Forwood, 1998); and (2) tensile testing methodologies, in which the specimen is stretched by an external force, while both the specimen deformation and the applied force are recorded (e.g. Butler *et al.*, 1978; Ker, 1992). The latter methodology seems to be the preferable one in many cases, mostly because it is considered to mimic adequately the way loading is imposed on tendons in real life.

A tensile testing machine is composed of an oscillating actuator and a load cell (Figure 2.3.4). The tendon specimen is gripped by two clamps, a static one mounted on the load cell and a moving one mounted on the actuator. The actuator is then set in motion while the load cell records the tension associated with the stretching applied. The tensile deformation of the specimen is taken from the displacement of the actuator.

A typical force–deformation plot of an isolated tendon is shown in Figure 2.3.5. Generally, in force–deformation curves, slopes relate to stiffness (N/mm), and areas to energy (J). In elongation-to-failure conditions, four different regions can be identified in the tendon force–deformation curve. Region I is the initial concave portion of the curve in which stiffness gradually increases; it is referred to as the tendon 'toe' region. Loads within the 'toe' region elongate the tendon by reducing the crimp angle of the collagen fibres at rest, but they do not cause further fibre stretching. Hence, loading within the 'toe' region does not exceed the tendon elastic limit; that is, subsequent unloading restores the tendon initial length. Further elongation brings the tendon into the 'linear' region II, in which stiffness remains constant as a function of elongation. In this region, elongation is the result of stretching imposed in the already aligned fibres by the load imposed in the preceding 'toe' region. At the end point of this region some fibres start to fail. Thus, (a) the tendon stiffness begins to drop and (b) unloading from this point does not restore the tendon initial length. Elongation beyond the 'linear' region brings the tendon into region III, where additional fibre failure occurs in an unpredictable way. Further elongation brings the tendon into region IV, where complete failure occurs.

Although regions I, II, III, and IV are apparent in tendon force–deformation curves during elongation-to-failure conditions, the shapes of the curves obtained differ between specimens. To a great extent these differences can be accounted for by inter-specimen dimensional differences. For example, tendons of equal length but different cross-sectional area exhibit different force–deformation properties; thicker tendons are stiffer. Similarly, different force–deformation curves are obtained from tendons of equal cross-sectional area but different initial length; shorter tendons are stiffer (see Butler *et al.*, 1978). To account for inter-specimen dimensional differences, tendon force is reduced to stress (MPa) by normalization to the tendon cross-sectional area, and tendon deformation is reduced to strain (%) by normalization to the tendon original length. The tendon stress–strain curve is similar in shape to the force–deformation curve, but reflects the intrinsic material properties rather than the structural properties of the specimen.

The most common material parameters taken from a stress–strain curve under elongation-to-failure conditions are the Young's modulus (GPa), the ultimate stress (MPa), and the ultimate strain (%). The Young's modulus is calculated as the product of the stiffness and the original length-to-cross-sectional area ratio of the specimen. Experiments on several tendons indicate that the Young's modulus reaches the level of 1–2 GPa at stresses exceeding 30 MPa, the ultimate tendon stress (i.e. stress at failure) ranges from 50 to 100 MPa, and the ultimate tendon strain (i.e. strain at failure) ranges from 4 to 10% (Bennett *et al.*, 1986; Butler *et al.*, 1978; Pollock and Shadwick, 1994).

2.3.3.2 *In vivo* testing

The examination of human tendon properties under *in vitro* conditions necessitates the use of donor specimens, which are not always readily available. Moreover, caution should be adopted when using the results of the *in vitro* test to infer *in vivo* function for the following reasons: (1) the forces exerted by maximal tendon loading under *in vivo* conditions may not reach the 'linear' region where stiffness and Young's modulus are measured under *in vitro* conditions; (2) clamping of an excised specimen in a testing rig is inevitably associated with some collagen fibre slippage and stress concentration that may result in premature rupture (Ker, 1992); (3) *in vitro* experiments have often been performed using preserved tendons, which may have altered properties (Matthews and Ellis, 1968; Smith, Young and Kearney, 1996).

Recently, however, a non-invasive method for assessing the mechanical properties of human tendons *in vivo* that circumvents the above problems was developed. This method is based on ultrasound scanning of a reference point along the tendon during an isometric contraction (Maganaris and Paul, 1999). The muscle forces generated by activation are measurable by dynamometry. They pull the tendon, causing a longitudinal deformation that can be quantified by measuring the displacement of the reference landmark on the scans recorded (Figure 2.3.6). The force–elongation plots obtained during loading–unloading can be transformed to the respective stress–strain plots by normalization to the dimensions of the tendon, which can also be measured using non-invasive imaging.

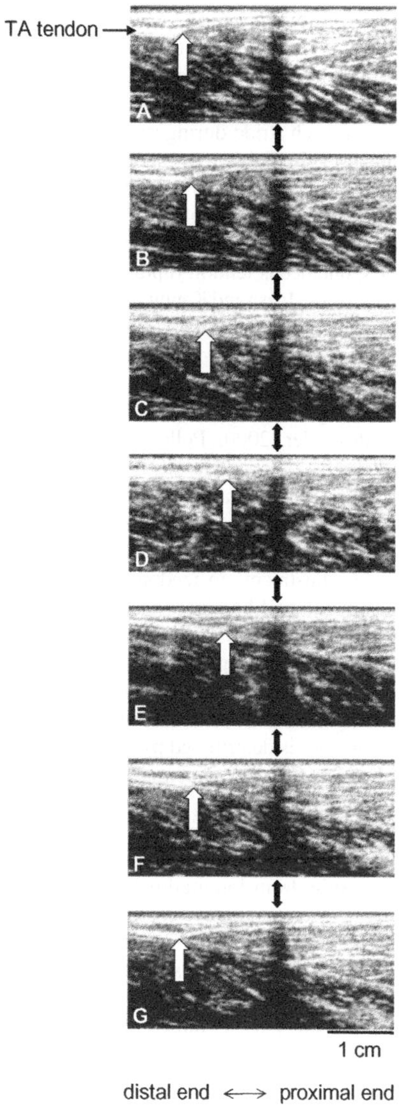

Figure 2.3.6 Typical *in vivo* sonographs of the human tibialis anterior (TA) tendon. A = resting state, B = 40% of maximal isometric contraction during activation, C = 80% of maximal isometric contraction during activation, D = 100% of maximal isometric contraction, E = 80% of maximal isometric contraction during relaxation, F = 40% of maximal isometric contraction during relaxation, G = 0% of maximal isometric contraction at the end of relaxation. The white arrow in each scan points to the TA tendon origin. The black double arrows point to the shadow generated by an echo-absorptive marker glued on the skin to identify any displacements of the scanning probe during muscle contraction–relaxation. The tendon origin displacement is larger during relaxation than with contraction at each loading level, indicating the presence of mechanical hysteresis in the tendon. Reproduced, with permission, from Maganaris, C. N. & Paul, J. P., 2000. 'Hysteresis Measurements in Intact Human Tendon'. J. Biomechanics. 33:12 © Elsevier

2.3.3.3 Tendon adaptations to altered mechanical loading

The general principles of *in vivo* tendon testing have often been applied to the study of the adaptations of human tendon to altered mechanical loading. The effect of increased or decreased chronic mechanical loading on the mechanical properties of human tendons *in vivo* has been examined in a number of cross-sectional and longitudinal design studies.

Cross-sectional studies

Cross-sectional studies have mainly compared (1) highly-stressed versus low-stressed tendons in a given sample, (2) tendons in sedentary versus athletic populations, (3) healthy versus disused tendons and (4) tendons in different age groups.

1. Tendons subjected to different habitual loading: when comparing the human Achilles/gastrocnemius and tibialis anterior tendons of young adults using the same methodology, it emerges that these two tendons have very similar Young's modulus (1.2 GPa) (Maganaris, 2002; Maganaris and Paul, 1999, 2002). This finding should be interpreted bearing in mind that the gastrocnemius and tibialis anterior tendons are subjected to different physiological forces. The gastrocnemius tendon is subjected to the high forces generated in late

stance and the tibialis anterior tendon is subjected to the lower forces generated by controlling ankle plantar flexion in the early stance phase of gait. *In vivo* measurements of tendon force indicate that the Achilles tendon may carry up to 110 MPa in each stride during running (Komi, Fukashiro and Jarvinen, 1992). This stress exceeds the average ultimate tensile tendon stress of 100 MPa (Butler *et al.*, 1978; Bennett *et al.*, 1986), which highlights the possibility of Achilles tendon rupture in a single pull in real life. Epidemiological studies of spontaneous tendon rupture verify these theoretical considerations (Jozsa and Kannus, 1997). Notwithstanding the above stress difference between the two tendons, the gastrocnemius tendon was not found to be intrinsically stiffer than the tibialis anterior tendon, which agrees with previous findings on several isolated animal tendons (Pike, Ker and Alexander, 2000; Pollock and Shadwick, 1994). Consistent with this result is the finding of Couppé *et al.* (2008) of increased patellar tendon cross-sectional area and stiffness, but not Young's modulus, in the stronger leg compared to the weaker leg of elite athletes. These findings indicate that adjustments in tendon structural properties to differences in habitual loading are accomplished by adding or removing material so that the tendon becomes thicker or thinner, rather than altering the intrinsic material properties. This notion is consistent with a theory developed by Ker, Alexander and Bennett (1988), according to which the thickness of the tendon is determined by the criterion of minimal combined mass for the tendon and its muscle, and the muscle-to-tendon-area ratio is rather constant. Cross-sectional data on humans (An, Linscheid and Brand, 1991) support the theory of Ker, Alexander and Bennett (1988), but recent results from longitudinal, interventional studies, which will be discussed below, challenge it and indicate that alterations in tendon size may not always be the sole mechanism underlying tendon stiffness adaptations to loading.

2. Tendons in sedentary versus athletic populations: The results of such comparative studies are insightful, but it should be considered that they may be confounded by self-selection bias and inter-subject variability in exercise training history.

   In one study, the cross-sectional area of the Achilles tendon has been shown to be greater along its length in long-distance runners than controls (Magnusson and Kjaer, 2003), with greater differences in the distal region of the tendon, close to the calcaneal attachment, suggesting that site-specific tendon hypertrophy may be associated with compressive rather than tensile tendon loading. Despite the differences in tendon size between the two groups and the reduced tendon stress in the runners, the stiffness of the Achilles tendon-aponeurosis was similar in runners and non-runners (Rosager *et al.*, 2002). A similar finding was reported by Arampatzis *et al.* (2007a), who included three subject groups in their study: endurance runners, sprinters, and controls. The stiffness of the Achilles tendon-aponeurosis was greater in the sprinters than in controls, but there were no differences between endurance runners and non-runners and there was a significant correlation between tendon stiffness and tendon force during maximum voluntary contraction. However, Kubo *et al.* (2000a) reported no differences in the Achilles tendon-aponeurosis stiffness between sprinters and controls. Taken together, the above data suggest that the Achilles tendon-aponeurosis stiffness does not show a response proportional to the intensity of the physical activity performed and there may be a threshold, corresponding to the intensity/frequency/volume of training, above which mechanical stimuli can evoke adaptive responses. In contrast to the Achilles tendon-aponeurosis, the quadriceps femoris tendon-aponeurosis in endurance runners has been shown to be stiffer than in controls (Kubo *et al.*, 2000b), indicating that the threshold value for adaptation in this tendon-aponeurosis may be lower than in the Achilles tendon-aponeurosis. In a study on tendon stiffness in relation to running economy (Arampatzis *et al.*, 2006) it was found that the most economical runners had lower quadriceps tendon-aponeurosis stiffness at low tendon forces, resulting in higher strains and a greater elastic energy storage and release, which could reduce the metabolic cost of running. A lower quadriceps tendon-aponeurosis stiffness was also reported for faster sprinters compared to slower sprinters and a correlation was found between the maximum tendon-aponeurosis strain and the 100 m time (Stafilidis and Arampatzis, 2007), suggesting greater energy storage and recovery, and faster contractile shortening in the faster sprinters.

3. Healthy versus disused tendons: in cross-sectional studies, pathology-induced immobilization has been used as a model of disuse. One study compared the mechanical properties of the patellar tendon in able-bodied men and SCI men, with durations of lesion ranging from 1.5 to 24 years (Maganaris *et al.*, 2006). It was found that the tendons of the SCI subjects had a reduced stiffness of 77% and a reduced Young's modulus of 59% compared with the able-bodied subjects (Figure 2.3.7). The reduced Young's modulus with SCI indicates that chronic disuse deteriorates the material of the tendon. However, the tendons of the SCI subjects also had smaller cross-sectional areas (by 17%, Figure 2.3.8), indicating that they may have undergone atrophy. The possibility that some of the differences in tendon cross-sectional area between the two subject groups relate to their anthropometric characteristics rather than atrophy cannot be excluded, although it should be noted that the length of the tendons was similar in the two groups. No apparent relation was observed between the deterioration of tendon and the duration of lesion, indicating that most of the deterioration of tendon occurred rapidly, within the first few months of immobilization. However, a proper longitudinal study with measurements at different time points after the lesion would be required to shed more light on the dose–response relation of adaptations in human tendon with immobilization.

   In a more recent study, the mechanical properties of the human patellar tendon were examined in individuals who had undergone surgical reconstruction of the anterior cruciate ligament using a patellar tendon graft between 1 and 10 years before the study (Reeves *et al.*, 2009). Cross-sectional area of the operated patellar tendons was 21% larger than

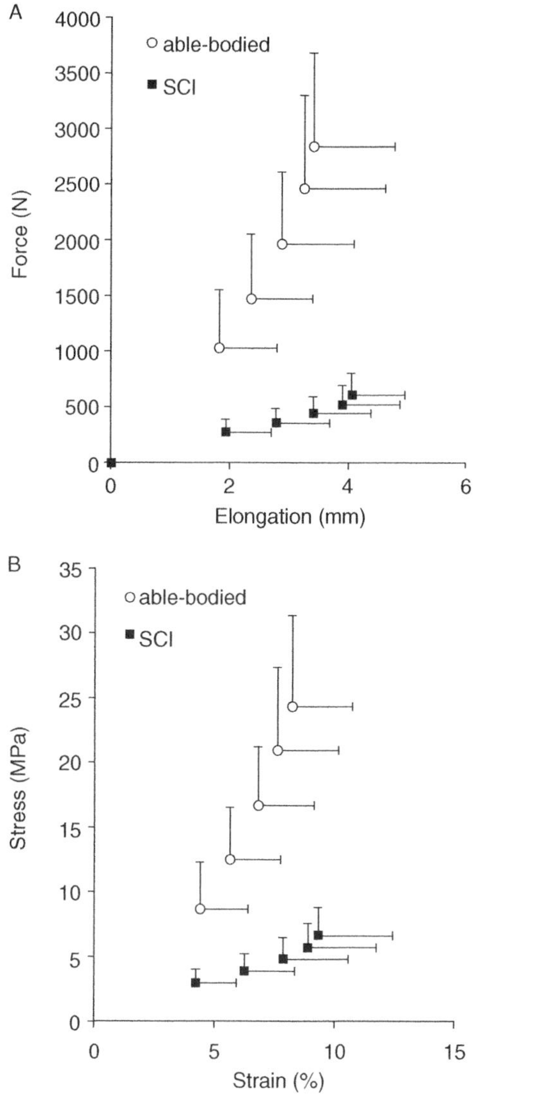

Figure 2.3.7 Patellar tendon (a) force–elongation and(b) stress–stain data for able-bodied (AB) and spinal cord-injured (SCI) subjects. Data are means and SD (n = 6 per group). Reproduced, with permission, from Maganaris, C. N., Reeves, N. D., Rittweger, J. *et al.* 2005. 'Adaptive Response of Human Tendon to Paralysis' in 'Muscle and Nerve' © John Wiley & Sons

the contralateral, control tendons. Although the Young's modulus was 24% lower in the operated tendons, the operated and control tendons had similar stiffness values. It seems that a compensatory enlargement of the patellar tendon, presumably due to scar-tissue formation, enabled a recovery of tendon stiffness in the operated tendons.

4. Tendons in different age groups: maturation and ageing are two contrasting biological processes which involve altered mechanical loading for the tendon and for the whole musculoskeletal system. This is because in pre-pubertal age the muscle forces that are exerted in order to support the smaller mass of the human body are also smaller, and older individuals are generally more sedentary than younger individuals. This factor alone could reduce the tendon mechanical properties in both pre-pubescence and old age compared with adolescence. Indeed, studies have shown that there is an increase in the stiffness (Kubo *et al.*, 2001a) and Young's modulus (O'Brien *et al.*, 2010) of tendons in adolescents compared with pre-pubertal children, whereas ageing has been shown to gradually deteriorate the tendon's mechanical properties (Karamanidis and Arampatzis, 2005; Onambele, Narici and Maganaris, 2006). However, it should be emphasized that there are age-driven processes taking place that affect the mechanical behaviour of the tendons independent of the loading conditions present; for example, in ageing there is an increase in collagen cross-linking, which increases the stiffness of the tendon (Kjaer, 2004). However, in one recent study there were no differences in the patellar tendon Young's modulus between younger and older individuals, despite the increased collagen cross-linking in the older group (Couppé *et al.*, 2009). This inter-study discrepancy may reflect lifestyle differences between the older individuals examined.

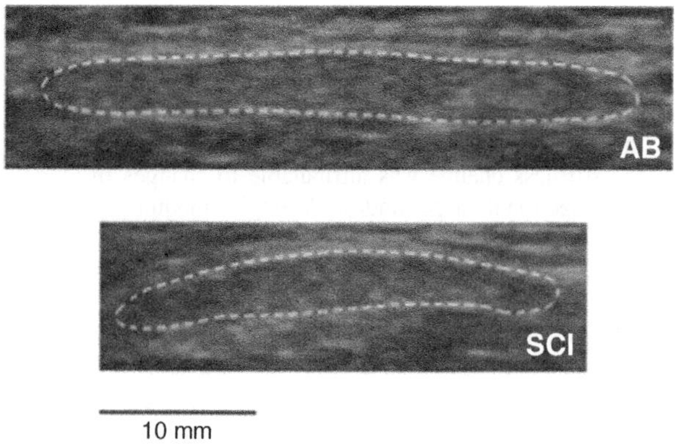

Figure 2.3.8 Axial-plane ultrasounds showing the cross-sectional area of the patellar tendon in its mid-region in an able-bodied (AB) subject (tendon length = 38 mm) and a spinal cord-injured (SCI) subject (tendon length = 46 mm). Reproduced, with permission, from Maganaris, C. N., Reeves, N. D., Rittweger, J. et al. 2005. 'Adaptive Response of Human Tendon to Paralysis' in 'Muscle and Nerve' © John Wiley & Sons

Longitudinal studies

Longitudinal studies have made pre- versus post-intervention comparisons where chronic loading has been manipulated by introducing exercise training or disuse.

1. Exercise training: most training studies have employed resistance exercise for several weeks in order to chronically overload the tendons of the exercised muscles.

Kubo *et al.* (2001b) examined the effects of isometric knee extensor muscle strength training for 12 weeks, four times a week, and found an increase in the stiffness of the quadriceps tendon-aponeurosis after training. No measurements of tendon size were taken, so it is not known whether the stiffness change was attributable to changes in tendon cross-sectional area and/or Young's modulus. Reeves, Maganaris and Narici (2003) examined the effects of dynamic knee extensor muscle strengthening for 14 weeks, three times a week, in older individuals, and found an increase in both patellar tendon stiffness and Young's modulus, but not in tendon cross-sectional area. It was concluded that the stiffness increase was caused solely by changes in the material of the tendon and not by tendon hypertrophy. Three recent studies, however, challenge this finding. Kongsgaard *et al.* (2007), Arampatzis, Karamanidis and Albracht (2007b), and Seynnes *et al.* (2009) all showed that there was regional tendon hypertrophy in response to resistance training between 9 and 14 weeks, indicating that in the earlier study of Reeves, Maganaris and Narici (2003) some regional tendon hypertrophy may actually have occurred but gone undetected due to the limited number of scans (three) recorded along the tendon. In contrast to Reeves, Maganaris and Narici (2003), Arampatzis, Karamanidis and Albracht (2007b), and Seynnes *et al.* (2009), it must be noted that Kongsgaard *et al.* (2007) reported no improvement in the tendon Young's modulus after training. However, it should also be noted that in the latter study the smallest tendon cross-sectional area was used in the calculation of tendon stress, while the former three studies used mean values. In the studies of Kubo *et al.* (2001b), Kongsgaard *et al.* (2007), and Arampatzis, Karamanidis and Albracht (2007b)), the contralateral leg was also trained. Kubo *et al.* used smaller durations of loading per repetition, Kongsgaard *et al.* used smaller resistance during contraction and therefore smaller tendon strains, and Arampatzis *et al.* used directly smaller tendon strains. In all three studies, the volume of training was the same for the two legs. However, the properties of tendon did not change in the legs trained using the lighter stimuli, indicating that there is a threshold above which mechanical loading can induce tendon hypertrophy and/or improvement in Young's modulus.

Studies have also examined the effects of stretching on tendon properties to investigate whether the resultant increases in joint range of movement and reduction in joint stiffness are associated with changes in tendon properties. Kubo, Kanehisa and Fukunaga (2002) applied ankle stretching for three weeks and found that there was no change in Achilles tendon stiffness. Similarly, Morse *et al.* (2008) found no changes in Achilles tendon stiffness after acute passive ankle stretches, and the measurements taken showed that the increased whole-muscle tendon length due to the greater joint range of movement post-stretching was attributed to an extension of the muscle belly, which became more compliant. However, the changes in muscle fascicle length corrected for the effect of pennation could not account for the extension of the muscle belly, indicating lengthening of the muscle's aponeuroses, rather than muscle extension by pivoting of the muscle fascicles around their insertion points in the aponeuroses alone.

2. Disuse: longitudinal studies have examined the effects of disuse, induced by means of immobilization and bed rest.

In a unilateral limb-suspension study, the patellar tendon of the unloaded leg of the participants was tested at base line and after 14 and 23 days of suspension. There was no change in the tendon's dimensions, but there was a gradual decrease in the tendon stiffness and Young's modulus, by about 10% and 30% on days 14 and 23 of the intervention respectively. The rate of tendon collagen synthesis, assessed by quantifying the incorporation of non-radioactive isotopes in tendon tissue samples obtained by biopsy, also fell gradually during the intervention, with most of the reduction occurring in the first 10 days. The above results indicate that the reduction in protein synthesis could be associated with a reduced collagen fibril density, rather than with atrophy at a whole-tendon level. In contrast to these findings, Christensen *et al.* (2008a) showed no changes in Achilles tendon collagen synthesis after two weeks of plaster-induced immobilization, by using microdialysis to detect markers of collagen synthesis in the peritendinous area. This discrepancy may indicate that: (i) the microdialysis and isotope tracing techniques do not have the same sensitivity; (ii) there is a difference in the level of disuse between the limb-suspension and plaster models for equal durations of application of the two interventions; (iii) different tendons require different durations/levels of disuse before their collagen synthesis starts to drop, depending on the habitual use and loading of the tendon. Somewhat surprising is the finding of Christensen *et al.* (2008b), who used the micodialysis technique and showed that seven weeks of plaster-induced immobilization, not for experimental purposes but for the treatment of unilateral ankle fractures, increased tendon collagen synthesis and degradation, while a remobilization period of similar length reduced collagen synthesis and degradation to baseline levels. However, it is possible that the ankle fracture close to the tendon sampling site affected the above data. There was no change in the Achilles tendon cross-sectional area throughout the entire study.

The results of bed-rest studies show substantial deterioration of tendon properties with disuse. Kubo *et al.* (2004) examined the quadriceps tendon-aponeurosis before and after 20 days of bed rest, and Reeves *et al.* (2005) examined the gastrocnemius tendon before and after 90 days of bed rest. In both studies there was a substantial reduction in tendon stiffness after bed rest. No tendon dimensions were measured in the former study, but in the latter there was no significant change in the tendon's cross-sectional area after bed rest and therefore the tendon stiffness reduction could almost entirely be attributed to deterioration in the tendon's material. However, as stated earlier, the possibility of some regional tendon atrophy going undetected due to the limited scans recorded cannot be excluded. The assessment of

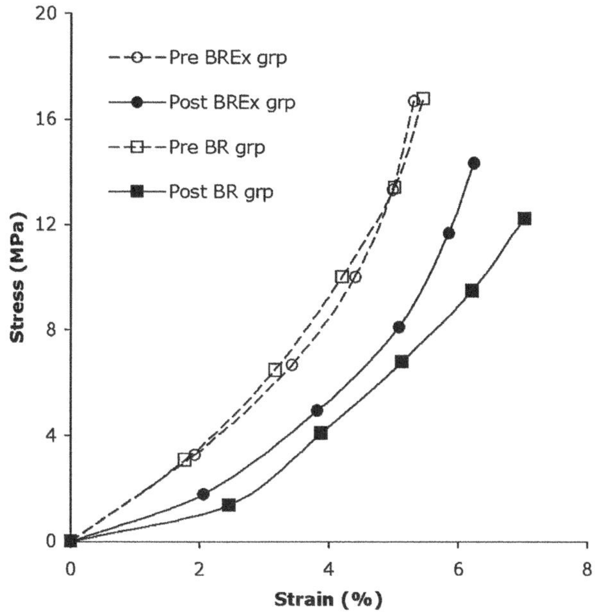

Figure 2.3.9 Tendon stress–stain data before and after 90 days of bed rest (preBR and postBR, respectively) and before and after bed rest plus exercise (preBREx and postBREx, respectively). Exercise partly attenuates the deteriorating effect of disuse. Data are means ($n = 9$ per group). Reproduced, with permission, from Reeves, N. D., Maganaris, C. N., Ferretti, G. and Narici, M. V. 2005. 'Influence of 90-day simulated microgravity on human tendon mechanical properties and the effect of resistive countermeasures'. J. Appl. Physiol. 98: 2278–2286. © Americal Physiological Society

tendon properties in the study of Reeves *et al.* (2005) was carried out using the same methodology and analysis as in the SCI study by Maganaris *et al.* (2006), and the deteriorations in tendon stiffness and Young's modulus were of similar magnitude, despite the two studies employing different durations of mechanical unloading: 90 days in the bed-rest study and up to 24 years of paralysis in the SCI study. This is consistent with the notion that that most of the deterioration of tendon during chronic disuse occurs rapidly, within the first few months. In the studies of both Kubo *et al.* (2004) and Reeves *et al.* (2005) there was an additional subject group, which underwent resistance exercise throughout the bed-rest period, and in both studies the addition of exercise attenuated, at least partly the deteriorating effect of disuse (Figure 2.3.9). This indicates that any exercises undertaken to strengthen and condition the skeletal muscles during prolonged periods of clinical disuse, for example confinement to a hospital bed, may also be beneficial for the tendons.

2.3.4 CONCLUSION

In conclusion, there is clear evidence that exercise has positive effects upon bone, leading to denser trabecular networks, greater cortical thickness, and thus an enhancement of whole-bone strength. These effects are most likely mechanically mediated, as they are site-specific, reversible, and strongest when the exercise involves large forces and impacts. Moreover, the effects are most pronounced during puberty, and they seem to diminish with increasing age.

However, despite the fact that exercise effects upon bone are beneficial, they are quite small and do not usually exceed 3% in interventional studies. Differences between athletes and sedentary people can amount to up to 20%, but this is probably partly due to self-selection bias (Wilks *et al.*, 2009a). Therefore, osteogenic exercise effects are small when compared to the rapid bone accrual in the early stages after bed rest (Figure 2.3.3a). The important question therefore arises as to what limits further accrual of bone mass, both in exercise and after bed rest (Figure 2.3.3b). In consideration of the enormous potential to recover lost bone (Rittweger and Felsenberg, 2009), it seems that this limitation is not imposed by bone itself. Rather, the soft tissues defining the mechanical interfaces to bone, such as tendinous structures (Rittweger *et al.*, 2005b) and effective joint size (Rittweger, 2008), are more likely to set effective limits to bone accrual.

Similar to bones, tendons deteriorate in response to mechanical unloading, and this is at least partly explained by an inferior tendon material as a whole. Unlike bones, however, tendons do show substantial positive adaptations and increase their mechanical stiffness in response to chronic exercise, as shown by a number of interventional and cross-sectional studies. Furthermore, we now understand a lot more about the mechanisms mediating the conversion of mechanical loading to biochemical changes that lead to tendon stiffness improvements. Using the microdialysis technique it has been shown that the rate of tendon collagen synthesis increases very rapidly in response to exercise, within six hours after a single bout of exercise, and by twice as much as at rest 24 hours post-exercise (Miller *et al.*, 2005). We also know that net collagen synthesis in the tendon increases in response to exercise training, and that this is associated with a dominance of collagen synthesis over degradation (Langberg, Rosendal and Kjaer, 2001). Collagen degradation is mediated by activation of matrix metaloproteinases (MMPs), while synthesis is mediated by over-expression of growth factors, such as TGF-a, IL-6, and IGF-I, some of which may also reduce collagen degradation by suppressing MMPs and stimulating their inhibitors (TIMPs) (Heinemeier *et al.*, 2003; Koskinen *et al.*, 2004; Kjaer, 2004; Langberg *et al.*, 2002; Olesen *et al.*, 2007). Interestingly, oral-contraceptive users have suppressed IGF-I and rate of tendon collagen synthesis after a single bout of exercise (Hansen *et al.*, 2008), and it remains to be investigated whether a similar effect occurs after long-term training, which may provide clues to the incidence of certain gender-related sports injuries. At a whole-tendon level, *in vivo* studies show that the adaptations to increased mechanical loading vary, and may include (1) regional increases in tendon cross-sectional area without changes in the tendon's material, (2) improvement in the tendon's material without any changes in its dimensions, or (3) changes in both material and cross-sectional area of the tendon.

But why are there size increases in some and not all, or in none of the tendon regions? Is the answer simply that the current *in vivo* imaging techniques do not have the sensitivity to allow the measurement of small tendon size changes, or is this a true effect, related, for example, to the anatomical location and biomechanical environment of the tendon? If the latter is the case, what is the molecular basis underlying the differentiation of tendon hypertrophy from a qualitative tendon change unrelated to the tendon's size? Addressing these important questions will allow us to unravel the mechanisms regulating tendon plasticity and help us optimize the effectiveness of exercise interventions for 'tendon strengthening' in different tendons and individuals according to specific needs, such as sporting competition and rehabilitation after injury.

References

Adami S., Gatti D., Braga V. *et al.* (1999) Site-specific effects of strength training on bone structure and geometry of ultradistal radius in postmenopausal women [see comments]. *J Bone Miner Res*, **14**, 120–124.

Alexander R.M.N. (1988) *Elastic Mechanisms in Animal Movement*, Cambridge University Press, Cambridge.

An K.N., Linscheid R.L. and Brand P.W. (1991) Correlation of physiological cross-sectional areas of muscle and tendon. *J Hand Surg Br*, **16**, 66–67.

Arampatzis A., De Monte G., Karamanidis K. *et al.* (2006) Influence of the muscle-tendon unit's mechanical and morphological properties on running economy. *J Exp Biol*, **209**, 3345–3357.

Arampatzis A., Karamanidis K., Morey-Klapsing G. *et al.* (2007a) Mechanical properties of the triceps surae tendon and aponeurosis in relation to intensity of sport activity. *J Biomech*, **40**, 1946–1952.

Arampatzis A., Karamanidis K. and Albracht K. (2007b) Adaptational responses of the human Achilles tendon by modulation of the applied cyclic strain magnitude. *J Exp Biol*, **210**, 2743–2753.

Baecker M., Tomic A., Mika C. *et al.* (2003) Bone resorption is induced on the second day of bed rest: results of a controlled crossover trial. *J Appl Physiol*, **95**, 977–982.

Baratta R., Solomonow M., Zhou B.H. *et al.* (1988) Muscular coactivation. The role of the antagonist musculature in maintaining knee stability. *Am J Sports Med*, **16**, 113–122.

Baron J.A., Karagas M., Barrett J. *et al.* (1996) Basic epidemiology of fractures of the upper and lower limb among Americans over 65 years of age. *Epidemiology*, **7**, 612–618.

Barry D.W. and Kohrt W.M. (2008) BMD decreases over the course of a year in competitive male cyclists. *J Bone Miner Res*, **23**, 484–491.

Bass S., Pearce G., Young N. and Seeman E. (1994) Bone mass during growth: the effects of exercise. Exercise and mineral accrual. *Acta Univ Carol [Med] (Praha)*, **40**, 3–6.

Bass S., Pearce G., Bradney M. *et al.* (1998) Exercise before puberty may confer residual benefits in bone density in adulthood: studies in active prepubertal and retired female gymnasts. *J Bone Miner Res*, **13**, 500–507.

Bass S.L., Saxon L., Daly R.M. *et al.* (2002) The effect of mechanical loading on the size and shape of bone in pre-, peri-, and postpubertal girls: a study in tennis players. *J Bone Miner Res*, **17**, 2274–2280.

Bass S.L., Naughton G., Saxon L. *et al.* (2007) Exercise and calcium combined results in a greater osteogenic effect than either factor alone: a blinded randomized placebo-controlled trial in boys. *J Bone Miner Res*, **22**, 458–464.

Bassey E.J. and Ramsdale S.J. (1994) Increase in femoral bone density in young women following high-impact exercise. *Osteoporos Int*, **4**, 72–75.

Bassey E.J., Rothwell M.C., Littlewood J.J. and Pye D.W. (1998) Pre- and postmenopausal women have different bone mineral density responses to the same high-impact exercise. *J Bone Miner Res*, **13**, 1805–1813.

Bennell K.L., Malcolm S.A., Khan K.M. *et al.* (1997) Bone mass and bone turnover in power athletes, endurance athletes, and controls: a 12-month longitudinal study. *Bone*, **20**, 477–484.

Bennett M.B., Ker R.F., Dimery N.J. and Alexander R.M.N. (1986) Mechanical properties of various mammalian tendons. *J Zool Lond (A)*, **209**, 537–548.

Biering-Sorensen F., Bohr H.H. and Schaadt O.P. (1990) Longitudinal study of bone mineral content in the lumbar spine, the forearm and the lower extremities after spinal cord injury. *Eur J Clin Invest*, **20**, 330–335.

Butler D.L., Goods E.S., Noyes F.R. and Zerniche R.F. (1978) Biomechanics of ligaments and tendons. *Exerc Sports Sci Rev*, **6**, 125–181.

Cardinale M. and Rittweger J. (2006) Vibration exercise makes your muscles and bones stronger: fact or fiction? *J Br Menopause Soc*, **12**, 12–18.

Christensen B., Dyrberg E., Aagaard P. *et al.* (2008a) Short-term immobilization and recovery affect skeletal muscle but not collagen tissue turnover in humans. *J Appl Physiol*, **105**, 1845–1851.

Christensen B., Dyrberg E., Aagaard P. *et al.* (2008b) Effects of long-term immobilization and recovery on human triceps surae and collagen turnover in the Achilles tendon in patients with healing ankle fracture. *J Appl Physiol*, **105**, 420–426.

Couppé C., Kongsgaard M., Aagaard P. *et al.* (2008) Habitual loading results in tendon hypertrophy and increased stiffness of the human patellar tendon. *J Appl Physiol*, **105**, 805–810.

Couppé C., Hansen P., Kongsgaard M. *et al.* (2009) Mechanical properties and collagen cross-linking of the patellar tendon in old and young men. *J Appl Physiol*, **107**, 880–886.

Currey J.D. (2002) *Bones: Structure and Mechanics*, Princeton University Press, Princeton.

Diab T., Condon K.W., Burr D.B. and Vashishth D. (2006) Age-related change in the damage morphology of human cortical bone and its role in bone fragility. *Bone*, **38**, 427.

Dickerman R.D., Pertusi R. and Smith G.H. (2000) The upper range of lumbar spine bone mineral density? An examination of the current world record holder in the squat lift. *Int J Sports Med*, **21**, 469–470.

Ebbesen E.N., Thomsen J.S. and Mosekilde L. (1997) Nondestructive determination of iliac crest cancellous bone strength by pQCT. *Bone*, **21**, 535–540.

Eser P., Frotzler A., Zehnder Y. *et al.* (2004) Relationship between the duration of paralysis and bone structure: a pQCT study of spinal cord injured individuals. *Bone*, **34**, 869–880.

Ettema G.J., Goh J.T. and Forwood M.R. (1998) A new method to measure elastic properties of plastic-viscoelastic connective tissue. *Med Eng Phys*, **20**, 308–314.

Frey-Rindova P., de Bruin E.D., Stussi E. *et al.* (2000) Bone mineral density in upper and lower extremities during 12 months after spinal cord injury measured by peripheral quantitative computed tomography. *Spinal Cord*, **38**, 26–32.

Friedlander A.L., Genant H.K., Sadowsky S. *et al.* (1995) A two-year program of aerobics and weight training enhances bone mineral density of young women. *J Bone Miner Res*, **10**, 574–585.

Gilsanz V., Wren T.A., Sanchez M. *et al.* (2006) Low-level, high-frequency mechanical signals enhance musculoskeletal development of young women with low BMD. *J Bone Miner Res*, **21**, 1464–1474.

Grigoriev A.I., Morukov B.V., Oganov V.S. *et al.* (1992) Effect of exercise and bisphosphonate on mineral balance and bone density during 360 day antiorthostatic hypokinesia. *J Bone Miner Res*, **7** (Suppl. 2), S449–S455.

Gullberg B., Johnell O. and Kanis J.A. (1997) World-wide projections for hip fracture. *Osteoporos Int*, **7**, 407–413.

Gusi N., Raimundo A. and Leal A. (2006) Low-frequency vibratory exercise reduces the risk of bone fracture more than walking: a randomized controlled trial. *BMC Musculoskelet Disord*, **7**, 92.

Haapasalo H., Kontulainen S., Sievanen H. *et al.* (2000) Exercise-induced bone gain is due to enlargement in bone size without a change in volumetric bone density: a peripheral quantitative computed tomography study of the upper arms of male tennis players. *Bone*, **27**, 351–357.

Hagino H., Raab D.M., Kimmel D.B. *et al.* (1993) Effect of ovariectomy on bone response to in vivo external loading. *J Bone Miner Res*, **8**, 347–357.

Hansen M., Koskinen S.O., Petersen S.G. *et al.* (2008) Ethinyl oestradiol administration in women suppresses synthesis of collagen in tendon in response to exercise. *J Physiol*, **586**, 3005–3016.

Heaney R.P., Abrams S., Dawson-Hughes B. *et al.* (2000) Peak bone mass. *Osteoporos Int*, **11**, 985–1009.

Heinemeier K., Langberg H., Olesen J.L. and Kjaer M. (2003) Role of TGF-beta1 in relation to exercise-induced type I collagen synthesis in human tendinous tissue. *J Appl Physiol*, **95**, 2390–2397.

Heinonen A., Kannus P., Sievanen H. *et al.* (1996) Randomised controlled trial of effect of high-impact exercise on selected risk factors for osteoporotic fractures. *Lancet*, **348**, 1343–1347.

Heinonen A., Oja P., Sievanen H. *et al.* (1998) Effect of two training regimens on bone mineral density in healthy perimenopausal women: a randomized controlled trial. *J Bone Miner Res*, **13**, 483–490.

Heinonen A., Sievanen H., Kannus P. *et al.* (2000) High-impact exercise and bones of growing girls: a 9-month controlled trial. *Osteoporos Int*, **11**, 1010–1017.

Hind K. and Burrows M. (2007) Weight-bearing exercise and bone mineral accrual in children and adolescents: a review of controlled trials. *Bone*, **40**, 14–27.

Huang R.P., Rubin C.T. and McLeod K.J. (1999) Changes in postural muscle dynamics as a function of age. *J Gerontol A Biol Sci Med Sci*, **54**, B352–B357.

Hughes J.M., Novotny S.A., Wetzsteon R.J. and Petit M.A. (2007) Lessons learned from school-based skeletal loading intervention trials: putting research into practice. *Med Sport Sci*, **51**, 137–158.

Johannsen N., Binkley T., Englert V. *et al.* (2003) Bone response to jumping is site-specific in children: a randomized trial. *Bone*, **33**, 533–539.

Jorgensen L., Crabtree N.J., Reeve J. and Jacobsen B.K. (2000) Ambulatory level and asymmetrical weight bearing after stroke affects bone loss in the upper and lower part of the femoral neck differently: bone adaptation after decreased mechanical loading. *Bone*, **27**, 701–707.

Jozsa L.G. and Kannus P. (1997) Spontaneous rupture of tendons, in *Human Tendons: Anatomy, Physiology and Pathology* (eds L.G. Jozsa and P. Kannus), Human Kinetics, Champaign, IL, pp. 254–325.

Kalkwarf H.J. and Specker B.L. (1995) Bone mineral loss during lactation and recovery after weaning. *Obstet Gynecol*, **86**, 26–32.

Kalkwarf H.J., Specker B.L., Heubi J.E. *et al.* (1996) Intestinal calcium absorption of women during lactation and after weaning. *Am J Clin Nutr*, **63**, 526–531.

Kalu D.N., Salerno E., Liu C.C. *et al.* (1991) A comparative study of the actions of tamoxifen, estrogen and progesterone in the ovariectomized rat. *Bone Miner*, **15**, 109–123.

Kanis J.A., Johnell O., Oden A. *et al.* (2001) Ten year probabilities of osteoporotic fractures according to BMD and diagnostic thresholds. *Osteoporos Int*, **12**, 989–995.

Kannus P., Haapasalo H., Sievanen H. *et al.* (1994) The site-specific effects of long-term unilateral activity on bone mineral density and content. *Bone*, **15**, 279–284.

Karamanidis K. and Arampatzis A. (2005) Mechanical and morphological properties of different muscle-tendon units in the lower extremity and running mechanics: effect of aging and physical activity. *J Exp Biol*, **208**, 3907–3923.

Karlsson M.K., Johnell O. and Obrant K.J. (1993) Bone mineral density in weight lifters. *Calcif Tissue Int*, **52**, 212–215.

Ker R.F. (1992) Tensile fibres: strings and straps, in *Biomechanics – Materials: A Practical Approach* (ed. J.F.V. Vincent), Oxford University Press, New York, NY, pp. 75–97.

Ker R.F., Alexander R.M.N. and Bennett M.B. (1988) Why are mammalian tendons so thick? *J Zool Lond*, **216**, 309–324.

Kjaer M. (2004) Role of extracellular matrix in adaptation of tendon and skeletal muscle to mechanical loading. *Physiol Rev*, **84**, 649–698.

Kohrt W.M., Bloomfield S.A., Little K.D. *et al.* (2004) American College of Sports Medicine Position Stand: physical activity and bone health. *Med Sci Sports Exerc*, **36**, 1985–1996.

Komi P.V., Fukashiro S. and Jarvinen M. (1992) Biomechanical loading of Achilles tendon during normal locomotion. *Clin Sports Med*, **11**, 521–531.

Kongsgaard M., Reitelseder S., Pedersen T.G. *et al.* (2007) Region specific patellar tendon hypertrophy in humans following resistance training. *Acta Physiol*, **191**, 111–121.

Kontulainen S., Kannus P., Haapasalo H. *et al.* (1999) Changes in bone mineral content with decreased training in competitive young adult tennis players and controls: a prospective 4-yr follow-up. *Med Sci Sports Exerc*, **31**, 646–652.

Kontulainen S., Heinonen A., Kannus P. *et al.* (2004) Former exercisers of an 18-month intervention display residual aBMD benefits compared with control women 3.5 years post-intervention: a follow-up of a randomized controlled high-impact trial. *Osteoporos Int*, **15**, 248–251.

Koskinen S.O., Heinemeier K.M., Olesen J.L. *et al.* (2004) Physical exercise can influence local levels of matrix metalloproteinases and their inhibitors in tendon-related connective tissue. *J Appl Physiol*, **96**, 861–864.

Kubo K., Kanehisa H., Kawakami Y. and Fukunaga T. (2000a) Elasticity of tendon structures of the lower limbs in sprinters. *Acta Physiol Scand*, **168**, 327–335.

Kubo K., Kanehisa H., Kawakami Y. and Fukunaga T. (2000b) Elastic properties of muscle-tendon complex in long-distance runners. *Eur J Appl Physiol*, **81**, 181–187.

Kubo K., Kanehisa H., Kawakami Y. and Fukanaga T. (2001a) Growth changes in the elastic properties of human tendon structures. *Int J Sports Med*, **22**, 138–143.

Kubo K., Kanehisa H., Ito M. and Fukunaga T. (2001b) Effects of isometric training on the elasticity of human tendon structures in vivo. *J Appl Physiol*, **91**, 26–32.

Kubo K., Kanehisa H. and Fukunaga T. (2002) Effect of stretching training on the viscoelastic properties of human tendon structures in vivo. *J Appl Physiol*, **92**, 595–601.

Kubo K., Akima H., Ushiyama J. *et al.* (2004) Effects of resistance training during bed rest on the viscoelastic properties of tendon structures in the lower limb. *Scand J Med Sci Sports*, **14**, 296–302.

Langberg H., Rosendal L. and Kjaer M. (2001) Training-induced changes in peritendinous type I collagen turnover determined by microdialysis in humans. *J Physiol*, **534**, 297–302.

Langberg H., Olesen J.L., Gemmer C. and Kjaer M. (2002) Substantial elevation of interleukin-6 concentration in peritendinous tissue, in contrast to muscle, following prolonged exercise in humans. *J Physiol*, **542**, 985–990.

LeBlanc A.D., Schneider V.S., Evans H.J. *et al.* (1990) Bone mineral loss and recovery after 17 weeks of bed rest. *J Bone Miner Res*, **5**, 843–850.

Leppala J., Kannus P., Natri A. *et al.* (1999) Effect of anterior cruciate ligament injury of the knee on bone mineral density of the spine and affected lower extremity: a prospective one-year follow-Up study. *Calcif Tissue Int*, **64**, 357–363.

Lohman T., Going S., Pamenter R. *et al.* (1995) Effects of resistance training on regional and total bone mineral density in premenopausal women: a randomized prospective study. *J Bone Miner Res*, **10**, 1015–1024.

Loram I.D., Maganaris C.N. and Lakie M. (2009) Paradoxical muscle movement during postural control. *Med Sci Sports Exerc*, **41**, 198–204.

Mack P.B., LaChange P.A., Vose G.P. and Vogt F.B. (1967) Bone demineralization of foot and hand of Gemini-Titan IV, V and VII astronauts during orbital space flight. *Am J Roentgenol Radium Ther Nucl Med*, **100**, 503–511.

Mackelvie K.J., McKay H.A., Khan K.M. and Crocker P.R. (2001) A school-based exercise intervention augments bone mineral accrual in early pubertal girls. *J Pediatr*, **139**, 501–508.

Maddalozzo G.F. and Snow C.M. (2000) High intensity resistance training: effects on bone in older men and women. *Calcif Tissue Int*, **66**, 399–404.

Maganaris C.N. (2002) Tensile behaviour of in vivo human tendinous tissue. *J Biomech*, **35**, 1019–1027.

Maganaris C.N. and Paul J.P. (1999) In vivo human tendon mechanical properties. *J Physiol*, **521**, 307–313.

Maganaris C.N. and Paul J.P. (2000) Hysteresis measurements in intact human tendon. *J Biomech*, **33**, 1723–1727.

Maganaris C.N. and Paul J.P. (2002) Tensile properties of the in vivo human gastrocnemius tendon. *J Biomech*, **35**, 1639–1946.

Maganaris C.N., Reeves N.D., Rittweger J. *et al.* (2006) Adaptive response of human tendon to paralysis. *Muscle Nerve*, **33**, 85–92.

Magnusson S.P. and Kjaer M. (2003) Region-specific differences in Achilles tendon cross-sectional area in runners and non-runners. *Eur J Appl Physiol*, **90**, 549–553.

Maimoun L., Couret I., Mariano-Goulart D. *et al.* (2005) Changes in osteoprotegerin/RANKL system, bone mineral density, and bone biochemicals markers in patients with recent spinal cord injury. *Calcif Tissue Int*, **76**, 404–411.

Martin R.B., Burr D.B. and Sharkey N.A. (1998) *Skeletal Tissue Mechanics*, Springer-Verlag, New York, NY.

Matthews B.L., Bennell K.L., McKay H.A. *et al.* (2006) Dancing for bone health: a 3-year longitudinal study of bone mineral accrual across puberty in female non-elite dancers and controls. *Osteoporos Int*, **17**, 1043–1054.

Matthews L.S. and Ellis D. (1968) Viscoelastic properties of cat tendon: Effects of time after death and preservation by freezing. *J Biomech*, **1**, 65–71.

Medelli J., Lounana J., Menuet J.J. *et al.* (2009) Is osteopenia a health risk in professional cyclists? *J Clin Densitom*, **12**, 28–34.

Miller B.F., Olesen J.L., Hansen M. *et al.* (2005) Coordinated collagen and muscle protein synthesis in human patella tendon and quadriceps muscle after exercise. *J Physiol*, **567**, 1021–1033.

Milliken L.A., Going S.B., Houtkooper L.B. *et al.* (2003) Effects of exercise training on bone remodeling, insulin-like growth factors, and bone mineral density in postmenopausal women with and without hormone replacement therapy. *Calcif Tissue Int*, **72**, 478–484.

Morris F.L., Naughton G.A., Gibbs J.L. *et al.* (1997) Prospective ten-month exercise intervention in premenarcheal girls: positive effects on bone and lean mass. *J Bone Miner Res*, **12**, 1453–1462.

Morse C.I., Degens H., Seynnes O.R. *et al.* (2008) The acute effect of stretching on the passive stiffness of the human gastrocnemius muscle tendon unit. *J Physiol*, **586**, 97–106.

Mosley J.R. and Lanyon L.E. (1998) Strain rate as a controlling influence on adaptive modeling in response to dynamic loading of the ulna in growing male rats. *Bone*, **23**, 313–318.

Nalla R.K., Kruzic J.J., Kinney J.H. and Ritchie R.O. (2004) Effect of aging on the toughness of human cortical bone: evaluation by R-curves. *Bone*, **35**, 1240–1246.

Nichols J.F., Palmer J.E. and Levy S.S. (2003) Low bone mineral density in highly trained male master cyclists. *Osteoporos Int*, **14**, 644–649.

Nikander R., Sievanen H., Heinonen A. and Kannus P. (2005) Femoral neck structure in adult female athletes subjected to different loading modalities. *J Bone Miner Res*, **20**, 520–528.

Nikander R., Sievanen H., Uusi-Rasi K. *et al.* (2006) Loading modalities and bone structures at nonweight-bearing upper extremity and weight-bearing lower extremity: a pQCT study of adult female athletes. *Bone*, **39**, 886–894.

O'Brien T.D., Baltzopoulos V., Reeves N.D. and Maganaris C.N. (2010) Mechanical properties of the patellar tendon in adults and pre-bupertal children. *J Biomech*, **19** (43), 1190–1195.

Oganov V.S., Grigoriev A.I., Voronin L.I. *et al.* (1992) [Bone mineral density in cosmonauts after flights lasting 4.5–6 months on the Mir orbital station]. *Aviakosm Ekolog Med*, **26**, 20–24.

Olesen J.L., Heinemeier K.M., Gemmer C. *et al.* (2007) Exercise-dependent IGF-I, IGFBPs, and type I collagen changes in human peritendinous connective tissue determined by microdialysis. *J Appl Physiol*, **102**, 214–220.

Onambele G.L., Narici M.V. and Maganaris C.N. (2006) Calf muscle-tendon properties and postural balance in old age. *J Appl Physiol*, **100**, 2048–2056.

Otis C.L., Drinkwater B., Johnson M. *et al.* (1997) American College of Sports Medicine position stand. The Female Athlete Triad. *Med Sci Sports Exerc*, **29**, i–ix.

Özkaya N. and Nordin M. (1998) *Fundamentals of Biomechanics*, Springer, New York, NY.

Pavy-Le Traon A., Heer M., Narici M.V. *et al.* (2007) From space to Earth: advances in human physiology from 20 years of bed rest studies (1986–2006). *Eur J Appl Physiol*, **101**, 143–194.

Petit M.A., McKay H.A., MacKelvie K.J. *et al.* (2002) A randomized school-based jumping intervention confers site and maturity-specific benefits on bone structural properties in girls: a hip structural analysis study. *J Bone Miner Res*, **17**, 363–372.

Pike A.V.L., Ker R.F. and Alexander R.M.N. (2000) The development of fatigue quality in high-and low-stressed tendons of sheep (Ovis aries). *J Exp Biol*, **203**, 2187–2193.

Pollock C.M. and Shadwick R.E. (1994) Relationship between body mass and biomechanical properties of limb tendons in adult mammals. *Am J Physiol*, **266**, R1016–R1021.

Rack P.M.H. and Ross H.F. (1984) The tendon of flexor pollicis longus: its effects on the muscular control of force and position at the human thumb. *J Physiol*, **351**, 99–110.

Rambaut P.C., Smith M.C., Mack P.B. and Vogel J.M. (1975) Skeletal response, in *Biomedical Results of Apollo* (eds R.S.D. Johnston and C.A. Berry), NASA, USA., pp. 303–322.

Recker R., Lappe J., Davies K. and Heaney R. (2000) Characterization of perimenopausal bone loss: a prospective study. *J Bone Miner Res*, **15**, 1965–1973.

Reeves N.D., Maganaris C.N. and Narici M.V. (2003) Effect of strength training on human patella tendon mechanical properties of older individuals. *J Physiol*, **548**, 971–981.

Reeves N.D., Maganaris C.N., Ferretti G. and Narici M.V. (2005) Influence of 90-day simulated microgravity on human tendon mechanical properties and the effect of resistive countermeasures. *J Appl Physiol*, **98**, 2278–2286.

Reeves N.D., Maganaris C.N., Maffulli N. and Rittweger J. (2009) Human patellar tendon stiffness is restored following graft harvest for anterior cruciate ligament surgery. *J Biomech*, **42**, 797–803.

Reiter A.L., Volk A., Vollmar J. *et al.* (2007) Changes of basic bone turnover parameters in short-term and long-term patients with spinal cord injury. *Eur Spine J*, **16**, 771–776.

Rhodes E.C., Martin A.D., Taunton J.E. *et al.* (2000) Effects of one year of resistance training on the relation between muscular strength and bone density in elderly women. *Br J Sports Med*, **34**, 18–22.

Riggs B.L., Melton I.L.J. III, Robb R.A. *et al.* (2004) Population-based study of age and sex differences in bone volumetric density, size, geometry, and structure at different skeletal sites. *J Bone Miner Res*, **19**, 1945–1954.

Rittweger J. (2006) Can exercise prevent osteoporosis? *J Musculoskelet Neuronal Interact*, **6**, 162–166.

Rittweger J. (2008) Ten years muscle-bone hypothesis: What have we learned so far? – Almost a Festschrift. *J Musculoskelet Neuronal Interact*, **8**, 174–178.

Rittweger J. and Felsenberg D. (2009) Recovery of muscle atrophy and bone loss from 90 days bed rest: Results from a one-year follow-up. *Bone*, **44**, 214–224.

Rittweger J., Beller G., Ehrig J. *et al.* (2000) Bone-muscle strength indices for the human lower leg. *Bone*, **27**, 319–326.

Rittweger J., Frost H.M., Schiessl H. *et al.* (2005a) Muscle atrophy and bone loss after 90 days of bed rest and the effects of flywheel resistive exercise and pamidronate: results from the LTBR study. *Bone*, **36**, 1019–1029.

Rittweger J., Maffulli N., Maganaris C.N. and Narici M.V. (2005b) Reconstruction of the anterior cruciate ligament with a patella-tendon-bone graft may lead to a permanent loss of bone mineral content due to decreased patellar tendon stiffness. *Med Hypotheses*, **64**, 1166–1169.

Rittweger J., Belavy D., Hunek P. *et al.* (2006a) Highly demanding resistive exercise program is tolerated during 56 days of strict bed rest. *Int J Sports Med*, **27**, 553–559.

Rittweger J., Gerrits K., Altenburg T. *et al.* (2006b) Bone adaptation to altered loading after spinal cord injury: a study of bone and muscle strength. *J Musculoskel Neuron Interact*, **6**, 269–276.

Rittweger J., Simunic B., Bilancio G. *et al.* (2009) Bone loss in the lower leg during 35 days of bed rest is predominantly from the cortical compartment. *Bone*, **44**, 612–618.

Rittweger J., Beller G., Armbrecht G. *et al.* (2010) Prevention of bone loss during 56 days of strict bed rest by side-alternating resistive vibration exercise. *Bone*, **46** (1), 137–147.

Rosager S., Aagaard P., Dyhre-Poulsen P. *et al.* (2002) Load-displacement properties of the human triceps surae aponeurosis and tendon in runners and non-runners. *Scand J Med Sci Sports*, **12**, 90–98.

Rubin C.T. and Lanyon L.E. (1987) Kappa Delta Award paper. Osteoregulatory nature of mechanical stimuli: function as a determinant for adaptive remodeling in bone. *J Orthop Res*, **5**, 300–310.

Rubin C., Recker R., Cullen D. *et al.* (2004) Prevention of postmenopausal bone loss by a low-magnitude, high-frequency mechanical stimuli: a clinical trial assessing compliance, efficacy, and safety. *J Bone Miner Res*, **19**, 343–351.

Rubin C.T., Sommerfeldt D.W., Judex S. and Qin Y. (2001) Inhibition of osteopenia by low magnitude, high-frequency mechanical stimuli. *Drug Discov Today*, **6**, 848–858.

Ruffoni D., Fratzl P., Roschger P. *et al.* (2007) The bone mineralization density distribution as a fingerprint of the mineralization process. *Bone*, **40**, 1308–1319.

Runge M., Rittweger J., Russo C.R. *et al.* (2004) Is muscle power output a key factor in the age-related decline in physical performance? A comparison of muscle cross section, chair-rising test and jumping power. *Clin Physiol Funct Imaging*, **24**, 335–340.

Schiessl H., Frost H.M. and Jee W.S. (1998) Estrogen and bone-muscle strength and mass relationships. *Bone*, **22**, 1–6.

Seynnes O.R., Erskine R.M., Maganaris C.N. *et al.* (2009) Training-induced changes in structural and mechanical properties of the patellar tendon are related to muscle hypertrophy but not to strength gains. *J Appl Physiol*, **107**, 523–530.

Shackelford L.C., LeBlanc A.D., Driscoll T.B. *et al.* (2004) Resistance exercise as a countermeasure to disuse-induced bone loss. *J Appl Physiol*, **97**, 119–129.

Shiraki M., Kuroda T., Tanaka S. *et al.* (2008) Nonenzymatic collagen cross-links induced by glycoxidation (pentosidine) predicts vertebral fractures. *J Bone Miner Metab*, **26**, 93–100.

Sievanen H., Heinonen A. and Kannus P. (1996) Adaptation of bone to altered loading environment: a biomechanical approach using X-ray absorptiometric data from the patella of a young woman. *Bone*, **19**, 55–59.

Smith C.W., Young I.S. and Kearney J.N. (1996) Mechanical properties of tendons: Changes with sterilization and preservation. *J Biomech Eng*, **118**, 56–61.

Snow C.M., Williams D.P., LaRiviere J. *et al.* (2001) Bone gains and losses follow seasonal training and detraining in gymnasts. *Calcif Tissue Int*, **69**, 7–12.

Snow-Harter C., Bouxsein M.L., Lewis B.T. *et al.* (1992) Effects of resistance and endurance exercise on bone mineral status of young women: a randomized exercise intervention trial. *J Bone Miner Res*, **7**, 761–769.

Specker B. and Binkley T. (2003) Randomized trial of physical activity and calcium supplementation on bone mineral content in 3- to 5-year-old children. *J Bone Miner Res*, **18**, 885–892.

Stafilidis S. and Arampatzis A. (2007) Muscle-tendon unit mechanical and morphological properties and sprint performance. *J Sports Sci*, **25**, 1035–1046.

Tsuzuku S., Ikegami Y. and Yabe K. (1998) Effects of high-intensity resistance training on bone mineral density in young male powerlifters. *Calcif Tissue Int*, **63**, 283–286.

Tsuzuku S., Ikegami Y. and Yabe K. (1999) Bone mineral density differences between paraplegic and quadriplegic patients: a cross-sectional study. *Spinal Cord*, **37**, 358–361.

Umemura Y., Ishiko T., Yamauchi T. *et al.* (1997) Five jumps per day increase bone mass and breaking force in rats. *J Bone Miner Res*, **12**, 1480–1485.

Verschueren S.M., Roelants M., Delecluse C. *et al.* (2004) Effect of 6-month whole body vibration training on hip density, muscle strength, and postural control in postmenopausal women: a randomized controlled pilot study. *J Bone Miner Res*, **19**, 352–359.

Vico L., Chappard D., Alexandre C. *et al.* (1987) Effects of a 120 day period of bed-rest on bone mass and bone cell activities in man: attempts at countermeasure. *Bone Miner*, **2**, 383–394.

Vincent K.R. and Braith R.W. (2002) Resistance exercise and bone turnover in elderly men and women. *Med Sci Sports Exerc*, **34**, 17–23.

Watanabe Y., Ohshima H., Mizuno K. *et al.* (2004) Intravenous pamidronate prevents femoral bone loss and renal stone formation during 90-day bed rest. *J Bone Miner Res*, **19**, 1771–1778.

Wilks D.C., Winwood K., Gilliver S.F. *et al.* (2009a) Bone mass and geometry of the tibia and the radius of Master sprinters, middle and long distance runners, race walkers, and sedentary control participants: A pQCT study. *Bone*, **45**, 91–97.

Wilks D.C., Gilliver S.F. and Rittweger J. (2009b) Forearm and tibial bone measures of distance- and sprint-trained master cyclists. *Med Sci Sports Exerc*, **41**, 566–573.

Wilks D.C., Winwood K., Gulliver S.F. *et al.* (2009c) Age-dependency of bone mass and geometry: A cross sectional pQCT study on male and female Master sprinters, middle distance runners, long distance runners, race-walkers, and sedentary people. *J Musculoskel Neuron Interact*, **9** (4), 236–246.

Wilson A.M. and Goodship A.E. (1994) Exercise-induced hyperthermia as a possible mechanism for tendon degeneration. *J Biomech*, **27**, 899–905.

Winters K.M. and Snow C.M. (2000) Detraining reverses positive effects of exercise on the musculoskeletal system in premenopausal women. *J Bone Miner Res*, **15**, 2495–2503.

Winters-Stone K.M. and Snow C.M. (2006) Site-specific response of bone to exercise in premenopausal women. *Bone*, **39**, 1203–1209.

Witzke K.A. and Snow C.M. (2000) Effects of plyometric jump training on bone mass in adolescent girls. *Med Sci Sports Exerc*, **32**, 1051–1057.

Zajac F.E. (1989) Muscle and tendon: properties, models, scaling, and application to biomechanics and motor control. *CRC Crit Rev Biomed Eng*, **17**, 359–411.

Zanchetta J.R., Plotkin H. and Alvarez-Filgueira M.L. (1995) Bone mass in children: normative values for the 2-20-year-old population. *Bone*, **16**, S393–S399.

Zehnder Y., Luthi M., Michel D. *et al.* (2004) Long-term changes in bone metabolism, bone mineral density, quantitative ultrasound parameters, and fracture incidence after spinal cord injury: a cross-sectional observational study in 100 paraplegic men. *Osteoporos Int*, **15**, 180–189.

2.4 Biochemical Markers and Resistance Training

Christian Cook[1,2,3] and Blair Crewther[3], [1]United Kingdom Sport Council, London, UK, [2]University of Bath, Sport, Health and Exercise Science, Bath, UK, [3]Imperial College, Institute of Biomedical Engineering, London, UK

2.4.1 INTRODUCTION

The mechanical and hormonal responses to resistance training or weight training have been proposed as necessary stimuli for training adaptations to occur (Crewther *et al.*, 2006; Häkkinen, 1989; Kraemer, 2000; Kraemer and Mazzetti, 2003). In particular, hormones play an important role in the resistance training process by regulating protein turnover and long-term changes in skeletal muscle growth. Since the cross-sectional area (CSA) of muscle is also a determinant of muscle force production (Bruce, Phillips and Woledge, 1997), the long-term effects of hormones upon muscle CSA can also influence both power and strength adaptation. Perhaps the most widely examined steroid hormones from this perspective are testosterone (T) and cortisol (C), often considered the primary anabolic and catabolic hormones, respectively, along with the polypeptide growth hormone (GH) (Crewther *et al.*, 2006; Häkkinen, 1989; Kraemer, 2000; Kraemer and Mazzetti, 2003).

Resistance training is known to stimulate acute changes in blood hormone concentrations, which are thought to mediate those cellular processors involved in muscle growth (Kraemer, 2000; Kraemer and Mazzetti, 2003). Thus, the design of individual workouts (e.g. load, volume, rest periods, etc.) plays an important role in mediating long-term adaptation with training. With this in mind, examining the acute responses of the aforementioned hormones to different workouts (e.g. hypertrophy, strength, explosive power) would provide a better understanding of how various training methods affect the hormonal milieu and hypertrophy, power, and strength adaptation thereafter. Discussion of the additional effects of age, gender, and training history would add to this analysis. Although this review will focus primarily on T, C, and GH, this in no way suggests the primacy of these three examples, but it does allow for a more detailed discussion of their relevance to the resistance training process.

2.4.2 TESTOSTERONE RESPONSES TO RESISTANCE TRAINING

The free or biologically active form of T accounts for ~1–2% of all T, with ~38–50% bound to albumin and the remaining portion bound to sex hormone-binding globulin (Dunn, Nisula and Rodbard, 1981; Loebel and Kraemer, 1998). The T portion bound to albumin is believed to be readily available for metabolism, due to enhanced dissociation (Pardridge, 1986), effectively contributing to the bioavailable hormone pool. The majority of evidence supports T as having an important anabolic effect upon muscle tissue growth (Crewther *et al.*, 2006; Herbst and Bhasin, 2004). In this role, T is thought to contribute directly to muscle protein synthesis, whilst also reducing muscle protein degradation (Herbst and Bhasin, 2004). Indirectly, T may also stimulate the release of other anabolic hormones (e.g. GH) important for muscle tissue growth (Crewther *et al.*, 2006).

2.4.2.1 Effects of workout design

Hypertrophy workouts generally result in large increases in T concentrations, whilst maximal strength workouts elicit much smaller or no hormonal changes, measured in blood or saliva (Consitt, Copeland and Tremblay, 2001; Crewther *et al.*, 2008; Gotshalk *et al.*, 1997; Häkkinen and Pakarinen, 1993; Kraemer *et al.*, 1991; Linnamo *et al.*, 2005; McCaulley *et al.*, 2009; Smilios *et al.*, 2003). For instance, a 72% increase in T concentrations was noted after a hypertrophy session (10 repetition maximum (RM) load, high volume, 1 min rest periods) in men, whilst a maximal strength session (5 RM, moderate volume, 3 min rest periods) elevated T by only 27% in the same group (Kraemer *et al.*, 1991). Explosive power workouts (<50% 1 RM, low–moderate volume, 1–3 min rest periods) can also

Strength and Conditioning – Biological Principles and Practical Applications Marco Cardinale, Rob Newton, and Kazunori Nosaka.

acutely increase T concentrations (Mero *et al.*, 1991; Volek *et al.*, 1997), although not to the same extent as those training sessions designed to increase muscle size. On the other hand, other studies have reported no changes in the acute T responses to explosive training protocols (Crewther *et al.*, 2008; Linnamo *et al.*, 2005; McCaulley *et al.*, 2009).

2.4.2.2 Effects of age and gender

Age is one factor influencing the acute hormonal responses to resistance exercise. Resistance exercise is known to increase T concentrations among younger males (Fry *et al.*, 1993; Kraemer *et al.*, 1992; Mero *et al.*, 1993), but to a much lower extent than in their older male counterparts (Mero *et al.*, 1993; Pullinen *et al.*, 2002). Resting T concentrations are also much greater in adult males (Mero *et al.*, 1993; Pullinen *et al.*, 2002). These differences are likely to provide adult males with a more beneficial response for adaptation and thereby explain muscle mass and strength differences between adult and younger males. The acute T responses to resistance exercise decrease in later life, as do basal concentrations of T (Häkkinen and Pakarinen, 1995; Kraemer *et al.*, 1998a, 1999). These changes in androgen activity may partly explain the reduction in muscle mass and strength seen with increasing age. Proposed mechanisms include failure of the hypothalamic-pituitary axis, changes in testicular function, an increase in SHBG levels, and/or increased sensitivity of gonadotropin secretion to androgen negative-feedback inhibition (Vermeulen, Rubens and Verdonck, 1972).

Gender also plays a role in modulating the hormonal responses, with resistance training sessions found to produce reliable T increases in males, but no T changes noted in females performing the same workouts (Kraemer *et al.*, 1991; Linnamo *et al.*, 2005). Possibly this relates to different mechanisms of secretion, with males producing much greater androgen concentrations overall, as well as larger acute responses via the testes. However, a broad range of protocols have not been explored and females have shown elevated T across different modes of exercise (Consitt, Copeland and Tremblay, 2001; Copeland, Consitt and Tremblay, 2002). The differential response patterns between males and females, along with the greater (5–10-fold) basal concentrations of T in males (Kraemer *et al.*, 1991; Linnamo *et al.*, 2005), support the general observation that men exhibit greater muscle mass and strength than females, as well as having a greater training capacity for inducing changes in these variables.

2.4.2.3 Effects of training history

Training history can also influence the acute hormonal responses. Resistance-trained athletes were found to produce a greater T response than endurance athletes (Tremblay, Copeland and Van Helder, 2004) and non-athletes (Ahtiainen *et al.*, 2004). Likewise, a nine-week training period markedly increased the acute T response in men (Kraemer *et al.*, 1998b), with other evidence showing that greater training experience is accompanied by greater T responses to resistance exercise (Kraemer *et al.*, 1988, 1992). Training can also affect resting hormones; for instance male sprinters have higher resting T than soccer players, who in turn have higher T than cross-country skiers (Bosco and Viru, 1998), while sprinters have higher T than handball players (Cardinale and Stone, 2006). Conversely, the T concentrations of endurance athletes were found to be less than more explosive athletes and/or sedentary or active controls (Grasso *et al.*, 1997; Hackney, Fahrner and Gulledge, 1998, Hackney, Szczepanowska and Viru, 2003; Izquierdo *et al.*, 2004). These T differences might be functionally important in meeting the force demands of athletes during exercise, training, and competition; alternatively, these findings could represent genetic predisposition and the natural selection of participants towards different sports.

2.4.3 CORTISOL RESPONSES TO RESISTANCE TRAINING

Most C is bound to plasma proteins in blood, with ~80% bound to corticosteroid-binding globulin and ~6% to albumin (Dunn, Nisula and Rodbard, 1981). Approximately 1–5% of C circulates freely and is biologically active, although the saturation of corticosteroid-binding globulin can also lead to dramatic increases in the free moiety (Dunn, Nisula and Rodbard, 1981; Rowbottom *et al.*, 1995). Traditionally, C has been regarded as the primary catabolic hormone, as it decreases protein synthesis and increases the breakdown of muscle protein (Crewther *et al.*, 2006; Viru and Viru, 2004). Likewise the anti-anabolic properties of C are linked to the attenuation of other anabolic hormones such as T and GH (Crewther *et al.*, 2006). However, a more correct interpretation may be to view acute changes in C as a prerequisite for the repartitioning of metabolic resources; an essential step in hypertrophy (Viru and Viru, 2004).

2.4.3.1 Effects of workout design

Most research concludes that hypertrophy workouts elicit a greater acute C response than maximal-strength workouts (Crewther *et al.*, 2008; Häkkinen and Pakarinen, 1993; Kraemer *et al.*, 1993; McCauley *et al.*, 2009; Smilios *et al.*, 2003; Zafeiridis *et al.*, 2003). Kraemer *et al.* (1993) reported a 125% increase in female C concentrations after a hypertrophy workout, but no hormonal changes were noted following a maximal-strength workout. Similarly, workouts designed to improve explosive power generally do not alter C concentrations (Crewther *et al.*, 2008; Linnamo *et al.*, 2005; McCaulley *et al.*, 2009; Mero *et al.*, 1991). The results of some studies have suggested that C concentrations may even decline across certain forms of resistance training (Beaven, Gill and Cook, 2008a; Crewther *et al.*, 2008; Smilios *et al.*, 2003). Theoretically, if the testosterone-to-cortisol ratio (T/C) is important, as some suggest, then the use of the T/C may be advantageous to evaluate the net hormonal effects imposed by different training workouts. However, sampling differences, the study population, and training status all need careful consideration when comparing studies.

2.4.3.2 Effects of age and gender

The acute C response to resistance exercise was found to be much greater in younger males than in adult males (Mero *et al.*, 1993; Pullinen *et al.*, 2002). In their study, Pullinen *et al.* (2002) compared the hormonal responses of men, women, and pubescent boys to a single exercise bout (5 sets x 10 knee extensions, 40% 1 RM) followed by two sets to exhaustion. Only the boys revealed an acute increase in C concentrations. Furthermore, peak epinephrine concentrations (another stress hormone) were found to be twice as high in this group, as compared to men and women. These data are indicative of a greater stress response among younger males. Differences in maturation, anxiety and adaptive capability to resistance exercise may be important factors in this respect (Pullinen *et al.*, 2002). Once adulthood has been reached, however, age appears to have little or no effect upon the acute C responses to a bout of resistance exercise (Häkkinen and Pakarinen, 1995; Kraemer *et al.*, 1999).

The responsiveness of C to resistance exercise appears to be similar between males and females, regardless of the type of workout performed (Häkkinen and Pakarinen, 1995; Kostka *et al.*, 2003; McGuigan, Egan and Foster, 2004; Pullinen *et al.*, 2002). Kraemer *et al.*, 1998b also found no significant differences in the acute C responses of males and females to a single exercise bout, performed before and after an eight-week period of heavy-resistance training. Subsequently, the stimulus of resistance exercise would seem to elicit similar adrenal stress responses for both males and females. The resting concentrations of C are also largely similar for males and females (Häkkinen and Pakarinen, 1995; Kostka *et al.*, 2003; McGuigan, Egan and Foster, 2004; Pullinen *et al.*, 2002). Given these findings, it may be speculated that gender differences in muscle mass and performance are primarily driven by the differential response patterns of the anabolic hormones.

2.4.3.3 Effects of training history

The effects of training history upon C secretory patterns are less clear. A similar C response was reported by resistance-trained (380%) and sedentary males (380%), although the loads utilized were vastly different (Tremblay, Copeland and Van Helder, 2004). Other research has reported either small or no changes in the acute C responses after periods of weight training (Kraemer *et al.*, 1998b; McCall *et al.*, 1999), or demonstrated that training experience does not influence these responses (Ahtiainen *et al.*, 2004; Kraemer *et al.*, 1992). However, one study reported a greater C response in untrained individuals following resistance exercise, compared to that seen in trained individuals (McMillan *et al.*, 1993), the possible result of greater perception of stress among those with no training experience, even though they exercised at the same relative load (McMillan *et al.*, 1993). In contrast to T, basal C concentrations were found to be no different between various athletic groups and controls (Ahtiainen *et al.*, 2004; Grasso *et al.*, 1997; Hackney, Fahrner and Gulledge, 1998; Izquierdo *et al.*, 2004).

2.4.4 DUAL ACTIONS OF TESTOSTERONE AND CORTISOL

T and C appear to have dual actions upon the neuromuscular system: first, acting in the traditional longer time frame through a genomic pathway; and second, acting on a much shorter time frame through a non-genomic pathway. As supporting evidence, both T and C have been shown to induce rapid (i.e. seconds to minutes) changes in neuronal activity within the animal brain (Chen *et al.*, 1991; Jansen *et al.*, 1993; Smith, Jones and Wilson, 2002; Zaki and Barrett-Jolley, 2002). These steroid hormones can also produce relatively rapid changes in mood, behaviour, and/or cognitive function in animals (Aikey *et al.*, 2002; James and Nyby, 2002; Schiml and Rissman, 1999) and humans (Aleman *et al.*, 2004; Hermans *et al.*, 2006; Hsu *et al.*, 2003). In addition, these hormones can both produce rapid changes in intracellular Ca^{2+} in muscle cells (Estrada *et al.*, 2000, 2003; Passaquin, Lhote and Rüegg, 1998). These findings are important because Ca^{2+} triggers the muscle contraction and regulates energy metabolism (Berchtold, Brinkmeier and Muntener, 2000).

Hormones can influence other neural pathways important to muscle function. Early research identified rapid C effects upon the electrophysiological properties of the diaphragm muscle in rats (Dlouhá and Vyskočil, 1979), and rapid T effects upon the reflex activity of rat penile muscle (Sachs and Leipheimer, 1988). More recently, an increase in T concentrations was found to be effective in decreasing the cortical motor threshold in humans (Bonifazi *et al.*, 2004), whereas C concentrations had a direct effect upon the motor cortex and electromyographic activity of hand muscle in humans (Sale, Ridding and Nordstorm, 2008). Collectively, these rapid hormonal effects have possible implications for muscle function. In fact, correlations have been reported between hormone concentrations and individual performance during speed, power, and strength tests (Bosco *et al.*, 1996a, 1996b; Crewther *et al.*, 2010; Jürimäe and Jürimäe, 2001; Słowińska-Lisowska and Witkowski, 2001). The rapid effects of T and C could also play a role in controlling muscle growth, since the expression of performance (e.g. forces produced, work done) during individual workouts is a major factor influencing protein metabolism (Crewther, Cronin and Keogh, 2005).

2.4.5 GROWTH HORMONE RESPONSES TO RESISTANCE TRAINING

GH is another anabolic hormone which influences muscle growth by increasing protein synthesis and reducing muscle protein degradation (Crewther *et al.*, 2006; Velloso, 2008). The release of GH may further enhance the training environment by stimulating the release of a family of polypeptides called the somatomedins from the liver (Velloso, 2008); this family is also known for its anabolic effects upon muscle tissue. Unlike the steroid hormones, human GH represents a family of proteins rather than a single hormone complex, with over 100 forms of GH currently identified within plasma fluid (Baumann, 1991). However, the function of the different variants of GH has not yet been fully established. The most dominant form of GH in circulation, and the most widely examined within research, is the 22kD variant.

2.4.5.1 Effects of workout design

Similar to the steroid hormones, GH is also acutely responsive to the design of different training workouts. Hypertrophy training sessions have, for example, been shown to produce large increases in circulating GH concentrations, and more so than maximal-strength training sessions (Häkkinen and Pakarinen, 1993; Kraemer *et al.*, 1990, 1991, 1993; Smilios *et al.*, 2003; Zafeiridis *et al.*, 2003). Much less research has investigated the hormonal responses to an explosive power workout in males and females (Linnamo *et al.*, 2005), with GH increasing after exercise, but only in men. It is interesting to note that the magnitude of the acute changes in GH (up to 170-fold) after different training sessions is generally much greater than that observed for T (up to 1-fold) and C (up to 2-fold), relative to baseline measurements (Crewther *et al.*, 2006).

2.4.5.2 Effects of age and gender

A reduction in the GH response to resistance exercise is seen during later life. Häkkinen and Pakarinen (1995) compared the GH responses of three groups of males (27 years, 47 years, 68 years) and females (25 years, 48 years, 68 years), each performing an identical weight-training session. For males, the 27-year-old group reported the greatest GH response (200-fold), followed by the 47-year-olds (19-fold), with no changes reported in the 68-year-olds. For females, the 48-year-old group produced a much larger increase in GH (20-fold) than the 25-year-olds (2-fold), while the oldest female group was non-responsive to the exercise protocol. Other research findings are in agreement (Häkkinen *et al.*, 1998, 2000; Kraemer *et al.*, 1998a, 1999; Pyka, Wiswell and Marcus, 1992) and confirm that the decline in muscle size and performance with age may be partially attributed to changes in GH activity. No significant differences in GH concentrations were found between men and boys (Pullinen *et al.*, 2002).

Males generally attain greater acute GH responses (relative changes) to resistance than females (Häkkinen and Pakarinen, 1995; Häkkinen *et al.*, 2000, 2002; Kraemer *et al.*, 1991; Pullinen *et al.*, 2002). On the other hand, females have been shown to produce greater GH responses (absolute values) after a single training session (Häkkinen *et al.*, 2000; Kraemer *et al.*, 1991, 1998b). These findings may be related to the oestrogen senitization of the somatotrophs, which are known to give an increased responsiveness to a variety of stimuli among females (Kraemer *et al.*, 1991), and/or baseline differences in GH concentrations, which are greater for females (Kraemer *et al.*, 1991; Pullinen *et al.*, 2002). The importance of elevated GH among females requires further investigation. It is noteworthy that most studies have examined females during the follicular phase of menstruation (Consitt, Copeland and Tremblay, 2001; Copeland, Consitt and Tremblay, 2002; Kraemer *et al.*, 1991, 1993, 1998b; Mulligan *et al.*, 1996; Pullinen *et al.*, 2002; Taylor *et al.*, 2000). Clearly, the effects of resistance exercise on hormone release in different phases of the menstrual cycle is another area warranting investigation.

2.4.5.3 Effects of training history

The GH response to a resistance training sessions does not appear to be influenced by training history. For instance, no differences in the acute GH responses were found between trained and untrained athletes (Ahtiainen *et al.*, 2004), or as a function of weight-training experience (Kraemer *et al.*, 1992). Likewise, resistance training of varying periods produced little or no change in the GH responses (Häkkinen *et al.*, 2001; Kraemer *et al.*, 1998b, 1999; McCall *et al.*, 1999; Nicklas *et al.*, 1995). Taylor *et al.* (2000) did report a greater GH response to resistance exercise among trained females (90%) when compared to untrained females (30%). This result may be partially attributed to the significantly lower resting GH values for the trained group. Disparate findings in this area may also be attributed to individual variability in the GH responses to exercise and different training methodologies, as well as possible interactions with age and gender.

2.4.6 OTHER BIOCHEMICAL MARKERS

The mammalian target of rapamycin (mTOR) is one of the more recently discovered pathways that appear important for hypertrophy (Drummond *et al.*, 2009a; Terzis *et al.*, 2008). Possibly the most influencial study in this area is that of Terzis *et al.* (2008); using an indirect measure of mTOR activation, via muscle biopsy tissue, a strong relationship ($r = 0.81–0.89$) was seen between the hypertrophy and strength gains achieved over a 14-week training period and mTOR activity. Taken together with other studies, in which rapamycin treatment can prevent resistance exercise-induced hypertrophy (Drummond *et al.*, 2009b), this is good evidence of an involvement of mTOR in muscle adaptation. However, biopsy assessments are somewhat longitudinally limited when compared to blood or saliva analyses and comprehensive data relating mTOR to training adaptation with different protocols remains to be collected. Likewise, very few systematic studies have investigated the modulating effects of age, gender, and training history upon mTOR.

Insulin is also believed to have a strong anabolic effect upon muscle growth, repair, and recovery (Crewther *et al.*, 2006; Kraemer, 2000). Whilst the actions of the other anabolic hormones would appear to directly influence cellular reactions and protein accretion, by either binding directly inside the cell (steroid hormones) or to the cell membrane (peptide hormones), the actions of insulin mediate the adaptive processors involved in tissue regeneration and growth (Crewther *et al.*, 2006; Kraemer, 2000). The catecholamines norepinephrine and epinephrine also help to mediate weight-training adaptation, although this role is probably related to neuromuscular function, energy mobilization, and utilization, as well as preparing individuals for the stress of resistance exercise (French *et al.*, 2007; Zouhal *et al.*, 2008). Few data are available concerning the acute response patterns of these hormonal markers to different resistance training workouts and long-term adaptation.

2.4.7 LIMITATIONS IN THE USE AND INTERPRETATION OF BIOCHEMICAL MARKERS

Biochemical markers can be regarded as simple markers; that is, their changes mirror or predict changes in performance or as causal agents (blocking their effect abolishes any expected gains in performance). They can also be regarded as permissive (as long as they are present in physiological quantities the performance changes can occur) or dose-dependent (the relationship to performance depends on the dose present). Three areas have been investigated. In the case of T, blocking its effect on androgen receptors prevents resistance training-induced hypertrophy and strength changes (Kvorning *et al.*, 2006; Inoue *et al.*, 1994). Whether this is permissive or has a dose-response nature is not as yet known. Clearly, supraphysiological levels of T are associated with extreme hypertrophy, and very low levels (hypogonadism) with loss of muscle mass and strength (alleviated by returning T to physiological levels).

The roles of GH and IGF-1 remain more equivocal. Although mechanical stimulation increases IGF-1, studies have shown that functional IGF receptors are not needed for hypertrophy to occur (Spangenburg *et al.*, 2008), in contrast to the apparent permissive necessity of T. Recent data suggest that IGF-1 may be of greater importance in inducing connective tissue adaption than in inducing muscle growth per se (Yakar *et al.*, 2009). Other biochemical pathways and hormones have not been investigated to the same extent as T, C, and GH, but may have overlapping observations with the above. It is highly likely that an orchestra of biochemical factors are needed for resistance training adaptation to occur.

The most obvious difficulty in trying to assess biochemical markers relative to resistance training adaptation is the disparity of methods and protocols across studies, as well as the use of non-uniform training-status populations. Many studies use a single-point assay when frequent longitudinal sampling is needed to adequately understand the dynamics of hormonal response to resistance exercise. Similarly, sampling methods – saliva, blood (venous or capillary), tissue – differ, as do protocols. Control data is generally inadequate and often fails to take into account such variables as time of day, given the circadian rhythm of hormones (Kraemer *et al.*, 2001; Nindl *et al.*, 2001a; Thuma *et al.*, 1995). This understanding can be further confounded by the pulsatile secretion of some hormones (e.g. GH) (Nindl *et al.*, 2001b). Plasma volume changes during resistance exercise, mainly due to an accumulation of lactate and other metabolites in the muscles (Durand *et al.*, 2003; Wallace, Moffatt and Hancock, 1990), can potentially change the measured concentrations of hormones, irrespective of actual changes in hormone production.

Saliva is a relatively new format for the measurement of hormones, compared to blood and muscle sampling, and offers an easy compliant method that can be applied frequently with relatively low stress (Cook, 2002; Vining and McGinley, 1986). Furthermore, saliva reflects the biologically active component of the steroid hormones, since only the free hormone partitions cross blood into saliva (Vining and McGinley, 1986; Viru and Viru, 2001). However, the contamination of saliva with blood and salivary protein from the mucosa can interfere with hormonal measurements (Tremblay, Chu and Mureika, 1995). The partitioning of hormones between blood and saliva (e.g. time lag, influence of flow rate, etc.), particularly under stress is also curretnly not well known. Assay selection adds another variable, with different antibodies and assays used by suppliers, producing different cross-reactivity and potentially different results (Tremblay, Chu and Mureika, 1995).

2.4.8 APPLICATIONS OF RESISTANCE TRAINING

Applications of the literature to resistance training are extremely difficult, in large part due to the lack of standardization of design. Sampling points, assay designs, and hormonal fraction examined all vary greatly, and in many cases subjects are naïve resistance trainers of variable age. Consistent pattern studies are needed in elite athletes in order to understand the usefulness, if any, of hormones in patterning of training adaptation. It seems likely that the endocrine response to resistance exercise plays an important role in facilitating changes in both maximal strength and power and, at least to some extent, through morphological adaptation. In conjunction with mechanical stimuli, stimulation of multiple pathways including mTOR hormones probably assists in the remodelling of muscle tissue (i.e. protein synthesis and degradation) post-exercise. Over time, a net increase in protein synthesis should increase muscle fibre area and thereby lead to greater CSA of the gross muscle. Resistance exercise is also suggested to be capable of modulating the 'quality' of muscle protein synthesized (Goldspink, 1992, 2003). However, it is likely that hormones also act to effectuate other changes that contribute to power and strength.

Resistive exercise programmes used for hypertrophy generally elicit the largest anabolic (e.g. T, GH) responses, which are theoretically conducive to the accretion of protein and muscle size. However, mTOR is also strongly stimulated by these types of workout, and antagonizing either T or mTOR, as discussed earlier, severely hampers hypertrophy. Hypertrophy workouts also elicit the largest C responses, traditionally regarded as catabolic. In breaking down muscle protein the catabolic actions of C may create an increased pool of amino acids, enabling protein synthesis to occur, or increase protein turnover rate in previously inactive muscles (Viru and Viru, 2001), aiding the remodelling of muscle protein. Typically, maximal-strength workouts produce smaller hormonal responses (and mTOR has yet to be explored outside standard hypertrophy models), suggesting that the remodelling of muscle tissue is less likely to occur. Strength athletes (i.e. bodybuilders and power lifters/ Olympic lifters) are also characterized by different morphological profiles (Alway *et al.*, 1992; Tesch, 1987; Tesch and Larsson, 1982; Tesch, Colliander and Kaiser, 1986), anecdotally supporting these underlying hormonal differences.

Power-loading workouts also elicit smaller hormonal responses, and do not produce as much apparent change in muscle CSA, compared to hypertrophy workouts. Endurance exercise can, under some circumstances, produce significant increases in circulating hormone levels, similar to those changes seen across hypertrophy workouts (Consitt, Copeland and Tremblay, 2001; Kindermann *et al.*, 1982; Thuma *et al.*, 1995; Zafeiridis *et al.*, 2003), but these do not generally result in any substantial changes in muscle size, which is perhaps attributable to a lack of the necessary mechanical stress. Thus, other mechanical factors (e.g. total tension, work done, time under tension) associated with the training stimulus may determine whether or not morphological adaptation will occur, and again mTOR mechanisms are not know for these types of training protocol.

Obviously the design of resistance training workouts plays an important role in determining the acute hormonal response. However, other factors such as training status, type of training experience, gender, age, nutrition, and genetic predisposition, all of which are highly individual, can also influence the hormonal responses. Where one pathway may become saturated or unavailable, another may alter its contribution to make the hormonal response a dynamic feature over time. The T/C has proven to be a sensitive tool for monitoring individual changes in power and/or strength during periods of training and detraining (Alén *et al.*, 1988; Fry *et al.*, 2000; Häkkinen *et al.*, 1985, 1988; Raastad *et al.*, 2001). Recently, Haff *et al.* (2008) demonstrated that the changes in the T/C could predict temporary over-reaching (or lack of recovery) in a group of elite female weightlifters and that these T/C changes matched well with transient deficits in force–time curves (Haff *et al.*, 2008).

Beaven *et al.* (2008b) presented a different opinion on the classical hypertrophy–power–maximal strength workouts. When they applied these workouts to well-trained professional rugby players, overall group data suggested little T response to any of the protocols and a fall in C across all protocols (Beaven, Gill and Cook, 2008a). After re-examining the data on an individual basis, there were marked and significant differences, with some of the athletes responding with large increases in T to hypertrophy, others to strength, and still others to power workouts. This individual-workout-dependent response appeared reasonably stable over a three-week period. Beaven and co-authors went on to show that utilizing these individual responses could produce greater hypertrophy and strength gains than any of the other protocols (Beaven, Cook and Gill, 2008b). There are limitations in this study, such as the small sample size used, but it a reasonably valid conclusion to suggest that much of the variability seen in the literature across resistance training protocols may be due to large consistent individual differences.

Biochemical monitoring has the potential to greatly improve our understanding of resistance training adaption and ultimately to assist in accelerating acquisition of its features of hypertrophy, power, and strength. However, we are still a considerable distance away from achieving this. Consistency in the study protocols, athletes used, and type of hormonal biochemical monitoring are needed. Limitations in the elite athlete world will include sampling; for example, while mTOR appears fascinating at present, it requires muscle biopsy for its assessment. Planning frequent longitudinal sampling based on biopsy is extremely difficult in elite athlete tracking. In contrast, saliva collection is extremely easy, but may suffer from the fact that we are only examining the free component of the hormonal pool, with time lags relative to other body pools that are currently poorly understood. The goals are tangible but will require good consistency across a number of studies to be achieved.

2.4.9 CONCLUSION

The contribution of biochemical markers to hypertrophy, strength, and power development, while tangible, are far from fully elucidated. Similarly, the actual roles of different resistance workouts in producing different muscle outcomes are also questionable. To provide further understanding in these areas, research needs to adopt a more systematic approach when evaluating biochemical changes. The differences between non-elite and elite athletes also need addressing and any approach made should have sufficient athlete compliance to collect a rich longitudinal base of both biochemical and related performance data. Collectively, this information would not only provide a better understanding of the roles of different biochemical processes in mediating resistance training adaptation, but would also lead to the effective prescription of training protocol on an individual-athlete basis.

References

Ahtiainen J.P. *et al.* (2004) Acute hormonal responses to heavy resistance exercise in strength athletes versus nonathletes. *Can J Appl Physiol*, **29** (5), 527–543.

Aikey J.L. *et al.* (2002) Testosterone rapidly reduces anxiety in male house mice (Mus musculus). *Horm Behav*, **42** (4), 448–460.

Aleman A. *et al.* (2004) A single administration of testosterone improves visuospatial ability in young women. *Psychoneuroendocrinology*, **29** (5), 612–617.

Alén M. *et al.* (1988) Responses of serum androgenic–anabolic and catabolic hormones to prolonged strength training. *Int J Sports Med*, **9** (3), 229–233.

Alway S.E. *et al.* (1992) Effects of resistance training on elbow flexors of highly competitive bodybuilders. *J Appl Physiol*, **72** (4), 1512–1521.

Baumann G. (1991) Growth hormone heterogeneity: genes, isohormones, variants, and binding proteins. *Endocr Rev*, **12** (4), 424–449.

Beaven M.C., Gill N.D. and Cook C.J. (2008a) Salivary testosterone and cortisol responses in professional rugby players after four resistance exercise protocols. *J Strength Cond Res*, **22** (2), 426–432.

Beaven M.C., Cook C.J. and Gill N.D. (2008b) Significant strength gains observed in rugby players after specific resistance exercise protocols based on individual salivary testosterone responses. *J Strength Cond Res*, **22** (2), 419–425.

Berchtold M.W., Brinkmeier H. and Muntener M. (2000) Calcium ion in skeletal muscle: its crucial role for muscle function, plasticity, and disease. *Physiol Rev*, **80** (3), 1215–1265.

Bonifazi M. *et al.* (2004) Effects of gonadal steroids on the input–output relationship of the corticospinal pathway in humans. *Brain Res*, **1011** (2), 187–194.

Bosco C. and Viru A. (1998) Testosterone and cortisol levels in blood of male sprinters, soccer players and cross–country skiers. *Biol Sport*, **15** (1), 3–8.

Bosco C. *et al.* (1996a) Hormonal responses in strenuous jumping effort. *Jpn J Physiol*, **46** (1), 93–98.

Bosco C., Tihanyi J. and Viru A. (1996b) Relationships between field fitness test and basal serum testosterone and cortisol levels in soccer players. *Clin Physiol*, **16** (3), 317–322.

Bruce S.A., Phillips S.K. and Woledge R.C. (1997) Interpreting the relation between force and cross–sectional area in human muscle. *Med Sci Sports Exerc*, **29** (5), 677–683.

Cardinale M. and Stone M.H. (2006) Is testosterone influencing explosive performance? *J Strength Cond Res*, **20** (1), 103–107.

Chen Y.Z. *et al.* (1991) An electrophysiological study on the membrane receptor–mediated action of glucocorticoids in mammalian neurons. *Neuroendocrinology*, **53** (Suppl. 1), S25–S30.

Consitt L.A., Copeland J.L. and Tremblay M.S. (2001) Hormone responses to resistance vs. endurance exercise in premenopausal females. *Can J Appl Physiol*, **26** (6), 574–587.

Cook C.J. (2002) Rapid non–invasive measurement of hormones in transdermal exudate and saliva. *Physiol Behav*, **75** (1–2), 169–181.

Copeland J.L., Consitt L.A. and Tremblay M.S. (2002) Hormonal responses to endurance and resistance exercise in females aged 19–69 years. *J Gerontol A Biol Sci Med Sci*, **57A** (4), B158–B165.

Crewther B., Cronin J. and Keogh J. (2005) Possible stimuli for strength and power adaptation: acute mechanical responses. *Sports Med*, **35** (11), 967–989.

Crewther B. *et al.* (2006) Possible stimuli for strength and power adaptation: acute hormonal responses. *Sports Med*, **36** (3), 215–238.

Crewther B. *et al.* (2008) The salivary testosterone and cortisol response to three loading schemes. *J Strength Cond Res*, **22** (1), 250–255.

Crewther B.T. *et al.* (2010) Neuromuscular performance of elite rugby union players and relationships with salivary hormones. *J Strength Cond Res*, in press.

Dlouhá H. and Vyskočil F. (1979) The effect of cortisol on the excitability of the rat muscle fibre membrane and neuromuscular transmission. *Physiol Bohemoslov*, **28** (6), 485–494.

Drummond M.J. *et al.* (2009a) Nutritional and contractile regulation of human skeletal muscle protein synthesis and mTORC1 signalling. *J Appl Physiol*, **106**, 1374–1384.

Drummond M.J. *et al.* (2009b) Rapamycin administration in humans blocks the contraction–induced increase in skeletal muscle protein synthesis. *J Physiol*, **587** (7), 1535–1546.

Dunn J.F., Nisula B.C. and Rodbard D. (1981) Transport of steroid hormones: binding of 21 endogenous steroids to both testosterone–binding globulin and corticosteroid–binding globulin in human plasma. *J Clin Endocrinol Metab*, **53** (1), 58–68.

Durand R.J. *et al.* (2003) Different effects of concentric and eccentric muscle actions on plasma volume. *J Strength Cond Res*, **17** (3), 541–548.

Estrada M. *et al.* (2000) Aldosterone- and testosterone-mediated intracellular calcium response in skeletal muscle cell cultures. *Am J Physiol Endocrinol Metab*, **279** (1), E132–E139.

Estrada M. *et al.* (2003) Testosterone stimulates intracellular calcium release and mitogen–activated protein kinases via a G protein–coupled receptor in skeletal muscle cells. *Endocrinology*, **144** (8), 3586–3597.

French D.N. *et al.* (2007) Anticipatory responses of catecholamines on muscle force production. *J Appl Physiol*, **102** (1), 94–102.

Fry A.C. *et al.* (1993) Endocrine and performance responses to high volume training and amino acid supplementation in elite junior weightlifters. *Int J Sport Nutr*, **3** (3), 306–322.

Fry A.C. *et al.* (2000) Relationships between serum testosterone, cortisol, and weightlifting performance. *J Strength Cond Res*, **14** (3), 338–343.

Goldspink G. (1992) Cellular and molecular aspects of adaptation in skeletal muscle, in *Strength and Power in Sport* (ed. P. Komi), Blackwell Scientific, Boston, MA, pp. 211–229.

Goldspink G. (2003) Gene expression in muscle in response to exercise. *J Muscle Res Cell Motil*, **24**, 121–126.

Gotshalk L.A. *et al.* (1997) Hormonal responses of multiset versus single–set heavy–resistance exercise protocols. *Can J Appl Physiol*, **22** (3), 244–255.

Grasso G. *et al.* (1997) Glucocorticoid receptors in human peripheral blood mononuclear cells in relation to age and to sport activity. *Life Sci*, **61** (3), 301–308.

Hackney A.C., Fahrner C.L. and Gulledge T.P. (1998) Basal reproductive hormonal profiles are altered in endurance trained men. *J Sports Med Phys Fitness*, **38** (2), 138–141.

Hackney A.C., Szczepanowska E. and Viru A.M. (2003) Basal testicular testosterone production in endurance–trained men is suppressed. *Eur J Appl Physiol*, **89** (2), 198–201.

Haff G.G. *et al.* (2008) Force–time curve characteristics and hormonal alterations during an eleven-week training period in elite women weightlifters. *J Strength Cond Res*, **22** (2), 433–446.

Häkkinen K. (1989) Neuromuscular and hormonal adaptations during strength and power training. *J Sports Med Phys Fitness*, **29** (1), 9–26.

Häkkinen K. and Pakarinen A. (1993) Acute hormonal responses to two different fatiguing heavy–resistance protocols in male athletes. *J Appl Physiol*, **74**, 882–887.

Häkkinen K. and Pakarinen A. (1995) Acute hormonal responses to heavy resistance exercise in men and women at different ages. *Int J Sports Med*, **16**, 507–513.

Häkkinen K. *et al.* (1985) Serum hormones during prolonged training of neuromuscular performance. *Eur J Appl Physiol Occup Physiol*, **53** (4), 287–293.

Häkkinen K. *et al.* (1988) Neuromuscular and hormonal adaptations in athletes to strength training in two years. *J Appl Physiol*, **65** (6), 2406–2412.

Häkkinen K. *et al.* (1998) Acute hormone responses to heavy resistance lower and upper extremity exercise in young versus old men. *Eur J Appl Physiol Occup Physiol*, **77** (4), 312–319.

Häkkinen K. *et al.* (2000) Basal concentrations and acute responses of serum hormones and strength development during heavy resistance training in middle-aged and elderly men and women. *J Gerontol Ser A Biol Sci Med Sci*, **55A** (2), B95–105.

Häkkinen K. *et al.* (2001) Selective muscle hypertrophy, changes in EMG and force, and serum hormones during strength training in older women. *J Appl Physiol*, **91**, 569–580.

Häkkinen K. *et al.* (2002) Effects of heavy resistance/power training on maximal strength, muscle morphology, and hormonal response patterns in 60–75-year-old men and women. *Can J Appl Physiol*, **27** (3), 213–231.

Herbst K.L. and Bhasin S. (2004) Testosterone action on skeletal muscle. *Curr Opin Clin Nutr Metab Care*, **7** (3), 271–277.

Hermans E.J. *et al.* (2006) A single administration of testosterone reduces fear–potentiated startle in humans. *Biol Psychiatry*, **59** (9), 872–874.

Hsu F.C. *et al.* (2003) Effects of a single dose of cortisol on the neural correlates of episodic memory and error processing in healthy volunteers. *Psychopharmacology*, **167** (4), 431–442.

Inoue K. *et al.* (1994) Androgen receptor antagonist suppresses exercise–induced hypertrophy of skeletal muscle. *Eur J Appl Physiol Occup Physiol*, **69**, 88–91.

Izquierdo M. *et al.* (2004) Maximal strength and power, muscle mass, endurance and serum hormones in weightlifters and road cyclists. *J Sports Sci*, **22** (5), 465–478.

James P.J. and Nyby J.G. (2002) Testosterone rapidly affects the expression of copulatory behavior in house mice (Mus musculus). *Physiol Behav*, **75** (3), 287–294.

Jansen H.T. *et al.* (1993) A re-evaluation of the effects of gonadal steroids on neuronal activity in the male rat. *Brain Res Bull*, **31** (1–2), 217–223.

Jürimäe J. and Jürimäe T. (2001) Responses of blood hormones to the maximal rowing ergometer test in college rowers. *J Sports Med Phys Fitness*, **41** (1), 73–77.

Kindermann W. *et al.* (1982) Catecholamines, growth hormone, cortisol, insulin and sex hormones in anaerobic and aerobic exercise. *Eur J Appl Physiol Occup Physiol*, **49**, 389–399.

Kostka T. *et al.* (2003) Anabolic and catabolic responses to experimental two–set low–volume resistance exercise in sendentary and active elderly people. *Aging Clin Exp Res*, **15** (2), 123–130.

Kraemer W.J. (2000) Endocrine responses to resistance exercise, in *Essentials of Strength Training and Conditioning* (eds T.R. Baechle and R.W. Earle), Human Kinetics, Champaign, IL, pp. 91–114.

Kraemer W.J. and Mazzetti S.A. (2003) Hormonal mechanisms related to the expression of muscular strength and power, in *Strength and Power in Sport* (ed. P.V. Komi), Blackwell Scientific Publishing, Boston, MA, pp. 73–95.

Kraemer W.J. *et al.* (1990) Hormonal and growth factor responses to heavy resistance exercise protocols. *J Appl Physiol*, **69**, 1442–1450.

Kraemer W.J. *et al.* (1991) Endogenous anabolic hormonal and growth factor responses to heavy resistance exercise in males and females. *Int J Sports Med*, **12** (2), 228–235.

Kraemer W.J. *et al.* (1992) Acute hormonal responses in elite junior weightlifters. *Int J Sports Med*, **13** (2), 103–109.

Kraemer W.J. *et al.* (1993) Changes in hormonal concentrations after different heavy–resistance exercise protocols in women. *J Appl Physiol*, **75**, 594–604.

Kraemer W.J. *et al.* (1998a) Acute hormonal responses to heavy resistance exercise in younger and older men. *Eur J Appl Physiol Occup Physiol*, **77** (3), 206–211.

Kraemer W.J. *et al.* (1998b) The effects of short-term resistance training on endocrine function in men and women. *Eur J Appl Physiol Occup Physiol*, **78** (1), 69–76.

Kraemer W.J. *et al.* (1999) Effects of heavy-resistance training on hormonal response patterns in younger vs. older men. *J Appl Physiol*, **87** (3), 982–992.

Kraemer W.J. *et al.* (2001) The effect of heavy resistance exercise on the circadian rhythm of salivary testosterone in men. *Eur J Appl Physiol*, **84**, 13–18.

Kvorning T. *et al.* (2006) Suppression of endogenous testosterone production attenuates the response to strength training: a randomized, placebo-controlled, and blinded intervention study. *Am J Physiol Endocrinol Metab*, **291** (6), E1325–E1332.

Linnamo V. *et al.* (2005) Acute hormonal responses to submaximal and maximal heavy resistance and explosive exercises in men and women. *J Strength Cond Res*, **19** (3), 566–571.

Loebel C.C. and Kraemer W.J. (1998) A brief review: testosterone and resistance exercise in men. *J Strength Cond Res*, **12** (1), 57–63.

McCall G.E. *et al.* (1999) Acute and chronic hormonal responses to resistance training designed to promote muscle hypertrophy. *Can J Appl Physiol*, **24** (1), 96–107.

McCaulley G.O. *et al.* (2009) Acute hormonal and neuromuscular responses to hypertrophy, strength and power type resistance exercise. *Eur J Appl Physiol*, **105**, 695–704.

McGuigan M.R., Egan A.D. and Foster C. (2004) Salivary cortisol responses and perceived exertion during high intensity and low intensity bouts of resistance exercise. *J Sports Sci Med*, **3**, 8–15.

McMillan J.L. *et al.* (1993) 20-hour physiological responses to a single weight-training session. *J Strength Cond Res*, **7** (1), 9–21.

Mero A. *et al.* (1991) Acute EMG, force and hormonal responses in male athletes to four strength exercise units. *J Biomech*, **25** (7), 287–288.

Mero A. *et al.* (1993) Biomechanical and hormonal responses to two high intensity strength exercise units with different recovery in pubertal and adult male athletes. XIVth Congress of the International Society of Biomechanics, Paris, France.

Mulligan S.E. *et al.* (1996) Influence of resistance exercise volume on serum growth hormone and cortisol concentrations in women. *J Strength Cond Res*, **10** (4), 256–262.

Nicklas B.J. *et al.* (1995) Testosterone, growth hormone and IGF-1 responses to acute and chronic resistive exercise in men aged 55–70 years. *Int J Sports Med*, **16** (7), 445–450.

Nindl B.C. *et al.* (2001a) LH secretion and testosterone concentrations are blunted after resistance exercise in men. *J Appl Physiol*, **91**, 1251–1258.

Nindl B.C. *et al.* (2001b) Growth hormone pulsatility profile characteristics following acute heavy resistance exercise. *J Appl Physiol*, **91**, 163–172.

Pardridge W.M. (1986) Serum bioavailability of sex steroid hormones. *Clin Endocrinol Metab*, **15** (2), 259–278.

Passaquin A.C., Lhote P. and Rüegg U.T. (1998) Calcium influx inhibition by steroids and analogs in C2C12 skeletal muscle cells. *Br J Pharmacol*, **124** (8), 1751–1759.

Pullinen T. *et al.* (2002) Resistance exercise-induced hormonal responses in men, women, and pubescent boys. *Med Sci Sports Exerc*, **34** (5), 806–813.

Pyka G., Wiswell R.A. and Marcus R. (1992) Age-dependent effect of resistance exercise on growth hormone secretion in people. *J Clin Endocrinol Metab*, **75** (2), 404–407.

Raastad T. *et al.* (2001) Changes in human skeletal muscle contractility and hormone status during 2 weeks of heavy strength training. *Eur J Appl Physiol*, **84**, 54–63.

Rowbottom D.G. *et al.* (1995) Serum free cortisol responses to a standard exercise test among elite triathletes. *Aust J Sci Med Sport*, **27** (4), 103–107.

Sachs B.D. and Leipheimer R.E. (1988) Rapid effect of testosterone on striated muscle activity in rats. *Neuroendocrinology*, **48** (5), 453–458.

Sale M.V., Ridding M.C. and Nordstorm M.A. (2008) Cortisol inhibits neuroplasticity induction in human motor cortex. *J Neurosci*, **28** (33), 8285–8293.

Schiml P.A. and Rissman E.F. (1999) Cortisol facilitates induction of sexual behavior in the female musk shrew (Suncus murinus). *Behav Neurosci*, **113** (1), 166–175.

Słowińska-Lisowska M. and Witkowski K. (2001) The influence of exercise on the functioning of the pituitary–gonadal axis in physically active older and younger men. *Aging Male*, **4** (3), 145–150.

Smilios I. *et al.* (2003) Hormonal responses after various resistance exercise protocols. *Med Sci Sports Exerc*, **35** (4), 644–654.

Smith M.D., Jones L.S. and Wilson M.A. (2002) Sex differences in hippocampal slice excitability: role of testosterone. *Neuroscience*, **109** (3), 517–530.

Spangenburg E.E. *et al.* (2008) A functional insulin-like growth factor receptor is not necessary for load-induced skeletal muscle hypertrophy. *J Physiol*, **586** (1), 283–291.

Taylor J.M. *et al.* (2000) Growth hormone response to an acute bout of resistance exercise in weight-trained and non-weight-trained women. *J Strength Cond Res*, **14** (2), 220–227.

Terzis G. *et al.* (2008) Resistance exercise–induced increase in muscle mass correlates with p70S6 kinase phosphorylation in human subjects. *Eur J Appl Physiol*, **102**, 145–152.

Tesch P. (1987) Acute and long-term metabolic changes consequent to heavy resistance exercise. *Med Sport Sci*, **26**, 67–89.

Tesch P.A. and Larsson L. (1982) Muscle hypertrophy in bodybuilders. *Eur J Appl Physiol Occup Physiol*, **49**, 301–306.

Tesch P.A., Colliander E.B. and Kaiser P. (1986) Muscle metabolism during intense, heavy–resistance exercise. *Eur J Appl Physiol Occup Physiol*, **55**, 362–366.

Thuma J.R. *et al.* (1995) Circadian rhythm of cortisol confounds cortisol responses to exercise: implications for future research. *J Appl Physiol*, **78** (5), 1657–1664.

Tremblay M.S., Chu S.Y. and Mureika R. (1995) Methodological and statistical considerations for exercise-related hormone evaluations. *Sports Med*, **20** (2), 90–108.

Tremblay M.S., Copeland J.L. and Van Helder W. (2004) Effect of training status and exercise mode on endogenous steroid hormones in men. *J Appl Physiol*, **96** (2), 531–539.

Velloso C.P. (2008) Regulation of muscle mass by growth hormone and IGF-1. *Br J Pharmacol*, **154**, 557–568.

Vermeulen A., Rubens R. and Verdonck L. (1972) Testosterone secretion and metabolism in male senescence. *J Clin Endocrinol*, **34**, 730–735.

Vining R.F. and McGinley R.A. (1986) Hormones in saliva. *Crit Rev Clin Lab Sci*, **23** (2), 95–146.

Viru A. and Viru M. (2001) Hormones as tools for training monitoring, in *Biochemical Monitoring of Sport Training*, Human Kinetics, Champaign, IL, pp. 73–112.

Viru A. and Viru M. (2004) Cortisol – essential adaptation hormone in exercise. *Int J Sports Med*, **25**, 461–464.

Volek J. *et al.* (1997) Testosterone and cortisol relationship to dietary nutrients and resistance exercise. *J Appl Physiol*, **82** (1), 49–54.

Wallace B.M., Moffatt R.J. and Hancock L.A. (1990) The delayed effects of heavy resistance exercise on plasma volume shifts. *J Strength Cond Res*, **4** (4), 154–159.

Yakar S. *et al.* (2009) Serum IGF-1 determines skeletal strength by regulating sub-periosteal expansion and trait interactions. *J Bone Miner Res*, **24** (8), 1481–1492.

Zafeiridis A. *et al.* (2003) Serum leptin responses after acute resistance exercise protocols. *J Appl Physiol*, **94**, 591–597.

Zaki A. and Barrett-Jolley R. (2002) Rapid neuromodulation by cortisol in the rat paraventricular nucleus: an in vitro study. *Br J Pharmacol*, **137** (1), 87.

Zouhal H. *et al.* (2008) Catecholamines and the effects of exercise, training and gender. *Sports Med*, **38** (5), 401–423.

2.5 Cardiovascular Adaptations to Strength and Conditioning

Andrew M. Jones and Fred J. DiMenna, School of Sport and Health Sciences, University of Exeter, Exeter, UK

2.5.1 INTRODUCTION

During a lifetime, the cardiovascular system of an average person will be responsible for the movement of over 200 million litres of blood just to sustain life at rest. Add in the extra work necessary to support all of our daily physical activities and it is obvious that the human cardiovascular system must operate as efficiently and effectively as any machine that man has ever made. The purpose of cardiovascular strength and conditioning training is to systematically overload cardiovascular function so that specific components of the system that might limit its capacity will improve their level of performance.

2.5.2 CARDIOVASCULAR FUNCTION

2.5.2.1 Oxygen uptake

The cardiovascular system plays a vital role in maintaining cellular homoeostasis, especially during exercise. For example, when exercise is initiated and muscles begin to contract vigorously, aerobic energy metabolism within active muscle cells increases in an attempt to support the higher level of energy turnover. This mandates a coordinated effort by the pulmonary and cardiovascular systems to ensure that sufficient atmospheric oxygen is delivered to the mitochondria of the involved fibres. The Fick equation indicates that the quantity of oxygen that will be consumed (oxygen uptake; VO_2) is a function of both the amount contained in the blood that leaves the heart and the quantity removed from that blood as it perfuses active tissues. Oxygen consumption is proportional to the intensity of the contractions being performed and in elite athletes can surpass 20 times the resting requirement at maximal levels of exertion. This means that critical cardiovascular adjustments must occur to facilitate delivery of the appropriate amount of oxygenated blood to the body's periphery during exercise.

2.5.2.2 Maximal oxygen uptake

Maximal oxygen uptake (VO_{2max}) represents the maximum rate at which oxygen can be consumed by a human at sea level. Therefore, VO_{2max} is often considered to reflect functional capacity and is typically used as an index to assess cardiovascular function. During exercise at sea level that requires more than approximately one third of an individual's total muscle mass, it is generally accepted that VO_{2max} is principally limited by the rate at which oxygen can be supplied to the muscles and not by the muscles' ability to extract oxygen from the blood they receive (Andersen and Saltin, 1985; Gonzalez-Alonso and Calbet, 2003; Knight *et al.*, 1993; Saltin and Strange, 1992; Wagner, 2000). Total blood volume for an average adult is approximately 5 l, and this blood can carry a maximum of 180–200 ml of oxygen per litre in a healthy individual at sea level (Brooks *et al.*, 2000). Given that the cardiovascular system is a closed system (additional blood cannot be added when circulatory demands are high), this means that prodigious VO_{2max} values can only be attained if the limited amount of blood that is available for distribution is both sent to and returned from the active musculature very rapidly.

2.5.2.3 Cardiac output

Cardiac output is the volume of blood ejected from the left ventricle of the heart each minute. Elite endurance athletes who can achieve VO_{2max} values of over 5–6 l/min must generate cardiac outputs well in excess of 30 l/min to do so (Jones and Poole, 2009). Cardiac output can be calculated as the product of heart rate (myocardial contractions per minute) and stroke volume (millilitres of blood ejected per contraction), and both of these variables increase during exercise (see Figure 2.5.1). Maximum heart rate is relatively fixed for a given individual at a given point in time, although it does decrease with age. Conversely, maximum stroke volume is a trainable

Strength and Conditioning – Biological Principles and Practical Applications Marco Cardinale, Rob Newton, and Kazunori Nosaka.

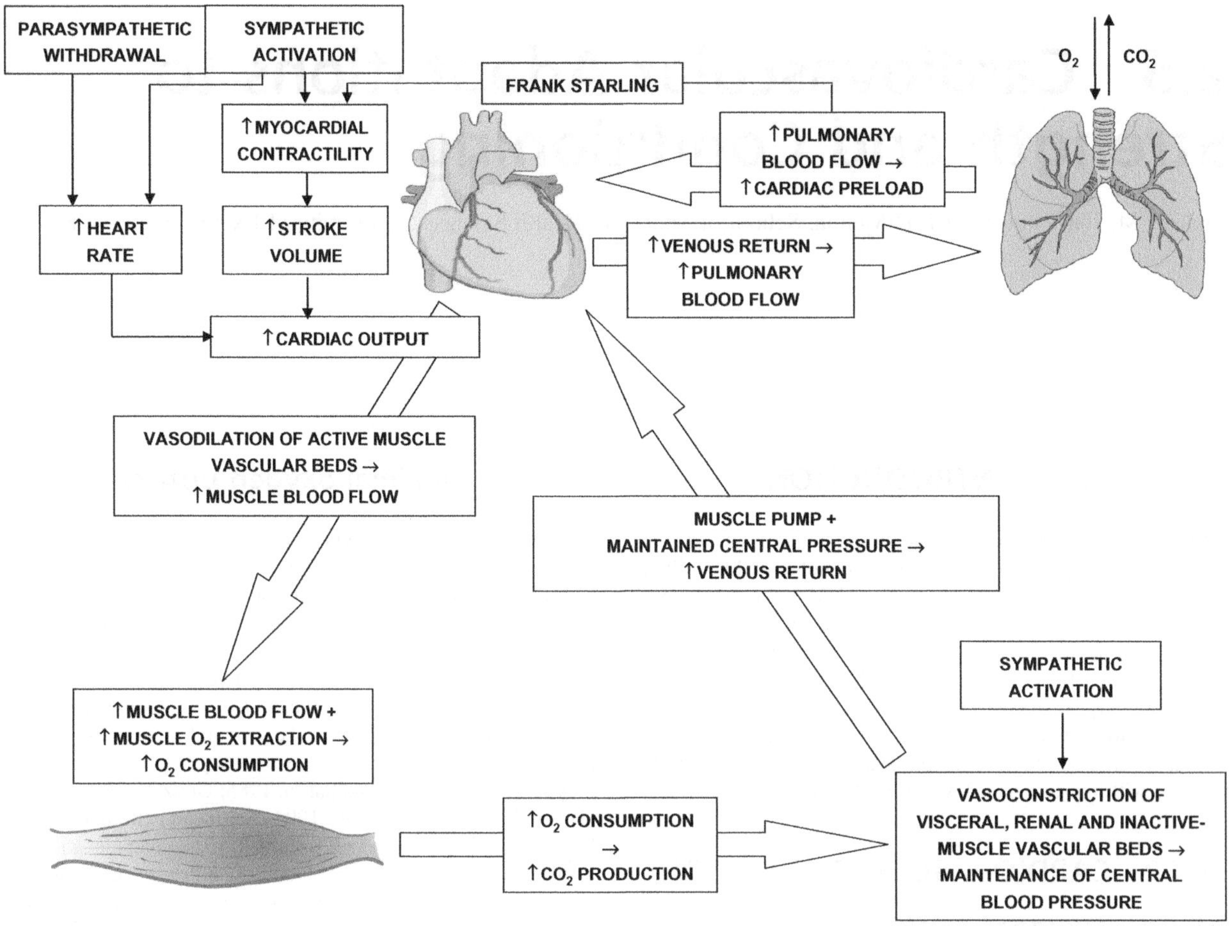

Figure 2.5.1 Endurance exercise provides the precise stimulus that is specific for inducing positive cardiovascular adaptations. When large muscle groups contract rhythmically at a sustainable percentage of the maximal voluntary contraction, aerobic metabolism predominates and the deliverance of atmospheric oxygen to the body's periphery must increase. Consequently, circulation is elevated and the pumping action of the muscles ensures that blood returning to the heart is similarly enhanced. This results in eccentric left-ventricular hypertrophy because the heart's chambers must expand from the inside out in order to accommodate more blood. Illustration by Jamie Blackwell

physiological parameter that is determined by the ability of the left ventricle to accommodate blood during diastole (the relaxation phase of the cardiac cycle) and to eject the blood it contains during systole (the cardiac contraction). For many years, exercise physiologists believed that stroke volume reached a plateau at approximately 40% of VO_{2max}, after which heart rate was the sole mechanism for increasing cardiac output (Åstrand *et al.*, 1964). However, more current research indicates that this is not the case, at least for highly-trained endurance athletes, because in these individuals stroke volume continues to rise throughout incremental exercise to exhaustion (Glehdill, Cox and Jamnik, 1994; Wiebe *et al.*, 1999; Zhou *et al.*, 2001). Regardless of this distinction, however, maximal heart rate and maximal stroke volume together determine maximal cardiac output, which has been estimated to account for 70–85% of the limitation to VO_{2max} (Bassett and Howley, 2000), and it is stroke volume that can be predominantly altered by training.

2.5.2.4 Cardiovascular overload

With regard to exercise training, the principle of specificity states that the structural and functional adaptations that can be expected to result from repeated application of a given overload exercise stressor will be unique and highly dependent on the particular stressor. Therefore, identifying the appropriate training stimulus requires knowledge of the system being targeted in order to determine the specific form of exercise that will maximally tax its principal function. Given that the cardiovascular system's primary purpose is to circulate blood, the only forms of exercise that will be optimal for inducing positive adaptations in cardiovascular function are those that require rapid movement of blood to and from the body's periphery. Unfortunately, there is often a lack of understanding in the strength and conditioning field regarding the precise overload that achieves this purpose.

2.5.2.5 The cardiovascular training stimulus

It has long been known that VO_2 and heart rate rise in an approximately linear manner as exercise intensity is increased (Åstrand and Saltin, 1961). This has led to the use of heart rate as a convenient gauge of VO_2 and therefore cardiovascular training intensity (Burke, 1998). However, it is important to recognize that an elevated heart rate per se is not the requisite stimulus that drives chronic cardiovascular adaptations. For example, during conventional resistance training (i.e. muscular contractions performed against a relatively high percentage of the maximal voluntary contraction), heart rate and VO_2 will be dissociated because elevation of the former will occur due to a drive for motor unit recruitment that is mediated by the sympathetic branch of the autonomic nervous system. Furthermore, the intense contractile effort that characterizes each repetition and the corresponding increase in intramuscular pressure will trap blood in peripheral vascular beds, thereby *reducing* blood flow until the 'sticking point' of each repetition is surpassed (Shepherd, 1987). This is *not* consistent with the specific overload that drives positive cardiovascular adaptation.

2.5.2.6 Endurance exercise training

Endurance exercise training is a broad term that is used quite liberally in the field of strength and conditioning. Generally speaking, endurance exercise has been suggested to include all of those (usually continuous) sports events or physical activities that rely predominantly on oxidative metabolism for energy supply, provided that they are sustained for a sufficiently long period of time (e.g. ≥90 seconds during high-intensity exercise and ≥10 minutes during submaximal exercise) (Jones and Poole, 2009). Endurance (aerobic) exercise is characterized by rhythmic contractions of a significant portion of the body's larger muscle groups at a sustainable percentage of the maximal voluntary contraction; this type of effort will also elevate heart rate. However, in this case, VO_2 is similarly increased, circulation is promoted, and an overload specific to circulatory function can be achieved (see Figure 2.5.1). This represents the precise stimulus that induces cardiovascular adaptations, which will be reflected in an increased functional capacity. Generally speaking, intense endurance training results in a VO_{2max} increase of approximately 20%; however, greater increases are possible if a participant's initial fitness level is low (Brooks *et al.*, 2000; Hickson *et al.*, 1981). The duration of the training programme and the intensity, duration, and frequency of individual training sessions are critical factors which determine the magnitude of the increase that can be expected (Wenger and Bell, 1986). The genetic proclivity for improvement and the age of the participant also play important roles.

2.5.3 CARDIOVASCULAR ADAPTATIONS TO TRAINING

Endurance exercise training facilitates numerous chronic positive changes that affect both central and peripheral aspects of cardiovascular system function. In addition, the heart itself operates much more efficiently in the trained state, with a reduced heart rate and requirement for oxygen at rest and during submaximal exercise at the same metabolic rate (Barnard *et al.*, 1977). These adaptations are responsible for marked changes in aerobic fitness, which typically manifest as an increased VO_{2max} and the ability to perform all submaximal exercise challenges with reduced cardiovascular stress (see Figures 2.5.2 and 2.5.3).

2.5.3.1 Myocardial adaptations to endurance training

Increased stroke volume/cardiac output

The most significant positive cardiovascular adaptation to chronic endurance training is a marked increase in stroke volume at rest and during submaximal and maximal exercise (Ekblom and Hermansen, 1968; Saltin *et al.*, 1968). In the trained state, cardiac output at rest and at any absolute submaximal work rate is either unchanged or decreased (Brooks *et al.*, 2000). This means that the increased capacity to move blood with each contraction of the myocardium allows any submaximal cardiac output, including the requirement at rest, to be established at a lower heart rate (see Figure 2.5.2). Furthermore, the maximal heart rate is also unchanged or slightly decreased after training (Ekblom *et al.*, 1968; Wilmore *et al.*, 2001), but maximal cardiac output is increased by 30% or more (Saltin and Rowell, 1980). This means that increased stroke volume provides the exclusive means by which maximal cardiac output is increased due to training (see Figure 2.5.3). This improvement also represents the predominant mechanism that facilitates an increased VO_{2max} because only a modest enhancement of arteriovenous oxygen difference (which reflects the ability to extract oxygen from the blood as it circulates through active tissues) at maximal exercise is typically present in the trained state (e.g. 16.5 compared to 16.2 ml of oxygen per 100 ml of blood) (Brooks *et al.*, 2000). Training-induced increases in stroke volume are achieved when the ventricle is able to accommodate more blood and/or when it can eject a greater percentage of the blood that it contains.

Increased left-ventricular end-diastolic volume

The increase in maximal stroke volume that occurs due to endurance training is predominantly attributable to an increased quantity of blood available for ejection (Brandao *et al.*, 1993; Ehsani, Hagberg and Hickson, 1978; Levy *et al.*, 1993). During

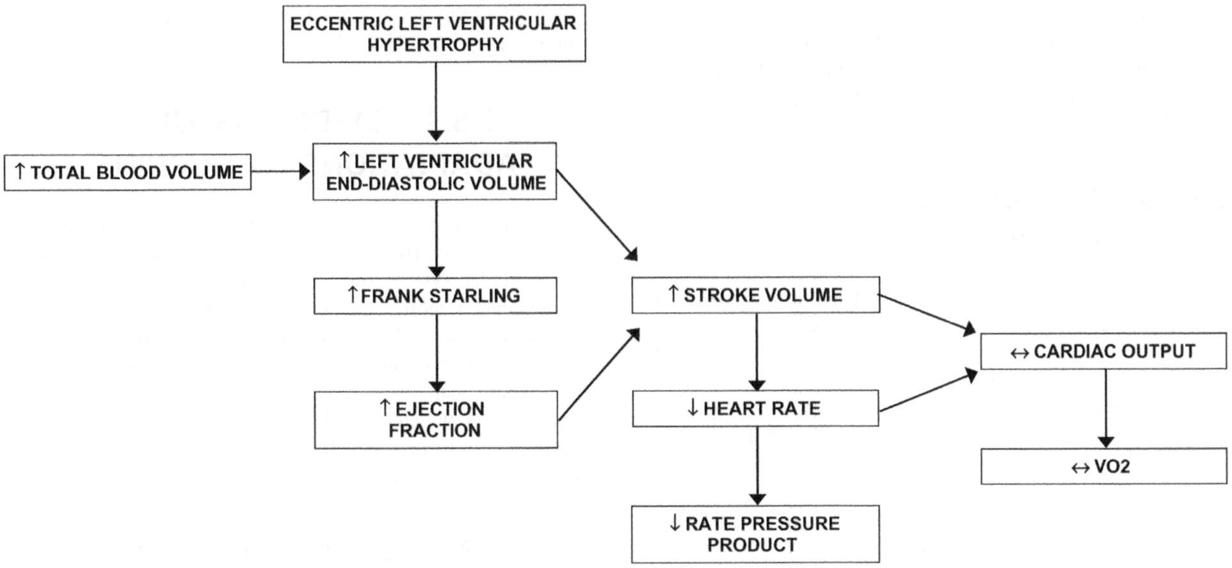

Figure 2.5.2 Endurance exercise training results in eccentric left-ventricular hypertrophy and an increase in total blood volume. This allows the same submaximal exercise work rate/VO_2 to be maintained at a lower heart rate and RPP

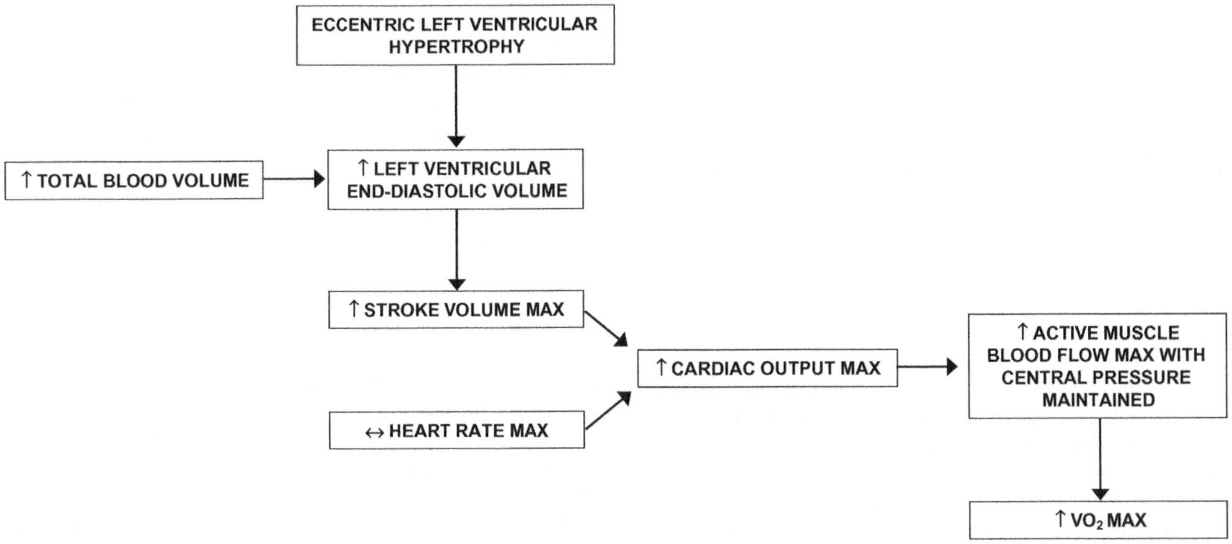

Figure 2.5.3 Endurance exercise training results in eccentric left-ventricular hypertrophy and an increase in total blood volume. This loosens the central circulatory restriction at maximal exercise, which allows a greater VO_{2max} to be achieved

diastole, the ventricles relax and the blood that will be expelled during the ensuing contraction phase fills the emptied chambers. Left-ventricular end-diastolic volume (LVEDV) or preload is the amount of blood contained in the left ventricle once this filling phase is complete. The amount of oxygenated blood available for filling is dependent upon the amount of deoxygenated blood that was sent to the lungs from the right ventricle of the heart during the preceding myocardial contraction. Consequently, the critical determinant of preload is venous return (the flow of deoxygenated blood back to the right side of the heart from the periphery). An endurance exercise session provides a prolonged period during which venous return is greatly increased due to the rhythmic contractions performed by the active musculature (the muscle pump) (see Figure 2.5.1).

Regular endurance exercise therefore mandates an adaptive change in ventricular volume to accommodate this oft-encountered overload.

Left-ventricular eccentric hypertrophy

Investigations that have used echocardiography or magnetic resonance imaging (MRI) to measure myocardial dimensions have revealed that athletes involved in endurance-training sports possess increased left-ventricular mass (Cohen and Segal, 1985; Henriksen *et al.*, 1996; Maron, 1986; Milliken *et al.*, 1988; Riley-Hagan *et al.*, 1992; Scharhag *et al.*, 2002). This indicates that when the heart is subjected to extended periods of elevated venous return on a regular basis, the chronic volume overload causes myocardial tissue to adapt. Specifically, the muscle cells (cardiocytes) which make up the myocardium grow larger and the myocardium as a whole grows bigger from the inside out (Morganroth *et al.*, 1975). This eccentric hypertrophy results in larger chambers that can accommodate more blood. Left-ventricular eccentric hypertrophy is the critical adaptation that allows for increased preload and it's important to note that this myocardial growth occurs exclusively in a longitudinal direction, without a corresponding change in the length of the muscle's contractile units (sarcomeres) (Moore, 2006). Consequently, the training-induced increase in preload that can be achieved due to left-ventricular eccentric hypertrophy causes myocardial sarcomeres to experience greater stretch during diastole. A training-induced increase in left-ventricular end-diastolic diameter can be achieved from as little as 10 consecutive days of cycle ergometer training (Mier *et al.*, 1997).

The Frank–Starling mechanism

The length–tension relationship of skeletal muscle indicates that there is a specific sarcomere length at which the contractile myofilaments (actin and myosin) are aligned to provide the greatest opportunity for cross-bridge interaction. This is the optimal length for tension development in a passive setting. However, a muscle fibre also contains series elastic components that absorb energy when the fibre is placed on stretch. Consequently, during active tension development, a contraction can benefit from this 'potential energy of elongation' through the stretch–shortening cycle (SSC). The Frank–Starling mechanism involves similar storage and subsequent use of elastic energy by the muscle fibres of the heart. Frank–Starling facilitation allows for strong myocardial contractions during endurance-training sessions when venous return and cardiac preload are high (i.e. when left ventricular myocardial tissue is placed on significant stretch) (see Figure 2.5.1) and the resultant eccentric left-ventricular hypertrophy, without a corresponding increase in myocardial sarcomere length, ensures that these stretch–shortening properties are enhanced in the trained state (Levine *et al.*, 1991).

Increased myocardial contractility

An endurance-trained heart consists of compliant tissue that can relax sufficiently to allow optimal filling of the enlarged left ventricle during diastole. However, an increased ability to accommodate blood will only prove beneficial if a significant portion of that blood can be expelled. The ejection fraction is the percentage of end-diastolic volume that is pumped from the ventricles each time the myocardium contracts. Enhancement of the Frank–Starling mechanism due to a training-induced increase in preload is one way that the trained heart achieves adequate emptying. However, there is evidence to suggest that architectural changes that facilitate a more forceful myocardial contraction also occur due to endurance training.

In addition to intrinsic growth of the heart's chambers, myocardial tissue can also become enlarged (hypertrophied) due to increased synthesis of actin and myosin. A greater presence of these contractile proteins will allow for a more forceful contraction; a heart that has thickened in this way due to exercise training is referred to as an 'athlete's heart'. However, this generic description is complicated by the fact that different athletes overload their hearts in different ways. For example, sports like weight lifting and wrestling involve transient periods of work against heavy resistance to blood flow (afterload), which is unlike the overload present during endurance exercise. Originally, this led to a distinction of two forms of athlete's heart: a strength-trained myocardium, characterized exclusively by increased wall thickness (concentric ventricular hypertrophy) and an endurance-trained heart with minimal wall growth relative to the increased chamber size (Mitchell *et al.*, 2005; Morganroth *et al.*, 1975). It is now believed that there is more cross-over between these two effects (Pluim *et al.*, 2000) and a considerable increase in wall thickness can also occur in endurance-trained athletes (Spirito *et al.*, 1994). However, any thickening of myocardial tissue due to endurance training is approximately proportional to the increase in chamber dimension that occurs concurrently. Furthermore, while a 12% disproportionate increase of wall thickness compared to chamber volume has been reported for strength-trained athletes (Fagard, 1997), the extent of this concentric ventricular hypertrophy is relatively modest compared to the pathological changes that occur when hypertrophic cardiomyopathies are present (Pelliccia and Maron, 1997).

Much like skeletal muscle, an increase of contractile proteins within the myocardium allows for a stronger muscular contraction. There is also evidence to suggest that the specific type of myocardial contractile protein is altered with training, as a myosin isoform that is more energetically active (myosin V_1) has been reported in the myocardial tissue of swim-trained rats (Pagani and Solaro, 1983). However, swim training can also enhance stroke volume and cardiac output in rats without this conversion so this is not an obligatory feature of a trained heart (Sharma, Tomanek and Bhalla, 1985). Trained rats also demonstrate greater calcium storage capacity in myocardial sarcoplasmic reticulum and increased sarcolemmal calcium flux (Penpargkul *et al.*, 1977; Tibbits *et al.*, 1981). These training-induced improvements in the ability to regulate calcium availability may improve cardiac performance, especially when pathological compromise is present (Moore, 2006).

Training-induced bradycardia

Bradycardia is defined as a heart rate that is abnormally low; for example, below 60 beats per minute at rest. In patient populations, bradycardia can be caused by chronotropic impairment due to diseases that compromise electrical conductivity within the myocardium. However, at the other end of the spectrum, a trained heart will also beat less frequently at rest and during submaximal exercise at the same absolute and relative exercise intensity because it can accommodate and eject more blood with each contraction (Blomqvist and Saltin, 1983; Wilmore *et al.*, 2001). Consequently, training-induced bradycardia is actually a beneficial cardiac dysrhythmia.

Myocardial work and the associated oxygen requirement of the heart during any physical activity can be determined from the rate–pressure product (RPP; the product of systolic blood pressure and heart rate). For example, symptoms of limited blood flow (i.e. insufficient oxygen supply) through coronary arteries (e.g. angina pectoris) typically manifest at a reproducible RPP (oxygen demand) (Clausen and Trap-Jensen, 1976). Training-induced bradycardia is therefore a beneficial cardiovascular adaptation because it allows the resting systemic circulatory requirements and those associated with any submaximal level of physical exertion to be met with less work on the part of the myocardium (i.e. at a lower RPP; see Figure 2.5.2).

A relatively low heart rate also provides a longer period of diastole before an ensuing cardiac contraction, which is necessary because more time is needed to completely fill the more compliant trained chambers (Brooks *et al.*, 2000). Consequently, training-induced bradycardia is an important adaptation as it contributes to the increased stroke volume that is present at rest and during submaximal exercise in the trained state. It is also interesting to note that in addition to resting and submaximal work, a number of investigations have reported that maximal heart rate is *reduced* after training (on average, six beats per minute; see Zavorsky, 2000 for review). There are numerous training-induced adaptations that might be responsible; however, regardless of the mechanistic basis, this effect indicates the importance of adequate ventricular filling time for the conditioned heart and further supports the critical role that stroke volume plays in establishing enhanced maximal cardiac output after training. Furthermore, an unchanged or reduced heart rate at maximal exercise in conjunction with systolic blood pressure that is unchanged or only slightly increased (Wilmore *et al.*, 2001) collectively ensures that the increase in maximal capacity in the trained state is achieved at an RPP (and, therefore, amount of myocardial work) that is no greater than that which was present prior to training (Secher, 2009).

2.5.3.2 Circulatory adaptations to endurance training

Blood redistribution during exercise

One of the biggest challenges that the cardiovascular system faces during endurance exercise is distributing its cardiac output to critical areas of need. At rest, the splanchnic organs (stomach, spleen, pancreas, intestines, and liver) receive more blood flow than any other region, while the skeletal muscles receive a modest provision because their metabolic demands are relatively low (Rowell, 1973). However, during exercise, peak blood flow in the active muscle mass can increase 100-fold above the resting value (Andersen and Saltin, 1985). This presents a problem because the cardiovascular system has a limited amount of blood available for distribution and also has the ability to accommodate much more than it actually contains. Simply opening up all of its vessels to allow more blood to flow to the active musculature without a corresponding sparing of blood from other areas would result in a catastrophic fall in central pressure/venous return if multiple muscles were activated. This means that blood distribution during exercise mandates precise circulatory adjustments that involve both active-muscle vasodilation (an increase in blood-vessel radius that reduces vascular resistance to blood flow) and visceral, renal, and inactive muscle vasoconstriction (a decrease in blood-vessel radius that presents an impediment to flow) (Clifford, 2007; Rowell, 1973; Shepherd, 1987).

During challenging exercise, the sympathetic branch of the autonomic nervous system plays a vital role in activating cardiovascular function to meet the increased demand for circulation. This influence is mediated through sympathetic nerve fibres and also via catecholamines (epinephrine and norepinephrine) released from the adrenal glands. Sympathetic activation causes an increase in both heart rate and myocardial contractility so that cardiac output is elevated to the appropriate level. Sympathetic activation also protects central pressure/venous return by promoting systemic vasoconstriction. For example, during exercise at or near VO_{2max}, splanchnic blood flow is reduced by almost 80% (Rowell, 1973). However, this drive to *reduce* flow to all areas of the body except the heart and brain is opposed by countering mechanisms in active muscles which allow for a 'functional sympatholysis' that results in local vasodilation and increased perfusion (hyperaemia) (Remensnyder, Mitchell and Sarnoff, 1962; Thomas, Hansen and Victor, 1994; Thomas and Segal, 2004). The collective effect is the ability to divert limited cardiac output from less active tissues to the areas of greatest need.

Despite the coordinated efforts of neural and hormonal mechanisms, peripheral feedback afferents, and metabolite-related circulating substances, the human body still cannot adequately redistribute blood to simultaneously protect central pressure and support peripheral requirements during challenging endurance exercise involving large/multiple muscle groups. In fact, there is evidence to suggest that some vasoconstriction in *active* muscles also occurs during intense exercise as a consequence of activating a greater quantity of muscle than the available cardiac output can fully support. For example, active muscle blood flow and oxygen uptake are less during two-compared to one-legged cycling and when the arms and legs are exercised together (Klausen *et al.*, 1982; Secher, 2009; Secher and Volianitis, 2006; Secher *et al.*, 1977). Furthermore, leg cycling after β_1-adrenergic blockade that reduces cardiac output is characterized by active muscle vasoconstriction (Pawelczyk *et al.*, 1992). These examples indicate that the quantity of blood available for distribution presents a critical limitation to aerobic exercise performance, so it is not surprising that other chronic cardiovascular adaptations to regular endurance exercise address this restriction.

Increased total blood volume

It is perhaps intuitive that architectural changes that allow for greater stroke volume per cardiac contraction will be accompanied by adaptations that increase the blood that is available for distribution. For example, a training-induced increase in blood volume within the cardiovascular network is responsible for greater ventricular filling pressure during diastole, which ensures that a larger, more compliant trained left ventricle is optimally filled when circulatory requirements are low. This allows a given cardiac output to be established with a greater stroke volume and reduced heart rate at rest and during submaximal exercise (see Figure 2.5.2). Furthermore, more blood in the system provides an increased quantity for distribution to active muscles during intense efforts. This loosens the restriction of circulation that limits VO_{2max} (see Figure 2.5.3). Therefore, adaptations that provide for an increased amount of blood in the circulatory system (hypervolemia) are critical because they facilitate both increased cardiovascular efficiency at rest/during submaximal exercise and the ability to surpass previous limitations when maximal levels of work are encountered. Highly-trained endurance athletes typically possess a 20–25% larger blood volume than sedentary subjects due to a number of training-related adaptations (Convertino, 1991). The primary one is expansion of plasma volume (Convertino, 1991; Kanstrup, Marving and Hoilund-Carlsen, 1992; Oscai, Williams and Hertig, 1968).

Plasma-volume expansion

The increase in stroke volume that occurs due to endurance training is approximately mirrored by an increase in plasma volume (Green, Jones and Painter, 1990). This is not surprising given the dramatic effect that plasma-volume expansion has on left-ventricular function. For example, a 9% increase in blood volume achieved via plasma infusion has been shown to facilitate a 10–14% increase in submaximal exercise stroke volume, which is largely attributable to increased LVEDV (Kanstrup, Marving and Hoilund-Carlsen, 1992). In addition to providing more blood for expulsion, this change also exacerbates the Frank–Starling effect (Goodman, Liu and Green, 2005). Furthermore, central venous pressure is significantly elevated during exercise after acute plasma-volume expansion and the fall that normally occurs with increasing workload is eliminated (Kanstrup, Marving and Hoilund-Carlsen, 1992). The functional significance of these hemodynamic alterations is reflected by the 6% increase in VO_{2peak} and 16% increase in time to exhaustion that has been reported for exhaustive constant-load cycle exercise after a 14% expansion of plasma volume via clinical plasma-expanding solution (Berger *et al.*, 2006).

There is evidence to suggest that a 10% expansion of plasma volume can occur within 24 hours of the first endurance exercise session (Gillen *et al.*, 1991), and an overall increase of 12% after eight days of training has been shown (Convertino *et al.*, 1980, 1983). Importantly, this training-induced hypervolemia is accompanied by a similar increase in VO_{2max}, and a 0.78 correlation between total blood volume and VO_{2max} expressed relative to body weight has been reported (Convertino, 1991; Convertino, Keil and Greenleaf, 1983; Goodman, Liu and Green, 2005). This confirms the importance of blood volume as a determinant of functional capacity. Another major benefit of a rapidly occurring plasma expansion is that it protects against acute fluid loss from the vascular space that occurs during extended endurance exercise sessions; specifically, although a similar exercise-induced plasma-volume shift occurs in the trained state, training-induced hypervolemia ensures that more blood remains for circulation once this shift has taken place (Convertino, 1983). Rapidly-occurring plasma-volume expansion is therefore an important adaptation that acclimatizes an athlete's cardiovascular system to prolonged heavy exercise in the heat (Green, Jones and Painter, 1990).

The mechanism that facilitates plasma-volume expansion after as little as one endurance exercise session appears to be related to the plasma protein albumin, which exerts an osmotic pull that draws fluid from the extracellular space (Gillen *et al.*, 1991). Plasma albumin content is elevated within one hour following a bout of intense upright cycle exercise (Nagashima *et al.*, 1999). Furthermore, after eight days of training, a ninefold elevation of plasma renin activity and vasopressin concentration during exercise promotes fluid and sodium retention, which also plays a role (Convertino *et al.*, 1980).

Increased red blood cells

For up to the initial 10 days of training, changes in plasma volume are primarily responsible for changes in blood volume, with little or no associated elevation of red blood cell mass (Convertino, 1991; Convertino *et al.*, 1980). However, the total increase in blood volume that eventually occurs due to training appears to be the result of both plasma-volume expansion and increased red blood cell number (Hoffman, 2002). A greater number of red blood cells increases the oxygen-carrying capacity of the blood; however, the increase in plasma volume outstrips the increased production of red blood cells so that haematocrit (the relative portion of the blood comprising red blood cells) is actually reduced in the trained state (Mier *et al.*, 1997). This is important because it results in decreased blood viscosity and facilitated flow.

Improved muscle blood-flow capacity

Together, the training-induced improvement in myocardial capacity and increased total blood volume would be of limited use if endurance training did not also improve the cardiovascular system's ability to perfuse the exercising musculature with blood. Once it leaves the left ventricle through the aorta, oxygenated blood travels through large arteries to small arteries and arterioles before accessing the interstitial space via capillaries. Training-induced remodelling of these structures must therefore occur on multiple levels as existing arterial vessels must be enlarged and new capillaries must be formed. These adaptations take place due to processes known as arteriogenesis and angiogenesis, respectively (Andersen and Henriksson, 1977; Lehoux, Tronc and Tedgui, 2002; Prior, Yang and Terjung, 2004). Furthermore, there is evidence to suggest that functional alterations in the vasomotor reactivity of arteries/arterioles also characterize the trained state. The degree to which these structural and functional adaptations contribute to the improved capacity for muscle blood flow after training is likely the result of muscle fibre-type composition, motor unit recruitment pattern, and the mode/duration/intensity of the exercise that was performed (e.g. interval sprint versus endurance training) as non-uniform adaptations between muscles of different fibre-type composition and within the same vascular network have been reported (Laughlin and Roseguini, 2008).

Arteriogenesis and angiogenesis

Arteriogenesis allows for greater bulk blood flow to the periphery, while angiogenesis provides a denser capillary network that increases red blood cell mean transit time and decreases diffusion distance. Increased transit time through active muscle might contribute to the higher oxygen extraction fraction that has been observed during submaximal exercise in the trained state (Kalliokoski *et al.*, 2001). It is interesting to note that strikingly different spatial patterns of adaptation can exist in arteries compared to capillaries within and between trained muscle, which suggests that the factors/signals promoting arteriogenesis and angiogenesis might be different (Laughlin and Roseguini, 2008). There is also evidence to suggest that the coronary arteries that supply blood to myocardial tissue are altered by endurance training; specifically, an increase in diameter (Currens and White, 1961; Wyatt and Mitchell, 1978) and an enhanced ability to dilate during exercise (Kozàkovà *et al.*, 2000) have been reported. Furthermore, the cardiac capillary network is enhanced and some cardiac capillaries are transformed into arterioles due to training (Brown, 2003).

Improved blood distribution

The trained state is characterized by reduced sympathetic nervous system activity at any specific absolute level of submaximal exertion because the relative stress associated with the effort is decreased (Peronnet *et al.*, 1981). Consequently, the reduction of blood flow to the splanchnic and renal regions during exercise is attenuated in the trained state. This has the potential to be beneficial because homoeostatic disturbances associated with decreased flow are reduced and the ability to metabolize glucose during prolonged exercise is enhanced (McAllister, 1998). There is also evidence to suggest that there is an altered control of vascular resistance in exercising muscle in the trained state. For example, in some circumstances, trained animal muscle exhibits increased endothelial-dependent dilation due to changes in endothelial nitric oxide synthase (Jasperse and Laughlin, 2006). This enzyme generates nitric oxide, which is an important vasodilator in the arterial network. However, it appears as if the presence/extent of any effect of training on endothelium-dependent dilation in the arterial network depends on numerous factors, including the duration of the training programme, the size and anatomical location of the artery/arteriole, and the health of the individual (e.g. greater adaptations typically occur in diseased than in healthy individuals) (Jasperse and Laughlin, 2006). It is also interesting to note that endurance training and interval sprint training induce non-uniform changes in smooth muscle and endothelium throughout the arteriolar network, which suggests a highly specific nature to these alterations in vasomotor control (Laughlin and Roseguini, 2008). Finally, heterogeneity of blood flow within active muscles is also reduced in the trained state (Kalliokoski *et al.*, 2001). This vascular adaptation may facilitate oxygen extraction by ensuring that blood perfusing a contracting muscle is more uniformly distributed to specific areas of need (Piiper, 2000).

Reduced blood pressure

Regular endurance training elicits only modest reductions in systolic and diastolic blood pressure at rest (e.g. <3 mmHg). However, during submaximal exercise at the same absolute work rate, greater reductions are present (Wilmore *et al.*, 2001). Arterial blood pressure is a function of blood flow in the cardiovascular system (cardiac output) and resistance to flow created by the peripheral vasculature (total peripheral resistance). Arteriogenesis/angiogenesis and reduced sympathetic nervous system activity at rest and in response to submaximal work are two training-induced alterations responsible for reducing total peripheral resistance and, by extension, blood pressure in the trained state (Fagard, 2006). At maximal exercise, total peripheral resistance is also reduced after training; however, cardiac output is increased. The collective effect is an unchanged or only modestly increased systolic, but decreased diastolic, blood pressure during maximal exercise after training (Wilmore *et al.*, 2001).

2.5.4 CARDIOVASCULAR-RELATED ADAPTATIONS TO TRAINING

Generally speaking, cardiovascular adaptations are directly responsible for the enhanced functional capacity that is achieved in the trained state. However, a training-induced increase in VO_{2max} is only possible if the ability to transfer oxygen to the blood is sufficient. Similarly, an increased capacity for circulation is of no benefit if oxygen cannot be extracted from the blood and utilized where it is required. Consequently, a summary of cardiovascular adaptations to strength and conditioning must include mention of both the pulmonary and the skeletal muscular systems.

2.5.4.1 The pulmonary system

Pulmonary diffusion capacity

In humans, the lungs provide the critical link that is necessary for transport of metabolic gases to and from the atmosphere by the cardiovascular system. Therefore, cardiovascular function is intimately related to pulmonary diffusion capacity. However, most evidence suggests that endurance training has little or no effect on the structure and function of the lungs and airways (Brown, 2000; Dempsey, Miller and Romer, 2006a; Wagner, 2005). For example, lung diffusion capacity and pulmonary capillary blood volume are not substantially different between endurance-trained and healthy untrained subjects at rest or during exercise, and static lung volumes and maximum flow-volume loops are also similar (Dempsey, Miller and Romer, 2006a). This has been interpreted as evidence that pulmonary diffusion capacity exceeds cardiovascular capacity by a sufficient margin that training-induced alterations in cardiovascular function can occur without associated pulmonary change. One reason for this discrepancy appears to be the degree to which each system can elevate its level of function when challenged. During maximal exercise, 4–6-fold increases in respiratory rate and 5–7-fold increases in tidal volume are typically observed (Brown, 2000). Consequently, minute ventilation (the product of the two) can increase by as much as 40-fold during all-out exercise. This far outstrips the 4–5-fold increase in cardiac output that is attainable under the same circumstances.

Ventilatory efficiency

The robust ability to increase minute ventilation during exercise means that training-induced cardiovascular adaptations that allow for an increased VO_{2max} need not be accompanied by adaptive changes in either respiratory rate or tidal volume. However, the trained state *is* characterized by a more efficient ventilatory exchange. At the same submaximal steady-state work rate, minute ventilation is less after training and the ventilatory equivalent for oxygen (V_E/VO_2 ratio) is reduced (Davis *et al.*, 1979). Furthermore, a given minute ventilation is achieved with greater tidal volume and reduced breathing frequency. This decreases the oxygen cost of ventilation and reduces ventilatory muscle fatigue, which can adversely affect performance (Dempsey *et al.*, 2006b).

Exercise-induced arterial hypoxaemia

Generally speaking, the alveolar partial pressure of oxygen rises during exercise and the partial pressure of oxygen in arterial blood is therefore maintained. However, an appreciable fall in arterial oxygen saturation (e.g. exercise-induced arterial hypoxaemia, EIAH) is often observed during near-maximal or maximal exercise in highly-trained individuals possessing high VO_{2max} values (Dempsey, Hanson and Henderson, 1984). Furthermore, inspiration of a hyperoxic gas mixture by subjects who experience EIAH improves both VO_{2peak} and performance (Wagner, 2005; Wilkerson, Berger and Jones, 2006). The cause of this pulmonary limitation has yet to be identified, but might relate to the prodigious maximal cardiac outputs that these athletes can generate. For example, the lung has a finite and relatively small capillary volume, so an extremely high cardiac output could result in an excessively short red blood cell transit time that restricts oxygen loading. Regardless of the mechanism, however, the presence of a ventilatory-related restriction to maximal exercise in some highly-trained individuals does raise the question as to why pulmonary diffusion capacity does not improve in healthy individuals as a consequence of endurance exercise training (Wagner, 2005).

2.5.4.2 The skeletal muscular system

Increased mitochondrial enzyme activity

A circulatory constraint of VO_{2max} does not mean that endurance exercise adaptations are limited to those that facilitate enhanced oxygen delivery. On the contrary, it is well documented that the specific sites where oxygen is consumed (skeletal muscle mitochondria) are also dramatically altered by training. Specifically, enzymatic activity is increased 200–300% in some mitochondrial enzymes and 30–60% in others (Holloszy and Coyle, 1984). While increases are not present in all mitochondrial enzymes, it is interesting to note that improvements of this magnitude far outstrip the increase in VO_{2max} that occurs due to training. This indicates that it is beneficial to improve mitochondrial capacity for reasons other than simply increasing the maximal oxidative capacity of the organism.

Increased mitochondrial volume/density

A four-week programme of daily sprint training can facilitate a VO_{2max} increase of similar magnitude to that elicited by endurance training. However, only the latter stimulates a synthesis of mitochondrial protein that results in greater mitochondrial content in the trained musculature (Davies, Packer and Brooks, 1981, 1982). This proliferation is directly responsible for the aforementioned training-induced increase in enzymatic activity because the specific activity of mitochondrial enzymes (i.e. enzymatic activity per unit of mitochondrial protein) remains unaltered in the trained state (Tonkonogi and Sahlin, 2002). It is also interesting to note that significant increases in mitochondrial content are present in all three principal fibre types (Howald *et al.*, 1985). Increased mitochondrial mass ensures that any submaximal VO_2 can be sustained with less metabolic stress per mitochondrial unit.

Improved respiratory control

When mitochondrial mass is increased after training, better respiratory control (the ability to sustain a given rate of oxidative flux with less free ADP concentration, that is at a higher ATP : ADP ratio) is achieved and feedback activation of glycolysis at the same absolute submaximal work rate is reduced (Bassett and Howley, 2000). Consequently, a shift in substrate utilization will occur that is supported by the increased presence of enzymes that mobilize and metabolize fat (Holloszy and Coyle, 1984). The end result is that fat catabolism is increased, glycogen is spared, and less lactate is formed during submaximal exercise in the trained state (Azevedo *et al.*, 1998; Coggan *et al.*, 1993). This means that challenging exercise can be sustained for longer periods with less fatigue (i.e. endurance capacity is improved). Furthermore, in accordance with the model of respiratory control proposed by Meyer in 1988, increased mitochondrial mass and tighter respiratory control should allow for a more rapid VO_2 response when a transition is made from a lower to a higher metabolic rate (i.e. faster VO_2 kinetics) (Meyer, 1988). This is beneficial because faster VO_2 kinetics during a transition to the same work rate will reduce the magnitude of the oxygen deficit, thereby sparing the associated fall in intramuscular phosphocreatine concentration and attenuating the production of lactic acid (i.e. allowing the same VO_2 to be attained with less perturbation of the phosphorylation and redox potentials).

2.5.5 CONCLUSION

As depicted in Figure 2.5.1, an endurance (aerobic) exercise training session mandates acute circulatory changes that provide the precise stimulus required to provoke a number of chronic positive adaptations in cardiovascular function. The changes in cardiovascular and related parameters that accompany these adaptations are summarized in Table 2.5.1. At rest and during submaximal exercise, cardiac output is unchanged after training; however, a given cardiac output is established via an increased stroke volume that is attributable to left-ventricular eccentric hypertrophy, exercise-induced bradycardia, and increased myocardial contractility. This means that heart rate and corresponding myocardial work is reduced at any submaximal level of exertion (see Figure 2.5.2). At maximal exercise, training facilitates a substantial increase in cardiac output that is exclusively attributable to a greater maximal stroke volume. This enhancement is mediated by a training-induced increase in blood volume that facilitates greater venous return, with central pressure maintained during exercise. Blood vessels are also altered by training, so that greater flow can be accommodated with less resistance and consequently the time available for gas exchange with active tissue is increased. As depicted in Figure 2.5.3, the end result is that the functional capacity of the cardiovascular system is improved; this is reflected in the ability to achieve a greater VO_{2max}.

Table 2.5.1 Endurance training-induced cardiovascular adaptations at rest and during submaximal (same absolute work rate) and maximal exercise

	Rest	Submaximal	Maximal
Cardiovascular adaptations			
Myocardial			
Venous return	↑	↑	↑
Left-ventricular end diastolic volume	↑	↑	↑
Left-ventricular end systolic volume	↔	↔	↔
Stroke volume	↑	↑	↑
Heart rate	↓	↓	↔↓
Cardiac output	↔↓	↔↓	↑
Rate–pressure product	↓	↓	↔
Coronary blood flow	↑	↑	↑
Peripheral			
Plasma volume	↑	NA	NA
Red blood cells	↑	NA	NA
Total blood volume	↑	NA	NA
Haematocrit	↓	NA	NA
Total peripheral resistance	↓	↓	↓
Systolic blood pressure	↓	↓	↔↑
Diastolic blood pressure	↓	↓	↓
Related			
Pulmonary diffusion capacity	↔	↔	↔
Mitochondrial volume/density	↑	NA	NA
Oxidative enzymes (specific activity)	↔	NA	NA
Oxidative enzymes (total activity)	↑	NA	NA
Lipid utilization	↔	↓	↔
Arterio-venous oxygen difference	↔↑	↑	↑
VO_2	↔	↔	↑

References

Andersen P. and Henriksson J. (1977) Capillary supply of the quadriceps femoris muscle of man: adaptive response to exercise. *J Physiol*, **270**, 677–690.

Andersen P. and Saltin B. (1985) Maximal perfusion of skeletal muscle in man. *J Physiol*, **366**, 233–249.

Åstrand P.O. and Saltin B. (1961) Oxygen uptake during the first minutes of heavy muscular exercise. *J Appl Physiol*, **16**, 971–976.

Åstrand P.O., Cuddy T.E., Saltin B. and Stenberg J. (1964) Cardiac output during submaximal and maximal work. *J Appl Physiol*, **19**, 268–274.

Azevedo J.L. Jr., Linderman K., Lehman S.L. and Brooks G.A. (1998) Training decreases muscle glycogen turnover during exercise. *Eur J Appl Physiol*, **78**, 479–486.

Barnard R.J., MacAlpin R., Kattus A.A. and Buckberg G.D. (1977) Effect of training on myocardial oxygen supply/demand balance. *Circulation*, **56**, 289–291.

Bassett D.R. and Howley E.T. (2000) Limiting factors for maximum oxygen uptake and determinants of endurance performance. *Med Sci Sports Exerc*, **32**, 70–84.

Berger N.J.A., Campbell I.T., Wilkerson D.P. and Jones A.M. (2006) Influence of acute plasma volume expansion on VO_2 kinetics, VO_{2peak}, and performance during high-intensity cycle exercise. *J Appl Physiol*, **101**, 707–714.

Blomqvist C.G. and Saltin B. (1983) Cardiovascular adaptations to physical training. *Annu Rev Physiol*, **45**, 169–189.

Brandao M.U., Wajngarten M., Rondon E. *et al.* (1993) Left ventricular function during dynamic exercise in untrained and moderately trained subjects. *J Appl Physiol*, **75**, 1989–1995.

Brooks G.A., Fahey T.D., White T.P. and Baldwin K.M. (2000) Cardiovascular dynamics during exercise, in *Exercise Physiology: Human Bioenergetics and Its Application* (eds G.A. Brooks, T.D. Fahey, T. White and K.M. Baldwin), McGraw-Hill, New York, NY, pp. 317–338.

Brown D.D. (2000) Pulmonary responses to exercise and training, in *Exercise and Sport Science* (eds W.E. Garrett Jr. and D.T. Kirkendall), Lippincott Williams & Wilkins, Philadelphia, PA, pp. 116–134.

Brown M.D. (2003) Exercise and coronary vascular remodelling in the healthy heart. *Exp Physiol*, **88**, 645–658.

Burke E.R. (1998) Heart rate monitoring and training, in *Precision Heart Rate Training* (ed. E.R. Burke), Human Kinetics, Champaign, IL, pp. 1–28.

Clausen J.P. and Trap-Jensen J. (1976) Heart rate and arterial blood pressure during exercise in patients with angina pectoris. Effect of training and nitroglycerin. *Circulation*, **53**, 436–442.

Clifford P.S. (2007) Skeletal muscle vasodilation at the onset of exercise. *J Physiol*, **583**, 825–833.

Coggan A.R., Habash D.L., Mendenhall L.A. *et al.* (1993) Isotopic estimation of CO2 production during exercise before and after endurance training. *J Appl Physiol*, **75**, 70–75.

Cohen J.L. and Segal K.R. (1985) Left ventricular hypertrophy in athletes: an exercise-echocardiographic study. *Med Sci Sports Exerc*, **17**, 695–700.

Convertino V.A. (1983) Heart rate and sweat rate responses associated with exercise-induced hypervolemia. *Med Sci Sports Exerc*, **15**, 77–82.

Convertino V.A. (1991) Blood volume: its adaptation to endurance training. *Med Sci Sports Exerc*, **23**, 1338–1348.

Convertino V.A., Brock P.J., Keil L.C. *et al.* (1980) Exercise training-induced hypervolemia: role of plasma albumin, rennin, and vasopressin. *J Appl Physiol*, **48**, 665–669.

Convertino V.A., Keil L.C. and Greenleaf J.E. (1983) Plasma volume, rennin, and vasopressin responses to graded exercise after training. *J Appl Physiol*, **54**, 508–514.

Currens J.H. and White P.D. (1961) Half a century of running. Clinical, physiologic and autopsy findings in the case of Clarence DeMar ('Mr. Marathon'). *N Engl J Med*, **265**, 988–993.

Davies K.J.A., Packer L. and Brooks G.A. (1981) Biochemical adaptation of mitochondria, muscle, and whole-animal respiration to endurance training. *Arch Biochem Biophys*, **209**, 539–554.

Davies K.J.A., Packer L. and Brooks G.A. (1982) Exercise bioenergetics following sprint training. *Arch Biochem Biophys*, **215**, 260–265.

Davis J.A., Frank M.H., Whipp B.J. and Wasserman K. (1979) Anaerobic threshold alterations caused by endurance training in middle-aged men. *J Appl Physiol*, **46**, 1039–1046.

Dempsey J.A., Hanson P.G. and Henderson K.S. (1984) Exercise-induced arterial hypoxaemia in healthy human subjects at sea level. *J Physiol*, **355**, 161–175.

Dempsey J.A., Miller J.D. and Romer L.M. (2006a) The respiratory system, in *Acsm's Advanced Exercise Physiology* (ed. C.M. Tipton), Lippincott Williams & Wilkins, Philadelphia, PA, pp. 246–299.

Dempsey J.A., Romer L., Rodman J. *et al.* (2006b) Consequences of exercise-induced respiratory muscle work. *Respir Physiol Neurobiol*, **151**, 242–250.

Ehsani A.A., Hagberg J.M. and Hickson R.C. (1978) Rapid changes in left ventricular dimensions and mass in response to physical conditioning and deconditioning. *Am J Cardiol*, **42**, 52–56.

Ekblom B. and Hermansen L. (1968) Cardiac output in athletes. *J Appl Physiol*, **25**, 619–625.

Ekblom B., Astrand P.O., Saltin B. *et al.* (1968) Effect of training on the circulatory response to exercise. *J Appl Physiol*, **24**, 518–528.

Fagard R.H. (1997) Impact of different sports and training on cardiac structure and function. *Cardiol Clin*, **15**, 397–312.

Fagard R.H. (2006) Exercise is good for your blood pressure: effects of endurance training and resistance training. *Clin Exp Pharmacol Physiol*, **33**, 853–856.

Gillen C.M., Lee R., Mack G.W. *et al.* (1991) Plasma volume expansion in humans after a single intense exercise protocol. *J Appl Physiol*, **71**, 1914–1920.

Glehdill N., Cox D. and Jamnik R. (1994) Endurance athlete's stroke volume does not plateau: major advantage in diastolic function. *Med Sci Sports Exerc*, **26**, 1116–1121.

Gonzalez-Alonso J. and Calbet J.A. (2003) Reductions in systemic and skeletal muscle blood flow and oxygen delivery limit maximal aerobic capacity in humans. *Circulation*, **107**, 824–830.

Goodman J.M., Liu P.P. and Green H.J. (2005) Left ventricular adaptations following short-term endurance training. *J Appl Physiol*, **98**, 454–460.

Green H.J., Jones L.L. and Painter D.C. (1990) Effects of short-term training on cardiac function during prolonged exercise. *Med Sci Sports Exerc*, **22**, 488–493.

Henriksen E., Landelius J., Wesslén L. *et al.* (1996) Echocardiographic right and left ventricular measurements in male elite endurance athletes. *Eur Heart J*, **17**, 1121–1128.

Hickson R.C., Hagberg J.M., Ehsani A.A. and Holloszy J.O. (1981) Time course of the adaptive responses of aerobic power and heart rate to training. *Med Sci Sports Exerc*, **13**, 17–20.

Hoffman J. (2002) Cardiovascular system and exercise, in *Physiological Aspects of Sport Training and Performance* (ed. J. Hoffman), Human Kinetics, Champaign, IL, pp. 39–56.

Holloszy J.O. and Coyle E.F. (1984) Adaptations of skeletal muscle to endurance exercise and their metabolic consequences. *J Appl Physiol*, **56**, 831–838.

Howald H., Hoppeler H., Claassen H. *et al.* (1985) Influences of endurance training on the ultrastructural composition of the different muscle fiber types in humans. *Pflugers Arch*, **403**, 369–376.

Jasperse J. and Laughlin M. (2006) Endothelial function and exercise training: evidence from studies using animal models. *Med Sci Sports Exerc*, **38**, 445–454.

Jones A.M. and Poole D.C. (2009) Physiological demands of endurance exercise, in *The Olympic Textbook of Science in Sport* (ed. R.J. Maughan), Blackwell Publishing Ltd., Oxford, UK, pp. 43–55.

Kalliokoski K.K., Oikonen V., Takala T.O. *et al.* (2001) Enhanced oxygen extraction and reduced flow heterogeneity in exercising muscle in endurance-trained men. *Am J Physiol Endocrinol Metab*, **280**, 1015–1021.

Kanstrup I.L., Marving J. and Hoilund-Carlsen P.F. (1992) Acute plasma expansion: left ventricular hemodynamics and endocrine function during exercise. *J Appl Physiol*, **73**, 1791–1796.

Klausen K., Secher N.H., Clausen J.P. *et al.* (1982) Central and regional circulatory adaptations to one-leg training. *J Appl Physiol*, **52**, 976–983.

Knight D.R., Schaffartzik W., Poole D.C. *et al.* (1993) Effects of hyperoxia in maximal leg O_2 supply and utilization in men. *J Appl Physiol*, **75**, 2586–2594.

Kozàkovà M., Galetta F., Gregorini L. *et al.* (2000) Coronary vasodilator capacity and epicardial vessel remodeling in physiological and hypertensive hypertrophy. *Hypertension*, **36**, 343–349.

Laughlin M.H. and Roseguini B. (2008) Mechanisms for exercise training-induced increases in skeletal muscle blood flow capacity: differences with interval sprint training versus aerobic endurance training. *J Physiol Pharmacol*, **59**, 71–88.

Lehoux S., Tronc F. and Tedgui A. (2002) Mechanisms of blood flow-induced vascular enlargement. *Biorheology*, **39**, 319–324.

Levine B.D., Lane L., Buckey J.C. *et al.* (1991) Left ventricular pressure-volume and Frank-Starling relations in endurance athletes. *Circulation*, **84**, 1016–1023.

Levy W.C., Cerqueira M.D., Abrass I.B. *et al.* (1993) Endurance exercise training augments diastolic filling at rest and during exercise in healthy young and older men. *Circulation*, **88**, 116–126.

McAllister R.M. (1998) Adaptations in control of blood flow with training: splanchnic and renal blood flows. *Med Sci Sports Exerc*, **30**, 375–381.

Maron B.J. (1986) Structural features of the athlete's heart as defined by echocardiography. *J Am Coll Cardiol*, **7**, 190–203.

Meyer R.A. (1988) A linear model of muscle respiration explains monoexponential phosphocreatine changes. *Am J Physiol*, **254**, C548–C553.

Mier C.M., Turner M.J., Ehsani A.A. and Spina R.J. (1997) Cardiovascular adaptations to 10 days of cycle exercise. *J Appl Physiol*, **83**, 1900–1906.

Milliken M.C., Stray-Gundersen J., Peshock R.M. *et al.* (1988) Left ventricular mass as determined by magnetic resonance imaging in male endurance athletes. *Am J Cardiol*, **62**, 301–305.

Mitchell J.H., Haskell W., Snell P. and Van Camp S.P. (2005) Task Force 8: classification of sports. *J Am Coll Cardiol*, **45**, 1364–1367.

Moore R.L. (2006) The cardiovascular system: cardiac function, in *Acsm's Advanced Exercise Physiology* (ed. C.M. Tipton), Lippincott Williams & Wilkins, Philadelphia, PA, pp. 326–342.

Morganroth J., Maron B.J., Henry W.L. and Epstein S.E. (1975) Comparative left ventricular dimensions in trained athletes. *Ann Intern Med*, **82**, 521–524.

Nagashima K., Mack G.W., Haskell A. *et al.* (1999) Mechanism for the posture-specific plasma volume increase after a single intense exercise protocol. *J Appl Physiol*, **86**, 867–873.

Oscai L.B., Williams B.T. and Hertig B.A. (1968) Effect of exercise on blood volume. *J Appl Physiol*, **24**, 622–624.

Pagani E.D. and Solaro R.J. (1983) Swimming exercise, thyroid state, and the distribution of myosin isoenzymes in rat heart. *Am J Physiol Heart Circ Physiol*, **245**, H713–H720.

Pawelczyk J.A., Hanel B., Pawelczyk R.A. *et al.* (1992) Leg vasoconstriction during dynamic exercise with reduced cardiac output. *J Appl Physiol*, **73**, 1838–1846.

Pelliccia A., Maron B.J., Spataro A. *et al.* (1991) The upper limit of physiological cardiac hypertrophy in highly trained elite athletes. *N Engl J Med*, **324**, 295–301.

Pelliccia A. and Maron B.J. (1997) Outer limits of the athlete's heart: the effect of gender and relevance to the differential diagnosis with primary cardiac diseases. *Cardiol Clin*, **15**, 381–396.

Penpargkul S., Repke D.I., Katz A.M. and Scheuer J. (1977) Effect of physical training on calcium transport by rat cardiac sarcoplasmic reticulum. *Circ Res*, **40**, 134–138.

Peronnet F., Cleroux J., Perrault H. *et al.* (1981) Plasma norepinephrine response to exercise before and after training in humans. *J Appl Physiol*, **51**, 812–815.

Piiper J. (2000) Perfusion, diffusion and their heterogeneities limiting blood-tissue O_2 transfer in muscle. *Acta Physiol Scand*, **168**, 603–607.

Pluim B.M., Zwinderman A.H., Van der Laarse A. and Van der Wall E.E. (2000) The athlete's heart: a meta-analysis of cardiac structure and function. *Circulation*, **101**, 336–344.

Prior B.M., Yang H.T. and Terjung R.L. (2004) What makes vessels grow with exercise training? *J Appl Physiol*, **97**, 1119–1128.

Remensnyder J.P., Mitchell J.H. and Sarnoff S.J. (1962) Functional sympatholysis during muscular activity. Observations on influence of carotid sinus on oxygen uptake. *Circ Res*, **11**, 370–380.

Riley-Hagan M., Peshock R.M., Stray-Gundersen J. *et al.* (1992) Left ventricular dimensions and mass using magnetic resonance imaging in female endurance athletes. *Am J Cardiol*, **69**, 1067–1074.

Rowell L.B. (1973) Regulation of splanchnic blood flow in man. *Physiologist*, **16**, 127–142.

Saltin B. and Rowell L.B. (1980) Functional adaptations to physical activity and inactivity. *Fed Proc*, **39**, 1506–1513.

Saltin B. and Strange S. (1992) Maximal oxygen uptake: 'old' and 'new' arguments for a cardiovascular limitation. *Med Sci Sports Exerc*, **24**, 43–55.

Saltin B., Blomqvist C.G., Mitchell J.H. *et al.* (1968) Response to submaximal and maximal exercise after bed rest and training. *Circulation*, **38** (Suppl. 7), 1–78.

Scharhag J., Schneider G., Urhausen A. *et al.* (2002) Athlete's heart: right and left ventricular mass and function in male endurance athletes and untrained individuals determined by magnetic resonance imaging. *J Am Coll Cardiol*, **40**, 1856–1863.

Secher N.H. (2009) Cardiorespiratory limitations to performance, in *The Olympic Textbook of Science in Sport* (ed. R.J. Maughan), Blackwell Publishing Ltd., Oxford, UK, pp. 307–323.

Secher N.H. and Volianitis S. (2006) Are the arms and legs in competition for cardiac output? *Med Sci Sports Exerc*, **38**, 1797–1803.

Secher N.H., Clausen J.P., Klausen K. *et al.* (1977) Central and regional circulatory effects of adding arm exercise to leg exercise. *Acta Physiol Scand*, **100**, 288–297.

Sharma R.V., Tomanek R.J. and Bhalla R.C. (1985) Effect of swimming training on cardiac function and myosin ATPase activity in SHR. *J Appl Physiol*, **59**, 758–765.

Shepherd J.T. (1987) Circulatory response to exercise in health. *Circulation*, **76**, VI3–V10.

Spirito P., Pelliccia A., Proschan M.A. *et al.* (1994) Morphology of the 'athlete's heart' assessed by echocardiography in 947 elite athletes representing 27 sports. *Am J Cardiol*, **74**, 802–806.

Thomas G.D. and Segal S.S. (2004) Neural control of muscle blood flow during exercise. *J Appl Physiol*, **97**, 731–738.

Thomas G.D., Hansen J. and Victor R.G. (1994) Inhibition of alpha 2-adrenergic vasoconstriction during contraction of glycolytic, not oxidative, rat hindlimb muscle. *Am J Physiol Heart Circ Physiol*, **266**, H920–H929.

Tibbits G.F., Barnard R.J., Baldwin K.M. *et al.* (1981) Influence of exercise on excitation-contraction-coupling in rat myocardium. *Am J Physiol Heart Circ Physiol*, **240**, H472–H480.

Tonkonogi M. and Sahlin K. (2002) Physical exercise and mitochondrial function in human skeletal muscle. *Exerc Sport Sci Rev*, **30**, 129–137.

Wagner P.D. (2000) New ideas on limitations to VO_{2max}. *Exerc Sport Sci Rev*, **28**, 10–14.

Wagner P.D. (2005) Why doesn't exercise grow the lungs when other factors do? *Exerc Sport Sci Rev*, **33**, 3–8.

Wenger H.A. and Bell G.J. (1986) The interactions of intensity, frequency and duration of exercise training in altering cardiorespiratory fitness. *Sports Med*, **3**, 346–356.

Wiebe C.G., Gledhill N., Jamnik V.K. and Ferguson S. (1999) Exercise cardiac function in young through elderly endurance trained women. *Med Sci Sports Exerc*, **31**, 684–691.

Wilkerson D.P., Berger N.J.A. and Jones A.M. (2006) Influence of hyperoxia on pulmonary O2 on-kinetics following the onset of exercise in humans. *Respir Physiol Neurobiol*, **153**, 92–106.

Wilmore J.H., Stanforth P.R., Gagnon J. *et al.* (2001) Heart rate and blood pressure changes with endurance training: the HERITAGE Family Study. *Med Sci Sports Exerc*, **33**, 107–116.

Wyatt H.L. and Mitchell J. (1978) Influences of physical conditioning and deconditioning on coronary vasculature of dogs. *J Appl Physiol*, **45**, 619–625.

Zavorsky G.S. (2000) Evidence and possible mechanisms of altered maximum heart rate with endurance training and tapering. *Sports Med*, **29**, 13–26.

Zhou B., Conlee R.K., Jensen R. *et al.* (2001) Stroke volume does not plateau during graded exercise in elite male distance runners. *Med Sci Sports Exerc*, **33**, 1849–1854.

2.6 Exercise-induced Muscle Damage and Delayed-onset Muscle Soreness (DOMS)

Kazunori Nosaka, Edith Cowan University, School of Exercise, Biomedical and Health Sciences, Joondalup, WA, Australia

2.6.1 INTRODUCTION

Injurious physical, chemical, or biological stressors damage skeletal muscles. The severity of muscle damage varies from micro injury of a small number of muscle fibres to disruption of a whole muscle, depending on the cause of damage. This chapter focuses on muscle damage indicated by delayed-onset muscle soreness (DOMS), which is the most common type of damage that we experience in our daily life and exercise. It is predominantly induced by lengthening contractions or isometric contractions at a long muscle length (Clarkson, Nosaka and Braun, 1992; Jones, Newham and Torgan, 1989). It is also known that isometric contractions evoked by electrical muscle stimulation (Aldayel *et al.*, 2009; Jubeau *et al.*, 2008) induce muscle damage, especially when they are performed for the first time or a long time after a previous bout.

Here we will describe the characteristics of the muscle damage induced by eccentric exercise of the elbow flexors. Although the elbow flexors are often used to study muscle damage, it should be noted that the muscle damage of the elbow flexors does not necessarily represent the muscle damage of other muscles. In resistance training, 'no pain, no gain' is often advocated, therefore we also discuss whether DOMS or muscle damage is necessary for maximising the effects of resistance training on muscle hypertrophy and strength gain.

2.6.2 SYMPTOMS AND MARKERS OF MUSCLE DAMAGE

2.6.2.1 Symptoms

One prominent symptom of muscle damage is sore muscle developing after exercise (Cheung, Hume and Maxwell, 2003; Clarkson, Nosaka and Braun, 1992). As shown in Figure 2.6.1, other symptoms of muscle damage include muscle weakness, fatigue, stiff muscle, and muscle swelling (Clarkson, Nosaka and Braun, 1992; Nosaka, 2008). In order to assess muscle damage, several direct and indirect markers, including the measures to quantify the symptoms, are used.

2.6.2.2 Histology

The direct indicator of muscle damage is histological abnormality observed under light (e.g. infiltration of inflammatory cells to muscle fibres, absence of dystrophin or desmin staining) and/or electron microscope (e.g. ultrastructural alternation of myofilaments and intermediate filaments) (Fridén and Lieber, 2001; Gibala *et al.*, 1995; Jones *et al.*, 1986; Stauber and Smith, 1998). To investigate such pathological changes, an invasive muscle-sampling technique (muscle biopsy) is required; this can provide only a small amount of muscle tissue, which might not reflect the muscle damage of a whole muscle.

It should be noted that methodological artefacts could produce pathological characteristics (Malm, 2001). Raastad *et al.* (2010) have reported that 36% of muscle fibres obtained from muscle biopsy samples show myofiblillar disruptions characterized by myofillament disorganization and loss of Z-disk integrity following 300 maximal voluntary eccentric contractions of the knee extensors. However, infiltration of leukocytes in muscle fibres is not seen extensively even after such exercise, and leucocyte accumulation is primarily in the endomysium and perimysium (Paulsen *et al.*, 2010). It appears that the extent of actual histological alternations is minor, even when sever DOMS is present (Jones *et al.*, 1986; Lieber and Fridén, 2002).

Crameri *et al.* (2007) reported that no histological changes in muscle fibres were observed following 210 maximal

Strength and Conditioning – Biological Principles and Practical Applications Marco Cardinale, Rob Newton, and Kazunori Nosaka.

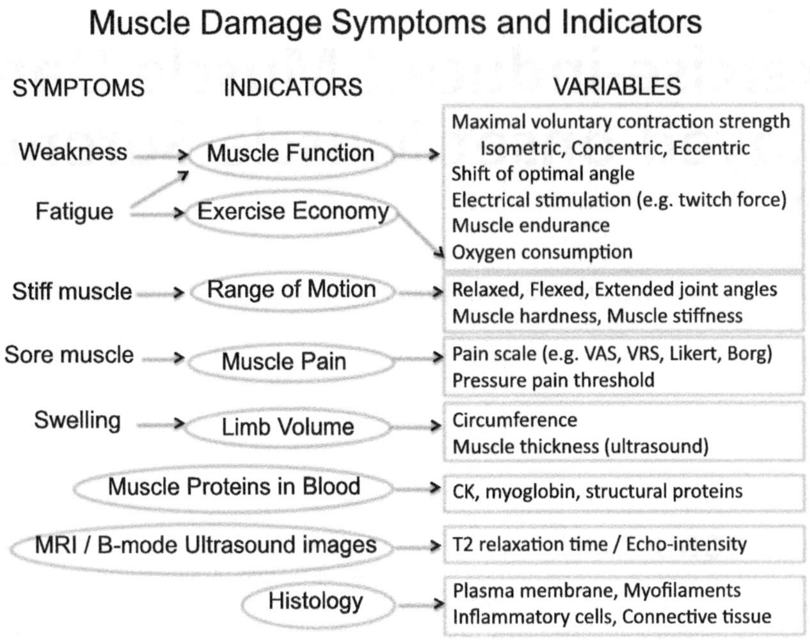

Figure 2.6.1 Symptoms, indicators, and measures of exercise-induced muscle damage

voluntary eccentric contractions of the knee extensors, although severe DOMS occurred and muscle strength was largely decreased. This suggests that histological abnormality in muscle fibres is not necessarily associated with DOMS, and it does not appear that many muscle fibres are actually degenerated following eccentric exercise resulting in DOMS. It has been documented that ultrastructual changes in muscle fibres indicate 'muscle remodelling' rather than muscle damage (Yu and Thornell, 2002; Yu, Malm and Thornell, 2002).

2.6.2.3 MRI and B-mode ultrasound images

Muscle damage can be visualized by magnetic resonance imaging (MRI) as an increased T2 relaxation time, or by B-mode ultrasound as an increased echo intensity (Foley *et al.*, 1999; Nosaka and Clarkson, 1996; Nosaka and Sakamoto, 2001; Nurenberg *et al.*, 1992). The increased T2 relaxation time is an indicative of increases in water in the area, probably due to inflammation. It has been shown that T2 relaxation time increases following eccentric exercise of the elbow flexors, peaks several days (e.g. six) post-exercise, and returns to the pre-exercise value in 4–8 weeks (Nosaka and Clarkson, 1996). As depicted in Figure 2.6.2, the region showing the increased echo intensity corresponds to the region showing the increased T2 relaxation time. It is not fully understood what causes the increases in echo intensity, although it appears to be associated with inflammation (Fujikake, Hart and Nosaka, 2009).

2.6.2.4 Blood markers

Muscle damage is represented by increases in muscle-specific proteins in the blood such as creatine kinase (CK) and myoglobin (Mb). When muscle fibres are severely damaged, their plasma membrane is disrupted and they become necrotic; soluble proteins located in the cytoplasm then leak out from the plasma membrane and get into the bloodstream. Small proteins can enter the bloodstream through capillaries, but large proteins appear to enter via lymph (Lindena *et al.*, 1979). For example, the molecular weights of CK and Mb are approximately 80 kD and 18 kD, respectively, and CK reaches the bloodstream via lymph, while Mb can get in through capillary (Sayers and Clarkson, 2003). This could explain the delayed increases in CK activity in the blood, which peaks 4–5 days following eccentric exercise of the elbow flexors (Figure 2.6.3), while Mb peaks earlier (e.g. 2–3 days following the eccentric exercise). The delayed large increases in CK or other enzyme activities (e.g. aldorase, lactate dehydrogenase, aspartate aminotransferase) in the blood are often seen following eccentric exercise of limb muscles (Nosaka and Clarkson, 1996). However, these enzymes already show large increases immediately after endurance events such as marathon or triathlon, and peak 1–2 days following exercise (Nosaka *et al.*, 2009; Suzuki *et al.*, 2006).

As shown in Figure 2.6.3, structural proteins such as troponin increase in the blood following eccentric exercise of the elbow flexors. Interestingly, only fast-twitch-type troponin I increases in a similar way to CK activity. This suggests that only fast-twitch fibres are damaged in the eccentric exercise. When the magnitude of CK activity in the blood is small (e.g.

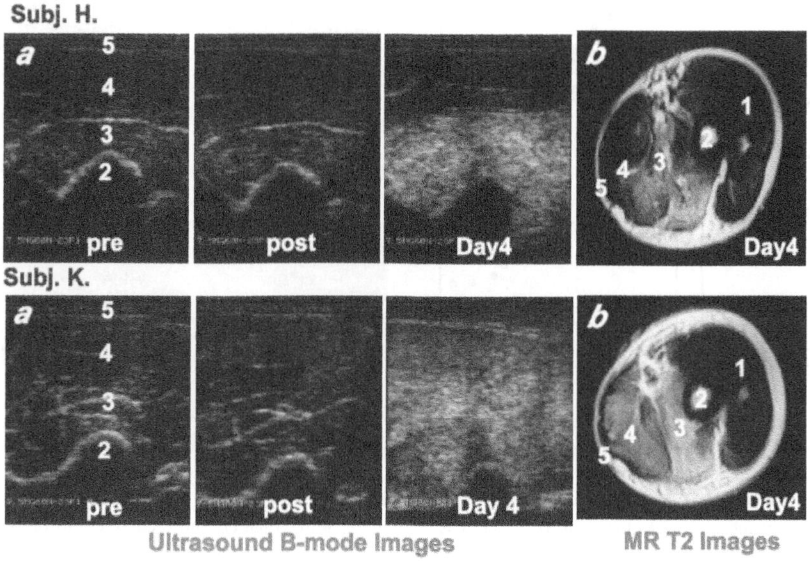

Figure 2.6.2 (a) B-mode ultrasound images of the upper arm taken before (pre), immediately after (post), and four days following eccentric exercise of the elbow flexors. (b) Magnetic resonance T2 images taken four days post-exercise for two subjects (Subj. H and Subj. K). 1 = triceps brachii, 2 = humerus, 3 = brachialis, 4 = biceps brachii, 5 = subcutaneous fat

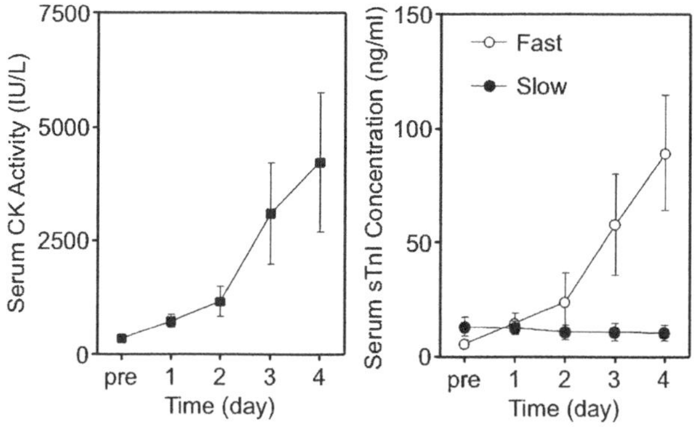

Figure 2.6.3 Changes in serum CK activity and serum skeletal muscle tropinin I concentration for its fast type and slow type before (pre) and 1–4 days following 210 (35 sets of 6) maximal eccentric contractions of the elbow flexors performed by eight non-resistance-trained men. From Nosaka *et al.*, unpublished data

<300 IU/l), it may not represent muscle damage, but rather increased permeability of the plasma membrane, or squeezing out of CK in the lymph (Nosaka, Sakamoto and Newton, 2002b).

2.6.2.5 Muscle function

A loss of muscle function lasting more than two days is a typical indicator of muscle damage, and muscle function measures such as maximal voluntary contraction strength of any contraction mode are considered to be the best tool for quantifying muscle damage (Warren, Lowe and Armstrong, 1999). A shift of optimum angle to a long muscle length has been reported after eccentric and isometric contractions (Philippou *et al.*, 2004), and this has been advocated as a sensitive marker of muscle damage (Proske and Morgan, 2001). In some cases, muscle strength does not return to the baseline for more than two months when maximal eccentric exercise of the elbow flexors is performed by 'untrained' individuals (Nosaka and Clarkson, 1996).

Using electrical muscle and/or nerve stimulation, it is possible to assess contractile property such as twitch torque and voluntary activation (Prasartwuth *et al.*, 2005, 2006), and give more information about the causes of strength decrement. It appears that some central inhibitory mechanisms are associated with the strength loss after eccentric exercise (Prasartwuth *et al.*, 2006), but maximal voluntary contraction strength provides a good picture of muscle damage and recovery. If muscle damage needs to be assessed in practice, a muscle-function measure in which the presumed 'damaged' muscle is involved (e.g. 1 RM test, jump test) should be performed.

Figure 2.6.4 shows the relationship between the magnitude of maximal voluntary isometric contractionstrength (MVC) and peak plasma CK activity. No significant correlation is seen between the MVC and CK at immediately after exercise, but the MVC and CK four days post-exercise are highly correlated. As explained in Section 2.6.2.4, the magnitude of increase in plasma CK activity is considered to represent the magnitude of muscle fibre necrosis. Thus, the high correlation seems to indicate that the inability to generate muscle force is associated with

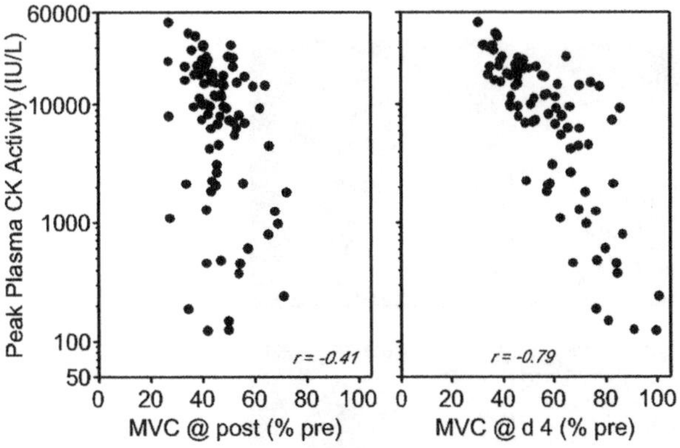

Figure 2.6.4 Relationship between the level of maximal voluntary isometric contraction strength (% of pre-exercise level) immediately after 24 maximal eccentric contractions of the elbow flexors (MVC @ post) or four days after the exercise (MVC @ d 4) and peak plasma CK activity. No significant correlation between MVC and CK was evident for post, but a high correlation is evident for d 4. Modified from Nosaka *et al.* (2006)

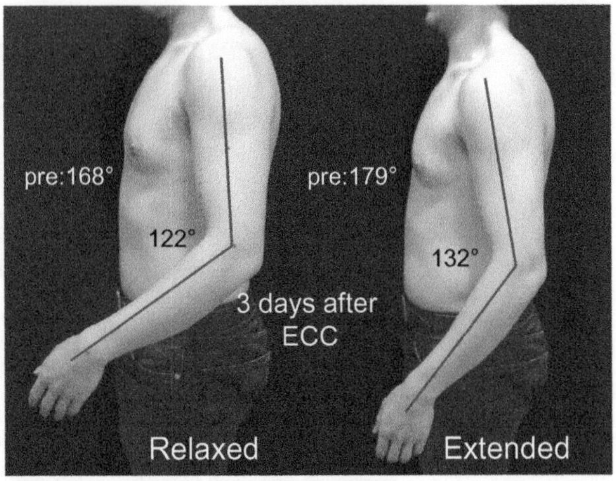

Figure 2.6.5 Relaxed and extended elbow joint angles at three days after 24 maximal eccentric contractions of the elbow flexors performed by a young man. The relaxed elbow joint was 168° before the exercise, but decreased to 122°. When he was asked to extend the elbow joint, he was able to extend to only 132°, although it was 179° previously

muscle-fibre damage. Raastad *et al.* (2010) have recently reported that the magnitude of decrease in MVC is highly correlated with the number of muscle fibres with myofibrillar disruptions. When muscle strength is affected by muscle damage, other performance related to muscle function, such as jump performance (Miyama and Nosaka, 2004), is also affected.

2.6.2.6 Exercise economy

Exercise economy is decreased when muscle damage is present. For example, Chen *et al.* (2007b) showed that running economy during submaximal treadmill running was significantly reduced for three days following downhill running. Chen *et al.* (2009b) reported that significant decreases in running economy during level running were evident at 80% and 90% VO_{2peak} intensities, but not at 70% VO_{2peak}. It is possible that the significant effect of muscle damage on running economy at high intensities (80% and 90%) reflects the greater number of muscle fibres that needed to be recruited, along with impairment of the ability to utilise elastic energy.

2.6.2.7 Range of motion (ROM)

Muscles become stiff with muscle damage (Clarkson, Nosaka and Braun, 1992; Howell, Chleboun and Conatser, 1993; Jones, Newham and Clarkson, 1987), and for some muscles this is reflected in a decrease in range of motion (ROM) around the joint at which they are located.

Figure 2.6.5 shows a typical elbow joint angle observed after eccentric exercise of the elbow flexors. The picture on the left shows the angle when the subject is relaxing so that the arm hangs down at his side. Before exercise, the angle is 168°, at which the elbow joint is nearly straightened. However, the angle gets smaller following exercise, and generally the largest decrease is observed at around three days post-exercise. In this situation, it is not possible for the subject to extend the elbow joint fully, and as shown in the right-hand picture, the subject is able to extend the elbow joint only 10° further than the relaxed angle. Following eccentric exercise of the elbow flexors, the subjects cannot fully flex the elbow joint angle. Because of a decrease in the extended elbow joint angle and an increase in the flexed elbow joint angle, the ROM of the elbow joint decreases. This can also happen to other joints, but the magnitude of the change does not appear to be as large as that seen in the elbow joint. Murayama *et al.* (2000) reported that muscle becomes harder after eccentric exercise, which could be associated with increases in muscle stiffness.

2.6.2.8 Swelling

Damaged muscles are swollen (Clarkson, Nosaka and Braun, 1992; Howell, Chleboun and Conatser, 1993; Nosaka and Clarkson, 1996), and this can be detected by an increase in muscle thickness or volume shown by B-mode ultrasound (Figure 2.6.4) or MRI, and/or an increase in circumference (Figure 2.6.6). In Figure 2.6.4, the distance between the subcutaneous fat layer and the humerus is greater at day 4 than before that point. Figure 2.6.6 shows an arm which performed eccentric exercise of the elbow flexors four days earlier. Compared to the pre-exercise value, the circumference is more than 40 mm larger at the upper arm and forearm. The circumference increases after exercise, but the increase is small (less than 10 mm) immediately after and one-day post-exercise, and peaks

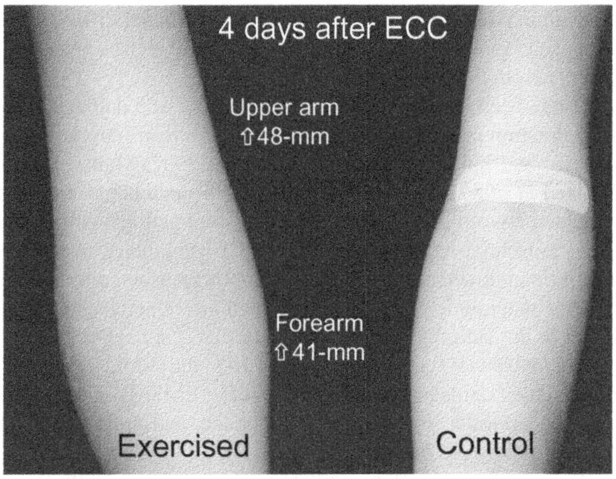

Figure 2.6.6 Swelling of the arm observed at four days after 24 maximal eccentric contractions of the elbow flexors performed by a young man. Compared to the control arm, the exercise arm is bigger, but their circumferences were similar before exercise. The upper arm and forearm increased 48 mm and 41 mm, respectively, compared to pre-exercise value

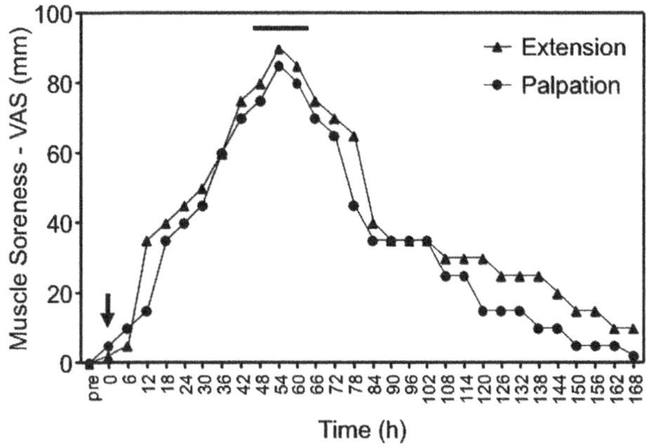

Figure 2.6.7 Changes in muscle soreness of a subject (48-year-old, non-resistance-trained man) evaluated by a 100 mm visual analogue scale (VAS: 0 = no pain, 100 = extremely painful) for extension of the elbow joint (extension) and palpating the upper arm before (pre), immediately after (0), and 6–168 hours after 100 maximal eccentric contractions (5 sets of 20 repetitions) of the elbow flexors on an isokinetic dynamometer. From Nosaka, unpublished data

4–6 days following exercise. Generally, the swelling of the forearm becomes more conspicuous several days later, because of the shift of fluid from the upper arm to the forearm due to gravity.

2.6.2.9 Muscle pain

DOMS is characterized by a sensation of dull, aching pain, usually felt during movement or palpation of the affected muscle (Clarkson, Nosaka and Braun, 1992; Miles and Clarkson, 1994). DOMS typically develops several hours after exercise, peaks at 1–3 days, and disappears by 7–10 days post-exercise (Cheung *et al.*, 2003). Because of its subjective nature, it is difficult to quantify the magnitude of muscle pain; however, several scales, including visual analogue scale (VAS), numerical rating scale, verbal rating scale (VRS), and descriptors differential scale, are often used (Nosaka, 2008).

Two popular scales that have been used in muscle-damage studies are VAS, which consists of a line (50–100 mm) with 'no pain' at the left and 'unbearable pain' at the right, and VRS, in which descriptors such as 'no pain' and 'unbearable pain' correspond to numbers (e.g. 1–10). Figure 2.6.7 shows changes in muscle pain of the biceps brachii on the 100 mm VAS when the muscle is palpated or extended following 100 maximal eccentric contractions of the elbow flexors by a middle-aged man. Muscle pain immediately post-exercise is minimal, but the pain level increases gradually for the next 48 hours, peaks between 48 and 60 hours post-exercise, and gradually subsides by seven days post-exercise. In contrast, Figure 2.6.8 shows changes in pain of four regions of the body following an iron-man triathlon race performed by an experienced triathlete

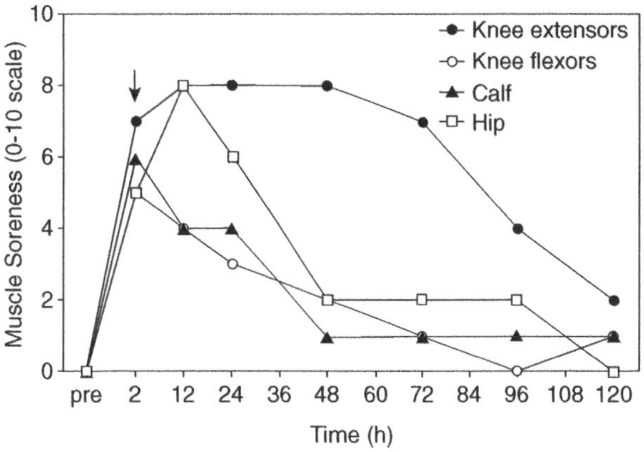

Figure 2.6.8 Changes in muscle soreness of a subject (38-year-old man, experienced triathlete) evaluated by a Borg scale (0 = no pain, 10 = worst pain imaginable) for palpating knee extensors, knee flexors, calf and hip muscles before (pre) and 2–120 hours following an iron-man triathlon race. Modified from Nosaka *et al.* (2009)

(38-year-old man), described using VRS. The athlete experienced severe muscle pain two hours post-race, and muscle pain peaked within 12 hours after the race. It is interesting that the time course of changes in muscle pain is different between different muscles, with the knee extensors showing most long-lasting pain, which subsides by five days post-race.

Pain is generally considered a warning signal, but this does not appear to be the case for DOMS (Nosaka, 2008). Exercise of sore muscles does not induce additional DOMS and muscle damage, nor does it retard recovery from the previous eccentric exercise bout (Chen and Hsieh, 2001; Nosaka and Newton, 2002a, 2002b). When DOMS is present, moving the sore muscles attenuates the magnitude of muscle soreness; however, this effect is temporary and the time course of changes in soreness is not affected (Zainuddin *et al.*, 2006).

2.6.3 RELATIONSHIP BETWEEN DOMS AND OTHER INDICATORS

It appears that each symptom or marker shows a different aspect of muscle damage. It is important to note that DOMS does not represent the time course of muscle damage, and the level of DOMS is poorly correlated with the magnitude of changes in other indicators of muscle damage (Nosaka, Newton and Sacco, 2002a; Rodenburg *et al.*, 1993). For example, after maximal eccentric exercise of the elbow flexors, muscle soreness does not develop immediately following exercise, but muscle strength shows its largest decrease at this time point. When muscle soreness subsides, swelling of the upper arm peaks, and abnormality in MRI and ultrasound images is greatest around this time period (Nosaka, 2008).

As shown in Figure 2.6.9, peak muscle soreness assessed by VAS does not relate to the magnitude of decrease in maximal voluntary isometric contraction strength and ROM at four days post-exercise, or to peak plasma CK activity. Severe DOMS develops with little or no indication of muscle damage, and severe muscle damage does not necessarily result in severe DOMS. DOMS does not reflect the magnitude of muscle damage even within the same individuals. For example, when the same subjects performed two different intensities of eccentric exercise of the elbow flexors, all of the indirect markers of muscle damage showed greater changes for maximal intensity than submaximal intensity; however, no significant differences in muscle soreness were seen between the exercises (Nosaka and Newton, 2002c).

It has been documented that connective tissue damage and inflammation is more responsible for DOMS than muscle-fibre damage and inflammation (Crameri *et al.*, 2007; Malm, 2001; Paulsen *et al.*, 2010). Crameri *et al.* (2007) compared muscle damage between 210 maximal eccentric contractions with electrical muscle stimulation (EMS) and 210 voluntary maximal eccentric contractions (VOL) of the knee extensors, and found that the magnitude of DOMS developed after exercise and the increase in staining of the intramuscular connective tissue (tenascin C) were similar between EMS and VOL; however muscle-fibre damage was evident only after EMS. This suggests that extracellular matrix (ECM) damage and inflammation plays a major role in DOMS.

Paulsen *et al.* (2010) recently reported that the magnitude of leucocyte accumulation in the endomysium and perimysium was negatively correlated with the magnitude of DOMS developed after 300 maximal voluntary eccentric contractions of the knee extensors, and stated that DOMS cannot be explained by the presence of leukocytes. The cause of DOMS is still not fully understood.

2.6.4 FACTORS INFLUENCING THE MAGNITUDE OF MUSCLE DAMAGE

The magnitude of eccentric exercise-induced muscle damage is influenced by factors such as intensity, number, velocity, and muscle length of eccentric contractions, as well as training status and previous use of the muscles in exercise and daily activities.

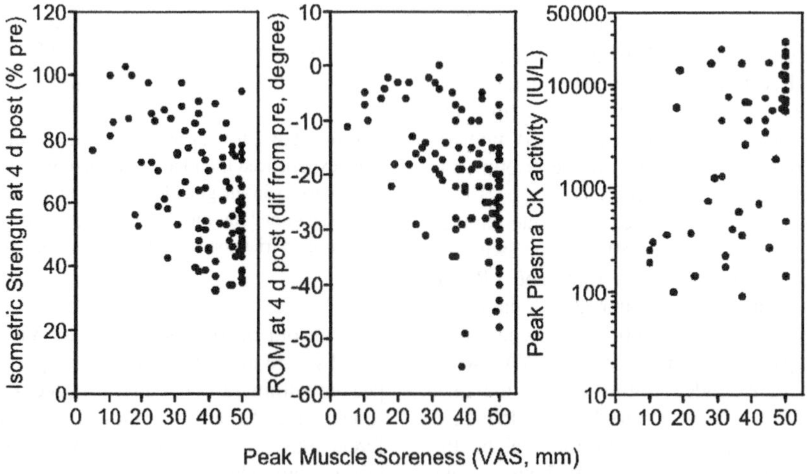

Figure 2.6.9 Relationship between peak muscle soreness evaluated by VAS (0 = no pain, 50 = extremely painful) and maximal voluntary isometric strength at four days post-exercise (% of pre-exercise level), ROM around the elbow joint at four days post-exercise (absolute difference from the pre-exercise value), and peak plasma CK activity following 24 maximal eccentric contractions of the elbow flexors performed by 89 non-resistance-trained men. Modified from Nosaka *et al.* (2002a)

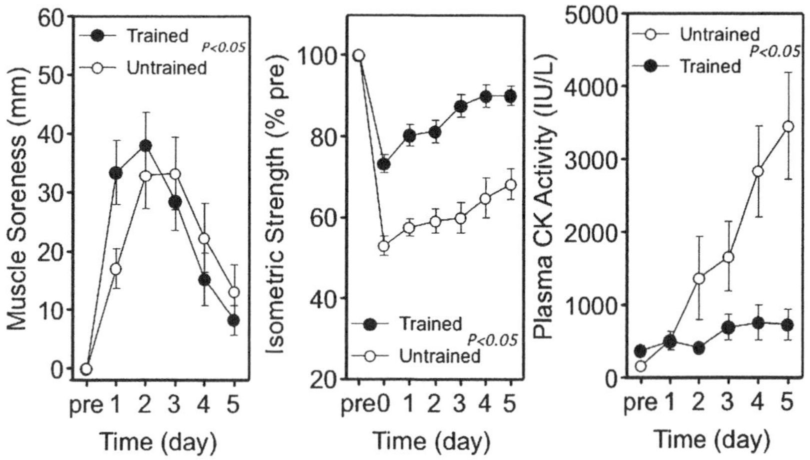

Figure 2.6.10 Changes in muscle soreness, maximal voluntary isometric strength, and plasma CK activity before (pre), immediately after (0), and 1–5 days after 60 (6 sets of 10) maximal eccentric contractions of the elbow flexors on an isokinetic dynamometer performed by resistance-trained men (Trained) and non-resistance-trained men (Untrained). Modified from Newton *et al.* (2008)

2.6.4.1 Contraction parameters

The magnitude of muscle damage induced by eccentric exercise is greater with higher intensity, larger numbers of contractions, faster velocity, and at longer muscle lengths (Nosaka, 2009). Chapman *et al.* (2008a) showed that in 30 eccentric contractions, the effect of velocity was minor, but when 210 eccentric contractions were performed, the fast-velocity (210 °/s) exercise resulted in significantly greater changes in most indirect markers of muscle damage than the slow-velocity (30 °/s) exercise. This suggests that the number of eccentric contractions is a stronger predictor of muscle damage, but contraction velocity also affects the magnitude of muscle damage.

Comparison between two maximal lengthening contraction regimens in which the elbow joint was extended from 50 to 130 ° (short condition) and from 100 to 180 ° (long condition) showed that changes in MVC, ROM, upper-arm circumference, muscle soreness, plasma CK activity, and abnormalities in B-mode ultrasound and MRI images were significantly greater for the long condition than the short condition (Nosaka and Sakamoto, 2001). This was confirmed in the subsequent study (Nosaka *et al.*, 2005b), in which a short (50–100 °) and a long (130–180 °) extension range were compared.

2.6.4.2 Training status

The characteristics of muscle damage in trained athletes are different from those of untrained subjects. Newton *et al.* (2008) compared trained (performing resistance training at least three days a week for at least a year) and untrained men for changes in some indirect markers of muscle damage following 10 sets of six maximal voluntary eccentric contractions of the elbow flexors of one arm on an isokinetic dynamometer, which was an unfamiliar exercise for all subjects in both groups. As shown in Figure 2.6.10, the trained group showed significantly smaller decreases and faster recovery of maximal voluntary isometric contraction strength compared with the untrained group. No significant increases in plasma CK activity were seen in the trained group, but the untrained group showed large increases. Interestingly, no significant difference between the groups is evident for muscle soreness. This is another example where muscle pain does not reflect the magnitude of muscle-fibre damage. It should be noted that the muscle strength of the trained group recovered to the baseline by three days post-exercise, when the untrained group showed approximately 40% lower strength than baseline. These results suggest that resistance-trained men are less susceptible to muscle damage induced by maximal eccentric exercise than untrained subjects.

2.6.4.3 Repeated-bout effect

The magnitude of muscle damage from the same exercise bout is never the same, even for untrained individuals. Compared to the initial bout of eccentric exercise, a subsequent bout of the same exercise performed within several weeks results in less indications of muscle damage. This phenomenon is known as the 'repeated-bout effect' (Clarkson, Nosaka and Braun, 1992; McHugh, 2003). As depicted in Figure 2.6.11, compared to the first bout of eccentric exercise of the elbow flexors, a second bout of the same exercise with the same arm performed four weeks later resulted in less DOMS, faster recovery of muscle strength, and smaller increases in plasma CK activity. It has also been reported that changes in ROM, upper-arm circumference (swelling), B-mode ultrasound and MRI are smaller after the second bout than after the first (Foley *et al.*, 1999; Nosaka and Clarkson, 1996). This protective effect is reported to last

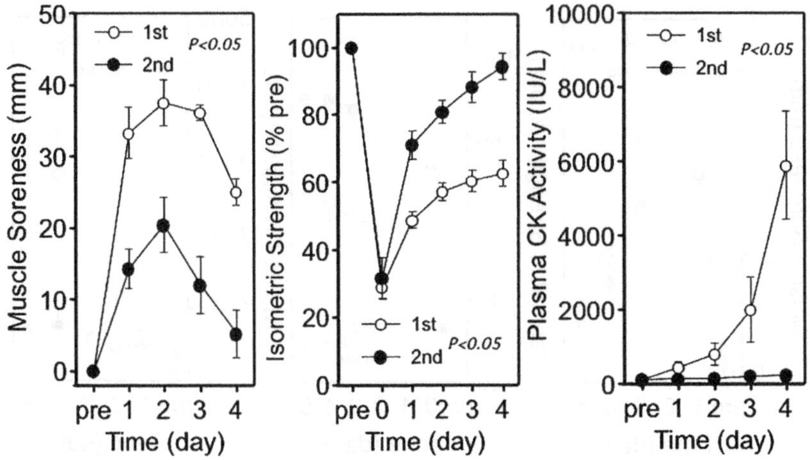

Figure 2.6.11 Changes in muscle soreness, maximal voluntary isometric strength, and plasma CK activity before (pre), immediately after (0), and 1–4 days after eccentric exercise of the elbow flexors (6 sets of 5 eccentric contractions with a dumbbell) for the first and second (performed four weeks after the first) bouts by 10 non-resistance-trained men. Modified from Hirose *et al.* (2004)

for at least several weeks, but the length of the repeated-bout effect is dependent on markers of muscle damage (Nosaka *et al.*, 2001a; Nosaka, Newton and Sacco, 2005a).

The initial eccentric bout with minor damage can still confer some protective effect. It has been reported that performing an initial eccentric bout with a relatively small number of eccentric contractions produced the repeated-bout effect (Chen and Nosaka, 2006; Nosaka *et al.*, 2001b), lower-intensity eccentric contractions (Chen *et al.*, 2007a, 2009b), or eccentric contractions at short muscle length, which produced minor muscle damage (Nosaka *et al.*, 2005b). Lavender and Nosaka (2008a) have shown that a light eccentric exercise which does not induce changes in any of the indirect markers of muscle damage confers protection against muscle damage after a more strenuous eccentric exercise performed two days later. The greatest adaptation of the muscle against muscle damage is induced after the first eccentric exercise bout, but some further protective adaptations are induced in subsequent bouts (Chen *et al.*, 2009a). This seems to explain the case of the 'resistance-trained individuals' shown in Section 2.6.4.2 (Figure 2.6.9). Using the repeated-bout effect, it is possible to minimize muscle damage.

2.6.4.4 Muscle

The susceptibility to muscle damage appears to be different among muscles. Jamurtas *et al.* (2005) reported that an eccentric exercise of the knee extensors resulted in smaller decreases and faster recovery of muscle strength, and smaller increases in plasma CK activity, compared with an eccentric exercise of the elbow flexors, when the two exercises had the same number of contractions (6 sets of 12 reps) performed at the same relative intensity (75% of maximal eccentric torque) using the same subjects (Figure 2.6.12). Interestingly, no significant difference in DOMS was seen in two exercises. This is another example where the level of muscle pain does not necessarily reflect the magnitude of muscle damage. The decreased muscle damage in the knee extensors seems to be due to the repeated-bout effect.

It has been reported that little muscle damage occurs to the wrist extensors (Slater *et al.*, 2010). This may be associated with how much change in muscle length occurs during eccentric contractions in the range of joint movement.

2.6.4.5 Other factors

Age could be a factor, as shown in animal studies reporting that old muscles are more susceptible to eccentric exercise-induced muscle damage than young muscles (e.g. Brooks and Faulkner, 1996). However, this has not been shown clearly in human studies, at least where voluntary eccentric contractions are performed. For example, a study comparing changes in indirect markers of muscle damage following eccentric exercise of the elbow flexors between young and middle-aged (50-year-old) men did not find significant differences between the groups for changes in indirect markers of muscle damage, except for muscle soreness (Lavender and Nosaka, 2008b). This was also the case for studies comparing young and old (>65-year-old) men (Chapman *et al.*, 2008b; Lavender and Nosaka, 2006). Interestingly, muscle soreness is less for the middle-aged and elderly individuals than for the young (Lavender and Nosaka, 2006, 2008b).

Controversy exists concerning the difference between men and women for their susceptibility to muscle damage (Nosaka, 2009). It does not appear that gender differences are greater than individual differences.

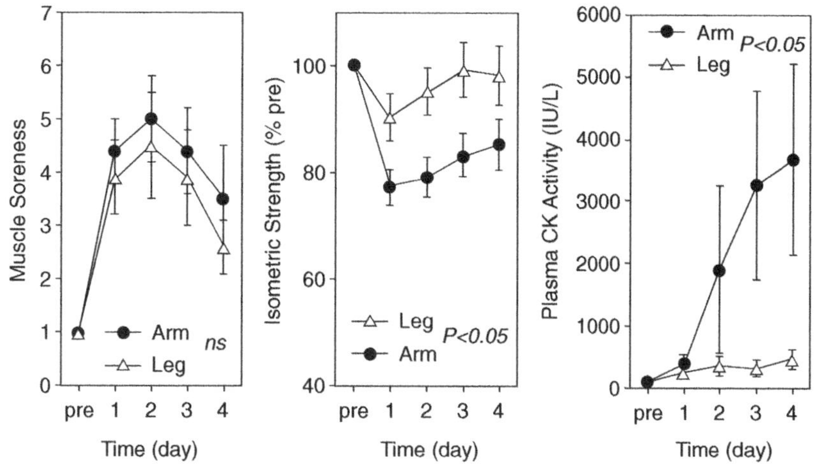

Figure 2.6.12 Changes in muscle soreness, maximal voluntary isometric strength, and plasma CK activity before (pre) and 1–4 days after eccentric exercise of the elbow flexors (Arm) and the knee extensors (Leg) performed by non-resistance-trained men. Modified from Jamurtas *et al.* (2005)

2.6.5 MUSCLE DAMAGE AND TRAINING

Skeletal muscles adapt structurally and physiologically to mechanical stimuli generated in muscle contractions, and two common consequences of such adaptations are strength gain and muscle hypertrophy. The third position stand of the American College of Sports Medicine on the guidelines for resistance training (Ratamess *et al.*, 2009) states that both concentric and eccentric contractions should be included to maximize muscle hypertrophy. As explained above, eccentric contractions could induce muscle damage and DOMS.

'Pain' occurs during or after training, and some people think that pain is necessary for a gain in muscle strength and size. The concept of 'no pain, no gain' is popular in sports culture. This section tries to clarify how much 'pain' is necessary for a 'gain' in muscle function and muscle volume.

2.6.5.1 The importance of eccentric contractions

Several studies have reported superiority of eccentric contractions over concentric contractions for muscle hypertrophy. For example, Higbie *et al.* (1996) showed that quadriceps cross-sectional area increased to a greater extent when training was eccentric (6.6%) than concentric (5.0%). Hortobagyi *et al.* (2000) reported that increases in muscle strength and type II muscle fibre size after 12 weeks of retraining after 3 weeks of immobilization were greater for eccentric training than for concentric training. Farthing and Chilibeck (2003) demonstrated that eccentric training of elbow flexors increased muscle thickness significantly more than concentric training. Furthermore, Vikne *et al.* (2006) showed that eccentric training (2–3 times a week over 12 weeks) increased cross-sectional area of the elbow flexors and their muscle fibres significantly more than concentric training. One of the reasons why eccentric training has the potential to induce greater muscle hypertrophy than concentric training is the higher force generated during eccentric contractions than during concentric contractions (Adams *et al.*, 2004; Farthing and Chilibeck, 2003). However, it is noted that concentric training can increases muscle size and strength in a similar magnitude to eccentric training (Blazevich *et al.*, 2007). Thus, it does not appear that muscle damage is a prerequisite for muscle hypertrophy.

The rate of protein synthesis is mediated through activations of protein kinase B (Akt), mammalian target of rapamycin (mTOR), and p70 S6 kinase ($p70^{s6k}$), and the Akt/mTOR/$p70^{s6k}$ pathway is known to be involved in exercise-induced muscle hypertrophy (Atherton *et al.*, 2005; Bolster *et al.*, 2003). Eliasson *et al.* (2006) reported that maximal eccentric contractions of the knee extensors activated $p70^{s6k}$ in the vastus lateralis, but maximal concentric contractions did not, suggesting that maximal eccentric contractions are more effective than maximal concentric contractions in stimulating protein synthesis. Therefore, it seems reasonable to state that eccentric contractions are more potent stimuli than concentric contractions.

2.6.5.2 The possible role of muscle damage in muscle hypertrophy and strength gain

Satellite cells, located in the muscle basal laminae, maintain the myonuclei-to-cytoplasmic volume ratio by adding myonuclei to the increasing area and length of muscle fibres (Kadi *et al.*,

2005), and play a major role in muscle hypertrophy (Adams, 2006). It is known that satellite cells within the injured muscle fibres are activated to proliferate and differentiate, forming myoblasts, which can fuse to form myotubes (Quintero *et al.*, 2009). Thus, it is possible that the activation of satellite cells by muscle damage will increase hypertrophic responses. In fact, Crameri *et al.* (2004) showed that satellite cells were proliferated after a single bout of eccentric exercise of the knee extensors. Since satellite cells are responsible for muscle regeneration, it seems reasonable to assume that a greater activation of satellite cells will be induced when severe muscle damage is induced. However, it may be that the satellites cells are used only to regenerate the necrotic regions of muscle fibres, and are not necessarily involved in muscle hypertrophy in severe muscle damage. It should also be noted that protein synthesis should exceed protein degradation in order for a muscle to be hypertrophied, and muscle damage is the process of protein degradation. If a muscle is damaged regularly, no muscle hypertrophy will occur. Therefore, damaging muscles does not necessarily result in muscle hypertrophy, and the benefits of eccentric contractions as a stimulus for muscle hypertrophy should be considered separately from the issue of muscle damage.

2.6.5.3 No pain, no gain?

As explained in the Section 2.6.4.3, muscles become less susceptible to muscle damage when the same or a similar exercise is repeated. Figure 2.6.13 shows changes in MVC, plasma CK activity, and muscle soreness when three sets of ten eccentric contractions of the elbow flexors with a dumbbell of 50% MVC were performed once a week for eight weeks (Nosaka and Newton, 2002d). The first training session induced more muscle damage than the other sessions, and each session resulted in a large decrease in MVC immediately after exercise, but the pre-exercise MVC gradually returned to the pre-training level, and MVC finally returned to the baseline at week 6. Plasma CK activity increased only after the first training session, indicating that muscle-fibre damage occurred only after the first session. Muscle soreness developed following all training sessions, but the magnitude was smaller following the second to eighth sessions compared with the first session. This may indicate that some minor connective tissue damage inflammation was induced in each session. In the study, no muscle hypertrophy measurement was made, but considering the fact that muscle hypertrophy is generally evident after several weeks of training, it seems reasonable to state that muscle damage is not the main factor for muscle hypertrophy.

The magnitude of the training effect seems to be greater for eccentric training, possibly due to the greater mechanical stress of eccentric contractions than of concentric or isometric contractions. It is important to note that eccentric training does not necessarily induce muscle damage, and muscle is capable of hypertrophy in the absence of muscle damage. However, some muscle damage indicators, such as muscle pain and weakness, are frequently accompanied by resistance training when intending to maximasing mechanical stress to muscles. This is why some people believe that 'muscle damage' is necessary for muscle hypertrophy and strength gain.

2.6.6 CONCLUSION

DOMS is one of the peculiar symptoms of muscle damage; however, the magnitude of DOMS does not reflect the magnitude of muscle damage, which can be indicated by the magnitude of loss of muscle strength. The magnitude of muscle

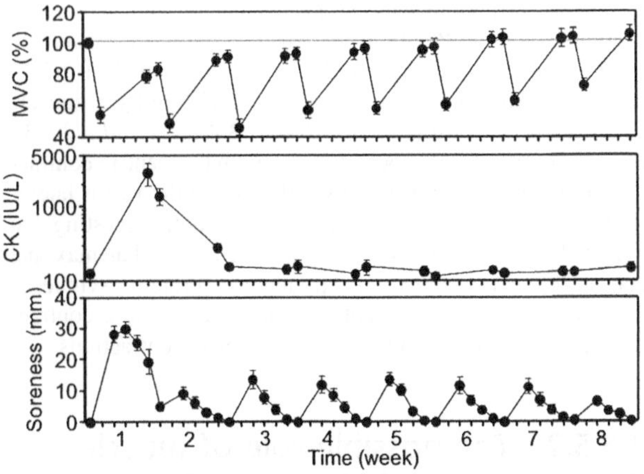

Figure 2.6.13 Changes in maximal voluntary isometric strength, plasma CK activity, and muscle soreness (VAS: 0 = no pain, 50 = extremely painful) over eight weeks in which a bout of eccentric exercise of the elbow flexors (3 sets of 10 eccentric contractions of the elbow flexors using a dumbbell) was performed once a week. Modified from Nosaka and Newton (2002d)

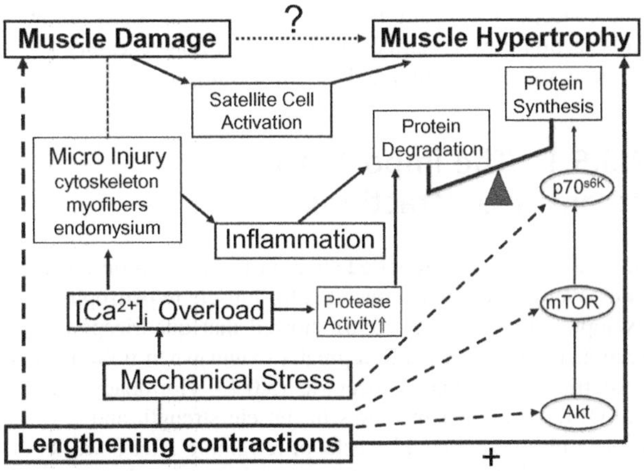

Figure 2.6.14 Relationship between muscle damage and muscle hypertrophy. Instead of muscle damage, lengthening contractions appear to be more potent stimuli for muscle hypertrophy

damage is affected by many factors, such as intensity, volume, muscle length of eccentric contractions, and individual characteristics such as experience and training. Muscles adapt rapidly to damaging exercise in order to attenuate the magnitude of muscle damage. This makes muscles less susceptible to muscle damage, and when resistance training is progressed, less muscle damage is induced. Thus, muscle hypertrophy and muscle-strength gain in resistance training do not necessarily relate to muscle damage. However, there is no doubt that eccentric contractions are important for muscle hypertrophy and muscle-strength gain. It is not muscle damage due to eccentric contractions but eccentric contractions per se that are responsible for the greater muscle hypertrophy and muscle-strength gain conferred by eccentric training (Figure 2.6.14).

References

Adams G.R. (2006) Satellite cell proliferation and skeletal muscle hypertrophy. *Appl Physiol Nutr Metab*, **31**, 782–790.

Adams G.R., Cheng D.C., Haddad F. and Baldwin K.M. (2004) Skeletal muscle hypertrophy in response to isometric, lengthening and shortening training bouts of equivalent duration. *J Appl Physiol*, **96**, 1613–1618.

Aldayel A., Jubeau M., McGuigan M.R. and Nosaka K. (2009) Less indication of muscle damage in the second than initial electrical muscle stimulation bout consisting of isometric contractions of the knee extensors. *Eur J Appl Physiol*, **108**, 709–717.

Atherton P.J., Babraj J.A., Smith K. *et al.* (2005) Selective activation of AMPK-PGC-1andalpha; or PKB-TSC2-mTOR signalling can explain specific adaptive responses to endurance or resistance training-like electrical muscle stimulation. *FASEB J*, **19**, 768–788.

Blazevich A.J., Gill N.D., Deans N. and Zhou S. (2007) Lack of human muscle architectural adaptation after short-term strength training. *Muscle Nerve*, **35**, 78–86.

Bolster D.R., Kubica N., Crozier S.J. *et al.* (2003) Immediate response of mammalian target of rapamycin (mTOR)-mediated signalling following acute resistance exercise in rat skeletal muscle. *J Physiol*, **553**, 213–220.

Brooks S.V. and Faulkner J.A. (1996) The magnitude of the initial injury induced by stretches of maximally activated muscle fibres of mice and rats increases in old age. *J Physiol*, **497** (Pt 2), 573–580.

Chapman D., Newton M., McGuigan M. and Nosaka K. (2008a) Effect of contraction velocity on muscle damage after maximal lengthening contractions. *Med Sci Sports Exerc*, **40**, 926–933.

Chapman D., Newton M., McGuigan M.R. and Nosaka K. (2008b) Comparison between old and young men for responses to fast velocity maximal lengthening contractions of the elbow flexors. *Eur J Appl Physiol*, **104**, 531–539.

Chen T.C. and Hsieh S.S. (2001) Effects of a 7-day eccentric training period on muscle damage and inflammation. *Med Sci Sports Exerc*, **33**, 1732–1738.

Chen T.C. and Nosaka K. (2006) Effects of number of eccentric muscle actions on first and second bouts of intensive eccentric exercise of the elbow flexors. *J Sci Med Sport*, **9**, 57–66.

Chen T.C., Nosaka K. and Sacco P. (2007a) Intensity of initial eccentric exercise and the magnitude of repeated bout effect. *J Appl Physiol*, **102**, 992–999.

Chen T.C., Nosaka K. and Tu J.H. (2007b) Changes in running economy following downhill running. *J Sports Sci*, **25**, 55–63.

Chen T.C., Chen H.-L., Lin M.-R. *et al.* (2009a) Responses to four eccentric exercise bouts of the elbow flexors performed every four weeks. *Eur J Appl Physiol*, **106**, 267–275.

Chen T.C., Nosaka K., Chen H.-L. *et al.* (2009b) Changes in running economy at different intensities following downhill running. *J Sports Sci*, **27**, 1–8.

Cheung K., Hume P.A. and Maxwell L. (2003) Delayed onset muscle soreness. Treatment strategies and performance factors. *Sports Med*, **33**, 145–164.

Clarkson P.M., Nosaka K. and Braun B. (1992) Muscle function after exercise-induced muscle damage and rapid adaptation. *Med Sci Sports Exerc*, **24**, 512–520.

Crameri R.M., Langberg H., Magnusson P. *et al.* (2004) Changes in satellite cells in human skeletal muscle after a single bout of high intensity exercise. *J Physiol*, **558**, 333–340.

Crameri R.M., Aagaard P., Qvortrup K. *et al.* (2007) Myofibre damage in human skeletal muscle: effects of electrical stimulation versus voluntary contraction. *J Physiol*, **583** (Pt 1), 365–380.

Eliasson J., Elfegoun T., Nilsson J. *et al.* (2006) Maximal lengthening contractions increase p70 S6 kinase phosphorylation in human skeletal muscle in the absence of nutritional supply. *Am J Physiol Endocrinol Metab*, **291**, 1197–1205.

Farthing J.P. and Chilibeck P.D. (2003) The effects of eccentric and concentric training at different velocities on muscle hypertrophy. *Eur J Appl Physiol*, **89**, 578–586.

Foley J.M., Jayaraman R.C., Prior B.M. *et al.* (1999) MR measurements of muscle damage and adaptation after eccentric exercise. *J Appl Physiol*, **87**, 2311–2318.

Fridén J. and Lieber R.L. (2001) Eccentric exercise-induced injuries to contractile and cytoskeletal muscle fibre components. *Acta Physiol Scand*, **171**, 321–326.

Fujikake T., Hart R. and Nosaka K. (2009) Changes in B-mode ultrasound echo intensity following injection of bupivacaine hydrochloride injection rat hind limb muscles in relation to histological changes. *Ultrasound Med Biol*, **35**, 687–696.

Gibala M.J., MacDougall J.D., Stauber W.T. and Elorriaga A. (1995) Changes in human skeletal muscle ultrastructure and force production after acute resistance exercise. *J Appl Physiol*, **78**, 702–708.

Higbie E.J., Cureton K.J., Warren G.L. and Prior B.M. (1996) Effects of concentric and eccentric training on muscle strength, cross-sectional area, and neural activation. *J Appl Physiol*, **81**, 2173–2181.

Hirose L., Nosaka N., Newton M. *et al.* (2004) Changes in inflammatory mediators following eccentric exercise of the elbow flexors. *Exerc Immunol Rev*, **10**, 75–90.

Hortobagyi T., Dempsey L., Fraser D. *et al.* (2000) Changes in muscle strength, muscle fibre size and myofibrillar gene expression after immobilization and retraining in humans. *J Physiol*, **524**, 293–304.

Howell J.N., Chleboun G. and Conatser R. (1993) Muscle stiffness, strength loss, swelling and soreness following exercise-induced injury in humans. *J Physiol*, **464**, 183–196.

Jamurtas A.Z., Theocharis V., Tofas T. *et al.* (2005) Comparison between leg and arm eccentric exercise of the same relative intensity on indices of muscle damage. *Eur J Appl Physiol*, **95**, 179–185.

Jones D.A., Newham D.J., Round J.M. and Tolfree S.E.J. (1986) Experimental human muscle damage: morphological changes in relation to other indices of damage. *J Physiol*, **375**, 435–448.

Jones D.A., Newham D.J. and Clarkson P.M. (1987) Skeletal muscle stiffness and pain following eccentric exercise of the elbow flexors. *Pain*, **30**, 233–242.

Jones D.A., Newham D.J. and Torgan C. (1989) Mechanical influences on long-lasting human muscle fatigue and delayed-onset pain. *J Physiol*, **412**, 415–427.

Jubeau M., Sartorio A., Marinone P.G. *et al.* (2008) Differences in growth hormone and neuromuscular responses between electrical stimulation and voluntary exercise of the quadriceps femoris. *J Appl Physiol*, **104**, 75–81.

Kadi F., Charifi N., Denis C. *et al.* (2005) The behaviour of satellite cells in response to exercise: what have we learned from human studies? *Pflugers Arch*, **451**, 319–327.

Lavender A.P. and Nosaka K. (2006) Comparison of changes in markers of muscle damage between old and young men following voluntary eccentric exercise of the elbow flexors. *Appl Physiol Nutr Metab*, **31**, 218–225.

Lavender A. and Nosaka K. (2008a) A light load eccentric exercise confers protection against a subsequent bout of more demanding eccentric exercise performed 2 days later. *J Sci Med Sport*, **11**, 292–298.

Lavender A. and Nosaka K. (2008b) Changes in markers of muscle damage of middle age men following eccentric exercise of the elbow flexors. *J Sci Med Sport*, **11**, 124–131.

Lieber R.L. and Fridén J. (2002) Morphologic and mechanical basis of delayed-onset muscle soreness. *J Am Acad Orthop Surg*, **10**, 67–73.

Lindena J., Küpper W., Friedel R. and Trautschold I. (1979) Lymphatic transport of cellular enzymes from muscle into the intravascular compartment. *Enzyme*, **24**, 120–131.

McHugh M.P. (2003) Recent advances in the understanding of the repeated bout effect: the protective effect against muscle damage from a single bout of eccentric exercise. *Scand J Med Sci Sports*, **13**, 88–97.

Malm C. (2001) Exercise-induced muscle damage and inflammation: fact or fiction? *Acta Physiol Scand*, **171**, 233–239.

Miles M.P. and Clarkson P.M. (1994) Exercise-induced muscle pain, soreness, and cramps. *J Sports Med Phys Fitness*, **34**, 203–216.

Miyama M. and Nosaka K. (2004) Muscle damage and soreness following repeated bouts of consecutive drop jumps. *Adv Exerc Sports Physiol*, **10**, 63–69.

Murayama M., Nosaka K., Yoneda T. and Minamitani K. (2000) Changes in hardness of the human elbow flexor muscles after eccentric exercise. *Eur J Appl Physiol*, **82**, 361–367.

Newton M., Morgan G.T., Sacco P. *et al.* (2008) Comparison between resistance trained and untrained men for responses to a bout of strenuous eccentric exercise of the elbow flexors. *J Strength Cond Res*, **22**, 597–607.

Nosaka K. (2008) Chapter 5. Muscle soreness, muscle damage and repeated bout effect, in *Skeletal Muscle Damage and Repair* (ed. P.M. Tiidus), Human Kinetics, Champaign, IL, pp. 59–76.

Nosaka K. (2009) Chapter 18. Muscle damage and adaptation induced by lengthening contractions, in *Advances in Neuromuscular Physiology of Motor Skills and Muscle Fatigue* (ed. M. Shinohara), Research Signpost, Kerala, India, pp. 415–435.

Nosaka K. and Clarkson P.M. (1996) Changes in indicators of inflammation after eccentric exercise of the elbow flexors. *Med Sci Sports Exerc*, **28**, 953–961.

Nosaka K. and Newton M. (2002a) Repeated eccentric exercise bouts do not exacerbate muscle damage and repair. *J Strength Cond Res*, **16**, 117–122.

Nosaka K. and Newton M. (2002b) Is recovery from muscle damage retarded by a subsequent bout of eccentric exercise inducing larger decreases in force? *J Sci Med Sport*, **5**, 204–208.

Nosaka K. and Newton M. (2002c) Differences in the magnitude of muscle damage between maximal and submaximal eccentric loading. *J Strength Cond Res*, **16**, 202–208.

Nosaka K. and Newton M. (2002d) Concentric or eccentric training on eccentric exercise-induced muscle damage. *Med Sci Sports Exerc*, **34**, 63–69.

Nosaka K. and Sakamoto K. (2001) Effect of elbow joint angle on the magnitude of muscle damage to the elbow flexors. *Med Sci Sports Exerc*, **33**, 22–29.

Nosaka K., Sakamoto K., Newton M. and Sacco P. (2001a) How long does the protective effect on eccentric-induced muscle damage last? *Med Sci Sports Exerc*, **33**, 1490–1495.

Nosaka K., Sakamoto K., Newton M. and Sacco P. (2001b) The repeated bout effect of reduced-load eccentric exercise on elbow flexor muscle damage. *Eur J Appl Physiol*, **85**, 34–40.

Nosaka K., Newton M. and Sacco P. (2002a) DOMS does not reflect the magnitude of eccentric exercise-induced muscle damage. *Scand J Med Sci Sports*, **12**, 337–346.

Nosaka K., Sakamoto K. and Newton M. (2002b) Influence of arm-cranking on changes in plasma CK activity after high force eccentric exercise of the elbow flexors. *Adv Exerc Sports Physiol*, **8**, 45–50.

Nosaka K., Newton M.J. and Sacco P. (2005a) Attenuation of protective effect against eccentric exercise induced muscle damage. *Can J Appl Physiol*, **30**, 529–542.

Nosaka K., Newton M., Sacco P. *et al.* (2005b) Partial protection against muscle damage by eccentric actions at short muscle lengths. *Med Sci Sports Exerc*, **37**, 746–753.

Nosaka K., Newton M., Chapman D. and Sacco P. (2006) Is isometric strength loss immediately after eccentric exercise related to changes in indirect markers of muscle damage? *Appl Physiol Nutr Metab*, **31**, 313–319.

Nosaka K., Abbiss C., Watson G. *et al.* (2009) Recovery following an Ironman triathlon – a case study. *Eur J Sport Sci*, **10**, 159–165.

Nurenberg P., Giddings C.J., Stray-Gundersen J. *et al.* (1992) MR imaging-guided muscle biopsy for correlation of increased signal intensity with ultrastructural change and delayed-onset muscle soreness after exercise. *Radiology*, **184**, 865–869.

Paulsen G., Crameri R., Benestad H.B. *et al.* (2010) Time course of leukocyte accumulation in human muscle after eccentric exercise. *Med Sci Sports Exerc*, **42**, 75–85.

Philippou A., Bogdanis G.C., Nevill A.M. and Maridaki M. (2004) Changes in the angle-force curve of human elbow flexors following eccentric and isometric exercise. *Eur J Appl Physiol*, **93**, 237–244.

Prasartwuth O., Taylor J.L. and Gandevia S.C. (2005) Maximal force, voluntary activation and muscle soreness after eccentric damage to human elbow flexor muscles. *J Physiol*, **567** (Pt 1), 337–348.

Prasartwuth O., Allen T.J., Butler J.E. *et al.* (2006) Length-dependent changes in voluntary activation, maximum voluntary torque and twitch responses after eccentric damage in humans. *J Physiol*, **571** (Pt 1), 243–252.

Proske U. and Morgan D.L. (2001) Muscle damage from eccentric exercise: mechanism, mechanical signs, and adaptation and clinical applications. *J Physiol*, **537**, 333–345.

Quintero A.J., Wright V.J., Fu F.H. and Huard J. (2009) Stem cells for the treatment of skeletal muscle injury. *Clin Sports Med*, **28**, 1–11.

Raastad T., Owe S.G., Paulsen G. *et al.* (2010) Changes in calpain activity, muscle structure, and function after eccentric exercise. *Med Sci Sports Exerc*, **42**, 86–95.

Ratamess N.A., Alvar B.A., Evetoch T.K. *et al.* (2009) American College of Sports Medicine position stand. Progression models in resistance training for healthy adults. *Med Sci Sports Exerc*, **41**, 687–708.

Rodenburg J.B., Bär P.R. and De Boer R.W. (1993) Relationship between muscle soreness and biochemical and functional outcomes of eccentric exercise. *J Appl Physiol*, **74**, 2976–2983.

Sayers S.P. and Clarkson P.M. (2003) Short-term immobilization after eccentric exercise. Part II: creatine kinase and myoglobin. *Med Sci Sports Exerc*, **35**, 762–768.

Slater H., Thériault E., Ronningen B.O. *et al.* (2010) Exercise-induced mechanical hyperalgesia in musculotendinous tissue of the lateral elbow. *Man Ther*, **15**, 66–73.

Stauber W.T. and Smith C.A. (1998) Cellular responses in exertion-induced skeletal muscle injury. *Mol Cell Biochem*, **179**, 189–196.

Suzuki K., Peake J., Nosaka K. *et al.* (2006) Alterations in markers of muscle damage, inflammation, HSP70 and clinical biochemical variables after an Ironman triathlon race. *Eur J Appl Physiol*, **98**, 525–534.

Vikne H., Refsnes P.E., Ekmark M. *et al.* (2006) Muscular performance after concentric and eccentric exercise in trained men. *Med Sci Sports Exerc*, **38**, 1770–1781.

Warren G.L., Lowe D.A. and Armstrong R.B. (1999) Measurement tools used in the study of eccentric contraction-induced injury. *Sports Med*, **27**, 43–59.

Yu J.-G. and Thornell L.-E. (2002) Desmin and actin alterations in human muscles affected by delayed onset muscle soreness: a high resolution immunocytochemical study. *Histochem Cell Biol*, **118**, 171–179.

Yu J.-G., Malm C. and Thornell L.-E. (2002) Eccentric contractions leading to DOMS do not cause loss of desmin nor fibre necrosis in human muscle. *Histochem Cell Biol*, **118**, 29–34.

Zainuddin Z., Sacco P., Newton M. and Nosaka K. (2006) Light concentric exercise has a temporarily analgesic effect on DOMS but no effect on recovery from eccentric exercise. *Appl Physiol Nutr Metab*, **31**, 126–134.

2.7 Alternative Modalities of Strength and Conditioning: Electrical Stimulation and Vibration

Nicola A. Maffiuletti[1] and Marco Cardinale[2,3], [1]Schulthess Clinic, Neuromuscular Research Laboratory, Zurich, Switzerland, [2]British Olympic Medical Institute, Institute of Sport Exercise and Health, University College London, London, UK, [3]University of Aberdeen, School of Medical Sciences, Aberdeen, UK

2.7.1 INTRODUCTION

Recent advancements in technology and the need to develop innovative alternative solutions to exercise in the last few years have determined an increase in the use of various unconventional systems able to stimulate skeletal muscles in a similar way to conventional resistance exercise. In particular, electrical stimulation and vibration have been proposed as alternative forms of exercise for various populations, ranging from patients in rehabilitation settings to elite athletes trying to maximize performance. Unfortunately, while research activities are still trying to explain and understand how such modalities could be used in various populations, there is an enormous interest in the commercial aspect, which has determined marketing strategies based on overrated and scientifically incorrect claims. The aim of this chapter is to introduce the reader to such modalities and explain the scientific principles of their applications. We also aim to provide useful guidelines to help the reader in designing effective programmes employing electrical stimulation and vibration in various populations.

2.7.2 ELECTRICAL-STIMULATION EXERCISE

Electrical stimulation (ES), which is also called neuromuscular electrical stimulation, involves artificially activating the muscle, with a protocol designed to minimize the discomfort associated with the stimulation. ES has long been used both to supplement for voluntary muscle activation in many rehabilitation settings, for example for maintenance of muscle mass and function during prolonged periods of disuse or immobilization (Lake, 1992), particularly for the quadriceps muscle, and to improve the muscle strength of healthy muscles (Currier and Mann, 1983). More recently, ES has been implemented in competitive athletes (Delitto *et al.*, 1989).

Typical settings of ES exercise involve the application of intermittent electrical stimuli (tetani) through surface electrodes, positioned over the muscle motor point, and preprogrammed stimulation units (Figure 2.7.1). Thanks to recent advances in ES technology, portable battery-powered and relatively low-cost stimulators can be purchased and used by a growing number of individuals. Therefore, in order to optimize ES use and minimize the possible risks for the user, it is important to know (1) the main stimulus parameters, (2) how a contraction is triggered by ES, and (3) the acute and chronic effects of ES use on neuromuscular function.

After a quick overview of the main methodological and physiological aspects of ES, we will try to answer the following important questions about the use of ES in sports training:

1. Does ES training improve muscle strength?
2. Could ES training improve sport performance?

Finally, some recommendations and practical suggestions for optimizing ES use will be presented.

2.7.2.1 Methodological aspects: ES parameters and settings

Depending upon the objectives established at the beginning of the treatment, different ES protocols and parameters can be used. Besides electrode characteristics (material, size, placement), the main stimulus parameters for ES, which are dictated

Strength and Conditioning – Biological Principles and Practical Applications Marco Cardinale, Rob Newton, and Kazunori Nosaka.

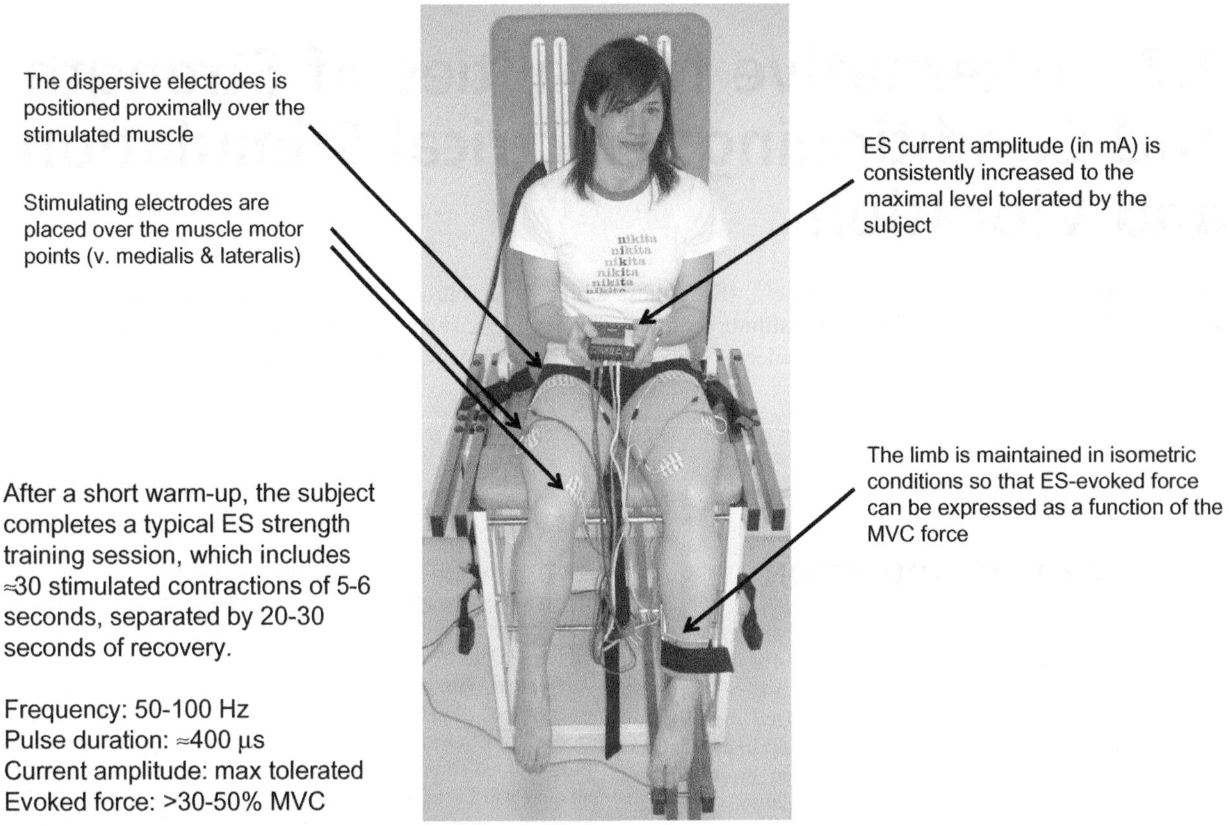
The dispersive electrodes is positioned proximally over the stimulated muscle
Stimulating electrodes are placed over the muscle motor points (v. medialis & lateralis)
ES current amplitude (in mA) is consistently increased to the maximal level tolerated by the subject
The limb is maintained in isometric conditions so that ES-evoked force can be expressed as a function of the MVC force
After a short warm-up, the subject completes a typical ES strength training session, which includes ≈30 stimulated contractions of 5-6 seconds, separated by 20-30 seconds of recovery.
Frequency: 50-100 Hz
Pulse duration: ≈400 μs
Current amplitude: max tolerated
Evoked force: >30-50% MVC

Figure 2.7.1 Typical settings of ES exercise for the quadriceps muscle

by the physiological characteristics of nerves and muscles, include:

1. frequency (number of pulses per second)
2. intensity, or current amplitude (which is probably the most important parameter)
3. pulse characteristics (shape and duration)
4. on/off cycle, or duty cycle (to minimize the occurrence of fatigue)
5. ramping (to reduce contraction abruptness and improve comfort).

There is no general consensus on the optimal stimulus parameters and stimulation conditions (isometric or not), but there is agreement on some current characteristics. For example, in order to maximize the level of evoked force (muscle tension), which is probably the key factor in ES effectiveness (Lieber and Kelly, 1991), ES strength training should be performed using pulse rates of 50–100 Hz (Vanderthommen and Duchateau, 2007) and the highest tolerated dose of current (Lake, 1992). Biphasic symmetrical rectangular pulses lasting 100–500 microseconds are commonly adopted. Short (4–6 second) current bursts are generally separated by rest periods of 20–30 seconds, and the stimulated muscle is maintained in isometric conditions to control the level of evoked force.

Even though the modulation of ES parameters may facilitate the effectiveness of this technique, practitioners agree that there is considerable subject variation in response to ES, and optimization may relate more to the characteristics of the subject than to the stimulus parameters themselves (Lloyd *et al.*, 1986). In the same way, Lieber and Kelly (1991) recently suggested that ES effectiveness would not depend on external controllable factors (such as electrode size or stimulation current) but rather on some intrinsic anatomical properties, such as individual motor nerve branching. We strongly support this notion, and concur with the idea that ES success is determined, at least in part, by uncontrollable factors.

2.7.2.2 Physiological aspects: motor unit recruitment and muscle fatigue

When skeletal muscles are artificially activated by ES, the involvement of motor units is different from that underlying voluntary activation. The main argument supporting such a difference is that large-diameter axons are more easily excited by electrical stimuli, which will alter the activation order during ES compared with voluntary contractions (Enoka, 2002). However, human experiments have yielded contradictory findings, with some studies suggesting preferential or selective activation of fast motor units with ES, and others demonstrating minimal or no difference between the two contraction modalities (for an overview of these studies see Gregory and Bickel, 2005). In a recent review paper, Gregory and Bickel (2005) suggested that motor unit recruitment during ES is non-selective or random (see also Jubeau *et al.*, 2007); that is, muscle fibres are recruited without obvious sequencing related to fibre types ('disorderly' recruitment), and therefore ES can be used to activate fast motor units (in addition to the slow ones) at relatively low force levels. The main differences in motor unit recruitment between voluntary and stimulated contractions are summarized in Table 2.7.1.

Table 2.7.1 Motor unit recruitment during voluntary and ES contractions

	Voluntary contraction	ES contraction
Temporal	Asynchronous	Synchronous
Spatial	Dispersed	Superficial (close to the electrodes)
	Rotation is possible	Spatially fixed
	Quasi-complete (even at the maximum)	Largely incomplete (even at the maximum)
Orderly	Yes, selective (slow to fast)	No, nonselective/ random (slow and fast)
Consequence	Partially fatiguing	Extremely fatiguing

The main consequence of such a unique motor unit recruitment pattern for ES is the exaggerated metabolic cost of an electrically evoked contraction (Vanderthommen *et al.*, 2003), which, compared to a voluntary action of the same intensity, provokes greater and earlier muscle fatigue (Theurel *et al.*, 2007). According to Vanderthommen and Duchateau (2007), these differences, in motor unit recruitment and thus in metabolic demand between evoked and voluntary contractions, constitute an argument in favour of the combination of these two modalities of activation in the context of sports training.

2.7.2.3 Does ES training improve muscle strength?

There is no doubt that it is possible to improve muscle strength by means of ES training programmes. However, it should be remembered that ES training-induced strength gains for healthy muscles are not greater than those that can be achieved with traditional voluntary training. In a recent systematic review of ES studies, Bax, Staes and Verhagen (2005) concluded that for unimpaired quadriceps the effectiveness of ES training is generally lower than with voluntary modalities, while for impaired (completely or partially immobilized) quadriceps, ES training could be more effective than voluntary training. This has important implications for the use of ES in the context of post-injury and/or postoperative rehabilitation. Training studies performed in the last 20 years have also demonstrated that it is possible to obtain significant improvements in muscle strength – particularly for the lower-extremity muscles – in amateur and competitive athletes of all levels (Table 2.7.2).

ES training-induced increases in muscle strength are largely mediated by neural adaptations, for example increased muscle activation (e.g. Maffiuletti *et al.*, 2002b), particularly in the case of short-term (3–4 weeks) training programmes. In contrast, only ES regimens of longer duration (6–8 weeks) can elicit

Table 2.7.2 Applications of ES strength training in competitive sport

Study (year)	Sport	Muscle	Weeks (x/wk)	Type of ES: settings; frequency	Main findings
Delitto *et al.* (1989)	Weightlifting	Q	6 (3)	I-LE; 2500 Hz	↑ weightlifting
Wolf *et al.* (1989)	Tennis	Q	3 (4)	C-S; 75 Hz	↑ strength, sprint, jump
Pichon *et al.* (1995)	Swimming	LD	3 (3)	I-OC; 80 Hz	↑ strength, swimming
Willoughby and Simpson (1996)	Basketball	BB	6 (3)	I-PC; 2500 Hz	↑ strength
Willoughby and Simpson (1998)	Track & field	Q	6 (3)	C/E-LE; 2500 Hz	↑ strength, jump
Maffiuletti *et al.* (2000)	Basketball	Q	4 (3)	I-LE; 100 Hz	↑ strength, jump
Malatesta *et al.* (2003)	Volleyball	Q+CA	4 (3)	I-S; 105–120 Hz	↑ strength, jump
Maffiuletti *et al.* (2002a)	Volleyball	Q+CA	4 (3)	I-LE/SC; 120 Hz	↑ strength, jump
Brocherie *et al.* (2005)	Ice hockey	Q	3 (3)	I-LE; 85 Hz	↑ strength, sprint
Babault *et al.* (2007)	Rugby	Q+CA+G	6 (1–3)	I-LE/CM; 100 Hz	↑ strength, jump
Maffiuletti *et al.* (2009)	Tennis	Q	3 (3)	I-LE; 85 Hz	↑ strength, sprint, jump

Q: quadriceps; LD: latissimus dorsi; BB: biceps brachii; CA: calf; G: gluteus; I: isometric; LE: leg extension; C: concentric; S: squat; OC: open chain; PC: preacher curl; E: eccentric; SC: standing calf; CM: calf machine; ↑: significant improvement.

morphological changes in the muscle (Gondin *et al.*, 2005). In their well-designed study, Gondin *et al.* (2005) demonstrated the time course of neuromuscular adaptations to ES strength training. After four weeks of training, strength increases were accompanied by increased muscle activation, while muscle cross-sectional area was not significantly modified. Interestingly, both neural and muscular adaptations mediated the strength improvements observed after eight weeks of ES, which is similar to the classical model proposed by Sale (1988) for neuromuscular adaptations to voluntary strength training.

2.7.2.4 Could ES training improve sport performance?

Several studies with individual (e.g. swimming, tennis, track-and-field, weightlifting) and team-sport (e.g. basketball, volleyball, ice hockey, rugby) athletes have reported a significant improvement in maximal strength, and in some cases even in anaerobic power production (vertical jump and sprint ability), after ES training programmes lasting three to six weeks (Table 2.7.2). Despite the lack of scientific evidence, these improvements are likely to affect field performance. However, since the stress is applied during nonspecific contractions (i.e. isometric in general), an excessive use of ES could impair motor coordination. Therefore, performance of complex movements requiring high levels of coordination can only be achieved if ES is used in conjunction with voluntary 'technical' exercise, such as plyometrics (Maffiuletti *et al.*, 2002a). If ES is adequately combined with sport-specific training and logically integrated into the yearly training season, improvements could be achieved in the following capabilities:

- jumping ability (both general and specific jumps)
- sprinting ability (including shuttle sprints)
- other sport-related performances (swimming, weightlifting).

As a practical recommendation for both individual and team sports, it is suggested that ES training could be used to enhance muscle strength and anaerobic performance without interfering excessively with sport-specific training; however, it would be best used early in the training season (i.e. at the beginning of the preparatory training season).

The main interest of using ES in high-level sport is that this modality represents a new form of stress from a neuromuscular, metabolic, and also psychological point of view, and therefore it could be viewed as a new stimulus to favour plasticity. ES could be particularly useful for athletes whose performance has plateaued after several years of training and competition, but it would be supplementary to, rather than a substitute for, more traditional forms of training. Another interest of ES for elite sportsmen is that a single ES bout is usually less time-consuming (12–15 minutes) than traditional volitional exercise sessions, and this is extremely appealing for athletes who have a limited amount of time for conditioning (e.g. tennis players).

2.7.2.5 Practical suggestions for ES use

Who should administer ES?

For the first one or two training sessions, ES should be administered by a physical therapist or physical trainer who is familiar with the methodological and physiological aspects of ES exercise presented here.

How to use ES within a single session

Each ES exercise session should include the following phases:

1. warm-up with and without ES
2. progressive increase in current amplitude to the maximal tolerated level (5–10 contractions)
3. first series of 10–15 contractions
4. 3–5 minutes of recovery
5. second series of 10–15 contractions
6. cool-down.

Recommendations:

1. Current amplitude should be consistently increased during the session.
2. Electrode position and joint angle should be modified between the two series.
3. Whenever possible, both current amplitude (in mA) and evoked force (as a percentage of the MVC force) should be controlled to allow ES intensity to be carefully quantified (Figure 2.7.1).

How to use ES in the context of a training programme

ES strength-training programmes should be designed as follows:

1. total duration: 2–8 weeks, depending on the objective
2. frequency: 2 then 3 sessions per week
3. volume: 2 then 3 series of 10–15 contractions
4. intensity: current amplitude (and thus evoked force) should be increased from session to session.

Recommendations:

1. It is desirable to attain an evoked force level of approximately 50% of the MVC force by the end of the first week of training.
2. Muscle palpation is recommended on a daily basis after the first few training sessions in order to estimate the degree of muscle soreness and accordingly adjust training volume/intensity.

Caution for acute and chronic use

Athletes are generally reluctant to use this technique as a supplement to training because of the discomfort associated with artificial activation. Additionally, post-exercise muscle soreness and damage produced by the electrical current (Jubeau *et al.*, 2008), particularly in the area beneath the stimulating electrodes, are associated with a high risk of peripheral overreaching/overtraining (Maffiuletti *et al.*, 2006; Zory, Jubeau and Maffiuletti, 2010). Despite large inter-individual differences in the acute and chronic response to ES exercise, these limitations should be made clear to the end user.

2.7.3 VIBRATION EXERCISE

Before discussing the possibility of implementing vibration in a training programme it is important to define what characterizes vibration and how it is possible to control its parameters in training prescription.

Vibration is a mechanical stimulus characterized by an oscillatory motion. The biomechanical parameters determining its intensity are the frequency and the extent of the oscillation. The extent of the oscillatory motion can be given as the amplitude (maximum displacement from equilibrium) or peak-to-peak displacement (p-to-p D) (the displacement from the lowest to the highest point). The repetition rate of the cycles of oscillation determines the frequency (F) of the vibration (measured in Hertz).

Consequently the magnitude of the vibration stimulus depends on the amplitude of the oscillations, the frequency, and the resulting acceleration transmitted to the body, or to parts of it (see Figures 2.7.2 and 2.7.3). A large number of scientific studies have been conducted with the aim of investigating the negative effects of vibrating tools and whole-body vibration (WBV) on workers exposed to such stimuli for long hours during their working tasks (Bovenzi, 2002; Cherniack *et al.*, 2008; Griffin, 2004; Mirbod *et al.*, 1999; Pope *et al.*, 2002; Sanya and Ogwumike, 2005). The possibility of using vibration as a form of exercise became a novel idea in the 1980s and 1990s thanks to the work of Nazarov (Nazarov and Spivak, 1985) and Issurin (Issurin and Tenenbaum, 1999; Issurin *et al.*, 1994), who suggested the use of vibrating pulley machines during strength and flexibility training, and that vibrating devices be applied directly to the body while the athlete was performing strengthening exercises. Later on, the use of vibrating plates to allow individuals to perform WBV exercises was introduced as an effective training modality in elite athletes (Bosco *et al.*, 1998). Previous work has suggested that vibration exposure using this and similar modalities elicits small but rapid changes in muscle length, producing reflex muscle activity in an attempt to dampen the vibratory waves (Cardinale and Wakeling, 2005). This reflex muscle activation is likely to be similar to the tonic vibration reflex (TVR) (Hagbarth *et al.*, 1976); however, there is currently no firm evidence to suggest that the neuromuscular responses observed with WBV exercise are in fact the expression of TVRs. Because muscle spindle primary endings are most responsive to vibration and thus responsible for the TVR, most studies have focused on understanding neuromuscular responses to WBV exercise.

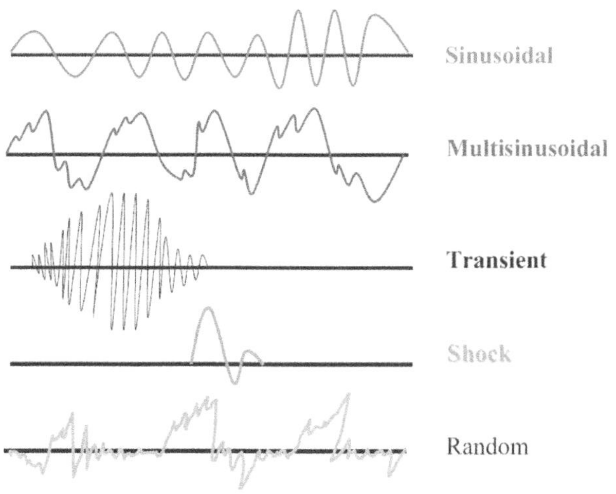

Figure 2.7.2 Types of vibration

Due to the potential for vibration to timulate the neuromuscular system both acutely and chronically, our group introduced the concept of WBV, hypothesizing the possibility of improving force and power production of the lower limbs by such an approach (Bosco *et al.*, 1998, 1999a, 1999b; Cardinale and Lim, 2003). Since then, numerous studies have followed, all of which have tried to analyse and understand how vibration could be used as a training intervention not only in elite athletes but as an alternative exercise modality in populations including disabled children (Ward *et al.*, 2004), aged individuals (Bautmans *et al.*, 2005), and adults with cerebral palsy (Ahlborg *et al.*, 2006). In recent years, different technologies have been developed to provide vibration-exercise devices similar to typical gym equipment and/or implementing mechanical and acoustic vibration modalities (Dessy *et al.*, 2008; Mischi and Kaashoek, 2007; Poston *et al.*, 2007).

There is no current consensus on the effectiveness of vibration exercise in improving force- and power-generating capacity in humans. Some authors are suggesting that there is no evidence for WBV (Nordlund and Thorstensson, 2007) improving strength and power in the lower limbs; others (in a systematic review of the literature) seem to suggest strong to moderate evidence for the positive effects of WBV exercise in untrained people and elderly women (Rehn *et al.*, 2007). A few more recent reviews of the literature support the suggestion that vibration exercise could be an effective modality in improving strength and power of almost all limbs in various populations (Cardinale and Rittweger, 2006; Cardinale and Wakeling, 2005; Issurin, 2005; Pavy-Le Traon *et al.*, 2007; Wilcock *et al.*, 2009).

Despite the clear need for further research, the aggressive and unscrupulous marketing of many companies has determined the unfortunate appearance of devices which are sold without proper and safe advice for the user and without any

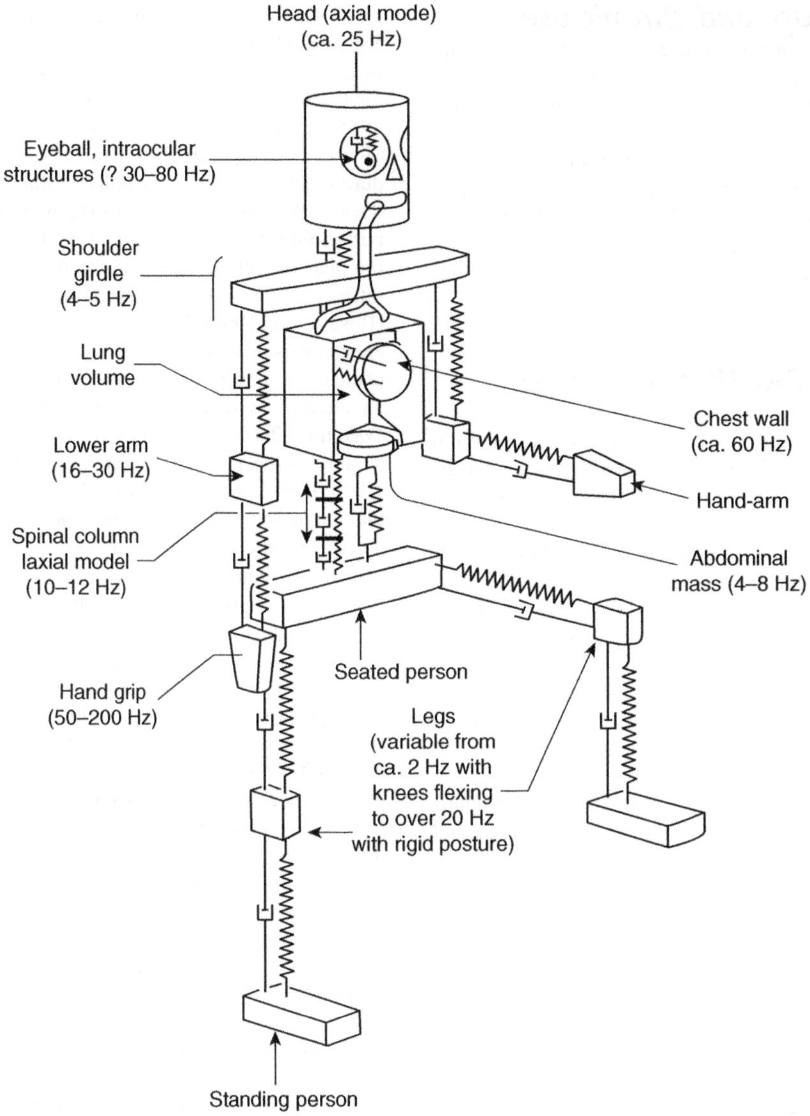

Figure 2.7.3 Biodynamic responses to vibration. From Rasmussen (1982)

scientific validation of their safety and effectiveness. This of course suggests the need for tighter regulatory control over the production and sale of such devices, which in extreme cases could result in harmful effects to users. The fitness market is fundamentally inundated by WBV devices of two types: vibrating platforms which provide a vertical oscillation of the whole plate, and vibrating platforms with a seesaw movement side-to-side of the platform (see Figure 2.7.4). Other vibration exercise devices include vibrating dumbbells (see Figure 2.7.5), gym devices capable of transmitting vibration to the user (see Figure 2.7.6), or other vibration exercise tools.

Most of the studies investigating the effectiveness of vibration training have been conducted on non-competitive athletes, sports science students, and/or injured and aged individuals and special populations. Furthermore, they have largely been conducted using only a few devices, while the market is full of hundreds that have never been investigated in any well-controlled study.

It is very important for the reader to understand that the effectiveness of vibration exercise is not correlated to the cost of the device being used, but rather to the correct use of the effective combination of the key vibration parameters: frequency and amplitude, in combination with appropriate levels of muscle tension.

Needless to say, devices which provide consistent vibration parameters and allow such parameters to be unaffected by

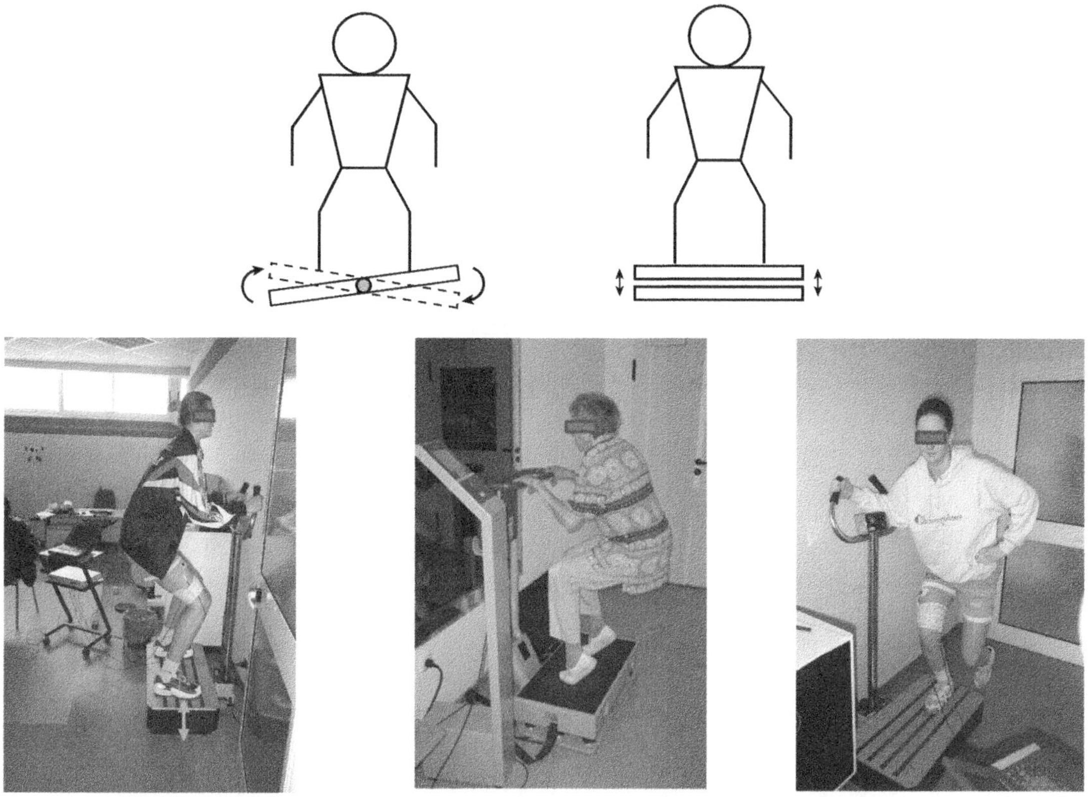

Figure 2.7.4 WBV exercise devices

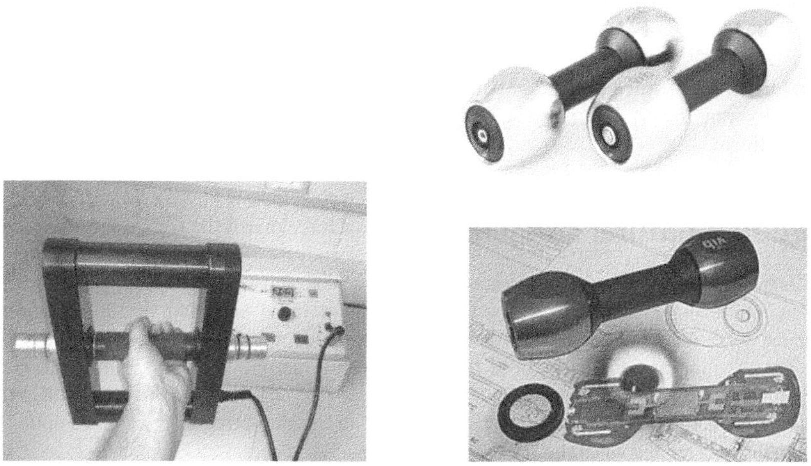

Figure 2.7.5 Vibrating dumbbells

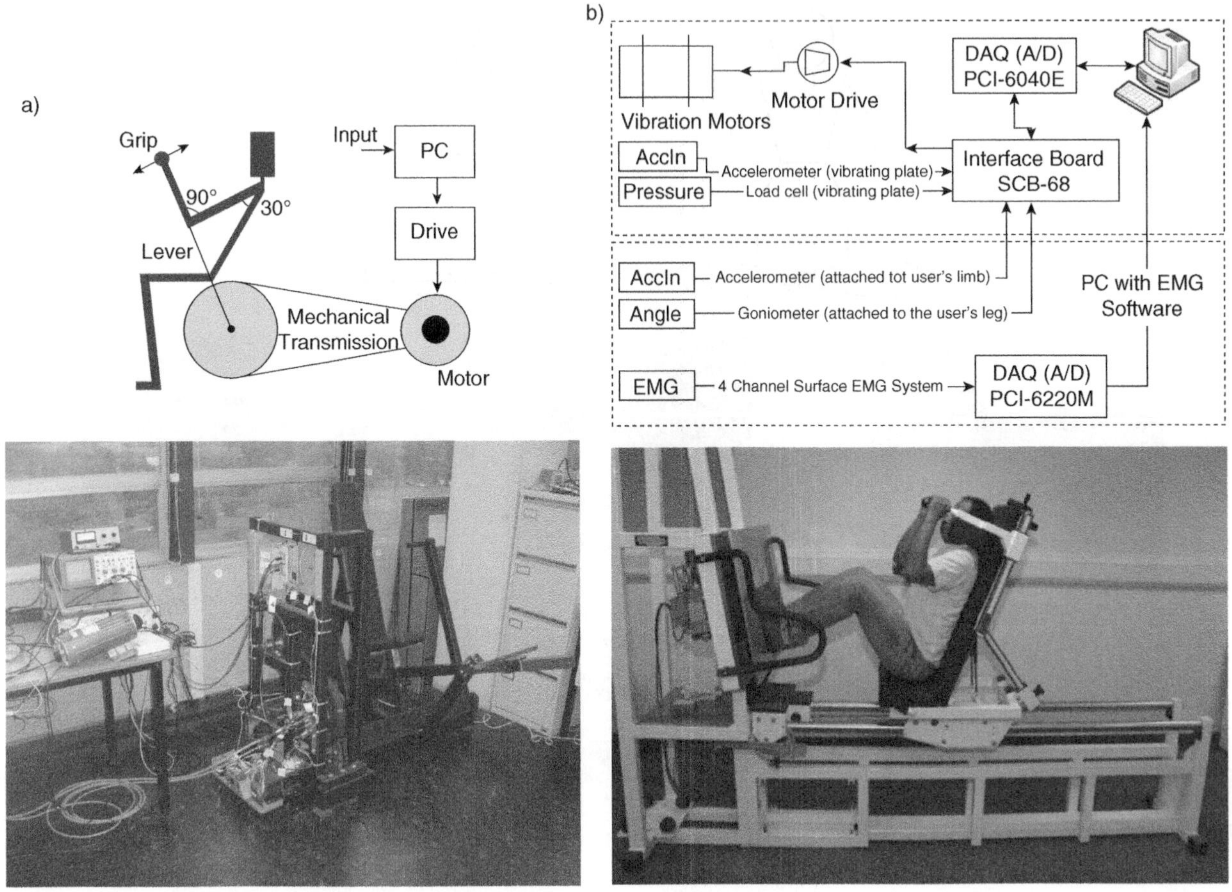

Figure 2.7.6 Vibrating gym devices for upper- and lower-body training. (a) Arm flexion/extension device used in Mischi and Cardinale (2009). (b) Leg press used in Pujari, Neilson and Cardinale (2009) [US Patent No. US7416518 (B2), EU Patent No. EP1524957 (B1)]

the user's body mass and the level of force applied is a must, particular when athletes/users want to perform loaded squatting exercises on a vibrating platform.

2.7.3.1 Is vibration a natural stimulus?

During all sporting and daily activities our bodies interact with the external environment and experience externally applied forces. These forces induce vibrations and oscillations within the tissues of the body. Tissue vibrations can be induced from impacts where either a part of the body or an item of sporting equipment in contact with the body collides with an object. Impact shocks experienced through the leg when the heel strikes the ground during each running stride are the most typical example. Another common one in sport is the impact shock that occurs when a racquet is used to hit a ball (Elliott *et al.*, 1980; Hatze, 1992; Li *et al.*, 2004; Timme and Morrison, 2009).

The initial impact causes vibrations within the soft tissues; after the impact has occurred, the tissues will continue to oscillate as a free vibration; that is, vibrating at their natural frequency, with the amplitude of these vibrations decaying due to damping within the tissues. Other examples of tissue vibrations are those experienced through the legs when skiing or through the arms when riding on a bicycle. A continuously oscillating input force drives the soft-tissue vibrations to occur at the same frequency as the input force, and allows resonance to occur; however, active damping from muscles can reduce the amplitude of these larger-amplitude vibrations. Therefore, we can state that we experience soft-tissue vibrations in almost all sporting activities and the amplitude and frequency of these vibrations and the ability to cope with them is partly determined by the natural frequency and damping characteristics of the tissues.

The body relies on a range of structures and mechanisms in order to regulate the transmission of impact shocks and vibrations through the body, including bone, cartilage, synovial fluids, soft tissues, joint kinematics, and muscular activity. The body is capable of adapting the response to external vibrations by producing changes in joint kinematics and muscle activity which can be controlled on a short time scale. A few recent studies have suggested that the body has a strategy of 'tuning' its muscle activity in order to reduce its soft-tissue vibrations in an attempt to reduce such deleterious effects (Nigg and Wakeling, 2001). The muscle-tuning hypothesis predicts that the level of muscle activity used for a particular movement task is, to some degree, dependent on the interaction between the body and the externally applied vibration forces.

A maximally activated muscle can damp free vibrations so that the tissue oscillations are eliminated after a couple of cycles (Boyer and Nigg, 2007; Wakeling and Liphardt, 2006). The fact that muscles are actively engaged in damping vibratory stimuli suggests that exercising with vibratory stimulation could provide an effective alternative to resistance exercise and improve strength and power capabilities not only in athletic populations, but in groups which cannot otherwise perform strength training. Recent work from our laboratory has shown that neuromuscular activity during vibration affects agonist and antagonist muscles involved in the task on which vibration is superimposed, suggesting that the real benefits of vibration stimulation might appear only when superimposed on high levels of muscle tension (Mischi and Cardinale, 2009).

2.7.3.2 Vibration training is not just about vibrating platforms

Vibration was used in the 1980s mainly by Russian scientists as an alternative way of improving strength and flexibility in gymnasts (Nazarov and Spivak, 1985). Recent work from Mischi (Mischi and Kaashoek, 2007) has shown greater gains in the use of vibration through combination of vibrating devices with typical gym strength-training equipment. Applying vibration to barbells has been shown to increase peak and average power output during bench-pressing exercises (Poston *et al.*, 2007), while gymnasts have been shown to improve flexibility using a novel vibrating tool which allows them to perform flexibility training with vibration, suggesting another means of using vibration as a training stimulus (see Figure 4.4.2) (Sands *et al.*, 2006). Direct application of vibration to muscles and tendons has also been suggested to produce acute and chronic effects in neuromuscular performance and will be discussed later in this chapter.

2.7.3.3 Acute effects of vibration in athletes

The acute effects of vibration exercise have been studied mainly with the aim of identifying the most appropriate combination of frequency and amplitude to affect strength and power performance. Furthermore, considering the observed effects on muscle perfusion (Kerschan-Schindl *et al.*, 2001), the interest in understanding the acute responses to such stimuli is increased with respect to using vibration as some form of training aid/warm-up/preconditioning procedure. Few studies have been conducted using athletes competing in national- and/or international-level competitions. Bosco *et al.* (1999b) showed an acute shift to the right of the force–velocity and power–velocity relationship in the vibrated leg of elite female volleyball players after 10 sets of 1 minute of static squatting on one leg performed on a tilting device (F = 26 Hz, p-t-p D = 10 mm), suggesting the possibility of such modality producing a potentiation effect. Elite female field hockey players were also shown to increase vertical jumping ability and hamstrings flexibility after five minutes of WBV (squat) performed on a tilting device (Novotec, Pforzheim, Germany; F = 26 Hz, p-to-p D = 6 mm). We have previously suggested that WBV exercise increased electromyographic (EMG) activity of the leg extensor muscles in elite female volleyball players by 34% as compared to just squatting in no-vibration condition (Cardinale and Lim, 2003), with the peak EMG reached at 30 Hz. For this reason, and due to the resemblance of such response to the tonic vibration reflex, we have suggested a 'muscle tuning' response aimed at controlling active damping in the soft tissues as one of the mechanisms leading to the observed performance enhancements (Cardinale and Wakeling, 2005). This increase in EMG activity has also been observed by other authors in various populations (Abercromby *et al.*, 2007; Delecluse *et al.*, 2003; Hazell *et al.*, 2007; Roelants *et al.*, 2006).

These data need to be also interpreted with caution, as vibration noise can contribute to higher EMG levels and appropriate filtering techniques will need to be used in future studies to understand the true contribution of vibration to muscle activation (Abercromby *et al.*, 2007; Fratini *et al.*, 2008; Mischi and Cardinale, 2009). Short-duration dynamic squatting (30 seconds) performed on a vibrating platform (F = 35 Hz, p-to-p D = 4 mm) has been suggested to allow college athletes to produce higher power output during a subsequent squat exercise with 75% of 1 RM when compared to just resting in between sets (Rhea and Kenn, 2009). Acute improvements in the ability to produce force during a plantar flexion task lasting up to eight minutes have been observed following WBV (F = 30 Hz, p-to-p D = 3.5 mm) but could not be related to improvement in motoneural excitability (McBride *et al.*, 2009). Similar results have been obtained in other studies (Bullock *et al.*, 2008; McBride *et al.*, 2009). So far, there seems to be a consensus that some forms of vibration stimulation superimposed on various levels of muscle activation might increase neuromuscular demands, suggesting the potential of such intervention to determine acute and chronic effects on force and power production.

Despite the accepted evidence of higher neuromuscular activation during WBV, the local metabolic demand in leg extensors while performing static squats on a whole-plate oscillating device (F = 30 Hz, p-to-p D = 4 mm) (Cardinale *et al.*, 2007) as compared to squatting without vibration does not seem to be higher; if anything it is actually less after the first 30–40 seconds. However, it seems obvious to suggest that a combination of increases in blood flow, and consequent muscle perfusion (Kerschan-Schindl *et al.*, 2001; Lohman *et al.*, 2007; Yamada *et al.*, 2005), might acutely favour power production due to a consequent increase in muscle temperature (Cochrane *et al.*, 2008b).

The aforementioned studies show that there is a paucity of data on elite/well-trained athletes and also that is it very difficult to recommend specific protocols. However, the following recommendations can be made when using WBV in warm-up or as a potentiation routine:

1. Frequencies ranging from 20 to 50 Hz seem to be effective.
2. Oscillations ranging from 3 to 10 mm (peak-to-peak D) seem to be effective.
3. Durations should not be longer than five minutes per set as they might cause fatigue rather than potentiation.
4. It is preferable if WBV is combined with loaded static and dynamic exercise to produce the best effects.
5. In order to personalize protocols to each athlete it is important to conduct some simple measurements in order to identify the dose–response relationship to various protocols (Adams *et al.*, 2009).

Vibrating dumbbells have also shown some promise in increasing power output acutely due to a large increase in EMG activity in elite boxers (Bosco *et al.*, 1999a; Issurin and Tenenbaum, 1999). Recent work has suggested that using a vibrating dumbbell requires the recruitment of higher-threshold motor units during an arm flexion task (McBride *et al.*, 2004). However, no acute gains were observed in climbers using such a modality of training (Cochrane and Hawke, 2007) and the potentiation effects were similar to arm-cranking exercise (Cochrane *et al.*, 2007).

Finally, performing stretching on vibrating devices has been shown to increase flexibility in gymnasts more than conventional stretching (Sands *et al.*, 2006), and to increase flexibility without affecting explosive abilities when combined with stretching (Kinser *et al.*, 2008).

To date, due to the paucity of data of using the dumbbells and flexibility protocols, it is almost impossible to advise specific protocols. Definitively more studies are needed in this field, though it is possible to suggest that relatively low frequency ranges (20–40 Hz) and small amplitudes (<5 mm) should be used to produce maximum gains.

Further applications could see WBV as a recovery modality (Edge *et al.*, 2009), but more work is needed in this area before specific guidelines can be provided.

2.7.3.4 Acute effects of vibration exercise in non-athletes

A wide range of physiological responses to WBV exercise have been studied so far in non-athletic populations. Aged individuals seem to be able to cope well with five minutes of WBV exercise and have also been shown to receive some benefits in the form of acute increases in circulating levels of IGF-1 (Cardinale *et al.*, 2008). The hormonal responses observed in an aged population have been similar to those previously observed in young healthy individuals (Bosco *et al.*, 2000). However, recent work does seem to indicate that WBV per se does not represent a demanding training stimulus able to challenge homoeostasis as measured by hormonal levels in saliva and/or blood (Cardinale *et al.*, 2006; Di Loreto *et al.*, 2004; Erskine *et al.*, 2007). A high intensity of neuromuscular activity has been suggested as the necessary pre-requisite to determine a marked alteration in hormonal levels, and when WBV is combined with conventional resistance exercise it seems that such an intensity could be high enough to trigger training adaptations modulated by hormonal responses (Kvorning *et al.*, 2006). Various authors have suggested the effectiveness of WBV exercise in improving force and power production in various populations with similar protocols (Abercromby *et al.*, 2007; Cochrane *et al.*, 2008a; Jacobs and Burns, 2009; Ronnestad, 2009).

Vibrating devices directly applied to the skin of the biceps brachii (F = 65 Hz, p-to-p D = 1.2 mm) do not seem to provide any acute benefit to power output (Moran *et al.*, 2007) in an arm flexion task. And vibration directly applied to the rectus femoris for prolonged periods (Jackson and Turner, 2003) can markedly reduce force-generating capacity not only in the limb where vibration is applied but also in the contralateral limb. Squatting on a vibrating platform (10 sets × 60 seconds at 26 Hz) seems to be able to acutely reduce arterial stiffness in healthy men (Otsuki *et al.*, 2008) and affects blood flow, further suggesting the acute use of vibration exercise as a warm-up preparation-to-exercise tool in various populations.

2.7.3.5 Chronic programmes of vibration training in athletes

Regular programmes of vibration training have been studied in a wide range of subjects, suggesting a lot of benefits. When we analyse the effectiveness of such programmes in athletes, the evidence does not seem to be very strong. A five-week training period characterized by progressive WBV training of frequencies of 35–40 Hz and amplitudes of 1.7–2.5 mm produced no significant improvements in vertical jumping ability, isometric and dynamic muscle strength, or maximal knee extension velocity in well-trained sprinters (Delecluse *et al.*, 2005). Young elite skiers were shown to benefit from a resistance exercise programme supplemented by WBV (F = 24–28 Hz, amplitude 2–4 mm), showing greater improvements in vertical jumping ability and isokinetic knee and ankle measures than the resistance exercise-only group (Mahieu *et al.*, 2006). Eight weeks of WBV training (5 × 40 seconds, with 1 minute rest; 30 Hz, 5 g magnitude) produced a significant improvement in vertical jumping ability and knee-extensor strength in ballerinas (Annino *et al.*, 2007).

In our opinion, it is unlikely that vibration exercise alone using the currently available technologies (vibrating platforms with frequencies ranging from 10 to 60 Hz and amplitudes ranging from <1 to 10 mm (peak-to-peak D)) can benefit athletes. The benefit of such technology would occur only if resistance exercise could be performed in combination with vibration stimulation and/or vibration applied to pre-existing high levels of muscle activation. Preliminary studies in our laboratory (Pujari, Neilson and Cardinale, 2009) on a novel patented exercise apparatus (US Patent No. US7416518 (B2) EU Patent No. EP1524957 (B1), see Figure 2.7.6) characterized by a vibrating leg press device strongly suggest the superimposition of vibration to specific levels of muscle tension and muscle actions to obtain maximal gains.

2.7.3.6 Chronic programmes of vibration training in non-athletes

Various studies have been conducted on the effects of short-, medium- and long-term programmes of vibration training on non-athletic populations. Delecluse *et al.* (2003) have initially suggested that WBV was superior to resistance exercise in improving strength of the lower limbs in healthy young women. It should be noted that the resistance-exercise protocol used by the resistance-exercise group in this experiment was of relatively low intensity. Roelants *et al.* (2004a) showed that WBV produced not only gains in strength but also increases in fat-free mass in young untrained women, with the effects being similar in magnitude to those produced by conventional fitness regimes characterized by a mixture of cardiovascular and strength-training activities. Four months of WBV training improved vertical jump in young men and women but was unable to improve balance in the same age group (Torvinen *et al.*, 2002). Eight months with a similar regime improved vertical jumping activity but did not have an effect on bone density in a young mixed age group (Torvinen *et al.*, 2003). Young healthy subjects seem to be able to improve strength and power performance if vibration is combined with resistance exercise (Kvorning *et al.*, 2006), but such improvements are not larger than that observed with an appropriate conventional resistance-exercise programme. Chronic low back pain (Rittweger *et al.*, 2002) and proprioception of the spine (Fontana *et al.*, 2005) seem to benefit from WBV exercise in untrained populations.

The elderly seem to benefit most from WBV exercise programmes. Numerous studies have in fact suggested that WBV exercise programmes can be effective in improving strength and power of the lower limbs in post-menopausal women (Roelants *et al.*, 2004b; Verschueren *et al.*, 2004). WBV exercise has also been shown to be effective in preventing osteoporosis in aged populations and in young populations with low bone density (Gilsanz *et al.*, 2006; Gusi *et al.*, 2006; Iwamoto *et al.*, 2005; Lindqvist, 2003), suggesting such modality as an effective alternative not only to conventional forms of exercise but potentially also to pharmacological agents for the prevention of osteoporosis.

2.7.3.7 Applications in rehabilitation

Vibration exercise has the potential of being an effective alternative and companion to conventional exercise modalities due to the potential to exercise bed-ridden patients. Recent studies conducted on bed-rest models have shown the beneficial effects of resistive exercise combined with vibration in preserving muscle form and function (Bleeker *et al.*, 2005; Blottner *et al.*, 2006; Mulder *et al.*, 2006, 2007, 2008; Rittweger *et al.*, 2006). These encouraging results strongly propose the efficacy of using such a modality in bed-ridden patients and patients for whom there is the need to preserve and/or restore muscle form and function. More studies are needed to verify the efficacy of such a modality on athletes recovering from various injuries. However, recent results are encouraging. Proprioception and balance around the knee and ankle joint seem to be improved following WBV exercise programmes in injured individuals (Moezy *et al.*, 2008; Trans *et al.*, 2009). Patients with Parkinson's disease benefit from WBV exercise, even if the effects are not superior to conventional physiotherapy and improved balance (Ebersbach *et al.*, 2008). Adults suffering from cerebral palsy seem to benefit from WBV exercise (Ahlborg *et al.*, 2006), with improved ability to produce strength and reduced spasticity and mobility (Semler *et al.*, 2007). Children and adolescents with osteogenesis imperfecta also benefit from WBV exercise by improving muscle strength and mobility (Semler *et al.*, 2008). Spinal-cord-injury patients gain benefit from using vibrating dumbbells by improving average power during arm-flexion tasks (Melchiorri *et al.*, 2007). These results are very encouraging as vibration could definitively help special populations due to the simplicity of use of such technology and the cost-effectiveness of performing training sessions at home, without the need for bulky exercise machines. It is important to state that combinations of various modalities (ES, vibration, and conventional exercise) should be sought in all cases to make sure maximum gains are obtained with the least effort in various populations.

2.7.3.8 Safety considerations

Performing resistance exercise on the currently marketed WBV plates can be a risky business. First, the surface limits the stance of the athlete (if squatting is the exercise of choice). Second, and most importantly, because of the poor manufacturing and the cheap choice of components, many vibrating plates cannot sustain heavy loads. In our experience, many commercially available plates vary frequencies, amplitudes (and hence magnitudes), and directions of vibration in a stochastic manner when someone tries to perform strengthening exercises on them.

When using WBV devices it is important to recognize that body posture and the type of vibrating plate used (Abercromby *et al.*, 2007) affect transmissibility of vibration to the spine and the head when exercises are performed standing on the plate. Lying and/or sitting on the plate should be discouraged as transmission to the head is quite high and would put the athlete at risk of spinal degeneration, visual and vestibular damage, hearing loss, and other health risks (Griffin, 2004). Handheld vibrating exercise devices and vibrating dumbbells, cables, and

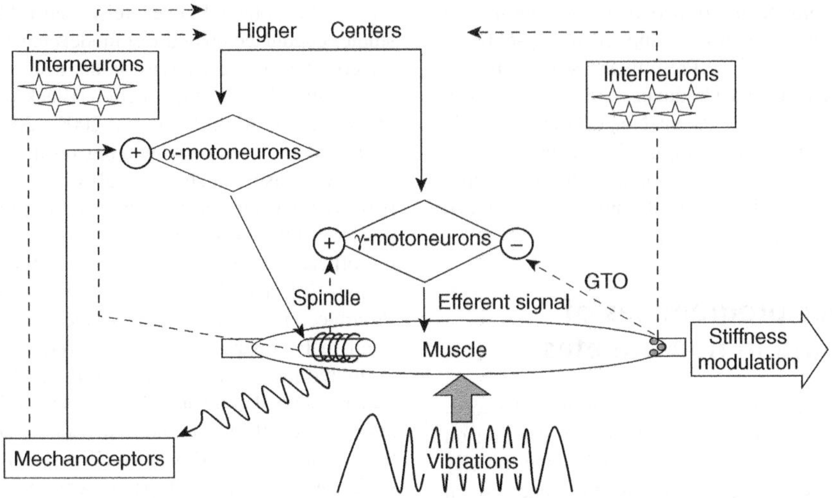

Figure 2.7.7 The effects of vibration stimulation on humans. From Cardinale and Bosco 2003

barbells should be used with caution as prolonged use of similar tools has been linked to degeneration of soft tissues, reduction in vascularization, and bone and articular damage (Griffin, 2004).

There is a need for more research to fully understand the potential of using vibratory stimulation to improve strength and power in various populations. However, considering the potential of vibration to stimulate various physiological structures and modulate muscle function (see Figure 2.7.7; Cardinale and Bosco, 2003), the aim of future research activities should be to understand appropriate dose–response relationships to vibration training.

References

Abercromby A.F. *et al.* (2007) Variation in neuromuscular responses during acute whole-body vibration exercise. *Med Sci Sports Exerc*, **39** (9), 1642–1650.

Adams J.B. *et al.* (2009) Optimal frequency, displacement, duration, and recovery patterns to maximize power output following acute whole-body vibration. *J Strength Cond Res*, **23** (1), 237–245.

Ahlborg L. *et al.* (2006) Whole-body vibration training compared with resistance training: effect on spasticity, muscle strength and motor performance in adults with cerebral palsy. *J Rehabil Med*, **38** (5), 302–308.

Annino G. *et al.* (2007) Effect of whole body vibration training on lower limb performance in selected high-level ballet students. *J Strength Cond Res*, **21** (4), 1072–1076.

Babault N., Cometti G., Bernardin M. *et al.* (2007) Effects of electromyostimulation training on muscle strength and power of elite rugby players. *J Strength Cond Res*, **21** (2), 431–437.

Bautmans I. *et al.* (2005) The feasibility of Whole Body Vibration in institutionalised elderly persons and its influence on muscle performance, balance and mobility: a randomised controlled trial [ISRCTN62535013]. *BMC Geriatr*, **5**, 17.

Bax L., Staes F. and Verhagen A. (2005) Does neuromuscular electrical stimulation strengthen the quadriceps femoris? A systematic review of randomised controlled trials. *Sports Med*, **35** (3), 191–212.

Bleeker M.W. *et al.* (2005) Vascular adaptation to deconditioning and the effect of an exercise countermeasure: results of the Berlin Bed Rest study. *J Appl Physiol*, **99** (4), 1293–1300.

Blottner D. *et al.* (2006) Human skeletal muscle structure and function preserved by vibration muscle exercise following 55 days of bed rest. *Eur J Appl Physiol*, **97** (3), 261–271.

Bosco C. *et al.* (1998) The influence of whole body vibration on jumping ability. *Biol Sport*, **15** (3), 157–164.

Bosco C. *et al.* (1999a) Influence of vibration on mechanical power and electromyogram activity in human arm flexor muscles. *Eur J Appl Physiol Occup Physiol*, **79** (4), 306–311.

Bosco C. *et al.* (1999b) Adaptive responses of human skeletal muscle to vibration exposure. *Clin Physiol*, **19** (2), 183–187.

Bosco C. *et al.* (2000) Hormonal responses to whole–body vibration in men. *Eur J Appl Physiol*, **81** (6), 449–454.

Bovenzi M. (2002) [The European Directive 2002/44/CE on the protection of workers from risks resulting from mechanical vibration]. *Med Lav*, **93** (6), 551–560.

Boyer K.A. and Nigg B.M. (2007) Changes in muscle activity in response to different impact forces affect soft tissue compartment mechanical properties. *J Biomech Eng*, **129** (4), 594–602.

Brocherie F., Babault N., Cometti G. *et al.* (2005) Electrostimulation training effects on the physical performance of ice hockey players. *Med Sci Sports Exerc*, **37** (3), 455–460.

Bullock N. *et al.* (2008) Acute effect of whole-body vibration on sprint and jumping performance in elite skeleton athletes. *J Strength Cond Res*, **22** (4), 1371–1374.

Cardinale M. and Bosco C. (2003) The use of vibration as an exercise intervention. *Exerc Sport Sci Rev*, **31** (1), 3–7.

Cardinale M. and Lim J. (2003) Electromyography activity of vastus lateralis muscle during whole–body vibrations of different frequencies. *J Strength Cond Res*, **17** (3), 621–624.

Cardinale M. and Rittweger J. (2006) Vibration exercise makes your muscles and bones stronger: fact or fiction? *J Br Menopause Soc*, **12** (1), 12–18.

Cardinale M. and Wakeling J. (2005) Whole body vibration exercise: are vibrations good for you? *Br J Sports Med*, **39** (9), 585–589, discussion 589.

Cardinale M. *et al.* (2006) The acute effects of different whole body vibration amplitudes on the endocrine system of young healthy men: a preliminary study. *Clin Physiol Funct Imaging*, **26** (6), 380–384.

Cardinale M. *et al.* (2007) Gastrocnemius medialis and vastus lateralis oxygenation during whole-body vibration exercise. *Med Sci Sports Exerc*, **39** (4), 694–700.

Cardinale M. *et al.* (2008) Hormonal responses to a single session of whole body vibration exercise in elderly individuals. *Br J Sports Med*, **3** (2), 232–239.

Cherniack M. *et al.* (2008) Syndromes from segmental vibration and nerve entrapment: observations on case definitions for carpal tunnel syndrome. Int Arch Occup Environ Health. *Int Arch Occup Environ Health*, **81** (5), 661–669.

Cochrane D.J. and Hawke E.J. (2007) Effects of acute upper–body vibration on strength and power variables in climbers. *J Strength Cond Res*, **21** (2), 527–531.

Cochrane D.J. *et al.* (2007) The acute effect of vibration exercise on concentric muscular characteristics. *J Sci Med Sport*, **21** (2), 527–531.

Cochrane D.J. *et al.* (2008a) A comparison of the physiologic effects of acute whole–body vibration exercise in young and older people. *Arch Phys Med Rehabil*, **89** (5), 815–821.

Cochrane D.J. *et al.* (2008b) The rate of muscle temperature increase during acute whole-body vibration exercise. *Eur J Appl Physiol*, **103** (4), 441–448.

Currier D.P. and Mann R. (1983) Muscular strength development by electrical stimulation in healthy individuals. *Phys Ther*, **63** (6), 915–921.

Delecluse C. *et al.* (2003) Strength increase after whole–body vibration compared with resistance training. *Med Sci Sports Exerc*, **35** (6), 1033–1041.

Delecluse C. *et al.* (2005) Effects of whole body vibration training on muscle strength and sprint performance in sprint-trained athletes. *Int J Sports Med*, **26** (8), 662–668.

Delitto A., Brown M., Strube M.J. *et al.* (1989) Electrical stimulation of quadriceps femoris in an elite weight lifter:

a single subject experiment. *Int J Sports Med*, **10** (3), 187–191.

Dessy L.A. *et al.* (2008) The use of mechanical acoustic vibrations to improve abdominal contour. *Aesthetic Plast Surg*, **32** (2), 339–345.

Di Loreto C. *et al.* (2004) Effects of whole-body vibration exercise on the endocrine system of healthy men. *J Endocrinol Invest*, **27** (4), 323–327.

Ebersbach G. *et al.* (2008) Whole body vibration versus conventional physiotherapy to improve balance and gait in Parkinson's disease. *Arch Phys Med Rehabil*, **89** (3), 399–403.

Edge J. *et al.* (2009) The effects of acute whole body vibration as a recovery modality following high–intensity interval training in well-trained, middle-aged runners. *Eur J Appl Physiol*, **105** (3), 421–428.

Elliott B.C. *et al.* (1980) Vibration and rebound velocity characteristics of conventional and oversized tennis rackets. *Res Q Exerc Sport*, **51** (4), 608–615.

Enoka R.M. (2002) Activation order of motor axons in electrically evoked contractions. *Muscle Nerve*, **25** (6), 763–764.

Erskine J. *et al.* (2007) Neuromuscular and hormonal responses to a single session of whole body vibration exercise in healthy young men. *Clin Physiol Funct Imaging*, **27** (4), 242–248.

Fontana T.L. *et al.* (2005) The effect of weight–bearing exercise with low frequency, whole body vibration on lumbosacral proprioception: a pilot study on normal subjects. *Aust J Physiother*, **51** (4), 259–263.

Fratini A. *et al.* (2008) Relevance of motion artifact in electromyography recordings during vibration treatment. *J Electromyogr Kinesiol*, **19** (4), 710–718.

Gilsanz V. *et al.* (2006) Low–level, high–frequency mechanical signals enhance musculoskeletal development of young women with low BMD. *J Bone Miner Res*, **21** (9), 1464–1474.

Gondin J., Guette M., Ballay Y. and Martin A. (2005) Electromyostimulation training effects on neural drive and muscle architecture. *Med Sci Sports Exerc*, **37** (8), 1291–1299.

Gregory C.M. and Bickel C.S. (2005) Recruitment patterns in human skeletal muscle during electrical stimulation. *Phys Ther*, **85** (4), 358–364.

Griffin M.J. (2004) Minimum health and safety requirements for workers exposed to hand–transmitted vibration and whole–body vibration in the European Union: a review. *Occup Environ Med*, **61** (5), 387–397.

Gusi N. *et al.* (2006) Low–frequency vibratory exercise reduces the risk of bone fracture more than walking: a randomized controlled trial. *BMC Musculoskelet Disord*, **7**, 92.

Hagbarth K. *et al.* (1976) Single unit spindle responses to muscle vibration in man. *Prog Brain Res*, **44**, 281–289.

Hatze H. (1992) The effectiveness of grip bands in reducing racquet vibration transfer and slipping. *Med Sci Sports Exerc*, **24** (2), 226–230.

Hazell T.J. *et al.* (2007) The effects of whole-body vibration on upper- and lower-body EMG during static and dynamic contractions. *Appl Physiol Nutr Metab*, **32** (6), 1156–1163.

Issurin V.B. (2005) Vibrations and their applications in sport: a review. *J Sports Med Phys Fitness*, **45** (3), 324–336.

Issurin V.B. and Tenenbaum G. (1999) Acute and residual effects of vibratory stimulation on explosive strength in elite and amateur athletes. *J Sports Sci*, **17** (3), 177–182.

Issurin V.B. *et al.* (1994) Effect of vibratory stimulation training on maximal force and flexibility. *J Sports Sci*, **12** (6), 561–566.

Iwamoto J. *et al.* (2005) Effect of whole-body vibration exercise on lumbar bone mineral density, bone turnover, and chronic back pain in post-menopausal osteoporotic women treated with alendronate. *Aging Clin Exp Res*, **17**, 157–163.

Jackson S.W. and Turner D.L. (2003) Prolonged muscle vibration reduces maximal voluntary knee extension performance in both the ipsilateral and the contralateral limb in man. *Eur J Appl Physiol*, **88** (4–5), 380–386.

Jacobs P.L. and Burns P. (2009) Acute enhancement of lower-extremity dynamic strength and flexibility with whole-body vibration. *J Strength Cond Res*, **23** (1), 51–57.

Jubeau M., Gondin J., Martin A. *et al.* (2007) Random motor unit activation by electrostimulation. *Int J Sports Med*, **28** (11), 901–904.

Jubeau M., Sartorio A., Marinone P.G. *et al.* (2008) Comparison between voluntary and stimulated contractions of the quadriceps femoris for growth hormone response and muscle damage. *J Appl Physiol*, **104** (1), 75–81.

Kerschan-Schindl K. *et al.* (2001) Whole–body vibration exercise leads to alterations in muscle blood volume. *Clin Physiol*, **21** (3), 377–382.

Kinser A.M. *et al.* (2008) Vibration and stretching effects on flexibility and explosive strength in young gymnasts. *Med Sci Sports Exerc*, **40** (1), 133–140.

Kvorning T. *et al.* (2006) Effects of vibration and resistance training on neuromuscular and hormonal measures. *Eur J Appl Physiol*, **96** (5), 615–625.

Lake D.A. (1992) Neuromuscular electrical stimulation. An overview and its application in the treatment of sports injuries. *Sports Med*, **13** (5), 320–336.

Li F.X. *et al.* (2004) String vibration dampers do not reduce racket frame vibration transfer to the forearm. *J Sports Sci*, **22** (11–12), 1041–1052.

Lieber R.L. and Kelly M.J. (1991) Factors influencing quadriceps femoris muscle torque using transcutaneous neuromuscular electrical stimulation. *Phys Ther*, **71** (10), 715–721, discussion 722–3.

Lindqvist K. (2003) [Purr as a cat – and avoid osteoporosis]. *Lakartidningen*, **100** (45), 3688–3690.

Lloyd T., De Domenico G., Strauss G. and Singer K. (1986) A review of the use of electro-motor stimulation in human muscles. *Austr J Physiother*, **32**, 18–30.

Lohman E.B. 3rd *et al.* (2007) The effect of whole body vibration on lower extremity skin blood flow in normal subjects. *Med Sci Monit*, **13** (2), CR71–CR76.

McBride J.M. *et al.* (2004) Effect of vibration during fatiguing resistance exercise on subsequent muscle activity during maximal voluntary isometric contractions. *J Strength Cond Res*, **18** (4), 777–781.

McBride J.M. *et al.* (2009) Effect of an acute bout of whole body vibration exercise on muscle force output and motor neuron excitability. *J Strength Cond Res*, **24** (1), 184–189.

Maffiuletti N.A., Cometti G., Amiridis I.G. *et al.* (2000) The effects of electromyostimulation training and basketball practice on muscle strength and jumping ability. *Int J Sports Med*, **21** (6), 437–443.

Maffiuletti N.A., Dugnani S., Folz M. *et al.* (2002a) Effect of combined electrostimulation and plyometric training on vertical jump height. *Med Sci Sports Exerc*, **34** (10), 1638–1644.

Maffiuletti N.A., Pensini M. and Martin A. (2002b) Activation of human plantar flexor muscles increases after electromyostimulation training. *J Appl Physiol*, **92** (4), 1383–1392.

Maffiuletti N.A., Zory R., Miotti D. *et al.* (2006) Neuromuscular adaptations to electrostimulation resistance training. *Am J Phys Med Rehabil*, **85** (2), 167–175.

Maffiuletti N.A., Bramanti J., Jubeau M. *et al.* (2009) Feasibility and efficacy of progressive electrostimulation strength training for competitive tennis players. *J Strength Cond Res*, **23** (2), 677–682.

Mahieu N.N. *et al.* (2006) Improving strength and postural control in young skiers: whole-body vibration versus equivalent resistance training. *J Athl Train*, **41** (3), 286–293.

Malatesta D., Cattaneo F., Dugnani S. and Maffiuletti N.A. (2003) Effects of electromyostimulation training and volleyball practice on jumping ability. *J Strength Cond Res*, **17** (3), 573–579.

Melchiorri G. *et al.* (2007) Use of vibration exercise in spinal cord injury patients who regularly practise sport. *Funct Neurol*, **22** (3), 151–154.

Mirbod S.M. *et al.* (1999) A four-year follow-up study on subjective symptoms and functional capacities in workers using hand-held grinders. *Ind Health*, **37**, 415–425.

Mischi M. and Cardinale M. (2009) The effects of a 28-Hz vibration on arm muscle activity during isometric exercise. *Med Sci Sports Exerc*, **41** (3), 645–653.

Mischi M. and Kaashoek I. (2007) Electromyographic hyperactivation of skeletal muscles by time–modulated mechanical stimulation. *Conf Proc IEEE Eng Med Biol Soc*, **1**, 5373–5376.

Moezy A. *et al.* (2008) A comparative study of whole body vibration training and conventional training on knee proprioception and postural stability after anterior cruciate ligament reconstruction. *Br J Sports Med*, **42** (5), 373–378.

Moran K. *et al.* (2007) Effect of vibration training in maximal effort (70% 1RM) dynamic bicep curls. *Med Sci Sports Exerc*, **39** (3), 526–533.

Mulder E.R. *et al.* (2006) Strength, size and activation of knee extensors followed during 8 weeks of horizontal bed rest and the influence of a countermeasure. *Eur J Appl Physiol*, **97** (6), 706–715.

Mulder E.R. *et al.* (2007) Knee extensor fatigability after bedrest for 8 weeks with and without countermeasure. *Muscle Nerve*, **36** (6), 798–806.

Mulder E.R. *et al.* (2008) Characteristics of fast voluntary and electrically evoked isometric knee extensions during 56 days of bed rest with and without exercise countermeasure. *Eur J Appl Physiol*, **103** (4), 431–440.

Nazarov V. and Spivak G. (1985) Development of athlete's strength abilities by means of biomechanical stimulation method. *Theory Pract Phys Cult*, **12**, 445–450.

Nigg B.M. and Wakeling J.M. (2001) Impact forces and muscle tuning: a new paradigm. *Exerc Sport Sci Rev*, **29**, 37–41.

Nordlund M.M. and Thorstensson A. (2007) Strength training effects of whole-body vibration? *Scand J Med Sci Sports*, **17** (1), 12–17.

Otsuki T. *et al.* (2008) Arterial stiffness acutely decreases after whole-body vibration in humans. *Acta Physiol (Oxf)*, **194** (3), 189–194.

Pavy-Le Traon A. *et al.* (2007) From space to Earth: advances in human physiology from 20 years of bed rest studies (1986–2006). *Eur J Appl Physiol*, **101** (2), 143–194.

Pichon F., Chatard J.C., Martin A. and Cometti G. (1995) Electrical stimulation and swimming performance. *Med Sci Sports Exerc*, **27** (12), 1671–1676.

Pope M.H. *et al.* (2002) Spine ergonomics. *Annu Rev Biomed Eng*, **4**, 49–68.

Poston B. *et al.* (2007) The acute effects of mechanical vibration on power output in the bench press. *J Strength Cond Res*, **21** (1), 199–203.

Pujari A.N., Neilson R.D. and Cardinale M. (2009) A novel vibration device for neuromuscular stimulation for sports and rehabilitation applications. *Conf Proc IEEE Eng Med Biol Soc*, 839–844.

Rasmussen, G. (1982) Human body vibration exposure and its measurement. *Bruel & Kjaer Techn. Rev.*, **1**, 3–31.

Rehn B. *et al.* (2007) Effects on leg muscular performance from whole-body vibration exercise: a systematic review. *Scand J Med Sci Sports*, **17** (1), 2–11.

Rhea M.R. and Kenn J.G. (2009) The effect of acute applications of whole-body vibration on the iTonic platform on subsequent lower-body power output during the back squat. *J Strength Cond Res*, **23** (1), 58–61.

Rittweger J. *et al.* (2002) Treatment of chronic lower back pain with lumbar extension and whole–body vibration exercise: a randomized controlled trial. *Spine*, **27** (17), 1829–1834.

Rittweger J. *et al.* (2006) Highly demanding resistive vibration exercise program is tolerated during 56 days of strict bed-rest 1. *Int J Sports Med*, **27** (7), 553–559.

Roelants M. *et al.* (2004a) Effects of 24 weeks of whole body vibration training on body composition and muscle strength in untrained females. *Int J Sports Med*, **25** (1), 1–5.

Roelants M. *et al.* (2004b) Whole-body-vibration training increases knee-extension strength and speed of movement in older women. *J Am Geriatr Soc*, **52** (6), 901–908.

Roelants M. *et al.* (2006) Whole-body-vibration-induced increase in leg muscle activity during different squat exercises. *J Strength Cond Res*, **20** (1), 124–129.

Ronnestad B.R. (2009) Acute effects of various whole-body vibration frequencies on lower-body power in trained and untrained subjects. *J Strength Cond Res*, **23** (4), 1309–1315.

Sale D.G. (1988) Neural adaptation to resistance training. *Med Sci Sports Exerc*, **20** (Suppl. 5), S135–S145.

Sands W.A. *et al.* (2006) Flexibility enhancement with vibration: acute and long–term. *Med Sci Sports Exerc*, **38** (4), 720–725.

Sanya A.O. and Ogwumike O.O. (2005) Low back pain prevalence amongst industrial workers in the private sector in Oyo State, Nigeria. *Afr J Med Med Sci*, **34** (3), 245–249.

Semler O. *et al.* (2007) Preliminary results on the mobility after whole body vibration in immobilized children and adolescents. *J Musculoskelet Neuronal Interact*, **7** (1), 77–81.

Semler O. *et al.* (2008) Results of a prospective pilot trial on mobility after whole body vibration in children and adolescents with osteogenesis imperfecta. *Clin Rehabil*, **22** (5), 387–394.

Theurel J., Lepers R., Pardon L. and Maffiuletti N.A. (2007) Differences in cardiorespiratory and neuromuscular responses between voluntary and stimulated contractions of the quadriceps femoris muscle. *Respir Physiol Neurobiol*, **157** (2–3), 341–347.

Timme N. and Morrison A. (2009) The mode shapes of a tennis racket and the effects of vibration dampers on those mode shapes. *J Acoust Soc Am*, **125** (6), 3650–3656.

Torvinen S. *et al.* (2002) Effect of four-month vertical whole body vibration on performance and balance. *Med Sci Sports Exerc*, **34** (9), 1523–1528.

Torvinen S. *et al.* (2003) Effect of 8-month vertical whole body vibration on bone, muscle performance, and body balance: a randomized controlled study. *J Bone Miner Res*, **18** (5), 876–884.

Trans T. *et al.* (2009) Effect of whole body vibration exercise on muscle strength and proprioception in females with knee osteoarthritis. *Knee*, **16** (4), 256–261.

Vanderthommen M. and Duchateau J. (2007) Electrical stimulation as a modality to improve performance of the neuromuscular system. *Exerc Sport Sci Rev*, **35** (4), 180–185.

Vanderthommen M., Duteil S., Wary C. *et al.* (2003) A comparison of voluntary and electrically induced contractions by interleaved 1H- and 31P-NMRS in humans. *J Appl Physiol*, **94** (3), 1012–1024.

Verschueren S.M. *et al.* (2004) Effect of 6-month whole body vibration training on hip density, muscle strength, and postural control in postmenopausal women: a randomized controlled pilot study. *J Bone Miner Res*, **19** (3), 352–359.

Wakeling J.M. and Liphardt A.M. (2006) Task–specific recruitment of motor units for vibration damping. *J Biomech*, **39** (7), 1342–1346.

Ward K. *et al.* (2004) Low magnitude mechanical loading is osteogenic in children with disabling conditions. *J Bone Miner Res*, **19** (3), 360–369.

Wilcock I.M. *et al.* (2009) Vibration training: could it enhance the strength, power, or speed of athletes? *J Strength Cond Res*, **23** (2), 593–603.

Willoughby D.S. and Simpson S. (1996) The effects of combined electromyostimulation and dynamic muscular contractions on the strength of college basketball players. *J Strength Cond Res*, **10** (1), 40–44.

Willoughby D.S. and Simpson S. (1998) Supplemental ES and dynamic weight training: effects on knee extensor strength and vertical jump of female college track and field athletes. *J Strength Cond Res*, **12** (3), 131–137.

Wolf S.L., Ariel G.B., Saar D. *et al.* (1989) The effect of muscle stimulation during resistive training on performance parameters. *Am J Sports Med*, **14** (1), 18–23.

Yamada E. *et al.* (2005) Vastus lateralis oxygenation and blood volume measured by near-infrared spectroscopy during whole body vibration. *Clin Physiol Funct Imaging*, **25** (4), 203–208.

Zory R., Jubeau M. and Maffiuletti N.A. (2010) Contractile impairment after quadriceps strength training via electrical stimulation. *J Strength Cond Res*, **24**, 458–464.

2.8 The Stretch–Shortening Cycle (SSC)

Anthony Blazevich, School of Exercise, Biomedical and Health Sciences, Edith Cowan University, Joondalup, WA, Australia

2.8.1 INTRODUCTION

Motor tasks requiring high movement speeds or economy are commonly performed with movement patterns that allow muscle–tendon units (MTUs) to be stretched before shortening rapidly, without a significant delay between the two phases. This is called the stretch–shortening cycle (SSC). Walking, running, hopping, jumping, stair climbing, throwing, hitting, and kicking are examples of activities that are typically performed with an SSC. It is well known that the SSC improves performance by a considerable amount when compared to a concentric-only movement. For example, vertical jump height has been shown to be improved by about 8% (Markovic *et al.*, 2004) and running economy by approximately 50% (Cavagna, Saibene and Margaria, 1964) through the use of the SSC. In this chapter, the mechanisms that underpin the performance enhancement gained through the SSC will be reviewed – in many cases the relative impact of each mechanism on performance has not been fully elucidated. In addition, some considerations on how to optimize SSC performance are discussed.

2.8.2 MECHANISMS RESPONSIBLE FOR PERFORMANCE ENHANCEMENT WITH THE SSC

There is still some debate as to the mechanisms responsible for the performance enhancement seen with an SSC. This is partly because the relative contribution of each mechanism varies as the force–time characteristics of a movement change. Here, we will examine the possible role of six elastic and contractile mechanisms.

2.8.2.1 Elastic mechanisms

Mechanism 1: recovery of elastic potential energy

In order to understand the role of elastic energy in the enhancement of performance by the SSC, it is important to first understand some fundamental principles. Many tissues deform (elongate or compress) when a force is applied to them and can therefore store elastic potential energy (Figure 2.8.1). The deformation of a tissue is dependent upon its stiffness and the force applied to it, according to Hooke's law: $F = kx$ (Hooke's law actually calculates the restoring force rather than the force required to deform the tissue, so is usually written $F = -kx$), where k is the stiffness of the tissue and x is the deformation distance of the tissue. The equation can be rewritten as $k = F/x$, which shows that a stiff tissue requires a high stretching force (F) to achieve a given stretch or deformation (x). Compliance is the inverse of stiffness ($1/k$), so a compliant tissue deforms easily under a small force.

Most tissues return to shape, or undergo restitution, when the force that caused the deformation is removed or reduced; such tissues are called 'elastic'. Much of the stored energy is returned as kinetic (movement) energy. The recoil speed of elastic tissues such as tendons is practically unlimited, so their shortening can occur at speeds far exceeding those of muscle shortening. Thus, tissues such as tendons can ensure that MTUs shorten at speeds exceeding those that use muscle contraction alone.

Elastic energy storage in muscle–tendon units (MTUs)

Both muscles and tendons can store elastic energy according to the equation: $E = \frac{1}{2}kx^2$, where E is the energy stored. The stored energy is thus equal to the area under the force–elongation curve, as shown in Figure 2.8.1. Muscle–tendon elongation is the dominant factor influencing energy storage, and this is reflected by the squared x term in the equation. Although it is clear that tendons will store energy as they are stretched, it should be remembered that muscles can store energy as they are stretched too, particularly over short stretch distances where cross-bridge cycling does not occur. At all but the longest muscle lengths this energy is stored in the series elastic components (SECs) within the muscle, including the cross-bridges and actin and myosin filaments. Studies show that muscle fibres can be stretched by approximately 3% before cross-bridge cycling occurs (Flitney and Hirst, 1978), although in pennate muscles, where fibre rotation occurs as the muscle is stretched,

Strength and Conditioning – Biological Principles and Practical Applications Marco Cardinale, Rob Newton, and Kazunori Nosaka.

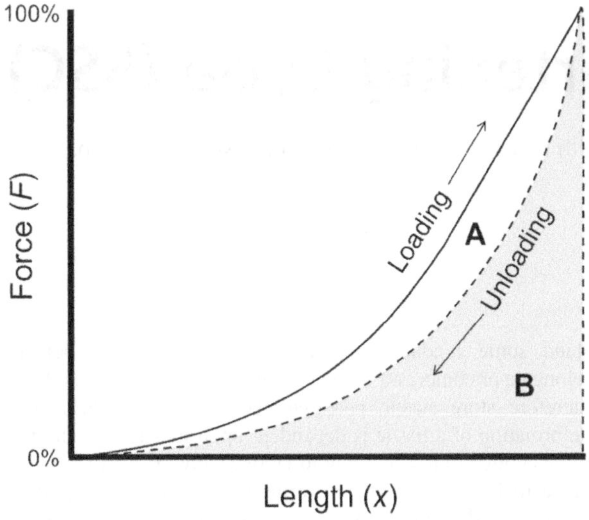

Figure 2.8.1 The force–length curve for viscoelastic materials, such as muscle and tendon. An applied force causes stretch in the tissue (loading). Stiffer tissues stretch less (x) for a given force (F) and therefore display a steeper curve. The elastic potential energy stored is equivalent to the area under the loading curve (areas A + B). The tissue shortens when force is reduced (unloading), but some energy is dissipated. The energy returned is equal to the area under the unloading curve (B). The efficiency of elastic energy recovery (%) is equal to $\frac{B-(A+B)}{(A+B)}\times 100$

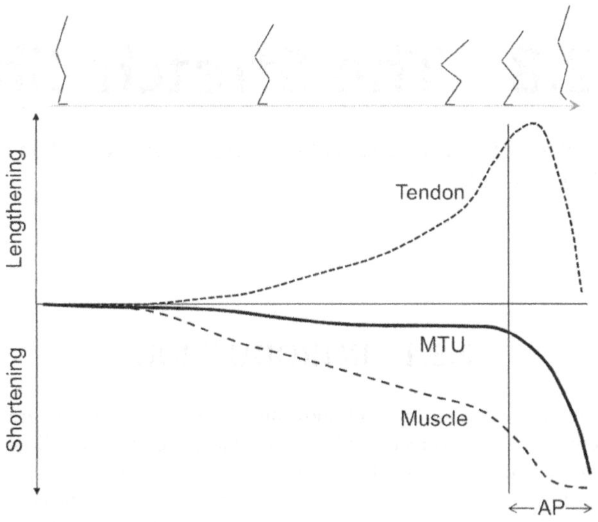

Figure 2.8.2 The muscle and tendon undergo shortening and lengthening at different times in many SSC actions. For example, research on countermovement jumping (e.g. Kurokawa *et al.*, 2003) shows that the Achilles tendon–gastrocnemius MTU stays at a relatively constant length until it shortens rapidly during ankle plantar flexion (AP) late in the jump sequence. However, the tendon is first stretched by the active muscles and then recoils at high speed late in the movement, whereas the muscle shortens slowly during the movement but remains quasi-isometric late in the movement. Thus, the high-speed phase of the jump is associated with fast recoil of the Achilles tendon but minimal shortening of the gastrocnemius muscle

the whole muscle may be able to stretch by ~5%. Thus, a long muscle such as the vastus lateralis (30–40cm) can stretch by 1.5–2.0cm before cross-bridge cycling occurs, which is similar to the elongation of the Achilles tendon during countermovement jumping (Kurokawa *et al.*, 2003) and hopping (Lichtwark and Wilson, 2005).

The question therefore arises as to which tissue is most appropriate for the storage of the energy. The answer lies not in which tissue can store the most energy, but in the efficiency and rate with which energy is returned during recoil. Many studies on human tendon indicate that >85% of the stored energy can be returned (e.g. Bennett *et al.*, 1986; Pollock and Shadwick, 1994), although *in vivo* studies using newly developed ultrasound imaging techniques estimate this to be ~65–90% (Lichtwark and Wilson, 2005; Maganaris and Paul, 2000). Whilst few data exist detailing the efficiency of muscle, one study on activated rabbit muscle indicates that perhaps only ~60% of the energy is returned (Best *et al.*, 1994), as the viscous properties of muscle ensure that energy is converted to heat and dissipated. These viscous effects also slow the rate of muscle shortening during recoil, so it is better to store as much energy as possible in the more efficient tendons in order to optimize movement performance.

In many SSC actions, muscles are activated to become highly stiff, and tendon elongation and recoil dominate the overall muscle–tendon length change. As shown in Figure 2.8.2, this results in the muscle and the tendon operating out of phase, with tendon recoil providing the greatest shortening during the high-speed propulsive phase of an SSC movement. Of course, the influence of tendon recoil versus muscle shortening will be different for the various MTUs within a limb, and for movements with different force–time characteristics or movement patterns.

Another question thus arises: how do we ensure that most of the elastic potential energy is stored in the tendons rather than the muscles? The answer lies in the fact that when two springs of different stiffness are placed in series, more energy is stored in the compliant spring. Therefore, more energy would be stored in a tendon if its associated muscle were relatively stiffer. Increases in muscle stiffness can result from both active and passive mechanisms:

1. Muscles that can produce a large contractile force will resist lengthening and clearly have a high stiffness (remember, $k = F/x$), so increasing a muscle's size and neural activation is important (active mechanism).
2. At a given level of muscle force, stretch reflex activation may increase muscle stiffness during rapid stretch above that which is developed in the absence of the reflex (Hoffer and

Andreassen, 1981; Sinkjaer *et al.*, 1988) and contributes within 50 ms of strertch onset (Sinkjaer *et al.*, 1988), so maximizing facilitatory (e.g. stretch reflex/Ia and II afferents) and minimizing inhibitory (Ib (Golgi), III, IV afferents) reflex input is important.

3. Muscles that possess stiffer intra-muscular connective tissue structures (both series and parallel elastic components) will have a higher passive stiffness (passive mechanism). Little research has examined intra- and inter-muscular variations in the stiffness of elastic structures – although, for example, aponeurosis stiffness is known to vary between muscles and between individuals (Bojsen-Møller *et al.*, 2004) – and the influence of force and stiffness *in vivo* is still unclear.
4. Shorter muscles, or those with shorter fascicles, will have a higher passive stiffness because compliance (the inverse of stiffness) increases proportionally with tissue length (passive mechanism); two springs of the same mechanical properties placed in series will stretch further than a single spring for a given force. Indeed, many distal-limb MTUs that store and release considerable energy in SSC actions have short muscles attaching to longer tendons (e.g. gastrocnemius–Achilles tendon complex), which optimizes them for SSC function.
5. Muscles with greater fluid mass per muscle volume (i.e. increased turgidity) are stiffer; fluid shifts induced by ion-dependent osmosis increase muscle stiffness (Grazi, 2008). Such fluid shifts occur in exercising muscle and may remain for hours or days after exercise.
6. Muscles, like all viscoelastic materials, will be stiffer when stretched more rapidly (Lamontagne, Malouin and Richards, 1997; Mutungi and Ranatunga, 1996). Viscous effects are caused by fluids in the muscle having to flow past intramuscular structures; the more viscous a muscle's fluid, the more resistance there is at high stretch speeds.

Thus, interventions that influence these factors might be expected to influence SSC performance, although in many cases more research is required to fully understand their relative impact.

Importance of stored elastic energy in SSC actions

Although the storage and release of elastic energy is often considered a primary mechanism affecting SSC performance (Alexander, 1987), there is some debate as to its true influence (Bobbert *et al.*, 1996; Voigt *et al.*, 1995). For example, Bobbert *et al.* (1996) suggested that the performance enhancement in countermovement jumping, when compared to squat (static) jumping, results almost entirely from the increased time for muscle activation afforded by the countermovement phase (see Mechanism 3). Nonetheless, other researchers have shown strong evidence for an important role of elastic energy in countermovement jump performance (e.g. Kawakami *et al.*, 2002; Kurokawa *et al.*, 2003). A point that might be considered is that an increase in muscle force is required in order to decelerate in the eccentric (downward) phase and then re-accelerate in the concentric (upward) phase in order to change the body's momentum, which will stretch the tendons further. This increases energy storage and the restoring force applied by the tendons (e.g. Finni *et al.*, 2003), according to Hooke's law. By this reasoning it is not possible for an increase in muscle force to be the sole contributor to performance enhancement because the muscle must stretch an elastic tendon. Nonetheless, it could be argued that the increased restoring force of the tendon is proportional to the increase in muscle force resulting from the countermovement, and thus the increased muscle force might ultimately be responsible for the greater performance.

Regardless, for SSC movements with different force–time characteristics, it is very likely that the recoil of elastic tissues contributes to both the velocity and force output of the MTU, as well as to movement economy. For example, Finni *et al.* (2003) showed that in the concentric phase of a drop jump the stretch and shortening occurred in the quadriceps tendon with little change in muscle length, whilst in the countermovement jump there was a greater elongation of the muscle. Subsequently it was shown that greater intensities of loading (i.e. increases in force with little or no increase in force production time) in the eccentric phase of drop-jump tasks increased the stretch–shortening distance of the vastus lateralis tendon and reduced that of the muscle, when the height of the jump was held constant (Ishikawa and Komi, 2004; Ishikawa *et al.*, 2006). Nonetheless, very high loading intensities in the eccentric phase might result in greater muscle elongation, and thus relatively less tendon elongation, as the muscle yields under the load (Ishikawa, Niemelä and Komi, 2005). Thus, the force–time characteristics of a movement appear to strongly influence the relative contribution of elastic recoil to velocity and force production.

With respect to increasing movement speed, it is well known that muscle power during the concentric phase of an SSC movement far exceeds that which can be produced by even the fastest muscles. Peak shortening speeds alone might be as fast as 7–8 fibre lengths/second in fast muscles (Close, 1972), but force, and thus power (F × v, where v is shortening velocity), will be low. However, tendons can recoil at very high speeds and with a large restoring force, so power output can be high. For example, de Graaf *et al.* (1987) estimated that 1400 W was generated by the plantar flexors in a one-leg countermovement jump. Mammalian limb muscles with a high proportion of fast-twitch fibres might produce ~500 W/kg (Brooks, Faulkner and McCubbrey, 1990; Josephson, 1993); for a plantar flexor mass of 950 g (based on a muscle density of 1.06 g/ml from Méndez and Keys, 1960, and a total volume of ~900 ml per leg from Fukunaga *et al.*, 1996) the peak power output of muscle alone would be ~475 W. This is about a third less of 1400 W reported by Graaf *et al.* (1987). Ankle power in two-legged countermovement (~1800 W) and drop jumps (~2400 W) also clearly exceeds the capacity of muscle. Although some other factors might act to augment muscle power in complex movements, much of this extra power output is thought to come from the re-use of stored elastic energy. Tendon recoil increases muscle power output largely because of the high recoil speeds; that is, the velocity component of the power equation (F × v) is improved. Power outputs of MTUs that exceed those that could be achieved by muscle alone are commonly seen in animal locomotion, so tendons are often referred to as 'power amplifiers'.

Power output can also be improved by increasing the force component of the equation. One limitation of skeletal muscle is that its ability to develop force decreases with increasing velocity, according to the force–velocity relationship. In fact, peak power production is typically produced at shortening speeds of about 1/3 of maximum, so faster muscle shortening will inevitably lead to force reduction. However, both modelled (Hof, van Zandwijk and Bobbert, 2002) and experimental data (Finni, 2001; Finni, Ikegawa and Komi, 2001; Finni *et al.*, 2003; Fukashiro and Komi, 1987) show that joint torque and MTU force at higher velocities are far greater than that predicted by the force–velocity relationship. As shown in Figure 2.8.3, the peak force attained at zero MTU velocity is within expected limits, but the force produced at higher concentric speeds is greater than expected. This force enhancement has been shown to occur without greater muscle activity or an altered working length of the fascicles (Finni, Ikegawa and Komi, 2001), so the likely reason for this phenomenon is that the restoring force of tendon recoil increases the total MTU force, according to Hooke's law. At these high MTU shortening speeds, the muscle itself will often shorten relatively slowly (see Section 2.8.2.2 and Figure 2.8.2), which allows for a higher force output than if it were shortening rapidly. The combination of this higher force and the tendon-restoring force results in a substantive increase in force at high MTU shortening speeds.

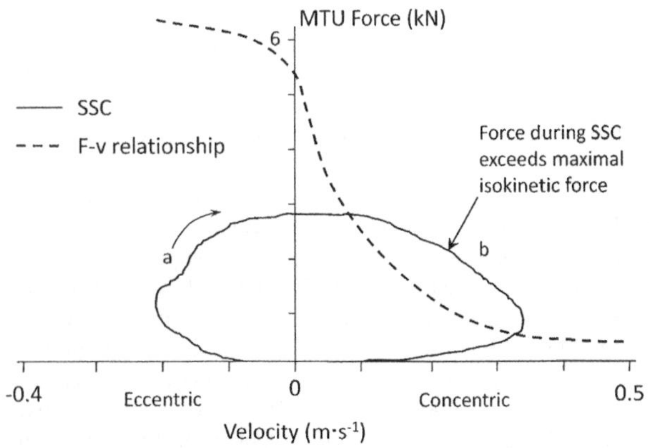

Figure 2.8.3 Example data of a force–velocity curve of a distal MTU measured during an SSC, compared to the force–velocity relationship for that MTU measured with isokinetic dynamometry (i.e. concentric-, isometric-, or eccentric-only force recordings measured at constant angular velocities). During an SSC, the MTU force rises as it is stretched eccentrically (a), and then falls as the MTU recoils in the concentric phase (b). The force at some concentric speeds is higher than that predicted from the force–velocity curve measured isokinetically (e.g. Finni *et al.*, 2003; Hof, van Zandwijk and Bobbert, 2002)

With respect to movement economy, it has long been known that significant energy savings are achieved through the re-use of stored elastic energy. In human running for example, Cavagna, Saibene and Margaria (1964) estimated that around 50% of the energy for running propulsion comes from the storage and release of elastic energy. This process was suggested to account largely for the efficiency of muscle (actually, the whole MTU) being increased from ~0.25 to ~0.50. This is because if the deceleration of the body during the early stance phase were performed only by muscles working eccentrically, the energy would be lost as heat and muscle work would be required to re-accelerate the body. But a portion of the kinetic energy of a body or limb performing a countermovement can be stored as elastic potential energy and then regained as kinetic energy in the concentric phase. Recent research suggests that similar (~50%) energy savings are made in movements such as vertical jumping (Bosco, Saggini and Viru, 1997). In fact, *in vivo* estimates of energy contributions from individual elastic tissues are significant, with ~17% of the energy of running coming from energy stored in the arch of the foot (Ker *et al.*, 1987), and ~16% (Lichtwark and Wilson, 2005) and ~6% (Maganaris and Paul, 2002) coming from energy stored in the Achilles tendon alone during one-legged hopping and walking, respectively. However, some researchers caution that the increased economy should not be confused with an increase in the efficiency of muscle contraction itself, which has been argued to decrease since some of the work done by the muscle on the tendon will be lost as heat (e.g. Ettema, 1996; Ingen Schenau, Bobbert and de Haan, 1997).

2.8.2.2 Contractile mechanisms

While the recovery of elastic potential energy (Mechanism 1) is a major contributor to the improved force/power production and economy of many human movements, the storage of this energy is ultimately dependent on the magnitude of the contractile force that stretches the series elastic components. Five contractile mechanisms (Mechanisms 2–6) are thought to contribute to the increased contractile force and ultimately to enhanced performance in SSC actions.

Mechanism 2: force potentiation

Although it is typically believed that only factors such as the length or velocity of a muscle dictate its force-production capacity, force production also has a history dependence (Abbott and Aubert, 1952). It is commonly observed that muscle or fibre force increases by up to twofold when an isometrically contracting muscle is stretched. This substantial increase is greater at high stretch velocities, but when the stretch is applied over a constant time period, it is relatively independent of stretch amplitude (Edman, Elzinga and Noble, 1978; Lombardi and Piazzesi, 1990). Experiments performed on muscle fibres and whole muscles show that the increase in force is transient when the stretch is imposed at shorter muscle lengths (ascending or plateau region of the force–length curve); that is, it is lost in approximately 20 ms if the muscle is held at constant length or allowed to shorten. Since many muscles involved in SSC actions, such as the human gastrocnemius, typically work on their ascending limb, it has been suggested that this mechanism is unlikely to be a factor influencing force production (Brown and Loeb, 2000). However, when stretch is imposed on muscles at longer lengths (on their descending limb), the force remains higher compared to an isometric contraction performed at the same muscle length (Edman, Elzinga and Noble, 1978; Rassier and Herzog, 2004). This increase in force remains for as long as the muscle is active. One caveat is that the force enhancement due to stretch is still rapidly lost if the muscle is allowed to shorten (Edman, Elzinga and Noble, 1978; Herzog and Leonard, 2000), so this mechanism is again thought not to be responsible for the increases in force observed when muscles undergo substantial shortening during contraction (e.g. Walshe, Wilson and Ettema, 1998).

Nonetheless, the increase in force of the muscle in the stretch phase could allow the SEC to store more elastic energy, and the subsequent drop in force as the muscle shortens provides the necessary conditions for recoil of the SEC (see Section 2.8.3). Also, viscoelastic materials such as muscle increase in stiffness when stretched rapidly (Mutungi and Ranatunga, 1996). Thus, the rapid stretch of activated muscles in the eccentric phase might ultimately be responsible for some of the increase in muscle power and efficiency in some SSC actions. Further research is required to assess the relative importance of this mechanism for SSC performance.

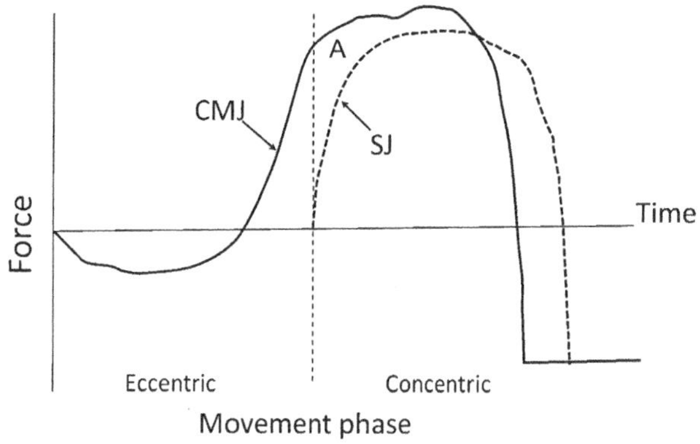

Figure 2.8.4 Representative GRF traces from both countermovement jump (CMJ) and squat (static) jumps (SJ). The higher propulsive force early in the concentric phase of a CMJ is purported to contribute substantially to the increased jump height compared to an SJ. This increased force, and thus work, is shown as area A

Mechanism 3: increased time for muscle activation

Near-maximal muscle force (>90% maximum) typically takes several hundred milliseconds to develop, although this can vary between about ~150 and ~900 ms depending on the architecture and voluntary activation rate of the muscle. In many human movements, the force application phase is considerably shorter than this (e.g. 100 ms for ground contact in sprint running), so near-maximal force will not normally be achievable. However, the concentric phase of an SSC action is preceded by an eccentric, or countermovement, phase where muscle force begins to increase. Thus, there is a higher level of force prior to the start of the concentric phase, which increases the amount of work done ($W = F \times d$, where F is the average force and d is the distance over which force is produced), as shown in Figure 2.8.4. Bobbert *et al.* (1996) found that joint moments at the beginning of the concentric phase of a countermovement jump were much higher than in the squat jump. Their modelling indicated that the increase in moment could account for most of the difference in jump height. This is good evidence that, at least in some instances such as vertical jump, the increased time for force development enhances movement performance.

However, several other arguments must be considered. First, an increase in muscle force will result in a greater elongation of elastic tissues and hence an increase in elastic energy contribution in the propulsive phase. While this might not always increase the work done, it can increase the rate of work (power), as tendon recoil velocity can be much higher than muscle shortening speeds. So, because greater muscle forces result in greater tendon stretch, an increased time for activation is unlikely to ever be a sole contributor. Second, it is likely that SSC actions performed very rapidly benefit more from the elastic energy mechanism (Finni *et al.*, 2003; Ishikawa, Finni and Komi, 2003), so the importance of the increased time of activation will decrease as movement time decreases. In some movements, such as countermovement jumping on a sledge apparatus, it has been shown that joint torque measured at the beginning of the concentric phase is not higher than that obtained in an isometric contraction at the same joint angle, although there is a greater joint torque later in the movement (Finni, Ikegawa and Komi, 2001). So the increased time for muscle activation does not always result in a greater force at the start of the concentric phase and cannot always be responsible for the increased movement performance. Regardless, the greater time for force development is a likely contributor to greater movement performance in some SSC movements.

Mechanism 4: pre-load effect

A significant drawback of concentric-only movements is that a muscle's ability to develop force is reduced as soon as concentric shortening speed increases, in accordance with the force–velocity relationship. So a mechanism that allows the muscle to shorten more slowly (or eccentrically/isometrically) will result in a higher overall muscle force. Many insects such as fleas and froghoppers (spittle bugs) use a catch-like mechanism to prevent joint extension until agonist muscle force is high, so the concentric phase begins when muscles have already developed high force levels and the elastic tendons are stretched (Bennet-Clark and Lucey, 1967; Burrows, 2003). Humans use a countermovement to similar effect. Once a countermovement is initiated, muscles are activated in order to slow the body or limb, i.e. to reduce its negative momentum, prior to the concentric phase. This allows muscle force to increase during the eccentric (countermovement) and isometric (transition) phases of the movement, when movement speed is relatively low, so the concentric phase begins with the muscle already having developed a high level of force. Thus, the countermovement preloads the body or limb so that muscle force can be developed well before the concentric movement speed increases; interestingly, it has been shown that preloading the body using an eccentric contraction is more beneficial than loading with an isometric contraction prior to a vertical jump (Walshe, Wilson and Ettema, 1998), which makes sense given that force potentiation (Mechanism 2) might also be greater when a muscle is actively stretched.

The result of both an increased time for muscle activation and this preload effect can be seen in the force traces obtained during maximal vertical jumps with and without a countermovement (see Figure 2.8.4). It is likely that for some SSC actions this preload effect, in combination with the greater time available for force accumulation, substantially improves SSC performance. However, in other movements such as a repeated drop-jump exercise (performed on a sledge apparatus; Finni, 2001; Finni, Ikegawa and Komi, 2001) there is clearly no effect of this mechanism. So other mechanisms must be of greater importance in some movements.

Mechanism 5: muscle–tendon interaction (concerted contraction)

A benefit of having tendons that stretch and then recoil during a movement is that the muscle lengthening and shortening distance can be less for a given joint angle change. This smaller muscle displacement, also known as concerted contraction (defined by Hof, Geelen and Berg, (1983) as 'a contraction in which the activation is matched to the load to the effect that the length of the contractile component remains constant'), allows the sarcomeres to work through shorter ranges and thus remain closer to their optimum length, should the muscle length be appropriate for it. Therefore, force production might be optimized from a force–length perspective. A smaller displacement would also result in a slower shortening speed, and thus a greater force potential according to the force–velocity relationship. The importance of this mechanism is underlined by the fact that muscles could not shorten at the speeds required for many high-speed human tasks, such as sprint running or jumping, so of course muscle force would be zero under these conditions. But even in movements where muscles could have completed the movement without the aid of tendons, the faster muscle shortening speed would seriously limit maximum force output. Importantly, isometric muscle contraction requires less energy per unit force than concentric shortening (Hill, 1938), so the ability to contract at zero or lower speeds reduces energy use and increases movement economy. Thus, the ability for the muscle to contract quasi-isometrically in many SSC activities has a substantial effect on force, power, and movement economy.

Mechanism 6: reflex contribution

An important role of muscle in SSC actions is to provide sufficient stiffness to allow elastic energy storage in the SEC, and in particular the tendon. In addition to α-motoneurone activity (Dyhre-Poulsen, Simonsen and Voigt, 1991), reflex activation (primarily the short-latency, monosynaptic response) is thought to contribute to muscle stiffness and force (Cordo and Rymer, 1982; Komi, 2003; Voigt, Dyhre-Poulsen and Simonsen, 1998). The stretch reflex response, resulting from muscle spindle afferent discharge, has been shown to increase muscle force and joint stiffness (Houk, 1979; Nichols and Houk, 1976). In fact, Hoffer and Andreassen (1981) showed that a muscle's resistance to stretch was greater when reflexes were intact than when they were not, given identical muscle force prior to stretch. This increased stiffness was not correlated with the amplitude of the early (presumably monosynaptic) EMG peak, which indicates a complex, non-activation-dependent influence of the reflex. So the stretch reflex appears to be an important mechanism for force application and stiffness regulation in SSC actions. Nonetheless, there are several arguments against the assertion that reflexes contribute to SSC performance:

1. The latency of the reflex's mechanical response (i.e. rise of force) may be too long to contribute to many SSC movements.
2. There is considerable evidence that important agonist muscles may not lengthen during many common SSC actions, so the stretch reflex cannot be present.
3. The overall increase in muscle activation, measured by the area under the EMG – time curve, is insufficient to substantially increase muscle force above that resulting from voluntary activation.

Given that these arguments have provided a framework around which research into the role of the stretch reflex has progressed, it is probably a good idea to address each in turn.

Latency of the stretch reflex mechanical response

One of the main concerns regarding the potential importance of the stretch reflex is that its mechanical response might be too slow to make a substantial contribution to muscle force or stiffness (e.g. Ingen Schenau, Bobbert and de Haan, 1997). This is based on the reasonable logic that the fastest stretch reflex component (the short-latency component) manifests approximately 40 ms after the onset of rapid muscle stretch, and then the electro-mechanical coupling process takes a further 90–100 ms (Ingen Schenau *et al.*, 1995; Vos, Harlaar and Ingen Schenau, 1991). This results in a minimum period of 130 ms between stretch onset and the resulting rise in force, although a further delay would be likely as the resulting muscle force is transferred through compliant tissues such as the tendons to the skeleton (i.e. there is a moderate rate of force development). By this argument, an increase in muscle force and stiffness might only be expected to occur in movements lasting considerably longer than 130 ms from the onset of stretch (Ingen Schenau, Bobbert and de Haan, 1997), which rules out human running, hopping, and bouncing movements. Furthermore, in movements such as countermovement jumping where there is probably sufficient time for the reflex response to impact on movement performance, the benefit of the SSC has previously been largely explained by the increase in muscle active state prior to the concentric phase (Bobbert *et al.*, 1996).

However, experimental evidence is not congruent with this argument. Nicol and Komi (1998) applied rapid stretches to the plantar flexor muscles and found a 13–15 ms delay between the reflex onset and force rise (measured in the Achilles tendon), giving a total mechanical response time of only ~55 ms; this response time includes the time required for force to be transmitted through the elastic components to the skeleton, although it could be considered that there would be an additional time required for the force to reach sufficient levels to impact on movement performance. Previous data from Melvill-Jones and Watt (1971) had already suggested that the longer-latency stretch reflex response was noticeable after 50–120 ms and was of sufficient time to influence the concentric phases of human hopping, although it was too slow to influence fast hopping or stepping down from a step. Also, Ishikawa and Komi (2007) showed that the short-latency stretch reflex occurred rapidly enough to affect force production in slow–moderate-speed (2–5 m/s), but not higher-speed (6.5 m/s), running. Thus, the evidence suggests that the mechanical response to the stretch reflex has a time course that fits within the force application times of many, but not the fastest, human movements.

Do agonist muscles stretch during SSC actions?

Another assumption that needs to be satisfied is that agonist muscles are stretched prior to the concentric movement phase. While it might be assumed that muscles must be stretched in the eccentric phase of an SSC, it is often the case that the muscles and tendons work out of phase (as shown in Figure 2.8.2), so muscle stretch–shortening should not be confused with whole-MTU stretch–shortening. In fact, because of the substantial compliance of tendons, modelling studies have predicted that the triceps surae (ankle plantar flexors) contract only isometrically and concentrically in human walking and running, while the tendon is stretched and recoils over a considerable range (Hof, van Zandwijk and Bobbert, 2002). Also, some studies, using ultrasound imaging to visualize the fascicle shortening–lengthening behaviour of the triceps surae, have not shown stretch of the fascicles in movements such as walking (Fukunaga *et al.*, 2001) and running (Lichtwark, Bougoulias and Wilson, 2007).

However, the recent use of high-frequency ultrasound video systems has allowed the detection of brief, high-speed stretches of muscles during some of these SSC movements. For example, Ishikawa and Komi (2007) showed that the medial gastrocnemius muscle underwent a brief stretch immediately after foot–ground contact in running at both slow and fast speeds. These stretches were followed by a detectable short-latency stretch reflex response. In all but the highest running speeds (6.5 m/s) it was assumed that the reflex was sufficiently timed to aid force production and improve muscle stiffness in propulsion. Importantly, fascicle stretch might also be different in the various muscles within a synergist group. Sousa *et al.* (2007) found that the contraction mode of soleus and gastrocnemius medialis was not the same during different phases of a drop-jump exercise; in fact, soleus underwent lengthening before shortening over a range of drop-jump intensities, whereas gastrocnemius tended to shorten at lower intensities but then to contract either isometrically or eccentrically in the eccentric (braking) phase of drop jumps performed at very high intensities. Thus, not only does fascicle behaviour differ between muscles, but it also changes according to the force–time characteristics of the task. Regardless, brief stretch of some muscles can be seen clearly in human SSC activities when appropriate techniques are used.

Is the stretch reflex activation of muscle sufficient to increase muscle force and stiffness?

One difficulty with ascribing increases in muscle force in SSC movements to the stretch reflex is that it needs to be shown that the brief reflex activation could have a substantial effect in addition to the large muscle activation usually present in these movements. This is a very difficult thing to do. While it is clear that the force production from a reflex that is induced in relaxed muscle is substantial, and easily measured, some researchers have questioned the potential benefits of stretch reflexes in SSC actions because the total quantity of EMG measured in the concentric phase of an SSC action has not been greater than that of a purely concentric action. For example, Bobbert *et al.* (1996) found no increase in EMG amplitude in lower-limb muscles in a countermovement jump compared to a squat jump, and Finni, Ikegawa and Komi (2001) found no difference in the EMG of vastus lateralis in submaximal sledge jumping exercises between SSC and concentric-only conditions. Thus, there is a question as to the magnitude of the benefit of the reflex stimulation.

However, novocaine (1%) injections to the motor point of vastus lateralis, which reduced the stretch reflex response substantially (58–87%) by blocking the gamma efferents to the muscle spindles, resulted in a significant decrease in countermovement jump height (12.5%) despite no decrease in voluntary (α-motoneurone) activation (Kilani *et al.*, 1989). This is very good evidence of an important role of (particularly short-latency, monosynaptic) stretch reflexes. Importantly, Hoffer and Andreassen (1981) showed that a muscle's resistance to stretch was greater when reflexes were intact than when they were abolished, when muscle force was the same prior to stretch. Further, Sinkjaer *et al.*, (1988) showed that reflexes in human dorsiflexor muscles influenced muscle stiffness within 50 ms of stretch onset and accounted for up to half of the increased stiffness resulting from rapid muscle stretch; reflex contributions to stiffness were maximal at intermediate levels of voluntary muscle force. Thus, a stretch reflex appears to be beneficial for muscle stiffness even if total muscle activation is not noticeably greater.

Another benefit of the stretch reflex is that it might optimize the use of the catch-like property of muscles (Burke, Rudomin and Zajac, 1970), which is an increase in force that occurs when a brief, high-frequency burst of stimulation, followed by 'normal' lower-frequency stimulation, is applied to a muscle. The high-frequency bursts are associated with increases in isometric and dynamic muscle force, with substantial gains found in some studies. For example, Binder-Macleod and Barrish (1992) reported a 20% increase in force and 50% reduction in time to a target force when a two-pulse rapid burst was applied prior to a lower-frequency train of stimulation in rat soleus muscle. Given that muscles that contribute largely to SSC movements (e.g. gastrocnemius) work on the ascending limb of their force–length curve (i.e. at relatively short lengths) and are required to produce high isometric and concentric forces during the propulsive phase, it is perhaps important to note that the potentiating effect of high-frequency bursts is greater when muscles are at a shorter length (Lee, Gerdom and Binder-Macleod, 1999; Mela *et al.*, 2002; Sandercock and Heckman, 1997), are contracting isometrically (Callister, Reinking and Stuart, 2003; Sandercock and Heckman, 1997) or concentrically (Binder-Macleod and Lee, 1996; Callister, Reinking and Stuart, 2003; Sandercock and Heckman, 1997), and when the muscle force is higher (Lee, Becker and Binder-Macleod, 2000). Nonetheless, performance enhancement during SSC actions has not been explicitly tested, and, although it is known that variable stimulation influences force production (Maladen *et al.*, 2007), it has not been determined whether high-frequency bursts delivered after muscle force has risen during normal human movements result in a force augmentation. Further research is required to determine whether the catch-like property of muscle can be exploited during SSC actions, and whether the brief high-frequency stretch reflex burst is appropriate to elicit this response.

It should also be remembered that in some SSC movements the stretch reflex is thought to contribute to a substantial increase in muscle activity (measured as an increased EMG amplitude) compared to the concentric-only condition (120–190% increase; Trimble, Kukulka and Thomas, 2000), although some of this increase might be associated with the greater time available for muscle activation and the preload effect of the countermovement phase. Indeed, Trimble, Kukulka and Thomas (2000) showed that brief high-frequency (100 Hz) nerve stimulations of the tibial nerve, of magnitudes that would evoke an H- but not an M-wave response (i.e. reflex, but not direct, muscle activation), produced negligible EMG activity at rest but significantly increased EMG values when imposed over an MVC. Thus, the addition of a reflex component appears to be able to substantially increase activity in contracting muscles. Regardless of whether an increase in muscle activity is seen, it is likely that stretch reflexes increase muscle force and stiffness, and therefore SSC performance.

2.8.2.3 Summary of mechanisms

The increase in concentric power output achieved when a rapid stretch of the MTU precedes it is probably a result of several mechanisms, which might include the contribution of elastic energy (i.e. recoil of the SEC), force potentiation from rapid muscle stretch, an increased time for muscle activation, the preload effect, force and stiffness augmentation from stretch reflexes, and a unique interaction between muscle and tendon that allows the muscle to operate at lower shortening speeds and over shorter distances. There is still considerable doubt over whether force potentiation can contribute to SSC performance, although no conclusive data negate the possibility. The relative contribution of the other mechanisms is debated, although cumulative evidence suggests that: (1) increased time for activation and preload effects play a more important role in larger-range-of-motion, slower movements such as the vertical jump; (2) stretch reflexes have their greatest influence in moderate-duration tasks such as hopping and slow running; and (3) the contribution of stored elastic energy increases with (particularly eccentric) movement speeds, at least until eccentric loading is great enough to cause the muscle to yield. Clearly, there is a need for considerable research into the mechanisms of performance enhancement with SSC use.

2.8.3 FORCE UNLOADING: A REQUIREMENT FOR ELASTIC RECOIL

It is clear that SSC actions involve the storage of elastic energy in elastic structures, which comes from the kinetic energy of the countermovement (in jumps, this kinetic energy results from the gravitational potential energy of the body prior to the jump) and the work done by the muscles. What is rarely considered is that, counter-intuitively, elastic energy can only be recovered when force decreases; it is worth looking at how this happens.

Muscle–tendon forces peak when their shortening velocity is close to zero, and maximum energy is stored in the tendons. However, muscle force decreases as movement velocity increases in the concentric phase, according to the force–velocity relationship. This allows the tendons to recoil with a force proportional to their stiffness, according to Hooke's law ($F = -kx$). This high tendon recoil speed allows the muscle to shorten slower, and thus muscle force can remain higher than it would have been if it were responsible for the total MTU shortening. In effect, the rapid shortening of tendons occurs because the increasing muscle shortening speed ensures that muscle force decreases.

This mechanism of tendon recoil is assisted by the changing moment arm about the distal joints (Carrier, Heglund and Earls, 1994; Roberts and Marsh, 2003). At the ankle joint during hopping, for example, the external moment arm of force is large (approximately the length of the foot) and the internal moment arm is small (the distance from the ankle's joint centre to the Achilles tendon). In this case there is a large gear ratio, as shown in Figure 2.8.5. As an example, an external force of, say, 1000 N would create a joint moment of ~200 Nm. This needs to be exceeded in order to jump, so the plantar flexor muscles must produce a force greater than ~3330 N, based on a moment arm of 6 cm (Maganaris, Baltzopoulos and Sargeant, 1998). This large force causes little joint angular acceleration (plantar flexion) because the joint torque only slightly exceeds the torque created by the ground-reaction force (GRF). However, the Achilles tendon is stretched considerably by the muscle force. As plantar flexion continues, the external moment arm

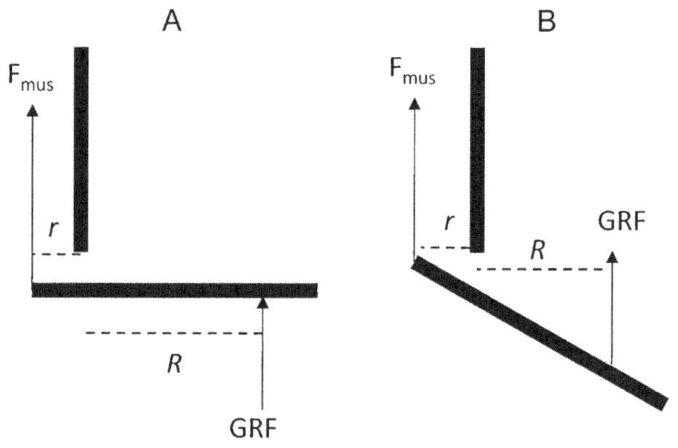

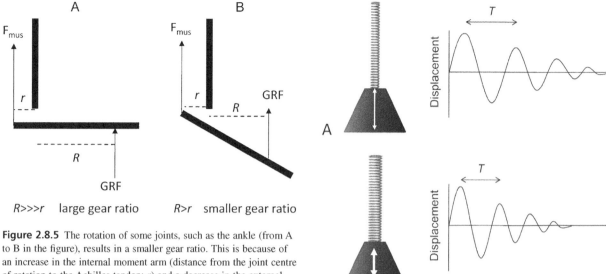

Figure 2.8.5 The rotation of some joints, such as the ankle (from A to B in the figure), results in a smaller gear ratio. This is because of an increase in the internal moment arm (distance from the joint centre of rotation to the Achilles tendon: *r*) and a decrease in the external moment arm (distance from the GRF to the joint centre: *R*). The unfavourable gear ratio in A ensures that a large muscle force causes little joint rotation and maximum tendon elongation, whereas the favourable gear ratio in B allows a faster muscle shortening speed and hence results in a decreased muscle force, allowing tendon recoil to occur

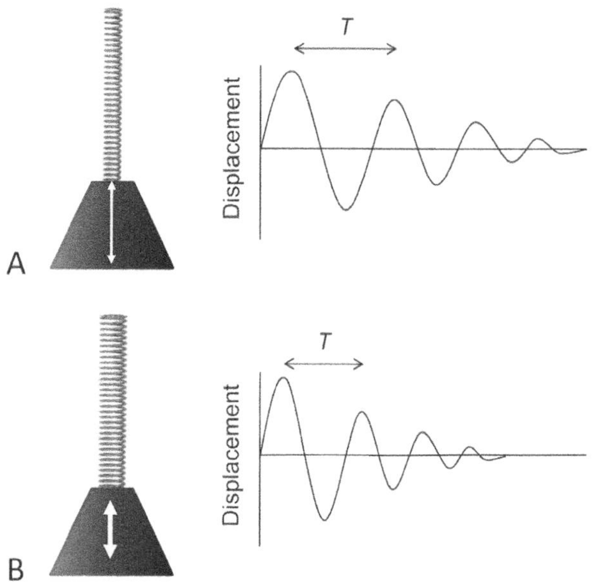

Figure 2.8.6 The natural frequency of oscillation of a spring system is a function of the spring stiffness. If the load on the system is the same, spring A will oscillate slower (longer period of oscillation, *T*) than spring B when perturbed. It should also be noted that the amplitude of oscillation of the stiffer spring (B) will be smaller, but the rate of energy dissipation (seen as an amplitude reduction) will be equal. Muscle–tendon systems operate according to the same principles

decreases and the internal moment arm increases. Thus, the relative mechanical advantage of the plantar flexors is increased (decreased gear ratio) and the larger joint moment results in substantial joint rotation. The increasing joint angular velocity necessitates an increase in the muscle shortening speed and hence muscle force decreases, allowing the tendon to recoil. So the changing moment arm about the ankle joint provides a condition for energy storage early in the movement and for tendon recoil later in the movement. In humans, moment arms at joints such as the ankle (Maganaris, Baltzopoulos and Sargeant, 1998) and elbow (Ettema, Styles and Kippers, 1998) increase with joint extension, which is ideal for optimizing tendon energy stretch and recoil.

2.8.4 OPTIMUM MTU PROPERTIES FOR SSC PERFORMANCE

Properties of muscles and tendons that would be optimum for a muscle–tendon system must vary as the force–time characteristics vary. It is well established that spring-like systems operate most efficiently when the forces driving them act in resonance (i.e. at the natural frequency) with the spring. This can be seen in a simple experiment where a small weight attached to a spring is perturbed so that is oscillates (Figure 2.8.6): if the same weight is attached to a stiffer spring, the oscillation frequency will increase; that is, the natural frequency increases with spring stiffness. Thus, SSC performance will be optimized when the natural frequency of the MTU system matches the movement frequency. This principle has been well demonstrated for SSC movements such as the bench-press (Wilson, Wood and Elliot, 1991b). It follows that movements requiring large forces to be produced in short time intervals, such as in the contact phase of sprint running, will likely benefit from MTUs being relatively stiff, whereas movements requiring smaller forces to be produced over longer time intervals, such as a vertical jump, might benefit from MTUs being relatively compliant. An aim of future research is to develop practical methods of measuring muscle–tendon stiffness in different muscle groups and then determining the optima for movements performed by different individuals and with different force–time characteristics.

Of course, regardless of the loads imposed on the system, faster recoil speeds will be realised with stiffer systems since the acceleration of a mass is dependent on the force applied (according to Newton's second law, $F = m \times a$) and the restoring force of a spring increases with spring stiffness (according to Hooke's law, $F = -kx$). The caveat is that a large enough muscle force must be applied to stretch the stiffer tendon in order to allow that force to be produced over sufficient time for acceleration to occur (according to the impulse–momentum relationship, $Ft = \Delta mv$). This might also require a longer activation time since muscle force increases relatively slowly.

Attention must be given to this because (1) fatigue is increased as the requirement for force production increases (Kram and Taylor, 1990) and (2) lower force production, which can be caused by the fatigue, will lead to reduced energy storage and a decrease in movement performance. For a sprint runner, the beneficial effects of an increased recoil speed from a stiffer MTU might outweigh the increased energy cost, but for other athletes this will not necessarily be the case. So an optimum stiffness must be found for each MTU, for each individual, and for movements with different force–time characteristics.

2.8.5 EFFECTS OF THE TRANSITION TIME BETWEEN STRETCH AND SHORTENING ON SSC PERFORMANCE

It is well documented that an SSC action is one in which MTU lengthening precedes rapid shortening; however, it is also often considered that a minimum delay between these phases is essential (e.g. Komi and Gollhofer, 1997). One reason for this is that imposing a delay between the phases substantially reduces concentric movement performance (Wilson, Elliot and Wood, 1991a) and efficiency (Thys, Faraggiana and Margaria, 1972; Henchoz *et al.*, 2006); both performance and efficiency decay exponentially with time, with a half-life of ~0.5–0.9 s (Thys, Faraggiana and Margaria, 1972; Wilson, Elliot and Wood, 1991a). While the mechanisms underpinning this phenomenon remain to be fully described, it is generally accepted that two are chiefly at fault: (1) decreased muscle force prior to the concentric phase; and (2) the dissipation of stored elastic energy.

The loss of muscle force is probably a consequence of several factors. First, the reflexes that might be initiated by the rapid muscle stretch in an SSC have very short response times (<200 ms), so delaying the concentric phase will limit the opportunities for the stretch reflex to augment force output. Second, the preload effect of a countermovement allows a muscle to build up very large forces, but relatively little force is required to maintain a posture in the absence of a preload; for example, it takes less force to hold a squatting position than to decelerate the body during the countermovement phase of a vertical jump. So muscle force necessarily drops when a pause is allowed between the eccentric and concentric phases. The loss of elastic potential energy is largely a consequence of this drop in muscle force because: (1) the pause allows for cross-bridge cycling to occur, so elastic energy stored is dissipated as heat; and (2) the decrease in muscle force allows the tendon to shorten and dissipate its energy slowly. Thus, the minimization of the transition time is essential for optimum SSC performance.

2.8.6 CONCLUSION

SSCs are commonly performed with the aim of improving movement speed, power, and/or movement economy. SSC actions are optimized when the natural frequency of the MTU system matches the movement frequency (i.e. stiffness is optimized) and when there is minimal delay between eccentric and concentric forces; however, it should be remembered that stiffer MTUs provide greater restoring forces and thus faster recoil speeds, but compliant MTUs store more energy for a given imposed muscle force. The eccentric, or countermovement, phase provides conditions for greater concentric performance through at least six main mechanisms, whose contributions vary according to the force–time characteristics of the SSC movement. Broadly speaking: (1) the contribution of elastic energy seems to increase with movement speed (i.e. shorter SSC duration) because total lengthening–shortening decreases for muscle but increases for tendon; (2) the benefit of having an increased time for muscle activation is probably greater for SSC movements with longer durations; (3) the importance of the preload effect is probably greater when there is a greater change in momentum between eccentric and concentric phases; (4) the beneficial effect of the stretch reflex seems to be greatest in movements of moderate SSC duration (i.e. it is inconsequential for fast movements lasting less than ~120 ms and in movements with very low rates of muscle stretch or slow eccentric–concentric transitions); (5) there is still debate as to whether the force potentiation phenomenon contributes to SSC performance, although rapid stretch can increase muscle force and stiffness because of muscle's viscous properties; and (6) the opportunity for muscles to contract over shorter ranges and at slower speeds contributes significantly to movement economy and (in addition to the restoring force of the SEC) to an increase in force development at high MTU shortening speeds, which increases MTU power production. Although more research is required to understand the relative importance of these mechanisms, there is probably an even greater need to develop tests to easily measure MTU properties (including their stiffness, etc.) and determine which properties suit SSC movements with specific force–time characteristics. Such tests would provide the practitioner with the opportunity of optimizing SSC performance in both clinical and athletic populations.

References

Abbott B.C. and Aubert X.M. (1952) The force exerted by active striated muscle during and after change of length. *J Physiol (Lond)*, **117**, 77–86.

Alexander R.M. (1987) The spring in your step. *New Sci*, **114**, 42–44.

Bennet-Clark H.C. and Lucey E.C.A. (1967) The jump of the flea: a study of the energetics and a model of the mechanism. *J Exp Biol*, **47**, 59–76.

Bennett M.B., Ker R.F., Dimery N.J. and Alexander R.M. (1986) Mechanical properties of various mammalian tendons. *J Zool Ser A*, **209**, 537–548.

Best T.M., McElhaney J., Garrett W.E. and Myers B.S. (1994) Characterization of the passive responses of live skeletal muscle using the quasi-linear theory of viscoelasticity. *J Biomech*, **27**, 413–419.

Binder-Macleod S.A. and Barrish W.J. (1992) Force response of rat soleus muscle to variable-frequency train stimulation. *J Neurophysiol*, **68**, 1068–1078.

Binder-Macleod S.A. and Lee S.C.K. (1996) Catchlike property of human muscle during isovelocity movements. *J Appl Physiol*, **80**, 2051–2059.

Bobbert M.F., Gerritsen K.G.M., Litjens M.C.A. and van Soest A.J. (1996) Why is countermovment jump height greater than squat jump height? *Med Sci Sports Exerc*, **28**, 1402–1412.

Bojsen-Møller J., Hansen P., Aagaard P. *et al.* (2004) Differential displacement of the human soleus and medial gastrocnemius aponeuroses during isometric plantar flexor contractions in vivo. *J Appl Physiol*, **97**, 1908–1914.

Bosco C., Saggini R. and Viru A. (1997) The influence of different floor stiffness on mechanical efficiency of leg extensor muscle. *Ergonomics*, **40**, 670–679.

Brooks S.V., Faulkner J.A. and McCubbrey D.A. (1990) Power outputs of slow and fast skeletal muscles of mice. *J Appl Physiol*, **68**, 1282–1285.

Brown I.E. and Loeb G.E. (2000) Measured and modeled properties of mammalian skeletal muscle: III. The effects of stimulus frequency on stretch-induced force enhancement and shortening-induced force depression. *J Muscle Res Cell Motil*, **21**, 21–31.

Burke R.E., Rudomin P. and Zajac F.E. (1970) Catch property in single mammalian motor units. *Science*, **168**, 122–124.

Burrows M. (2003) Froghopper insects leap to new heights. *Nature*, **424**, 509.

Callister R.J., Reinking R.M. and Stuart D.G. (2003) Effects of fatigue on the catchlike property in turtle hindlimb muscle. *J Comput Physiol A*, **189**, 857–866.

Carrier D.R., Heglund N.C. and Earls K.D. (1994) Variable gearing during locomotion in the human musculoskeletal system. *Science*, **265**, 651–653.

Cavagna G.A., Saibene F.P. and Margaria R. (1964) Mechanical work in running. *J Appl Physiol*, **19**, 249–256.

Close R.I. (1972) Dynamic properties of mammalian skeletal muscles. *Physiol Rev*, **52**, 129–197.

Cordo P.J. and Rymer W.Z. (1982) Motor-unit activation patterns in lengthening and isometric contractions of hindlimb extensor muscles in the decerebrate cat. *J Neurophysiol*, **47**, 782–796.

de Graaf J.B., Bobbert M.F., Tetteroo W.E. and van Ingen Schenau G.J. (1987) Mechanical output about the ankle in countermovement jumps and jumps with extended knee. *Hum Mov Sci*, **6**, 333–347.

Dyhre-Poulsen P., Simonsen E.B. and Voigt M. (1991) Dynamic control of muscle stiffness and H reflex modulation during hopping and jumping in man. *J Physiol (Lond)*, **437**, 287–304.

Edman K.A.P., Elzinga G. and Noble M.I.M. (1978) Enhancement of mechanical performance by stretch during tetanic contractions of vertebrate skeletal muscle fibres. *J Physiol (Lond)*, **281**, 139–155.

Ettema G.J.C. (1996) Mechanical efficiency and efficiency of storage and release of series elastic energy in skeletal muscle during stretch–shorten cycles. *J Exp Biol*, **199**, 1983–1997.

Ettema G.J.C., Styles G. and Kippers V. (1998) The moment arms of 23 muscle segments of the upper limb with varying elbow and forearm positions: implications for motor control. *Hum Mov Sci*, **17**, 201–220.

Finni T. (2001) Muscle mechanics during human movement revealed by in vivo measurements of tendon force and muscle length. *Studies in Sport, Physical Education and Health*, **78**. PhD Thesis, University of Jyväskylä.

Finni T., Ikegawa S. and Komi P.V. (2001) Concentric force enhancement during human movement. *Acta Physiol Scand*, **173**, 369–377.

Finni T., Ikegawa S., Lepola V. and Komi P.V. (2003) Comparison of force–velocity relationships of vastus lateralis muscle in isokinetic and in stretch-shortening cycle exercises. *Acta Physiol Scand*, **177**, 483–491.

Flitney F.W. and Hirst D.G. (1978) Cross-bridge detachment and sarcomere 'give' during stretch of active frog's muscle. *J Physiol (Lond)*, **276**, 449–465.

Fukashiro S. and Komi P.V. (1987) Joint moment and mechanical power flow of the lower limb during vertical jump. *Int J Sports Med*, **8**, 15–21.

Fukunaga T., Roy R.R., Shellock J.A. *et al.* (1996) Specific tension of human plantar flexors and dorsiflexors. *J Appl Physiol*, **80**, 158–165.

Fukunaga T., Kubo K., Kawakami Y. *et al.* (2001) In vivo behaviour of human muscle tendon during walking. *Proc R Soc B Biol Sci*, **268**, 229–233.

Grazi E. (2008) Water and muscle contraction. *Int J Mol Sci*, **9**, 1435–1452.

Henchoz Y., Malatesta D., Gremion G. and Belli A. (2006) Effects of the transition time between muscle–tendon stretch and shortening on mechanical efficiency. *Eur J Appl Physiol*, **96**, 665–671.

Herzog W. and Leonard T.R. (2000) The history dependence of force production in mammalian skeletal muscle following stretch-shortening and shortening-stretch cycles. *J Biomech*, **33**, 531–542.

Hill A.V. (1938) The heat of shortening and the dynamic constants of muscle. *Proc R Soc B Biol Sci*, **138**, 136–195.

Hof A.L., Geelen B.A. and van den Berg J.W. (1983) Calf muscle moment, work and efficiency in level walking; role of series elasticity. *J Biomech*, **16**, 523–537.

Hof A.L., van Zandwijk J.P. and Bobbert M.F. (2002) Mechanics of human triceps surae muscle in walking, running and jumping. *Acta Physiol Scand*, **174**, 17–30.

Hoffer J.A. and Andreassen S. (1981) Regulation of soleus muscle stiffness in premammillary cats: Intrinsic and reflex components. *J Neurophysiol*, **45**, 267–285.

Houk J.C. (1979) Regulation of stiffness by skeletomotor reflexes. *Annu Rev Physiol*, **41**, 99–114.

Ingen Schenau G.J., van Dorssers W.M.M., Welter T.G. *et al.* (1995) The control of mono-articular muscles in multi-joint leg extensions in man. *J Physiol (Lond)*, **484**, 247–254.

Ingen Schenau G.J., van Bobbert M.F. and de Haan A. (1997) Does elastic energy enhance work and efficiency in the stretch-shortening cycle? *J Appl Biomech*, **13**, 389–415.

Ishikawa M. and Komi P.V. (2004) Effects of different dropping intensities on fascicle and tendinous tissue behaviour during stretch-shortening cycle exercise. *J Appl Physiol*, **96**, 848–852.

Ishikawa M. and Komi P.V. (2007) The role of the stretch reflex in the gastrocnemius muscle during human locomotion at various speeds. *J Appl Physiol*, **103**, 1030–1036.

Ishikawa M., Finni T. and Komi P.V. (2003) Behaviour of vastus lateralis muscle-tendon during high intensity SSC exercises in vivo. *Acta Physiol Scand*, **178**, 205–213.

Ishikawa M., Niemelä E. and Komi P.V. (2005) Interaction between fascicle and tendinous tissues in short-contact stretch-shortening cycle exercise with varying eccentric intensities. *J Appl Physiol*, **99**, 217–223.

Ishikawa M., Komi P.V., Finni T. and Kuitunen S. (2006) Contribution of the tendinous tissue to force enhancement during stretch-shortening cycle exercise depends on the prestretch and concentric phase intensities. *J Electromyogr Kinesiol*, **16**, 423–431.

Josephson R.K. (1993) Contraction dynamics and power output of skeletal muscle. *Annu Rev Physiol*, **55**, 527–546.

Kawakami Y., Muraoka T., Ito S. *et al.* (2002) In vivo muscle fibre behaviour during counter-movement exercise in humans reveals a significant role for tendon elasticity. *J Physiol*, **540**, 635–646.

Ker R.F., Bennett M.B., Bibby S.R. *et al.* (1987) The spring in the arch of the human foot. *Nature*, **325**, 147–149.

Kilani H.A., Palmer, S.S., Adrian, M.J. and Gapsis J.J. (1989) Block of the stretch reflex of vastus lateralis during vertical jumps. *Hum Mov Sci*, **8**, 247–269.

Komi P.V. (2003) Stretch-shortening cycle, in *Strength and Power in Sport* (ed. P.V. Komi), Blackwell Science, Oxford, UK, pp. 184–202.

Komi P.V. and Gollhofer A. (2007) Stretch reflexes can have an important role in force enhancement during SSC exercise. *J Appl Biomech*, **13**, 451–460.

Kram R. and Taylor C.R. (1990) Energetics of running: a new perspective. *Nature*, **346**, 265–267.

Kurokawa S., Fukunaga T., Nagano A. and Fukahiro S. (2003) Interaction between fascicles and tendinous structures during counter movement jumping investigated in vivo. *J Appl Physiol*, **95**, 2306–2314.

Lamontagne A., Malouin F. and Richards C.L. (1997) Viscoelastic behavior of plantar flexor muscle-tendon unit at rest. *J Occup Sports Phys Ther*, **26**, 244–252.

Lee S.C., Gerdom M.L. and Binder-Macleod S.A. (1999) Effects of length on the catchlike property of human quadriceps femoris muscle. *Phys Ther*, **79**, 738–748.

Lee S.C., Becker C.N. and Binder-Macleod S.A. (2000) Activation of human quadriceps femoris muscle during dynamic contractions: effects of load on fatigue. *J Appl Physiol*, **89**, 926–936.

Lichtwark G.A. and Wilson A.M. (2005) In vivo mechanical properties of the human Achilles tendon during one-legged hopping. *J Exp Biol*, **208**, 4715–4725.

Lichtwark G.A., Bougoulias K. and Wilson A.M. (2007) Muscle fascicle and series elastic element length changes along the length of the human gastrocnemius during walking and running. *J Biomech*, **40**, 157–164.

Lombardi V. and Piazzesi G. (1990) The contractile response during lengthening of stimulated frog muscle fibres. *J Physiol (Lond)*, **431**, 141–171.

Maganaris C.N. and Paul J.P. (2000) Hysteresis measurements in intact human tendon. *J Biomech*, **33**, 1723–1727.

Maganaris C.N. and Paul J.P. (2002) Tensile properties of the in vivo human gastrocnemius tendon. *J Biomech*, **35**, 1639–1646.

Maganaris C.N., Baltzopoulos V. and Sargeant A.J. (1998) Changes in Achilles tendon moment arm from rest to maximum isometric plantarflexion: in vivo observations in man. *J Physiol*, **510**, 977–985.

Maladen R.D., Perumal R., Wexler A.S. and Binder-Macleod S.A. (2007) Effects of activation pattern on nonisometric human skeletal muscle performance. *J Appl Physiol*, **102**, 1985–1991.

Markovic G., Dizdar D., Jukic I. and Cardinale M. (2004) Reliability and factorial validity of squat and countermovement jump tests. *J Strength Cond Res*, **18**, 551–555.

Mela P., Veltink P.H., Huijing P.A. *et al.* (2002) The optimal stimulation pattern for skeletal muscle is dependent on length. *IEEE Trans Neural Syst Rehabil Eng*, **10**, 85–93.

Melvill-Jones G. and Watt D.G.D. (1971) Observations on the control of stepping and hopping movements in man. *J Physiol (Lond)*, **219**, 709–727.

Méndez J. and Keys A. (1960) Density and composition of mammalian muscle. *Metabolism*, **9**, 184–188.

Mutungi G. and Ranatunga K.W. (1996) The viscous, viscoelastic and elastic characteristics of resting fast and slow mammalian (rat) muscle fibres. *J Physiol (Lond)*, **496**, 827–836.

Nichols T.R. and Houk J.C. (1976) Improvement in linearity and regulation of stiffness that results from actions of stretch reflex. *J Neurophysiol*, **39**, 119–142.

Nicol C. and Komi P.V. (1998) Significance of passively induced stretch reflexes on Achilles tendon force enhancement. *Muscle Nerve*, **21**, 1546–1548.

Pollock C.M. and Shadwick R.E. (1994) Relationship between body mass and biomechanical properties of limb tendons in adult mammals. *Am J Physiol Regul Integr Comp Physiol*, **266**, R1016–R1021.

Rassier D.E. and Herzog W. (2004) Considerations on the history dependence of muscle contraction. *J Appl Physiol*, **96**, 419–427.

Roberts T.J. and Marsh R.L. (2003) Probing the limits to muscle-powered accelerations: lessons from jumping bullfrogs. *J Exp Biol*, **206**, 2567–2580.

Sandercock T.G. and Heckman C.J. (1997) Doublet potentiation during eccentric and concentric contractions of cat soleus muscle. *J Appl Physiol*, **82**, 1219–1228.

Sinkjaer T., Toft E., Andreassen S. and Hornemann B.C. (1998) Muscle stiffness in human dorsiflexors: Intrinsic and reflex components. *J Neurophysiol*, **60**, 1110–1120.

Sousa F., Ishikawa M., Vilas-Boas J.P. and Komi P.V. (2007) Intensity- and muscle-specific fascicle behaviour during human drop jumps. *J Appl Physiol*, **102**, 382–389.

Thys H., Faraggiana T. and Margaria R. (1972) Utilization of muscle elasticity in exercise. *J Appl Physiol*, **32**, 491–494.

Trimble M.H., Kukulka C.G. and Thomas R.S. (2000) Reflex facilitation during the stretch-shortening cycle. *J Electromyogr Kinesiol*, **10**, 179–187.

Voigt M., Simonsen E.B., Dyhre-Poulsen P. and Klausen K. (1995) Mechanical and muscular factors influencing the performance in maximal vertical jumping after different prestretch loads. *J Biomech*, **28**, 293–307.

Voigt M., Dyhre-Poulsen P. and Simonsen E.B. (1998) Modulation of short latency stretch reflexes during human hopping. *Acta Physiol Scand*, **163**, 181–194.

Vos E.J., Harlaar J. and van Ingen Schenau G.J. (1991) Electromechanical delay during knee extensor contractions. *Med Sci Sports Exerc*, **23**, 1187–1193.

Walshe A.D., Wilson G.J. and Ettema G.J.C. (1998) Stretch-shorten cycle compared with isometric preload: contributions to enhanced muscular performance. *J Appl Physiol*, **84**, 97–106.

Wilson G.J., Elliot B.C. and Wood G.A. (1991a) The effect on performance of imposing a delay during a stretch-shorten cycle movement. *Med Sci Sports Exerc*, **23**, 364–370.

Wilson G.J., Wood G.A. and Elliot B.C. (1991b) Optimal stiffness of series elastic component in a stretch-shorten cycle activity. *J Appl Physiol*, **70**, 825–833.

2.9 Repeated-sprint Ability (RSA)

David Bishop[1] and Olivier Girard[2], [1]Institute of Sport, Exercise and Active Living (ISEAL), School of Sport and Exercise Science, Victoria University Melbourne, Australia, [2]ASPETAR, Qatar Orthopaedic and Sports Medicine Hospital, Doha, Qatar

2.9.1 INTRODUCTION

High-intensity sprints of short duration, interspersed with brief recoveries, are common during most team sports (Spencer *et al.*, 2005). The ability to recover and reproduce performance in subsequent sprints is therefore important for team-sport athletes and has been termed repeated-sprint ability (RSA). As RSA has not been strongly correlated with either aerobic power or anaerobic capacity (Bishop, Lawrence and Spencer, 2003; Wadley and Le Rossignol, 1998), this suggests that RSA is a specific quality and that specific tests are required to evaluate this fitness component (see Chapter 3.3). Tests of RSA predict both the distance of high-intensity running (>19.8 km/hour) and the total sprint distance during a professional soccer match (Rampinini *et al.*, 2007). RSA is therefore a specific fitness requirement of team-sport athletes and it is important to better understand the factors which can limit and improve it.

2.9.1.1 Definitions

The main feature of repeated-sprint exercise (RSE) is the alteration of brief, maximal-intensity sprint bouts with periods of incomplete recovery (consisting of complete rest or low- to moderate-intensity activity). However, there is potential for confusion as some authors have also used the word 'sprint' to describe exercise lasting 30 seconds or more (Bogdanis *et al.*, 1996; Sharp *et al.*, 1986). For the purposes of this chapter, the definition of 'sprint' activity will be limited to brief exercise bouts, in general ≤10 seconds, where peak intensity (power/velocity) can be maintained until the end of the entire bout. Longer-duration, maximal-intensity exercise, where there is a considerable decrease in performance, will be referred to as 'all-out' exercise (Figure 2.9.1).

When sprints are repeated, it is also useful to define two different types of exercise: intermittent-sprint exercise and RSE. Intermittent-sprint exercise can be characterized by short-duration sprints (≤10 seconds), interspersed with recovery periods long enough (60–300 seconds) to allow near-complete recovery of sprint performance (Balsom *et al.*, 1992a). In comparison, RSE is characterized by short-duration sprints (≤10 seconds) interspersed with brief recovery periods (usually ≤60 seconds). The main difference is that during intermittent-sprint exercise there is little or no performance decrement (Balsom *et al.*, 1992b; Bishop and Claudius, 2005), whereas during RSE there is a marked performance decrement (Bishop *et al.*, 2004) (Figure 2.9.2).

2.9.1.2 Indices of RSA

During RSE, fatigue typically manifests as a decline in maximal sprint speed (running), or a decrease in peak power or total work (cycling), over sprint repetitions (Figure 2.9.2). To quantify the amount of fatigue experienced during RSE, researchers have tended to use one of two terms: the fatigue index (FI) or the percentage decrement score (S_{dec}) (Glaister *et al.*, 2008). These terms are described in more detail in Chapter 3.3. Other fatigue indices, such as changes in the subsequent maximal voluntary contraction torque and decreases in the pedalling rate or moment, have also been used to describe performance decrement (Billaut, Basset and Falgairette, 2005; Hautier *et al.*, 2000; Racinais *et al.*, 2007). These fatigue-induced modifications are generally inversely related to the initial sprint performance (Hamilton *et al.*, 1991; Mendez-Villanueva, Hamer and Bishop, 2008).

2.9.2 LIMITING FACTORS

The degree of fatigue experienced during RSE is largely influenced by the particulars of the task: the intensity and duration of each exercise bout, the recovery time between sprint bouts, the nature of the recovery, the preceding number of high-intensity exercise bouts (Glaister, 2005). Other factors such as time of day (Giacomoni, Billaut and Falgairette, 2006; Racinais *et al.*, 2005), the sex, age and training status of subjects (Abrantes, Macas and Sampaio, 2004; Ratel *et al.*, 2002, 2005), and whether or not subjects are sickle cell trait carriers (Connes *et al.*, 2006) also influence the ability to resist fatigue during RSE.

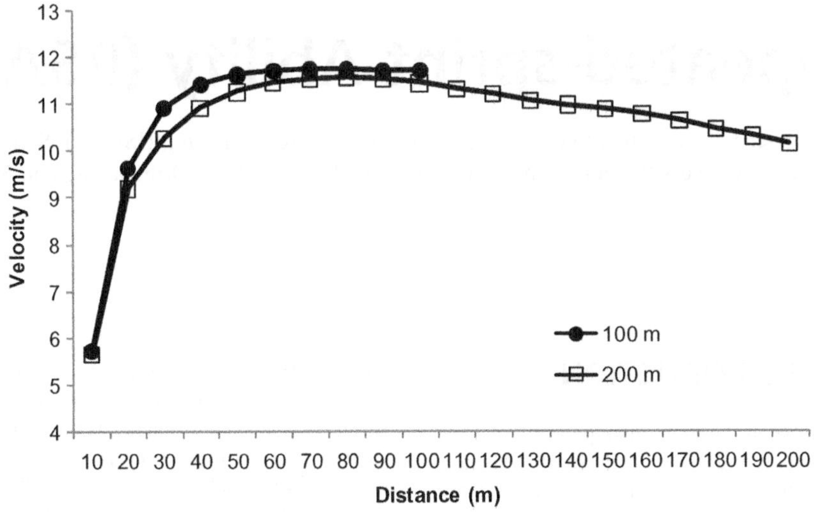

Figure 2.9.1 Sprint profiles of 100 (circle) and 200 m (square) world records for men. The 100 m performance would be defined as a 'sprint' exercise, while the 200 m performance would be defined as an 'all-out' exercise

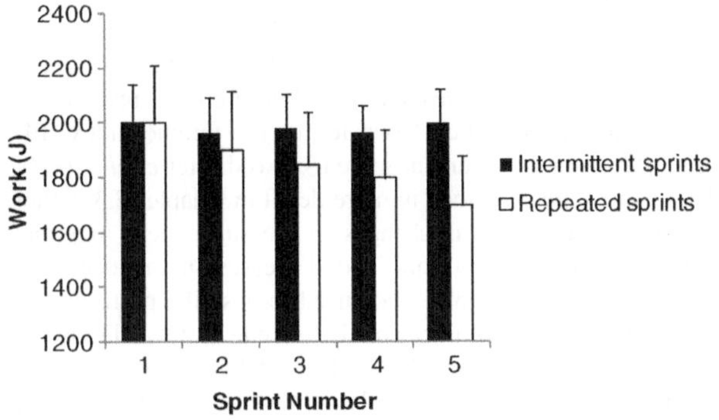

Figure 2.9.2 Graph showing the effect of rest duration on maximal four-second, bicycle-ergometer sprint performance. Intermittent sprints were performed every two minutes (Bishop and Claudius, 2005), whereas repeated sprints were executed every 30 seconds (Bishop *et al.*, 2004)

As a consequence of the unpredictable changes that occur during field testing, most of our knowledge concerning fatigue-induced adaptations during RSE has been gained from cycle-based repeated sprinting performed in the laboratory environment. The strength of this approach is that most of the influencing variables can be accurately controlled and manipulated. However, the applicability of findings arising from laboratory settings has been questioned (Glaister *et al.*, 2006).

The complex nature of muscle fatigue is also highlighted by the considerable number of approaches, models, and indices (see Section 2.9.1) that have been used to account for the decline in muscular performance. It is now accepted that rather than a single factor, the decline in performance during RSE might involve changes within the muscle cell itself (muscular factors) and/or neural adjustments (neuromechanical factors) (Figure 2.9.3). An improved understanding of the factors precipitating fatigue might allow for the better design of interventions to improve RSE.

2.9.2.1 Muscular factors

Muscle excitability

At the skeletal muscle level, dramatic changes in transmembrane distribution of Na^+ and K^+ have been observed following high-intensity exercise (Allen, Lamb and Westerblad, 2008). During intense contractions, the Na^+–K^+ pump cannot readily reaccumulate the K^+ into the muscles cells, inducing up to a doubling of the extracellular K^+ concentration (Clausen *et al.*, 1998). Such marked cellular K^+ efflux may cause muscle membrane depolarization and in turn reduce muscle excitability (Clausen *et al.*, 1998). By applying an electrical stimulus to peripheral nerves, the study of the muscle compound action potential (M-wave) characteristics has been used to determine whether fatigue alters muscle excitability during multiple-sprint activities (e.g. tennis; Girard *et al.*, 2008). Decreased M-wave amplitude, but not duration, was reported after a RSE, suggesting that action potential synaptic transmission, rather than propagation (impulse-conduction velocity along the sarcolemma), may be impaired during such exercise (Girard *et al.*, 2007). This suggests that interventions which can increase the content and/or activity of the Na^+–K^+ pump may improve RSA. However, whether a loss of membrane excitability contributes to muscle fatigue is debatable since a potentiation of the M-wave response has also been reported following RSE (Racinais *et al.*, 2007).

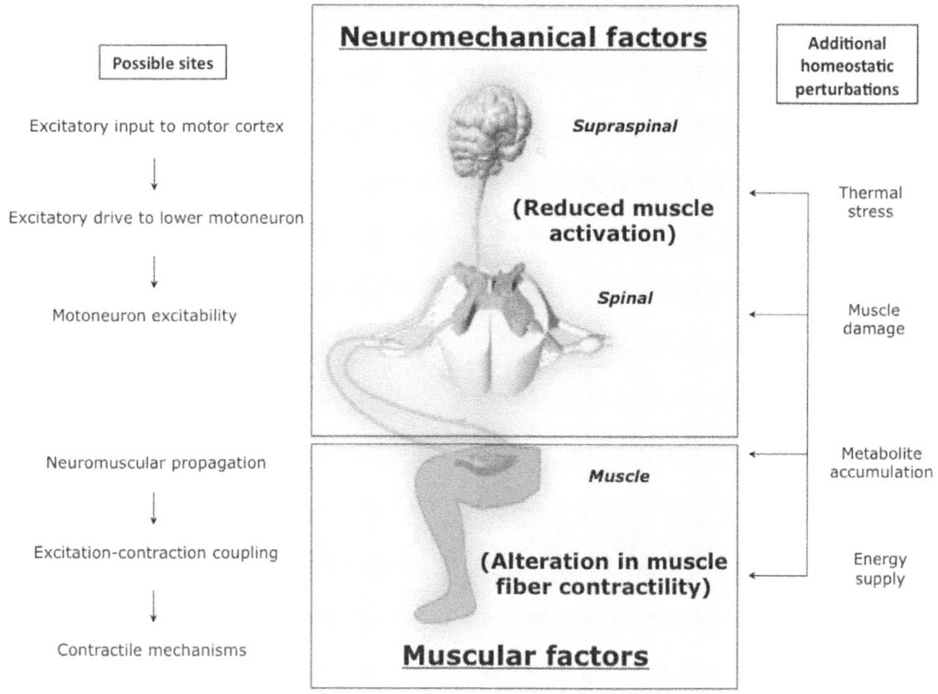

Figure 2.9.3 Potential neuromechanical and muscular sites contributing to fatigue

Limitations in energy supply

Phosphocreatine availability

Phosphocreatine (PCr) depletion after maximal sprinting has been reported to be around 35–55% of resting values (Dawson *et al.*, 1997; Gaitanos *et al.*, 1993), and the complete recovery of PCr stores can last more than five minutes (Tomlin and Wenger, 2001). As recovery times during RSE tests generally do not exceed 30 seconds, this will lead to only a partial restoration of PCr stores between two successive sprints (Bogdanis *et al.*, 1996; Dawson *et al.*, 1997). Moreover, these stores will progressively deplete with the repetition of high-intensity efforts (Gaitanos *et al.*, 1993; Yoshida and Watari, 1993). Coupled with the fact that the recovery of power output occurs in parallel with the resynthesis of PCr, several authors have proposed that performance during such an exercise mode may become increasingly limited by PCr availability; that is, there will be a decrease in the absolute contribution of PCr to the total ATP production with each subsequent sprint (Bogdanis *et al.*, 1995; Sahlin *et al.*, 1976) (Figure 2.9.4). In line with this proposition, strong relationships have been reported between the percentage of PCr resynthesis and the recovery of performance during repeated, all-out exercise bouts (Bogdanis *et al.*, 1995, 1996). Thus, performance during multiple-sprint activities may be improved by interventions (training and/or ergogenic aids) that can increase resting concentrations of PCr stores and/or the rate of PCr resynthesis.

Anaerobic glycolysis

Anaerobic glycolysis supplies approximately 40% of the total energy during a single six-second sprint (Figure 2.9.5) (Boobis, Williams and Wootton, 1982; Gaitanos *et al.*, 1993). As muscle glycogen content has been reported to range from 250 to 650 mmol/kg dw, and maximal ATP production from anaerobic glycolysis is 6–9 mmol ATP/kg dw/s (Hultman and Sjoholm, 1983; Jones *et al.*, 1985; Parolin *et al.*, 1999), this suggests that normal glycogen stores are unlikely to represent a limiting factor during RSE, especially as glycogen use decreases with subsequent sprints (Figure 2.9.6). However, reductions in dietary carbohydrate intake, and consequently large decreases in resting muscular glycogen, have been associated with a greater decline in end-pedalling frequency (last three seconds) during the last four six-second bouts of an RSE consisting of 15 bouts (Balsom *et al.*, 1999). Thus, the maintenance of appropriate muscle glycogen stores during training (e.g. with a carbohydrate supplementation for subjects on a normal diet) appears to be important in maximizing RSA (Hultman *et al.*, 1990). Interventions that are able to increase the contribution of anaerobic glycolysis during individual sprints may also improve RSA.

Oxidative metabolism

The contribution of oxidative phosphorylation to the total energy expenditure during a single short sprint is limited (<10%; McGawley and Bishop, 2008). As sprints are repeated,

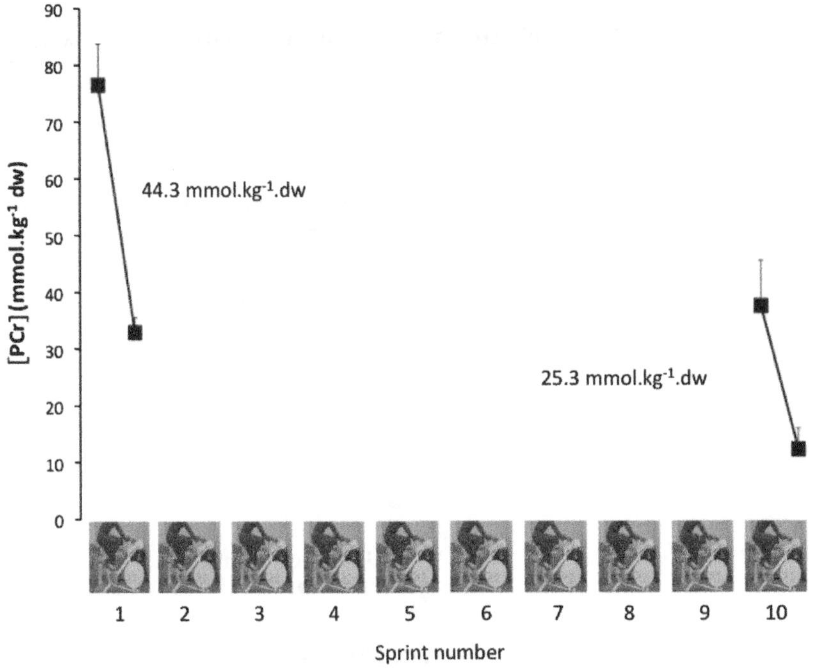

Figure 2.9.4 Changes in PCr utilization before and after the first and last sprint of a 10 × six-second repeated-sprint test (with 30 seconds of recovery between sprints) on a cycle ergometer (Gaitanos *et al.*, 1993)

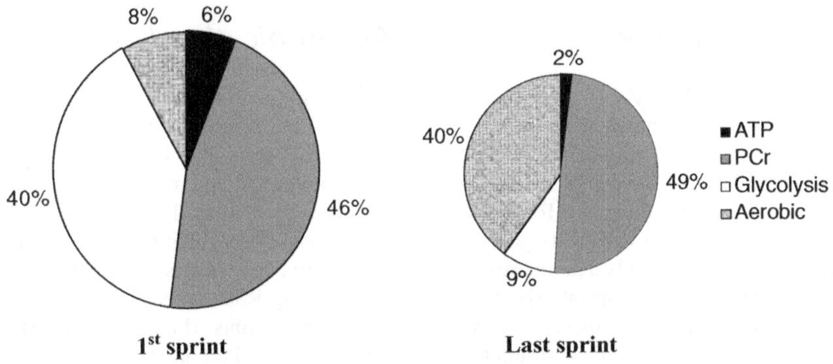

Figure 2.9.5 Changes in metabolism during the first and last sprint of a repeated-sprint exercise (Gaitanos *et al.*, 1993; McGawley and Bishop, 2008; Mendez-Villanueva, Hamer and Bishop, 2007). Note the area of each paragraph represents the total absolute energy used during each sprint

however, aerobic participation increases with time and may contribute as much as 40% of the total energy supply during the final repetitions of RSE (McGawley and Bishop, 2008). This may explain why oxygen availability, which has the potential to influence the magnitude of the aerobic contribution to ATP resynthesis (work periods) and the rate of PCr resynthesis (rest periods), has been associated with performance decrements during multiple-sprint activities (Balsom *et al.*, 1994b; Buchheit *et al.*, 2009; Haseler, Hogan and Richardson, 1999). In line with this proposition, research has shown that subjects with greater aerobic fitness (Bishop and Edge, 2006) or faster oxygen uptake (VO_2) kinetics (Dupont *et al.*, 2005) have a superior ability to resist fatigue during RSE. As subjects can reach their VO_{2max} during the final stage of a RSE (McGawley and Bishop, 2008), it is not surprising that training (Edge, Bishop and Goodman, 2005) and ergogenic aids (Balsom, Ekblom and Sjodin, 1994a) which increase VO_{2max} have also been found to improve fatigue resistance. Increasing VO_{2max} may therefore allow for a greater aerobic contribution during the latter sprints, potentially improving RSA.

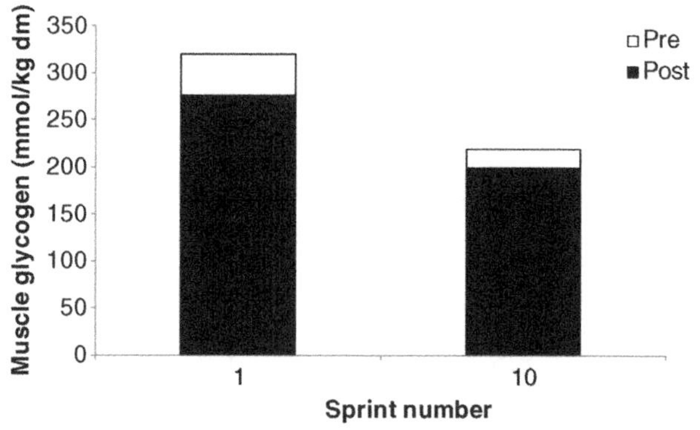

Figure 2.9.6 Muscle glycogen levels before and after the first and last sprint of a 10 × six-second repeated-sprint test (with 30 seconds of recovery between sprints) performed on a cycle ergometer (Gaitanos *et al.*, 1993)

Metabolite accumulation

Acidosis

Historically, it has been proposed that reduced pH (accumulation of H^+) in a vigorously contracting muscle, as typically observed during RSE (Bishop *et al.*, 2004), interferes with the contractile machinery and/or the rate of glycolytic reactions (Allen, Lamb and Westerblad, 2008). In support of this suggestion, correlations have been observed between RSA and both muscle buffer capacity (ßm) and changes in blood pH (Bishop and Edge, 2006; Bishop *et al.*, 2003, 2004). It has been argued that the considerable increases in muscle (Bishop and Edge, 2006; Spencer *et al.*, 2008) and blood (Bishop, Lawrence and Spencer, 2003; Ratel *et al.*, 2005) H^+ accumulation may affect sprinting performance through the inhibition of ATP generation derived from glycolysis, possibly via negative effects on phosphofructokinase and glycogen phosphorylase enzymes (Spriet *et al.*, 1989). Thus, repeated-sprint performance may be improved by interventions that can reduce the accumulation of H^+. At physiological temperatures, however, acidification as a direct cause of muscle fatigue during RSE has been challenged for at least three reasons: (1) the time course of the recovery of force/power following a bout of intense/maximal work is much faster than that of pH; (2) high power outputs have been obtained under acidic conditions; (3) the ingestion of sodium bicarbonate (known to increase extracellular buffering capacity) has, in some cases, no effect on RSE performance (Glaister, 2005). Other studies have even proposed that lactic acid might preserve muscle excitability (Pedersen *et al.*, 2004), opening new research avenues to determine whether the link between acidosis (muscle buffering) and fatigue is just coincidental.

Inorganic phosphate

As a result of rapid breakdown of ATP and PCr during sprinting, there is an increased intramuscular accumulation of Mg^{2+}, ADP, and inorganic phosphate (P_i), which may in turn impair RSA. Increasing evidence indicates that at high P_i concentrations, there could be a net influx of P_i into the sarcoplasmic reticulum, which could result in Ca^{2+}–P_i precipitation limiting Ca^{2+} available for release (Fryer *et al.*, 1995). In examining modifications of single-twitch contractile properties pre- to post-fatigue, RSE-based studies have confirmed that a failure of the excitation–contraction coupling can occur (Girard *et al.*, 2007; Racinais *et al.*, 2007). Low-frequency fatigue (i.e. a decrease in the ratio between mechanical responses to tetanus at low- and high- frequency stimulations) has also been detected after run-based repeated sprinting (Girard *et al.*, 2007), which further suggests that RSA can be improved through interventions that can limit sarcoplasmic reticulum dysfunction under fatigue (e.g. reduction of the amount of free Ca^{2+} available for release due to Ca^{2+}–P_i precipitation in the sarcoplasmic reticulum; Allen, Lamb and Westerblad, 2008). However, corroborative scientific evidence for the specific role of P_i in performance decrement during multiple-sprint activities is far from substantive.

2.9.2.2 Neuromechanical factors

Neural drive

As maximal sprint exercise demands high levels of neural drive, failure to fully activate the contracting musculature will theoretically decrease force production and therefore reduce repeated-sprint performance (Ross and Leveritt, 2001). While not a universal finding (Billaut and Basset, 2007; Girard *et al.*, 2007; Hautier *et al.*, 2000; Matsuura *et al.*, 2007), a concurrent decline in sprint performance and the amplitude of EMG signals (e.g. root mean square and integrated EMG values) has been reported in several studies (Mendez-Villanueva, Hamer and Bishop, 2007, 2008; Racinais *et al.*, 2007) (Table 2.9.1). This suboptimal motor unit activity (i.e. a decrease in recruitment, firing rate, or both) – which seems to be related to fatigue resistance (Mendez-Villanueva, Hamer and Bishop, 2008) – has also been highlighted via the MRI technique (Kinugasa *et al.*, 2004) and interpolated-twitch results have been obtained during post-RSE assessment of neuromuscular function (Girard *et al.*, 2007; Racinais *et al.*, 2007). Such neural adjustments may be caused by both pre- and post-synaptic inhibitory mechanisms arising from peripheral receptors (e.g. muscle spindles, Golgi tendon organs, free endings of group III and IV nerves). Preliminary evidence, however, suggests that decreases in neural drive during RSE are not caused by decreases in motoneuron excitability (as assessed by the normalized H-reflex amplitude) (Girard *et al.*, 2007). In addition, adaptions in neural function can result in a reduced efficiency in the generation of the motor command, possibly due to disturbances in brain neurotransmitter (e.g. serotonin, dopamine, acetylcholine) concentration (Gandevia, 2001). Under this view, it has been suggested that an increased concentration of free tryptophan (the serotonin precursor) in the brain, at least during prolonged intermittent exercise (four hours of tennis singles; Struder *et al.*, 1995), may reduce the rate of central drive. Although scientific evidence is lacking, it is likely that ergogenic aids (CHO ingestion), or training that has the potential to attenuate the rise in the free

Table 2.9.1 A summary of the characteristics and results of studies that have investigated changes in muscle activation parameters during repeated-sprint exercise

	Repeated-sprint exercise				Fatigue effects	
Study	Subjects	Repetitions	Sprint duration	Recovery duration	Mechanical changes	Muscle activation changes
Billaut and Basset (2007)	13 PA	10	6s	30s	↘ PPO & MPO (~9–13%) from sp1 to sp8 ↘ MVC_{post} (~10%) = MVC_{+5min} (~−8%) & MVC_{+10min} (~−6%)	= VL RMS_{MVC} (~+10%) = MF_{MVC} (~−8%)
Billaut, Basset and Falgairette (2005)	12 PA	10	6s	30s	↘ PPO (11%), PR (7%) & PT (14%) from sp1 to sp10	= $iEMG_{sprint}$ during RSE ↘ activation time delays (~90 ms) between VL and BF EMG onsets from sp1 to sp10
Hautier *et al.* (2000)	10 PA	15	5s	25s	↘ PPO (~11%), moment produced (~6%) & PR (~5%) from sp1 to sp13	↘ RMS_{sprint} in BF and GL (~13 and 17%) from sp1 to sp13 = RMS_{sprint} in GM, VL and RF from sp1 to sp13
Billaut *et al.* (2006)	12 PA	10	6s	30s	↘ PPO (8%, 10 and 11% after sp8, 9, and 10, respectively) ↘ MVC_{post} (13%) & MVC_{+5min} (10%)	↗ VL RMS_{MVC} (~15%) post-RSE = VM RMS_{MVC} (~+10%) post-RSE ↘ VL and VM MF_{MVC} (~15%) post-RSE
Racinais *et al.* (2007)	9 PA	10	6s	30s	↘ PPO (~10%) from sp1 to sp10 ↘ MVC_{post} (16.5%)	↘ VL RMS_{sprint} (~14%) ↘ VL RMS/M_{MVC} (14.5%) & VA (2.5%)
Mendez-Villanueva, Hamer and Bishop (2008)	8 PA	10	6s	30s	↘ PPO & MPO from sp1 to sp5 (~14 and ~17%) and from sp1 to sp10 (~25 and ~28%), respectively	↘ VL RMS_{sprint} from s1 to s5 (~9%) and from sp1 to sp10 (~14%)
Mendez-Villanueva, Hamer and Bishop (2007)	8 PA	10	6s	30s	↘ PPO (~24%) from sp1 to sp10 ↘ TW (~27%) from sp1 to sp10	↘ VL RMS_{sprint} (~14%) from s1 to s10 ↘ VL MF_{sprint} (~11%) from s1 to s10
Matsuura *et al.* (2006)	8 T	10	10s	35s	↘ PPO (~17%) from sp1 to sp10	↘ VL and RF $iEMG_{sprint}$ (~12–15 and 20%) from sp1 to sp10 = VL and RF MPF_{sprint} from s1 to s10
Matsuura *et al.* (2007)	8 T	10	10s	30s	↘ PPO & MPO (~25%) after sp8	= VL RMS_{sprint} during RSE ↘ VL MPF_{sprint} (~5–10%) after sp3 and sp7 only
Giacomoni, Billaut and Falgairette (2006)	12 T	10	6s	30s	↘ PPO (~10 and 11%), PT (~2 and 9%) & TW (~16%) from sp1 to sp10 in the morning and evening, respectively ↘ MVC_{post} (~15 and 13%) & MVC_{+5min} (~10 and 11%) in the morning and evening, respectively	↗ VL RMS_{MVC} (~7.5 and 10% in the morning and evening) after RSE

↘, decrease; ↗, increase; =, no significant change; PA, physically active; T, trained; RSE, repeated-sprint exercise; sp1, sprint 1. PPO, peak power output; MPO, mean power output; MVC, maximal voluntary contraction torque; PR, maximal pedalling rate; PT, peak torque applied on the crank; TW, total work.

RMS, root mean square; RMS/M, normalized RMS activity; iEMG, integrated EMG; MF, median frequency; MPF, mean power frequency; MVC_{+5min}, MVC torque measured +5min after RSE; RMS_{sprint}, RMS measured during sprinting; RMS_{MVC}, RMS measured during MVC; VA, voluntary activation. VL, vastus lateralis; VM, vastus medialis; RF, rectus femoris; BF, biceps femoris; GL, gastrocnemius lateralis; GM, gluteus maximus.

All data were collected during cycle-based repeated-sprinting protocols using a passive recovery mode between exercise bouts.

tryptophan-branched-chain amino acids ratio in the circulation (by lowering the free fatty acid levels during exercise), could theoretically improve repeated-exercise performance. Further studies are needed to better understand the role of decreases in neural drive (and the mechanisms of these adaptations) on the aetiology of muscle fatigue during RSE.

Muscle recruitment strategies

Billaut, Basset and Falgairette (2005) have reported that the time delay between the knee extensor and the flexor EMG activation onsets was reduced during the last sprint of an RSE, owing to an earlier antagonist activation with fatigue occurrence. Using fifteen 5-second cycle sprints separated by 25 seconds of recovery, Hautier *et al.* (2000) observed a significant reduction in RMS for the knee flexor muscles in the thirteenth compared to the first sprint, without changes in the knee extensor muscles, highlighting a possible decrease in muscle coactivation with fatigue. In addition, a shift of median frequency values (obtained during MVC) toward lower frequencies has been reported during post-RSE tests (Billaut *et al.*, 2006). This has been interpreted as a modification in the pattern of muscle fibre recruitment and a decrease in the conduction velocity of active fibres. More tellingly, it is probable that the relative contribution of type I muscle fibres involved in force generation increases during RSE protocols as a result of the greater fatigability of type II fibres, highly solicited during this exercise mode (Casey *et al.*, 1996). Matsuura *et al.* (2006) added further support to this notion by showing that during cycle-based repeated sprinting, mean power frequency is lower with 35-second compared with 350-second recovery periods. This would ultimately suggest that greater fatigue is linked to a preferential recruitment of slow-twitch motor units. Thus, training that can attenuate or delay the fast-to-slow shift in muscle fibre recruitment that occurs with fatigue could potentially improve repeated-sprint performance.

Stiffness regulation

Although not as extensively studied (Ross and Leveritt, 2001), changes in mechanical behaviour (stiffness regulation) may also contribute to performance decrements during maximal efforts such as sprinting. It is generally believed that a stiffer system allows for more efficient elastic energy contribution, potentially enhancing force production during the concentric phase of the movement (Farley *et al.*, 1991). Supported by the close relationship between leg stiffness and sprint performance (Chelly and Denis, 2001), it has been argued that stiffness regulation is a vital component of setting stride frequency (Farley *et al.*, 1991). In line with this statement, decreased stride frequencies have been shown to accompany fatigue development during run-based repeated sprinting (Buchheit *et al.*, 2009; Ratel *et al.*, 2005). Although stiffness has never been systematically calculated during each sprint repetition of an RSE, it is interesting to note that impairment in the spring-mass model properties of the runner's lower limbs (e.g. vertical stiffness) and performance decrement induced by the repetition of all-out efforts (four × 100 m interspersed with two minutes of recovery) have been correlated (Morin *et al.*, 2006). This finding may support the view that the ability to maintain a high-level stiffness condition improves fatigue resistance during RSE. As joint-stiffness regulation has been related to a subject's fitness level (Clark, 2008), this provides another mechanism by which improving aerobic fitness may improve RSE-based performance.

Homoeostatic perturbations affecting fatigue resistance

Fatigue resistance is likely to be compromised as a result of several homoeostatic perturbations, for example hypoxia, hyperthermia, dehydration, low glycogen concentration, a reduced cerebral function (decreased neurotransmitter concentration), and/or the occurrence of muscular damage. Investigating how hyperthermia affects repeated-sprint performance, Drust *et al.* (2005) have reported that the ability to produce power during such an exercise mode is impaired when core and muscle temperatures are simultaneously elevated. In the absence of metabolic changes (muscle lactate, extracellular potassium), these authors have associated the reduced performance in heat with the negative influence of high core temperature on the function of the central nervous system (e.g. alterations in brain activity, reductions in cerebral blood flow, increases in whole-brain energy turnover, reduced muscle activation). While more research is required to establish the mechanisms of these responses, interventions that can mitigate the development of homoeostatic-induced perturbations during RSE may improve fatigue resistance in hostile conditions (see Section 2.9.3).

2.9.2.3 Summary

The inability to reproduce performance in subsequent sprints (fatigue) during RSE is manifested by a decline in sprint speed or peak/mean power output. Proposed factors responsible for these performance decrements include limitations in energy supply, metabolic byproduct accumulation (e.g. P_i, H^+), and reduced excitation of the sarcolemma (increases in extracellular K^+). Although not as extensively studied, failure to fully activate the contracting muscle may also limit repeated-sprint performance. Moreover, the details of a task (e.g. changes in the nature of the work/recovery bouts) will determine the relative contribution of the underlying mechanisms (task dependency) to fatigue. Additional homoeostatic perturbations (e.g. hypoglycaemia, muscle damage, hyperthermia) are likely to further compromise fatigue resistance during RSE. Interventions (e.g. ergogenic aids or training) which are able to ameliorate the influence of these limiting factors should improve RSA.

2.9.3 ERGOGENIC AIDS AND RSA

The use of ergogenic aids provides a valuable means of experimentally assessing and confirming the possible determinants of fatigue during repeated-sprint tasks (see Section 2.9.2). The term 'ergogenic' is derived from the Greek words *ergon* (work) and *gennan* (to produce). Hence, an ergogenic aid usually refers to something that enhances work. With such a broad definition, ergogenic aids could include nutritional aids (e.g. supplements), physiological aids (e.g. training or taper techniques), psychological aids (e.g. imagery), and biochemical aids (e.g. factors that modify technique). For the purpose of this chapter, the discussion of ergogenic aids that can improve repeated-sprint performance will be limited to ingestible substances. Furthermore, this chapter will only discuss substances that are not currently banned by the International Olympic Committee (IOC). While the legal implications regarding the use of ergogenic aids are generally well defined, the moral/ethical implications are less clear. The following section is concerned with how ergogenic aids can help us better understand the mechanisms contributing to fatigue during RSE and does not constitute an endorsement or recommendation of any of the ergogenic aids discussed.

2.9.3.1 Creatine

As repeated sprints produce a severe reduction in intramuscular PCr concentration (Figure 2.9.4), it has been proposed that increasing muscle PCr stores may improve RSA via a reduced ATP degradation and a faster resynthesis of PCr between sprints (Yquel *et al.*, 2002). However, while creatine supplementation (typically ~ 20 g/day for 5–7 days) has been reported to improve intermittent-sprint performance (Preen *et al.*, 2001; Skare, Skadberg and Wisnes, 2001; van Loon *et al.*, 2003; Wiroth *et al.*, 2001), it has not generally been reported to improve repeated-sprint performance (Cornish, Chilibeck and Burke, 2006; Delecluse, Diels and Goris, 2003; Glaister *et al.*, 2006; Kinugasa et al., 2004; McKenna *et al.*, 1999). For example, creatine supplementation (~ 20 g/day for 5 days) has been reported to improve the performance of repeated 10-second cycle sprints interspersed with 60 seconds of recovery (Wiroth *et al.*, 2001), but not with 30 seconds of recovery (Cornish, Chilibeck and Burke, 2006). More tellingly, within one study, creatine supplementation was reported to improve the performance of repeated six-second cycle sprints with recovery intervals of 54 or 84 seconds, but not 24 seconds (Preen *et al.*, 2001). A possible explanation for these conflicting findings is the proposal that creatine supplementation improves intermittent-sprint performance via a faster resynthesis of PCr between sprints (Yquel *et al.*, 2002). Indeed, improved intermittent-sprint performance has been associated with an improved PCr replenishment rate (Preen *et al.*, 2001). However, as creatine supplementation has been reported to increase PCr resynthesis after 60 and 120 seconds, but not 20 seconds (Greenhaff *et al.*, 1994) (Figure 2.9.7), this may explain why creatine supplementation generally does not improve repeated-sprint performance (i.e. repeated sprints with recovery periods of ~30 seconds or less).

2.9.3.2 Carbohydrates

As the body's carbohydrate stores are limited, it has been suggested that maintaining muscle glycogen stores could potentially be important in attenuating fatigue during prolonged, multiple-sprint exercise (Balsom *et al.*, 1999). However, while many studies have demonstrated that carbohydrate ingestion is ergogenic for the performance of prolonged endurance exercise (for review see Coggan and Coyle, 1991), there is limited research that has investigated the effects of carbohydrate ingestion on repeated-sprint performance. In one of the few studies to date, it was reported that increasing muscle glycogen stores resulted in better maintenance of power output over 15 six-second sprints (with 30 seconds of rest between sprints) (Balsom *et al.*, 1999). These results suggest that carbohydrate loading (e.g. 8–10 grams of carbohydrate per kilogram body mass over the 24-hour period preceding exercise) can improve RSA.

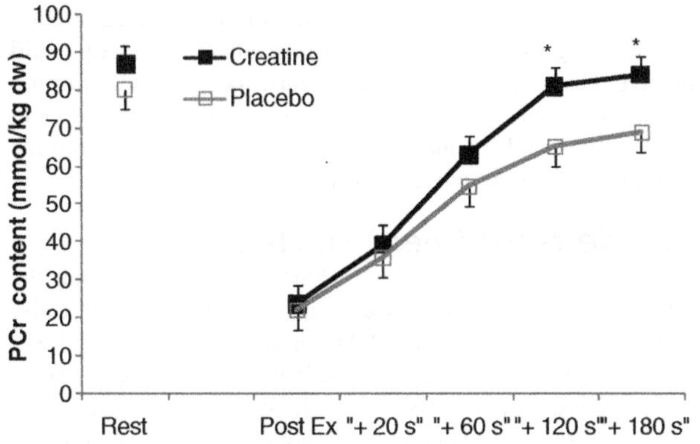

Figure 2.9.7 Changes in resting and post-exercise PCr content following short-term creatine supplementation (20 g/day for five days); * $P < 0.05$ (Greenhaff *et al.*, 1994; Preen *et al.*, 2001)

H^+ + HCO_3^- (Weak base) ⟶ H_2CO_3 (Weak acid)

↓

H_2O + CO_2 ⟶ Exhaled

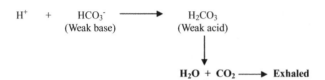

Figure 2.9.8 The bicarbonate buffer system is present in both the intracellular and extracellular fluids and operates to resist changes in H^+ concentration when a strong acid or base is added. When a strong acid is added to the fluid, the bicarbonate ions (HCO_3^-) act as weak bases to tie up the H^+ released by the stronger acid, forming carbonic acid (H_2CO_3). The [HCO_3^-] in the extracellular fluid is normally around 25 mmol/l at rest, and this has been reported to increase by an average of 5.3 mmol/l after the ingestion 0.3 g/kg of body mass of sodium bicarbonate ($NaHCO_3$)

2.9.3.3 Alkalizing agents

The ingestion of alkalizing agents (e.g. $NaHCO_3$) prior to RSE should improve performance by enhancing the efflux of H^+ from the muscle into the blood, and thus maintaining muscle

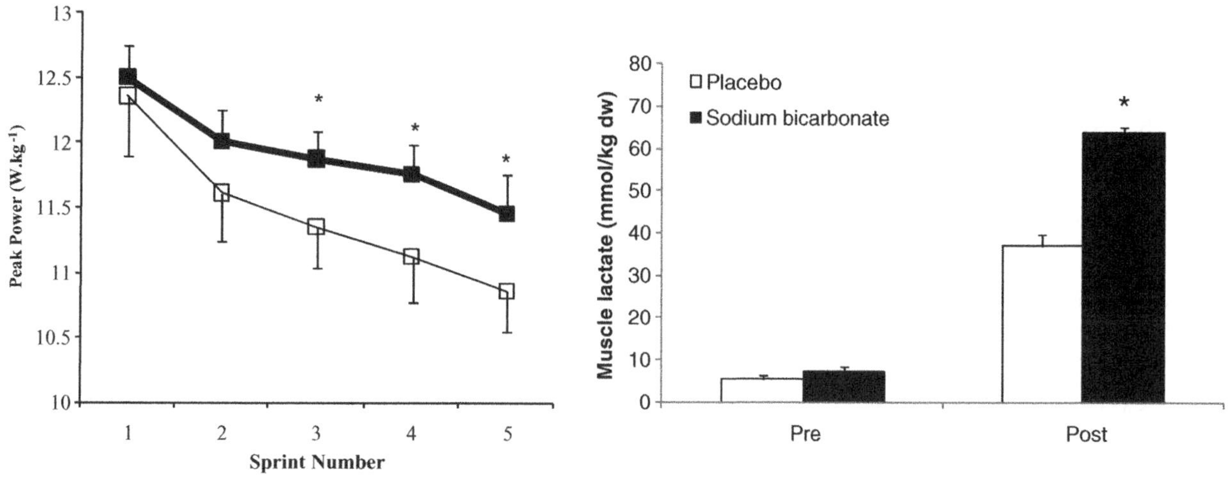

Figure 2.9.9 (a) Peak power output during a five × six-second repeated-sprint exercise (RSE), following the ingestion of either sodium bicarbonate or a placebo (NaCl). (b) Muscle lactate values before and after the RSE. Values are mean ± SE (N = 10). * denotes significant difference from placebo ($P < 0.05$) (Bishop *et al.*, 2004)

pH levels closer to normal during RSE (Hirche *et al.*, 1976; Mainwood and Worseley-Brown, 1975). Although the ergogenic benefits of alkaline ingestion have been largely attributed to an enhanced extracellular buffer capacity, it has recently been demonstrated that the ingestion of alkalizing agents (either sodium bicarbonate or sodium citrate) can also reduce the exercise-induced increase in extracellular K^+ (Sostaric *et al.*, 2006; Street *et al.*, 2005).

Many studies have investigated the effects of alkaline ingestion on high-intensity exercise performance (McNaughton, Ford and Newbold, 1997; Spriet *et al.*, 1986; Stephens *et al.*, 2002). There is however a paucity of studies that have investigated the effects of alkaline ingestion on RSA. Bishop *et al.* (2004) reported a better power maintenance during sprints 3, 4, and 5 of a repeated-sprint test (five six-second sprints performed every 30 seconds) following the ingestion of 0.3 g/kg of $NaHCO_3$ (Figure 2.9.9). In support of this finding, Lavender and Bird (1989) also reported $NaHCO_3$ ingestion to be ergogenic for the performance of ten 10-second cycle sprints with 50 seconds of recovery in between. In contrast, $NaHCO_3$ ingestion produced only a small (~2%), non-significant improvement in the performance of ten six-second running sprints (on a non-motorized treadmill), separated by 30-second recovery periods (Gaitanos *et al.*, 1991). While it is possible that these contrasting findings are due to differing effects of $NaHCO_3$ ingestion on running and cycling repeated-sprint performance, the more likely explanation is the relatively small change in blood pH reported in the final study (7.38–7.43, possibly due to the greater time delay (150 min) between ingestion and exercise). Thus, while confirmatory research is required, it appears that alkaline ingestion (e.g. 0.3 g/kg of $NaHCO_3$ 90 minutes before exercise), leading to a large increase in pH (~0.1 of a pH unit) and $[HCO_3^-]$ (~5.0 mmol/l), is likely to improve repeated-sprint performance. Furthermore, while improved K^+ regulation may contribute to the improved performance, the greater production of lactate suggests that reduced inhibition of anaerobic glycolysis also plays a role (Figure 2.9.9). Such studies suggest that H^+ accumulation is indeed a limiting factor during RSE.

2.9.3.4 Caffeine

Caffeine (1,3,7-trimethylxanthine) is the most commonly consumed drug in the world and is found in coffee, tea, cola, chocolate, and various 'energy' drinks. While there is good evidence that caffeine is also a potent ergogenic aid for many athletic pursuits (Billaut and Basset, 2007; Billaut, Basset and Falgairette, 2005), it is of limited value in providing better understanding of the mechanisms contributing to fatigue during RSE. This is because the ergogenic effects of caffeine have been attributed to a number of possible mechanisms, including: the blocking of adenosine receptors (Fredholm *et al.*, 1999), central nervous system facilitation (Williams, 1991), increased Na^+/K^+ATPase activity (Lindinger, Graham and Spriet, 1993), mobilization of intracellular calcium (Sinclair and Geiger, 2000), and increased plasma catecholamine concentration (Mazzeo, 1991).

Nonetheless, research has investigated the effects of caffeine (6 mg/kg of body mass) on multiple-sprint performance. Stuart *et al.* (2005) reported a negligible effect of caffeine ingestion on RSA (10 × 20-minute sprints performed every 10 seconds), while Schneiker *et al.* (2006) reported a 7% increase in mean power when 10 male team-sport athletes performed an intermittent-sprint test consisting of two 36-minute 'halves', each half comprising 18 four-second sprints with two minutes of active recovery at 35% VO_{2peak} between each sprint. Thus, while further research is certainly warranted, these results suggest that caffeine ingestion is likely to improve intermittent, but not repeated, sprint performance.

2.9.3.5 Summary

To date, few studies have investigated the effects of ergogenic aids on RSA, and contradictory findings have often been reported (possibly due to the effects of task dependency on fatigue). Further research is required as such studies provide important insights into possible determinants of fatigue during RSE. For example, reports of improved repeated-sprint performance following alkaline ingestion provide support for the hypothesis that H^+ accumulation is an important limiting factor during RSA. Similarly, decreases in RSA following the lowering of muscle glycogen stores suggest that, under certain conditions, muscle glycogen content can limit RSA. While creatine-loading studies do not support the hypothesis that PCr availability is an important limiting factor for RSA, this can probably be attributed to the observation that creatine loading does not significantly accelerate short-term (<60 seconds) PCr resynthesis.

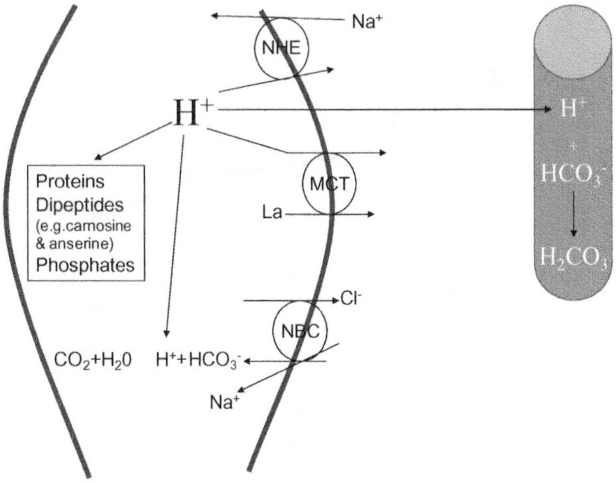

Figure 2.9.10 Muscle pH-regulating systems. NHE = sodium–hydrogen exchanger, MCT = monocarboxylate transporters, NBC = sodium bicarbonate cotransporter

2.9.4 EFFECTS OF TRAINING ON RSA

2.9.4.1 Introduction

Recently, there has been an increase in scientific research regarding the importance of RSA for team- and racket-sport athletes (Bravo *et al.*, 2008; Girard *et al.*, 2007; Impellizzeri *et al.*, 2008; Rampinini *et al.*, 2007; Spencer *et al.*, 2004, 2005). Surprisingly, however, there has been little research about the best training methods to improve this component. In the absence of strong scientific evidence, one concept that has emerged is the need to train RSA by performing RSE. While such a concept appeals to common sense, the scientific evidence in support of this approach is lacking. In this section, it will be argued that the best way to improve RSA is to improve the underlying factors responsible for fatigue (see Section 2.9.2).

2.9.4.2 Training the limiting factors

Ion regulation

The removal of H^+ during intense skeletal muscle contractions (such as repeated sprints) occurs via intracellular buffering ($\beta m_{in\ vitro}$) and a number of different membrane transport systems, especially the monocarboxylate transporters (MCTs) (Juel, 1998) (Figure 2.9.10). The muscle membrane also contains Na^+–K^+ pumps, which are largely responsible for the maintenance of extracellular [K^+] (Clausen, 2003).

A large increase in muscle H^+ and/or lactate during exercise has been proposed to be an important stimulus for adaptations of the muscle pH-regulating systems (Weston *et al.*, 1997). This is supported by increases in $\beta m_{in\ vitro}$ in response to high-intensity, but not moderate-intensity, endurance training (Edge, Bishop and Goodman, 2006). However, greater accumulation of lactate and H^+ during training has not always been associated with greater increases in MCT (Juel *et al.*, 2004b; Mohr *et al.*, 2007) or $\beta m_{in\ vitro}$ (Bishop *et al.*, 2008). Furthermore, research suggests that too large an accumulation of H^+ during training may have a detrimental effect on adaptations to the pH regulatory systems within the muscle (Bishop *et al.*, 2008; Thomas *et al.*, 2007). Thus, while further research is required, it appears that intramuscular accumulation of H^+ and/or lactate provides an important stimulus to improvement of the muscle pH-regulating systems; however, maximizing H^+ accumulation during training does not maximize these adaptations.

The above considerations have important implications for the design of training programmes to improve the muscle pH-regulating systems and hence RSA. To increase $\beta m_{in\ vitro}$, it appears important to employ high-intensity interval training (80–90% VO_{2max}), interspersed with rest periods that are shorter than the work periods, so that the muscle is required to contract while experiencing a reduced pH. The use of intensities $\geq VO_{2max}$ does not appear to provide additional benefits, and has the potential to actually decrease $\beta m_{in\ vitro}$ (Bishop *et al.*, 2008). In addition, the use of rest periods that are longer than the work periods allows greater removal of lactate and H^+ prior to subsequent intervals (Sahlin *et al.*, 1976) and typically does not result in a significant increase in $\beta m_{in\ vitro}$ (Harmer *et al.*, 2000; Nevill *et al.*, 1989). While an optimal training volume for improvement of $\beta m_{in\ vitro}$ has not yet been established, it appears that training at the above-mentioned intensities, two to three times per week, for three to five weeks, can result in significant increases in $\beta m_{in\ vitro}$ (Edge, Bishop and Goodman, 2006; Gibala *et al.*, 2006; Weston *et al.*, 1997).

It is more difficult to recommend the ideal training programme for increasing the MCTs as significant increases have been reported following low- (Juel, Holten and Dela, 2004a), moderate- (Bonen *et al.*, 1998; Dubouchaud *et al.*, 2000) and high- (Burgomaster *et al.*, 2007; Juel *et al.*, 2004b) intensity training (Table 2.9.2). However, one factor that these training programmes tend to have in common is that they are associated with only modest increases in blood lactate concentration (~4–8 mmol/l). When high-intensity training has been employed, the rest periods between high-intensity intervals have ranged from 90 to 180 seconds (e.g. a work-to-rest ratio of ≤1 : 2) (Juel *et al.*, 2004b; Pilegaard *et al.*, 1999), allowing substantial removal of lactate and H^+ prior to subsequent intervals (Sahlin *et al.*, 1976). Thus, in contrast to the high-intensity training required to increase $\beta m_{in\ vitro}$, it appears that both low- and high-intensity training can increase the MCTs, but that training should be structured so as to provoke only a modest increase in blood lactate concentration. Significant changes in MCT content appear more likely when training is performed two to three times per week for six to eight weeks.

There is less agreement about the training stimulus required to increase the Na^+–K^+ pump content. However, it has been suggested that greater adaptations may be induced by training that invokes greater Na^+–K^+ pump activity (Mohr *et al.*, 2007). This is supported by larger increases in Na^+–K^+ pump content following high-intensity interval training (8–12 × 30 seconds all-out separated by 1.5–3 minutes of recovery), compared with either intermittent-sprint training (15 × 6-second sprint separated by 1 minute of recovery) (Mohr *et al.*, 2007) or endurance training (continuous or interval training at 70–80% VO_{2max}) (Aughey *et al.*, 2007; Iaia *et al.*, 2008). However, high-volume, moderate-intensity, continuous training (2 hours at 60–65% VO_{2max}) has also been reported to increase Na^+–K^+ pump protein content in untrained subjects (Green *et al.*, 1993, 2004). Thus, despite the paucity of studies, it appears that high-intensity (>VO_{2max}) interval training may be required to increase Na^+–K^+ pump content in moderately-trained athletes (e.g. team-sport athletes). Furthermore, it appears that these adaptations can occur quite quickly (after less than two weeks of training).

Energy supply

PCr resynthesis

The oxidative metabolism pathways are essential for PCr resynthesis during recovery from exercise (Haseler, Hogan and Richardson, 1999). This suggests that individuals with an elevated aerobic fitness should be able to more rapidly resynthesize PCr between repeated sprints. Indeed, cross-sectional

Table 2.9.2 A summary of the characteristics and results of training studies that have investigated changes in MCT relative abundance in humans

	Subjects		Training		Adaptations			
Study	Type	VO_{2max}	Programme	Blood La	MCT1	MCT4	VO_{2max}	CS
Bonen *et al.*(1998)	7 UT ♂	45.1 ± 2.5	1 wk: 2h/d @ 65% VO_{2max}	<4 mM	↗18%	nr	—	nr
Pilegaard *et al.* (1999)	4 UT ♂	nr	8 wk: 3–5 × (5 × 30–60 s @ 150–200% VO_{2max}: 2 min rest), 3–5 d/wk	5–6 mM	↗70±85%	↗ 33±26%	nr	↗ 18%^{ns}
Dubouchaud *et al.* (2000)	9 UT ♂	43.5 ± 3.9	9 wk: 1 h @ 75% VO_{2max}, 6d/wk	—	↗ 60±24%	↗ 47±24%	↗ 15%	↗ 75%
Eversten *et al.* (2001)	20 E ♂♀	58–73	20 wk: 10–16 h/wk @ 60–70% VO_{2max} 20 wk: 10–16 h/wk @ 60–70% VO_{2max}	<1.5 mM 3–4 mM	↘12±9% —	— —	— —	↘ 6%^{ns} ↘6%^{ns}
Juel *et al.* (2004b)	6 MT ♂	50.2 ± 3.0	7–8 wk: 15 × (1 min @ 150% VO2max: 3 min rest), 3–5 d/wk	>8 mM	↗ 15±12%	↗11±27%^{ns}	nr	nr
Juel, Holten and Dela (2004a)	7 UT ♂	nr	6 wk: 3–4 × (8–12 reps: 2 min rest), 3d/wk	<4–5 mM	↗ 48±58%	↗ 32±56%	nr	nr
Mohr *et al.* (2007)	6 MT ♂ 7 MT ♂	50.2 ± 3.7 49.0 ± 4.2	8 wk: 15 × (6 s sprint: 1 min @ 95% max), 3–5 d/wk 8 wk: 8 × (30 s @ 130% max: 90 s rest), 3–5 d/wk	~9 mM ~16 mM	↗ 28±32% ↗30±24%	— —	nr nr	nr
Burgomaster *et al.* (2007)	8 MT ♂	50.0 ± 5.6	1 wk: 4–6 × (30 s all-out: 240 s rest), 3 d/wk 6 wk: 4–6 × (30 s all-out: 240 s rest), 3 d/wk	—	↗ 60±65%^{ns} ↗120±230%	↗40±55% ↗ 55±70%	nr	nr
Bishop *et al.* (2008)	6 UT ♀	43.2 ± 4.9	5 wk: 6–12 × (2 min @ 95–115% VO_{2max}: 1 min rest), 3 d/wk	~16 mM	↘ 4±43%^{ns}	↗20±53%^{ns}	↗ 10%	↗ 6%^{ns}

VO_{2max}: mL·kg^{-1}·min^{-1}; UT: untrained; MT: moderately trained; E: elite; ♂: Males; ♀: Females; ↘: *decrease*; ↗: *increase*; nr: not reported; —: no change; ns: not significant; CS: citrate synthase; La: lactate, MCT: monocarboxylate transporters.

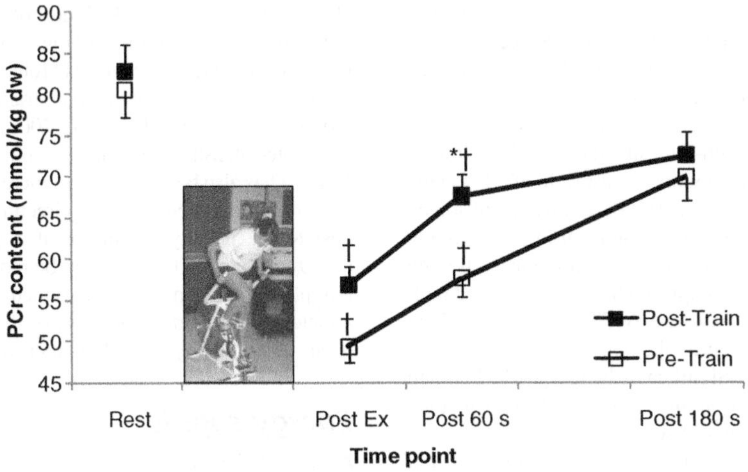

Figure 2.9.11 Changes in resting and post-exercise PCr content following high-intensity interval training (Bishop *et al.*, 2008, and unpublished research). * = significantly different from pre-train, † = significantly different from rest

research (Yoshida and Watari, 1993) and one training study (McCully *et al.*, 1991) support the hypothesis that endurance training enhances PCr resynthesis following low-intensity exercise. Recently, it has also been reported that high-intensity interval training (80–90% VO_{2max}), interspersed with rest periods that are shorter than the work periods (e.g. similar to that described above for improving $\beta m_{in\ vitro}$), can significantly improve brief (60-second) PCr resynthesis following high-intensity exercise (Figure 2.9.11) (Bishop *et al.*, 2008). In contrast, no changes in the rate of PCr resynthesis have been reported following speed-endurance or intermittent-sprint training (Mohr *et al.*, 2007), or training involving repeated 30-second all-out efforts (Stathis *et al.*, 1994). These results can probably be attributed to the absence of changes in muscle oxidative capacity with these types of training. Thus, while the optimal training intensity has not yet been established, it appears that improvements in muscle oxidative capacity may be required to improve PCr resynthesis. To date, no research has investigated the optimal volume of training for improving PCr resynthesis. However, as improvements in aerobic fitness have been associated with training volume, this suggests that training-induced increases in the rate of PCr resynthesis may be greater with higher volumes of training.

Anaerobic glycolysis

The maximal amount of ATP that can be produced via anaerobic metabolism (e.g. PCr breakdown and anaerobic glycolysis) has been defined as an athlete's anaerobic capacity (Medbo and Burgers, 1990). As training typically does not increase the amount of PCr breakdown (Harmer *et al.*, 2000), changes in the ability to produce ATP via anaerobic glycolysis are likely to be well reflected by training-induced changes in the anaerobic capacity. A high rate of anaerobic energy release during exercise has been proposed to be an important stimulus for adaptations of the muscle's anaerobic capacity (Medbo and Burgers, 1990). This is supported by increases in the anaerobic capacity in response to high-intensity (20–120-second intervals at 100–170% VO_{2max}) (Medbo and Burgers, 1990; Tabata *et al.*, 1996; Weber and Schneider, 2002), but not moderate-intensity (60-minute intervals at 70% VO_{2max}) (Tabata *et al.*, 1996), endurance training. Furthermore, greater changes in anaerobic capacity have been reported in response to interval training that produces larger changes in blood lactate concentration (Medbo and Burgers, 1990; Tabata *et al.*, 1996). These results are consistent with the observation that training-induced changes in enzymes important for anaerobic glycolysis (e.g. phosphofructokinase and phosphorylase) are greater following training that involves repeated 30-second bouts, than following repeated six-second bouts or continuous training (Daussin *et al.*, 2008; Jacobs *et al.*, 1987). Interestingly, greater increases in glycolytic enzymes have also been reported when high-intensity intervals are separated by short (10-second) rather than long (3–5-minute) rest periods. Thus, to improve anaerobic glycolysis, it is recommended that short (20–30-second), high-intensity (all-out) intervals separated by relatively short rest periods (<1 minute) are utilized.

Aerobic fitness

While further research is required, many physiologists believe that it is the reduced oxygen levels in the muscle (hypoxia) during training that provide the stimulus to increase aerobic fitness (Daussin *et al.*, 2008). As the oxygen level in the muscle decreases as the exercise intensity increases up to 100% VO_{2max}, but does not decrease further once the exercise intensity exceeds 100% VO_{2max} (Figure 2.9.12) (MacDougall and Sale, 1981), interval training at intensities which approximate VO_{2max} may be most effective. This is supported by previous studies which have reported greater improvements in aerobic fitness after interval training compared with continuous training, despite total work being matched (Daussin *et al.*, 2008; Eversten,

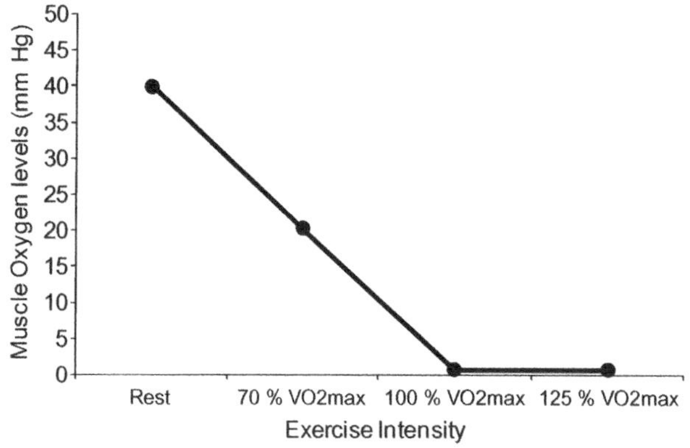

Figure 2.9.12 Muscle oxygen levels during rest and different intensities of exercise (MacDougall and Sale, 1981)

Medbo and Bonen, 2001; Gorostiaga *et al.*, 1991; Helgerud *et al.*, 2007). It should be noted, however, that most of these studies performed their continuous training at very low intensities (≤56% of the power at VO_{2max}) (Daussin *et al.*, 2008; Gorostiaga *et al.*, 1991; Helgerud *et al.*, 2007). When continuous training has been performed at intensities >60% the power at VO_{2max}, interval and continuous training have been reported to produce similar improvement in VO_{2max} (Cunningham, McCrimmon and Vlach, 1979; Eddy, Sparks and Adelizi, 1977; Edge, Bishop and Goodman, 2006; Poole and Gaesser, 1985). Furthermore, other studies appear to demonstrate a strong relationship between total workload of training and changes in VO_{2max} (Laursen *et al.*, 2002; Stepto *et al.*, 1999). These results suggest that if a minimum training intensity is exceeded (>60% the power at VO_{2max}), interval and continuous training matched for total work can result in similar improvements in VO_{2max} and the lactate threshold. Furthermore, as mentioned above, improvements in aerobic fitness appear to be greater with higher volumes of training.

Muscle activation

As previously discussed (Section 2.9.2.1), RSE is associated with dramatic metabolic disturbances within the muscle cell. Despite the direct effects that selected metabolites have on muscle contractility, it is also believed that changes in their concentration may have an indirect action via centrally controlled mechanisms (Gandevia, 2001). Limiting metabolite accumulation, but also reducing the extent of muscle damage, with proper training or ergogenic aids might theoretically prevent a precocious central fatigue. Evidence of training-induced neural plasticity (increases in EMG activity) has been provided through the use of electromyostimulation (Gondin, Duclay and Martin, 2006), and eccentric (Duclay *et al.*, 2008) or isometric (Del Balso and Cafarelli, 2007) resistance training. This elevated motor neuronal output, which may involve increased firing rates, increased motoneuron excitability, decreased presynaptic inhibition, downregulation of inhibitory neural pathways, and increased levels of central descending motor drive, has the potential to increase RSE performance requiring near-maximal levels of neural activation (Hautier *et al.*, 2000). However, while the theoretical basis for such reasoning is compelling, corroborative research is far from substantive. Future research should therefore investigate whether training regimens known to improve muscle activation can also improve RSE variables and delay fatigue of neural origin. Over time, research into fatigue/training-related mechanisms should determine the relative involvement and functional significance of these neural factors.

2.9.4.3 Putting it all together

The above discussion highlights the fact that there is not one single type of training that can be recommended to improve all of the factors responsible for performance decrements during repeated-sprint tasks. It appears, however, that to increase $\beta m_{in\ vitro}$ it is important to include some high-intensity (80–90% VO_{2max}) interval training, interspersed with rest periods that are shorter than the work periods. This type of training can also be used to improve aerobic fitness and PCr resynthesis. Higher-intensity (>VO_{2max}) interval training (e.g. repeated 30-second, all-out efforts separated by 1.5–3 minutes of recovery), invoking larger increase in Na^+–K^+ pump activity, is probably required to increase Na^+–K^+ pump content, especially in moderately- to well-trained athletes. Similar training can also be employed to improve anaerobic glycolysis, although it is recommended that the intervals be separated by shorter rest periods (<1 minute). In contrast, to improve the MCTs, training (either continuous or interval) should be structured so as to provoke only a modest increase in blood lactate concentration. For most athletes, it is probably impossible to perform all of the above-described training concurrently. It is therefore important that a periodized training programme, designed to improve RSA, is structured such that different aspects are emphasized, at different times, in accordance with the strengths and weaknesses of the individual athlete.

2.9.5 CONCLUSION

RSA is an important component of many popular sports. Although a great number of issues remain unresolved, there has been an exponential growth in research into adaptive changes in neuromuscular function (e.g. fatigue, ergogenic aids, training) over recent years. The aetiology of muscle fatigue during RSE is a complex phenomenon that involves, depending on the task being performed, ionic disturbances and the accumulation of metabolites within the muscle fibres, and a reduced neural activation to the contracting musculature. Of the strategies used to try and combat the factors limiting RSA, most of the research to date has focused on the effects of various ergogenic aids (e.g. creatine, sodium bicarbonate, caffeine, CHO) and training regimens. Interventions that can increase the ability to buffer H^+ and the rate of PCr resynthesis are likely to improve RSA performance.

Success in RSAs is also likely to depend on an athlete's ability to generate explosive power within a few seconds. Future studies should therefore explore the effect of fatigue/training on explosive muscle strength (the rate of force development) to establish its functional significance. More studies are also needed to assess the effects of manipulating homoeostatic perturbations (e.g. environmental conditions (hypoxia, hot/humid), dehydration status, intake of carbohydrate, creatine, or caffeine) on mechanisms underlying changes in the neuromuscular function during RSE.

The physiological demands imposed on athletes during sports which consist of intermittent load profiles are difficult to duplicate under laboratory conditions. The increasing interest in multiple-sprint work must therefore be expanded to sport-specific test settings, in parallel with a high level of standardization and reliability of measures.

References

Abrantes C., Macas V. and Sampaio J. (2004) Variation in football players' sprint test performance across different ages and levels of competition. *JSSM*, **3** (YISI 1), 44–49.

Allen D.G., Lamb G.D. and Westerblad H. (2008) Skeletal muscle fatigue: cellular mechanisms. *Physiol Rev*, **88**, 287–332.

Aughey R.J., Murphy K.T., Clark S.A. *et al.* (2007) Muscle Na+-K+-ATPase activity and isoform adaptations to intense interval exercise and training in well-trained athletes. *J Appl Physiol*, **103**, 39–47.

Balsom P.D., Seger J.Y., Sjodin B. and Ekblom B. (1992a) Maximal-intensity intermittent exercise: effect of recovery duration. *Int J Sports Med*, **13** (7), 528–533.

Balsom P.D., Seger J.Y., Sjodin B. and Ekblom B. (1992b) Physiological responses to maximal intensity intermittent exercise. *Eur J Appl Physiol*, **65**, 144–149.

Balsom P., Ekblom B. and Sjodin B. (1994a) Enhanced oxygen availablility during high intensity intermittent exercise decreases anaerobic metabolite concentration in blood. *Acta Physiol Scand*, **150**, 455–456.

Balsom P.D., Gaitanos G.C., Ekblom B. and Sjodin B. (1994b) Reduced oxygen availability during high intensity intermittent exercise impairs performance. *Acta Physiol Scand*, **152**, 279–285.

Balsom P.D., Gaitanos G.C., Soderlund K. and Ekblom B. (1999) High-intensity exercise and muscle glycogen availability in humans. *Acta Physiol Scand*, **165**, 337–345.

Billaut F. and Basset F.A. (2007) Effect of different recovery patterns on repeated-sprint ability and neuromuscular responses. *J Sports Sci*, **25**, 905–913.

Billaut F., Basset F.A. and Falgairette G. (2005) Muscle coordination changes during intermittent cycling sprints. *Neurosci Lett*, **380**, 265–269.

Billaut F., Basset F.A., Giacomoni M. *et al.* (2006) Effect of high-intensity intermittent cycling sprints on neuromuscular activity. *Int J Sports Med*, **27**, 25–30.

Bishop D. and Claudius B. (2005) Effects of induced metabolic alkalosis on prolonged intermittent-sprint performance. *Med Sci Sports Exerc*, **37**, 759–767.

Bishop D. and Edge J. (2006) Determinants of repeated-sprint ability in females matched for single-sprint performance. *Eur. J. Appl. Physiol*, **97**, 373–379.

Bishop D., Lawrence S. and Spencer M. (2003) Predictors of repeated-sprint ability in elite female hockey players. *J Sci Med Sport*, **6**, 199–209.

Bishop D., Edge J., Davis C. and Goodman C. (2004) Induced metabolic alkalosis affects muscle metabolism and repeated-sprint ability. *Med Sci Sports Exerc*, **36**, 807–813.

Bishop D., Edge J., Thomas C. and Mercier J. (2008) Effects of high-intensity training on muscle lactate transporters and postexercise recovery of muscle lactate and hydrogen ions in women. *Am J Physiol Regul Integr Comp Physiol*, **295**, R1991–R1998.

Bogdanis G.C., Nevill M.E., Boobis L.H. *et al.* (1995) Recovery of power output and muscle metabolites following 30 s of maximal sprint cycling in man. *J Physiol (Lond)*, **482**, 467–480.

Bogdanis G.C., Nevill M.E., Boobis L.H. and Lakomy H.K.A. (1996) Contribution of phosphocreatine and aerobic metabolism to energy supply during repeated sprint exercise. *J Appl Physiol*, **80** (3), 876–884.

Bonen A., McCullagh K.J.A., Putman C.T. *et al.* (1998) Short-term training increases human muscle MCT1 and femoral venous lactate in relation to muscle lactate. *Am J Physiol Endocrinol Metab*, **274**, E102–E107.

Boobis L., Williams C. and Wootton S. (1982) Human muscle metabolism during brief maximal exercise. *J Physiol (Lond)*, **338**, P22.

Bravo D.F., Impellizzeri F.M., Rampinini E. *et al.* (2008) Sprint vs. interval training in football. *Int J Sports Med*, **29**, 668–674.

Buchheit M., Cormie P., Abbis C.R. *et al.* (2009) Muscle deoxygenation during repeated sprint running: effect of active vs. passive recovery. *Int J Sports Med*, **30** (6), 418–425.

Burgomaster K.A., Cermak N.M., Phillips S.M. *et al.* (2007) Divergent response of metabolite transport proteins in human skeletal muscle after sprint interval training and detraining. *Am J Physiol Regul Integr Comp Physiol*, **292**, R1970–R1976.

Casey A., Constantin-Teodosiu D., Howell S. *et al.* (1996) Metabolic responses of type I and II muscle fibres during repeated bouts of maximal exercise in humans. *Am J Physiol*, **271**, E38–E43.

Chelly S.M. and Denis C. (2001) Leg power and hopping stiffness: relationship with sprint running performance. *Med Sci Sports Exerc*, **33**, 326–333.

Clark R.A. (2008) The effect of training status on inter-limb joint stiffness regulation during repeated maximal sprints. *J Sci Med Sport*, **12** (3), 406–410.

Clausen T. (2003) Na+-K+ pump regulation and skeletal muscle contractility. *Physiol Rev*, **83**, 1269–1324.

Clausen T., Nielsen O.B., Harrison A.P. *et al.* (1998) The Na+,K+ pump and muscle excitability. *Acta Physiol Scand*, **162**, 183–190.

Coggan A.R. and Coyle E.F. (1991) Carbohydrate ingestion during prolonged exercise: effects on metabolism and performance. *Exerc Sport Sci Rev*, **19**, 1–40.

Connes P., Racinais S., Sra F. *et al.* (2006) Does the pattern of repeated sprint ability differ between sickle cell trait carriers and healthy subjects. *Int J Sports Med*, **27**, 937–942.

Cornish S.M., Chilibeck P.D. and Burke D.G. (2006) The effect of creatine monohydrate supplementation on sprint skating in ice-hockey players. *J Sports Med Phys Fitness*, **46**, 90–98.

Cunningham D.A., McCrimmon D. and Vlach L.F. (1979) Cardiovascular response to interval and continuous training in women. *Eur J Appl Physiol*, **41**, 187–197.

Daussin F.N., Zoll J., Dufour S.P. *et al.* (2008) Effect of interval versus continuous training on cardiorespiratory and mitochondrial functions: relationship to aerobic performance improvements in sedentary subjects. *Am J Physiol Regul Integr Comp Physiol*, **295**, R264–R272.

Dawson B., Goodman C., Lawrence S. *et al.* (1997) Muscle phosphocreatine repletion following single and repeated short sprint efforts. *Scand J Med Sci Sports*, **7**, 206–213.

Del Balso C. and Cafarelli E. (2007) Adaptations in the activation of human skeletal muscle induced by short-term isometric resistance training. *J Appl Physiol*, **103**, 402–411.

Delecluse C., Diels R. and Goris M. (2003) Effect of creatine supplementation on intermittent sprint running performance in highly trained athletes. *J Strength Cond Res*, **17**, 446–454.

Drust B., Rasmussen P., Mohr M. *et al.* (2005) Elevations in core and muscle temperature impairs repeated sprint performance. *Acta Physiol Scand*, **183**, 181–190.

Dubouchaud H., Butterfield G.E., Wolfel E.E. *et al.* (2000) Endurance training, expression and physiology of LDH, MCT1 and MCT4 in human skeletal muscle. *Am J Physiol Endocrinol Metab*, **278**, E571–E579.

Duclay J., Martin A., Robbe A. and Pousson M. (2008) Spinal reflex plasticity during maximal dynamic contractions after eccentric training. *Med Sci Sports Exerc*, **40**, 722–734.

Dupont G., Millet G.P., Guinhouya C. and Berthoin S. (2005) Relationship between oxygen uptake kinetics and performance in repeated running sprints. *Eur J Appl Physiol*, **95**, 27–34.

Eddy D.O., Sparks K.L. and Adelizi D.A. (1977) The effects of continuous and interval training in women and men. *Eur J Appl Physiol*, **37**, 83–92.

Edge J., Bishop D. and Goodman C. (2005) Effects of high- and moderate-intensity training on metabolism and repeated sprints. *Med Sci Sports Exerc*, **37**, 1975–1982.

Edge J., Bishop D. and Goodman C. (2006) The effects of training intensity on muscle buffer capacity in females. *Eur J Appl Physiol*, **96**, 97–105.

Eversten F., Medbo J.I. and Bonen A. (2001) Effect of training intensity on muscle lactate transporters and lactate threshold of cross-country skiers. *Acta Physiol Scand*, **173**, 195–205.

Farley C.T., Blickhan R., Saito J. and Taylor C.R. (1991) Hopping frequency in humans: a test of how springs set stride frequency in bouncing gaits. *J Appl Physiol*, **71**, 2127–2132.

Fredholm B.B., Bättig K., Holmén J. *et al.* (1999) Actions of caffeine in the brain with special reference to factors that contribute to its widespread use. *Pharmacol Rev*, **51**, 83–133.

Fryer M.W., Owen V.J., Lamb G.D. and Stephenson D.G. (1995) Effects of creatine phosphate and P(i) on Ca2+ movements and tension development in rat skinned skeletal muscle fibres. *J Physiol*, **482** (Pt 1), 123–140.

Gaitanos G.C., Nevill M.E., Brooks S. and Williams C. (1991) Repeated bouts of sprint running after induced alkalosis. *J Sports Sci*, **9**, 355–369.

Gaitanos G.C., Williams C., Boobis L.H. and Brooks S. (1993) Human muscle metabolism during intermittent maximal exercise. *J Appl Physiol*, **75** (2), 712–719.

Gandevia S.C. (2001) Spinal and supraspinal factors in human muscle fatigue. *Physiol Rev*, **81**, 1725–1789.

Giacomoni M., Billaut F. and Falgairette G. (2006) Effects of the time of day on repeated all-out cycle performance and short-term recovery patterns. *Int J Sports Med*, **27**, 468–474.

Gibala M.J., Little J.P., van Essen M. *et al.* (2006) Short-term sprint interval versus traditional endurance training: similar initial adaptations in human skeletal muscle and exercise performance. *J Physiol (Lond)*, **575**, 901–911.

Girard O., Racinais S., Thedon T. *et al.* (2007) Alteration of neuromuscular function after repeated running sprints, in *12th Congress of the ACAPS* (eds R. Buy, M. Buekers and D. Daly), Leuven, Belgium, pp. 289–290.

Girard O., Lattier G., Maffiuletti N.A. *et al.* (2008) Neuromuscular fatigue during a prolonged intermittent exercise: application to tennis. *J Electromyogr Kinesiol*, **18**, 1038–1046.

Glaister M. (2005) Multiple sprint work : physiological responses, mechanisms of fatigue and the influence of aerobic fitness. *Sports Med*, **35**, 757–777.

Glaister M., Lockey R.A., Abraham C.S. *et al.* (2006) Creatine supplementation and multiple sprint running performance. *J Strength Cond Res*, **20**, 273–277.

Glaister M., Howatson G., Pattison J.R. and McInnes G. (2008) The reliability and validity of fatigue measures during multiple-sprint work: an issue revisited. *J Strength Cond Res*, **22**, 1597–1601.

Gondin J., Duclay J. and Martin A. (2006) Soleus- and gastrocnemii-evoked V-wave responses increase after neuromuscular electrical stimulation training. *J Neurophysiol*, **95**, 3328–3335.

Gorostiaga E.M., Walter C.B., Foster A. and Hickson R.C. (1991) Uniqueness of interval and continuous training at the same maintained exercise intensity. *Eur J Appl Physiol*, **63**, 101–107.

Green H.J., Chin E.R., Ball-Burnett M. and Ranney D. (1993) Increases in human skeletal muscle Na(+)-K(+)-ATPase concentration with short-term training. *Am J Physiol*, **264**, C1538–C1541.

Green H.J., Barr D.J., Fowles J.R. *et al.* (2004) Malleability of human skeletal muscle Na(+)-K(+)-ATPase pump with short-term training. *J Appl Physiol*, **97**, 143–148.

Greenhaff P.L., Bodin K., Soderlund K. and Hultman E. (1994) Effect of oral creatine supplementation on skeletal muscle phosphocreatine resynthesis. *Am J Physiol*, **266**, E725–E730.

Hamilton A.L., Nevill M.E., Brooks S. and Williams C. (1991) Physiological responses to maximal intermittent exercise: Differences between endurance trained runners and game players. *J Sport Sci*, **9**, 371–382.

Harmer A.R., McKenna M.J., Sutton J.R. *et al.* (2000) Skeletal muscle metabolic and ionic adaptations during intense exercise following sprint training in humans. *J Appl Physiol*, **89**, 1793–1803.

Haseler L.J., Hogan M.C. and Richardson R.S. (1999) Skeletal muscle phosphocreatine recovery in exercise-trained humans is dependent on O2 availability. *J Appl Physiol*, **86** (6), 2013–2018.

Hautier C.A., Arsac L.M., Deghdegh K. *et al.* (2000) Influence of fatigue on EMG/force ratio and cocontraction in cycling. *Med Sci Sports Exerc*, **32**, 839–843.

Helgerud J., Hoydal K., Wang E. *et al.* (2007) Aerobic high-intensity intervals improve VO2max more than moderate training. *Med Sci Sports Exerc*, **39**, 665–671.

Hirche H., Hornbach V., Langohr H.D. *et al.* (1976) Lactic acid permeation rate in working gastrocnemii of dogs during metabolic alkalosis and acidosis. *Pflugers Archiv*, **356**, 209–222.

Hultman E. and Sjoholm H. (1983) Energy metabolism and contraction force of human skeletal muscle in situ during electrical stimulation. *J Physiol*, **345**, 525–532.

Hultman E., Bergström M., Spriet L.L. and Soderlund K. (1990) Energy metabolism and fatigue, in *Biochemistry of Sport and Exercise VII* (eds A. Taylor, P. Gollnick and H. Green), Human Kinetics, Champaign, IL, pp. 73–92.

Iaia F.M., Thomassen M., Kolding H. *et al.* (2008) Reduced volume but increased training intensity elevates muscle Na+-K+ pump 1-subunit and NHE1 expression as well as short-term work capacity in humans. *Am J Physiol Regul Integr Comp Physiol*, **294**, R966–R974.

Impellizzeri F.M., Rampinini E., Castagna C. *et al.* (2008) Validity of a repeated-sprint test for football. *Int J Sports Med*, **29**, 899–905.

Jacobs I., Esbjornsson M., Sylven C. *et al.* (1987) Sprint training effects on muscle myoglobin, enzymes, fibre types and blood lactate. *Med Sci Sports Exerc*, **19** (4), 368–374.

Jones N.L., McCartney N., Graham T. *et al.* (1985) Muscle performance and metabolism in maximal isokinetic cycling at slow and fast speeds. *J Appl Physiol*, **59**, 132–136.

Juel C. (1998) Muscle pH regulation: role of training. *Acta Physiol Scand*, **162**, 359–366.

Juel C., Holten M.K. and Dela F. (2004a) Effects of strength training on muscle lactate release and MCT1 and MCT4 content in healthy and type 2 diabetic humans. *J Physiol (Lond)*, **556** (1), 297–304.

Juel C., Klarskov C., Nielsen J.J. *et al.* (2004b) Effect of high-intensity intermittent training on lactate and H+ release from human skeletal muscle. *Am J Physiol Endocrinol Metab*, **286**, E245–E251.

Kinugasa R., Akima H., Ota A. *et al.* (2004) Short-term creatine supplementation does not improve muscle activation or sprint performance in humans. *Eur J Appl Physiol*, **91**, 230–237.

Laursen P.B., Shing C.M., Peake J.M. *et al.* (2002) Interval training program optimisation in highly trained endurance cyclists. *Med Sci Sports Exerc*, **34** (11), 1801–1807.

Lavender G. and Bird S.R. (1989) Effect of sodium bicarbonate ingestion upon repeated sprints. *Br J Sports Med*, **23** (1), 41–45.

Lindinger M.I., Graham T.E. and Spriet L.L. (1993) Caffeine attenuates the exercise-induced increase in plasma [K+] in humans. *J Appl Physiol*, **74**, 1149–1155.

van Loon L.J.C., Oosterlaar A.M., Hartgens F. *et al.* (2003) Effects of creatine loading and prolonged creatine supplementation on body composition, fuel selection, sprint and endurance performance in humans. *Clin Sci*, **104**, 153–162.

McCully K.K., Kakihira H., Vandenborne K. and Kent-Braun J. (1991) Noninvasive measurements of activity-induced changes in muscle metabolism. *J Biomech*, **21** (Suppl. 1), 153–161.

MacDougall D. and Sale D. (1981) Continuous vs. interval training: a review for the athlete and the coach. *Can J Appl Sport Sci*, **6**, 93–97.

McGawley K. and Bishop D. (2008) Anaerobic and aerobic contribution to two, 5 × 6-s repeated-sprint bouts. *Coaching Sport Sci J*, **3**, 52.

McKenna M.J., Morton J., Selig S.E. and Snow R.J. (1999) Creatine supplementation increases muscle total creatine but not maximal intermittent exercise performance. *J Appl Physiol*, **87**, 2244–2252.

McNaughton L.R., Ford S. and Newbold C. (1997) Effect of sodium bicarbonate ingestion on high intensity exercise in moderately trained women. *J Strength Cond Res*, **11** (2): 98–102.

Mainwood G.W. and Worseley-Brown P. (1975) The effect of extracellular pH and buffer concentration on the efflux of lactate from frog sartorius muscle. *J Physiol (Lond)*, **250**, 1–22.

Matsuura R., Ogata H., Yunoki T. *et al.* (2006) Effect of blood lactate concentration and the level of oxygen uptake immediately before a cycling sprint on neuromuscular activation during repeated cycling sprints. *J Physiol Anthropol*, **25**, 267–273.

Matsuura R., Arimitsu T., Kimura T. *et al.* (2007) Effect of oral administration of sodium bicarbonate on surface EMG activity during repeated cycling sprints.

Mazzeo R.S. (1991) Catecholamine responses to acute and chronic exercise. *Med Sci Sports Exerc*, **23**, 839–845.

Medbo J.I. and Burgers S. (1990) Effect of training on the anaerobic capacity. *Med Sci Sports Exerc*, **22**, 501–507.

Mendez-Villanueva A., Hamer P. and Bishop D. (2007) Fatigue responses during repeated sprints matched for initial mechanical output. *Med Sci Sports Exerc*, **39**, 2219–2225.

Mendez-Villanueva A., Hamer P. and Bishop D. (2008) Fatigue in repeated-sprint exercise is related to muscle power factors and reduced neuromuscular activity. *Eur J Appl Physiol*, **103** (4), 411–419.

Mohr M., Krustrup P., Nielsen J.J. *et al.* (2007) Effect of two different intense training regimens on skeletal muscle ion transport proteins and fatigue development. *Am J Physiol Regul Integr Comp Physiol*, **292**, R1594–R1602.

Morin J.B., Jeannin T., Chevallier B. and Belli A. (2006) Spring-mass model characteristics during sprint running: correlation with performance and fatigue-induced changes. *Int J Sports Med*, **27**, 158–165.

Nevill M.E., Boobis L.H., Brooks S.T. and Williams C. (1989) Effect of training on muscle metabolism during treadmill sprinting. *J Appl Physiol*, **67**, 2376–2382.

Parolin M.L., Chesley A., Matsos M.P. *et al.* (1999) Regulation of skeletal muscle glycogen phosphorylase and PDH during maximal intermittent exercise. *Am J Physiol*, **277**, E890–E900.

Pedersen T.H., Nielsen O.B., Lamb G.D. and Stephenson D.G. (2004) Intracellular acidosis enhances the excitability of working muscle. *Science*, **305**, 1144–1147.

Pilegaard H., Domino K., Noland T. *et al.* (1999) Effect of high-intensity exercise training on lactate/hydrogen ion transport capacity in human skeletal muscle. *Am J Physiol*, **276**, E255–E261.

Poole D.C. and Gaesser G.A. (1985) Response of ventilatory and lactate thresholds to continuous and interval training. *J Appl Physiol*, **58** (4), 1115–1121.

Preen D., Dawson B., Goodman C. *et al.* (2001) The effect of oral creatine supplementation on 80 minutes of repeated-sprint exercise. *Med Sci Sports Exerc*, **33**, 814–825.

Racinais S., Connes P., Bishop D. *et al.* (2005) Morning versus evening power output and repeated-sprint ability. *Chronobiol Int*, **22**, 1029–1039.

Racinais S., Bishop D., Denis R. *et al.* (2007) Muscle deoxygenation and neural drive to the muscle during repeated sprint cycling. *Med Sci Sports Exerc*, **39**, 268–274.

Rampinini E., Bishop D., Marcora S.M. *et al.* (2007) Validity of simple field tests as indicators of match-related physical performance in top-level professional soccer players. *Int J Sports Med*, **28**, 228–235.

Ratel S., Bedu M., Hennegrave A. *et al.* (2002) Effects of age and recovery duration on peak power output during repeated cycling bouts. *Int J Sports Med*, **23**, 397–402.

Ratel S., Williams C.A., Oliver J. and Armstrong N. (2005) Effects of age and recovery duration on performance during multiple treadmill sprints. *Int J Sports Med*, **26**, 1–8.

Ross A. and Leveritt M. (2001) Long-term metabolic and skeletal muscle adaptations to short-sprint training. *Sports Med*, **31** (15), 1063–1082.

Sahlin K., Harris R.C., Nylind B. and Hultman E. (1976) Lactate content and pH in muscle samples obtained after dynamic exercise. *Pflugers Arch*, **367**, 143–149.

Schneiker K.T., Bishop D., Dawson B. and Hackett L.P. (2006) Effects of caffeine on prolonged intermittent-sprint ability in team-sport athletes. *Med Sci Sports Exerc*, **38**, 578–585.

Sharp R.L., Costill D.L., Fink W.J. and King D.S. (1986) Effects of eight weeks of bicycle ergometer sprint training on human muscle buffer capacity. *Int J Sports Med*, **7**, 13–17.

Sinclair G.I. and Geiger J.D. (2000) Caffeine use in sports: A pharmacological review. *J Sports Med Phys Fitness*, **40**, 71–79.

Skare O.C., Skadberg O. and Wisnes A.R. (2001) Creatine supplementation improves sprint performance in male sprinters. *Scand J Med Sci Sports*, **11**, 96–102.

Sostaric S.M., Skinner S.L., Brown M.J. *et al.* (2006) Alkalosis increases muscle K+ release, but lowers plasma [K+] and delays fatigue during dynamic forearm exercise. *J Physiol*, **570**, 185–205.

Spencer M., Bishop D. and Lawrence S. (2004) Longitudinal assessment of the effects of field-hockey training on repeated sprint ability. *J Sci Med Sport*, **7**, 323–334.

Spencer M., Bishop D., Dawson B. and Goodman C. (2005) Physiological and metabolic responses of repeated-sprint activities: specific to field-based team sports. *Sports Med*, **35**, 1025–1044.

Spencer M., Dawson B., Goodman C. *et al.* (2008) Performance and metabolism in repeated sprint exercise: effect of recovery intensity. *Eur J Appl Physiol*, **103**, 545–552.

Spriet L.L., Lindinger M.I., Heigenhauser G.J. and Jones N.L. (1986) Effects of alkalosis on skeletal muscle metabolism and performance during exercise. *Am J Physiol*, **251**, R833–R839.

Spriet L.L., Lindinger M.I., Mckelvie R.S. *et al.* (1989) Muscle glycogenolysis and H+ concentration during maximal intermittent cycling. *J Appl Physiol*, **66** (1), 8–13.

Stathis C.G., Febbraio M.A., Carey M.F. and Snow R.J. (1994) Influence of sprint training on human skeletal muscle purine nucleotide metabolism. *J Appl Physiol*, **76** (4), 1802–1809.

Stephens T.J., McKenna M.J., Canny B.J. *et al.* (2002) Effect of sodium bicarbonate on muscle metabolism during intense endurance cycling. *Med Sci Sports Exerc*, **34** (4), 614–621.

Stepto N.K., Hawley J.A., Dennis S.C. and Hopkins W.G. (1999) Effects of different interval-training programs on cycling time-trial performance. *Med Sci Sports Exerc*, **31** (5), 736–741.

Street D., Nielsen J.-J., Bangsbo J. and Juel C. (2005) Metabolic alkalosis reduces exercise-induced acidosis and potassium accumulation in human skeletal muscle interstitium. *J Physiol*, **566**, 481–489.

Struder H.K., Hollmann W., Duperly J. and Weber K. (1995) Amino acid metabolism in tennis and its possible influence on the neuroendocrine system. *Br J Sports Med*, **29**, 28–30.

Stuart G.R., Hopkins W.G., Cook C. and Cairns S.P. (2005) Multiple effects of caffeine on simulated high-intensity team-sport performance. *Med Sci Sports Exerc*, **37**, 1998–2005.

Tabata I., Nishimura K., Kouzaki M. *et al.* (1996) Effects of moderate-intensity endurance and high-intensity intermittent training on anaerobic capacity and VO2max. *Med Sci Sports Exerc*, **28**, 1327–1330.

Thomas C., Bishop D., Moore-Morris T. and Mercier J. (2007) Effects of high-intensity training on MCT1, MCT4 and NBC expressions in rat skeletal muscles: influence of chronic metabolic alkalosis. *Am J Physiol Endocrinol Metab*, **293**, E916–E922.

Tomlin D.L. and Wenger H.A. (2001) The relationship between aerobic fitness and recovery from high intensity intermittent exercise. *Sports Med*, **31** (1), 1–11.

Wadley G. and Le Rossignol P. (1998) The relationship between repeated sprint ability and the aerobic and anaerobic energy systems. *J Sci Med Sport*, **1** (2), 100–110.

Weber C.L. and Schneider D.A. (2002) Increases in maximal accumulated oxygen deficit after high-intensity interval training are not gender dependent. *J Appl Physiol*, **92**, 1795–1801.

Weston A.R., Myburgh K.H., Lindsay F.H. *et al.* (1997) Skeletal muscle buffering capacity and endurance performance after high-intensity interval training by well-trained cyclists. *Eur J Appl Physiol*, **75**, 7–13.

Williams J.H. (1991) Caffeine, neuromuscular function and high-intensity exercise performance. *J Sports Med Phys Fitness*, **31**, 481–489.

Wiroth J., Bermon S., Andrei S. *et al.* (2001) Effects of oral creatine supplementation on maximal pedalling performance in older adults. *Eur J Appl Physiol*, **84**, 533–539.

Yoshida T. and Watari H. (1993) 31P-Nuclear magnetic resonance spectroscopy study of the time course of energy metabolism during exercise and recovery. *Eur J Appl Physiol*, **66**, 494–499.

Yquel R.J., Arsac L.M., Thiaudiere E. *et al.* (2002) Effect of creatine supplementation on phosphocreatine resynthesis, inorganic phosphate accumulation and pH during intermittent maximal exercise. *J Sports Sci*, **20**, 427–437.

2.10 The Overtraining Syndrome (OTS)[1]

Romain Meeusen and Kevin De Pauw, Department of Human Physiology and Sports Medicine, Faculteit LK, Vrije Universiteit Brussel, Brussels, Belgium

2.10.1 INTRODUCTION

Training can be defined as a process of overload that is used to disturb homoeostasis which results in acute fatigue leading to an improvement in performance. When prolonged excessive training is concurrent with other stressors and insufficient recovery, performance decrements can result in chronic maladaptations that can lead to the overtraining syndrome (OTS). By using the expression 'syndrome' we emphasize the multifactorial aetiology and acknowledge that exercise (training) is not necessarily the sole causative factor.

Many sports require considerable contributions from both aerobic and anaerobic energy systems and use heavy-resistance exercise for supplemental training, but almost all scientific literature on overtraining is based on endurance and aerobic activities.

In this chapter we will not only present the current state of the art on the OTS, but also highlight the difficulties in detecting the possible underlying mechanisms that make an athlete evolve from acute fatigue to a state of overreaching (OR) and eventually OTS.

2.10.2 DEFINITIONS

Athletes often increase their training load in an attempt to enhance performance. As a consequence, they may experience acute feelings of fatigue and decreases in performance as a result of a single intense training session, or an intense training period. The resultant acute fatigue, in combination with adequate rest, can be followed by a positive adaptation or improvement in performance (supercompensation) and is the basis of effective training programmes. Several authors consider overtraining a status that evolves from normal training, through OR, and finally ends in an OTS. Probably these states (OR/OTS) show different defining characteristics, and the 'overtraining continuum' may be an oversimplification, since it emphasizes training characteristics, while the features of OTS consist of more than training errors, and coincide with other stressors (Meeusen *et al.*, 2006). However, as stated in the recent 'consensus statement' of the European College of Sport Science (Meeusen *et al.*, 2006), these definitions indicate that the difference between overtraining and OR is the amount of time needed for performance restoration and not the type or duration of training stress or degree of impairment. As it is possible to recover from a state of OR within a two-week period (Halson *et al.*, 2002; Jeukendrup *et al.*, 1992; Kreider, Fry and O'Toole, 1998; Lehmann *et al.*, 1999a; Steinacker *et al.*, 2000), it may be argued that this condition is a relatively normal and harmless stage of the training process. However, athletes who are suffering from OTS may take months or sometimes years to completely recover.

When looking at the current literature it seems that several papers use 'overtraining' as a verb and therefore indicate the *process* of more intensive or prolonged training that might lead to training (mal)adaptations (Armstrong and VanHeest, 2002; Halson and Jeukendrup, 2004; Meeusen *et al.*, 2006). In many studies 'overtraining' is used to describe both the process of training excessively and the fatigue states that may develop as a consequence (Callister *et al.*, 1990; Kuipers and Keizer, 1988; Morgan *et al.*, 1987; Meeusen *et al.*, 2006).

2.10.2.1 Functional overreaching (FO)

OR is often utilized by athletes during a typical training cycle to enhance performance. Intensified training can result in a decline in performance, but when appropriate periods of recovery are provided, a supercompensation effect may occur, with the athlete exhibiting an enhanced performance or functional overreaching (FO) compared to baseline levels.

Steinacker *et al.* (2000) made a distinction between OR and OTS. While following rowers preparing for the world

[1] This chapter is based on the 'Position Statement' of the European College of Sport Science (Meeusen *et al.*, 2006).

Strength and Conditioning – Biological Principles and Practical Applications Marco Cardinale, Rob Newton, and Kazunori Nosaka.

championships, they found clear signs of OR after 18 days of intense training. These signs were a decrease in performance, gonadal and hypothalamic hormone disturbances, and deterioration of recovery in the psychological questionnaire. The authors called these athletes 'overreached' because, after a tapering period, the values returned to normal. This study was a typical example of FO because OR was used as an integral part of successful training, although during the intensive training period some markers already showed disturbances.

2.10.2.2 Nonfunctional overreaching (NFO)

FO has no negative consequences for the athlete in the long term. When 'intensified training' continues, and performance does not improve and feelings of fatigue do not disappear after the recovery period, OR has not been functional and this is thus called nonfunctional overreaching (NFO). Full recovery does not take place within the pre-planned period of time. This is undesirable for two reasons: first, a recovery period that takes longer than planned might interfere with competitions; second, there can be deconditioning due to the longer recovery period (Nederhof *et al.*, 2006).

Both FO and NFO athletes will be able to fully recover after sufficient rest. It seems from the literature that in NFO the evolution on the overtraining continuum is not only quantitatively determined (i.e. by the increase in training volume) but that qualitative changes also occur (e.g. signs and symptoms of psychological and/or endocrine distress). This is in line with current neuroendocrine findings using a double-exercise test (Meeusen *et al.*, 2004, 2010; Urhausen, Gabriel and Kindermann, 1998a; Urhausen *et al.*, 1998b).

2.10.2.3 The overtraining syndrome (OTS)

While it is possible to recover from FO within a period of two weeks, recovery from the NFO state is less clear. This is probably because not many studies have tried to define the difference between extreme OR, which needs several weeks or even months of recovery (Meeusen *et al.*, 2006), and OTS. Athletes who suffer from the OTS may need months or even years to completely recover, leading frequently to cessation of a sports career.

The difficulty lies in the subtle differences that might exist between extremely OR athletes and those with OTS.

Reports on athletes suffering from OTS are mostly case descriptions, since it is not only unethical, but probably also impossible to train an athlete with a high training load while at the same time including other stressors, especially since the symptoms of OTS differ per individual.

Meeusen *et al.* (2004) reported on differences in normal training status and FO (after a training camp), and compared the endocrinological results to a double-exercise test with an OTS athlete. Athletes were tested in a double-exercise protocol (two exercise tests with four hours' rest in between) in order to register the recovery capacity of the athletes. Performance was measured as the time to voluntary exhaustion. They compared the first and the second exercise tests in order to verify whether the athletes were able to maintain the same performance. The training camp reduced exercise capacity in the athletes. There was a 3% decrease between performances in the first versus the second test, while in the FO condition there was a 6% performance decrease. The OTS subjects showed an 11% decrease in time to exhaustion. The OTS athletes also showed clear psychological and endocrinological disturbances. In a follow-up study (Meeusen *et al.*, 2010) it was clear that some hormonal differences could discriminate between NFO and OTS, providing a possible tool for detecting the 'real' overtraining status of the athletes (see below).

2.10.2.4 Summary

The consensus statement of the European College of Sport Science (Meeusen *et al.*, 2006) indicates that the difference between NFO and OTS is the amount of time needed for performance restoration and not the type or duration of training stress or the degree of impairment.

When athletes deliberately use a short-term period (e.g. training camp) to increase their training load they can experience short-term performance decrement, without severe psychological or other lasting negative symptoms. This FO will eventually lead to an improvement in performance after recovery.

However, when athletes do not sufficiently respect the balance between training and recovery, NFO can occur. At this stage, the first signs and symptoms of prolonged training distress, such as performance decrements, psychological disturbance (decreased vigour, increased fatigue), and hormonal disturbances, will occur and the athletes will need weeks or months to recover. Several confounding factors such as inadequate nutrition (energy and/or carbohydrate intake), illness (most commonly upper respiratory tract infections, URTIs), psychosocial stressors (work-, team-, coach-, family- related), and sleep disorders may be present. At this stage, the distinction between NFO and OTS is very difficult, and will depend on the clinical outcome and exclusion diagnosis. The athlete will often show the same clinical, hormonal, and other signs and symptoms. Therefore the diagnosis of OTS can often only be made retrospectively, when the time course can be overseen. A keyword in the recognition of OTS might be 'prolonged maladaptation', not only of the athlete, but also of several biological, neurochemical, and hormonal regulation mechanisms.

2.10.3 PREVALENCE

The borderline between optimal performance and performance impairment due to OTS is subtle. This applies especially to physiological and biochemical factors. The apparent vagueness surrounding OTS are further complicated by the fact that the clinical features are varied between individuals, nonspecific, anecdotal, and numerous.

Probably because of the differences in the definitions used, prevalence data on overtrained athletes are dispersed. Studies have reported that up to 60% of distance runners show signs of overtraining during their careers, while data on swimmers vary between 3 and 30% (Hooper, MacKinnon and Hanrahan, 1997; Lehmann *et al.*, 1993a; Morgan *et al.*, 1987; O'Connor *et al.*, 1989; Raglin and Morgan, 1994). If the definition of OTS given above is used, the incidence figures will probably be less high. We suggest that a distinction be made between NFO and OTS and that athletes are defined as suffering from OTS only when a clinical exclusion diagnosis (see below) establishes this.

2.10.4 MECHANISMS AND DIAGNOSIS

Probably because of the difficulty in detecting straightforward mechanisms responsible for OTS, many speculations have been made as to the 'real' reason for the genesis of OTS. This has led to many papers that present a possible hypothesis on its origins.

Although in recent years our knowledge of the central pathomechanisms of OTS has significantly increased, there is still a strong demand for relevant tools for its early diagnosis. OTS is characterized by a sports-specific decrease in performance, together with persistent fatigue and disturbances in mood state (Armstrong and VanHeest, 2002; Halson and Jeukendrup, 2004; Meeusen *et al.*, 2006; Urhausen and Kindermann, 2002). This underperformance persists, despite a period of recovery lasting several weeks or months. Importantly, as there is no diagnostic tool to identify an athlete as suffering from OTS, diagnosis can only be made by excluding all other possible influences on changes in performance and mood state. The definitive diagnosis of OTS always requires the exclusion of an organic disease, e.g. endocrinological disorders (thyroid or adrenal gland, diabetes), iron deficiency with anaemia, and infectious diseases (Meeusen *et al.*, 2006). Other major disorders or eating disorders, such as anorexia nervosa and bulimia, should also be excluded. However, it should be emphasized that many endocrinological and clinical findings due to NFO and OTS can mimic other diseases. The line between under- and over-diagnosis is very difficult to judge (Meeusen *et al.*, 2006).

Usable markers for NFO and/or OTS should fulfil six criteria. They should be: (1) objective; (2) non-manipulable; (3) applicable in training practice; (4) not too demanding for athletes; (5) affordable for the majority of athletes; and (6) based on a sound theoretical framework (Nederhof *et al.*, 2006).

2.10.4.1 Hypothetical mechanisms

Increased training loads, as well as other persistent stresses, can influence the human body chronically. This disturbance of homeostasis will be compensated by re-regulating mechanisms, but when the stress becomes excessive, a permanent disorder can occur. Several hypotheses include the sympathetic and parasympathetic nervous systems' imbalances (Lehmann *et al.*, 1998a), glutamine (Rowbottom *et al.*, 1995, 1996), and other amino acids (Gastmann and Lehmann, 1998). Smith (2000) proposed a hypothesis in which excessive muscular stress (Seene, Umnova and Kaasik, 1999) induces a local inflammatory response that can evolve into chronic inflammation and possibly end up in a systemic inflammation. Inflammatory agents, such as cytokines (Smith, 2000, 2003, 2004), IL-6 (Robson, 2003), might act on the central nervous system (CNS), leading to a sickness behaviour, and creating physiological, biochemical, neuroendocrine, and psychological disturbances. Many studies report changes in endocrine functioning in overtrained athletes, and this has led to hypotheses ranging from catabolic/anabolic imbalances indicated by a decrease in the testosterone/cortisol ratio (Adlercreutz *et al.*, 1986) and subsequent neuroendocrine disturbances (Keizer, 1998) to CNS imbalances in neurotransmitters which create similarities with depression (Armstrong and VanHeest, 2002; Kreider, 1998; Meeusen, 1999).

Although these theories have potential, until more prospective studies are carried out with a longitudinal follow-up of athletes (who may develop OTS), or specific diagnostic tools are developed, they remain speculative. In the following sections, we will briefly explain some of these proposed hypothetical mechanisms and will extract those measures that might give an indication of the training (or overtraining) status of the individual athlete.

2.10.4.2 Biochemistry

During prolonged training, glycogen stores get close to full depletion, glycogenolysis and glucose transport are downregulated in muscle and liver, as along with the liver's production of IGF-I, and catabolism is induced. This catabolic state could be a possible trigger for several disturbances of HOMEOSTASIS of blood parameters, and measurements of selected enzyme activities and blood markers are in line with these hypotheses; however, the validity of these variables is overestimated when it comes to being a diagnostic tool for OTS (Meeusen *et al.*, 2006; Urhausen and Kindermann, 2002).

Although most of the blood parameters (e.g. blood count, CRP, SR, CK, urea, creatinine, liver enzymes, glucose, ferritin, sodium, potassium, etc.) are not capable of detecting OR or OTS, they are helpful in providing information on the actual health status of an athlete, and are therefore useful in the exclusion diagnosis (Meeusen *et al.*, 2006).

2.10.4.3 Physiology

There have been several proposals as to which physiological measures might be indicative of OR or OTS. Reduced maximal heart rates after increased training may be the result of reduced sympathetic nervous system activity, decreased tissue responsiveness to catecholamines, changes in adrenergic receptor activity, or simply a reduced power output achieved with maximal effort. Several other reductions in maximal physiological measures (oxygen uptake, heart rate, etc.) might be the consequence of a reduction in exercise time and not related to abnormalities per se, and it should be noted that changes of resting heart rate are not consistently found in athletes suffering from OTS (Meeusen *et al.*, 2006).

Even when heart rate is relatively stable, the time between two beats (R–R interval) can differ substantially. The variation in time between beats is defined as heart-rate variability (HRV). Numerous studies have examined the effects of training on indices of HRV, but to date few have investigated HRV in OR or OTS athletes, with existing studies showing either no change (Achten and Jeukendrup, 2003; Hedelin *et al.*, 2000; Uusitalo *et al.*, 1998a, 1998b), inconsistent changes (Uusitalo, Uusitalo and Rusko, 2000), or changes in parasympathetic modulation (Hedelin *et al.*, 2000; Pichot *et al.*, 2000). However, much more research is necessary before HRV can be considered as a diagnostic measurement for OTS. It might be an indication of the actual training status of the individual, and therefore be part of the exclusion diagnosis as a marker that needs attention when examining an athlete suspected of having OTS.

2.10.4.4 The immune system

There are many reports of URTIs due to increased training, and in OR and OTS athletes. It seems feasible that intensified training (leading to OR or OTS) might increase both the duration of the so-called 'open window' and the degree of the resultant immunodepression. The amount of scientific information available to substantiate this argument is, however, limited. It might just be that the increased URTI incidence reflects the increase in training, regardless of the response of the athlete to the increased physical stress (Meeusen *et al.*, 2006).

Infection might be one of the 'triggering' factors leading to the induction of OTS, or in some cases the diagnosis of OTS might not be capable of being differentiated from a state of post-viral fatigue such as that observed with episodes of glandular fever.

The current information regarding the immune system and OR confirms that periods of intensified training result in depressed immune cell functions with little or no alteration in circulating cell numbers. However, although immune parameters change in response to increased training load, these changes do not distinguish between those athletes who successfully adapt to OR and those who maladapt and develop symptoms of OTS.

It is clear that the immune system is extremely sensitive to stress – both physiological and psychological – and thus, potentially, immune variables could be used as an index of stress in relation to exercise training.

Furthermore, at present it seems that measures of immune function cannot really distinguish OTS from infection or post-viral fatigue states (Gleeson, 2007).

2.10.4.5 Hormones

For several years it has been hypothesized that a hormone-mediated central deregulation occurs during the pathogenesis of OTS, and that measurements of blood hormones could help to detect OTS (Fry and Kraemer, 1997; Fry, Steinacker and Meeusen, 2006; Keizer, 1998; Kuipers and Keizer, 1988; Urhausen, Gabriel and Kindermann, 1995, 1998a). The results of the research devoted to this subject are far from unanimous, mostly because of the difference in measuring methods and/or detection limits of the analytical equipment used.

For a long time the plasma testosterone/cortisol ratio was considered a good indicator of the overtraining state, but this ratio only indicates the actual physiological strain of training and cannot be used for diagnosis of OR or OTS (Lehmann *et al.*, 1998a, 1999b, 2001; Meeusen, 1999; Urhausen, Gabriel and Kindermann, 1995).

Most of the literature agrees that OR and OTS must be viewed on a continuum with a disturbance, an adaptation, and finally a maladaptation of the hypothalamic pituitary adrenal (HPA) axis and all the other hypothalamic axes (Keizer, 1998; Lehmann *et al.*, 1993b, 1998a, 1999b, 2001; Meeusen, 1998, 1999; Meeusen *et al.*, 2004; Urhausen, Gabriel and Kindermann, 1995, Urhausen *et al.*, 1998b). However, it should be emphasized that depending on the training status, when the hormone measures are taken (diurnal variation), urinary, blood, and salivary measures create a great variation in the interpretation of the results. In OTS, a decreased rise in pituitary hormones (ACTH, growth hormone (GH), luteinizing hormone (LH), and follicle-stimulating hormone (FSH)) in response to a stressful stimulus is reported (Barron *et al.*, 1985; Lehmann *et al.*, 1993a, 1998a, 1998b, 1999a, 1999b; Meeusen *et al.*, 2004; Wittert *et al.*, 1996; Urhausen, Gabriel and Kindermann, 1995, 1998a).

This indicates that hormonal markers are potent parameters for registering disturbances of homoeostasis, but the literature is still very diffuse because of a lack of standardization in test methods.

2.10.4.6 Is the brain involved?

Over the last decade, there has been significant interest in determining specific peripheral markers for the metabolic, physiological, and psychological responses to exercise that have been suggested to be associated with OTS. To date, relatively little attention has been given to the role of the CNS in OTS (Meeusen, 1999). The neuroendocrine and CNS hypotheses, as well as the neuro-immunological and psychometric data, indicate that OTS occurs with a major disturbance of regulatory mechanisms including the 'brain–periphery' interaction.

The hypothalamus is under the control of several higher brain centres and several neurotransmitters (Meeusen and De Meirleir, 1995). Amongst these transmitters, serotonin (5-HT) is known to play a major role in various neuroendocrine and behavioural functions, such as activation of the HPA axis, feeding, and locomotion (Wilckens, Schweiger and Pirke,

1992). The possibility that impaired neurotransmission at the various central aminergic synapses is associated with major disturbances of the CNS, such as depression (and possibly OTS), has received increasing attention over the past several years. It has been suggested that exercise exerts its putative psychological effects via the same neurochemical substrate (the monoamines) as antidepressant drugs, which are known to increase the synaptic availability of transmitters (Armstrong and Vanheest, 2002; Meeusen *et al.*, 1996, 1997; Uusitalo *et al.*, 2004). In pathological situations, such as in major depression (Dishman, 1997), post-traumatic stress disorders (PTSDs) (Liberzon *et al.*, 1999; Porter *et al.*, 2004), and probably OTS, the glucocorticoids and the brain monoaminergic systems apparently fail to restrain the HPA response to stress (Meeusen *et al.*, 2004).

In OTS the neuroendocrine disorder is a hypothalamic dysfunction rather than a malfunction of the peripheral hormonal organs (Kuipers and Keizer, 1988). The interactive features of the periphery and the brain could be translated into possible immunological, psychological, and endocrinological disturbances. However, since OTS is athlete-specific, generalization of the signs and symptoms is at present not possible.

2.10.4.7 Training status

A hallmark feature of the OTS is its inability to sustain intense exercise, and a decreased sports-specific performance capacity when the training load is maintained or even increased (Meeusen *et al.*, 2004; Urhausen, Gabriel and Kindermann, 1995). Athletes suffering from NFO and OTS are usually able to start a normal training sequence or a race at their normal training pace, but are not able to complete the training load they are given, or to race as usual. The key indicator of the OTS can be considered an unexplainable decrease in performance. Therefore an exercise/performance test is considered to be essential for the diagnosis of OTS (Budgett *et al.*, 2000; Lehmann *et al.*, 1999a; Urhausen, Gabriel and Kindermann, 1995).

Urhausen, Gabriel and Kindermann (1998a) and Meeusen *et al.* (2004) have shown that multiple tests carried out on different days (Urhausen, Gabriel and Kindermann, 1998a), or the two-maximal-incremental-exercise test separated by four hours, can be valuable tools for assessing the performance decrements usually seen in OTS athletes. A decrease in exercise time of at least 10% is considered to be significant. Furthermore, this decrease in performance needs to be confirmed by specific changes in hormone concentrations (Meeusen *et al.*, 2004). In one follow-up study a two-bout exercise protocol was used to make an objective, immediately available distinction between NFO and OTS Meeusen *et al.* (2010). The authors studied 10 underperforming athletes who were diagnosed with the suspicion of NFO or OTS. The athletes' recovery was monitored by a sports physician in order to retrospectively distinguish NFO from OTS. Five athletes were retrospectively diagnosed with NFO and five were diagnosed with OTS, and stress-induced hormonal reactions were registered. Maximal blood lactate concentration was lower in OTS compared to NFO, while resting concentrations of cortisol, ACTH, and prolactin were higher. However, sensitivity of these measures was low. The ACTH and prolactin reactions to the second exercise bout were much higher in NFO athletes than in those with OTS and showed the highest sensitivity for making the distinction.

2.10.4.8 Psychometric measures

There is general agreement that OTS is characterized by psychological disturbances and negative affective states. When athletes suffer from OTS, they typically experience chronic fatigue, poor sleep patterns, a drop in motivation, episodes of depression, and helplessness (Lemyre, 2005). Not surprisingly, their performance is considerably impaired. Full recovery from OTS represents a complex process that may necessitate many months, or even years, of rest and removal from sport (Kellmann, 2002; Kentta and Hassmen, 1998).

Several questionnaires, such as the Profile of Mood State (POMS) (Morgan *et al.*, 1988; O'Connor, 1997; O'Connor *et al.*, 1989; Raglin, Morgan and O'Connor, 1991; Rietjens *et al.*, 2005), the Recovery–Stress Questionnaire (RestQ-Sport) (Kellmann, 2002), the Daily Analysis of Life Demands of Athletes (DALDA) (Halson *et al.*, 2002), and the 'self-condition scale' (Urhausen *et al.*, 1998b), have been used to monitor psychological parameters in athletes. Other tests, such as attention tests (finger pre-cuing tasks) (Rietjens *et al.*, 2005) and neurocognitive tests (Kubesch *et al.*, 2003), also serve as promising tools for detecting subtle neurocognitive disturbances registered in OR or OTS athletes. It is important to register the current state of stress and recovery, and to prospectively follow the evolution for each athlete individually (Morgan *et al.*, 1988; Kellmann, 2002). The great advantage of psychometric instruments is the quick availability of information (Kellmann, 2002), especially since psychological disturbances coincide with physiological and performance changes and are generally the precursors of neuroendocrine disturbances. In OTS the depressive component is more expressed than in OR (Armstrong and VanHeest, 2002). Changes in mood state may be a useful indicator of OR and OTS; however, it is necessary to combine mood disturbances with measures of performance.

2.10.4.9 Psychomotor speed

Psychomotor speed is the amount of time it takes a person to process a signal, prepare a response, and execute that response. It is measured by reaction time. It is well known that psychomotor speed is reduced in patients with many different pathologies, such as type 2 diabetes mellitus, Parkinson's disease, Alzheimer's disease, and depression. Most of these pathologies have symptoms in common with OTS; mainly neuropsychological symptoms such as inability to concentrate, depression, irritability, and difficulty thinking, but also sleep disturbances (Nederhof *et al.*, 2006).

Considering the similarities between major depression, chronic fatigue syndrome, and OTS, it is hypothesized that psychomotor speed is also impaired in overtrained athletes.

Computerized tests of psychomotor speed fulfil all six criteria mentioned above (Section 2.10.4) for early markers of NFO and OTS. Furthermore, this is a task that can be performed in a quiet room on a laptop and can easily be integrated into a training programme.

2.10.4.10 Are there definitive diagnostic criteria?

The need for definitive diagnostic criteria for OTS is reflected in much of the OR and overtraining research by a lack of consistent findings. None of the currently available or suggested markers meets all of the above-mentioned criteria for a reliable marker for the onset of OTS (see Section 2.10.4; Meeusen *et al.*, 2006). When choosing markers that might give an indication of the training or overtraining status of an athlete, one needs to take into account several potential problems that might influence decision-making.

When testing the athlete's performance, the intensity and reproducibility of the test should be sufficient to detect differences (max test, time trial, two max test). Baseline measures are often not available, and therefore the degree of performance limitation may not be exactly determined. Many of the performance tests are not sports-specific. HRV seems a promising tool in theory, but needs to be standardized when tested, and at present does not provide consistent results. Biochemical markers, such as lactate or urea, and immunological markers do not have sufficiently consistent reports in the literature to be considered as absolute indicators for OTS.

Many factors affect blood hormone concentrations; these include factors linked to sampling conditions and/or conservation of the sampling: stress of the sampling, intra- and inter-assay coefficient of variability. Other factors, such as food intake (nutrient composition and/or pre- versus post-meal sampling), can significantly modify either the basal concentration of some hormones (cortisol, DHEA-S, total testosterone) or their concentration change in response to exercise (cortisol, GH).

Diurnal and seasonal variations of the hormones are important factors that need to be considered. In female athletes, the hormonal response will depend on the phase of the menstrual cycle. Hormone concentrations at rest and following stimulation (exercise = acute stimulus) respond differently. Stress-induced measures (exercise, pro-hormones, etc.) need to be compared with baseline measures from the same individual. Poor reproducibility and feasibility of some techniques used to measure certain hormones can make the comparison of results difficult. Therefore the use of two maximal performance (or time trial) tests separated by four hours could help in comparing the individual results.

Psychometric data always need to be compared with the baseline status of the athlete. The lack of success induced by a long-term decrement of performance could be explained by the depression in OTS. The differences between self-assessment and questionnaires given by an independent experimenter, and the timing of the mood-state assessment, are important. Questionnaires should be used in standardized conditions. Other psychological parameters other than mood state (attention-focusing, anxiety) might also be influenced.

NFO and OTS could be prevented using early markers, which should be (1) objective, (2) non-manipulable, (3) applicable in training practice, (4) not too demanding, (5) affordable, and (6) based on a sound theoretical framework. No such markers exist today, although it is possible that psychomotor speed might be one.

Athletes and the field of sports medicine in general would benefit greatly if a specific, sensitive, simple diagnostic test existed for the identification of OTS. At present no such test meets these criteria. There is a need for a combination of diagnostic aids to pinpoint possible markers, and in particular, for a detection mechanism for early triggering factors.

2.10.5 PREVENTION

One general confounding factor when reviewing the literature on OTS is that the definition and diagnosis of OR and OTS is not standardized. One can even question whether in most of the studies subjects were actually suffering from OTS. Because OTS is difficult to diagnose, authors agree that it is important to prevent it (Foster *et al.*, 1988; Kuipers, 1996; Uusitalo, 2001). Moreover, because OTS is mainly due to an imbalance in the training/recovery ratio (too much training and competitions, and too little recovery), it is of the utmost importance that athletes record their daily training load, using a daily training diary or training log (Foster, 1998; Foster *et al.*, 1988). Psychological screening of athletes (Berglund and Safstrom, 1994; Hooper and McKinnon, 1995; Hooper *et al.*, 1995; Kellmann, 2002; McKenzie, 1999; Morgan *et al.*, 1988; Raglin, Morgan and O'Connor, 1991; Steinacker and Lehmann, 2002; Urhausen *et al.*, 1998b) and the Ratings of Perceived Exertion (RPE) (Acevedo, Rinehardt and Kraemer, 1994; Callister *et al.*, 1990; Foster, 1998; Foster *et al.*, 1996; Hooper and McKinnon, 1995; Hooper *et al.*, 1995; Kentta and Hassmen, 1998; Snyder *et al.*, 1993) have received more and more attention in recent years (Table 2.10.1).

Foster (1998) and Foster *et al.* (1996) have determined training load as the product of the subjective intensity of a training session using 'session RPE' and the total duration of the training session expressed in minutes. With this method of monitoring training they have demonstrated the utility of evaluating experimental alterations in training and have successfully related training load to performance (Foster *et al.*, 1996). However, training load is clearly not the only training-related variable contributing to the genesis of OTS. When athletes are suspected of suffering from OTS, a double-exercise test separated by four hours where ACTH and prolactin are measured could be helpful in distinguishing between NFO and OTS.

2.10.6 CONCLUSION

One difficulty with recognizing and conducting research into athletes with OTS is defining the point at which it develops. Many studies claim to have induced OTS, but it is more likely that they have induced a state of OR in their subjects. Consequently, the majority of studies aimed at identifying markers of ensuing OTS are actually reporting markers of excessive exercise stress resulting in the acute condition of OR and not the chronic condition of OTS.

The mechanism of OTS might be difficult to examine in detail, because the stress caused by excessive training load, in combination with other stressors, could possibly trigger different defence mechanisms, such as the immunological, neuroendocrine, and other physiological systems, which all interact and therefore probably cannot be pinpointed as being the 'sole' cause of OTS. It might be that, as in other syndromes (e.g. chronic fatigue syndrome, or burnout), the psychoneuroimmunology (study of brain–behaviour–immune inter-relationships) can shed light on the possible mechanisms of OTS. But until there is a definite diagnostic tool, it is of the utmost importance to standardize the measures that are now thought to provide a good inventory of the training status of an athlete. It is very important to emphasize the need to distinguish OTS from OR and other potential causes of temporary underperformance, such as anaemia, acute infection, muscle damage, and insufficient carbohydrate intake.

The physical demands of intensified training are not the only elements in the development of OTS. It seems that a complex set of psychological factors are also important, including excessive expectations from a coach or family members, competitive stress, personality structure, social environment, relationships with family and friends, monotony in training, personal or emotional problems, and school- or work-related demands. While no single marker can be taken as an indicator of impending OTS, the regular monitoring of a combination of performance, physiological, biochemical, immunological, and psychological variables would seem to be the best strategy for identifying athletes who are failing to cope with the stress of training.

Much more research is necessary to get a clear-cut answer as to the origin and detection of OTS. We therefore encourage researchers and clinicians to report as much as possible on individual cases where athletes are underperforming, and by following the exclusion diagnosis, to discover whether they might possibly be suffering from OTS. Until a definitive diagnostic tool for OTS is created, coaches and physicians need to rely on performance decrements as verification that OTS exists. If sophisticated laboratory techniques are not available, the considerations shown in Table 2.10.1 may be useful.

Table 2.10.1 Considerations for coaches and physicians

Maintain accurate records of performance during training and competition. Be willing to adjust daily training intensity/volume, or allow a day of complete rest, when performance declines or the athlete complains of excessive fatigue
Avoid excessive monotony of training
Always individualize the intensity of training
Encourage and regularly reinforce optimal nutrition, hydration status, and sleep
Be aware that multiple stressors such as sleep loss or sleep disturbance (e.g. jet lag), exposure to environmental stressors, occupational pressures, change of residence, and interpersonal or family difficulties may add to the stress of physical training
Treat OTS with rest! Reduced training may be sufficient for recovery in some cases of OR
Individualize resumption of training on the basis of the signs and symptoms, since there is no definitive indicator of recovery
Communicate with athletes (perhaps through an online training diary) about their physical, mental, and emotional concerns
Include regular psychological questionnaires to evaluate the emotional and psychological state of the athlete
Maintain confidentiality regarding each athlete's condition (physical, clinical, and mental)
Enforce regular health checks by a multidisciplinary team (physician, nutritionist, psychologist, etc.)
Allow the athlete time to recover after illness/injury
Note the occurrence of URTIs and other infectious episodes; the athlete should be encouraged to suspend training or reduce their training intensity when suffering from an infection
Always rule out an organic disease in cases of performance decrement
Unresolved viral infections are not routinely assessed in elite athletes, but it may be worth investigating this in individuals experiencing fatigue and underperformance in training and competition

References

Acevedo E., Rinehardt K. and Kraemer R. (1994) Perceived exertion and affect at varying intensities of running. *Res Q Exer Sport*, **65** (4), 372–376.

Achten J. and Jeukendrup A.E. (2003) Heart rate monitoring: applications and limitations. *Sports Med*, **33**, 517–538.

Adlercreutz H., Harkonen M., Kuoppasalmi K. *et al.* (1986) Effect of training on plasma anabolic and catabolic steroid hormones and their response during physical exercise. *Int J Sports Med*, **7**, S27–S28.

Armstrong L. and VanHeest J. (2002) The unknown mechanisms of the overtraining syndrome. Clues from depression and psychoneuroimmunology. *Sports Med*, **32**, 185–209.

Barron G., Noakes T., Levy W. *et al.* (1985) Hypothalamic dysfunction in overtrained athletes. *J Clin Endocrinol Metab*, **60**, 803–806.

Berglund B. and Safstrom H. (1994) Psychological monitoring and modulation of training load of world-class canoeists. *Med Sci Sports Exerc*, **26**, 1036–1040.

Budgett R., Newsholme E., Lehmann M. *et al.* (2000) Redefining the overtraining syndrome as the unexplained underperformance syndrome. *Br J Sports Med*, **34**, 67–68.

Callister R., Callister R., Fleck S. and Dudley G. (1990) Physiological and performance responses to overtraining in elite judo athletes. *Med Sci Sports Exerc*, **22** (6), 816–824.

Dishman R. (1997) The norepinephrine hypothesis, in *Physical Activity and Mental Health* (ed. W.P. Morgan), Taylor and Francis, Washington, DC, pp. 199–212.

Foster C. (1998) Monitoring training in athletes with reference to overtraining syndrome. *Med Sci Sports Exerc*, **30** (7), 1164–1168.

Foster C., Snyder A., Thompson N. and Kuettel K. (1988) Normalisation of the blood lactate profile. *Int J Sports Med*, **9**, 198–200.

Foster C., Daines E., Hector L. *et al.* (1996) Athletic performance in relation to training load. *Wis Med J*, **95**, 370–374.

Fry A. and Kraemer W. (1997) Resistance exercise overtraining and overreaching. *Sports Med*, **23**, 106–129.

Fry A., Steinacker J. and Meeusen R. (2006) Central mechanisms of overtraining endocrinology, in *The Endocrine System in Sports and Exercise,* volume XI of the *Encyclopaedia of Sports Medicine* (eds W. Kraemer and A. Rogol), Blackwell Publishing, USA, pp. 578–599.

Gastmann U. and Lehmann M. (1998) Overtraining and the BCAA hypothesis. *Med Sci Sports Exerc*, **30**, 1173–1178.

Gleeson M. (2007) Immune function in sport and exercise. *J Appl Physiol*, **103** (2), 693–699.

Halson S. and Jeukendrup A. (2004) Does overtraining exist? An analysis of overreaching and overtraining research. *Sports Med*, **34**, 967–981.

Halson S.L., Bridge M.W., Meeusen R. *et al.* (2002) Time course of performance changes and fatigue markers during intensified training in trained cyclists. *J Appl Physiol*, **93** (3), 947–956.

Hedelin R., Kentta G., Wiklund U. *et al.* (2000) Short-term overtraining: effects on performance, circulatory responses, and heart rate variability. *Med Sci Sports Exerc*, **32**, 1480–1484.

Hooper S. and Mackinnon L. (1995) Monitoring overtraining in athletes. Recommendations. *Sports Med*, **20** (5), 321–327.

Hooper S.L., Mackinnon L.T., Howard A. *et al.* (1995) Markers for monitoring overtraining and recovery. *Med Sci Sports Exerc*, **27** (1), 106–112.

Hooper S., MacKinnon L. and Hanrahan S. (1997) Mood states as an indication of staleness and recovery. *Int J Sport Psychol*, **28**, 1–12.

Jeukendrup A.E., Hesselink M.K., Snyder A.C. *et al.* (1992) Physiological changes in male competitive cyclists after two weeks of intensified training. *Int J Sports Med*, **13**, 534–541.

Keizer H. (1998) Neuroendicrine aspects of overtraining, in *Overtraining in Sport* (eds R. Kreider, A.C. Fry and M. O'Toole), Human Kinetics, Champaign, IL., pp. 145–168.

Kellmann M. (2002) *Enhancing Recovery: Preventing Underperformance in Athletes*, Human Kinetics, Chapmaign, IL.

Kentta G. and Hassmen P. (1998) Overtraining and recovery. *Sports Med*, **26** (1), 1–16.

Kreider R., Fry A.C. and O'Toole M. (1998) Overtraining in sport: terms, definitions, and prevalence, in *Overtraining in Sport* (eds R. Kreider, A.C. Fry and M. O'Toole), Human Kinetics, Champaign, IL, pp. vii–vix.

Kubesch S., Bretschneider V., Freudenmann R. *et al.* (2003) Aerobic endurance exercise improves executive functions in depressed patients. *J Clin Psychiatry*, **64** (9), 1005–1012.

Kuipers H. (1996) How much is too much? Performance aspects of overtraining. *Res Q Exer Sport*, **67** (Suppl. 3), 65–69.

Kuipers H. and Keizer H. (1988) Overtraining in elite athletes. *Sports Med*, **6**, 79–92.

Lehmann M., Foster C. and Keul J. (1993a) Overtraining in endurance athletes: a brief review. *Med Sci Sports Exerc*, **25** (7), 854–862.

Lehmann M., Knizia K., Gastmann U. *et al.* (1993b) Influence of 6-week, 6 days per week, training on pituitary function in recreational athletes. *Br J Sports Med*, **27**, 186–192.

Lehmann M., Foster C., Dickhuth H.H. and Gastmann U. (1998a) Autonomic imbalance hypothesis and overtraining syndrome. *Med Sci Sports Exerc*, **30**, 1140–1145.

Lehmann M., Foster C., Netzer N. *et al.* (1998b) Physiological responses to short- and long-term overtraining in endurance athletes, in *Overtraining in Sport* (eds R. Kreider, A.C. Fry and M. O'Toole), Human Kinetics, Champaign, IL, pp. 19–46.

Lehmann M., Foster C., Gastmann U. *et al.* (1999a) Definitions, types, symptoms, findings, underlying mechanisms, and frequency of overtraining and overtraining syndrome, in *Overload, Performance Incompetence, and Regeneration in Sport* (eds M. Lehmann, C. Foster, U. Gastmann *et al.*), Kluwer Academic/Plenum Publishers, New York, NY, pp. 1–6.

Lehmann M., Gastmann U., Baur S. *et al.* (1999b) Selected parameters and mechanisms of peripheral and central fatigue and regeneration in overtrained athletes, in *Overload, Performance Incompetence, and Regeneration in Sport* (eds M. Lehmann, C. Foster, U. Gastmann *et al.*), Kluwer Academic/Plenum Publishers, New York, NY, pp. 7–25.

Lehmann M., Petersen K.G., Liu Y. *et al.* (2001) Chronische und erschöpfende Belastungen im Sport - Einfluss von Leptin und Inhibin [Chronic and exhausting training in sports – influence of leptin and inhibin]. *Deutsch Zeitschrift Sportmedicine*, **51**, 234–243.

Lemyre N. (2005) Determinants of burnout in elite athletes. PhD Thesis, Norwegian University of Sport and Physical Education.

Liberzon I., Taylor S., Amdur R. *et al.* (1999) Brain activation in PTSD in response to trauma-related stimuli. *Biol Psychiatry*, **45**, 817–826.

McKenzie D.C. (1999) Markers of excessive exercise. *Can J Appl Physiol*, **24** (1), 66–73.

Meeusen R. (1998) Overtraining, indoor and outdoor. *Vlaams tijdschrift voor Sportgeneeskunde Sportwetenschappen*, **19**, 8–19.

Meeusen R. (1999) Overtraining and the central nervous system, the missing link? in *Overload, Performance Incompetence, and Regeneration in Sport* (eds M. Lehmann, C. Foster, U. Gastmann *et al.*), Kluwer Academic/Plenum Publishers, New York, NY, pp. 187–202.

Meeusen R. and De Meirleir K. (1995) Exercise and brain neurotransmission. *Sports Med*, **20** (3), 160–188.

Meeusen R., Chaouloff F., Thorré K. *et al.* (1996) Effects of tryptophan and/or acute running on extracellular 5-HT and 5-HIAA levels in the hippocampus of food-deprived rats. *Brain Res*, **740**, 245–252.

Meeusen R., Smolders I., Sarre S. *et al.* (1997) Endurance training effects on striatal neurotransmitter release, an 'in vivo' microdialysis study. *Acta Physiol Scand*, **159**, 335–341.

Meeusen R., Piacentini M.F., Busschaert B. *et al.* (2004) Hormonal responses in athletes: the use of a two bout exercise protocol to detect subtle differences in (over) training status. *Eur J Appl Physiol*, **91**, 140–146.

Meeusen R., Duclos M., Gleeson M. *et al.* (2006) Prevention, diagnosis and the treatment of the overtraining syndrome. *Eur J Sport Sci*, **6** (1), 1–14.

Meeusen R., Nederhof E., Buyse L. *et al.* (2010) Diagnosing overtraining in athletes using the two bout exercise protocol. *Br J Sports Med*, **44** (9), 642–648.

Morgan W.P., Brown D., Raglin J. *et al.* (1987) Psychological monitoring of overtraining and staleness. *Br J Sports Med*, **21**, 107–114.

Morgan W., Costill D., Flynn M. *et al.* (1988) Mood disturbance following increased training in swimmers. *Med Sci Sports Exerc*, **20**, 408–414.

Nederhof E., Lemmink K.A.P., Visscher C. *et al.* (2006) Psychomotor speed. Possibly a new marker for overtraining syndrome. *Sports Med*, **36** (10), 817–828.

O'Connor P. (1997) Overtraining and staleness, in *Physical Activity and Mental Health* (ed W.P. Morgan), Taylor & Francis, Washington, DC, pp. 145–160.

O'Connor P., Morgan W., Raglin J. *et al.* (1989) Mood state and salivary cortisol levels following overtraining in female swimmers. *Psychoneuroendocrinology*, **14**, 303–310.

Pichot V., Roche F., Gaspoz J.-M. *et al.* (2000) Relation between heart rate variability and training load in middle-distance runners. *Med Sci Sports Exerc*, **32** (10), 1729–1736.

Porter R., Gallagher P., Watson S. and Young A. (2004) Corticosteriod–serotonin interactions in depression: a review of the human evidence. *Psychopharmacology*, **173**, 1–17.

Raglin J.S. and Morgan W.P. (1994) Development of a scale for use in monitoring training–induced distress in athletes. *Int J Sports Med*, **15** (2), 84–88.

Raglin J.S., Morgan W.P. and O'Connor P.J. (1991) Changes in mood states during training in female and male college swimmers. *Int J Sports Med*, **12** (6), 585–589.

Rietjens G., Kuipers H., Adam J. *et al.* (2005) Physiological, biochemical and psychologival markers of strenuous training-induced fatigue. *Int J Sports Med*, **26**, 16–26.

Robson P. (2003) Elucidationg the unexplained underperformance syndrome in endurance athletes, the interleukin-6 hypothesis. *Sports Med*, **33** (10), 771–781.

Rowbottom D., Keast D., Goodman C. and Morton A. (1995) The haematological, biochemical and immunological profile of athletes suffering from the overtraining syndrome. *Eur J Appl Physiol*, **70** (6), 502–509.

Rowbottom D., Keast D. and Morton A. (1996) The emerging role of glutamine as an indicator of exercise stress and overtraining. *Sports Med*, **21** (2), 80–97.

Seene T., Umnova M. and Kaasik P. (1999) The exercise myopathy, in *Overload, Performance Incompetence, and Regeneration in Sport* (eds M. Lehmann, C. Foster, U. Gastmann *et al.*), Kluwer Academic/Plenum Publishers, New York, NY, pp. 119–130.

Smith L. (2000) Cytokine hypothesis of overtraining: a physiological adaptation to excessive stress? *Med Sci Sports Exerc*, **32**, 317–331.

Smith L. (2003) Overtraining, excessive exercise, and altered immunity. Is this a T helper-1 versus T helper-2 lymphocyte response? *Sports Med*, **33** (5), 347–364.

Smith L. (2004) Tissue trauma: the underlying cause of overtraining syndrome? *J Strength Cond Res*, **18** (1), 184–191.

Snyder A., Jeukendrup A., Hesselink M. *et al.* (1993) A physiological/psychological indicator of overreaching during intensive training. *Int J Sports Med*, **14**, 29–32.

Steinacker J.M. and Lehmann M. (2002) Clinical findings and mechanisms of stress and recovery in athletes, in *Enhancing Recovery: Preventing Underperformance in Athletes* (ed. M. Kellmann), Human Kinetics, Champaign, IL, pp. 103–118.

Steinacker J.M., Lormes W., Liu Y. *et al.* (2000) Training of junior rowers before world championships. Effects on performance, mood state and selected hormonal and metabolic responses. *J Phys Fitness Sports Med*, **40**, 327–335.

Urhausen A. and Kindermann W. (2002) Diagnosis of overtraining – what tools do we have? *Sports Med*, **32**, 95–102.

Urhausen A., Gabriel H. and Kindermann W. (1995) Blood hormones as markers of training stress and overtraining. *Sports Med*, **20**, 251–276.

Urhausen A., Gabriel H. and Kindermann W. (1998a) Impaired pituitary hormonal response to exhaustive exercise in overtrained endurance athletes. *Med Sci Sports Exerc*, **30**, 407–414.

Urhausen A., Gabriel H., Weiler B. and Kindermann W. (1998b) Ergometric and psychological findings during overtraining: a long-term follow-up study in endurance athletes. *Int J Sports Med*, **19**, 114–120.

Uusitalo A.L.T. (2001) Overtraining. Making a difficult diagnosis and implementing targeted treatment. *Phys Sportsmed*, **29**, 35–50.

Uusitalo A.L.T., Huttunen P., Hanin Y. *et al.* (1998a) Hormonal responses to endurance training and overtraining in female athletes. *Clin J Sports Med*, **8**, 178–186.

Uusitalo A.L.T., Uusitalo A.J. and Rusko H.K. (1998b) Exhaustive endurance training for 6–9 weeks did not induce changes in intrinsic heart rate and cardiac autonomic modulation in female athletes. *Int J Sports Med*, **19**, 532–540.

Uusitalo A.L.T., Uusitalo A.J. and Rusko H.K. (2000) Heart rate and blood pressure variability during heavy training and overtraining in the female athlete. *Int J Sports Med*, **21**, 45–53.

Uusitalo A.L.T., Valkonen-Korhonen M., Helenius P. *et al.* (2004) Abnormal serotonin reuptake in an overtrained insomnic and depressed team athlete. *Int J Sports Med*, **25**, 150–153.

Wilckens T., Schweiger U. and Pirke K. (1992) Activation of 5-HT1C-receptors suppresses excessive wheel running induced by semi-starvation in the rat. *Psychopharmacology*, **109**, 77–84.

Wittert G., Livesey J., Espiner E. and Donald R. (1996) Adaptation of the hypothalamopituitary adrenal axis to chronic exercise stress in humans. *Med Sci Sports Exerc*, **28**, 1015–1019.

Kreider R. (1998) Central fatigue hypothesis and ivertraining, in *Overtraining in Sport* (eds R. Kreider, A.C. Fry and M. O'Toole), Human Kinetics, Champaign, IL, pp. 309–334.

Section 3
Monitoring strength and conditioning progress

3.1 Principles of Athlete Testing

Robert U. Newton[1], Prue Cormie[1] and Marco Cardinale[2,3], [1]School of Exercise, Biomedical and Health Sciences, Edith Cowan University, Joondalup, WA, Australia, [2]British Olympic Medical Institute, Institute of Sport, Exercise and Health, University College London, London (UK), [3]School of Medical Sciences, University of Aberdeen, Aberdeen, Scotland (UK)

3.1.1 INTRODUCTION

A programme of ongoing testing for the assessment of any athlete is essential to optimizing training programme design, reducing injury or illness risk, increasing career longevity, and maximizing sports performance. The adage 'you can't manage what you can't measure' applies equally to athletes as to business. A second important function of athlete testing is to provide feedback to the athlete, which increases motivation as well as the athlete's understanding of their responses and adaptations to different training manipulations. A quality programme of athlete testing can also help build trust and respect between the strength and conditioning specialist and the other coaching staff as they come to appreciate the role and impact that strength and conditioning has in the overall athlete preparation.

3.1.2 GENERAL PRINCIPLES OF TESTING ATHLETES

Principles of measurement are described in many reference texts (e.g. Thomas and Nelson, 1990) and the interested reader is advised to make themselves familiar with the key issues prior to developing and implementing a programme of athlete assessment. Here we will summarize the key aspects.

3.1.2.1 Validity

To have any meaning a test must (1) actually measure what it is purported to measure and (2) be reliable; the latter point will be discussed in the next section. An example where the former is important is the use of bioimpedence to measure body fat mass in the athlete. This technique actually estimates total body water based on electrical impedance to current flow and then makes inferences about fat content. While a useful technique in some instances, changes in hydration state, for example, can impact markedly on the result despite actual fat content not changing. This challenges the validity of this test. As another example, the 6 RM is often used as a measure of maximal strength, but a true 1 RM or even isometric test is a much more valid measure of the neuromuscular system's maximal force capacity; we can argue that a 6 RM test provides more of an indication of strength endurance, with the initial five repetitions serving to fatigue the system such that the weight lifted is in fact indicative of what the athlete can lift for the sixth repetition.

3.1.2.2 Reliability

For any test result to be meaningful there must be sufficient reliability for differences between athletes and changes in a given athlete to be effectively detected above the noise of the measurement. How much a given value changes with repeated measurements due to variation in equipment or methods or effects of environment is termed the 'reliability'. The measure is usually expressed as a coefficient of variation, or as intraclass correlation (ICC) coefficients. An in-depth discussion of the statistics of reliability and methods for determination is beyond the scope of this chapter and the interested reader is referred to the excellent resources provided online by Dr Will Hopkins (http://www.sportsci.org/resource/stats/).

3.1.2.3 Specificity

Tests should be selected based on their ability to accurately assess key performance components useful for improving the quality of programme design in a specific sport. It is pointless to assess aerobic capacity in sprint runners because the physiological capacity being measured has no relevance to performance in the target event. Time is valuable to the athlete and coach, and testing can be seen as an imposition if it is not efficient and does not directly assess the key aspects that require monitoring and feedback. Unless the test is instructive in terms of informing training programme design or assessing injury risk then it is of no use.

Strength and Conditioning – Biological Principles and Practical Applications Marco Cardinale, Rob Newton, and Kazunori Nosaka.

Test control

To achieve good reliability of test measures, the procedures must be precisely described and followed consistently for all test sessions and for trials within sessions. This is termed test 'control' because you are controlling the extraneous variables that could influence the parameter you are measuring. For example, in a 1 RM back squat test it is very important to enforce consistent depth of squat and foot placement. To achieve this, strength and conditioning specialists often go to extensive lengths, marking out foot position, measuring joint angles, calibrating equipment, and ensuring repeat testing is performed at the same time of day.

Equipment calibration and standardization

Reliable and valid measurement of performance is dependent on solid test control and ensuring that the equipment is accurate. Modern electronics and computer interfaces usually ensure accuracy of measurement, but only if they are used correctly and are properly calibrated and zeroed. The simplest example is ensuring that a barbell and weight plates are accurately measured before using them for strength testing. Most weightlifting equipment is calibrated well, but typical fitness barbells and plates can vary by 5% or more. If electronic measurement systems are to be used it is important to calibrate these systems and check the calibration frequently during the test session. For force platforms, this involves placing known masses on the plate. For displacement transducers, the system must be moved through a known distance to ensure that it is measuring correctly. For strain gauges, this involves attaching known masses to the load cell. Calibration is a very important aspect of testing as results can be invalidated by the lack of accuracy of the system used. It is very important to calibrate the equipment before each testing session. Strength and conditioning specialists should also avoid the use of devices which do not allow the ability to check their own calibration.

3.1.2.4 Scheduling

Testing must be performed when the athlete is 'fresh' and their performance capacity is maximized. Generally this requires 48 hours' abstinence from high-intensity training or any competition. This is not always feasible but it certainly is desirable in order to get the best possible scores. In some instances it is the goal of testing to assess levels of fatigue, overtraining, and injury, and so this principle does not apply. For example, athletes are now routinely tested in the days following a competitive event to assess recovery strategies and readiness to perform in the subsequent competition. Planning of the test schedule must be given careful consideration as test order and rest periods can have a large effect on results. For example, a maximum-strength test can be performed before an endurance test in a single day, but they cannot be performed in the opposite order. Any tests which are fatiguing and whose effects are persistent must be scheduled towards the end of the testing session. In many cases the programme will have to be split over several days to avoid previous tests impacting later ones.

3.1.2.5 Selection of movement tested

The greatest specificity of testing to target performance will be achieved when the movement used as the test most closely approximates that of the sport or event being assessed. Often generic tests of upper- and lower-body strength are used to achieve a global picture of an athlete's strength; these might include bench press, seated row, and a lower-body movement such as squat or leg press. For the purposes of informing training programme design more specific testing is required and strength should be evaluated in movements important to the target sport or event. For example, very high leg extensor strength is required in jumping sports, so back squat is commonly used as the test movement for these athletes. In rugby and Australian rules football, the ability to grapple with the opponent and pull them to the ground is critical, so upper-body pulling tests such as seated row, high pull, and lat pulldown are commonly used. Selection of the appropriate test movement is best carried out through a biomechanical analysis of the movements in the sport or event. It is also very important to make sure the athletes are accustomed to the movement used in testing and that they have a good technique in order to avoid the likelihood of injury and make sure a 'true' maximal value is obtained. Familiarizing athletes with new testing movements before the actual testing sessions is crucial to obtaining good valid and reliable data for assessing their status and applying specific changes to their training programme.

3.1.2.6 Stretching and other preparation for the test

Better performance can be achieved in strength tests if some submaximal efforts are completed in preparation. However, the number of trials must not produce any amount of fatigue that could reduce strength effort. Generally three to five maximal efforts are sufficient. Rest between trials is also important; sufficient recovery requires three to five minutes' rest in order to avoid fatigue having an influence on the effort. It is advisable to standardize a warm-up procedure and make sure the athletes follow it to the letter for each testing session in order to make sure all preparation is kept standard. It is well known that muscle temperature affects power output, hence trying to keep consistent warm-up will improve the quality of data gathered. For this reason, environmental aspects should also be considered. In particular, performing tests in a very cold gym or outdoors in a cold climate can negatively affect the ability of the athletes to produce 'true' maximal scores. A final note on preparation is to avoid prolonged static stretching because this can acutely reduce force and power output of the muscle (Marek *et al.*, 2005).

3.1.3 MAXIMUM STRENGTH

Maximum strength can be determined using several methods, and each has advantages and disadvantages. The principle modes are isoinertial, isokinetic, and isometric. With all strength testing the force–velocity and length–tension relationships for muscle must be considered, as well as the effects of muscle angle of pull and mechanical advantage influences on the amount of force that can be generated. Briefly, the faster the velocity of muscle shortening, the lower the amount of force that can be generated. As the athlete moves through the range of motion for a particular movement, the amount of torque that can be generated about the joints and the overall force output produced will change. For example, much less force can be generated from a deep squat position than from a half squat where the knee angle is 90°. These effects must be considered in the following discussion as they have a large influence on the strength measure recorded.

3.1.3.1 Isoinertial strength testing

'Isoinertial' refers to the constant mass of the resistance. This is perhaps the most simple, inexpensive, and accessible form of strength testing. It involves the use of the gravitational force acting on a mass such as a barbell, dumbbell, or weight-resistance machine, and the athlete must overcome this force and move the mass through the range of motion. Common examples are the bench press, squat, and deadlift exercises, but some of the weightlifting movements and derivatives, such as hang clean, high pull, and push press, are increasingly being used.

Free-weight testing, as described above, has high transference to the sporting environment because almost all sports involve manipulating a freely moving mass against gravity. Also, experienced athletes should have a long training history with the use of free weights and so be very familiar with movements such as bench press, back squat, and hang clean. A disadvantage of free weights is the considerable component of skill required. To improve test control, possibly reduce injury risk, and limit the familiarization required, various forms of resistance machines may be used. The designs of such machines are myriad and many involve cams or levers to alter the resistance force profile. This makes comparison between research studies and athletes tested on different equipment difficult, while free-weight testing can be more easily standardized. For example, bench press strength measured with standard Olympic barbell and bench is very repeatable in any weight-training facility anywhere in the world.

Rationale for 1 RM or repetition maximum testing

For isoinertial testing, the maximum weight that can be lifted is determined ideally within three to five attempts. The two methods are to determine the maximum weight that can be lifted once, termed the one repetition maximum (1 RM) or to lift lighter weights three to ten times (3–10 RM). The 1RM test is a more direct measure of maximal strength as either the athlete has sufficient strength capability to lift the weight or they do not. The principle behind multiple-repetition testing is to fatigue the neuromuscular system by prior repetitions to such a degree that the last repetition they can complete is a maximal effort. Multiple-repetition testing is preferred by some coaches and scientists because they believe that there is lower risk of injury since a lighter weight is being lifted, but this has neither been confirmed nor refuted by research. An equally persuasive argument can be made that 1 RM testing involves only a single effort whereas multiple-repetition testing involves repeated events that can lead to possible injury and thus increases risk. Pre-fatiguing the athlete in an attempt to elicit a maximal effort may compromise technique, resulting in injury, and also increases the likelihood of muscle damage through the repeated eccentric actions when lowering the weight. Finally, a 6 RM weight for the bench press is approximately 80% of the 1 RM weight, which when lifted six times represents a much higher total work done. Also, due to the fact that the 6 RM load can be accelerated faster, the peak forces applied to the musculoskeletal system are not appreciably different.

1RM test protocol

A typical protocol for 1 RM testing (McBride *et al.*, 1999) is outlined in Table 3.1.1. Experienced athletes can be asked to estimate their 1 RM and this weight is used as a starting point. If the athlete does not know their approximate 1 RM then it must be estimated by the personnel performing the test based on the athlete's body weight, age, gender, and lifting experience. This can then be adjusted up or down depending on their performance in the warm-up sets. It should be noted that for 1 RM testing, as for any performance test, familiarization can have considerable impact, and consideration should be given to employing at least two sessions for accurate determination of 1 RM.

Multiple-repetition test protocol

There are two approaches to multiple-repetition isoinertial strength testing. The first involves selection of a definitive repetition maximum, say 6 RM, and then execution of a protocol to determine the maximum weight the athlete can lift six times. This protocol is similar to 1 RM testing with a submaximal warm-up set of 10 repetitions completed first, then an estimated 10 RM load lifted for six repetitions, with between 1.25 and 10 kg added until the 6 RM load is obtained. As for the 1 RM protocol, the true 6 RM load should be determined in three to

Table 3.1.1 Protocol for determination of 1 RM; 3–5 minutes' rest between attempts

Warm-up of 10 repetitions at 50% of 1 RM
5 repetitions at 70% of 1 RM
3 repetitions at 80% of 1 RM
1 repetition at 90% of 1 RM, followed by 3 attempts to determine actual 1 RM

five attempts. The other approach is to accurately determine between a 6 RM and a 10 RM load and then apply a regression equation to estimate 1 RM strength based on weight lifted and number of repetitions completed. This second method is appealing because an estimate of 1 RM strength is obtained, but there are limitations in that the regression equations decrease in accuracy of prediction with increasing difference between the characteristics of the athlete being tested and the population sample from which the equation was derived. An extensive discussion of protocols and the accuracy of these methods is provided by LeSuer *et al.* (1997).

The range of motion for isoinertial testing must be closely controlled because of the effect on load lifted. For example, the lower the depth in squat testing, the less the weight the athlete is able to lift. Multiple-repetition test protocols have the advantage of requiring virtually no equipment apart from the obvious weight machines and barbells. This method is advisable with young and development athletes, but due to the limitations in determining the actual maximal capabilities, data gathered should be used with caution when testing elite performers with the aim of determining 1 RM.

3.1.3.2 Isometric strength testing

Isometric strength testing involves the athlete performing maximal contractions against an immovable resistance. As no movement occurs, this testing is termed static or isometric. Isometric testing is appealing because the measures are not confounded by the issues of movement velocity and changing joint angle already discussed. However, there is also strong argument that isometric testing lacks specificity to the dynamic movements predominant in sports. In fact, research has found isometric measures to be quite poor at predicting dynamic performance (Murphy and Wilson, 1996b). Some sports and events involve isometric muscle contractions and such testing may be quite applicable in this instance. One of the advantages of isometric testing is that measures of rate of force development (RFD) can be obtained which have been suggested (Häkkinen *et al.*, 1985) to accurately represent the athlete's ability to rapidly and forcefully contact their muscles, an important aspect of power production which will be discussed shortly.

Isometric strength testing can be performed for single (e.g. elbow flexion, knee extension) or multiple (e.g. squat, bench, mid-thigh pull) joint techniques. Force-measuring equipment is required, such as a transducer, an amplifier, and a computer to collect, store, and analyse the force produced. The protocol consists of a warm-up of submaximal efforts followed by a series of maximal efforts. These are continued until no further improvement in maximal force output can be achieved. The instruction to the athlete should be to 'push (or pull) against the resistance as hard and fast as possible' for three to five seconds. The force will rise quickly to a peak within this time period and the athlete can relax once the peak is attained.

One test which is commonly used to assess functional isometric strength is the mid-thigh pull (Figure 3.1.1). A barbell is supported in a power rack with the stops adjusted to a height providing the correct body position (barbell at mid thigh), with the athlete standing on a force plate (FP). The barbell is then loaded with sufficient weight plates that the athlete cannot move the load. When ready, the athlete is instructed to perform a maximal effort to pull for three to five seconds and the force exerted through the feet is recorded. Measures of peak force, time to peak force, and RFD can be recorded. These parameters provide information on the maximum force the athlete can produce and the time course of force production (e.g. how fast can they reach the maximum force). It has to be said that despite the fact that information on maximum force and to some extent speed can be gathered by such a method, it is virtually impossible to use the data to precisely prescribe the training load to be used for strength training. This is due to the fact that isometric force production is very specific to the joint angle and body posture used and has a very limited relationship to the ability to produce force dynamically throughout the full range of motion of the joint of interest. However, isometric testing that has been well performed can provide information on the maximal voluntary muscle force an athlete can produce and can help the coach in determining the complete force–velocity and power–velocity relationships in specific muscle groups in order to properly assess the status of the athlete and/or the effectiveness of a training programme.

Figure 3.1.1 Isometric mid-thigh pull from a force plate

3.1.3.3 Isokinetic strength testing

Isokinetic testing involves the use of relatively sophisticated equipment, usually electromechanical, which provides resistance by limiting the speed of movement to a preset linear or angular velocity (Figure 3.1.2). The earlier models were limited to concentric-only muscle actions, but more contemporary systems incorporate both concentric and eccentric modes as well as zero-velocity (isometric) testing. The advantages of isokinetic strength testing are that the issues of force–velocity effects and changing force capability through the movement are well controlled and quantifiable. Most isokinetic systems are designed primarily for unilateral, single-joint movements. For this reason strength measures from isokinetic dynamometers are usually expressed in torque units such as Nm. Torque is the angular equivalent of linear force and can be thought of as a 'twisting force' (see Section 1.8.2.2).

Single-joint measurement lacks specificity to most sporting motions, which generally involve a coordinated movement of several joints and muscle groups. To address this possible deficiency, manufacturers such as Biodex have developed attachments which provide linear as well as rotary isokinetic resistance, thus enabling multi-joint movements such as leg press to be tested.

Perhaps the greatest application of isokinetic testing has been in assessing hamstrings and quadriceps torque production about the knee. Consequently, much research has been directed toward assessing the ratio between these two muscle groups, as well as left-to-right asymmetries, in an effort to understand the mechanisms and screen and rehabilitate injuries such as hamstring strain and ACL injury of the knee.

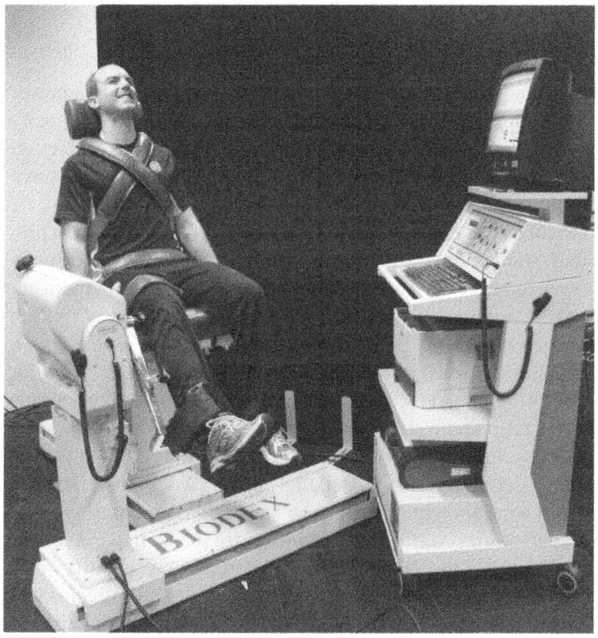

Figure 3.1.2 Biodex isokinetic dynamometer

By adjusting the resistance (accommodating) continuously to the torque capacity of the musculoskeletal system a profile throughout the range of movement can be obtained. This is a useful feature of the isokinetic mode as weak portions of the range can be identified, as well as shifts in the position at which the peak in torque occurs. Such information has been used to quantify the optimal angle, and some researchers have related shifts in this optimal angle to injury events such as hamstring strains (Brockett *et al.*, 2004).

Two issues that clinicians should be aware of when performing isokinetic testing are the torque overshoot phenomenon and gravity correction. At the beginning of each repetition there will be a period when the limb is accelerated from zero to the preset velocity of the dynamometer. This results in an impact when the limb attains this velocity and a spike and subsequent damped oscillation occurs in the torque signal. The faster the set speed of the dynamometer for the trial, the later into the movement this impact occurs and the larger the spike. Manufacturers have tried to address this problem with electronic and digital filters as well as by ramping the dynamometer speed, but measurements in the early phase of a movement are problematic regardless. Gravity correction is used to account for the fact that when testing, gravity will assist (increase) torque measured during downward-direction movements and impede (decrease) torque measured during upward-direction movements. This is easily accounted for contemporary systems by performing a weighing procedure for the limb and dynamometer arm prior to strength measurement; the torque measurement is then corrected throughout the range of movement. The main drawbacks of isokinetic testing in the applied setting reside in the relatively slow speeds at which the athlete can be tested and in the fact that isokinetic data are pretty much useless to the coach in determining the appropriate lifting load to be used in dynamic isoinertial exercises.

3.1.4 BALLISTIC TESTING

The majority of sports involve striking, kicking, throwing, or projecting the body into free space. The kinetic and kinematic profiles of isoinertial, isometric, and isokinetic testing methods are very different to the accelerative, high-power output characteristic of sport movements (Newton and Kraemer, 1994). As a result, measurement of force, velocity, and power during ballistic movements such as countermovement jumps (CMJs), jump squats (CMJs with additional load to body weight), squat jumps (concentric-only jumps with body weight or load), bench throws, bench pulls, and various weightlifting movements is becoming increasingly used for athlete testing.

It is instructive for the athlete to be tested over a range of loads depending on their experience, target sport, and task. A spectrum is useful so that an impression of the athlete's performance under heavy and light loads can be ascertained (i.e. high-load speed–strength or low-load speed–strength). Most of all, this approach will allow the definition of the force–velocity (F/V) and power–velocity (P/V) relationships and will help the strength and conditioning coach to identify the F/V and P/V characteristics of the athlete as well the most appropriate training load to improve specific aspects of force- and power-generating capacity. One scheme is to use loads of 30, 55, and 80% 1 RM (McBride *et al.*, 2001), while another is to use set loads such as 40, 70, and 100 kg in the jump squat. Such a loading protocol is particularly useful when the athlete has to move absolute loads in their sport, such as tackling in rugby. Other methods seek to determine the optimal load for power production (Baker *et al.*, 2001; Wilson *et al.*, 1993).

When determining the optimal load for power production, the athlete first performs the ballistic movement with a preselected load. After each trial the load is adjusted up or down and, due to the relationship between force and velocity, the power output changes. Only one load will produce the highest power output (Wilson *et al.*, 1993) and this is termed the optimal load for developing power output in the specific exercise tested. This may be expressed as an absolute force (or mass), or as a percentage of maximum isometric force or 1 RM. We know that in order to enhance explosive power performance, the load at which the maximal power is obtained must be used for training and the velocity of execution of the exercise used must always be maximal. This approach to strength training has been shown to produce remarkable enhancement of mechanical power (Berger, 1963; Hakkinen, 1994; Kaneko *et al.*, 1983; Moritani *et al.*, 1987).

The determination of F/V and P/V relationship is very useful in identifying the progress of the athlete as a consequence of the training programme used and should be routinely performed to ascertain whether the training is going in the right direction, improving maximal force, maximal power, or both.

Figure 3.1.3 Purpose-built power cage for training and assessment of upper- and lower-body power. Bottom stops reduce injury risk if the athlete falls or misses the barbell. A braking mechanism can be used to reduce landing impact or eccentric loading. A force plate and linear transducer are used to record performance characteristics for feedback to the athlete and performance diagnosis

3.1.4.1 Jump squats

Both upper- and lower-body ballistic performance can be assessed with the movement selected based on greatest specificity to the target sport. For the lower body the most common test is the jump squat, where the athlete holds a barbell in the back squat position, rapidly dips down, and then jumps for maximum height. These can be performed with a free weight or in a Smith machine to limit the movement to the vertical plane. For free-weight testing, the jump should be performed within a power rack with bottom stops in case the athlete falls (see Figure 3.1.3). In most cases, the athlete can perform a single jump, dipping down and then jumping upward for maximal height.

Nowadays, portable and sophisticated systems are available to measure force output and bar kinematics. From these systems, data such as height, velocity, force, impulse, and power output can be calculated for the jump squat. When creating an athlete profile or assessing the effectiveness of training programmes on power, assessing power across a spectrum of loads as discussed above (Figure 3.1.4) can provide valuable information regarding the strengths and weaknesses of the athlete or programme. For example, for the individual athlete, deficiencies in power with lighter loads may indicate a need for more high-velocity training (i.e. low-load speed–strength), while deficiencies in power at heavier loads may indicate a need for more high-force training (i.e. high-load speed–strength). Such data can be extremely valuable for subsequent training prescriptions as previous studies have found that different portions of the load–power relationship are trainable based on the loading intensity used in training (Cormie *et al.*, 2007b, Kaneko *et al.*, 1983; Moss *et al.*, 1997; Toji and Kaneko, 2004; Toji *et al.*, 1997). Furthermore, such measures may provide a useful tool for monitoring the effects of changing emphasis in programme periodization as well as for detecting overtraining, illness, and staleness. In the past, Bosco *et al.* (1984) showed that it is possible to determine the F/V relationship and assess the effectiveness of a training programme by using flight time-based systems and performing vertical jumps with extra loads. Individual athletes' data should be analysed to appropriately identify individual progression. On the other hand, groups of athletes can be averaged together to assess pre- and post-training power output, and these data can inform strength and conditioning coaches about the overall strengths and weaknesses of their

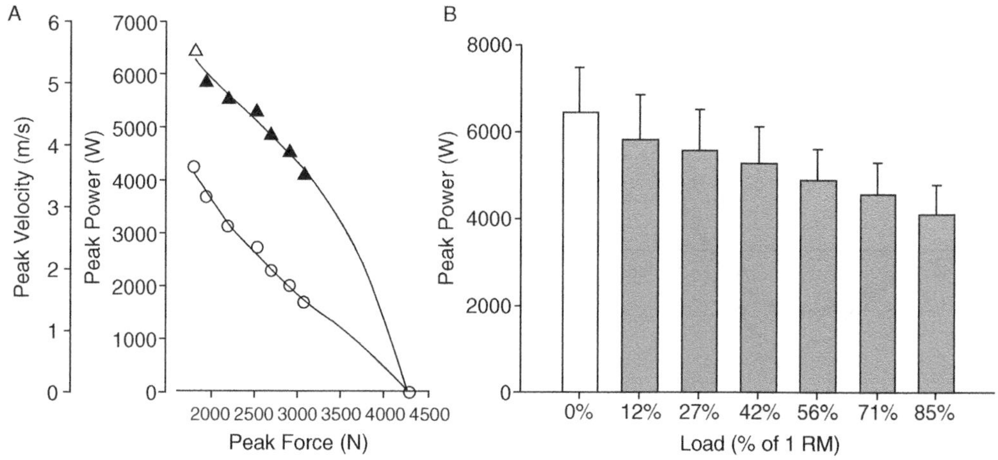

Figure 3.1.4 The jump-squat force–velocity (circles) and force–power (triangles) relationships (a) and load–power relationship (b). To assess these relationships, the athlete performs jump squats with a range of different loads. Due to the relationship between force and velocity, the power output will change and a certain load will produce the highest power output (i.e. the 'optimal' load, designated by the white triangle/bar). This may be expressed as an absolute force (or mass) or as a percentage of maximum isometric force or 1 RM. Adapted from Cormie *et al.* (2007c)

training programmes. It is important to remind the reader that when comparisons between athletes are performed, it is advisable to normalize force and power output data to each individual's body mass (and ideally fat-free mass) in order to make overall comparisons between athletes of difference body sizes.

3.1.4.2 Rate of force development (RFD)

There are several options to choose from for testing RFD. One common protocol uses isometric squats, but this incorporates many of the same advantages and disadvantages as testing maximal strength with isometrics (Murphy and Wilson, 1996a). RFD can also be determined during both concentric and eccentric phases of dynamic tests, which may have greater relevance to task performance (Wilson *et al.*, 1995; Murphy and Wilson, 1996a, 1997; Nuzzo *et al.*, 2008); this has not been well researched to date, however.

Two dynamic tests that are often utilized are the concentric-only jump and the concentric-only jump squat. For the concentric-only jump, the athlete squats down to a self-selected depth and holds that position for three to four seconds, then attempts to jump for maximum height without a preparatory movement. This can be difficult to perform and may require several trials to yield accurate data. Concentric-only jump squats are performed in a similar fashion to standard jump squats. For this variation, mechanical stops are positioned in the squat rack or Smith machine at the appropriate angle for the bottom position of the jump, and these become the starting position for the jump squat. Ground-reaction force and bar displacement can be recorded, as well as derived variables such as jump height, power output, and peak force developed.

The highest force produced during a concentric-only movement has been termed maximal dynamic strength (MDS) (Young, 1995): a strength and power quality with good predictive and discriminatory capability between athletes of different levels. Heavier external loads will result in greater MDS values, and greater test specificity will be obtained if the selected load is similar to the target task. Dynamic concentric-only tests (as with isometric tests) allow calculation of several measures of the ability to rapidly develop force, such as maximum RFD or the impulse (F × t) over the initial 100 ms or other epochs. The interested reader may consult the extensive discussions in other texts on the subject (see Zatsiorsky and Kraemer, 2006).

3.1.4.3 Temporal phase analysis

To date, most analyses conducted to examine power production capacity have reported either peak or mean power, force, velocity, or displacement. Peak power, for example, represents the single greatest data point of power production on the power–time curve, while mean or average power represents the total work done divided by the time which it took to perform that work (alternatively, the sum of the power measures divided by the number of samples). Typically mean power is calculated over the concentric phase of an exercise. While these variables are important indicators of power output, they are limited in their ability to delineate the exact nature and timing of changes following training interventions, differences between subject populations, and/or loading conditions *throughout the entire movement*. The most recent advancement in the analysis of power performance characteristics has been temporal phase analysis of power output throughout entire movements (Cormie

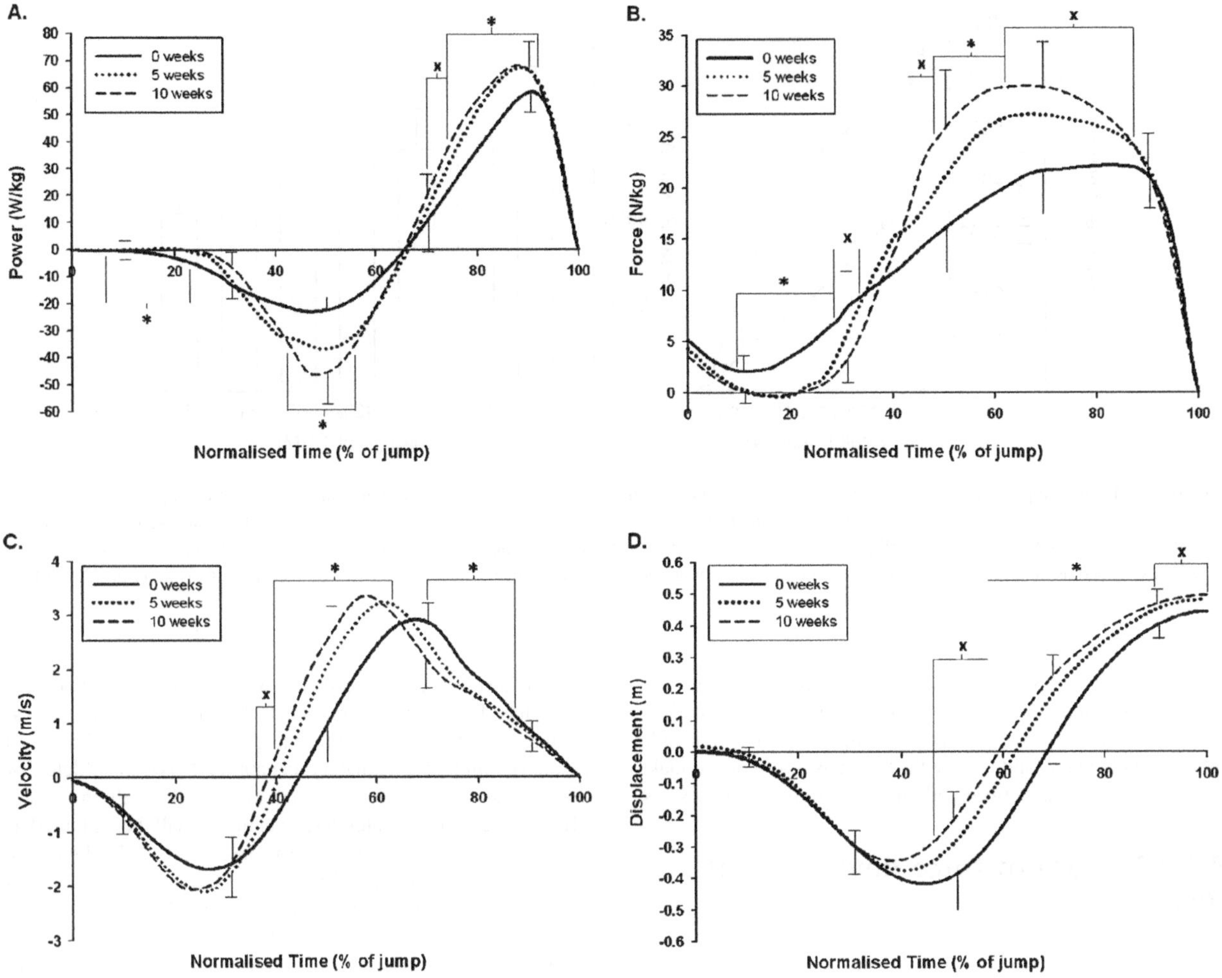

Figure 3.1.5 Changes to average ($n = 8$) power–time (a), force–time (b), velocity–time (c), and displacement_time (d) curves during a 0% 1 RM jump squat in response to 10 weeks of ballistic power training. Significant difference between 0 and 5 weeks of training (x) and 0 and 10 weeks of training (*). Adapted from Cormie *et al.*, (2010b)

et al., 2008, 2010a,b,c). This technique uses custom-designed computer programs to combine all individual power–time curves into one representative average power–time curve. Typically, the absolute time it takes for a group of individuals to complete a jump is different, and thus their power–time curves cannot be added together and later expressed as an average. However, through a re-sampling procedure each individual's power–time curve can be normalized to time so that the data can be pooled. For example, in a jump-squat movement, the time from the start of the eccentric phase to the point at which the individual leaves the ground is expressed on the same relative time scale (0–100% of time to complete the movement). Subsequently, since all individual power–time curves are on the same relative time scale, they can be added together and averaged to produce a single average curve for that group of individuals. Statistical analysis can then be used to detect significant differences in power in specific portions of the movement. Comparisons can be made before and after training interventions (see Figure 3.1.5), between loading conditions, and across different groups of subjects (Cormie *et al.*, 2008, 2010a,b,c). Research using average curve analysis in the field of strength and conditioning is limited, but preliminary evidence indicates that analysis of power output throughout the entire movement may provide novel insights into the differences between athletic populations, the nature of adaptations to power training, and the mechanisms involved in improving power output (Cormie *et al.*, 2008, 2010a,b,c).

For example, research has commonly attributed improvements in jump performance following strength and power training to alterations in the maximal neural activation, changes to

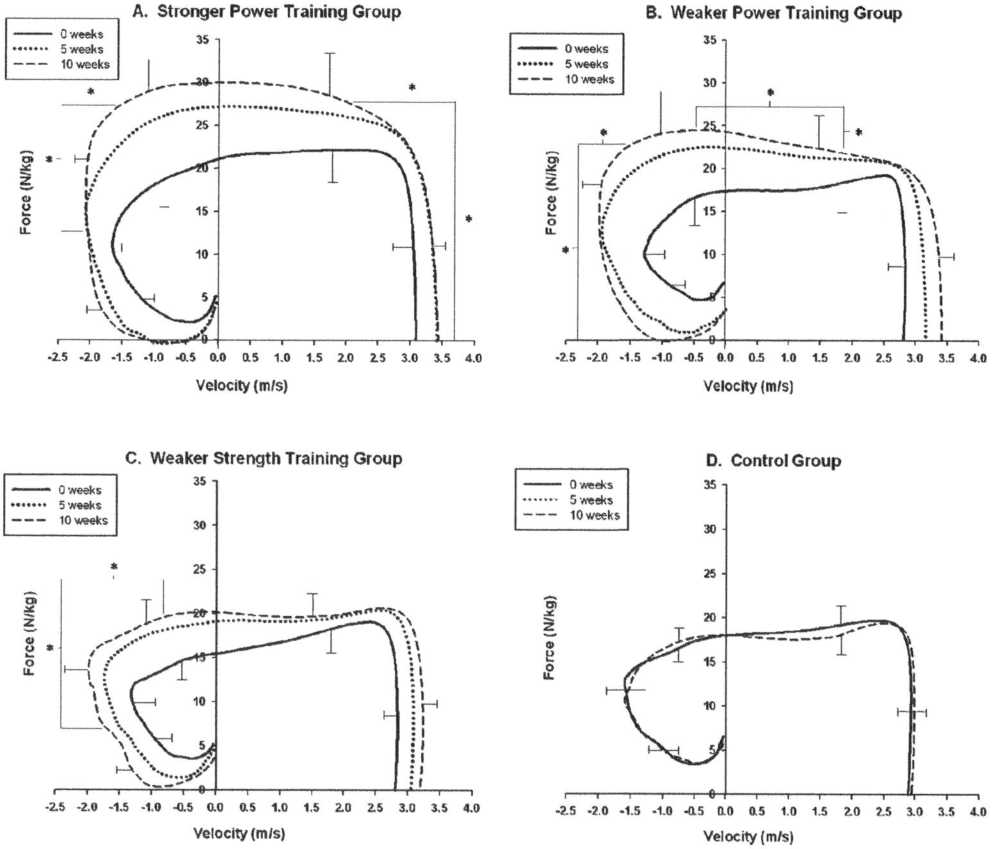

Figure 3.1.6 Changes to average ($n = 8$) force–velocity loops during a 0% 1 RM jump squat in response to 10 weeks of ballistic power training in relatively trained (a) and untrained (b) subjects; heavy strength training in relatively untrained subjects (c) or no training (d). These curves indicate the group average of force and velocity at every data point throughout the entire movement (i.e. from the initiation of the countermovement until take off). Because velocity is plotted along the x-axis, the force–velocity loops allow for a clear delineation between the eccentric phase (i.e. negative velocity) and the concentric phase (i.e. positive velocity). * denotes a significant difference in both force and velocity at a specific time point between 0 and 10 weeks. Adapted from Cormie *et al.*, (2010c)

neural activation patterns, and/or enhanced contractile capacity of the lower-limb musculature (Chimera *et al.*, 2004; Kraemer and Newton, 2000; McBride *et al.*, 2002; Schmidtbleicher *et al.*, 1988; Wilson *et al.*, 1993). However, results of a training study involving the use of temporal phase analyses indicated that an additional mechanism driving performance improvements was the optimization of SSC function (Cormie, 2010c). Specifically, improvements in jump performance were theorized to be driven by the development of a strategy to better utilize the eccentric phase during jumping. This is presented in Figure 3.1.6, illustrating changes to force and velocity throughout a 0% 1 RM jump squat following 10 weeks of ballistic power training in untrained (weaker) and trained (stronger) subjects, as well as heavy strength training in untrained subjects. This study indicated that greater unloading allowed for increased negative acceleration and therefore velocity during the countermovement. Increased musculotendinous stiffness resulted in an enhanced ability to translate the momentum developed during the countermovement into force, ultimately leading to improved concentric performance (i.e. force, velocity, power, jump height). Furthermore, these changes were theorized to positively influence the mechanisms involved with SSC (i.e. development of force prior to concentric phase, the interactions between contractile and elastic elements, potentiation of contractile elements, storage and utilization of elastic energy, as well as activation of stretch reflexes), which in turn contributed to the improved concentric performance. Interestingly, the changes to a variety of parameters during the eccentric phase (i.e. force, velocity, power) correlated significantly to the improvements in concentric jump performance. Thus, these data provide evidence that training-induced alterations in SSC function during the eccentric phase contribute to improvements in performance of SSC movements following both ballistic power training and heavy strength training

(Cormie, 2010c). The examination of such information would not be possible without using temporal phase analyses to investigate the changes that occur throughout the entire movement.

3.1.4.4 Equipment and analysis methods for ballistic testing

The methods commonly used in strength and power research to measure power performance characteristics involve: (1) an FP which directly measures the force the athlete produces; (2) a linear position transducer (LPT) which can provide highly accurate measurements of displacement; or (3) the combination of an FP and LPT(s).

As the athlete pushes against an FP during the leg press or squat, force–time data can be obtained. Alternatively, if the equipment is fixed so that no movement can occur, isometric measures of peak force and RFD can be gathered. During dynamic movements, it is possible to obtain measures of dynamic strength such as the highest force produced. Vertical jumps performed on an FP provide a rich array of performance data. If the athlete is isolated on the plate (i.e. does not touch any other surface), the impulse–momentum relationship can be used to derive velocity–time and displacement–time datasets from the force–time recording through a forward dynamics approach. Combining force and velocity data allows instantaneous power measurements to be derived throughout the jump, and summary variables such as peak and mean power can also be calculated. Calculating power from vertical ground reaction forces via FP is commonly utilized in strength and power research to compare various types of body weight jumps or to monitor vertical jump performance following an intervention.

Previous research has also utilized equipment that measures displacement to determine power output, the most common example being the LPT, but there are also rotary encoders, chronoscopic light systems, infrared light systems (all referred to as 'LPT' for the purpose of this chapter). The LPTs have an extendable cable that can be attached to an athlete or an implement such as a barbell. As the person or object moves, the displacement is measured and recorded with a computer system. Such systems are particularly useful for measuring performance during jumping with a barbell, upper-body movements such as bench throws, and the weightlifting movements. The LPT systems utilize inverse dynamics in order to determine power output solely from displacement data. Specifically, velocity is calculated through the differentiation of displacement data with respect to time (velocity = displacement / time). Double differentiation of the displacement data allows for the determination of acceleration (acceleration = velocity / time), which is in turn utilized to determine force (force = system mass / (acceleration + acceleration due to gravity)).

The combination of an FP and LPT(s) is commonly used throughout the literature to examine power output during jumping movements, weightlifting movements, and bench-press movements. The majority of research involving the combination of an FP and LPT(s) incorporates information obtained from an FP and a single LPT, but more recently an FP and two LPTs have been utilized to assess both vertical and horizontal displacement during multi-dimensional movements (Cormie *et al.*, 2007a, 2007b, 2007c). This measurement system is typically arranged to collect displacement and force data simultaneously at equal frequencies, allowing for the determination of power output (power = force × (displacement / time)).

Several investigations have compared the validity of these primary data collection and analysis methodologies (Chiu *et al.*, 2004; Cormie *et al.*, 2007a, 2007d; Cronin *et al.*, 2004; Hori *et al.*, 2007) and their impact on the load–power relationship in various movements (Cormie *et al.*, 2007a). Some comparisons between displacement-based modalities (i.e. LPT) and an FP have reported no difference in force data during jumping movements (Chiu *et al.*, 2004; Cronin *et al.*, 2004) or power data during concentric half squats (Rahmani *et al.*, 2001). However, these investigations are in contrast to more comprehensive analyses which have examined differences throughout a loading spectrum and compared the LPT and the FP methodologies to the FP + LPT technique (Cormie *et al.*, 2007a, 2007d; Hori *et al.*, 2005). The most comprehensive analysis to date has established that significant and meaningful differences in power values (both peak and mean power) exist depending on the methodological procedures utilized, and importantly that these differences affect the load–power relationship in the squat, jump squat, and power clean (Cormie *et al.*, 2007a) (Figure 3.1.7). Specifically, LPT methodologies relying solely on kinematic data consistently elevate power output across various loads in the squat, jump squat, and power clean in comparison to the FP + LPT methodologies (Cormie *et al.*, 2007a). Furthermore, when compared to the FP + LPT method, the FP technique under-represents velocity and power output during the squat, jump squat, and more prominently during movements involving the bar travelling independently of the body, such as the power clean (Cormie *et al.*, 2007a). The practical importance of this research is that for the exact same trial, different power values have been obtained through the various data collection techniques, and as a result, the load–power relationships in the squat and the power clean were defined incorrectly. Therefore, care must be taken when choosing what system to assess power performance characteristics.

The large discrepancies in the identification of the load that maximizes power in resistance exercises such as the jump squat have been reported to range from 0% 1 RM (Cormie *et al.*, 2007c) to 59% 1 RM (Baker *et al.*, 2001) and the impact of various collection and analysis techniques on the load–power relationship highlight the need for the standardization of methodological procedures used to assess power output. In applied strength and power research data collection and analysis procedures which measure both kinematic and kinetic components (i.e. FP + LPT) are the most appropriate for determining power output in exercises that involve synchronous movement of the bar and the body (Cormie *et al.*, 2007a; Dugan *et al.*, 2004). This is based on the fact that the production of muscular power involves both kinematic (velocity of shortening) and kinetic (force of concentric contraction) components. It is important to

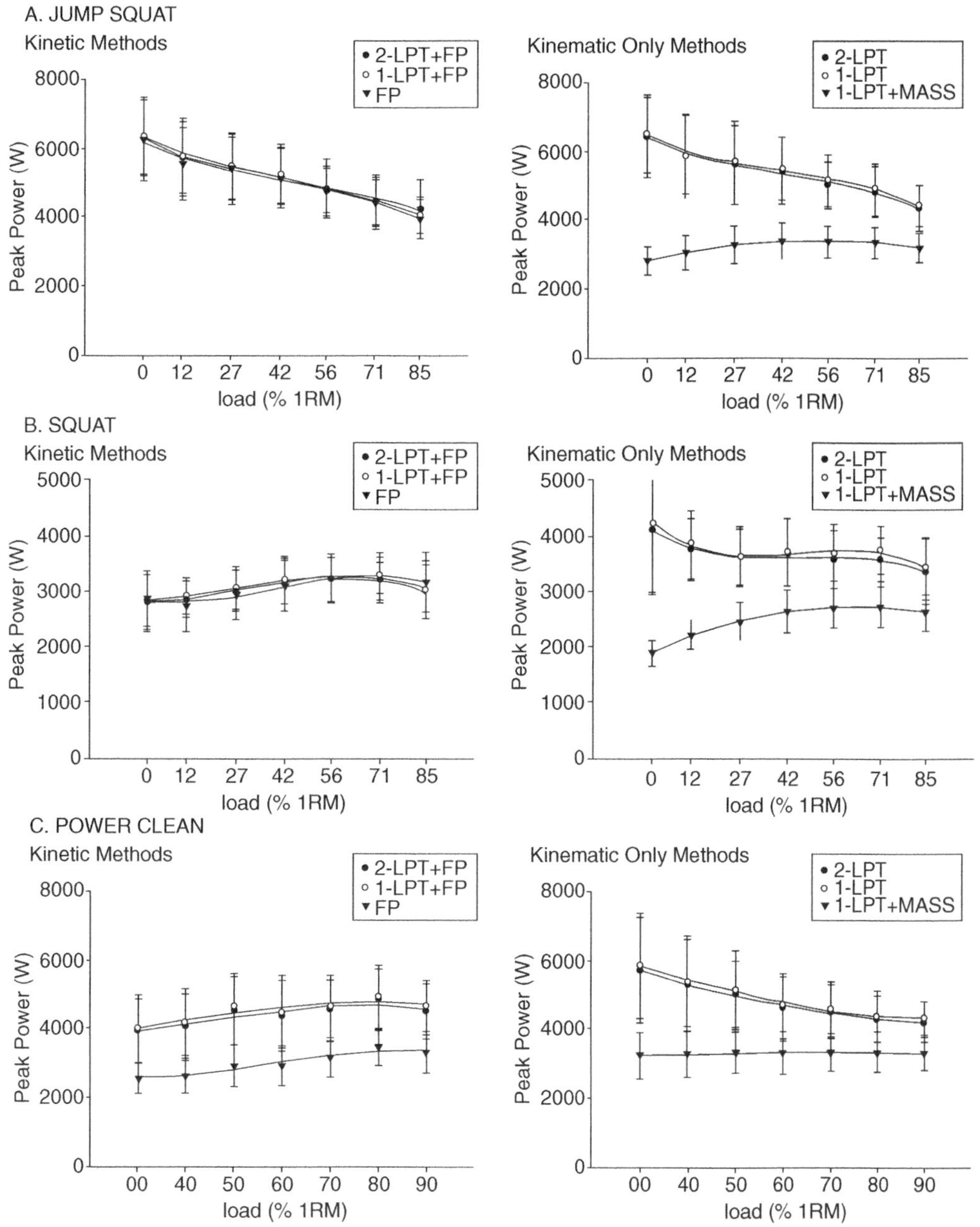

Figure 3.1.7 Load–power relationship in the jump squat (a), squat (b), and power clean (c) as derived by the six different data collection and analysis methodologies. Adapted from Cormie *et al.* (2007a)

note that care must be taken when considering movements involving the bar moving independently of the body (e.g. weightlifting movements), and researchers and practitioners need to consider the aspect of the movements they are most interested in examining and/or monitoring when examining these exercises (e.g. velocity of the barbell versus velocity of the system centre of mass in the power clean).

3.1.5 REACTIVE STRENGTH TESTS

The most common test for assessing an athlete's reactive strength is the depth jump. The athlete drops down from a box, lands, and then jumps upward for maximum height (Figure 3.1.8). A contact mat system or FP can be used to record the characteristics of the performance. It has been reported that

Figure 3.1.8 Athlete performing a drop jump on to a force plate with linear transducer attached to track displacement

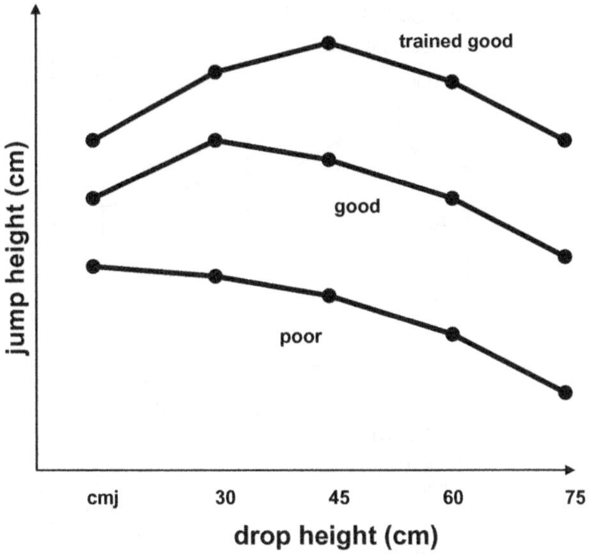

Figure 3.1.9 Plot of reactive strength index against drop height. The best score corresponds to the optimal drop height

the instructions given to the athlete affect the results and that athletes should attempt maximum jump height while minimizing ground contact time (Young *et al.*, 1995). Trials completed at increasing drop heights provide insight into how the athlete responds to increasing stretch loads. A common progression is to employ 0.30, 0.45, 0.60, and 0.75 m (12, 18, 24, and 30 inch) drop heights. Calculated variables include jump height, flight time, contact time, and flight time divided by contact time (which is also termed the reactive strength index or RSI). The 'best' drop height can be determined as the one that elicits the highest RSI (Figure 3.1.9). An athlete with a reasonable level of reactive strength should be able to produce a better jump height following a drop than they could from a countermovement jump (which effectively has a drop height of zero).

3.1.6 ECCENTRIC STRENGTH TESTS

Many sports involve movements with accentuated eccentric muscle actions. For example, it has been demonstrated that isokinetic eccentric strength of the knee extensors can discriminate elite and sub-elite downhill skiers (Abe *et al.*, 1992), which is a reasonable outcome given the repeated eccentric actions experienced by skiers. Therefore, it may be instructive to include tests of eccentric strength and power in the testing programme. The most common method of eccentric testing is the use of isokinetic dynamometers with this capability (Abe *et al.*, 1992), normally of the hamstrings and quadriceps muscle groups. Eccentric testing of the hamstrings is also performed for the purpose of assessing injury risk and recovery from injury, since eccentric strength and endurance have been implicated as possible risk factors. For detailed explanation the reader should consult the many extensive texts on the use of isokinetic dynamometry.

However, for more functional eccentric strength measures the options are much more limited. In one study an electromechanical device was developed which implemented an isokinetic squat movement (Wilson *et al.*, 1997) and it was reported that measures of strength from this device were more highly correlated with cycling performance than were single joint tests of the knee. It is possible to use isoinertial loads of greater than 1 RM to assess eccentric strength by measuring the forces exerted as a subject attempts to slow the rate of decent of the load (Murphy *et al.*, 1994). Essentially a weight such as a barbell is suspended over the subject in the squat or bench-press position and then released using an electromechanical device such as that shown in Figure 3.1.3. The loads used are between 130 and 150% of maximum isometric strength and so the subject cannot hold the load but attempts to slow the decent. Researchers have recorded measures of eccentric force, power absorption, and RFD from such tests and presented evidence that these measures are more informative for sports which rely on high eccentric performance.

3.1.6.1 Specific tests of skill performance

It is useful to include a test that is highly task-specific and which incorporates several strength/power qualities. We may refer to this as the 'gold standard' for the target task. For example, in volleyball the approach jump and reach is commonly used as a 'gold standard' test (Newton *et al.*, 1999). In basketball, the athlete could use an approach run onto a contact mat and then jump for maximal height while performing a jump shot action, landing back on the contact mat. In athletics, the actual field event performance (e.g. long jump or shot put distance) can be used.

Sometimes it is necessary to design highly specific tests that assess particular aspects of a sport. For example, power output produced under fatigue or following repeated impacts (or both) is important to the sport of rugby. In this instance, an obstacle course could be developed simulating a game, with outcome measures including time to complete certain sections, as well as performance in a power test such as a vertical jump.

3.1.6.2 Relative or absolute measures

Whether the results are expressed as absolute measures or normalized relative to body weight depends on the task and athlete; both methods have application. When the athlete must move body weight against gravity (e.g. high jump), relative measures may be more important. However, when momentum or total strength is key (e.g. rugby, American football), absolute measures may be more instructive. Relative measures allow for better comparison of athletes with different body mass, and most variables can be expressed in relative terms. For example, a 1 RM squat can be expressed as the number of body masses lifted (i.e. 1 RM/body mass), and power output during jumping can be expressed as watts per kilogram body mass. This approach is advisable when comparisons are made between different members of a team or large groups of athletes.

3.1.7 CONCLUSION

Testing of athlete performance is becoming more professional and sophisticated in response to demands by athletes, coaches, and clubs for further scientific approaches to training programme design and athlete monitoring. There is now a plethora of accurate, reliable, and relatively easy-to-use equipment for recording and comparing a myriad of measurements of displacement, velocity, force, impulse, and power parameters which can be used to describe the athlete's strength and power capabilities. The key to implementing a successful testing programme for strength and power is careful planning, thoughtful selection of test protocols, and well-controlled methods to ensure the resulting data are accurate, reliable, and instructive. Ideally, the testing programme should cover all areas which affect performance and help the strength and conditioning coach in building a clear profile of the athlete. Finally, once the profile is built, it is possible to verify how it is changing and whether the training programme is moving the athlete in the correct direction (e.g. max strength versus power versus body mass changes versus endurance changes;

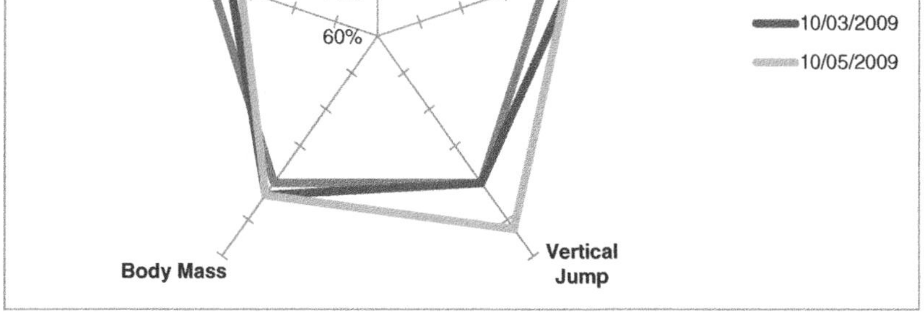

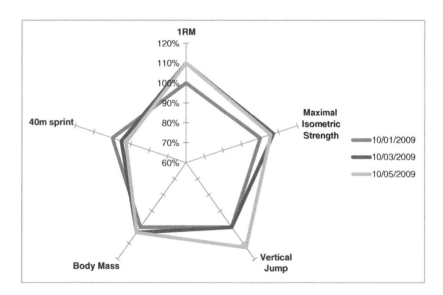

Figure 3.1.10 Example of a spider chart presenting various testing data of an athlete expressed as a percentage of the first test performed in the season. Improvements in specific tests were sought in the training programme used

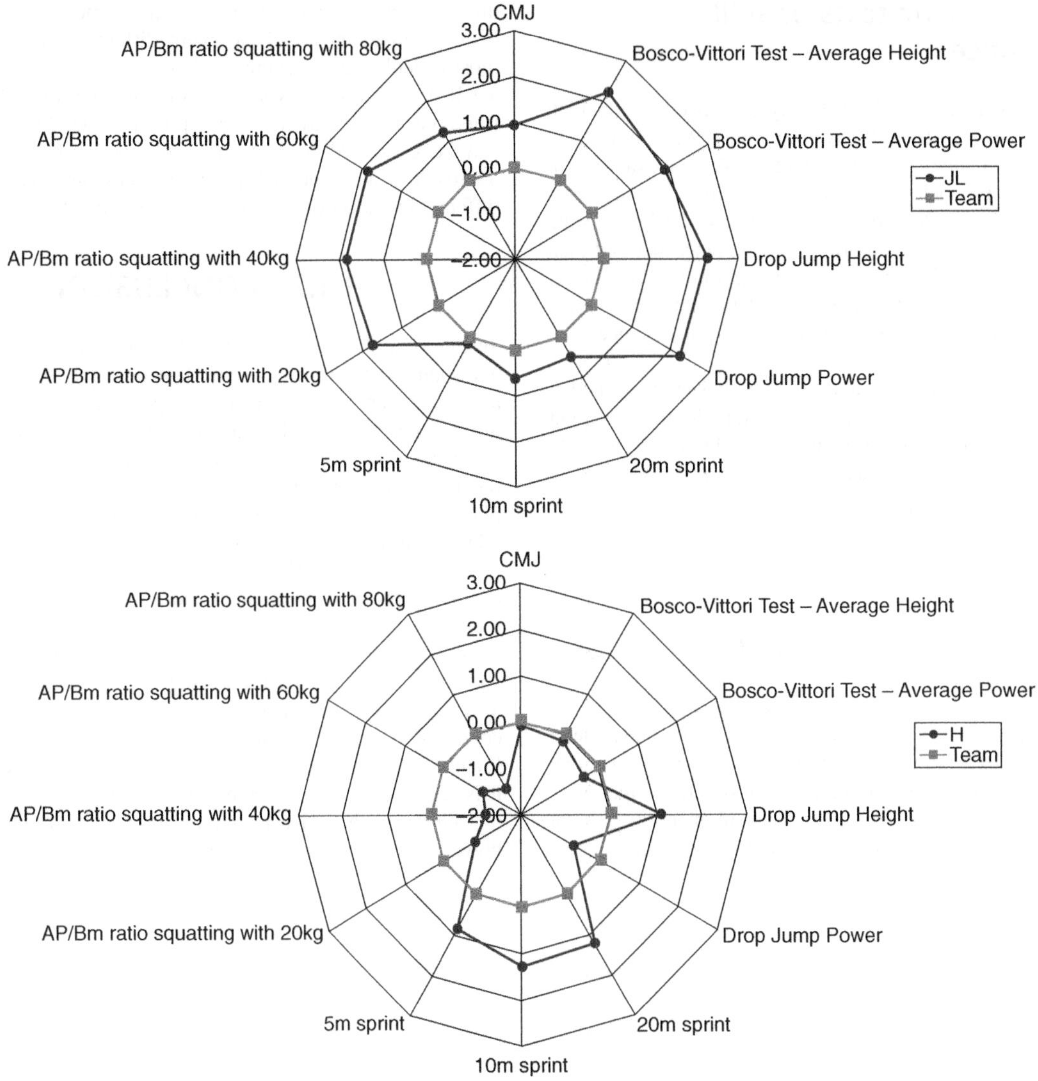

Figure 3.1.11 Example of a series of tests of an individual athlete presented as Z-scores to compare the individual's data to his team mates' data. (a) An individual who outscores the team average values in all tests. (b) An individual who outscores team results only in sprints. AP = Average Power; Bm = Body Mass

see Figure 3.1.10). Furthermore, when testing groups/teams, it is important to be able to benchmark each individual athlete with their 'peers' in order to gauge more information about their qualities in relation to the overall qualities of the team/group in which they train. This has some useful implications for team selection as well as for motivating athletes to improve. In this scope, it is advisable for coaches to familiarize themselves with simple statistical techniques such as the use of Z-scores to present and analyse each athlete's score (see Figure 3.1.11).

References

Abe T., Kawakami Y., Ikegawa S., Kanehisa H. and Fukunaga T. (1992) Isometric and isokinetic knee joint performance in Japanese alpine ski racers. *The Journal of Sports Medicine And Physical Fitness*, **32** (4), 353–357.

Baker D., Nance S. and Moore M. (2001) The load that maximizes the average mechanical power output during jump squats in power-trained athletes. *Journal of Strength and Conditioning Research*, **15** (1), 92–97.

Baker D., Nance S. and Moore M. (2001) The load that maximizes the average mechanical power output during explosive bench press throws in highly trained athletes. *Journal of Strength and Conditioning Research*, **15** (1), 20–24.

Berger R. (1963) Effect of dynamic and static training on vertical jumping. *Res Q*, **34**, 419–424.

Bosco C. *et al.* (1984) The influence of extra load on the mechanical behavior of skeletal muscle. *Eur J Appl Physiol*, **53**, 149–154.

Brockett C.L., Morgan D.L. and Proske U. (2004) Predicting hamstring strain injury in elite athletes. *Medicine and Science in Sports and Exercise*, **36** (3 (Print)), 379–387.

Chimera N.J., Swanik K.A., Swanik C.B. and Straub S.J. (2004) Effects of plyometric training on muscle-activation strategies and performance in female athletes. *J Athl Train*, **39** (1), 24–31.

Chiu L.Z.F., Schilling B.K., Fry A.C. and Weiss L.W. (2004) Measurement of resistance exercise force expression. *J Appl Biomech*, **20**, 204–212.

Cormie P., McBride J.M. and McCaulley G.O. (2007a) Validation of power measurement techniques in dynamic lower body resistance exercises. *J Appl Biomech*, **23** (2), 103–118.

Cormie P., McCaulley G.O. and McBride J.M. (2007) Power versus strength-power jump squat training: influence on the load-power relationship. *Med Sci Sports Exerc*, **39** (6), 996–1003.

Cormie P., McCaulley G.O., Triplett N.T. and McBride J.M. (2007c) Optimal loading for maximal power output during lower-body resistance exercises. *Med Sci Sports Exerc*, **39** (2), 340–349.

Cormie P., Deane R. and McBride J.M. (2007d) Methodological concerns for determining power output in the jump squat. *J Strength Cond Res*, **21** (2), 424–430.

Cormie P., McBride J.M. and McCaulley G.O. (2008) Power-time, force-time, and velocity-time curve analysis during the jump squat: impact of load. *J Appl Biomech*, **24** (2), 112–120.

Cormie P., McBride J.M. and McCaulley G.O. (2009) Power-time, force-time, and velocity-time curve analysis of the countermovement jump: impact of training. *J Strength Cond Res*, **23** (1), 177–186.

Cormie P., McGuigan M.R. and Newton R.U. (2010a) Influence of strength on magnitude and mechanisms of adaptation to power training. *Med Sci Sports Exerc*, **42** (8), 1566–1581.

Cormie P., McGuigan M.R. and Newton R.U. (2010b) Adaptations in athletic performance after ballistic power versus strength training. *Med Sci Sports Exerc*, **42** (8), 1582–1598.

Cormie P., McGuigan M.R. and Newton R.U. (2010c) Changes in the Eccentric Phase Contribute To Improved SSC Performance after Training. *Med Sci Sports Exerc*; E-Pub 4th February ahead of print. DOI 10.1249/MSS.0b013e3181d392e8.

Cronin J.B., Hing R.D. and McNair P.J. (2004) Reliability and validity of a linear position transducer for measuring jump performance. *J Strength Cond Res*, **18**, 590–593.

Dugan E.L., Doyle T.L., Humphries B., Hasson C.J. and Newton R.U. (2004) Determining the optimal load for jump squats: a review of methods and calculations. *J Strength Cond Res*, **18** (3), 668–674.

Hakkinen K. (1994) Neuromuscular adaptation during strength training, aging, detraining and immobilization. *Crit Rev Phys Rehab Med*, **6**, 161–198.

Häkkinen K., Alen M. and Komi P.V. (1985) Changes in isometric force- and relaxation-time, electromyographic and muscle fibre characteristics of human skeletal muscle during strength training and detraining. *Acta Physiol Scand*, **125** (4), 573–585.

Hori N., Newton R.U., Nosaka K. and McGuigan M.R. (2005) Comparison of system versus barbell force, velocity and power in the hang snatch. *J Strength Cond Res*, **19**, e19.

Hori N., Newton R.U., Andrews W.A., Kawamori N., McGuigan M.R. and Nosaka K. (2007) Comparison of four different methods to measure power output during the hang power clean and the weighted jump squat. *J Strength Cond Res*, **21** (2), 314–320.

Kaneko M., Fuchimoto T., Toji H. and Suei K. (1983) Training effect of different loads on the force-velocity relationship and mechanical power output in human muscle. *Scand J Med Sci Sports*, **5** (2), 50–55.

Kraemer W.J. and Newton R.U. (2000) Training for muscular power. *Phys Med Rehabil Clin N Am*, **11** (2), 341–368, vii.

LeSuer D.A., McCormick J.H., Mayhew J.L., Wasserstein R.L. and Arnold M.D. (1997) The accuracy of prediction equations for estimating 1-RM performance in the bench press, squat, and deadlift. *Journal of strength and conditioning research (Champaign, Ill.)*, **11** (4), 211–213.

Marek S.M., Cramer J.T., Fincher A.L., Massey L.L., Dangelmaier S.M., Purkayastha S., Fitz K.A. and Culbertson J.Y. (2005) Acute Effects of Static and Proprioceptive Neuromuscular Facilitation Stretching on Muscle Strength and Power Output. *Journal of athletic training (Dallas, Tex.)*, **40** (2), 94–103.

McBride J.M., Triplett-McBride T., Davie A. and Newton R.U. (1999) A comparison of strength and power characteristics between power lifters, Olympic lifters, and sprinters. *Journal of Strength and Conditioning Research*, **13** (1), 58–66.

McBride J.M., Triplett-McBride T., Davie A. and Newton R.U. (2002) The effect of heavy- vs. light-load jump squats on the development of strength, power, and speed. *J Strength Cond Res*, **16** (1), 75–82.

Moritani T., Muro M., Ishida K. and Taguchi S. (1987) Electrophysiological analyses of the effects of muscle power training. *Res J Phys Ed*, **1**, 23–32.

Moss B.M., Refsnes P.E., Abildgaard A., Nicolaysen K. and Jensen J. (1997) Effects of maximal effort strength training with different loads on dynamic strength, cross-sectional area, load-power and load-velocity relationships. *Eur J Appl Physiol Occup Physiol*, **75** (3), 193–199.

Murphy A.J., Wilson G.J. and Pryor J.F. (1994) Use of the iso-inertial force mass relationship in the prediction of dynamic human performance. *Eur J Appl Physiol Occup Physiol*, **69** (3), 250–257.

Murphy A.J. and Wilson G.J. (1996a) The assessment of human dynamic muscular function: A comparison of isoinertial and isokinetic tests. *J Sports Med Phys Fitness*, **36** (3), 169–177.

Murphy A.J. and Wilson G.J. (1996b) Poor correlations between isometric tests and dynamic performance: relationship to muscle activation. *European Journal of Applied Physiology and Occupational Physiology*, **73** (3–4), 353–357.

Murphy A.J. and Wilson G.J. (1997) The ability of tests of muscular function to reflect training–induced changes in performance. *J Sports Sci*, **15** (2), 191–200.

Newton R.U. and Kraemer W.J. (1994) Developing explosive muscular power: implications for a mixed methods training strategy. *Strength and Conditioning Journal*, **16** (5), 20–31.

Newton R.U. *et al.* (1999) Effects of ballistic training on preseason preparation of elite volleyball players. *Med Sci Sports Exerc*, **31** (2), 323–330.

Nuzzo J.L., McBride J.M., Cormie P. and McCaulley G.O. (2008) Relationship between countermovement jump performance and multijoint isometric and dynamic tests of strength. *J Strength Cond Res*, **22** (3), 699–707.

Rahmani A., Viale F., Dalleau G. and Lacour J.R. (2001) Force/velocity and power/velocity relationships in squat exercise. *Eur J Appl Physiol*, **84** (3), 227–232.

Schmidtbleicher D., Gollhofer A. and Frick U. (1988) *Effects of a stretch-shortening typed training on the performance capability and innervation characteristics of leg extensor muscles*. In: de Groot G., Hollander A.P., Huijing P.A. and van Ingen Schenau G.J. editors. Biomechanics XI-A. Amsterdam: Free University Press; pp. 185–189.

Thomas J.R. and Nelson J.K. (1990) *Research methods in physical activity (ed 2nd)*. Champaign, Ill., Human Kinetics.

Toji H. and Kaneko M. (2004) Effect of multiple-load training on the force–velocity relationship. *J Strength Cond Res*, **18** (4), 792–795.

Toji H., Suei K. and Kaneko M. (1997) Effects of combined training loads on relations among force, velocity, and power development. *Can J Appl Physiol*, **22** (4), 328–336.

Wilson G.J., Newton R.U., Murphy A.J. and Humphries B.J. (1993) The optimal training load for the development of dynamic athletic performance. *Medicine and Science in Sports and Exercise*, **25** (11), 1279–1286.

Wilson G.J., Walshe A.D. and Fisher M.R. (1997) The development of an isokinetic squat device: reliability and relationship to functional performance. *European Journal of Applied Physiology and Occupational Physiology*, **75** (5 (Print)), 455–461.

Wilson G.J., Lyttle A.D., Ostrowski K.J. and Murphy A.J. (1995) Assessing dynamic performance: A comparison of rate of force development tests. *J Strength Cond Res*, **9**(3), 176–181.

Young W.B. (1995) Laboratory strength assessment of athletes. *New Studies in Athletics*, **10** (1), 89–96.

Young W., Pryor J.F. and Wilson G. (1995) Effect of instructions on characteristics of countermovement and drop jump performance. *J Strength Cond Res*, **9**(4), 232–236.

Zatsiorsky V.M. and Kraemer W.J. (2006) *Science and Practice of Strength Training*, Human Kinetics, Champaign, IL.

3.2 Speed and Agility Assessment

Warren Young[1] and Jeremy Sheppard[2], [1]University of Ballarat, Victoria, Australia, [2]Queensland Academy of Sport, Australia

3.2.1 SPEED

3.2.1.1 Introduction

Speed is an important quality for many sports. The scientific definition of speed is distance divided by time, and it can refer to the movement of a body part such as the hand in boxing or to whole-body movement such as in running or cycling. For the purposes of this section, speed refers to total body running speed.

There are three forms of running speed:

1. Acceleration speed. This relates to situations where velocity increases rapidly and is important for sports played in a relatively confined space, for example tennis, basketball. Acceleration refers to the rate of change in velocity and is expressed in m/s/s. It is not usual for acceleration to be directly measured; it is more common to assess the time taken to sprint relatively short distances from a stationary start, for example 5–20 m.
2. Maximum speed. This is the peak speed reached in a sprint, and typically occurs at about 30–60 m in a maximum-effort sprint from a stationary start. Maximum speed may occur earlier in a maximum effort if the sprint is commenced from a running start (Duthie *et al.*, 2006a). Since speed is the distance covered divided by the time taken, it can be measured in m/s or km/h. Some equipment is available for the measurement of this speed quality (see Section 3.2.1.3), but the most common method is to report times taken to cover short distances, for example 10–20 m, from a 'flying' start.
3. Speed endurance. This is the ability to maintain relatively high running speeds, such as 90% of maximum or greater, and is epitomized by the 200 or 400 m running events in track and field. In many sports played on a field or court, speed endurance is important because repeated sprints are required, with relatively short recoveries between efforts. This is often termed 'repeat-sprint ability' (RSA).

Research has shown that the inter-correlations between these speed qualities are not high (Delecluse *et al.*, 1995), indicating that an athlete can be relatively good at one but not another. This may be explained by differences in the biomechanics and metabolic demands of these qualities. The practical application of this is that it is critical to conduct a needs analysis of the sport of interest to help identify the important speed qualities to assess.

3.2.1.2 Testing acceleration speed

This is usually assessed by reporting times taken to reach distances from 5 to 20 m in a maximum-effort sprint from a stationary start. Times should preferably be recorded by an electronic timing system, where 'gates' are positioned so that a beam is broken to start and stop the timing. Since timing should reflect the instant when the trunk passes through the beam, a single-beam device should use software to prevent triggering of the timing by a different part of the body, such as the hand. Dual-beam or triple-beam systems can also be used to ensure accuracy of timing. The timing resolution of the system should be at least 0.01 s.

If it is not possible to assess sprints with an electronic timing system, stop watches may be used, but care should be taken to maximize accuracy. For example, the timer could view the sprint by standing next to the finish line and stop the watch at the instant the trunk is perceived to cross the line. The watch should be started at the instant the back foot is perceived to leave the ground. Another potentially useful procedure would be to have two or three timers observing a sprint, and use the average of the times to represent the athlete's performance. A study by Moore *et al.* (2007) showed that hand timing with stop watches produced good inter-trial reliability but also produced significantly different times to an electronic timing gate system. Each procedure produces somewhat different times and therefore the results of the two methods should not be compared.

When testing relatively short sprints from a stationary position, the start can be expected to have a significant impact on the time recorded. Therefore it is crucial to standardize the starting technique. If sprinters from the sport of track and field are being assessed, they should be permitted to use starting blocks and commence the sprint from a starting signal, similar to that used in competition. For other athletes, reaction time from a starting signal is not of interest, and the acceleration sprint should begin from a stationary position so that timing commences at the instant the athlete initiates the first forward movement. Various starting techniques can be employed, such as:

- Standing start. The athlete adopts a forward leaning position with one foot forward and one foot back, and the opposite arms forward and back. The toe of the front foot is usually placed on a line, which may be directly under the start beam or a small distance behind it (e.g. 30 cm). There is no evidence regarding the optimum toe-to-start line distance. Two acceptable approaches are to use a standardized distance for all athletes, or to allow a self-selected distance to accommodate varying body dimensions. Regardless of this factor, the trunk should be positioned just behind the start beam for an electronic-timing light system. The athlete is not permitted to 'rock back' prior to the forward movement, as this would allow forward momentum to influence the time.
- Foot start. This involves a contact switch on the floor, such as a mat, on which the front foot rests in the set position. At the instant the foot leaves the mat, the switch opens and timing commences.
- Thumb start. This technique requires a 'three-point' start position with the hand (and thumb) on the same side as the back foot, touching a contact switch on the ground. Timing commences at the instant the hand comes off the switch.

Research comparing the above three starting techniques has shown that they all have acceptable and similar test-retest reliability (Duthie *et al.*, 2006b). However, the foot start was found to be the fastest and the thumb start was the slowest in a 10 m sprint. This indicates that the starting technique can significantly influence times, and therefore athletes should use a standardized technique that is most similar to the way they commence starts in their sport.

3.2.1.3 Testing maximum speed

One approach to assessing maximum speed or peak speed in a sprint is to record the instantaneous speed of the athlete when sprinting. In this situation the starting technique is not relevant; the important requirement is that the athlete runs far enough to have obtained maximum speed. This may be expected to be 30–60 m from a stationary start. One way to record instantaneous speed directly is with a laser or radar device that uses the Doppler principle. For example, the beam can be pointed at the back of the athlete so that they run directly away from it, and the change in frequency between the emitted and reflected signal is used to measure speed. The sampling rate and smoothing of the signal may be expected to influence the peak value obtained, and therefore should be taken into consideration.

Another more recent approach is to use a portable global positioning system (GPS). These have become commercially available and can accurately track the position of an athlete over time. A potential drawback of this technology for precise measurement of peak speed is the limited sampling rates, for example <10 Hz. Also, this assessment is presently limited to outdoor settings where sufficient satellite signals can be clearly received. An advantage of GPS tracking is that speeds can be obtained in the natural training or competition setting of the athlete.

With the use of a light gate system, maximum speed can be assessed in conjunction with acceleration speed. For example, timing gates can be placed at the start and at 10 m intervals up to 60 m. Acceleration capacity can be determined from the time to 10 or 20 m (from a stationary start), and maximum speed can be estimated from the smallest time interval between each 10 m segment of the sprint. Since this requires seven light gates, a simpler but cruder procedure would be to use light gates at the start, at 10 or 20 m, and at 50 and 60 m (four gates). Although this is a simple approach, the 50–60 m time is only an estimate of the true maximum speed; for example, if the time recorded were 1.10 s, the estimated maximum speed would be 9.09 m/s. A potential advantage of reporting speed or velocity in m/s is that velocity can be multiplied by body mass to yield the running momentum developed by the athlete. This approach has been used for rugby league (Baker and Newton, 2008) because momentum may be more important than velocity in impact situations in collision sports.

Whatever sprint distance is used, it is important to provide at least 15 m distance beyond the finish line to allow the athlete to decelerate safely. This suggests that at least 75 m is required to administer this test. Since it is preferable to conduct the test indoors to eliminate the influence of wind (see Section 3.2.1.5), it may be challenging to find a suitable indoor venue with enough space; a sprint that is less than 60 m may therefore be necessary. Research assessing elite Australian rules footballers (Young *et al.*, 2008) showed that in a 30 m sprint from a standing start, the correlation between 10 m time and the time between 20 and 30 m was $r = 0.65$, representing a common variance of 42%. The uniqueness of these two measures suggests that the 20–30 m split assesses a different speed quality to acceleration, interpreted by the authors to be a reasonable estimate of maximum speed. Therefore it is recommended that a sprint of at least 30 m and up to 60 m be used to estimate maximum speed capabilities.

3.2.1.4 Testing speed endurance

For sports such as track and field requiring 'long' continuous sprints (≥60 m), speed endurance is an important speed quality. It can be assessed simply by the time taken to reach appropriate distances. However, many sports require intermittent sprints over relatively short distances with insufficient recovery between efforts, resulting in sprinting in a fatigued state. There are many test protocols to assess RSA but no one protocol is universally accepted. Attempts have been made to determine sport-specific RSA protocols based on time-motion analyses (Spencer *et al.*, 2004; Wragg, Maxwell and Doust, 2000). Usually sprints are performed in a straight line, but a change of direction may be included in an attempt to increase specificity to field-based sports (Impellizzeri *et al.*, 2008; Wragg, Maxwell and Doust, 2000). Variables such as the sprint distance or duration, number of repetitions, recovery duration, and recovery activity will influence the precise physiological demands of an RSA test (Spencer *et al.*, 2005).

Regardless of the test protocol, sprint efforts should be performed with maximum effort each repetition so that 'pacing' is avoided. The test results can be reported as either the mean of the times for each repetition or as a calculated percentage decrement. The mean value is the recommended outcome measure for RSA testing, because the percentage decrement has been found to have poor test-retest reliability (Fitzsimons *et al.*, 1993; Impellizzeri *et al.*, 2008). For further details on RSA, see Chapter 3.3.

3.2.1.5 Standardizing speed-testing protocols

There are many variables that can influence speed-test results which should be controlled or standardized to maximize reliability. The starting procedure has already been outlined, but others are discussed below.

1. Environmental conditions. Temperature, humidity, and especially wind can all have a significant influence on sprint-test results. These variables can be controlled by conducting tests in a suitable indoor facility. However, for sports played outdoors, ecological validity may be compromised by conducting sprints indoors with relatively unfamiliar footwear and surfaces. If sprints are conducted outdoors and wind is present, the sprints should be aligned across the wind direction to minimize any assisting or resisting effect. If available, a wind gauge may also be used to report wind velocity recordings in order to assist the interpretation of results.
2. Floor surface and footwear. The type of surface and the footwear worn by athletes during sprint tests can have a profound effect on results. The combined surface and shoes should allow enough friction to prevent slipping, which is most likely to occur near the start of a sprint, when acceleration is high. If testing is conducted indoors, the floor should be clean and free of dust. If testing is conducted on natural grass, which may vary in nature, the ground condition should be noted to assist with interpretation of results.
3. Warm-up. Optimizing the warm-up is important in order to maximize performance and minimize the risk of injury. Maintaining warm muscles is especially important for sprint performance (Mohr *et al.*, 2004). One method of warm-up when testing a group of athletes is to administer a standardized procedure where all individuals perform the same activities with the same duration and intensity. An alternative approach is to allow each individual to perform their own preferred warm-up. This is advisable because it is more likely to cater for individual differences in fitness. However, for this approach to be effective, the athletes must be educated and experienced enough to determine their own optimum protocol. Regardless of the approach, it is recommended that the warm-up progresses in intensity and concludes with some practice of the specific skill about to be assessed (the starting procedure).
4. Number of trials. Since the objective of testing is to extract the best possible performance of the athlete, there should be no restrictions on the number of sprint trials allowed. However, if the warm-up is effective, one or two trials should be adequate. For fatiguing tests (e.g. tests for RSA), only one trial is appropriate.

3.2.2 AGILITY

Agility is an important quality in many sports played on a court or field. In team sports, agility can be important in evading a defender while attacking or putting pressure on an opponent when defending. Unlike other physical qualities, there is no universally accepted definition of agility. It has traditionally been thought of as the ability to accelerate, decelerate, and change direction (Brown and Ferrigno, 2005).

However more recently, agility has been described as containing a change-of-direction-speed component involving pre-planned movements, as well as a perceptual and decision-making component (Young, James and Montgomery, 2002). This acknowledges that in most sports, changes of direction are performed in response to a stimulus. The definition of agility that is adopted for this section is 'a rapid whole-body movement with change of velocity or direction in response to a stimulus' (Sheppard and Young, 2006).

Ellis *et al.* (2000) recognized the presence of the perceptual and decision-making factors but stated that the purpose of most agility tests is simply to measure the ability to rapidly change body direction and position. Such tests will be referred to here as change-of-direction (COD) speed tests (Young, James and Montgomery, 2002). These tests typically involve a maximum-effort sprint over a pre-determined course with the use of obstacles such as poles or cones to indicate directional changes. The test result is simply determined by the time taken to complete the course. In such tests, the issues discussed in relation to speed assessment, such as the standardization of protocols, are also applicable. With COD speed and agility tests, the ability to avoid slipping is particularly important.

Many COD speed tests have been developed, probably in an attempt to simulate sport-specific movement patterns. For example, some sports such as basketball, volleyball, and tennis involve sideways 'shuffling' and backwards running, and therefore tests have been devised which contain these patterns, such as SEMO and the 'T' test (Buckeridge *et al.*, 2000; Semenick, 1994). Time-motion studies should be consulted to determine the typical movement patterns that occur in the sport of interest.

Tests that contain a COD movement in response to a stimulus have been developed relatively recently, and are much less common. There are two categories of these 'reactive' agility tests. The first involves tests that require a directional change following a generic stimulus such as a flashing light (Oliver and Meyers, 2009). The second involves tests that require the athlete to react to a stimulus provided by a person performing a COD. This stimulus may be a video recording of an attacker in possession of a ball (Farrow, Young and Bruce, 2005; Williams and Davids, 1998) or live action of an 'opponent' changing direction (Gabbett, Kelly and Sheppard, 2008; Sheppard *et al.*, 2006) (Figures 3.2.1 and 3.2.2). Given the multitude of tests available, it would be fruitless to attempt to evaluate and recommend particular tests. Therefore the following section will discuss relevant issues that should assist the sport scientist or coach in selecting an appropriate COD speed or agility test.

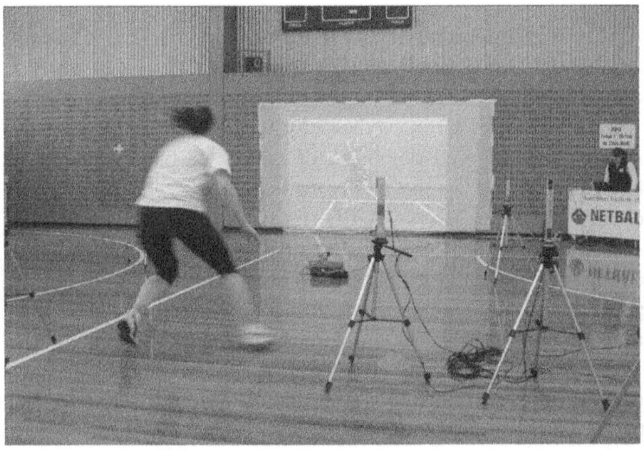

Figure 3.2.1 Reactive agility test for netball

Figure 3.2.2 Reactive agility test using a live tester (closest to camera)

One way to validate a test is to demonstrate that it can distinguish between athletes of different performance levels. For example, if an elite group of athletes performs significantly better than a sub-elite group, the test may be considered to be related to performance, and therefore to assess a relevant quality. For COD speed tests, such evidence has been reported for the hexagon test and side-shuffle in tennis (Roetert *et al.*, 1996) and a 40 m test in soccer (Reilly *et al.*, 2000). However, there is also evidence to the contrary, for example in volleyball (Barnes *et al.*, 2007) and rugby league (Baker and Newton, 2008). The reasons for this discrepancy are likely to be related to the variety of tests used, the specificity of the tests to each sport, and the samples studied.

Some research has assessed athletes with both COD speed tests and reactive agility tests. In all cases, the reactive tests were able to discriminate between athletes of different performance levels while the COD speed tests were not (Farrow, Young and Bruce, 2005; Gabbett, Kelly and Sheppard, 2008; Sheppard *et al.*, 2006) (Figure 3.2.2). It has also been shown that COD speed tests and reactive tests have modest correlations, which indicates they assess unique characteristics (Farrow, Young and Bruce, 2005; Gabbett, Kelly and Sheppard, 2008; Sheppard *et al.*, 2006). These results suggest that the reaction to an opponent's movements in netball, rugby league, and Australian rules football is an important component of agility in skilled performers. Indeed, decision-making time was found to be faster for a higher performance group of athletes (Farrow, Young and Bruce, 2005; Gabbett and Benton, 2009). Analysis of perceptual skills in soccer players has also revealed that better players are able to achieve faster and more accurate responses to an attacker dribbling a ball (Reilly *et al.*, 2000; Williams and Davids, 1998).

The above discussion indicates that in sports requiring a reaction to a directional change, the concept of 'reactive' agility is important, and tests will be more relevant if they contain this element. In some reports, the participants have been filmed with high-speed video (Gabbett and Benton, 2009; Gabbett, Kelly and Sheppard, 2008; Young and Willey, 2010). This procedure allows the analyst to determine the 'decision time', operationally defined as the time from the instant when the 'opponent' plants the foot to change direction to the instant when the reacting athlete plants his or her foot to change direction in pursuit. When this time is separated from the movement time of the athlete, the test can be used to determine whether decision time or movement time is a relative strength or weakness for the individual. This information can then be used to prescribe appropriate training.

There is little research available to provide evidence for the utility of reactive tests using a non-sport-specific stimulus. One study reported a relationship between a simple COD speed test and a test involving the same movement pattern following a COD directed by a flashing light (Oliver and Meyers, 2009). There was a high correlation, producing a common variance of 87%, indicating that the inclusion of a generic stimulus to change direction did not change the nature of the test. This result is consistent with the notion that the perceptual component of agility must be sport-specific.

Research investigating visual search strategies in soccer showed that more experienced players had a higher search rate, involving more fixations of shorter duration, and were fixated for longer on the hip region, indicating that this area was important in anticipating an opponent's movements (Williams and Davids, 1998). Further, research on rugby players demonstrated that highly skilled players were better than novices in detecting a deceptive movement such as 'dummy' side-step in a COD situation (Jackson, Warren and Abernethy, 2006). Therefore, agility tests should contain a stimulus to change direction that is as sport-specific as possible. This might involve an attacker in possession of a ball viewed from various angles to simulate the visual information displayed in the competition setting, for example.

As mentioned earlier, the stimulus to change direction may be either a video recording or a live tester acting as an 'opponent'. The main advantage of using various video recordings is that the time taken to produce the stimulus for the athlete to respond to is the same for all athletes tested. A possible disadvantage is that normal video displays are presented in two dimensions, which is somewhat different to the visual information seen in competition. A potential disadvantage of a live tester is that the time taken to execute the COD movement will vary for different athletes, and this time contributes to the total time recorded as the athlete's test score. Although this test has been shown to have adequate test-retest and inter-tester reliability (Sheppard *et al.*, 2006), recent research has shown that the 'tester time' can influence the score recorded for each athlete (Young and Willey, 2010). Therefore, when using this field test of reactive agility, it is advisable to use high-speed video analysis (e.g. sampling at 200 Hz) and to use a frame counter to isolate the various components of the test. For example, tester time, decision time, and the movement time of the athlete can be determined from the total time, which is recorded by an electronic light-gate system (Young and Willey, 2010).

3.2.3 CONCLUSION

Acceleration speed, maximum speed, and speed endurance are three independent speed qualities that can be assessed with various protocols. Practitioners designing tests batteries for athletes should conduct a needs analysis to identify the most important qualities to test in different athletes. Standardized protocols can then be developed to address the many factors that can influence test results. There are many tests available that assess the ability to change direction quickly and accurately, using obstacles to direct the movement pattern. These COD speed tests should mimic the specific movements of the sport of interest. If the sport involves reacting to a stimulus such as an opponent producing a side-step, evidence suggests that tests that include this element are specific and useful. The development of such reactive agility testing is evolving and should be encouraged. The challenge is to further develop tests based on sport-specific attacking and defending scenarios while maintaining reliability and validity.

References

Baker D.G. and Newton R.U. (2008) Comparison of lower body strength, power, acceleration, speed, agility, and sprint momentum to describe and compare playing rank among professional rugby league players. *J Strength Cond Res*, **22**, 153–158.

Barnes J.L., Schilling B.K., Favlo M.J. *et al.* (2007) Relationship of jumping and agility performance in female volleyball athletes. *J Strength Cond Res*, **21**, 1192–1196.

Brown L.R. and Ferrigno V.A. (2005) *Training for Speed, Agility and Quickness*, Human Kinetics, Champaign, IL.

Buckeridge A., Farrow D., Gastin P. *et al.* (2000) Protocols for the physiological assessment of high-performance tennis players, in *Physiological Tests for Elite Athletes, Australian Sports Commission* (ed. C. Gore), Human Kinetics, Champaign, IL, pp. 383–403.

Delecluse C., Van Coppenolle H., Willems E. *et al.* (1995) Influence of high-resistance and high-velocity training on sprint performance. *Med Sci Sports Exerc*, **27**, 1203–1209.

Duthie G.M., Pyne D.B., Marsh D.J. *et al.* (2006a) Sprint patterns in rugby union players during competition. *J Strength Cond Res*, **20**, 208–214.

Duthie G.M., Pyne D.B., Ross A.A. *et al.* (2006b) The reliability of ten-meter sprint time using different starting techniques. *J Strength Cond Res*, **20**, 246–251.

Ellis L., Gastin P., Lawrence S. *et al.* (2000) Protocols for the physiological assessment of team sport players, in *Physiological Tests for Elite Athletes, Australian Sports Commission* (ed. C. Gore), Human Kinetics, Champaign, IL, pp. 128–144.

Farrow D., Young W. and Bruce L. (2005) The development of a test of reactive agility for netball: a new methodology. *J Sci Med Sport*, **8**, 52–60.

Fitzsimons M., Dawson B.T., Ward D. *et al.* (1993) Cycling and running tests of repeated sprint ability. *Aust J Sci Med Sport*, **25**, 82–87.

Gabbett T. and Benton D. (2009) Reactive agility of rugby league players. *J Sci Med Sport*, **12**, 212–214.

Gabbett T., Kelly J. and Sheppard J. (2008) Speed, change of direction speed and reactive agility of rugby league players. *J Strength Cond Res*, **22**, 174–181.

Impellizzeri F.M., Rampinini E., Castagna C. *et al.* (2008) Validity of a repeated sprint test for football. *Int J Sports Med*, **11**, 899–905.

Jackson R., Warren S. and Abernethy B. (2006) Anticipation skill and susceptibility to deceptive movement. *Acta Psychol*, **123**, 355–371.

Mohr M., Krustrup P., Nybo L. *et al.* (2004) Muscle temperature and sprint performance during soccer matches – beneficial effect of re-warm-up at half-time. *Scand J Med Sci Sports*, **14**, 156–162.

Moore A.N., Decker A.J., Baarts J.N. *et al.* (2007) Effect of competitiveness on forty-yard dash performance in college men and women. *J Strength Cond Res*, **21**, 385–388.

Oliver J. and Meyers R. (2009) Reliability and generality of measures of acceleration, planned and reactive agility. *Int J Sports Physiol Perform*, **4**, 345–354.

Reilly T., Williams A.M., Nevill A. *et al.* (2000) A multidisciplinary approach to talent identification in soccer. *J Sports Sci*, **18**, 695–702.

Roetert E.P., Brown S.W., Piorkowski P.A. *et al.* (1996) Fitness comparisons among three different levels of elite tennis players. *J Strength Cond Res*, **10**, 139–143.

Semenick D. (1994) Testing protocols and procedures, in *Essentials of Strength and Conditioning* (ed. T.R. Baechle), Human Kinetics, Champaign, IL, pp. 258–273.

Sheppard J. and Young W. (2006) Agility literature review: classifications, training and testing. *J Sports Sci*, **24**, 919–932.

Sheppard J.M., Young W.B., Doyle T.A. *et al.* (2006) An evaluation of a new test of reactive agility, and its relationship to sprint speed and change of direction speed. *J Sci Med Sport*, **9**, 342–349.

Spencer M., Lawrence S., Rechichi C. *et al.* (2004) Time-motion analysis of elite field hockey, with special reference to repeated-sprint activity. *J Sports Sci*, **22**, 843–850.

Spencer M., Bishop D., Dawson B. *et al.* (2005) Physiological and metabolic responses of repeated-sprint activities. *Sports Med*, **35**, 1025–1044.

Williams M. and Davids K. (1998) Visual search strategy, selective attention, and expertise in soccer. *Res Q Exerc Sport*, **69**, 111–128.

Wragg C.B., Maxwell N.S. and Doust J.H. (2000) Evaluation of the reliability and validity of a soccer-specific field test of repeated sprint ability. *Eur J Appl Physiol*, **83**, 77–83.

Young W. and Willey B. (2010) Analysis of a reactive agility field test. *J Sci Med Sport*, **13**, 376–378.

Young W., James R. and Montgomery I. (2002) Is muscle power related to running speed with changes of direction? *J Sports Med Phys Fitness*, **42**, 282–288.

Young W., Russell A., Burge P. *et al.* (2008) The use of sprint tests for assessment of speed qualities of elite Australian Rules footballers. *Int J Sports Physiol Perform*, **3**, 199–206.

3.3 Testing Anaerobic Capacity and Repeated-sprint Ability

David Bishop[1] and Matt Spencer[2], [1]Institute of Sport, Exercise and Active Living (ISEAL), School of Sport and Exercise Science, Victoria University Melbourne, Australia, [2]Physiology Unit, Sport Science Department, ASPIRE, Academy for Sports Excellence, Doha, Qatar

3.3.1 INTRODUCTION

3.3.1.1 Energy systems

Adenosine triphosphate (ATP) is the immediate source of chemical energy for muscle contraction and subsequent movement. As intramuscular stores of ATP are limited (20–25 mmol/kg dry muscle (dm)), three well-regulated and closely integrated chemical pathways exist for the regeneration of ATP (see Figure 3.3.1). While often described as sequential pathways, it is important to emphasize that all three pathways are activated, to varying degrees, at the onset of muscle contraction.

The first process involves the splitting of phosphocreatine (PCr), which together with the stored ATP in the cell, provides the immediate energy in the initial stages of intense exercise. The second process involves the non-aerobic breakdown of carbohydrate (either glycogen or glucose) to lactate. These pathways are capable of regenerating large amounts of ATP per unit of time and are the predominate suppliers of energy during single or repeated bouts of brief intense exercise (Figure 3.3.2). It is important to recall, however, that the anaerobic contribution to repeated bouts of exercise will depend on the duration of the bouts, the number of bouts, and the recovery between bouts (see Figure 2.9.5). In particular, the contribution of the third chemical pathway, aerobic metabolism, has been shown to increase during repeated-sprint bouts (McGawley and Bishop, 2008).

3.3.1.2 The importance of anaerobic capacity and RSA

Given the important contribution of the anaerobic energy systems to brief intense exercise, it is not surprising that a well-developed anaerobic capacity (AC) appears important for the performance of such activities (generally consisting of <300 seconds of all-out exercise). For example, despite being more than 50% aerobic, both 500 m kayak (Bishop, 2000) and 4000 m individual cycle pursuit (Craig *et al.*, 1993) performance have been reported to be related to AC. While not all studies have reported significant relationships between AC and performance (Craig and Morgan, 1998; Olesen *et al.*, 1994), the reported positive correlations suggest that a complete assessment of athletes who participate in brief intense activities should include the assessment of AC in order to detect potential deficiencies and to inform the prescription of training. However, while the direct measurement of aerobic metabolism is relatively easy during most sporting activities, direct measurement of anaerobic metabolism is more complex and relies on various assumptions, described in Section 3.3.3.

Anaerobic metabolism is also important for the performance of high-intensity sprints of short duration, interspersed with brief recoveries. This ability to recover and to reproduce performance in subsequent sprints is an important fitness requirement of team-sport athletes (Rampinini *et al.*, 2007) and has been termed repeated-sprint ability (RSA). However, despite the important contribution of anaerobic metabolism to repeated sprints, AC does not seem to be well correlated with RSA (Wadley and Le Rossignol, 1998); additional testing may therefore be required for team-sport athletes. Recently, a number of tests have been devised to assess RSA. Such tests are useful for detecting potential weaknesses in individual athletes, helping inform training prescription, and allowing assessment of changes in RSA following different training or dietary interventions (see Chapter 2.9).

3.3.2 TESTING ANAEROBIC CAPACITY

3.3.2.1 Definition

Exhaustive exercise lasting a few minutes is reliant on the energy release of both the aerobic and the anaerobic energy systems (Figure 3.3.2). The energy for ATP formation, above that supplied by the measured VO_2, is extracted from several sources. Aerobic utilization of O_2 stores in the body is one source; this consists of O_2 bound to haemoglobin in the blood and myoglobin in the muscle, O_2 dissolved in body fluids, and O_2 present in the lungs (Astrand *et al.*, 1964). The contribution

Strength and Conditioning – Biological Principles and Practical Applications Marco Cardinale, Rob Newton, and Kazunori Nosaka.

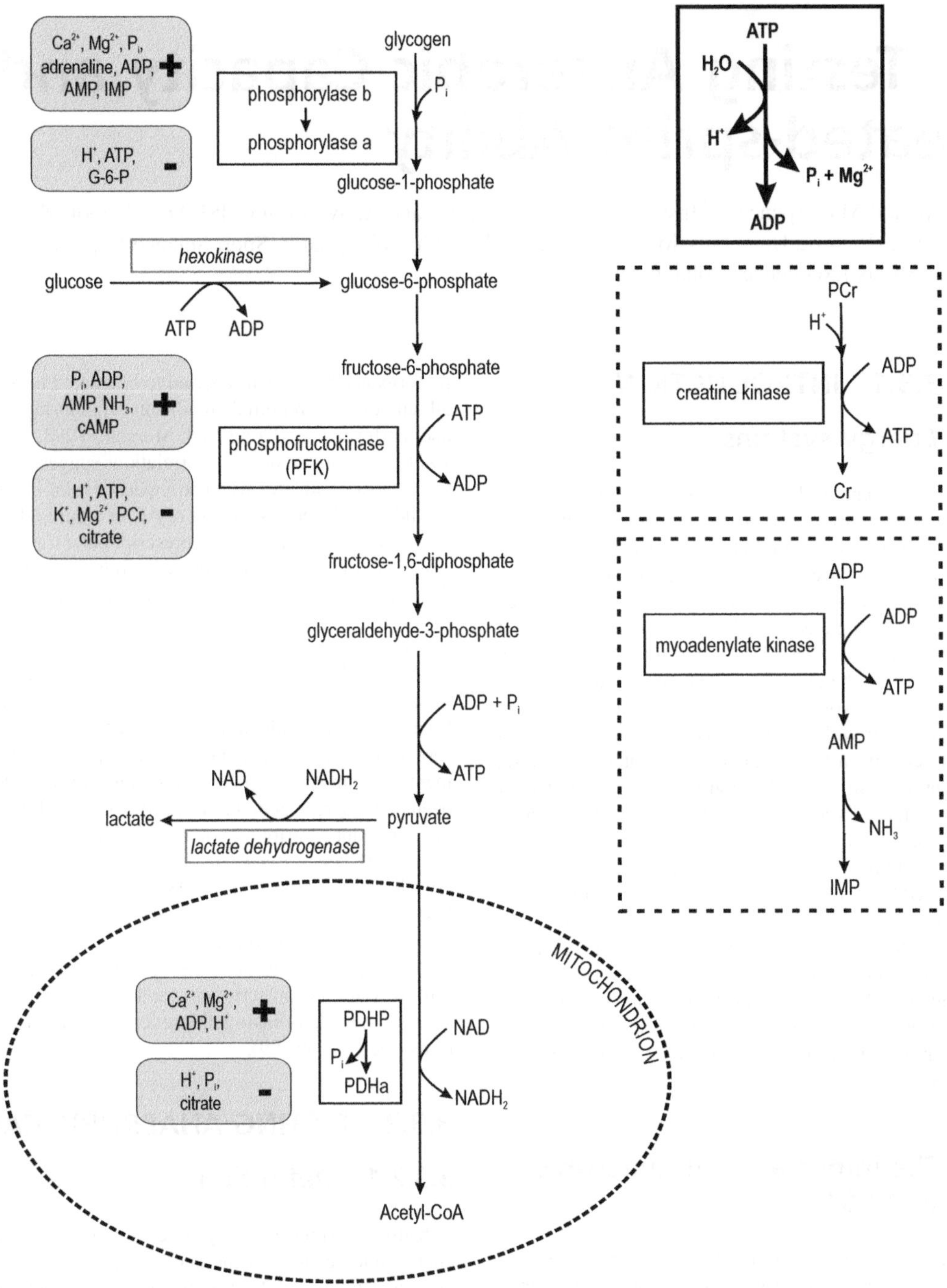

Figure 3.3.1 Schematic representation of the major metabolic pathways and a number of potential regulators

of the oxygen stores amounts to <10% (Medbo *et al.*, 1988). The second source of ATP consists of the breakdown of phosphocreatine stored in the exercising muscles (Hultman and Sjoholm, 1983; Sahlin, Harris and Hultman, 1979); this is also known as the alactic capacity. The third source is the breakdown of glycogen to lactate (sometimes termed lactic acid) (Hermansen and Vaage, 1977; Hultman and Sjoholm, 1983); this is often termed the lactic capacity. The total anaerobic energy release, termed the anaerobic capacity, is defined as the maximal amount of energy that can be released anaerobically

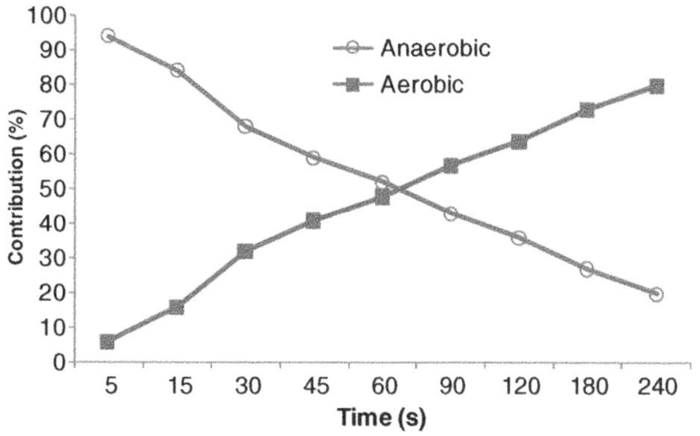

Figure 3.3.2 Contribution of the aerobic and anaerobic energy systems during periods of exhaustive exercise (Bangsbo *et al.*, 1993; Bishop, 2000; Gastin and Lawson, 1994; Gastin *et al.*, 1995; Spencer, Gastin and Payne, 1997)

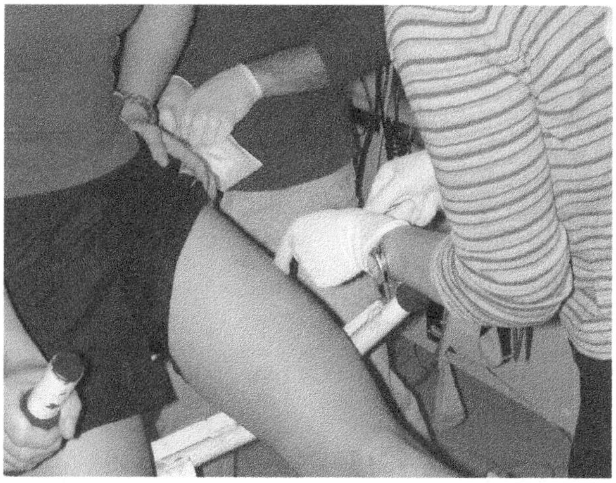

Figure 3.3.3 Post-exercise muscle biopsy performed while the subject remains seated on the cycle ergometer

during a specific bout of short-duration, exhaustive exercise (Medbo *et al.*, 1988). The anaerobic capacity is often abbreviated as AC. AC should be distinguished from its mechanical counterpart (anaerobic work capacity or AWC), which may be defined as the total amount of external (mechanical) work during a specific bout of short-duration, exhaustive exercise of sufficient duration to incur a near-maximal anaerobic ATP yield (Green, 1994). AWC is usually expressed in Joules.

3.3.2.2 Assessment

As the supply of anaerobic energy is dependent on intracellular mechanisms, with little reliance on central mechanisms, it has proven difficult to precisely quantify the AC. Furthermore, to date, there is no direct method for the validation of this parameter. Thus, unlike the aerobic energy system and the maximal oxygen uptake (VO_{2max}; see Chapter 3.4), a universally accepted method does not exist to measure an individual's AC. However, several methods have been proposed, including the invasive measures of muscle metabolites and blood lactate concentration, the oxygen deficit, and the critical power (CP) model. Each of these methods is discussed and evaluated below. For a discussion on less common methods for estimating AC, the reader is referred to Heck, Schulz and Bartmus (2003).

Muscle metabolites

The direct measurement of the AC is only possible via the measurement of changes in ATP, PCr, and lactate in muscle biopsy samples (Figure 3.3.3). ATP production is assumed to be equal to the $\Delta ATP + \Delta PCr + 1.5 \times \Delta lactate$ (Bangsbo, 1998). In addition, the net lactate release into the blood also needs to be considered. However, this simple calculation relies on the validity of a number of assumptions: first, that the biopsy sample is representative of the entire exercising muscle mass, and that all muscles are involved to the same extent as that sampled; second, that there is very little change in muscle metabolites from the cessation of exercise to the attainment of the biopsy sample; third, that there is an accurate estimation of the active muscle mass involved in the exercise. Using this method, it has been estimated that the maximal AC of the knee extensors is ~90 mmol ATP/kg/min (Bangsbo *et al.*, 1990).

Despite the difficulties and assumptions associated with quantifying the anaerobic energy release from changes in muscle metabolites, the procedure does offer a direct method of evaluating the anaerobic system. However, due to limited practicality of the regular use of such an invasive technique, indirect methods are also required. Where possible, the measurement of muscle metabolites should be used as a standard to which other indirect measures are compared.

Maximal accumulated oxygen deficit (MAOD)

Theoretical basis

The term 'oxygen deficit' was introduced by Krogh and Lindhard (1920). It is defined as the difference between the predicted oxygen demand and the actual oxygen uptake during muscular work (Figure 3.3.4). Medbo *et al.* (1988) described a testing procedure that allowed the determination of oxygen deficit as a measure of maximal AC. The maximum accumulated oxygen deficit (MAOD) methodology is calculated on the basis of the following three assumptions:

1. The release of anaerobic energy is the total energy released minus the aerobic release taken as the accumulated VO_2.
2. The O_2 demand increases linearly with the intensity of exercise.
3. The O_2 demand is constant from the beginning of exercise if the intensity of exercise remains constant.

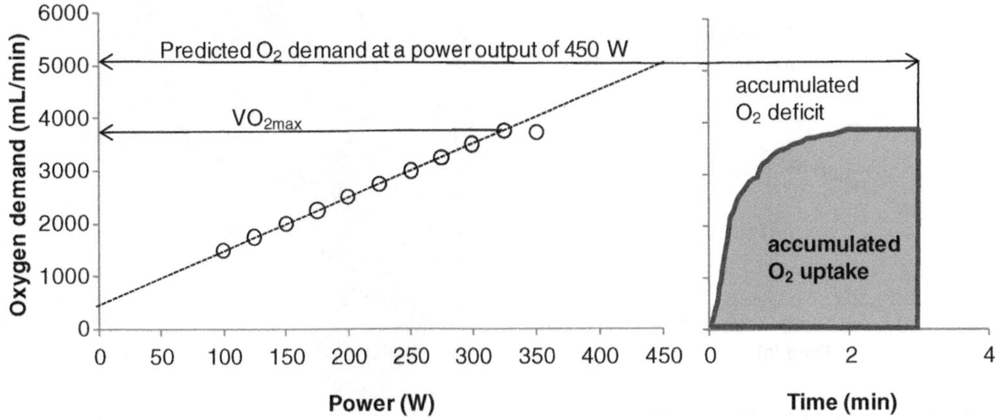

Figure 3.3.4 Example of MAOD assessment using the original methods of Medbo *et al.* (1988)

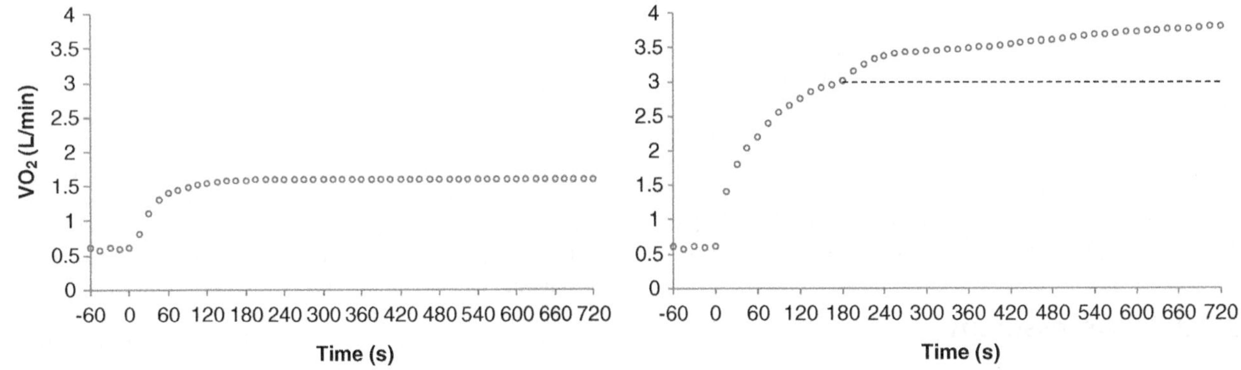

Figure 3.3.5 Typical examples of the VO_2 response to exercise performed at a power output below (left panel) and above (right panel) the lactate threshold. The circles represent the actual VO_2 values, while the dashed line represents the predicted VO_2 value

If the above-stated assumptions are justified then anaerobic energy release can be accurately estimated using the MAOD methodology. However, in recent years, some of these assumptions have been questioned. While good linearity of the power output–oxygen relationship has typically been reported (Buck and McNaughton, 1999a; Green and Dawson, 1996; Green *et al.*, 1996), others have reported that oxygen uptake does not always increase linearly with power output (Medbo, 1996; Medbo *et al.*, 1988; Zoladz, Rademaker and Sargeant, 1995). In particular, a disproportionately small increase in VO_2 has sometimes been reported at low power outputs. It has been argued that this may be attributable to a disproportionately greater increase in the oxygen demands of ventilation at higher power outputs (Heck, Schulz and Bartmus, 2003). Alterations in either mechanical or biochemical efficiency could also contribute to a nonlinear increase in O_2 demand with the intensity of exercise. This is supported by data obtained by assessing energy production during knee-extensor exercise (Bangsbo *et al.*, 2001). The authors reported that the energy turnover per unit of work (determined *in vitro* via muscle biopsy) was considerably greater during the initial 15 seconds than during the remaining duration of exercise. These findings suggest that in this particular mode of exercise, energy production was not constant, and that energy demand and work efficiency can change during the course of an exercise bout. Medbo (1996) suggested that these findings may have been influenced by the mechanical limitations of knee-extensor exercise used in this particular study and may therefore not apply to whole-body exercise (Medbo, 1996).

Another important assumption of the MAOD methodology is that, for a given workload, O_2 demand remains constant during the whole exercise duration. This appears to be a robust assumption for light-to-moderate exercise intensities as steady-state VO_2 values are usually achieved within three to four minutes (Figure 3.3.5, left panel) (Green and Dawson, 1995). However, during heavy exercise (intensity greater than the lactate threshold) there is a rapid monoexponential increase in VO_2 (the 'fundamental' or phase II component of the kinetics), followed by a further sustained increase in VO_2 ('slow' or phase III component) (Gaesser and Poole, 1996). This slow additional rise in VO_2 projects above the VO_2 predicted from exercise performed below the lactate threshold until either a delayed steady state is achieved or VO_{2max} is reached (Figure 3.3.5, right panel).

Methodology

The basic procedures of the accumulated oxygen deficit (AOD) methodology are as follows (see Figure 3.3.4). First, a series of sub-maximal exercise bouts are conducted throughout a range of power outputs. Steady-state VO_2 values are obtained from these exercise bouts and the relationship with the power output is then extrapolated and used to estimate energy demand during supra-maximal exercise. A supra-maximal exercise test is then performed and the accumulated VO_2 is calculated and subtracted from the estimated O_2 demand. The difference between the estimated O_2 demand and the actual VO_2 is termed the AOD. The AOD is often expressed in O_2 equivalents (i.e. ml/kg/min or L/min) as these are the units from which it is indirectly estimated.

The procedures for calculating the AOD recommended by Medbo (1996) are very extensive and time-consuming. Medbo *et al.* (1988) had subjects complete 10-minute steady-state exercise efforts over a three-week period at 20 different sub-maximal intensities ranging from 35 to 100% VO_2 max. A large variation in the range of regression slopes from the relationship between sub-maximal treadmill velocities and VO_2 (16% between subjects) was evidence of the need for the individual establishment of the exercise economy relationship and discredited the use of an assumed constant mechanical efficiency. Based on these results, Medbo recommended the use of 10 sub-maximal exercise bouts, each of 10 minutes' duration, to accurately determine the MAOD. If the stated procedures were used then a total sub-maximal exercise time of approximately 100 minutes would be required for each subject, often spread over several days. This level of testing is impractical for most applied situations and a number of modifications have been suggested and trialled. Such modifications are based on changes to the number of steady-state, sub-maximal exercise tests, the duration of the steady-state tests and the range (%VO_2 max) of the sub-maximal intensities used.

Number of steady-state, sub-maximal tests

Beyond the initial research, there has been little systematic investigation of the minimum number of bouts required to construct a suitable regression to predict MAOD. There is however one study which evaluated the effect of systematic removal of sub-maximal VO_2 values from the standard 10-point regression line (Buck and McNaughton, 1999a). This reported that inaccuracies occur in the measurement of MAOD when less than 10 points are used in the calculation. However, closer analysis of the results appears to suggest that there was very little difference from the MAOD value calculated using the standard 10-point regression line when only the four highest values were used. This is similar to an alternative procedure later adopted by Medbo and Burgers (1990) (four 10-minute efforts at intensities equivalent to 70–95% VO_2 max). It has also been suggested that the MAOD may be accurately determined using a common y-intercept (5 ml/kg/min) and two (Medbo *et al.*, 1988) or three (Scott *et al.*, 1991) measures of VO_2 at intensities of approximately 85–100% VO_2 max.

Sub-maximal test duration

Test durations of 4 (Green *et al.*, 1996), 5 (Bishop, 2000; Weyand *et al.*, 1994), 6 (Bangsbo, Michalsik and Petersen, 1993), 7 (Casaburi *et al.*, 1987), 8 (Bangsbo *et al.*, 1990), 10 (Medbo *et al.*, 1988), and 15 (Green and Dawson, 1996) minutes have all been used to construct the power–VO_2 regression. While reductions in the test duration from the standard 10 minutes have been made to make the test more feasible in applied testing situations, this has been shown to result in large differences in MAOD from the standard value. Research has shown that calculating the power–VO_2 regression with VO_2 values calculated at 2–4, 4–6, and 6–8 minutes of each test results in decreases in MAOD of 25, 10, and 5% compared with those calculated using 8–10-minute bouts (Buck and McNaughton, 1999b). Increasing the test duration to 15 minutes has not been reported to significantly alter the MAOD estimate (Green and Dawson, 1996).

Duration of the supra-maximal test

Hermansen and Medbo (1984) found that a supra-maximal test duration of at least two minutes is necessary to attain an MAOD (Figure 3.3.6). A shorter maximal exercise duration leads to a reduction in MAOD, which appears to be due to sub-maximal use of the anaerobic lactic capacity. This assumption is supported by the observation of higher post-exercise lactate values when supra-maximal exercise duration is extended (Renoux *et al.*, 1999). Renoux *et al.* (1999) also reported increases in MAOD with increases in exercise duration. It has been suggested however that sprint-trained athletes may benefit from a slightly shorter test duration (~70 seconds; Craig *et al.*, 1995). Increases in the duration of the supra-maximal test beyond two

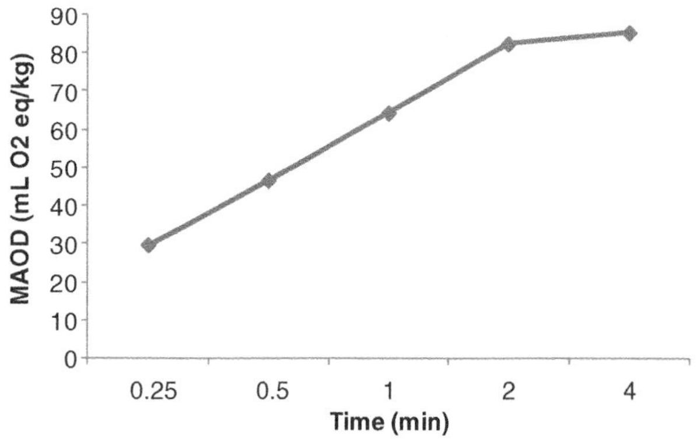

Figure 3.3.6 MAOD as a function of supra-maximal, constant-intensity test duration (Hermansen and Medbo, 1984)

minutes result in very little change in MAOD. The mean absolute difference in the MAOD between exhausting bouts lasting 2–5 minutes was <3 ml O_2/kg (Craig *et al.*, 1995; Hermansen and Medbo, 1984; Medbo *et al.*, 1988). It has also been suggested that the maximal accumulated oxygen deficit can also be determined from a 60–90-second all-out test (i.e. maximal, variable intensity) (Gastin, 1994). The proposed advantage of using an all-out protocol is that the shorter duration of exhausting exercise limits the aerobic contribution to the energy supply. However, as the initial power output is well in excess of the maximal oxygen uptake, and varies throughout the test, the estimation of energy demand by the linear extrapolation method may be less valid.

Reliability

Surprisingly, very few studies have investigated the reliability of the MAOD method. However, Green and Dawson (1995) have reported good reliability of both the power–VO_2 regression (r = 0.93) and the O_2 demand at supra-maximal intensities (r = 0.90) for trained subjects. Jacobs, Bleue and Goodman (1997) have also reported good reliability of the MAOD method for moderately-trained subjects (r = 0.97), while Green and Dawson (1995) have reported poor reliability of the MAOD test in untrained subjects. Thus it appears that MAOD is a reliable test for moderately- to well-trained athletes, but that more familiarization sessions may be required if it is performed on individuals with less experience of all-out, supra-maximal exercise.

Validity

More than 90% of MAOD (see Section 3.3.2.1) has been attributed to the alactic and lactic capacities. Therefore, the MAOD and results from muscle biopsy should be highly correlated. Indeed, a number of studies have reported MAOD calculations to be quantitatively similar to estimates of anaerobic ATP production determined from muscle biopsies (Bangsbo *et al.*, 1990; Withers *et al.*, 1991). In contrast, Green *et al.* (1996) reported a non-significant correlation coefficient between anaerobic energy contribution (measured in ATP equivalents) and MAOD (r = –0.38) in 10 cyclists. It is possible that the group of subjects was too homogeneous to reveal a significant correlation between MAOD and anaerobic energy contributions. It should also be noted that this final study did not include the energy associated with lactate release into the blood in its calculations.

As described above, the determination of AC from changes in muscle metabolites has many associated assumptions. Nonetheless, the validity of the MAOD method is supported by reductions in MAOD with a reduction in muscle glycogen stores (Lacombe *et al.*, 1999), and the decrease in VO_{2max} but not MAOD when subjects inspire air with a reduced O_2 content (Medbo *et al.*, 1988). In addition, creatine supplementation has been reported to increase MAOD (Jacobs, Bleue and Goodman, 1997). It therefore appears that MAOD changes in parallel with changes in AC, but further research is required to determine how accurately it provides a quantitative measure of maximal AC.

Interpretation of AC scores

The magnitude of the maximal accumulated oxygen deficit appears to depend on the genetic make-up and anaerobic training status of an individual, as well as the method used to estimate MAOD (see above). It would appear however that little difference exists between the anaerobic capacities of untrained and endurance-trained subjects (MAOD = 50–65 ml O_2/kg). Sprint-trained individuals have significantly higher maximal accumulated oxygen deficits, often in excess of 75 ml O_2/kg. Women have been reported to have a 20–30% lower MAOD than men (Medbo and Burgers, 1990; Weber and Schneider, 2002).

Summary

While not without its critics, the MAOD method appears to provide a valid, quantitative measure of AC. Further research is required to test some of the assumptions required of this procedure and to establish firm testing guidelines. Based on available data, at least four 10-minute efforts, at intensities between 70 and 90% of VO_2 max, are required to accurately and reliably determine MAOD. Research to date would also suggest that it is best not to perform these four bouts during one continuous trial. The length of the constant-intensity supra-maximal tests should be approximately 2–3 minutes. While MAOD appears to be a reliable method when using trained subject, more familiarization sessions of both the sub-maximal and the supra-maximal exercise bouts may be required to achieve acceptable reliability with untrained subjects.

Critical power

Theoretical basis

Originally described by Monod and Scherrer (Craig and Morgan, 1998), there is a hyperbolic relationship between power output and time to exhaustion (Figure 3.3.7, left panel), which can be described by the following formula:

$$t = W'/(P - CP) \tag{3.3.1}$$

where

t = duration of exercise (s)

W = AWC (J)

P = exercise intensity (W)

CP = critical power (aerobic power; W).

Multiplication of the equation with the term (P-CP) results in:

$$P \cdot t = W' + CP \cdot t \tag{3.3.2}$$

This equation describes a linear relation between exercise duration and work done. The gradient of the straight line (CP) thus corresponds to the aerobic power, while the y-intercept corresponds with W′ (AWC; Figure 3.3.7, right panel).

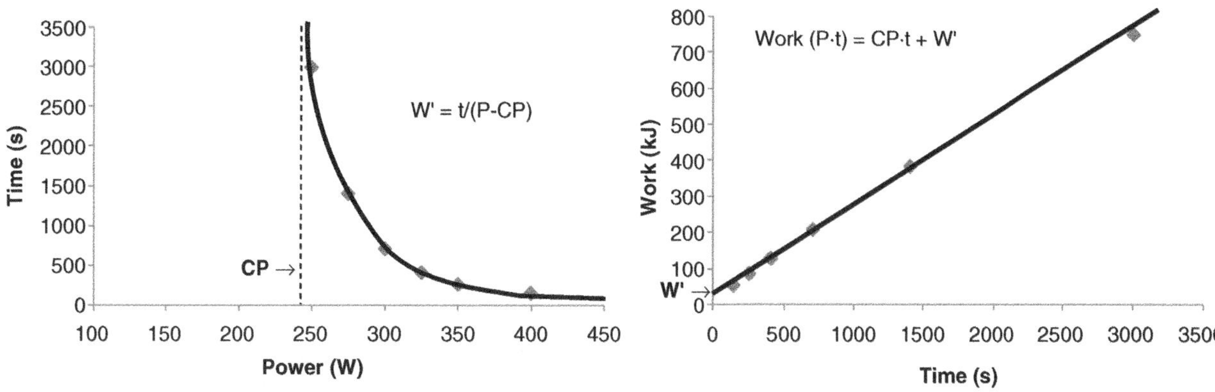

Figure 3.3.7 The critical power model describes the hyperbolic relationship between power and time to exhaustion (left panel). This relationship can be transformed to a linear relationship (right panel), where CP is the slope of the linear regression and W′ is the intercept with the y-axis

The critical power model is based on the following assumptions:

1. AC is constant and is completely used in every test.
2. Mechanical efficiency of muscular work is constant for the entire test duration.
3. Maximal aerobic power can be used completely from the beginning of exercise.

All three of these assumptions have been challenged by various authors. In particular, as shown before (Figure 3.3.5), there is a delayed increase in VO_2 at the beginning of exercise, which means that maximal aerobic power is never attained from the beginning of exercise.

Methodology

In order to estimate the parameters of the critical power function (i.e. CP and W′) it is necessary to complete at least two separate time-to-exhaustion trials. These may be performed for any exercise (e.g. cycling, running, swimming, rowing, kayak (Brickley and Doust, 1997; Clingeleffer, McNaughton and Davoren, 1994; Faff *et al.*, 1993; Hill, Steward and Lane, 1995; Housh *et al.*, 1992)), but must be performed at different but constant power outputs. Normally, at least three trials are performed so as to be able to construct the hyperbolic relationship (Figure 3.3.7, left panel). While it has been shown that critical power can be determined from two exercise trials (Clingeleffer, McNaughton and Davoren, 1994), it seems prudent to include at least three predictive trials so that one erroneous value does not have too great an influence on subsequent calculations. If subjects are unaccustomed to all-out exercise, one or two prior familiarization sessions are also recommended.

There is less agreement regarding the choice of predictive trials. A widely followed recommendation is that predictive tests should be designed to exhaust subjects within the 1–10 minute range (using power outputs ranging from approximately 120% to 95% of the power at VO_{2max}) (Poole, 1986). However, the use of short predictive trials (<5 min) has been shown to result in an underestimation of W′ and an overestimation of CP (Bishop, Jenkins and Howard, 1998). This is probably due to the delayed increase in VO_2 having a greater effect on the shorter than the longer predictive trials. The use of predictive trials which are too long (>15 min) may also result in poor estimation of the CP parameters due to the influence of motivation on time to exhaustion. Until further research is conducted, it is therefore recommended that three or four predictive trials designed to exhaust subjects within the 5–12 minute range be chosen. In addition, it is important to ensure that there is sufficient rest following predictive trials for subjects to completely recover prior to performing any subsequent trials (Bishop and Jenkins, 1995).

During each of the predictive tests, power and time (and hence work (power/time)) are the only variables that need to be recorded. Any one of six equations (including 3.3.1 and 3.3.2) can then be fitted, using any standard statistical curve-fitting software, to determine the critical power parameters (see Figure 3.3.8). While technically equivalent, in practice the choice of mathematical equation can have a small effect on the estimation of the critical power parameters (Gaesser *et al.*, 1995; Hill, 2004; Hill, Rose and Smith, 1993).

Reliability

In a review of studies that have investigated the reliability of critical power tests, it was reported that the mean coefficients of variation (CV) for the determination of the CP and the W′ were 4.3 and 9.2%, respectively (Hopkins, Schabort and Hawley, 2001). This indicates good reliability, although these CVs are 1.3–2.8 times larger than the CV of mean power for constant-work tests.

Validity

A major assumption of the critical power model is that CP is aerobic in nature, while W′ corresponds with the AC. Evidence

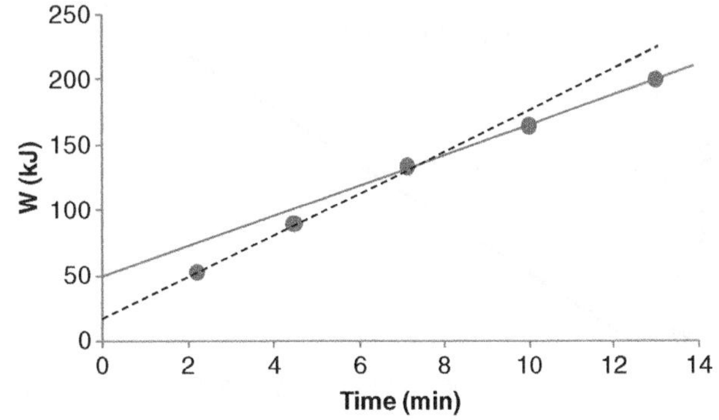

Figure 3.3.8 Relationship between duration of the time-to-exhaustion trials (min) and work completed, highlighting the effect of choice of test duration on the estimation of the critical power parameters (Bishop, Jenkins and Howard, 1998). Note that W′ is lower and CP is higher when the linear CP function is constructed using the three shortest trials (dashed line) than when the three longest trials are used (solid line)

supporting the aerobic nature of CP has been provided by its strong correlation with the ventilatory threshold (Moritani *et al.*, 1981) and observed increases in CP with predominately aerobic training (Gaesser and Wilson, 1988; Jenkins and Quigley, 1992). More importantly for this chapter, support for W′ being associated with the AC has been provided by its lack of response to ischaemia, hypoxia, and hyperoxia (Moritani *et al.*, 1981), its decrease in response to muscle glycogen depletion (Miura *et al.*, 2000), and its increase in response to both high-intensity training (Jenkins and Quigley, 1993) and creatine supplementation (Miura *et al.*, 1999).

Summary

The critical power model is based on the hyperbolic relationship between power output and time to exhaustion. While it has been shown that the critical power parameters (i.e. CP and W′) can be determined from two exhaustive trials, it is recommended that three or four predictive trials designed to exhaust the subject within the 5–12 minute range are used. It is important to ensure that there is sufficient rest between each of the predictive trials. The critical power test provides a reliable measure that reflects the AWC. However, its strong dependence on the protocol employed makes it less suitable, at this point in time, as a routine test to evaluate AC in athletes.

3.3.3 TESTING REPEATED-SPRINT ABILITY

3.3.3.1 Definition

The main feature of repeated-sprint exercise is the alternation of brief, maximal-intensity sprint bouts with periods of recovery (consisting of complete rest or moderate- to low-intensity activity). However, there is potential for confusion as some authors have also used the word ‘sprint’ to describe exercise lasting 30 seconds or more (Bogdanis *et al.*, 1996; Sharp *et al.*, 1986). For the purposes of this chapter, the definition of ‘sprint’ activity will be limited to brief exercise bouts, in general ≤10 seconds, in which peak intensity (power/velocity) can be maintained until the end of the entire exercise period. Longer-duration, maximal-intensity exercise in which there is a considerable decrease in performance will be referred to as all-out exercise (see Figure 2.8.1).

When sprints are repeated, it is also useful to define two different types of exercise: intermittent-sprint exercise and repeated-sprint exercise. Intermittent-sprint exercise can be characterized by short-duration sprints (≤10 s) interspersed with recovery periods long enough to allow near-complete recovery of sprint performance (>60 s) (Balsom *et al.*, 1992a). In comparison, repeated-sprint exercise is characterized by short-duration sprints (≤10 s) interspersed with brief recovery periods (usually ≤60 s). The main difference is that during intermittent-sprint exercise there is little or no performance decrement (Balsom *et al.*, 1992b; Bishop and Claudius, 2005), whereas there is a marked performance decrement in repeated-sprint exercise (Bishop *et al.*, 2004) (see Figure 2.9.2).

3.3.3.2 Assessment

Calculations

To quantify the amount of fatigue experienced during repeated-sprint exercise, researchers have tended to use one of two measures, the fatigue index (FI) or the sprint decrement (S_{dec}). The FI has generally been calculated as the drop-off in performance from the best to the worst performance during a set of repeated cycle sprints (3.3.3).

$$\mathrm{FI}(\mathrm{cycle}) = 100 \times \frac{(\mathrm{S1} - \mathrm{S}_{\mathrm{final}})}{S1} \quad (3.3.3)$$

S refers to sprint performance on a cycle ergometer and can be calculated for either work or power scores. Note that if these calculations are performed for running (where there is an *increase* in time as subjects fatigue) then S1 and S_{final} should be replaced with the slowest and fastest times, respectively (3.3.4).

$$\mathrm{FI}(\mathrm{running}) = 100 \times \frac{(\mathrm{S}_{\mathrm{slowest}} - \mathrm{S}_{\mathrm{fastest}})}{\mathrm{S}_{\mathrm{fastest}}} \quad (3.3.4)$$

Using the data in Figure 2.9.2, obtained from sprints performed on a cycle ergometer, the fatigue index would be calculated as:

$$\begin{aligned}\mathrm{FI} &= 100 \times (2000\,\mathrm{J} - 1700\,\mathrm{J})/2000\,\mathrm{J} \\ &= 15\%\end{aligned}$$

In comparison, the S_{dec} attempts to quantify fatigue by comparing actual performance with an imagined ‘ideal performance’ (where the best effort would be replicated in each sprint)

Table 3.3.1 Test protocols used to evaluate RSA for different team sports

Reference	Sport	RSA test		
		Sprint number	Sprint duration	Recovery between sprints
Castagna *et al.*, 2007	Basketball	10	15 m	30 s
Wadley and Le Rossignol, 1998	Australian rules football	12	20 m	~17 s
Spencer *et al.*, 2006b	Field hockey	6	30 m	~21 s
Rampinini *et al.*, 2007	Soccer	6	40 m	20 s

(Bishop *et al.*, 2001; Spencer *et al.*, 2006a, 2006b). A possible advantage of the S_{dec} is that it takes into consideration all sprints, is less influenced by a particularly good or bad first or last sprint.

$$S_{dec}(\%) = 1 - \frac{(S1 + S2 + S3 + \ldots + S_{final})}{S1 \times \text{number of sprints}} \times 100 \quad (3.3.5)$$

A slight modification of the formula is required for sprint running performance (as times will *increase* as subjects fatigue) (3.3.6):

$$S_{dec}(\%) = \frac{(S1 + S2 + S3 + \ldots + S_{final})}{S1 \times \text{number of sprints}} - 1 \times 100 \quad (3.3.6)$$

Using the same data from Figure 2.9.2, obtained from repeated sprints performed on a cycle ergometer, the S_{dec} would be calculated as:

$$S_{dec}(\%) = 1 - \frac{(2000 + 1900 + 1850 + 1800 + 1700)}{5 \times 200} \times 100$$
$$= 7.5\%$$

Methods

RSA is normally assessed by having athletes reproduce a number of all-out sprints separated by short recovery periods. Unlike other tests, where a fixed protocol is often recommended, it is difficult to provide a precise test prescription for tests of RSA. Ideally a repeated-sprint test should be based on a motion analysis of the sport to be evaluated (Table 3.3.1) and reflect the typical sprint and recovery durations required by the athlete. Altering the sprint duration (Balsom *et al.*, 1992b) or rest period between sprints (Balsom *et al.*, 1992a) will affect RSA. It is therefore important that once a test is chosen it does not change for subsequent testing. RSA may be tested using an ergometer (e.g. cycle) or a running protocol. In the case of running, it is preferable to use timing gates in order to more accurately record the time of each sprint.

While it is difficult to provide a precise test prescription for tests of RSA, a few general guidelines are important. To improve test reliability, as with most physiological tests, it is recommended that athletes don't perform any fatiguing training in the 24 hours prior to testing and that they do not ingest any food or liquids (except water) within two hours prior to testing.

A test of RSA should begin with a thorough warm-up. A typical warm-up consists of 10 minutes of moderate-intensity activity followed by three to five short-duration (i.e. 2–3 s) sprints. Following a short rest (~2 min), athletes should then perform a maximal single sprint ('criterion sprint'). While this single sprint is not included in many RSA test protocols, it is essential to ensure that the subsequent RSA test begins with a maximal sprint and that athletes do not pace themselves. Upon completion of the maximal single criterion sprint, subjects are required to rest for five minutes before starting the RSA test.

An RSA test consisting of the chosen sprint duration, sprint number, and rest duration between sprints is then begun. During the first sprint, subjects are required to achieve at least 95% of their criterion score, as a check on pacing. If 95% of the criterion score is not achieved, athletes are asked to rest for a further five minutes and then recommence the RSA test. If, however, 95% of the criterion score is achieved then athletes continue the RSA test until completion. In our experience, the threat of having to restart the test is sufficient motivation for nearly all athletes to achieve or exceed their criterion score on the first sprint.

The actual execution of the RSA test is relatively straightforward. If performed on an ergometer then five seconds before starting each sprint, athletes are asked to assume the ready position and await the start signal. If a running RSA test is performed then athletes commence each sprint a standard distance behind the start line (often 30–40 cm). This distance should be far enough to ensure that athletes don't break the start beam with their hands, and close enough to ensure that there is not a 'flying' start. It is also preferable to use a standardized starting position to minimize any variation in sprint start technique between subjects and between tests.

Following the completion of each sprint, subjects should decelerate to a walk and return to the start position. The type of recovery between sprints will affect performance (Spencer *et al.*, 2006a) and may be active or passive. An active recovery is often recommended as motion analysis of team sports has revealed that the recovery during repeated-sprint bouts is typically active (Spencer *et al.*, 2004). If an active recovery is used between sprints, it is important to control the pace of the recovery as accurately as possible (especially as the athletes

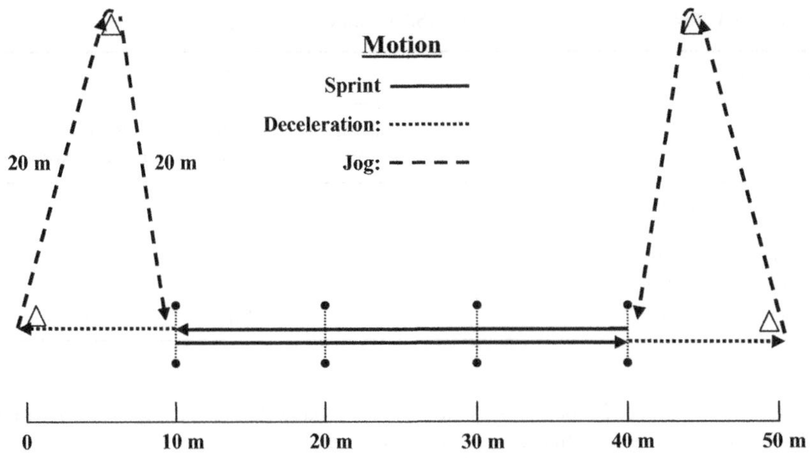

Figure 3.3.9 Schematic of the 6 × 30 m overground, repeated-sprint test with an active recovery (Spencer *et al.*, 2006b)

begin to fatigue). An example of a running RSA test protocol is shown in Figure 3.3.9.

Reliability of fatigue scores

Both single and total (or mean) sprint performance have been reported to have good reliability (e.g. CV < 4.0%) (Glaister *et al.*, 2008; McGawley and Bishop, 2006; Mendez-Villanueva, Bishop and Hamer, 2007; Oliver, 2009; Spencer *et al.*, 2006b; Wragg, Maxwell and Doust, 2000). In contrast, both fatigue indices are much less reliable, with CVs ranging from 11 to 50% (McGawley and Bishop, 2006; Oliver, 2009; Spencer *et al.*, 2006b). Consequently, several authors have concluded that any calculated fatigue scores should be viewed with caution (McGawley and Bishop, 2006) and that, where possible, changes in repeated-sprint performance should be discussed with respect to changes in mean or total sprint time (Oliver, 2009).

Interpretation of fatigue scores

The interpretation of RSA is complicated by the positive correlation between initial sprint performance and the various fatigue indices ($0.57 < r < 0.89$, $P < 0.05$) (Bishop, Lawrence and Spencer, 2003; Bishop and Spencer, 2004; Yanagiya *et al.*, 2003). That is, subjects with a better first sprint tend to experience greater fatigue during a repeated-sprint test. Thus, enhanced initial sprint performance may lead to greater fatigue during the latter stages of exercise, possibly due to a larger increase in the by-products of anaerobic metabolism, or other factors associated with the improved initial effort. In this instance, changes in mean or total work may not provide an accurate reflection of changes in repeated-sprint performance and it will be important to investigate changes in one of the fatigue indices discussed above.

Summary

The ability to recover and to reproduce performance in subsequent sprints is an important fitness requirement of team-sport athletes. RSA is normally assessed by having athletes reproduce a number of all-out sprints separated by short recovery periods. An active recovery is often recommended as motion analysis of team sports has revealed that the recovery during repeated-sprint bouts is typically active. Repeated-sprint performance should be discussed with respect to changes in mean or total sprint time, rather than the various fatigue indices, as these values have been shown to be more reliable.

3.3.4 CONCLUSION

While a subject's maximal aerobic power can quite easily and accurately be determined, the estimation of the maximal AC is much more difficult. Despite the difficulties and assumptions associated with quantifying the anaerobic energy release from changes in muscle metabolites, this procedure offers the only direct method of evaluating the anaerobic system. However, due to the limited practicality of such an invasive technique with elite athletes, indirect methods are also required. Two of the most common include MAOD and the critical power test. The MAOD method appears to provide a valid, quantitative measure of AC, although further standardization of the test methodology is required. The critical power test also appears to provide a measure that reflects the AC, but its strong dependence on the protocol employed makes it less suitable, at this point in time, as a routine test for the evaluation of AC in athletes. While anaerobic metabolism is also important for the performance of repeated sprints, AC is not well correlated with RSA and therefore specific tests to assess RSA have been devised. RSA is assessed by having athletes reproduce a number of all-out sprints separated by short recovery periods, with the exact protocol depending on the sport to be evaluated. It is recommended that the results of RSA tests be discussed with respect to changes in mean or total sprint time, rather than the various, less-reliable, fatigue indices.

References

Astrand P.O., Cuddy T.E., Saltin B. and Stenberg J. (1964) Cardiac output during submaximal and maximal work. *J Appl Physiol*, **19**, 268–274.

Balsom P.D., Seger J.Y., Sjodin B. and Ekblom B. (1992a) Maximal-intensity intermittent exercise: effect of recovery duration. *Int J Sports Med*, **13** (7), 528–533.

Balsom P.D., Seger J.Y., Sjodin B. and Ekblom B. (1992b) Physiological responses to maximal intensity intermittent exercise. *Eur J Appl Physiol*, **65**, 144–149.

Bangsbo J. (1998) Quantification of anaerobic energy production during intense exercise. *Med Sci Sports Exerc*, **30** (1), 47–52.

Bangsbo J., Gollnick P.D., Graham T.E. *et al.* (1990) Anaerobic energy production and O_2 deficit–debt relationship during exhaustive exercise in humans. *J Physiol (Lond)*, **422**, 539–559.

Bangsbo J., Krustrup P., Gonzalez-Alonso J. and Saltin B. (2001) ATP production and efficiency of human skeletal muscle during intense exercise: effect of previous exercise. *Am J Physiol Endocrinol Metab*, **280**, E956–E964.

Bangsbo J., Michalsik L. and Petersen A. (1993) Accumulated O_2 deficit during intense exercise and muscle characteristics of elite athletes. *Int J Sports Med*, **14** (4), 207–213.

Bishop D. (2000) Physiological predictors of flat-water kayak performance in women. *Eur J Appl Physiol*, **82** (1/2), 91–97.

Bishop D. and Claudius B. (2005) Effects of induced metabolic alkalosis on prolonged intermittent-sprint performance. *Med Sci Sports Exerc*, **37**, 759–767.

Bishop D. and Jenkins D.G. (1995) The influence of recovery duration between periods of exercise on the critical power function. *Eur J Appl Physiol*, **72**, 115–120.

Bishop D. and Spencer M. (2004) Determinants of repeated–sprint ability in well-trained team-sport athletes and endurance-trained athletes. *J Sports Med Phys Fitness*, **44**, 1–7.

Bishop D., Jenkins D.G. and Howard A. (1998) The critical power function is dependent on the durations of the predictive exercise tests chosen. *Int J Sports Med*, **19** (2), 125–129.

Bishop D., Spencer M., Duffield R. and Lawrence S. (2001) The validity of a repeated sprint ability test. *J Sci Med Sport*, **4** (1), 19–29.

Bishop D., Lawrence S. and Spencer M. (2003) Predictors of repeated-sprint ability in elite female hockey players. *J Sci Med Sport*, **6**, 199–209.

Bishop D., Edge J., Davis C. and Goodman C. (2004) Induced metabolic alkalosis affects muscle metabolism and repeated-sprint ability. *Med Sci Sports Exerc*, **36**, 807–813.

Bogdanis G.C., Nevill M.E., Boobis L.H. and Lakomy H.K.A. (1996) Contribution of phosphocreatine and aerobic metabolism to energy supply during repeated sprint exercise. *J Appl Physiol*, **80** (3), 876–884.

Brickley G. and Doust J. (1997) Protocol for the estimation of the critical power concept. *J Sports Sci*, **15**, 39.

Buck D. and McNaughton L. (1999a) Changing the number of submaximal exercise bouts effects calculation of MAOD. *Int J Sports Med*, **20**, 28–33.

Buck D. and McNaughton L. (1999b) Maximal accumulated oxygen deficit must be calculated using 10-min time periods. *Med Sci Sports Exerc*, **31**, 1346–1349.

Casaburi R., Storer T.W., Ben-Dov I. and Wasserman K. (1987) Effect of endurance training on possible determinants of VO_2 during heavy exercise. *J Appl Physiol*, **62** (1), 199–207.

Castagna C., Manzi V., D'Ottavio S. *et al.* (2007) Relation between maximal aerobic power and the ability to repeat sprints in young basketball players. *J Strength Cond Res*, **21**, 1172–1176.

Clingeleffer A., McNaughton L. and Davoren B. (1994) Critical power may be determined from two tests in elite kayakers. *Eur J Appl Physiol*, **68**, 36–40.

Craig I.S. and Morgan D.W. (1998) Relationship between 800-m running performance and accumulated oxygen deficit in middle-distance runners. *Med Sci Sports Exerc*, **30**, 1631–1636.

Craig N.P., Norton K.I., Bourdon P.C. *et al.* (1993) Aerobic and anaerobic indices contributing to track endurance cycling performance. *Eur J Appl Physiol*, **67**, 150–158.

Craig N.P., Norton K.I., Conyers R.A.J. *et al.* (1995) Influence of test duration and event specificity on maximal accumulated oxygen deficit of high performance track cyclists. *Int J Sports Med*, **16** (8), 534–540.

Faff J., Bienko A., Burkhard-Jagodzinska K. and Borkowski L. (1993) Diagnostic value of indices derived from the Critical Power Test in assessing the anaerobic work capacity of rowers. *Biol Sport*, **10**, 9–14.

Gaesser G.A. and Poole D.C. (1996) The slow component of oxygen uptake kinetics in humans. *Ex Sport Sci Rev*, **24**, 35–70.

Gaesser G.A. and Wilson L.A. (1988) Effects of continuous and interval training on the parameters of the power–endurance time relationship for high-intensity exercise. *Int J Sports Med*, **9**, 417–421.

Gaesser G.A., Carnevale T.J., Garfinkel A. *et al.* (1995) Estimation of critical power with nonlinear and linear models. *Med Sci Sports Exerc*, **27** (10), 1430–1438.

Gastin P. (1994) Quantification of anaerobic capacity. *Scand J Med Sci Sports*, **4**, 91–112.

Gastin P.B. and Lawson D.L. (1994) Influence of training status on maximal accumulated oxygen deficit during all-out cycle exercise. *Eur J Appl Physiol*, **69**, 321–330.

Gastin P.B., Costill D.L., Lawson D.L. *et al.* (1995) Accumulated oxygen deficit during supramaximal all-out and constant intensity exercise. *Med Sci Sports Exerc*, **27** (2), 255–263.

Glaister M., Howatson G., Pattison J.R. and McInnes G. (2008) The reliability and validity of fatigue measures during multiple-sprint work: an issue revisited. *J Strength Cond Res*, **22**, 1597–1601.

Green S. (1994) A definition and systems view of anaerobic capacity. *Eur J Appl Physiol*, **69**, 168–173.

Green S. and Dawson B.T. (1995) The oxygen uptake–power regression in cyclists and untrained men: implications for the accumulated O_2 deficit. *Eur J Appl Physiol*, **70**, 351–359.

Green S. and Dawson B.T. (1996) Methodological effects on the VO_2-power regression and the accumulated O_2 deficit. *Med Sci Sports Exerc*, **28** (3), 392–397.

Green S., Dawson B.T., Goodman C. and Carey M.F. (1996) Anaerobic ATP production and accumulated O_2 deficit in cyclists. *Med Sci Sports Exerc*, **28**, 315–321.

Heck H., Schulz H. and Bartmus U. (2003) Diagnostics of anaerobic power and capacity. *Eur J Sport Sci*, **3**, 1–23.

Hermansen L. and Medbo J. (1984) The relative significance of aerobic and anaerobic processes during maximal exercise of short duration. *Med Sci Sports Exerc*, **17**, 56–67.

Hermansen L. and Vaage O. (1977) Lactate disappearance and glycogen synthesis in human muscle after maximal exercise. *Am J Physiol*, **233** (5), E422–E429.

Hill D.W. (2004) The relationship between power and time to fatigue in cycle ergometer exercise. *Int J Sports Med*, **25**, 357–361.

Hill D.W., Rose L.E. and Smith J.C. (1993) Estimates of anaerobic capacity derived using different models of the power–time relationship. *Med Sci Sports Exerc*, **25** (5), S108 #606.

Hill D.W., Steward R.P. and Lane C.J. (1995) Application of the critical power concept to young swimmers. *Pediatr Exerc Sci*, **7**, 281–293.

Hopkins W.G., Schabort E.J. and Hawley J.A. (2001) Reliability of power in physical performance tests. *Sports Med*, **31**, 211–234.

Housh T.J., Johnson G.O., McDowell S.L. *et al.* (1992) The relationship between anaerobic running capacity and peak plasma lactate. *J Sports Med Phys Fitness*, **32**, 117–122.

Hultman E. and Sjoholm H. (1983) Energy metabolism and contraction force of human skeletal muscle in situ during electrical stimulation. *J Physiol*, **345**, 525–532.

Jacobs I., Bleue S. and Goodman J. (1997) Creatine ingestion increases anaerobic capacity and maximum accumulated oxygen debt. *Can J Appl Physiol*, **22** (3), 231–243.

Jenkins D.G. and Quigley B.M. (1992) Endurance training enhances critical power. *Med Sci Sports Exerc*, **24**, 1283–1289.

Jenkins D.G. and Quigley B.M. (1993) The influence of high-intensity exercise training on the W–T relationship. *Med Sci Sports Exerc*, **25**, 275–282.

Krogh A. and Lindhard J. (1920) The changes in respiration at the transition from work to rest. *J Physiol*, **53**, 431–439.

Lacombe V., Hinchcliff K.W., Geor R.J. and Lauderdale M.A. (1999) Exercise that induces substantial muscle glycogen depletion impairs subsequent anaerobic capacity. *Equine Vet J Suppl*, **30**, 293–297.

McGawley K. and Bishop D. (2006) Reliability of a 5 × 6-s maximal cycling repeated-sprint test in trained female team-sport athletes. *European Journal of Applied Physiology*, **98**, 383–393.

McGawley K. and Bishop D. (2008) Anaerobic and aerobic contribution to two, 5 × 6-s repeated-sprint bouts. *Coaching Sport Sci J*, **3**, 52.

Medbo J.I. (1996) Is the maximal accumulated oxygen deficit an adequate measure of the anaerobic capacity? *Can J Appl Physiol*, **21**, 370–383, discussion 384–8.

Medbo J.I. and Burgers S. (1990) Effect of training on the anaerobic capacity. *Med Sci Sports Exerc*, **22**, 501–507.

Medbo J.I., Mohn A.C., Tabata I. *et al.* (1988) Anaerobic capacity determined by maximal accumulated O_2 deficit. *J Appl Physiol*, **64**, 50–60.

Mendez-Villanueva A., Bishop D.J. and Hamer P. (2007) Reproducibility of a 6-s maximal cycling sprint test. *J Sci Med Sport*, **10**, 323–326.

Miura A., Kino F., Kajitani S. *et al.* (1999) The effect of oral creatine supplementation on the curvature constant parameter of the power–duration curve for cycle ergometry in humans. *Jpn J Physiol*, **49**, 169–174.

Miura A., Sato H., Sato H. *et al.* (2000) The effect of glycogen depletion on the curvature constant parameter of the power–duration curve for cycle ergometry. *Ergonomics*, **43**, 133–141.

Moritani T., Nagata A., deVries H.A. and Muro M. (1981) Critical power as a measure of physical work capacity and anaerobic threshold. *Ergonomics*, **24** (5), 339–350.

Olesen H.L., Raabo E., Bangsbo J. and Secher N.H. (1994) Maximal oxygen deficit of sprint and middle distance runners. *Eur J Appl Physiol Occup Physiol*, **69**, 140–146.

Oliver J.L. (2009) Is a fatigue index a worthwhile measure of repeated sprint ability? *J Sci Med Sport*, **12**, 20–23.

Poole D.C. (1986) Letter to the editor in chief. *Med Sci Sports Exerc*, **18**, 703–704.

Rampinini E., Bishop D., Marcora S.M. *et al.* (2007) Validity of simple field tests as indicators of match-related physical performance in top-level professional soccer players. *Int J Sports Med*, **28**, 228–235.

Renoux J.C., Petit B., Billat V. and Koralsztein J.P. (1999) Oxygen deficit is related to the exercise time to exhaustion at maximal aerobic speed in middle distance runners. *Arch Physiol Biochem*, **107**, 280–285.

Sahlin K., Harris R.C. and Hultman E. (1979) Resynthesis of creatine phosphate in human muscle after exercise in relation to intramuscular ph and availability of oxygen. *Scand J Clin Lab Invest*, **39**, 551–558.

Scott C.B., Roby F.B., Lohman T.G. and Bunt J.C. (1991) The maximally accumulated oxygen deficit as an indicator of anaerobic capacity. *Med Sci Sports Exerc*, **23**, 618–624.

Sharp R.L., Costill D.L., Fink W.J. and King D.S. (1986) Effects of eight weeks of bicycle ergometer sprint training on human muscle buffer capacity. *Int J Sports Med*, **7**, 13–17.

Spencer M., Bishop D., Dawson B. *et al.* (2006a) Metabolism and performance in repeated cycle sprints: active versus passive recovery. *Med Sci Sports Exerc*, **38**, 1492–1499.

Spencer M., Fitzsimons M., Dawson B. *et al.* (2006b) Reliability of a repeated-sprint test for field-hockey. *J Sci Med Sport*, **9**, 181–184.

Spencer M.R., Gastin P.B. and Payne W.R. (1997) Energy system contribution during 200 to 1500m running in elite athletes. The Australian Conference of Science and Medicine in Sport, 302–303.

Spencer M., Lawrence S., Rechichi C. *et al.* (2004) Time–motion analysis of elite field hockey, with special reference to repeated-sprint activity. *J Sports Sci*, **22**, 843–850.

Wadley G. and Le Rossignol P. (1998) The relationship between repeated sprint ability and the aerobic and anaerobic energy systems. *J Sci Med Sport*, **1** (2), 100–110.

Weber C.L. and Schneider D.A. (2002) Increases in maximal accumulated oxygen deficit after high-intensity interval training are not gender dependent. *J Appl Physiol*, **92**, 1795–1801.

Weyand P.G., Cureton K.J., Conley D.S. *et al.* (1994) Peak oxygen deficit predicts sprint and middle-distance track performance. *Med Sci Sports Exerc*, **26** (9), 1174–1180.

Withers R.T., Sherman W.M., Clark D.G. *et al.* (1991) Muscle metabolism during 30, 60 and 90 s of maximal cycling on an air-braked ergometer. *Eur J Appl Physiol*, **63**, 354–362.

Wragg C.B., Maxwell N.S. and Doust J.H. (2000) Evaluation of the reliability and validity of a soccer-specific field test of repeated sprint ability. *Eur J Appl Physiol*, **83**, 77–83.

Yanagiya T., Kanehisa H., Kouzaki M. *et al.* (2003) Effect of gender on mechanical power output during repeated bouts of maximal running in trained teenagers. *Int J Sports Med*, **24**, 304–310.

Zoladz J.A., Rademaker A.C. and Sargeant A.J. (1995) Non-linear relationship between O_2 uptake and power output at high intensities of exercise in humans. *J Physiol*, **488** (Pt 1), 211–217.

3.4 Cardiovascular Assessment and Aerobic Training Prescription

Andrew M. Jones and Fred J. DiMenna, School of Sport and Health Sciences, University of Exeter, Exeter, UK

3.4.1 INTRODUCTION

If you have an ailment and visit your doctor for treatment, he might suggest a course of action that involves ingestion of specific medications. On the surface, this issuance of a prescription appears relatively straightforward, but in reality it involves much more than just some illegible scribbling on a piece of paper and the deliverance of a bill. For example, before the physician is qualified to stipulate a pharmaceutical treatment, he must earn his credentials by studying the precise mechanisms by which each of the medications he can prescribe performs its function. Furthermore, even with a diploma certifying this knowledge displayed prominently on his wall, he must thoroughly assess each patient before deciding which of the drugs is appropriate in their particular case.

Aerobic exercise cannot be administered orally, transdermally, or via subcutaneous or intravenous injection. However, as an intervention, this type of treatment is every bit as powerful as any manmade pharmaceutical. Regular aerobic training causes numerous adaptations in cardiovascular function that are associated with an improved health profile. Furthermore, most athletic activities involve a significant contribution from aerobic energy transfer; therefore, aerobic training is required in a number of different sports. These powerful effects imply that just like a drug, aerobic exercise prescriptions must be calculated precisely by individuals with a sufficient knowledge base. The process of aerobic training prescription begins with cardiovascular assessment.

3.4.2 CARDIOVASCULAR ASSESSMENT

3.4.2.1 Health screening and risk stratification

The ACSM's Guidelines for Exercise Testing and Prescription (American College of Sports Medicine, 2009) state that it is important to provide an initial screening of prospective exercise participants relative to risk factors and/or symptoms for various cardiovascular, pulmonary, and metabolic diseases. The purpose of this screening is to identify individuals with medical contraindications to exercise, with clinically significant diseases that mandate a medically supervised programme, with increased risk for disease (due, for example, to age, symptoms, and/or risk factors), and with other special needs. This is necessary both for safety purposes and also to ensure that the programme that is prescribed is appropriate for the individual's specific circumstances. The Physical Activity Readiness Questionnaire (PAR-Q) is typically used as a minimal standard for this process.

Once the initial health screening is complete, the ACSM's Guidelines for Exercise Testing and Prescription (American College of Sports Medicine, 2009) suggest that prospective exercisers should be stratified based on the presence or absence of known cardiovascular, pulmonary, and/or metabolic disease, the presence or absence of signs and symptoms suggestive of cardiovascular, pulmonary, and/or metabolic disease, and the presence or absence of cardiovascular disease risk factors. This classification will affect both the specific aspects of the programme that is prescribed and also any preliminary testing performed to assess the participant's level of fitness. According to ACSM standards, initial risk stratification should delineate prospective exercisers as low, moderate, or high risk (American College of Sports Medicine, 2009). Decisions regarding the level of medical examination prior to testing and the need for medical supervision during testing are made according to this stratification.

3.4.2.2 Cardiorespiratory fitness and VO_{2max}

In the truest sense of the term, cardiovascular assessment involves the use of technologies such as electrocardiography, echocardiography, and radiography to examine the structure and function of the heart. These are important diagnostic tools, but in the typical exercise setting with apparently healthy

Strength and Conditioning – Biological Principles and Practical Applications Marco Cardinale, Rob Newton, and Kazunori Nosaka.

individuals, such examinations are often not warranted. In these cases, and specifically when prescribing exercise for apparently healthy competitive athletes, it is more useful to assess the integrated function of the pulmonary, cardiovascular, and neuromuscular systems by evaluating cardiorespiratory fitness. The criterion measure used to assess cardiorespiratory fitness is maximal oxygen uptake (VO_{2max}).

VO_{2max} represents the maximum rate at which oxygen can be consumed by a human at sea level. Consequently, VO_{2max} can be measured directly by maximal exercise testing. Furthermore, many submaximal testing protocols have been established to estimate VO_{2max}. According to the ACSM's Guidelines for Exercise Testing and Prescription (American College of Sports Medicine, 2009), the decision to use a maximal or submaximal test depends largely on the reasons for the assessment, the type of subject to be tested, and the availability of the appropriate equipment and personnel. Physician supervision is recommended for all (i.e. both submaximal and maximal) testing of high-risk subjects; however, for moderate-risk individuals, no supervision is required for submaximal tests. Low-risk subjects can be tested without physician supervision regardless of the type of test to be performed. Maximal exercise testing is typically not feasible for the vast majority of prospective exercisers; however, a direct measurement of VO_{2max} is beneficial for athletes. Numerous protocols have been used for both types of determination.

3.4.2.3 Submaximal exercise testing

Submaximal exercise testing to predict maximal cardiorespiratory capacity can be classified depending upon whether the testing protocol involves single or multiple stages and also upon whether VO_2 is measured directly via pulmonary gas exchange or estimated according to the work rate(s) being performed. Regardless of these distinctions, however, all submaximal tests base their estimation of VO_{2max} on the heart-rate response to submaximal exercise and a prediction of the maximal heart rate that can be achieved by the subject. This is problematic because maximal heart rate is highly variable between individuals and this inherent variability is not accounted for by age alone. Consequently, estimates of maximal heart rate derived from the age-based prediction formulae that are typically used (e.g. (220 − age) (Fox *et al.*, 1971) or (208 − (0.7 × age)) (Tanaka, Monahan and Seals, 2001)) exhibit relatively large standard deviations (e.g. ≥10 beats per minute; see Robergs and Landwehr 2002 for review). This variability is predominantly responsible for the error associated with the prediction of VO_{2max} from multistage submaximal incremental work-rate tests, which is typically 10–15%, but can be as high as 25% (Cooper and Storer, 2001).

One of the most frequently used submaximal exercise tests of VO_{2max} is the multistage YMCA cycle ergometer test (Golding, Myers and Sinning, 1989). This assessment requires subjects to cycle for three minutes at increasing work-rate stages with increments determined by heart-rate responses measured during the test. The test is terminated after four stages, or once 85% of the predicted maximal heart rate or 70% of the heart-rate reserve has been reached. During the test, a heart-rate value representative of the steady state of each stage is recorded, and once the test has been completed a plot of these values against work rate is made. This plot allows the maximal work rate to be predicted by extrapolation of the heart rate/work rate line of best fit, based upon the assumption that the relationship is linear. The VO_2 requirement of the predicted maximal work rate can then be determined using established regression equations (e.g. see American College of Sports Medicine, 2009, Table 7.2). The direct measurement of VO_2 during a similar test eliminates the need for the latter prediction as it allows for an actual heart-rate/VO_2 plot that can be extrapolated to the predicted maximal heart rate. This removes the error associated with prediction of the VO_2 cost of work (and the inherent flawed assumption of constant mechanical efficiency across subjects upon which that prediction is based); however, the error associated with the estimation of maximal heart rate remains.

There are a number of advantages associated with submaximal exercise testing (Cooper and Storer, 2001). For example, these tests are safer, require less effort to be expended by the subject, and are less time-consuming. Furthermore, unlike maximal exercise testing, physician supervision is not required for moderate-risk individuals and the results are less dependent on the motivation level of the subject. Finally, in addition to predicting the maximal work rate, the heart rate/work rate relationship derived from these tests can be used to monitor progress once the exercise programme has commenced, since a reduced heart rate at the same submaximal work rate is an important indicator of improved cardiorespiratory fitness. However, in addition to the questionable assumptions upon which VO_{2max} predictions are based, there are a number of other disadvantages with submaximal tests of functional capacity. For example, they only provide an estimate, as opposed to an actual measurement, of maximal aerobic capacity and therefore they do not afford any information regarding abnormal responses (e.g. angina pectoris) that might occur at work rates above the termination point. As a consequence of this latter limitation, for safety purposes, exercise prescriptions for intensity that are formulated according to the results of these tests should not exceed the highest work rate achieved on the test (Cooper and Storer, 2001). A maximal exercise test also provides information about other parameters of aerobic function that can be of particular use for prescribing and monitoring the training of athletes.

3.4.2.4 Maximal exercise testing

Step/ramp incremental exercise testing

In the early part of the 20th century, A.V. Hill and colleagues reported that during running, there was a specific rate of oxygen uptake beyond which no further increase could occur, regardless of how much running speed was increased. They termed this the maximal oxygen intake (Hill and Lupton, 1923). This concept suggests that during an incremental exercise test to exhaustion (i.e. a maximal graded exercise test), a plot of VO_2 against work rate will reveal a VO_2 plateau prior to exhaustion that reflects this VO_{2max}. However, Hill also recognized the absence of a plateau in some subjects (hence, the distinction of VO_{2peak}, as opposed to VO_{2max}) and to this day this phenomenon continues to provoke debate regarding both the criteria for validating the attainment of VO_{2max} (Poole, Wilkerson and Jones, 2008) and the physiological significance of the parameter (e.g. Noakes, 1988; see Hale, 2008 for review). Nevertheless, maximal exercise tests to determine VO_{2max}, either directly via concurrent gas exchange analysis or indirectly as a function of the predicted VO_2 at the maximum work rate achieved, have become the gold standard for assessing cardiorespiratory fitness.

The initial work of Hill and other research teams involved the determination of VO_{2max} via pulmonary gas exchange data collected during a discontinuous series of constant work-rate tests performed over several days or even weeks (Hale, 2008). The time between bouts was subsequently reduced to minutes (Mitchell, Sproule and Chapman, 1958), and eventually continuous maximal exercise tests became the most popular method for establishing key parameters of aerobic function. In addition to VO_{2max}, these include exercise economy and the lactate threshold (Whipp *et al.*, 1981). Continuous maximal exercise tests typically involve the work rate (cycle ergometer) or speed/incline (treadmill) being increased systematically as a function of time until the subject is unable to continue. Theoretically, at this limit of tolerance, a sufficiently motivated subject will have achieved their mode-specific VO_{2max}, as well as their maximal heart rate and cardiac output (Poole, Wilkerson and Jones, 2008). Further research suggested that during continuous incremental exercise to exhaustion, a work-rate increment that brought a subject to their limit of tolerance in approximately 10 minutes was optimal for evaluating cardiopulmonary function (Buchfuhrer *et al.*, 1983). The increments forming such a protocol can be applied in either a step (i.e. with discrete stages) or a ramp (i.e. continuously progressing) fashion. An example of the latter that is typically employed is leg cycle ergometry with a 1 W increase in work rate every two seconds (i.e. an increase of 30W/min) until volitional exhaustion. Figure 3.4.1 depicts a VO_2 profile in response to such a protocol for which VO_{2max} is clearly indicated by a definitive plateau.

Exercise economy

Exercise economy is the oxygen uptake required at a given absolute work rate (Jones and Carter, 2000). This parameter is influenced by anthropometric, physiological, biomechanical, and technical factors (Pate *et al.*, 1992), and there is considerable inter-individual variability in exercise economy, even in individuals with similar VO_{2max} values or performance capabilities (Conley and Krahenbuhl, 1980; Coyle *et al.*, 1991; Morgan and Craib, 1992). However, there is evidence to suggest that for some physical activities (e.g. running) trained athletes possess better exercise economy, which is advantageous for submaximal exercise (i.e. endurance) performance (Morgan *et al.*, 1995). Furthermore, exercise economy might explain why relatively low VO_{2max} values have been observed in some elite endurance athletes (Londeree, 1986; Morgan *et al.*, 1995). During an incremental maximal exercise test that brings a subject to exhaustion within approximately 10 minutes, the slope of the regression line describing the VO_2–work rate relationship is an index of the exercise economy for that form of exercise (e.g. see Figure 3.4.1).

The lactate/gas-exchange threshold

The lactate threshold (LT) is the metabolic rate at which blood lactate concentration begins to increase above the baseline

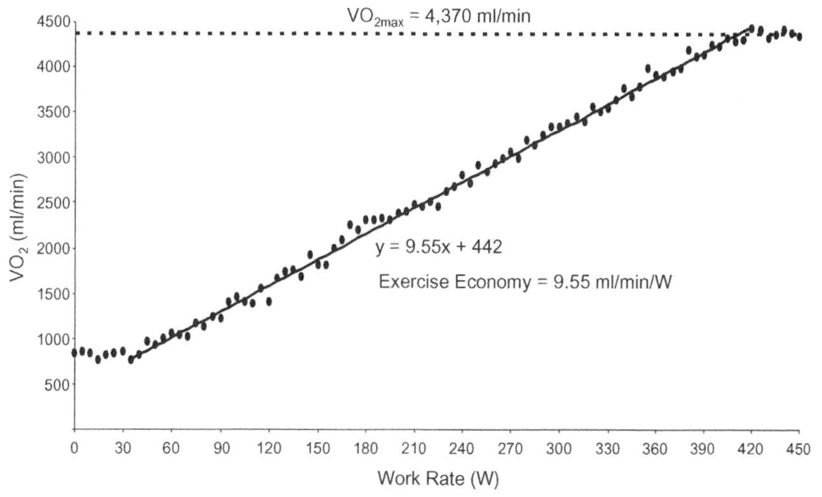

Figure 3.4.1 A typical VO_2 profile in response to a 1 W increase in work rate every two seconds (i.e. an increase of 30W/min) until volitional exhaustion during leg cycle ergometer exercise. The definitive VO_2 plateau prior to test termination confirms that a mode-specific VO_{2max} has been attained. This data also reveals exercise economy, which can be quantified as the slope of the regression line describing the VO_2–work rate relationship during the ascending portion of the response (in ml/min/W)

value of approximately 1.0 mM during incremental exercise (Wasserman *et al.*, 1973). This parameter is routinely determined in laboratory-based physiological assessments of endurance capacity. For example, as subjects perform a maximal incremental test involving multiple 3–5 minute stages of increasing intensity, blood samples can be taken upon completion of each stage and a plot will reveal the stage at which the increase of blood lactate above baseline occurs (see Figure 3.4.2). The LT can also be estimated non-invasively at the mouth during incremental or ramp exercise by determination of the gas-exchange threshold (GET) because lactic acidosis requires bicarbonate buffering that results in the generation of non-metabolic CO_2 and a consequent disproportionate increase in VCO_2 and ventilation. The GET is often determined from a cluster of measures, including: (1) the first disproportionate increase in VCO_2 from visual inspection of individual plots of VCO_2 versus VO_2 (i.e. the V-slope method; Beaver, Wasserman and Whipp, 1986); (2) an increase in V_E/VO_2 (V_E = expiratory ventilation) with no increase in V_E/VCO_2; and (3) an increase in end-tidal oxygen tension with no fall in end-tidal carbon dioxide tension. Figure 3.4.3 depicts a V-slope plot and identification of GET for the ramp incremental cycle ergometer maximal test depicted in Figure 3.4.1.

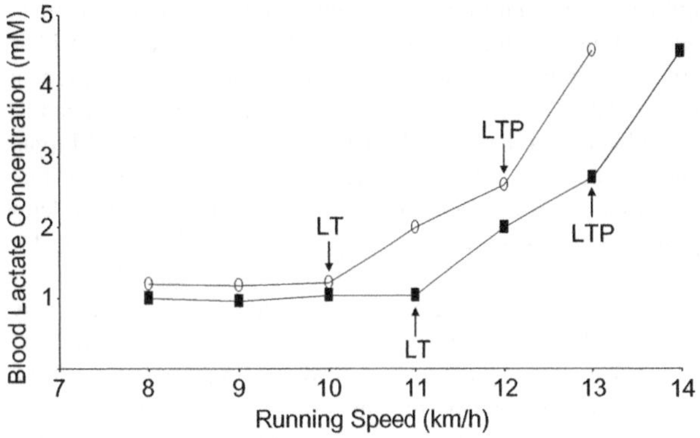

Figure 3.4.2 Plot of blood lactate concentration measured from blood samples taken upon completion of each stage during a maximal incremental treadmill test involving multiple 3–5 minute stages of increasing speed prior to initiation of an aerobic exercise training programme (open circles). From left to right, the two clear breakpoints indicate the lactate threshold (LT) and lactate turnpoint (LTP), respectively. After training (closed squares), a reduction in the degree of lactate accumulation for each running speed is apparent. This shifts both LT and LTP to the right (i.e. to a faster running speed)

The LT/GET is a powerful predictor of endurance performance (Farrell *et al.*, 1979; Tanaka *et al.*, 1983) that shifts to a higher power output (cycling) or speed (running) as a chronic adaptation to endurance training (e.g. see Figure 3.4.2) (Carter, Jones and Doust, 1999; Jones and Carter, 2000). This is an important physiological alteration because exercise above LT/GET is associated with a nonlinear increase in metabolic, respiratory, and perceptual stress and more rapid fatigue (Katch *et al.*, 1978; Sahlin, 1992). Furthermore, for constant work-rate exercise, the LT/GET divides exercise intensity domains within which a VO_2 slow component is not present (below LT/GET; i.e. the moderate-intensity domain) or is present (above LT/GET; i.e. the heavy- and severe-intensity domains). The VO_2 slow component reflects an 'excess' VO_2 that either elevates the steady-state requirement above that which would be predicted for the work rate and delays steady-state attainment (heavy exercise) or causes an inexorable rise that results in the achievement of VO_{2max} (severe exercise) (Poole *et al.*, 1988; Whipp and Wasserman, 1972). Consequently, the VO_2 slow component is also responsible for greater metabolic perturbation above LT/GET.

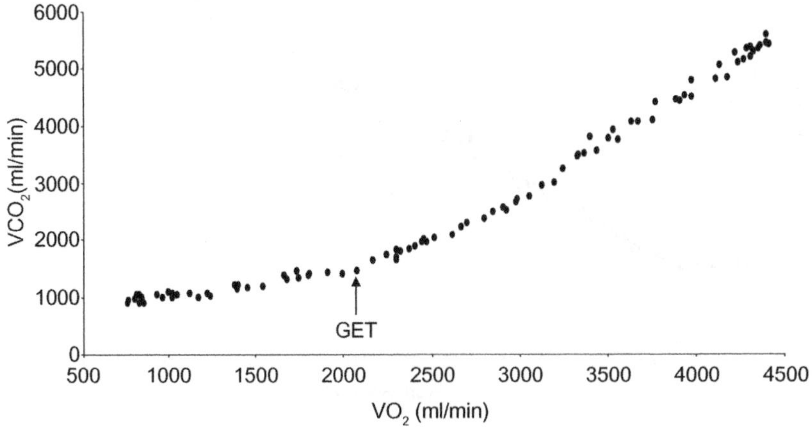

Figure 3.4.3 A typical plot of VCO_2 versus VO_2 (i.e. the V-slope method) depicting a response to a 1 W increase in work rate every two seconds until volitional exhaustion during leg cycle ergometer exercise. The clear breakpoint indicates the gas exchange threshold (GET) that approximates LT

Lactate turnpoint/maximal lactate steady state

During maximal incremental exercise, a second threshold beyond which the profile of lactate accumulation in the blood once again changes is generally observed (Davis *et al.*, 1983). This 'lactate turnpoint' (LTP) typically occurs at around 2.5–4.0 mM (e.g. see Figure 3.4.2) and, compared with LT, may provide a more robust parameter for assessing the endurance performance capacity of highly-trained subjects (Ribeiro *et al.*, 1986). For example, it has been suggested that LTP provides an estimate of the highest work rate that allows blood lactate to stabilize during prolonged constant-intensity exercise (i.e. the maximal lactate steady state, MLSS) (Aunola and Rusko, 1992; Smith and Jones, 2001). MLSS also represents the highest work rate that can be performed within the heavy-intensity domain (i.e. the highest work rate where a steady-state VO_2 can still be achieved); therefore, it designates the maximal pace that can be sustained during endurance exercise of up to approximately 60 minutes, and reflects an important predictor of endurance-exercise performance (Jones and Doust, 1998).

Despite its obvious utility for competitive athletes, MLSS testing is typically not suitable for routine diagnostic use because it requires subjects to perform a series of prolonged constant work-rate exercise bouts on separate days, with multiple blood samples drawn during each test, and this is both time-consuming and fraught with interpretational ambiguity (e.g. establishing a criterion for what constitutes blood lactate 'stabilization' when inherent measurement variability is present). However, if a single maximal incremental test involves stage durations that are sufficiently long and intensity increments that are sufficiently small to allow measured blood lactate concentrations to precisely reflect the accumulation associated with each stage, LTP can be relatively easily determined (e.g. see Figure 3.4.2) and used to provide a reasonable estimate of the MLSS work rate (Kilding and Jones, 2006). Furthermore, it has been suggested that LTP can be non-invasively determined by identifying the 'second ventilatory threshold' (i.e. the work rate during incremental exercise at which the onset of hyperventilation relative to VCO_2 occurs, which has been termed the respiratory compensation threshold, RCP) during fast incremental or ramp maximal exercise. However, confirmation of the hypothesized coincidence between RCP and MLSS has, thus far, proven elusive (e.g. see Laplaud *et al.*, 2006; Simon *et al.*, 1983).

'Maximal' constant-load exercise testing

LT/GET typically occurs at 45–60% VO_{2max}, but can be as low as 35% VO_{2max} in patients and as high as 80% VO_{2max} in athletes (Jones and Poole, 2005). The delineation point between the heavy and severe domains (for cycling, the point typically termed the 'critical power' or CP; see for example Monod and Scherrer, 1965; Moritani *et al.*, 1981), equivalent to the maximal lactate steady state, usually occurs approximately halfway between LT and VO_{2max} (Jones and Poole, 2005; Poole *et al.*, 1988); therefore, in general, a range of constant-intensity work rates from ~75 to 100% VO_{2max} should all be considered 'maximal' because if they are maintained until exhaustion, VO_{2max} will be attained. This means that it is possible to use constant work-rate severe exercise to determine VO_{2max}, which might be of particular use in confirming VO_{2max} when an initial maximal incremental exercise test has not revealed a discernible VO_2 plateau at exhaustion (Poole, Wilkerson and Jones, 2008). In these circumstances, secondary criteria are typically employed to provide confirmation (e.g. end-exercise respiratory exchange ratio ≥1.10 or 1.15; end-exercise heart rate = predicted maximum +/– 10 beats/minute; end-exercise blood lactate concentration ≥8 mM); however, these have been shown to underestimate VO_{2max} by as much as 27% (Poole, Wilkerson and Jones, 2008). Conversely, performance of a constant-intensity test to exhaustion at a work rate above the highest achieved on the incremental test will successfully confirm that the end-exercise VO_2 is the mode-specific maximal value that can be achieved, so long as the work rate during the constant-intensity bout can be sustained long enough for VO_2 to achieve this value before exhaustion occurs (e.g. at a work rate at or below ~170% CP; see Hill, Poole and Smith, 2002).

CP testing

The use of any of the aforementioned parameters of aerobic function to assess cardiorespiratory fitness is based upon the assumption that the measurement of a physiological variable during the response to exercise in some way reflects functional capacity. Conversely, the measurement of CP involves a direct determination of the ability to perform mechanical work at a level that might, from a practical standpoint, be considered the 'usable' portion of the metabolic reserve (i.e. the highest intensity that can be sustained for long periods without fatigue; see Monod and Scherrer, 1965; Moritani *et al.*, 1981). Consequently, it can be argued that CP provides a better indicator of the functional capacity of humans. However, determination of CP requires that subjects perform multiple (e.g. 4–6) constant-intensity exercise bouts to exhaustion on separate days to establish their power/time hyperbola and associated asymptote that represents CP. Heretofore, this has precluded the routine use of CP as a measurement of cardiorespiratory function. However, it has recently been confirmed that a single, three-minute all-out cycling test against a fixed resistance can be used to accurately predict CP by expending the finite capacity for work above CP prior to end-exercise (Vanhatalo, Doust and Burnley, 2007, 2008). Interestingly, it has also been shown that the VO_{2peak} measured during this all-out test is not different from that determined from a conventional ramp test, which suggests that this test can be used to assess both traditional (VO_{2max}) and practical (CP) measures of functional capacity (Burnley, Doust and Vanhatalo, 2006). However, determination of the fixed resistance against which subjects must cycle requires an initial maximal ramp test to determine GET and VO_{2max}; therefore, this test would not assuage the need for conventional VO_{2max} testing. A similar all-out testing protocol to determine CP and VO_{2max} during other forms of exercise (specifically, critical velocity during running, which would have the most widespread applicability) has yet to be established. Future research will likely address these limitations.

VO_2 kinetics testing

During most athletic endeavours, VO_{2max} is rarely attained, and it is even less likely for this metabolic rate to be encountered during typical activities of daily living. Furthermore, other than when sleeping or being completely immobile, humans are rarely in a metabolic steady state. However, the vast majority of sporting events and daily activities comprise countless transitions across a wide range of metabolic requirements and when muscular work changes in this manner, VO_2 must also adapt to support aerobic muscle metabolism. It has been shown that the VO_2 response to an increased work rate can be quite rapid, especially in endurance-trained athletes (Jones and Koppo, 2005); however, regardless of conditioning level or the magnitude of the increase that is required, the new requisite level (i.e. the steady-state VO_2 requirement for the work rate) cannot be achieved instantaneously. Therefore, every transition to a higher metabolic rate is characterized by an 'oxygen deficit', the magnitude of which is an important determinant of exercise tolerance. For example, a small oxygen deficit can be accounted for quite easily by hydrolysis of intramuscular phosphocreatine reserves and utilization of oxygen stores. Consequently, little metabolic disruption will result. However, a large deficit will exhaust these finite reserves and necessitate a significant contribution from the incomplete breakdown of glycogen (i.e. anaerobic glycolysis), which results in hydrogen/lactate-ion formation. This can result in metabolic disruption due to both substrate depletion and metabolite accumulation, especially when transitions that engender large oxygen deficits are made repeatedly in proximity (e.g. during start-and-stop athletic activities such as football, basketball, and rugby). For a given VO_2 increment, the size of the oxygen deficit is determined by the rapidity with which VO_2 adapts (e.g. see Figure 3.4.4); therefore, given the importance of the magnitude of the oxygen deficit during metabolic transitions, it could be argued that the 'truest' assessment of pulmonary, cardiovascular, and neuromuscular integrated function would consider the time course of the VO_2 response to exercise (i.e. VO_2 kinetics).

VO_2 kinetics testing involves determination of the rate at which VO_2 adapts when a transition is made between constant work rates. This requires breath-by-breath VO_2 measurements that provide the temporal resolution necessary to precisely characterize the response, and also mandates data collection for multiple repetitions of the same protocol, because breath-to-breath variability is present in the pulmonary signal. Ensemble averaging of like transitions allows this inherent 'noise' to be smoothed (Whipp *et al.*, 1982). Oxygen-uptake kinetics researchers typically assess the VO_2 time course during these step transitions by determining the time constant (τ) that describes the exponential VO_2 change. The time constant represents the time taken for 63% of the response to reach fruition (i.e. $\tau = 1.44 \times$ the response half-time).

VO_2 kinetics is generally faster in fitter humans (Jones and Koppo, 2005) and a correlation between τ and VO_{2max} has been reported over a very broad range of absolute and mass-specific VO_{2max} values in species with extremely diverse aerobic capacities (e.g. the lungless salamander and the thoroughbred racehorse) (Kindig, Behnke and Poole, 2005). Furthermore, the speeding of VO_2 kinetics that occurs after endurance training in humans is associated with an increased VO_{2max} (Jones and

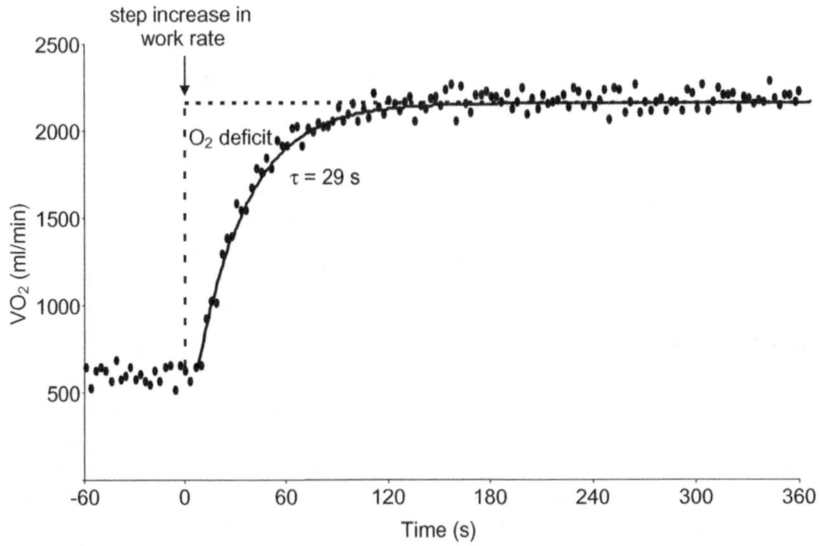

Figure 3.4.4 In response to a step increase in muscular work, VO_2 does not immediately achieve the requisite (steady-state) level. This creates an oxygen deficit that can profoundly influence exercise tolerance. The rapidity with which VO_2 adapts is an important indicator of 'functional fitness' and this speed can be quantified by determining the time constant (τ) that describes the exponential response

Koppo, 2005). This suggests that in addition to its functional significance, the determination of τ during VO_2 kinetics testing might also provide a correlate measure of the widely recognized criterion index of cardiorespiratory fitness (VO_{2max}) that can be obtained without either maximal tests, exhaustive 'submaximal' tests, or non-exhaustive submaximal tests based on potentially erroneous assumptions. Endurance training also results in faster *overall* VO_2 kinetics during exercise above LT by decreasing the amplitude of the VO_2 slow component (Casaburi *et al.*, 1987; Poole, Ward and Whipp, 1990; Womack *et al.*, 1995).

3.4.3 AEROBIC TRAINING PRESCRIPTION

Once health screening, risk stratification, and cardiorespiratory assessment have been completed, an aerobic exercise programme that is both safe and effective can be prescribed. Specific aspects of this programme should be customized according to the requirements of the aspiring exerciser, with consideration given to the subject's age, gender, physical fitness/health status, prior exercise history, and long-term objectives. There are two important principles upon which all training regimes should be based: specificity and overload. To ensure that the requirements associated with each of these principles are satisfied, training variables including mode, frequency, duration, and intensity must be prescribed appropriately and the concepts of progression and reversibility must be understood.

3.4.3.1 Specificity and aerobic overload

In 1951, Canadian endocrinologist Hans Selye reported that living organisms display a basic reaction pattern in response to stress (Selye, 1951). Selye termed this response the general adaptation syndrome (GAS) and although he formulated his views based upon what happened when he injected mice with damaging substances, the basic model he forwarded also applies to exercise training. GAS states that when stressors are introduced to a living organism, the organism experiences three response stages: alarm, resistance, and exhaustion. When applied to exercise training, this means that a training overload must be sufficient to provoke alarm (i.e. in excess of that to which the organism is accustomed), but not so intense that exhaustion occurs (i.e. not in excess of that to which the organism can adequately adapt). The principle of training specificity suggests that the structural and functional adaptations that can be expected to result from repeat application of an appropriate overload (i.e. the increased resistance displayed by the organism) will be unique and highly dependent on the particular stressor that has been encountered. For aerobic exercise prescription, this implies that only exercise that overloads the integrated capacities of the pulmonary, cardiovascular, and neuromuscular systems to facilitate aerobic energy transfer in the exercising musculature will be appropriate.

3.4.3.2 Mode

Generally speaking, aerobic (endurance) exercise includes all activities that require rhythmic contractions of a significant portion of the body's larger muscle groups at a sustainable percentage of the maximal voluntary contraction. Traditional activities that fit this description include walking, jogging, running, cycling, rowing, swimming, and cross-country skiing. Furthermore, there are many alternative forms of exercise that are typically done in the health club setting which also satisfy the principle of specificity with respect to an aerobic overload. These include treadmill walking/jogging/running, stationary leg cycling, arm cranking, stair stepping, elliptical stepping, aerobic dance, and spinning.

As is apparent, there are many different exercise modalities to choose from when prescribing aerobic exercise. Factors that should be considered when making a selection include subject preference, geographic location (e.g. climate, urban/rural, topography), equipment availability (e.g. public health club proximity, private home gym feasibility), economic concerns, and any orthopaedic limitations that might be present. Furthermore, certain aerobic activities require skilful execution that precludes their use by some subjects. For example, cross-country skiing (either outdoors or on an exercise apparatus designed to simulate the movement) is difficult to master and therefore is not suitable for everyone. Similarly, jogging is inappropriate for many, simply because it cannot be done at a low work rate; even subjects who express a desire to jog and have no orthopaedic contraindications to that form of exercise might not be able to do so until they develop sufficient aerobic capacity to sustain the requisite pace. For these individuals, fast-paced walking might provide an adequate stimulus overload upon initiation of the programme, and once aerobic capacity improves, sessions that involve walk/jog intervals can be carried out to bridge the gap between continuous full sessions of walking and jogging.

In addition to choosing exercise that is aerobic in nature, the principle of specificity also indicates that athletes in training for a particular competition should devote most of their training time to the performance of the specific mode(s) of exercise that their sport entails. However, 'cross-training', utilizing different aerobic exercises, can also be useful for these individuals, especially in the off-season and during periods of high-volume training, when orthopaedic stress is considerable. Cross-training is also beneficial for non-specific aerobic training (i.e. training to improve general aerobic fitness) because it can help foster programme adherence by reducing boredom and because it lowers the potential for overuse injuries that result from chronic repetitive physical stress (e.g. tendonitis).

3.4.3.3 Frequency

For many years, exercise physiologists have been trying to precisely define the optimal dose–response relationship for aerobic exercise (i.e. the volume of training that will yield maximal benefit with minimal risk of injury) (e.g. see Pollock *et al.*, 1977). With respect to the minimal stimulus required to invoke a training effect, when VO_{2max} is used to indicate aerobic capacity, it has been shown that frequencies of as low as two training sessions per week can result in improvements in less-fit subjects. When a greater initial fitness level is present (e.g. $VO_{2max} > 50$ ml/kg/min), at least three weekly sessions are required (Wenger and Bell, 1986). At the opposite end of the spectrum, the 1998 ACSM position stand suggests that minimal if any additional benefit can be gleaned from more than five sessions per week, while the incidence of injury increases disproportionately when exercise is performed this often (American College of Sports Medicine, 1998). Consequently, this position stand recommends three to five days per week of aerobic training to improve cardiorespiratory fitness. However, it is now apparent that frequency should not be considered as a standalone variable because the *total volume* of physical activity represents an important determinant of the training response. This volume reflects both duration and intensity, in addition to frequency.

3.4.3.4 Total volume

The influence of the total volume of training on the aerobic training effect means that once a minimal threshold has been exceeded for duration and intensity, the interaction between frequency, duration, and intensity has to be considered when prescribing these variables. This has been accounted for in the most recent position stand issued by the American College of Sports Medicine in conjunction with the American Heart Association (Haskell *et al.*, 2007). The revised recommendation for frequency of training suggests that to promote and maintain health, all healthy adults aged 18–65 require moderate-intensity aerobic activity for a minimum of 30 minutes five days a week, or vigorous-intensity aerobic activity for a minimum of 20 minutes three days a week. This position stand also states that moderate- and vigorous-intensity training sessions can be combined (e.g. two days of moderate-intensity exercise and two days of vigorous-intensity exercise) to meet these requirements. Other research has shown that the placement of aerobic training sessions during the week (e.g. three weekly sessions on consecutive compared to non-consecutive days) does not influence the improvements in VO_{2max} that will occur (Moffatt, Stamford and Neill, 1977); consequently, non-consecutive-day formats that provide for rest days interspersed between training sessions are typically preferred.

The 2007 recommendations from the ACSM and AHA also clarified that with respect to duration, the time commitment suggested (i.e. 30 and 20 minutes for moderate- and vigorous-intensity exercise, respectively) does not have to be accumulated in one continuous session. However, the position stand does define a minimum duration (10 minutes) for each bout of exercise making up the daily total. The revised document also emphasizes that the physical activity guidelines are based on the minimal amount of activity recommended to achieve substantial health benefits over and above the routine light-intensity activities of daily living, and that larger amounts of physical activity (including more activity at higher intensities) provide additional health benefits, but the nature of the relationship (amount compared to benefit) likely varies with health outcome. Consequently, the optimal dose–response relationship has yet to be determined (Haskell *et al.*, 2007).

3.4.3.5 Intensity

Intensity is perhaps the most challenging training variable to prescribe; it is also quite likely the most important. The 1998 ACSM position stand cites evidence that implicates total energy expenditure during the aerobic training session as the critical determinant of aerobic fitness development (American College of Sports Medicine, 1998). This suggests that once a minimum intensity threshold is exceeded, the aerobic training effect will be similar for lower- compared to higher-intensity activities so long as they are performed for long enough to engender a similar caloric outlay. The position stand recommends a minimum intensity of 55–65% of the maximum heart rate (40–50% VO_2 reserve) and suggests that this intensity is appropriate for most subjects because a high proportion of the adult population is both sedentary and has at least one risk factor for cardiovascular disease. Therefore, the ACSM recommends longer duration sessions at a moderate intensity to establish the requisite amount of energy expenditure. The position stand suggests that the upper limit for aerobic training is 90% of the maximum heart rate (85% VO_2 reserve).

3.4.3.6 Total volume vs volume per unit time

Unlike the original position stand, the revised version by the ACSM and AHA in 2007 recognizes that 'vigorous activity' (defined as >6 METS; i.e. activities like jogging that are characterized by rapid breathing and a substantial increase in heart rate) can also be an integral part of the physical activity recommendation. Furthermore, recent research indicates that in previously sedentary men, high-intensity aerobic training (cycling at 80% VO_{2max}) is *more effective* in increasing VO_{2max} compared with moderate-intensity exercise (60% VO_{2max}) at the same energy cost (three 400 kcal aerobic training sessions per week) (O'Donovan *et al.*, 2005). This contradicts the 1998 ACSM recommendation (American College of Sports Medicine, 1998) because it means that changes in cardiorespiratory fitness are not independent of aerobic exercise intensity (i.e. dependent exclusively on overall energy expenditure), which suggests that it is the total volume of activity per unit time as opposed to the total volume per se that is important. This is reflected in the most recent ACSM Guidelines for Exercise Testing and Prescription (American College of Sports Medicine, 2009), which suggests a minimum intensity threshold for health/fitness benefits for most people of 40 to <60% VO_2 reserve (i.e. 'moderate exercise'), but also recognizes a 'positive continuum of health/fitness benefits that exists with increasing exercise intensity' (American College of Sports Medicine, 2009). Furthermore, a combination of moderate and 'vigorous' (i.e. ≥60% VO_2 reserve) exercise is suggested to represent the ideal formula for the attainment of improvements in health/fitness in most adults.

3.4.3.7 Progression

In addition to improving the 'resistance' (which, in the case of aerobic function, is equivalent to the functional capacity) of the organism, GAS also suggests that the same stressor that initially elicited the alarm response will not instigate a similar reaction once resistance has increased. This is essential for survival of the species, but complicates exercise prescription because it means that adjustments in the training programme must be routinely made in order to facilitate further change. Progression can be achieved by increasing any of the aforementioned training variables (i.e. frequency, duration, and/or intensity); however, the ACSM's Guidelines for Exercise Testing and Prescription (American College of Sports Medicine, 2009) recommend increasing only duration during the initial phase of the programme (e.g. an increase of 5–10 minutes per session every 1–2 weeks over the first 4–6 weeks). On the other hand, the relative importance of exercise intensity for improving cardiorespiratory fitness (see above) suggests that this variable might provide the most viable means for progressing the aerobic overload stimulus for long-term training.

3.4.3.8 Reversibility

As far as exercise training is concerned, the antithesis of progression is reversibility. This concept implies that positive adaptations that have occurred due to repeat application of an exercise stimulus overload (i.e. the improved resistance of the organism) will be lost if a minimal volume of training is not maintained. Generally speaking, this volume is considerably less than that which is required to elicit further improvement; however, the effect of complete removal of physical activity on cardiorespiratory function is profound. For example, in a follow-up to the landmark study by Saltin *et al.* (1968), the five subjects who were evaluated at baseline and after three weeks of bed rest 30 years previously experienced less deterioration in cardiovascular and physical work capacity from 30 years of ageing than they did due to the forced inactivity that was imposed during the initial investigation (McGuire *et al.*, 2001). In that classic study, a 27% decrease in treadmill VO_{2max} and an approximately equal reduction of maximal cardiac output were observed (Saltin *et al.*, 1968). Of course, in that investigation, the supine posture associated with complete bed rest created orthostatic intolerance that removal of the cardiorespiratory training stimulus under 'normal circumstances' (e.g. cessation of training, but maintenance of activities of daily living) would not. However, it has also been shown that after a 10–12 month standardized training regimen, aerobically-conditioned subjects (VO_{2max}, 62 ml/kg/min; range, 53.1–75.5) who stopped training but were not restricted to bed rest and still performed the minimal physical activity required by their sedentary jobs experienced a 7% decrease in VO_{2max}, an 11% decrease in maximal stroke volume, and an 8% decrease in maximal cardiac output during the first two to three weeks of detraining, and a further decline in VO_{2max} of 9% from week three to week eight (Coyle *et al.*, 1984). Further research showed that the decreased stroke volume observed in these circumstances was largely due to a reduction in circulating blood volume (Coyle, Hemmert and Coggan, 1986). Furthermore, a 40% decrease in mitochondrial enzyme levels has also been reported after eight weeks of the cessation of endurance training, and this change was associated with a reduced maximal arterio-venous oxygen difference, which indicates that both central and peripheral factors are responsible for the loss of cardiorespiratory capacity due to detraining (Coyle *et al.*, 1984).

3.4.3.9 Parameter-specific cardiorespiratory enhancement

Generally speaking, enhanced cardiorespiratory fitness due to aerobic training should be reflected in improvements in each of the parameters of aerobic function. However, the capacity for aerobic energy metabolism in the cell and the systemic factors that are influenced by it both depend on the integrated function of multiple body systems, which could present unique limitations under different circumstances. Consequently, the principle of training specificity can also be applied to enhancement of the specific cardiorespiratory physiological parameters that have previously been mentioned (e.g. VO_{2max}, exercise economy, LT/LTP, τ/VO_2 slow component). This is of particular importance to athletes, whose competitive performance might be predominantly associated with one of these variables.

3.4.3.10 VO_{2max} improvement

Most studies show some increase of VO_{2max} due to endurance training; therefore, the optimal frequency, duration, and intensity of training designed specifically to improve this parameter are difficult to determine. However, a review that grouped many studies on different populations with different protocols revealed that maximal gains were elicited with exercise sessions of 35–45 minutes performed four times per week at intensities of 90–100% VO_{2max} (Wenger and Bell, 1986). It also appears that most of the improvement in VO_{2max} during long-term training programmes occurs quite early during the regimen. For example, during a nine-week endurance training programme divided into two constant-intensity phases with an increase in intensity after week four, VO_{2max} increased during both phases for the first three weeks and remained constant thereafter (Hickson *et al.*, 1981). These results also indicated a response half-time of less than 11 days, which exemplifies the rapid but fleeting nature of the capacity to improve VO_{2max}.

3.4.3.11 Exercise economy improvement

Compared to improvements in VO_{2max}, it appears that enhancement of exercise economy has a much longer time course. For example, a case study of an elite female distance runner revealed significant improvements with each year of training over a five-year period (Jones, 1998). This protracted response time might explain why investigations that have examined the effects of endurance training on exercise economy have produced equivocal findings (e.g. Conley *et al.*, 1984; Lake and Cavanagh, 1996; Overend, Paterson and Cunningham, 1992; Wilcox and Bulbulian, 1984). For example, these studies have typically involved 6–12 weeks of training, which might not be sufficiently long to produce measurable improvement, especially in subjects who are already trained (Jones and Carter, 2000). There are examples in the literature of improved exercise economy from short-term training. For example, six weeks of endurance training decreased VO_2 at a running speed of 12.0 km/hour from approximately 39 ml/kg/min to 36 ml/kg/min in recreationally active individuals (Jones, Carter and Doust, 1999), and an approximate 3% improvement in running economy was observed in recreational runners after six weeks of exhaustive distance or long-interval training (Franch *et al.*, 1998). Interestingly, short-interval training in the same study did not elicit any effect.

Endurance training has the potential to improve exercise economy in a number of ways. These include improved muscle oxidative capacity and associated changes in motor unit recruitment (Coyle *et al.*, 1992), reductions in ventilation and heart rate at the same submaximal exercise intensity (Franch *et al.*, 1998), and improved technique (Williams and Cavanagh, 1987). Generally speaking, athletes' most economical velocities or power outputs tend to be those at which they habitually train. In conjunction with the longer time course for adaptation, this might indicate that good exercise economy is a function of the total volume of endurance training performed. It has also been suggested that increasing maximal strength by supplementing endurance training with resistance training might improve this parameter by reducing the tension development required for each muscular contraction involved in the exercise (e.g. for each pedal thrust during cycling) (Hickson *et al.*, 1988). According to the well-established size principle of motor-unit recruitment (Henneman and Mendell, 1981), this would imply a reduced drive through the motor-unit recruitment hierarchy and less high-order (e.g. type II) fibre involvement for each contraction. In mouse/rat muscle, there is evidence to suggest that type II fibres are less energetically efficient (i.e. yield less ATP per volume oxygen consumed and/or require more ATP per contractile tension developed) (Crow and Kushmerick, 1982; Reggiani *et al.*, 1997; Wendt and Gibbs, 1973), and in humans the percentage of type I muscle fibres has been shown to be positively correlated with efficiency during both cycling and a novel exercise task (two-legged knee-extension exercise) in highly endurance-trained cyclists (Coyle *et al.*, 1992). This is why a reduced drive through the recruitment hierarchy should improve efficiency.

3.4.3.12 LT/LTP improvement

During exercise, blood lactate concentration increases above baseline when lactate appearance exceeds clearance. Consequently, blood lactate concentration ultimately reflects the dynamic balance between formation via glycolysis and removal via oxidation and gluconeogenesis. Endurance training is associated with a reduction in the degree of lactate accumulation for any given absolute or relative exercise intensity (see Figure 3.4.2) that is primarily attributable to improved clearance (Donovan and Brooks, 1983). Generally speaking, it appears that sustained 'tempo' endurance exercise at and above LTP might provide the ideal exercise stimulus to improve the body's ability to clear lactate. This would reduce its appearance at any submaximal work rate and, therefore, shift each specific breakpoint to a higher exercise work rate.

3.4.3.13 VO_2 kinetics improvement

Endurance training improves (i.e. accelerates) VO_2 kinetics by reducing τ, which describes the initial (i.e. phase II) VO_2 response (Berger *et al.*, 2006; Norris and Petersen, 1998), and/or by reducing the amplitude of the VO_2 slow component (Berger *et al.*, 2006; Carter *et al.*, 2000; Casaburi *et al.*, 1987; Womack *et al.*, 1995). While the slow-component phenomenon is still not completely understood, there is ample evidence to suggest that phase II VO_2 kinetics is controlled by mechanisms within active muscle mitochondria (i.e. an intrinsic slowness of oxidative metabolism to adjust to the new level of ATP turnover), at least under 'normal' circumstances where oxygen delivery is not substantially impaired (see Poole *et al.*, 2008 for a thorough review of this topic). Furthermore, the shortest phase II τ values are typically found in longer-distance specialists (Jones and Koppo, 2005), which suggests that long endurance training methods that increase muscle mitochondrial mass might be most useful for accelerating VO_2 kinetics. Total training volume as the critical overload stimulus for this adaptation is also supported by the observation that six weeks of low-intensity continuous cycle training is as effective in reducing τ and decreasing the amplitude of the VO_2 slow component as six weeks of high-intensity interval training (Berger *et al.*, 2006). However, more recent research indicates that for shorter training programmes (six training sessions performed over a two-week period), all-out 30-second cycle sprint interval training significantly improves VO_2 kinetics, while work-matched continuous moderate-intensity training does not (Bailey *et al.*, 2009). Consequently, volume per unit time, as opposed to volume per se, appears to play a role, at least during the earlier stages of training. It is also interesting to note that the faster phase II VO_2 response observed in this investigation was associated with faster deoxyhemoglobin kinetics, which suggests that muscle fractional oxygen extraction was enhanced by training (Bailey *et al.*, 2009). This confirms that changes to intracellular mechanisms that influence the rapidity with which available oxygen can be used are responsible for accelerating VO_2 kinetics in the trained state, at least during the phase II portion of the response. More research is clearly needed to elucidate the best training method(s) for accelerating the phase II and overall VO_2 response to exercise.

3.4.3.14 Prescribing exercise for competitive athletes

Given that physiological testing of athletes is so widespread, it is perhaps surprising that relatively little is known about the optimal training that might be prescribed to maximize improvements in the physiological parameters alluded to above (Midgley, McNaughton and Jones, 2007). This is related, in part, to the difficulty in recruiting already well-trained athletes for participation in studies that involve manipulation of their training. However, as reviewed by Midgley, McNaughton and Jones (2007), it remains possible for physiologists to use knowledge of the physiological responses within the various exercise-intensity domains to furnish coaches with practical advice about 'training zones' that can be used in the construction of exercise programmes. For example, for a distance runner, continuous training in the moderate domain (i.e. below his or her LT) which requires a relatively low fraction of the maximal heart rate and VO_2 can be sustained for long periods without fatigue. For this reason, easy, recovery runs of 20–40 minutes and long, steady runs of 60–120 minutes are typically performed in this domain. By extension, training in the heavy domain (between the LT and the LTP), while still submaximal, is physiologically more taxing. In this domain, the VO_2 slow component delays the attainment of steady state, blood lactate concentration is elevated, but eventually stabilizes (at perhaps 2–5 mM), and there is a greater rate of muscle glycogen utilization. The bulk of a distance runner's training, comprising steady but high-quality continuous running for periods of 30–90 minutes, will likely be performed in this domain. It has been speculated that such training might, over the longer term, enhance running economy and stimulate a right-shift of the LT and LTP to higher intensities (e.g. see Figure 3.4.2) (Jones, 2007; Midgley, McNaughton and Jones, 2007).

In contrast, to exercise in the moderate and heavy domains, exercise duration in the severe domain (≥LTP) will be compromised by a continuous rise in VO_2 towards its maximum and a continuous reduction in muscle energy charge (Jones *et al.*, 2008). While physically stressful, continuous exercise in this domain will require sustained high rates of pulmonary gas exchange and ventilation, which closely simulates the demands of shorter-distance competition. 'Tempo' sessions such as these are used regularly by the most successful endurance athletes and have been suggested to provide the stimulus for adaptations that collectively reduce the rate of blood lactate accumulation during high-intensity exercise (Jones, 2007; Midgley, McNaughton and Jones, 2007). Finally, with the assumption that VO_{2max} is limited by muscle O_2 delivery during exercise that engages more than approximately one-third of an individual's total muscle mass (Saltin and Strange, 1992; Wagner, 2000), training sessions which require the athlete to attain close to maximal heart rates and to sustain them for reasonable periods of time might be expected to facilitate improvements in VO_{2max} (Jones, 2007; Midgley, McNaughton and Jones, 2007). Interval training is most commonly used for this purpose.

Given the specific training zones mentioned above, measurements of the heart rates and running speeds that correspond to LT, LTP, and VO_{2max} during a physiological test can assist coaches in designing training sessions/programmes which provide the necessary stimuli to evoke physiological adaptations (e.g. increases in mitochondrial volume, capillary density, maximal cardiac output, etc.) that impact upon the relevant parameters dictating performance. However, prescribing the appropriate total volume and relative proportion of training to be performed in the easy, steady, tempo, and interval zones remains at least as much an art as it is a science. Therefore, more research is needed to determine the precise aspects of exercise prescription that are specific for competitive athletes.

3.4.4 CONCLUSION

Aerobic exercise training is a potent intervention for improving both general cardiovascular health and specific aspects of athletic performance. However, optimal return on the investment of time and effort in exercise can only be realised if the training programme is formulated according to the well-established principles of specificity and overload. Satisfying the requirements of these tenets mandates an initial comprehensive evaluation of the prospective exerciser, which should include health screening, risk assessment, and some form(s) of cardiorespiratory and/or cardiovascular testing. Health screening and risk assessment are necessary to ensure that all subsequent testing and training can be done with minimal risk of untoward incident. Furthermore, this evaluative process provides important information about the health and fitness status of the exercise aspirant that must be considered in conjunction with the results of testing when formulating the aerobic exercise prescription. Once all initial evaluations have been completed, prescription involves precise specification of training programme variables including the mode, frequency, duration, and intensity of exercise. In addition, the malleability of the aerobically trained state must be recognized by respecting the concepts of progression and reversibility. This is essential because exercise prescription is a dynamic process that must be ongoing in order for the changes that occur as the programme progresses to be duly accounted for.

References

American College of Sports Medicine (1998) The recommended quality and quantity of exercise for developing and maintaining cardiorespiratory and muscular fitness, and flexibility in adults. *Med Sci Sports Exerc*, **30**, 975–991.

American College of Sports Medicine (2009) *ACSM's Guidelines for Exercise Testing and Prescription*, 8th edn, Lippincott Williams and Wilkins, Philadelphia, PA.

Aunola S. and Rusko H. (1992) Does anaerobic threshold correlate with maximal lactate steady-state. *J Sports Sci*, **10**, 309–323.

Bailey S.J., Wilkerson D.P., DiMenna F.J. and Jones A.M. (2009) Influence of repeated sprint training on pulmonary O_2 uptake and muscle deoxygenation kinetics in humans. *J Appl Physiol*, **106**, 1875–1887.

Beaver W.L., Wasserman K. and Whipp B.J. (1986) A new method for detecting anaerobic threshold by gas exchange. *J Appl Physiol*, **60**, 2020–2027.

Berger N.J.A., Tolfrey K., Williams A.G. and Jones A.M. (2006) Influence of continuous and interval training on oxygen uptake on-kinetics. *Med Sci Sports Exerc*, **38**, 504–512.

Buchfuhrer M.J., Hansen J.E., Robinson T.E. *et al.* (1983) Optimizing the exercise protocol for cardiopulmonary assessment. *J Appl Physiol*, **55**, 1558–1564.

Burnley M., Doust J.H. and Vanhatalo A. (2006) A 3-min all-out test to determine peak oxygen uptake and the maximal steady state. *Med Sci Sports Exerc*, **38**, 1995–2003.

Carter H., Jones A.M. and Doust J.H. (1999) Effect of six weeks of endurance training on the lactate minimum speed. *J Sport Sci*, **17**, 957–967.

Carter H., Jones A.M., Barstow T.J. *et al.* (2000) Effect of endurance training on oxygen uptake kinetics during treadmill running. *J Appl Physiol*, **89**, 1744–1752.

Casaburi R., Storer T.W., Ben-Dov I. and Wasserman K. (1987) Effect of endurance training on possible determinants of VO_2 during heavy exercise. *J Appl Physiol*, **62**, 199–207.

Conley D.L. and Krahenbuhl G. (1980) Running economy and distance running performance of highly trained athletes. *Med Sci Sports*, **12**, 357–360.

Conley D.L., Krahenbuhl G.S., Burkett L.N. and Millar A.L. (1984) Following Steve Scott: physiological changes accompanying training. *Physician Sports Med*, **12**, 103–106.

Cooper C.B. and Storer T.W. (2001) *Exercise Testing and Interpretation: A Practical Approach*, Cambridge University Press, Cambridge.

Coyle E.F., Martin W.H., III, Sinacore D.R. *et al.* (1984) Time course of loss of adaptations after stopping prolonged intense endurance training. *J Appl Physiol*, **57**, 1857–1864.

Coyle E.F., Hemmert M.K. and Coggan A.R. (1986) Effects of detraining on cardiovascular responses to exercise: role of blood volume. *J Appl Physiol*, **60**, 95–99.

Coyle E.F., Feltner M.E., Kautz S.A. *et al.* (1991) Physiological and biomechanical factors associated with elite endurance cycling performance. *Med Sci Sports Exerc*, **23**, 93–107.

Coyle E.F., Sidossis L.S., Horowitz J.F. and Beltz J.D. (1992) Cycling efficiency is related to the percentage of type I muscle fibers. *Med Sci Sports Exerc*, **24**, 782–788.

Crow M.T. and Kushmerick M.J. (1982) Chemical energetics of slow- and fast-twitch muscles of the mouse. *J Gen Physiol*, **79**, 147–166.

Davis H., Bassett J., Hughes P. and Gass G. (1983) Anaerobic threshold and lactate turnpoint. *Eur J Appl Physiol*, **50**, 383–392.

Donovan C.M. and Brooks G.A. (1983) Endurance training affects lactate clearance, not lactate production. *Am J Physiol Endocrinol Metab*, **244**, E83–E92.

Farrell P.A., Wilmore J.H., Coyle E.F. *et al.* (1979) Plasma lactate accumulation and distance running performance. *Med Sci Sports Exerc*, **11**, 338–344.

Fox S.M., III, Naughton J.P. and Haskell W.L. (1971) Physical activity and the prevention of coronary heart disease. *Ann Clin Res*, **3**, 404–432.

Franch J., Madsen K., Djurhuus M.S. and Pedersen P.K. (1998) Improved running economy following intensified training correlates with reduced ventilatory demands. *Med Sci Sports Exerc*, **30**, 1250–1256.

Golding L., Myers C. and Sinning W.E. (1989) *Y's Way to Physical Fitness*, 3rd edn, Human Kinetics, Champaign, IL.

Hale T. (2008) History of developments in sport and exercise physiology: A.V. Hill, maximal oxygen uptake, and oxygen debt. *J Sport Sci*, **26**, 365–400.

Haskell W.L., Lee I.M., Pate R.R. *et al.* (2007) Physical activity and public health: updated recommendation for adults from the American College of Sports Medicine and the American Heart Association. *Med Sci Sports Exerc*, **39**, 1423–1434.

Henneman E. and Mendell L.M. (1981) Functional organisation of motoneuron pool and its inputs, in *Handbook of Physiology I* II (ed. V.B. Brooks), American Physiological Society, Bethesda, MA., pp. 423–507.

Hickson R.C., Hagberg J.M., Ehsani A.A. and Holloszy J.O. (1981) Time course of the adaptive responses of aerobic power and heart rate to training. *Med Sci Sports Exerc*, **13**, 17–20.

Hickson R.C., Dvorak B.A., Gorostiaga E.M. *et al.* (1988) Potential for strength and endurance training to amplify endurance performance. *J Appl Physiol*, **65**, 2285–2290.

Hill A.V. and Lupton H. (1923) Muscular exercise, lactic acid, and the supply and utilization of oxygen. *Q J Med*, **16**, 135–171.

Hill D.W., Poole D.C. and Smith J.C. (2002) The relationship between power and the time to achieve VO2 max. *Med Sci Sports Exerc*, **34**, 709–714.

Jones A.M. (1998) A 5-year physiological case study of an Olympic runner. *Br J Sports Med*, **32**, 39–43.

Jones A.M. (2007) Middle and long distance running, in *Sport and Exercise Science Testing Guidelines: The British Association of Sport and Exercise Sciences Guide* (eds E.M. Winter, A.M. Jones, R.R.C. Davison *et al.*), Routledge, London and New York, pp. 147–154.

Jones A.M. and Carter H. (2000) The effect of endurance training on parameters of aerobic fitness. *Sports Med*, **29**, 373–386.

Jones A.M. and Doust J.H. (1998) The validity of the lactate minimum test for determination of the maximal lactate steady state. *Med Sci Sports Exerc*, **30**, 1304–1313.

Jones A.M. and Koppo K. (2005) Effect of training on VO_2 kinetics and performance, in *Oxygen Uptake Kinetics in Sport, Exercise and Medicine* (eds A.M. Jones and D.C. Poole), Routledge, London and New York, pp. 373–398.

Jones A.M. and Poole D.C. (2005) Introduction to oxygen uptake kinetics and historical development of the discipline, in *Oxygen Uptake Kinetics in Sport, Exercise and Medicine* (eds A.M. Jones and D.C. Poole), Routledge, London and New York, pp. 3–35.

Jones A.M., Carter H. and Doust J.H. (1999) Effect of six weeks of endurance training on parameters of aerobic fitness. *Med Sci Sports Exerc*, **31**, S280.

Jones A.M., Wilkerson D.P., DiMenna F. *et al.* (2008) Muscle metabolic responses to exercise above and below the 'critical power' assessed using 31P-MRS. *Am J Physiol Regul Integr Comp Physiol*, **294**, R585–R593.

Katch V., Weltman A., Sady S. and Freedson P. (1978) Validity of the relative percent concept for equating training intensity. *Eur J Appl Physiol*, **39**, 219–227.

Kilding A.E. and Jones A.M. (2006) Validity of a single-visit protocol to estimate the maximum lactate steady state. *Med Sci Sports Exerc*, **39**, 219–227.

Kindig C.A., Behnke B.J. and Poole D.C. (2005) VO_2 dynamics in different species, in *Oxygen Uptake Kinetics in Sport, Exercise and Medicine* (eds A.M. Jones and D.C. Poole), Routledge, London and New York, pp. 115–137.

Lake M. and Cavanagh P. (1996) Six weeks of training does not change running mechanics or improve running economy. *Med Sci Sports Exerc*, **28**, 860–869.

Laplaud D., Guinot M., Favre-Juvin A. and Flore P. (2006) Maximal lactate steady state determination with a single incremental test exercise. *Eur J Appl Physiol*, **96**, 446–452.

Londeree B.R. (1986) The use of laboratory test results with long distance runners. *Sports Med*, **3**, 201–213.

McGuire D.K., Levine B.D., Williamson J.W. *et al.* (2001) A 30-year follow-up of the Dallas bed rest and training study: I. Effect of age on the cardiovascular response to exercise. *Circulation*, **104**, 1350–1357.

Midgley A.W., McNaughton L.R. and Jones A.M. (2007) Training to enhance the physiological determinants of long-distance running performance: can valid recommendations be given to runners and coaches based on current scientific knowledge? *Sports Med*, **37**, 857–880.

Mitchell J.H., Sproule B.J. and Chapman C.B. (1958) The physiological meaning of the maximal oxygen uptake test. *J Clin Invest*, **37**, 538–547.

Moffatt R.J., Stamford B.A. and Neill R.D. (1977) Placement of tri-weekly training sessions: importance regarding enhancement of aerobic capacity. *Res Q*, **48**, 583–591.

Monod H. and Scherrer J. (1965) The work capacity of a synergic muscle group. *Ergonomics*, **8**, 329–338.

Morgan D. and Craib M. (1992) Physiological aspects of running economy. *Med Sci Sports Exerc*, **24**, 456–461.

Morgan D.W., Bransford D.R., Costill D.L. *et al.* (1995) Variation in the aerobic demand of running among trained and untrained subjects. *Med Sci Sports Exerc*, **27**, 404–409.

Moritani T., Nagata A., de Vries H.A. and Muro M. (1981) Critical power as a measure of critical work capacity and anaerobic threshold. *Ergonomics*, **24**, 339–350.

Noakes T.D. (1988) Implications of exercise testing for prediction of athletic performance: a contemporary perspective. *Med Sci Sports Exerc*, **20**, 319–330.

Norris S.R. and Petersen S.R. (1998) Effects of endurance training on transient oxygen uptake responses in cyclists. *J Sports Sci*, **16**, 733–738.

O'Donovan G., Owen A., Bird S.R. *et al.* (2005) Changes in cardiorespiratory fitness and coronary heart disease risk factors following 24 wk of moderate- or high-intensity exercise of equal energy cost. *J Appl Physiol*, **98**, 1619–1625.

Overend T.J., Paterson D.H. and Cunningham D.A. (1992) The effect of interval and continuous training on the aerobic parameters. *Can J Sport Sci*, **17**, 129–134.

Pate R.R., Macera C.A., Bailey S.P. *et al.* (1992) Physiological, anthropometric, and training correlates of running economy. *Med Sci Sports Exerc*, **24**, 1128–1133.

Pollock M.L., Gettman R., Milesis C.A. *et al.* (1977) Effects of frequency and duration of training on attrition and incidence of injury. *Med Sci Sports Exerc*, **9**, 31–36.

Poole D.C., Ward S.A., Gardner G.W. and Whipp B.J. (1988) Metabolic and respiratory profile of the upper limit for prolonged exercise in man. *Ergonomics*, **31**, 1265–1279.

Poole D.C., Ward S.A. and Whipp B.J. (1990) The effects of training on the metabolic and respiratory profile of high-intensity cycle ergometer exercise. *Eur J Appl Physiol*, **59**, 421–429.

Poole D.C., Barstow T.J., McDonough P. and Jones A.M. (2008) Control of oxygen uptake during exercise. *Med Sci Sports Exerc*, **40**, 462–474.

Poole D.C., Wilkerson D.P. and Jones A.M. (2008) Validity of criteria for establishing maximal O_2 uptake during ramp exercise tests. *Eur J Appl Physiol*, **102**, 403–410.

Reggiani C., Potma E.J., Bottinelli R. *et al.* (1997) Chemo-mechanical energy transduction in relation to myosin isoform composition in skeletal muscle fibres of the rat. *J Physiol*, **502**, 449–460.

Ribeiro J.P., Hughes V., Fielding R.A. *et al.* (1986) Metabolic and ventilatory responses to steady state exercise relative to lactate thresholds. *Eur J Appl Physiol*, **55**, 215–221.

Robergs R. and Landwehr R. (2002) The surprising history of the 'HRmax = 220-age' equation. *J Exerc Physiol Online*, **5**, 1–10.

Sahlin K. (1992) Metabolic factors in fatigue. *Sports Med*, **13**, 99–107.

Saltin B. and Strange S. (1992) Maximal oxygen uptake: 'old' and 'new' arguments for a cardiovascular limitation. *Med Sci Sports Exerc*, **24**, 43–55.

Saltin B., Blomqvist C.G., Mitchell J.H. *et al.* (1968) Response to submaximal and maximal exercise after bed rest and training. *Circulation*, **38** (Suppl. 7), 1–78.

Selye H. (1951) The general-adaptation-syndrome. *Annual Review Medicine*, **2**, 327–342.

Simon J., Young J.L., Gutin B. *et al.* (1983) Lactate accumulation relative to the anaerobic and respiratory compensation thresholds. *J Appl Physiol*, **54**, 13–17.

Smith C.G.M. and Jones A.M. (2001) The relationship between critical velocity, maximal lactate steady-state velocity and lactate turnpoint velocity in runners. *Eur J Appl Physiol*, **85**, 19–26.

Tanaka K., Matsuura Y., Kumagai S. *et al.* (1983) Relationship of anaerobic threshold and onset of blood lactate accumulation with endurance performance. *Eur J Appl Physiol*, **52**, 51–56.

Tanaka H., Monahan K.D. and Seals D.R. (2001) Age-predicted maximal heart rate revisited. *J Am Coll Cardiol*, **37**, 153–156.

Vanhatalo A., Doust J.H. and Burnley M. (2007) Determination of critical power using a 3-min all-out cycling test. *Med Sci Sports Exerc*, **39**, 548–555.

Vanhatalo A., Doust J.H. and Burnley M. (2008) A 3-min all-out cycling test is sensitive to a change in critical power. *Med Sci Sports Exerc*, **40**, 1693–1699.

Wagner P.D. (2000) New ideas on limitations to VO_{2max}. *Exerc Sport Sci Rev*, **28**, 10–14.

Wasserman K., Whipp B.J., Koyl S.N. and Beaver W.L. (1973) Anaerobic threshold and respiratory gas exchange during exercise. *J Appl Physiol*, **35**, 236–243.

Wendt I.R. and Gibbs C.L. (1973) Energy production of rat extensor digitorum longus muscle. *Am J Physiol*, **224**, 1081–1086.

Wenger H.A. and Bell G.J. (1986) The interactions of intensity, frequency and duration of exercise training in altering cardiorespiratory fitness. *Sports Med*, **3**, 346–356.

Whipp B.J. and Wasserman K. (1972) Oxygen uptake kinetics for various intensities of constant-load work. *J Appl Physiol*, **33**, 351–356.

Whipp B.J., Davis J.A., Torres F. and Wasserman K. (1981) A test to determine parameters of aerobic function during exercise. *J Appl Physiol*, **50**, 217–221.

Whipp B.J., Ward S.A., Lamarra N. *et al.* (1982) Parameters of ventilatory and gas exchange dynamics during exercise. *J Appl Physiol*, **52**, 1506–1513.

Wilcox A. and Bulbulian R. (1984) Changes in running economy relative to VO_{2max} during a cross-country season. *J Sports Med*, **24**, 321–326.

Williams K.R. and Cavanagh P.R. (1987) Relationship between distance running mechanics, running economy, and performance. *J Appl Physiol*, **63**, 1236–1245.

Womack C.J., Davis S.E., Blumer J.L. *et al.* (1995) Slow component of O_2 uptake during heavy exercise: adaptation to endurance training. *J Appl Physiol*, **79**, 838–845.

3.5 Biochemical Monitoring in Strength and Conditioning

Michael R. McGuigan[1,2] and Stuart J. Cormack[3], [1]New Zealand Academy of Sport North Island, Auckland, New Zealand, [2]Auckland University of Technology, Auckland, New Zealand, [3]Essendon Football Club, Melbourne, Australia

3.5.1 INTRODUCTION

Strength and conditioning coaches and sports scientists are interested in reliable and valid methods of evaluating the effects of training, assessing training workload, and monitoring fatigue. Advances in technology have enabled the development of a variety of lab-based biochemical assays that can provide interesting insights into the effects of training perturbations on athletes. There are now a variety of metabolic, hormonal, and immunological markers that can be assessed in a relatively cost-effective manner and using a variety of different types of sample, such as blood, saliva, and urine. As with any measure that is made in sport and exercise, reliability and validity are important considerations. To date, no single reliable diagnostic marker has been found for biochemical monitoring.

In this chapter we outline some of the most common biochemical assays used in monitoring athletes in training and/or performance setting. In addition we discuss some of the research that has been conducted on the variety of methods of biochemical monitoring. We will also discuss how these measures can be implemented into the practical setting by providing an example from the sporting environment.

3.5.2 HORMONAL MONITORING

A number of hormones have been proposed as effective markers of training stress in athletes (Urhausen, Gabriel and Kindermann, 1995; Viru and Viru, 2001). Each has specific functions and its own pattern of exercise-induced responses, which must be considered when assessing its usefulness for monitoring purposes.

3.5.2.1 Cortisol

The influence of exercise on the levels of various hormones, particularly cortisol (C), testosterone (T), and their ratio value T:C, has been studied extensively in both laboratory (Bosco *et al.*, 1996, 2000; Urhausen, Kullmer and Kindermann, 1987) and field settings (Hoffman *et al.*, 2002, 2005; Maso *et al.*, 2005; Cormack, McGuigan and Newton, 2008; Cormack *et al.*, 2008). C is a major glucocorticoid and has an important role in both metabolism and immune function (Borer, 2003). Release of C is stimulated by adrenocorticotrophic hormone (ACTH), which is over-secreted as a response to the increased sensitivity of the hypthalamo–pituitary axis (HPA) to stress; the rise in C occurs approximately 15–30 minutes after ACTH release (Borer, 2003). C stimulates gluconeogenesis, which results in the sparing of blood glucose and protein stores. In metabolism, C causes increases in protein degradation in muscle and connective tissue, amino acid transport into the liver, liver glycogen synthesis, gluconeogenesis, and lipolysis (Borer, 2003). In contrast, reductions are caused in muscle protein synthesis, amino acid transport to muscle, and glucose uptake and utilisation. C is considered an important stress hormone and its presence provides an indication of the neuroendocrine system's response to exercise (Viru and Viru, 2001). The C response is also likely to be related to exercise duration and intensity, and is increased by some modes of training (Hill *et al.*, 2008; Jacks *et al.*, 2002; Kraemer and Ratamess, 2005). Like other hormones, C is influenced by circadian rhythms, with values highest in the morning and decreasing across the day (Aubets and Segura, 1995). This is an important consideration when using these measures for monitoring purposes in athletes.

3.5.2.2 Testosterone

T is anabolic in nature and has been shown to be important in muscle hypertrophy and increasing muscle glycogen synthesis. It has also been implicated in aggressive behaviour (Borer, 2003). T synthesis and secretion increase due to the effects of catecholamines, and plasma concentrations of T increase in response to acute exercise of a moderate or higher intensity (Kraemer and Ratamess, 2005; Tremblay, Copeland and Van

Strength and Conditioning – Biological Principles and Practical Applications Marco Cardinale, Rob Newton, and Kazunori Nosaka.

Helder, 2005). The T response to exercise appears to be inverse to exercise duration, with levels declining with increases in duration, whilst increases in intensity beyond 60–80% VO_{2max} do not lead to increased T concentration (Viru and Viru, 2003). Like C, T follows a diurnal pattern, with levels decreasing by 30–40% between early morning and late evening (Dabbs, 1990). The exercise responses of C, T and T:C will be discussed in more detail later in the chapter.

3.5.2.3 Catecholamines

Catecholamines (epinephrine and norepinephrine) are released in response to stress and reflect the acute demands of exercise. They are important for force production, energy availability, and muscle contraction, in addition to augmenting the effects of other hormones such as T (Kraemer and Ratamess, 2005). The exercise increases in catecholamines appear to be related to intensity (Bush *et al.*, 1999; Kraemer *et al.*, 2004); less is known about the chronic responses to exercise.

Researchers have examined changes in urinary and plasma catecholamine levels during periods of heavy training that resulted in overreaching and/or overtraining (Lehmann *et al.*, 1991). For example, Lehmann *et al.* (1991) reported decreased urinary norepinephrine and epinephrine excretion at night and increased submaximal plasma norepinephrine concentration after an increase in training volume. They also showed that submaximal and maximal heart rates significantly decreased in line with changes in catecholamines. It should be noted that findings of unaltered catecholamine levels and decreased maximal heart rates have also been reported (Urhausen, Gabriel and Kindermann, 1998). Unchanged resting, submaximal, and maximal free epinephrine and norepinephrine concentrations were described by Urhausen, Gabriel and Kindermann (1998) in underperforming cyclists and triathletes over a 15-month period. It has also been shown that epinephrine and norepinephrine have regulatory properties that modulate homoeostasis to meet the elevated demands of muscle force production, both before and during exercise (French *et al.*, 2007). In that study, subjects who were able to maintain force production throughout the exercise protocol had higher catecholamine concentrations than those whose performance decreased.

3.5.2.4 Growth hormone

The response of the growth hormone–insulin-like growth factor I (GH–IGF-I) axis has also been identified as potentially useful for monitoring purposes in athletes. Growth hormone (GH) has a number of important physiological effects, including promoting muscle growth and hypertrophy by facilitating amino acid transport and directly stimulating lipolysis (Borer, 2003). It is known that exercise acutely stimulates GH secretion (Kraemer and Ratamess, 2005); strength and power exercises have been shown to acutely increase GH, for example (Kraemer and Ratamess, 2005; Viru and Viru, 2001).

The use of GH for biochemical monitoring highlights the important issue of the measurement of hormones being complicated by the type of assay used to measure levels of GH (Nindl, 2007). A variety of immunoassays are commercially available to detect circulating GH concentrations. GH (and most hormones) exists as a family of related proteins of different molecular weights and structures. Traditional commercial assays simply measure one form and therefore neglect many others. There are more than 100 molecular isoforms of circulating GH, but the traditional measurement approach in the exercise literature has only focused on the main one (22 kDa).

One study by Hymer *et al.* (2001) shows the different responses of the different variants of GH. It is possible that these different forms have varying roles during exercise responses. Unfortunately, at the present time, the relationship between GH concentrations in the serum, GH signalling pathways, and longer-term changes in body composition or exercise performance is not well understood. It is hoped that, as this relationship becomes clearer, the role of exercise-induced GH release will be better defined and its use in biochemical monitoring can be made more useful.

3.5.2.5 IGF

Resting insulin-like growth factor I (IGF-I) and its binding proteins (e.g. IGFBP-3) concentrations have been studied as markers of stress (Elloumi *et al.*, 2005). The significance of monitoring IGF-I relates to its essential role in stimulating protein synthesis and maintaining muscle mass. IGF-I is a polypeptide produced by the liver, and plays a key role in mediating metabolic and anabolic responses during altered energy states (Kraemer and Ratamess, 2005). It also has a role in many physiological processes, particularly in bone and muscle anabolism. About 80% of IGF-I occurs in a ternary 150 kDa complex including IGFBP-3 and a protein called acid-labile subunit; less than 1% of IGF-I is free. Elloumi *et al.* (2005) have proposed that a reduction in resting IGFBP-3 may be used as a marker of overtraining. The IGFBPs serve as IGF carriers in circulation and act as regulators of their biological actions. The evaluation of possible changes in the concentrations of total IGF-I and its binding proteins may be of interest, because it is believed that they impact performance and potentially reflect the physical overload state of athletes (Elloumi *et al.*, 2005; Rosendal *et al.*, 2002).

Several studies have shown that intensive training stimulates both circulating IGF-I and IGFBP-3 (Borst *et al.*, 2001; Elloumi *et al.*, 2005; Koziris *et al.*, 1999). However, an associated fall in IGFBP-3 related to overtraining has been observed. A strong negative correlation has been shown between IGF-I and IGFBP-3 levels and overtraining. In particular, levels of IGFBP-3 after exercise decreased in fatigued subjects but were increased in fit subjects (Elloumi *et al.*, 2005). Borst *et al.* (2001) demonstrated a reduction in IGFBP-3 concentrations that paralleled an elevation in IGF-I concentrations after a 25-week training period.

Recent research has questioned the significance of circulating IGF-I and indicates that there is not a strong association between IGF-I alterations induced by a short period of increased activity and longer-term adaptations (Nindl *et al.*, 2007). Nindl *et al.* (2007) showed that seven days of increased physical activity resulted in declines in circulating IGF-I and that these response patterns were not altered by fitness level, dietary protein intake, or energy balance. The response of IGF-I to chronic training is even less clear. For example, short-term resistance-training studies have reported no change in resting concentration of IGF-I (McCall *et al.*, 1999), whereas long-term studies in men and women have reported significant elevations in resting IGF-I (Borst *et al.*, 2001; Koziris *et al.*, 1999). Acute overreaching, resulting from a dramatic increase in volume and/or intensity of training, has been shown to reduce IGF-I concentrations by 11% (Raastad *et al.*, 2003); concentrations returned to baseline when normal training resumed over the next cycle.

3.5.2.6 Leptin

Leptin is a protein hormone believed to relay satiety signals to the hypothalamus in order to regulate appetite and energy balance, as well as having roles with thermogenesis and metabolism. There has been some research into its potential response to exercise, with Simsch *et al.* (2002) reporting reductions in resting leptin concentrations in rowers following high-intensity resistance training and Maestu, Jurimae and Jurimae (2003) showing decreases in response to high-volume, low-intensity rowing training. Conversely, Nindl *et al.* (2002) showed no acute decrease in leptin concentration following high-volume resistance exercise, but rather a delayed decrease that may have reflected the large disruption in metabolic homoeostasis. Jurimae, Maestu and Jurimae (2003) have shown a relationship between training volume and plasma leptin. As with all hormonal measures, it is important to control for nutrient intake and diurnal variations, and the differences that have been seen in different studies may be accounted for by these factors. However, it has been suggested that monitoring of hormones such as leptin following exhaustive exercise bouts may be indicative of the hypothalamic down-regulation that occurs during overreaching (Maestu, Jurimae and Jurimae, 2005).

3.5.2.7 Research on hormones in sporting environments

Despite providing an important platform for the study of hormonal responses to exercise, results obtained in laboratory studies can be limited in terms of their applicability to elite training and competition environments. Numerous researchers have attempted to address the limitations associated with laboratory-based projects by examining hormonal response in athletic settings (Cormack, McGuigan and Newton, 2008; Cormack *et al.*, 2008; Coutts *et al.*, 2007a, 2007b; Coutts, Slattery and Wallace, 2007; Doan *et al.*, 2007; Elloumi *et al.*, 2005, 2008; Filaire *et al.*, 2001; Filaire, Lac and Pequignot, 2003; Hoffman *et al.*, 2002, 2005; Kraemer *et al.*, 2004; Maso *et al.*, 2005), for example in swimming, rowing, and running (Jurimae, Maestu and Jurimae, 2003; Lac and Berthon, 2000; Lehmann *et al.*, 1993; Maestu, Jurimae and Jurimae, 2003, 2005; Mujika *et al.*, 1996, 2000; Ramson *et al.*, 2009; Steinacker *et al.*, 2000; Urhausen, Kullmer and Kindermann, 1987). The response of hormonal measures to this type of exercise is not universal. A large body of research raises doubts about the efficacy of hormonal variables such as T, C, and T:C, while other studies indicate potential value in regular sampling of hormonal variables.

Changes in T:C have been shown to mirror performance changes in several studies, for example in swimming (Mujika *et al.*, 1996) and rugby league (Coutts *et al.*, 2007a). As a result, T:C may be a useful measure of the relationship between hormonal status and performance. In another study using swimmers, there was no difference between levels of C when comparing stale and non-stale athletes (Hooper *et al.*, 1995). In this study, a period of tapering was insufficient to allow stale swimmers to recover their level of performance. From the results of a further swimming study where performance, psychological, and blood markers were analysed, it was concluded that T and C were not reliable markers of overreaching or overtraining (Hooper, Mackinnon and Howard, 1999). A study by Coutts, Slattery and Wallace (2007) examined hormonal response to intensified training in experienced male triathletes. They found no change in T or C between the normal training and intensified training groups during the four weeks of the study. However, T:C ratio significantly increased and C significantly decreased in the intensified group during the taper period compared to the post-training measure, indicating an increase in anabolic:catabolic balance. The researchers concluded that performance measures are the only definitive measure of overreaching in endurance athletes and that a decrease in physiological state occurs with increased training load. The response of endocrine measures in the different studies suggest they may be capable of providing an early indication of relative anabolic:catabolic status which in turn could allow manipulation of the training stimulus to avoid unplanned fatigue.

There have also been an increasing number of studies in team sports such as Australian rules football (ARF), American football, soccer, rugby league, and rugby union (Cormack, McGuigan and Newton, 2008; Cormack *et al.*, 2008; Coutts *et al.*, 2007a, 2007b; Elloumi *et al.*, 2008; Filaire *et al.*, 2001; Filaire, Lac and Pequignot, 2003; Hoffman *et al.*, 2002, 2005; Moreira *et al.*, 2008, 2009). The results of longer-term studies have been equivocal; one study saw a rise in C during a soccer season and return to baseline two months post-season, with no change evident in T or T : C (Filaire, Lac and Pequignot, 2003), whilst a reduction in T was the only change noted during 15 weeks of American football training (Hoffman *et al.*, 2005) and a 14-week rugby season (Elloumi *et al.*, 2008). A competitive season of elite ARF competition has been shown to elicit fluctuations in endocrine responses (Cormack *et al.*, 2008); the pattern of response seen in C and T:C suggested subjects were unlikely to have been in a catabolic state as a result of the training and competition load and that C has a small relationship to performance. This study provided some evidence that C may be a useful variable for monitoring the response to elite ARF competition and training.

3.5.2.8 Methodological considerations

Traditionally a potential limitation to analysis of the hormonal response to exercise has been the need for a blood sample. Venipuncture is the primary method used to acquire blood samples for laboratory testing. However there are now a number of salivary kits available for analysis of different hormones. In an examination of the efficacy of obtaining C values via salivary samples, Neary, Malbon and McKenzie (2002) compared blood and saliva values and reported an ICC of r = 0.995 between serum and salivary C values. This suggests that salivary C accurately reflects free (unbound) C. This result has been confirmed in other work, reporting a strong relationship between salivary and serum C both at rest (r = 0.93) and during exercise (r = 0.90) (O'Connor and Corrigan, 1987). The use of saliva samples in monitoring hormonal responses to exercise greatly increases the opportunity to obtain data in athletic populations as the procedure is less invasive than the alternatives (Riad-Fahmy, Read and Walker, 1983).

Urine is easy to study and analyse, and urine analysis is non-invasive. Adaptations in the HPA axis can be analysed by looking at urine steroid hormone excretion (Gouarne *et al.*, 2005). Using reference values from an athlete's biochemical profile, it has also been suggested that urinary analysis could be used to quantify training loads and control workloads (Timon *et al.*, 2008). However, the hormone levels in urine will only provide general information, and the analysis can be time-consuming. Also, factors such as blood flow and rate of diuresis need to be taken into account (Viru and Viru, 2001).

Other methods are also currently being used, such as electrosonophoresis (Cook, 2002), which is used to collect transdermal samples of interstitial fluid; some recent studies have used this technology, particularly relating to sports such as rugby union (Gill, Beaven and Cook, 2006; Smart *et al.*, 2008).

Whatever method is used, the storage of the specimen should also be considered. Many hormones will be affected by temperature and should be stored in cool conditions as soon as possible. Some types of blood analysis also require separation of serum and plasma. Repeated freezing and thawing of samples should be avoided.

Advantages of hormonal analysis

- Measures can be undertaken non-invasively using saliva and urine.
- Regular monitoring of hormones throughout the season may allow implementation of appropriate interventions such as reduced training loads or periods of rest aimed at recovering hormonal status.

Problems with hormonal analysis

- Many factors affect blood hormone concentrations, such as factors linked to sampling conditions and storage of the sample.
- Food intake can modify significantly either the resting concentration of some hormones (T, C) or their concentration change in response to exercise (C, GH).
- In female athletes the hormonal response will depend on the phase of the menstrual cycle.
- Aerobic and resistance exercise protocols will typically result in different endocrinological responses.
- Hormone concentrations at rest and following exercise respond differently.
- There can be diurnal variations.
- Reproducibility of some hormonal analysis can be poor.
- Analysis can be time-consuming.

3.5.3 METABOLIC MONITORING

A number of different metabolites have been studied with the aim of establishing their usefulness for monitoring of training and performance (Viru and Viru, 2001). Here we discuss some of these metabolites, specifically looking at measures using muscle and blood. As with hormonal analysis, it is important for the coach to gain an understanding of the roles of the different metabolites and their significance in the various metabolic pathways.

3.5.3.1 Muscle biopsy

In order to investigate the various fibre and enzymatic features of skeletal muscle, it is first necessary to obtain a muscle sample. The percutaneous muscle biopsy technique described by Bergström *et al.* (1967) has commonly been used for research in exercise biochemistry. This involves the insertion of a hollow-bored needle under local anaesthetic and sterile conditions to obtain muscle specimens, and is useful for biochemical, histochemical, and immunohistochemical analyses. The muscle groups that are most commonly sampled are the vastus lateralis, triceps brachii, deltoid, and gastrocnemius muscles; this allows analysis of the cellular and molecular responses to exercise interventions. Due to the invasive nature of this approach, it can be difficult to use in more applied settings such as sport. However, it can reveal some very interesting information, such as muscle fibre type and area, hormone receptors, and enzyme levels.

Advantages of muscle biopsy analysis

- Direct measurement of enzymes and muscle characteristics, for example fibre type.
- Allows for more advanced analysis of molecular and cellular mechanisms such as receptor function.

Problems with muscle biopsy analysis

- It is invasive and expensive.
- Technical expertise is required in its use.
- It is laboratory-based.

3.5.3.2 Biochemical testing

Creatine kinase

Creatine kinase (CK) is an enzyme found in high concentration within muscle fibres. Increases in plasma and serum CK levels have been associated with muscle damage, but large variations between individuals have been reported (Clarkson, 1997). Plasma CK efflux is the most common indicator of loss of cellular homoeostasis, though investigators have also examined the release of L-aspartate aminotransferase, lactate dehydrogenase, CK isoforms, myoglobin, heart fatty acid binding protein, carbonic anhydrase isoenzyme III, contractile and regulatory proteins, and troponins (Warren, Lowe and Armstrong, 1999). Of these intracellular enzymes, myoglobin activity peaks in a similar time frame to CK efflux at 3–5 days post-exercise (Nosaka, Clarkson and Apple, 1992; Toft *et al.*, 2002). Individually increased blood concentrations of CK measured under standardized conditions at rest may provide information concerning an elevated muscular and/or metabolic strain, but cannot indicate an overreached or overtrained state (Brancaccio, Maffulli and Limongelli, 2007; Urhausen, Gabriel and Kindermann, 1998).

Cytokines

There has been a lot of interest recently in the response of cytokines to exercise. Cytokines such as tumour necrosis factor α (TNF-α) and interleukin 6 (IL-6) have multiple biological effects and are released from a number of different tissues, including skeletal muscle. Studies in adults have demonstrated large increases in inflammatory cytokines with heavy, prolonged exercise (Pedersen and Febbraio, 2008; Robson-Ansley, Blannin and Gleeson, 2007; Toft *et al.*, 2002). Their main role appears to be as cell–cell communicators. Contracting muscle fibres produce and release IL-6, which induces several metabolic effects (Pedersen and Febbraio, 2008), including lipolysis and fat oxidation; it is also involved in glucose homoeostasis during exercise. Very few studies have investigated long-term changes in cytokines in athlete populations. It is possible to measure some cytokines, for example IL-6, in saliva (Minetto *et al.*, 2007), but valid and cost-effective methods still need to be developed and more thoroughly researched.

Glutamine

Glutamine is a widely studied amino acid that is known to play an important role in many metabolic pathways. The concentration of plasma glutamine has been suggested as a potential indicator of high training load (Halson *et al.*, 2003; Smith and Norris, 2000). Reduced plasma glutamine and elevated plasma glutamate levels have been observed following periods of intense training and have been implicated in impaired immune system function (Filaire *et al.*, 2003; Halson *et al.*, 2003). In the study of rugby league players undergoing intensified training, the glutamine : glutamate ratio was shown to be one of the useful markers of training intolerance (Coutts *et al.*, 2007b).

Lactate

Blood lactate measurements are widely used to measure exercise responses. These measurements are dependent on the training status of the athlete and their usefulness in monitoring is questionable (Meeusen *et al.*, 2006). However, there is a reduced maximal lactate concentration in endurance athletes who are underperforming/overreached (Meeusen *et al.*, 2006).

There has been interest recently in the use of lactate dehydrogenase (LDH) in monitoring muscle adaptations to training (Brancaccio *et al.*, 2008; Butova and Masalov, 2009).

Advantages of biochemistry testing

- A number of different analyses can be performed using the blood samples of athletes.
- The glutamine : glutamate ratio is an indicator of excessive training stress.

Problems with biochemistry testing

- Analysis is time-consuming and expensive, particularly for cytokines.
- Measures can be highly variable and likely only provide a general indication of stress.

3.5.4 IMMUNOLOGICAL AND HAEMATOLOGICAL MONITORING

3.5.4.1 Immunological markers

Injury and/or illness can occur when physical demands outweigh the body's ability to fully recover between training sessions and competitions (Foster, 1998). There has been extensive research in this area, with differing conclusions being reached (Fry *et al.*, 1994; Mackinnon, 1997; Putlur *et al.*, 2004; Pyne and Gleeson, 1998). Intense exercise has been shown to change and suppress a number of immune markers including the circulating leukocytes, plasma cytokine concentrations, salivary immunoglobin A (S-IgA) secretion rate, and neutrophil and macrophage activity (Mackinnon, 1997). It has been shown that moderate exercise may stimulate the immune system, but heavy training loads suppress it, increasing the risk of infection. This could be particularly relevant for most elite-level sports in both the pre-season period when training loads are high and during the in-season when competition games are being played and physical demands on the body are at a peak.

Research has looked at S-IgA levels in elite athletes (Gleeson *et al.*, 1999; Moreira *et al.*, 2008; Neville, Gleeson and Folland,

2008; Pyne and Gleeson, 1998), specifically at the relationship with upper respiratory-tract infections (URTIs). Despite this, there is little evidence to confirm a direct link between depressed immune function and incidence of illness in athletes. A recent study by Neville, Gleeson and Folland (2008) in yacht-racing athletes showed that relative S-IgA determined a substantial proportion of the variability in weekly URTI incidence. Putlur *et al.* (2004) examined the alteration of immune function in women collegiate soccer players over a nine-week competitive season and compared these results with a control group of college students. They reported that illness occurred at a higher rate among the soccer players, and that the incidence of illness amongst this group was at its lowest in weeks when the training load was reduced. They also showed that 55% of the soccer players' illnesses were associated with a preceding spike in training load and that 82% of illnesses were associated with a preceding decrease in S-IgA levels. Foster (1998) showed that with experienced athletes, 84% of illnesses could be explained by a preceding increase in training load. This provides some evidence that an increase in training load can lead to an increase in illness amongst athletes and that monitoring immune status in athletes could be a useful strategy for avoiding loss of training and competition time. However, a number of studies indicate large variability in the responses of immune measures such as S-IgA (Moreira *et al.*, 2008; Neville, Gleeson and Folland, 2008).

3.5.4.2 Haematological markers

Haematological changes are sometimes reported in overreached or overtrained athletes (Halson *et al.*, 2002, 2003; Lehmann *et al.*, 1991, 1993; Rietjens *et al.*, 2005; Smith and Myburgh, 2005) where others have failed to find any changes (Coutts *et al.*, 2007b; Rowbottom *et al.*, 1995). The body of research would suggest that blood parameters (e.g. blood count, C-reactive protein, urea, creatinine, liver enzymes, glucose, ferritin, sodium, potassium, etc.) are not capable of detecting overreaching or overtraining in athletes (Rietjens *et al.*, 2005). Also, it has been shown that a lot of biochemical parameters do not sensitively reflect the physiological difference between athletes and control subjects. Many of these markers do not accurately represent physiological changes of athletes before and after training. Despite these limitations, these markers are helpful in providing general information on the health status of an athlete (Meeusen *et al.*, 2006).

It is clear that the immune system is extremely sensitive to physiological stress. Potentially, immune variables could be used as an index of stress in relation to exercise training. The current information regarding the immune system and training suggests that periods of intensified training result in depressed immune cell functions, with little or no alteration in circulating cell numbers. What is problematic is that even though these immune markers change in response to increased training loads, these changes do not appear to distinguish between those athletes who successfully adapt to high training loads and those who develop overtraining syndrome.

Advantages of immunological testing

- The decline in an individual's relative S-IgA over the two/three week period before a URTI appears to precede and contribute to URTI risk.
- General information on the health status of an athlete can be obtained.

Problems with immunological testing

- There is a high variability with measures such as S-IgA.
- Results can be affected by time of day and previous exercise.
- Analysis is time-consuming and expensive.

3.5.5 PRACTICAL APPLICATION

Strength and conditioning coaches can use biochemical monitoring for a number of different purposes, such as evaluation of training programmes, assessment of training and competition workload, and monitoring of fatigue. However, it is important to emphasize that there does not seem to be a single marker that can be used to inform all these different areas. It is therefore important to include various biochemical markers as part of an overall monitoring system that will incorporate other areas such as neuromuscular and psychological measures. The responses are directly influenced by the regulatory elements, including the exercise prescription used, environment (e.g. temperature, age, gender), nutritional status, and psychology (e.g. arousal level).

Despite previous studies of biochemical and endocrine responses in athletes, weekly variations in elite-level sport are poorly understood. There is likely to be some modification to biochemical and hormonal status during a sporting season, including various magnitudes of suppression or elevation that occur independently or in parallel. It is possible that change in these variables are related to workload or performance, and in the case of hormonal measures, may reflect modifications to total body anabolic : catabolic balance.

New technologies are developing all the time which have potential applications for the monitoring of athletes. For example, the use of metabolomics has recently been applied to athletes (Yan *et al.*, 2009). Metabolomics is the quantitative measurement of metabolic responses to stimuli or genetic modification and involves the determination of comprehensive metabolite profiles in biological markers from blood samples. However, the practicality of such analysis is often limited by the expense and time-consuming nature of these methods.

3.5.5.1 Evaluating the effects of training

An imbalance between training stress and recovery can lead to overreaching or overtraining syndrome (Meeusen *et al.*, 2006) and it appears that various hormonal measures may assist in assessing the response to training. C and T vary in opposite directions in response to exercise; this may represent an imbalance between anabolic and catabolic hormones resulting in a decreased T:C when training and performance loads are increased. However, the results of studies analysing the response of T, C, and T:C in relation to training, competition, and performance are varied (McCall *et al.*, 1999). What appears to be critical in evaluating the effects of training is regular, serial monitoring of concentrations. Although the acute response of these hormones has been investigated to a large extent, periods of substantially increased volume or intensity have been shown to elicit alterations (e.g. reduced total T, elevated C), thereby indicating these hormones are useful markers of chronic training stress.

Resting salivary hormone concentrations may also be important for workout performance, especially for individuals, potentially moderating training adaptation (Crewther *et al.*, 2009). The acute responses in the endocrine system during stressful training sessions are related to the intensity and duration of the specific exercise stimulus and also to the condition of the athlete (Simsch *et al.*, 2002). Hormones also play an important role in mediating adaptive changes in elite athletes, especially those with strength-training experience (Häkkinen, 1989; Häkkinen *et al.*, 1988). Adaptive changes in endogenous hormones during training have increasing importance in inducing performance changes in elite strength athletes (Häkkinen, 1989). However, more research is required to determine these relationships and clearly establish the role of hormonal monitoring for both predicting and tracking the effects of training programmes.

3.5.5.2 Assessment of training workload

Regular monitoring of hormones throughout the training cycle may allow instigation of appropriate interventions such as reduced training loads or periods of rest aimed at recovering neuromuscular and hormonal status. Responses should be examined in comparison to individual baseline measures on a weekly basis to specifically monitor training manipulations and competition load in an effort to maximize performance.

The measurement of exercise-induced hormones has an advantage in studying the adaptive state of athletes (Ramson *et al.*, 2009; Simsch *et al.*, 2002) and appears to be more indicative than basal levels of these hormones. Resting salivary hormone concentrations appear important for workout performance (Crewther *et al.*, 2009). Their study shows a significant relationship between resting salivary hormones and power and strength across all workouts in athletes. This finding can be attributed to individual differences in the target-tissue response to hormones (because saliva reflects the biologically active hormone) and the result of training and/or genetic factors (e.g. muscle fibre distribution, binding capacity, number of receptors).

3.5.5.3 Monitoring of fatigue

Fatigue is often a consequence of physical training and the effective management of fatigue by the coach is essential in optimizing adaptation and performance. There are a range of different methods of fatigue management for athletes (Robson-Ansley, Gleeson and Ansley, 2009); the expected hormonal and biochemical response and therefore usefulness of these measures for early detection of overreaching or overtraining in team sport athletes over extended periods is unclear. Responses may be sport-specific and cyclic across longer periods, necessitating profiling of individual sports. As stated previously there is no universal method available for the diagnosis and monitoring of fatigue.

It is arguable that the major conclusion to be drawn from previous work examining endocrine response in individual athletes is that a large variation in response is possible. This may be due to the variety of training-load interventions utilized and the length of study periods. It is also possible that changes in the endocrine measures studied do not reflect modifications to training load or performance. Therefore, it may be critical to evaluate the usefulness of these measures via analysis of results from sports similar to the one of interest, and ultimately research in specific environments may be the only method of obtaining valid and practically meaningful results in order to reach definitive conclusions regarding the value of endocrine measures for monitoring training and performance.

3.5.5.4 Conclusions and specific advice for implementation of a biochemical monitoring programme

In terms of practical application, the variability of the hormonal and biochemical responses amongst athletes leads us to suggest that there is a need to individually analyse the results with any sport. However, the coach does need to be aware that there is no single reliable, specific, and sensitive measure that will establish whether an athlete is overtrained or overreached. There is enough evidence to suggest that several measures can provide some useful information to coaches about the training status, adaptation to workload, and level of fatigue in their athletes.

Using a team-sport model, the following recommendations can be made in terms of a biochemical monitoring programme. These assume a relatively healthy budget for this aspect of the programme and access to a commercial or university biochemistry laboratory.

- Saliva collection each week for monitoring of C, T, and T:C. By establishing reliable baseline results for each individual, it may be possible to determine when workload in the athlete is becoming excessive; this can be reflected by an increase in C levels, for example (examined in conjunction with other monitoring data such as RPE reported training load). Ideally the sample should be collected in the morning, preferably after one day of rest.

- Blood collection once a month for routine biochemical monitoring; specifically, glutamine : glutamate for metabolic and immune function. Alteration of glutamine : glutamate could indicate that an athlete is overreached with the current training load and has suppressed immune function.
- Saliva collection once a month to evaluate levels of S-IgA. A suppressed level of S-IgA could be indicative of URTI risk.
- A maximal testing battery once every three months for athletes indentified as being at high risk of developing overtraining syndrome (Meeusen *et al.*, 2004), in order to assess hormones such as C, T, ACTH, and GH. A suppressed C and GH response to exercise could be indicative of overreaching, for example.

Some important considerations for the monitoring programme:

1. Exercise-induced measures need to be compared with baseline measures from the same individual.
2. Diurnal variation needs to be taken into account (samples need to be collected at the same time of day).
3. Comparisons with previously published results must be made with care, as assays can produce very different data (e.g. GH, glutamine).
4. These measures should be analysed in conjunction with other physiological and psychological measures, such as exercise test, wellness questionnaire, rating of perceived exertion.
5. Circulating levels of hormones and other markers, particularly at rest, are not always indicative of the molecular and cellular responses.
6. A single measure of a hormone or biochemical marker will not necessarily provide accurate information about the training status of an individual, and the effects of other factors such as diet, previous levels of exercise, stress, menstrual cycle, hormone secretion, and so on need to be taken into account.
7. Perhaps most importantly, the biochemical monitoring should provide coaches with information that can be used to make objective decisions about the training process of their athletes. Testing just for the sake of generating data is not useful.

References

Aubets J. and Segura J. (1995) Salivary cortisol as a marker of competition related stress. *Sci Sport*, **10**, 149–154.

Bergström J., Hermansen L., Hultman E. and Saltin B. (1967) Diet muscle glycogen and physical performance. *Acta Physiol Scand*, **71**, 140–150.

Borer K.T. (2003) *Exercise Endocrinology*, Human Kinetics, Champaign, IL.

Borst S.E., De Hoyos D.V., Garzarella L. *et al.* (2001) Effects of resistance training on insuline-like growth factor-I and IGF binding proteins. *Med Sci Sports Exerc*, **33**, 648–653.

Bosco C., Tihanyi J., Rivalta L. *et al.* (1996) Hormonal responses in strenuous jumping effort. *Jap J Physiol*, **46**, 93–98.

Bosco C., Colli R., Bonomi R. *et al.* (2000) Monitoring strength training: neuromuscular and hormonal profile. *Med Sci Sports Exerc*, **32**, 202–208.

Brancaccio P., Maffulli N. and Limongelli F.M. (2007) Creatine kinase monitoring in sport medicine. *Br Med Bull*, **81–2**, 209–230.

Brancaccio P., Maffulli N., Buonauro R. and Limongelli F.M. (2008) Serum enzyme monitoring in sports medicine. *Clin Sports Med*, **27**, 1–18.

Bush J.A., Kraemer W.J., Mastro A.M., Triplett-McBride N.T., Volek J.S., Putukian M., Sebastianelli W.J. and Knuttgen H.G. (1999) Exercise and recovery responses of adrenal medullary neurohormones to heavy resistance exercise. *Med Sci Sports Exerc*, **31** (4), 554–559.

Butova O.A. and Masalov S.V. (2009) Lactate dehydrogenase activity as an index of muscle tissue metabolism in highly trained athletes. *Human Phys*, **35**, 141–144.

Clarkson P.M. (1997) Eccentric exercise and muscle damage. *Int J Sports Med*, **18**, S314–S317.

Cook C.J. (2002) Rapid noninvasive measurement of hormones in transdermal exudate and saliva. *Physiol Behav*, **75**, 169–181.

Cormack S., McGuigan M.R. and Newton R.U. (2008) Neuromuscular and endocrine responses of elite players to an Australian Rules Football match. *Int J Sports Physiol Perform*, **3**, 359–374.

Cormack S., Newton R.U., McGuigan M.R. and Cormie P. (2008) Neuromuscular and endocrine responses of elite players during an Australian Rules season. *Int J Sports Physiol Perform*, **3** (4), 439–453.

Coutts A., Reaburn P., Piva T.J. and Murphy A. (2007a) Changes in selected biochemical, muscular strength, power and endurance measures during deliberate overreaching and tapering in rugby league players. *Int J Sports Med*, **28**, 116–124.

Coutts A.J., Reaburn P., Piva T.J. and Rowsell G.J. (2007b) Monitoring for overreaching in rugby league players. *Eur J Appl Physiol*, **99**, 313–324.

Coutts A.J., Slattery K.M. and Wallace L.K. (2007) Practical tests for monitoring performance, fatigue and recovery in triathletes. *J Sci Med Sport*, **10**, 372–381.

Crewther B.T., Lowe T., Weatherby R.P. and Gill N. (2009) Prior sprint cycling did not enhance training adaptation, but resting salivary hormones were related to workout power and strength. *Eur J Appl Physiol*, **105**, 919–927.

Dabbs J.M.J. (1990) Salivary testosterone measurements: reliability across hours, days and weeks. *Physiol Behav*, **48**, 83–86.

Doan B.K., Newton R.U., Kraemer W.J. *et al.* (2007) Salivary cortisol, testosterone and T/C ratio responses during a 36-hole golf competition. *Int J Sports Med*, **28**, 470–479.

Elloumi M., El Elj N., Zaouali M. *et al.* (2005) IGFBP-3, a sensitive marker of physical training and overtraining. *Br J Sports Med*, **39**, 604–610.

Elloumi M., Ben Ounis O., Tabka Z. *et al.* (2008) Psychoendocrine and physical performance responses in male Tunisian rugby players during an international competitive season. *Aggress Behav*, **34**, 623–632.

Filaire E., Bernain X., Sagnol M. and Lac G. (2001) Preliminary results on mood state, salivary testosterone:cortisol ratio and team performance in a professional soccer team. *Eur J Appl Physiol*, **86**, 179–184.

Filaire E., Lac G. and Pequignot J.-M. (2003) Biological, hormonal and psychological parameters in professional soccer players throughout a competitive season. *Percep Mot Skills*, **97**, 1061–1072.

Foster C. (1998) Monitoring training in athletes with reference to overtraining syndrome. *Med Sci Sports Exerc*, **30** (7), 1164–1168.

French D.N., Kraemer W.J., Volek J.S. *et al.* (2007) Anticipatory responses of catecholamines on muscle force production. *J Appl Physiol*, **102**, 94–102.

Fry R.W., Grove J.R., Morton A.R. *et al.* (1994) Psychological and immunological correlates of acute overtraining. *Br J Sports Med*, **28**, 241–246.

Gill N., Beaven C. and Cook C. (2006) Effectiveness of post-match recovery strategies in rugby players. *Br J Sports Med*, **40**, 260–263.

Gleeson M., McDonald W.A., Pyne D.B. *et al.* (1999) Salivary IgA levels and infection risk in elite swimmers. *Med Sci Sports Exerc*, **31**, 67–73.

Gouarne C., Groussard C., Gratas-Delamarche A. *et al.* (2005) Overnight urinary cortisol and cortisone add new insights into adaptation to training. *Med Sci Sports Exerc*, **37**, 1157–1167.

Häkkinen K. (1989) Neuromuscular and hormonal adaptations during strength and power training. *J Sports Med Phys Fitness*, **29**, 9–26.

Häkkinen K., Pakarinen A., Alén M. *et al.* (1988) Daily hormonal and neuromuscular responses to intensive strength training in 1 week. *Int J Sports Med*, **9**, 422–428.

Halson S.L., Bridge M.W., Meeusen R. *et al.* (2002) Time course of performance changes and fatigue markers during intensified training in trained cyclists. *J Appl Physiol*, **93**, 947–956.

Halson S.L., Lancaster G.I., Jeukendrup A.E. and Gleeson M. (2003) Immunological responses to overreaching in cyclists. *Med Sci Sports Exerc*, **35**, 854–861.

Hill E.E., Zack E., Battaglini C. *et al.* (2008) Exercise and circulating cortisol levels: the intensity threshold effect. *J Endocrinol Invest*, **31**, 587–591.

Hoffman J.R., Maresh C.M., Newton R.U. *et al.* (2002) Performance, biochemical and endocrine changes during a competitive football game. *Med Sci Sports Exerc*, **34**, 1845–1853.

Hoffman J.R., Kang J., Ratamess N.A. and Faigenbaum A.D. (2005) Biochemical and hormonal responses during an intercollegiate football season. *Med Sci Sports Exerc*, **37**, 1237–1241.

Hooper S.L., Mackinnon L.T., Howard A. *et al.* (1995) Markers for monitoring overtraining and recovery. *Med Sci Sports Exerc*, **27**, 106–112.

Hooper S.L., Mackinnon L.T. and Howard A. (1999) Physiological and psychometric variables for monitoring recovery during tapering for major competition. *Med Sci Sports Exerc*, **31**, 1205–1210.

Hymer W.C., Kraemer W.J., Nindl B.C., Marx J.O., Benson D.E., Welsch J.R., Mazzetti S.A., Volek J.S. and Deaver D.R. (2001) Characteristics of circulating growth hormone in women after acute heavy resistance exercise. *Am J Physiol Endocrinol Metab*, **281** (4), E878–E887.

Jacks D.E., Sowash J., Anning J. *et al.* (2002) Effect of exercise at three exercise intensities on salivary cortisol. *J Strength Cond Res*, **16**, 286–289.

Jurimae J., Maestu J. and Jurimae T. (2003) Leptin as a marker of training in highly trained male rowers? *Eur J Appl Physiol*, **90**, 533–538.

Koziris L.P., Hickson R.C., Chatterton R.T. *et al.* (1999) Serum levels of total and free IGF-1 and IGFBP-3 are increased and maintained in long-term training. *J Appl Physiol*, **86**, 1436–1442.

Kraemer W.J. and Ratamess N.A. (2005) Hormonal responses and adaptations to resistance exercise and training. *Sports Med*, **35**, 339–361.

Kraemer W.J., French D.N., Paxton N.J. *et al.* (2004) Changes in exercise performance and hormonal concentrations over a Big Ten soccer season in starters and nonstarters. *J Strength Cond Res*, **18**, 121–128.

Lac G. and Berthon P. (2000) Changes in cortisol testosterone and T/C ratio during and endurance competition and recovery. *J Sports Med Phys Fitness*, **40**, 139–144.

Lehmann M., Dickhuth H.H., Gendrisch G. *et al.* (1991) Training – overtraining. A prospective, experimental study with experienced middle- and long-distance runners. *Int J Sports Med*, **12**, 444–452.

Lehmann M., Knizia K., Gastmann U. *et al.* (1993) Influence of 6-week, 6 days per week, training on pituitary function in recreational athletes. *Br J Sports Med*, **27**, 186–192.

McCall G.E., Byrnes W.C., Fleck S.J. *et al.* (1999) Acute and chronic hormonal responses to resistance training designed to promote muscle hypertrophy. *Can J Appl Physiol*, **24**, 96–107.

Mackinnon L.T. (1997) Immunity in athletes. *Int J Sports Med*, Suppl 1, S62–S68.

Maestu J., Jurimae J. and Jurimae T. (2003) Effect of heavy increase in the training stress on the plasma leptin concentration in highly trained male rowers. *Horm Res*, **59**, 91–94.

Maestu J., Jurimae J. and Jurimae T. (2005) Monitoring of performance and training in rowing. *Sports Med*, **35**, 597–617.

Maso F., Lac G., Filaire E. *et al.* (2005) Salivary testosterone and cortisol in rugby players: correlation with psychological overtraining items. *Br J Sports Med*, **38**, 260–263.

Meeusen R., Piacentini M.F., Busschaert B. *et al.* (2004) Hormonal responses in athletes: the use of a two bout exercise protocol to detect subtle differences in (over) training status. *Eur J Appl Physiol*, **91**, 140–146.

Meeusen R., Duclos M., Gleeson M. *et al.* (2006) Prevention, diagnosis and the treatment of the overtraining syndrome. *Eur J Sport Sci*, **6**, 1–14.

Minetto M.A., Gazzoni M., Lanfranco F. *et al.* (2007) Influence of the sample collection method on salivary interleukin-6 levels in resting and post-exercise conditions. *Eur J Appl Physiol*, **101**, 249–256.

Moreira A., Arsati F., Cury P.R. *et al.* (2008) The impact of a 17-day training period for an international championship on mucosal immune parameters in top-level basketball players and staff members. *Eur J Oral Sci*, **116**, 431–437.

Moreira A., Arsati F., Arsati Y. *et al.* (2009) Salivary cortisol in top-level professional soccer players. *Eur J Appl Physiol*, **106**, 25–30.

Mujika I., Chatard J.-C., Padilla S. *et al.* (1996) Hormonal responses to training and its tapering off in competitive swimmers: relationships with performance. *Eur J Appl Physiol*, **74**, 361–366.

Mujika I., Goya A., Padilla S. *et al.* (2000) Physiological responses to a 6-d taper in middle-distance runners: influence of training intensity and volume. *Med Sci Sports Exerc*, **32**, 511–517.

Neary J.P., Malbon L. and McKenzie D.C. (2002) Relationship between serum, saliva and urinary cortisol and its implication during recovery from training. *J Sci Med Sport*, **5**, 108–114.

Neville V., Gleeson M. and Folland J.P. (2008) Salivary IgA as a risk factor for upper respiratory infections in elite professional athletes. *Med Sci Sports Exerc*, **40**, 1228–1236.

Nindl B.C. (2007) Exercise modulation of growth hormone isoforms: current knowledge and future directions for the exercise endocrinologist. *Br J Sports Med*, **41**, 346–348.

Nindl B.C., Kraemer W.J., Arciero P.J. *et al.* (2002) Leptin concentrations experience a delayed reduction after resistance exercise in men. *Med Sci Sports Exerc*, **34**, 608–613.

Nindl B.C., Alemany J.A., Kellogg M.D. *et al.* (2007) Utility of circulating IGF-I as a biomarker for assessing body composition changes in men during periods of high physical activity superimposed upon energy and sleep restriction. *J Appl Physiol*, **103**, 340–346.

Nosaka K., Clarkson P.M. and Apple F.S. (1992) Time course of serum protein changes after strenuous exercise of the forearm flexors. *J Lab Clin Med*, **119**, 183–188.

O'Connor P. and Corrigan D. (1987) Influence of short-term cycling on salivary cortisol levels. *Med Sci Sports Exerc*, **19**, 224–228.

Pedersen B.K. and Febbraio M.A. (2008) Muscle as an endocrine organ – focus on muscle-derived IL-6. *Physiol Rev*, **88**, 1379–1406.

Pedersen B.K., Akerstrom T.C., Nielsen A.R. and Fischer C.P. (2007) Role of myokines in exercise and metabolism. *J Appl Physiol*, **103**, 1093–1098.

Putlur P., Foster C., Miskowski J.A. *et al.* (2004) Alteration of immune function in women collegiate soccer players and collegiate students. *J Sports Sci Med*, **3**, 234–243.

Pyne D.B. and Gleeson M. (1998) Effects of intensive exercise on immunity in athletes. *Int J Sports Med*, **19**, S183–S194.

Raastad T., Glomsheller T., Bjøro T. and Hallén J. (2003) Recovery of skeletal muscle contractility and hormonal responses to strength exercise after two weeks of high-volume strength training. *Scand J Med Sci Sports*, **13** (3), 159–168.

Ramson R., Jürimäe J., Jürimäe T. and Mäestu J. (2009) Behavior of testosterone and cortisol during an intensity-controlled high-volume training period measured by a training task-specific test in men rowers. *J Strength Cond Res*, **23**, 645–651.

Riad-Fahmy D., Read G.F. and Walker R.F. (1983) Salivary steroid assays for assessing variation in endocrine activity. *J Steroid Biochem*, **19**, 265–272.

Rietjens G.J., Kuipers H., Adam J.J. *et al.* (2005) Physiological, biochemical and psychological markers of strenuous training-induced fatigue. *Int J Sports Med*, **26**, 16–26.

Robson-Ansley P.J., Blannin A. and Gleeson M. (2007) Elevated plasma interleukin-6 levels in trained male triathletes following an acute period of intense interval training. *Eur J Appl Physiol*, **99**, 353–360.

Robson-Ansley P.J., Gleeson M. and Ansley L. (2009) Fatigue management in the preparation of Olympic athletes. *J Sports Sci*, **27**, 1409–1420.

Rosendal L., Langberg H., Flyvbjerg A. *et al.* (2002) Physical capacity influences the response of insulin-like growth factor and its binding proteins to training. *J Appl Physiol*, **93**, 1669–1675.

Rowbottom D.G., Keast D., Goodman C. and Morton A.R. (1995) The haematological, biochemical and immunological profile of athletes suffering from the overtraining syndrome. *Eur J Appl Physiol*, **70**, 502–509.

Simsch C., Lormes W., Petersen K.G. *et al.* (2002) Training intensity influences leptin and thyroid hormones in highly trained rowers. *Int J Sports Med*, **23**, 422–427.

Smart D.J., Gill N.D., Beaven C.M. *et al.* (2008) The relationship between changes in interstitial creatine kinase and game-related impacts in rugby union. *Br J Sports Med*, **42**, 198–201.

Smith C. and Myburgh K.H. (2006) Are the relationships between early activation of lymphocytes and cortisol or testosterone influenced by intensified cycling training in men? *Appl Physiol Nutr Metab*, **31**, 226–234.

Smith D.J. and Norris S.R. (2000) Changes in glutamine and glutamate concentrations for tracking training tolerance. *Med Sci Sports Exerc*, **32**, 684–689.

Steinacker J.M., Lormes W., Kellmann M. *et al.* (2000) Training of junior rowers before world championships: effects on performance, mood state and selected hormonal and metabolic responses. *J Sports Med Phys Fitness*, **40**, 327–335.

Timon R., Olcina G., Maynar M. *et al.* (2008) Evaluation of urinary steroid profile in highly trained cyclists. *J Sports Med Phys Fitness*, **48**, 530–534.

Toft A.D., Jensen L.B., Bruunsgaard H. *et al.* (2002) Cytokine response to eccentric exercise in young and elderly humans. *Am J Physiol Cell Physiol*, **283**, C289–C295.

Tremblay M.S., Copeland J.L. and Van Helder W. (2005) Influence of exercise duration on post-exercise steroid hormone responses in trained males. *Eur J Appl Physiol*, **94**, 505–513.

Urhausen A., Kullmer T. and Kindermann W. (1987) A 7-week follow-up study of the behaviour of testosterone and cortisol during the competition period in rowers. *Eur J Appl Physiol*, **56**, 528–533.

Urhausen A., Gabriel H. and Kindermann W. (1995) Blood hormones as markers of training stress and overtraining. *Sports Med*, **20**, 251–276.

Urhausen A., Gabriel H. and Kindermann W. (1998) Impaired pituitary hormonal response to exhaustive exercise in overtrained endurance athletes. *Med Sci Sports Exerc*, **30**, 407–414.

Viru A. and Viru M. (2001) *Biochemical Monitoring of Sport Training*, Human Kinetics, Champaign, IL.

Viru A. and Viru M. (2003) Hormones in short-term exercises: anaerobic events. *Strength Cond J*, **25**, 31–37.

Warren G.L., Lowe D.A. and Armstrong R.B. (1999) Measurement tools used in the study of eccentric contraction-induced injury. *Sports Med*, **27**, 43–59.

Yan B., Wang G., Lu H. *et al.* (2009) Metabolomic investigation into variation of endogenous metabolites in professional athletes subject to strength-endurance training. *J App Physiol*, **106**, 531–538.

3.6 Body Composition: Laboratory and Field Methods of Assessment

Arthur Stewart[1] and Tim Ackland[2], [1]Centre for Obesity Research and Epidemiology, Robert Gordon University, Aberdeen, UK, [2]School of Sport Science, Exercise and Health, University of Western Australia, WA, Australia

3.6.1 INTRODUCTION

The concept of body composition – the constituents of the human body and their relative proportions – relates to both appearance (or phenotype) and functional capacity in terms of sports performance, as depicted in Figure 3.6.1. These attributes co-exist in an inter-dependent fashion, where change in one aspect invariably alters the others for athletes in most sports. As a result, some coaches have traditionally identified talented athletes by performance and appearance alone.

3.6.2 HISTORY OF BODY COMPOSITION METHODS

Understanding the body's make-up is not a new concept, and was championed by several ancient civilizations, most notably that of Greece. Hippocrates (c. 460–370 BC), the father of modern medicine, identified two different physique types (*physicus* and *apoplecticus*) and four bodily fluids (blood, dark bile, yellow bile, and phlegm) which were required to be in balance for optimum health. Archimedes (c. 287–212 BC) pioneered the science of densitometry, and correctly identified King Hieron's crown as an alloy and not pure gold. Polykleitos (c. 4th century BC) and other ancient sculptors generated the archetypal 'perfect physique' in works such as the 'Spear Bearer' by amassing anatomical data on several individuals, inadvertently founding the discipline of anthropometry. Much later, during the Renaissance period, Andreas Vesalius (1514–1564 AD) published artistic anatomical drawings of the muscular system, *De Corporis Humani Fabrica* (*Concerning the Structure of the Human Body*), which conveyed considerable detail from dissection studies. This followed the work of Leonardo da Vinci (1452–1519) and others at the commencement of the scientific revolution.

Against such a historic background, scientific enquiry in the 20th century posed military, health, nutritional, and more recently sports performance questions, and successfully developed these disciplines into a topic area which became established as 'body composition' in the latter half of the century. Medical, physiological, nutritional, and developmental specialists were to contribute to this new field. Jindrich Matiegka (1862–1962), a medic and lecturer in physical anthropology, used cadaver dissection to validate a tissue-specific model of mass fractionation, identifying skeletal, skin and adipose, and skeletal muscle masses (Matiegka, 1921). Albert Behnke (1903–1996) was a medic with the US Navy, and pioneered the densitometry method for body composition in humans, seeking to quantify nitrogen retention in adipose tissue. He also provided informative paradigms for future research efforts in the form of 'reference man' and 'reference woman', generic descriptors of dimensional and tissue proportions based on thousands of college and military subjects. During the 1980s, the Brussels Cadaver Study enabled revision of Matiegka's formulae for predicting tissue masses (Drinkwater *et al.*, 1986), representing a significant advance in quantifying the different tissues of interest.

It was not until the 1970s and 80s that body composition research was to focus specifically on athletes, after several landmark studies using generalized populations to validate techniques. At a time when strength and conditioning research was in its infancy, training methods themselves became more specialized and sophisticated, capable of conferring greater tissue adaptation. These circumstances were to generate difficulties for the scientist who professed body composition assessment methods were more precise or accurate than they are, or who used generalized body composition predictions which were not valid for athletes.

3.6.3 FRACTIONATION MODELS FOR BODY COMPOSITION

Different approaches are used in subdividing total body mass, from the whole body, to the tissues (and organs), the cells, the molecules, and finally the atomic level of the 44 chemical

Strength and Conditioning – Biological Principles and Practical Applications Marco Cardinale, Rob Newton, and Kazunori Nosaka.

elements in the human body (Wang, Pierson and Heymsfield, 1992). Classical approaches to body composition based on dissection have used tissues, whereas chemical component models have generally used molecular or atomic compartments (see Figure 3.6.2). What is important to appreciate is that all the different approaches to measuring body composition are assessing some parameter, but may not be at the same level of analysis. This can create problems when interpreting data from different methods. In addition, some methods can assess body composition by specific region, as well as across the whole, which enables much greater utility.

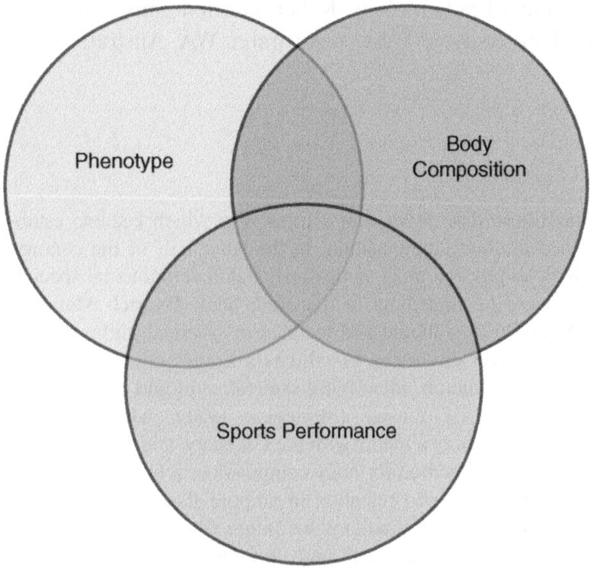

Figure 3.6.1 Inter-relationships of body composition and sport

3.6.4 BIOMECHANICAL IMPERATIVES FOR SPORTS PERFORMANCE

In biomechanical terms, athletes attempt to vary only a small number of imperatives: to maximize speed, force, or power; to minimize energy expenditure; and to optimize body moments in complex movements. To achieve these aims, the body seeks to optimize the forces of a movement, as, in a purely mechanical model, a perfect performance will be supported by a perfect physique, used optimally to the requirements of the task. As recently as the mid 1980s, there remained a somewhat simplistic view of body composition. Carter (1985) described the body mass as comprising 'productive mass' and 'ballast'; a concept which struck a chord with many athletes, who sought to minimize excess body fat and maximize muscle. It would be easy to believe that in strength sports, a large muscle mass results in a better performance, as this assumption underpins all weight-category sports. However, where weight is necessarily restricted, the biomechanics of the movement dictate whether a shorter, more muscular physique is advantageous, as in weightlifting (Gilbert and Lees, 2003), or whether a more slender physique with

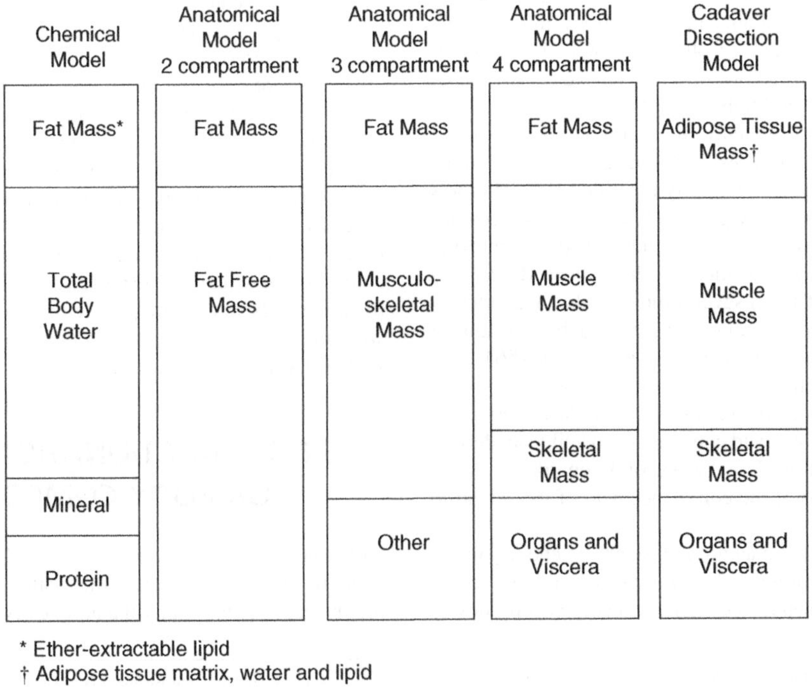

Figure 3.6.2 Theoretical models of mass fractionation

tall stature and long arm span is preferable, as in lightweight rowing (Stewart, 2001).

Coaches may assume that increased muscle mass will enhance force and power production, thereby justifying resistance training for athletes. However, more is not necessarily better when it comes to strength and body bulk, as added muscle must bear the burden of its additional weight, in consideration of the precise nature of the performance task. The desire to gain extra muscle mass must be counterpoised against the adverse effect on performance that additional weight might have. This is especially so given that a portion of weight gained through diet and training will comprise tissues other than muscle, including adipose tissue. Similarly, an athlete seeking to reduce total body mass to meet a weight category in their sport will lose a portion of their lean tissues in addition to adipose tissue, reducing strength accordingly. Contrary to what many athletes and coaches believed until recently, weight change is not a simple matter of gaining pure muscle and shedding pure adipose tissue, but involves a variety of tissues dependent upon initial body composition, rate of weight loss, and other factors (Hall, 2007).

In addition, the body requires a certain amount of essential fat for normal functioning. Adipose tissue provides several key functions beyond energy storage in the body, and historically its contribution to homoeostasis has perhaps been under-appreciated. Adipose tissue manufactures key chemicals, provides impact protection, affects temperature regulation, and effectively functions as a complex and essential endocrine organ (Kershaw and Flier, 2004).

Training paradigms operated by most coaches of elite athletes employ the cyclic concept of periodization, where the training programme is divided into units of differing magnitude and intensity, to provide structured overload, and sufficient rest and recovery to enable supercompensation. This is considered in detail in previous chapters. Such a programme in elite athletes generally has a four-year cycle, where the object is to maximize performance on the days that matter (trials and international competitions), rather than to maintain an unrealistic peak fitness for long periods. The anatomical expression of this was first described by Hawes and Sovak (1994) as the 'morphological prototype', where the ratio of adipose tissue to muscle or fat-free tissue alters with the proximity to the competition.

Recognizing that physique is a dynamic concept, and that a child's or adult's physique trajectory can be altered within certain limits, has enabled athletes to survive the rigors of increasingly ambitious and sophisticated training, which have conferred greater adaptation than ever before. Body composition analysis contributes to the knowledge base for effective performance management by establishing a necessary morphological platform for profiling, comparison with reference data, and tagging composition data to sports performance.

3.6.5 METHODS OF ASSESSMENT

There are well over 20 different methods of assessment for body composition, and whilst the majority of these have been used with athletes in specific cases, only a few are really practicable and worthy of recommendation for routine use for those on conditioning programmes. Of those not considered here, several are expensive medical imaging techniques used as diagnostic tools for health disorders, the funding for which can be justified on clinical grounds. For example, magnetic resonance imaging (MRI) is the 'gold standard' in terms of providing pin-sharp detailed pictures which quantify distances, areas, and volumes of specific tissues and structures. It achieves this non-invasively, and installations are becoming more widely available, but they remain beyond the scope of most exercise studies. Computed tomography (CT) also provides 3D images in cross-section, but at the expense of a significant radiation dose. This can be justified if patients have sufficient clinical need, but again, is not appropriate for routine use. Similarly, total body potassium (which uses the body's own naturally occurring decay of the K^{40} isotope) and neutron-activation analysis (where the body is bombarded by a neutron stream, inducing gamma rays and other phenomena which enable several chemical elements to be identified) are not practical for athlete monitoring. Facilities for these two approaches are rare, and are becoming limited to the nuclear power industry. Isotope dilution is a common technique in nutrition studies, involving the administration of a tracer (commonly tritium, deuterium, or oxygen-18) and calculation of total body water from its subsequent dilution. This approach, forming part of multi-component models, relies on the assumption that the tracer is not metabolized, is distributed equally throughout all water compartments of the body, and does not penetrate other tissues. The higher water flux incurred by exercising subjects mean that this method may be problematic, though still feasible, under some circumstances. Readers are referred to Heymsfield *et al.* (2005) for detail on this and the other methods not addressed here.

3.6.5.1 Laboratory methods

Densitometry

Densitometry involves measuring body mass and volume, calculating their ratio by underwater weighing, and relating this to the fat percentage (Brozek *et al.*, 1963; Siri, 1956). This partitions the body into fat-free mass (FFM) and fat mass (FM). Because fat is the only body constituent whose specific gravity is less than water (1.0), its buoyant force is counterpoised against the sinking force exerted by all other tissues. Fat content can thus be estimated, assuming the FM and FFM densities are constant and known. These are assumed to be approximately 0.90 g/ml and 1.10 g/ml respectively. While fat is constant in density, the same is not true for FFM, whose constituents may vary in density between individuals, as well as between their respective proportions of bone, muscle, and other tissues.

In practice, measuring human buoyancy in water is pivotally affected by the ability of the subject to exhale fully, and assessment of underwater weight must be made in conjunction with an assessment of entrapped air. This requires a tank of water that is sterilized, mixed, and heated to between 25 and 35 °C, with a submersible seat suspended from an autopsy scale or load cell above. Wearing swimwear, the participant is weighted in air to 0.01 kg and enters the water, rubs the skin surface to

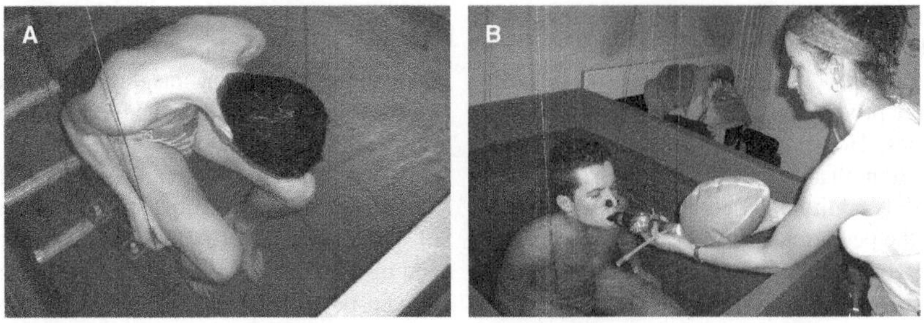

Figure 3.6.3 (a) Underwater weighing measurement. (b) Residual volume measurement

dislodge any air bubbles, and sits on the seat before exhaling fully and leaning forward, as depicted in Figure 3.6.3. The underwater weight is recorded to 0.01 kg as soon as the reading stabilizes (in practice about 5–10 seconds) and the individual is signalled to return to the surface. The procedure is repeated a minimum of three times and the tare weight is subtracted from the mean underwater weight.

Calculating body density (Db) relies on dividing mass in air by the measured volume V, which is:

$$V = [(Ma - Mw)/Dw] - A$$

where Ma is the body mass in air (kg), Mw is the body mass in water (kg), Dw is the density of water (g/ml) at the water temperature,[1] and A is the volume of entrapped air (ml).

Estimating the entrapped air volume involves measuring residual lung volume (RV) – the quantity of air remaining in the lungs following maximal expiration. Although RV can be predicted from mass, stature, age, and gender, this is unlikely to typify athletes who differ from norms in terms of thoracic dimensions and compliance of the chest wall. Thus RV should be measured at the time of underwater weighing, with the participant sitting in the tank, to simulate the hydrostatic pressure on the thorax. It can be measured directly by gas dilution using an inert gas such as helium, but the closed-circuit nitrogen/oxygen dilution method is more common (Wilmore *et al.*, 1980). This involves the participant wearing a nose clip and breathing pure oxygen from a spirometer, or more commonly an anaesthetic bag, within a closed and nitrogen-free spirometry system until nitrogen equilibrium is reached. The participant exhales maximally before connecting the system, breathing normally for five respiration cycles and exhaling maximally on the sixth. The mouthpiece is removed and the participant continues to breathe normally, and the closed-system gas composition is analysed for CO_2 and O_2 using a calibrated gas analyser. Then RV is calculated by measuring the nitrogen concentration (calculated by deducting CO_2 and O_2 from 100%) and the volumes of O_2 gases involved at the commencement and cessation of the test, using the following equation:

$$RV = \frac{VO_2 \times FEN_2}{0.798 - FEN_2} - DS \times BTPS\ \text{factor}$$

[1]Dw is 0.997 at 25 °C, 0.996 at 28 °C, 0.995 at 31 °C and 0.994 at 35 °C.

where RV is residual volume, VO_2 is the initial volume of O_2 in the anaesthetic bag, FEN_2 is the fraction of N_2 at equilibrium, DS is the dead space of mouthpiece and valve, and BTPS factor is the body temperature pressure saturated conversion factor.

Adjustment for gas in the gastrointestinal tract is necessary, and a constant 100 ml is normally added to RV in order to determine the entrapped air total (A). Density can be related to percentage fat via either of these equations:

$$\%\ \text{fat} = [(4.95/d) - 4.5] \times 100\ (\text{Siri, 1956})$$

$$\%\ \text{fat} = [(4.57/d) - 4.142] \times 100\ (\text{Brozek } et\ al., 1963)$$

While these equations produce similar results, predicted percentage fat is affected greatly by residual air (100 ml variation being equivalent to about 0.7% fat), and less so by errors in the recorded weight (100 g difference being equivalent to about 1% fat). By contrast, water temperature introduces a negligible effect.

In practice, densitometry presents problems for those subjects who have difficulty in exhaling fully or are not water-confident. Larger individuals need to take care not to contact the sides or bottom of the tank. The measuring apparatus can be customized for use in swimming pools, which can be effective provided water turbulence is not an issue.

Densitometry enjoyed the status of being the established reference method of choice during the landmark studies of the 1960s and 70s using skinfolds (Durnin and Womersley, 1974; Jackson and Pollock, 1978; Sloan, 1967; Wilmore and Behnke, 1969). Unfortunately for sports scientists today, many of these studies did not have athletes included in the sample, nor were the activity ranges of participants profiled. This presented a problem for their subsequent use with athletes, which was slow to be appreciated. Because of the known effects of exercise on bone and muscle, affecting their density and proportional mass, the assumption of constant density of the FFM is violated. This is elegantly illustrated in a key study of professional football players predicted to have −12% fat (Adams *et al.*, 1982), despite having significant and measurable superficial adipose tissue. A closer look at the participant sample reveals these individuals were highly strength-trained, and many were of Afro-Caribbean origin. Whereas such individuals' race might inflate bone density (Cote and Adams, 1993), resulting in greater density of the FFM (Wagner and Heyward, 2000) and thus reducing densitometry-predicted percentage fat, it has been suggested that muscle-mass increase would have the opposite effect, as the density of muscle is approximately 1.066 g/ml (Prior *et al.*, 2001). A study of 111 athletes suggested systematic deviations in the assumed FFM density, resulting in mean prediction errors in percentage fat of 2–5%. This work arrived at similar conclusions to a study of bone mineral density (Evans *et al.*, 2001), that the mass and density of FFM are not related to bone mineral mass or bone mineral density.

Air-displacement plethysmography (ADP)

Air-displacement plethysmography (ADP) is a relatively recent addition to available laboratory methods. The BOD POD body composition system (Life Measurement Inc., Concord, CA) has becoming established as an industry standard in nutritional and related studies of obese, military, and athletic groups. The BOD POD and more recently PEA POD (for infants up to 8 kg) have a similar theoretical framework to densitometry, in assessing body volume and relating the resultant density to percentage fat via a two-compartment model.

In ADP, the volunteer, wearing swimwear and a bathing cap, sits inside a sealed chamber (illustrated in Figure 3.6.4), linked to an adjacent chamber via a sealed diaphragm. Introduced perturbations trigger pressure fluctuations resulting from the volume changes. These establish the volume of the test chamber, both empty and with the participant inside. Because air behaves differently under isothermal and adiabatic conditions, scaling constants are applied to take account of the effects of hair, clothing, and body surface area. As with densitometry, lung volumes can be estimated, although athletes may not be typical of such reference values and so should be measured if possible. ADP calculates lung volumes from pressure–volume differences induced by tidal breathing inside the chamber. Once a normal breathing pattern is established, the airway is occluded at mid-exhalation, while the subject performs a gentle puffing manoeuvre. This induces pressure fluctuations, which are detected and related to volume by applying simple gas laws. Conformity with the procedure is addressed by calculating a score, referred to as the 'figure of merit', relating airway pressure to chamber pressure (Dempster and Aitkens, 1995). Values of <1 are considered acceptable for valid thoracic gas measurements.

Dempster and Aitkens (1995) used the BOD POD system to provide validation data between 25 and 150 l, a range encompassing all but the largest athletes. Precision errors for ADP have been shown to be slightly lower than for densitometry (1.7 versus 2.3 %CV for within-day reproducibility) (McCrory *et al.*, 1995). Predicted percentage fat in a sample of 69 football players was lower by ADP than by DXA or a three-compartment model using fat, bone mineral, and residual mass (Collins *et al.*, 1999). Factors including mixed ethnicity, the absence of a forced expiratory manoeuvre in ADP, and the DXA soft-tissue software not being benchmarked against such large muscular subjects may explain this difference. By contrast, a study comparing the body composition of 47 female athletes and 24 controls using ADP and DXA failed to detect significant differences (Ballard, Fafara and Vukovich, 2004) and the authors

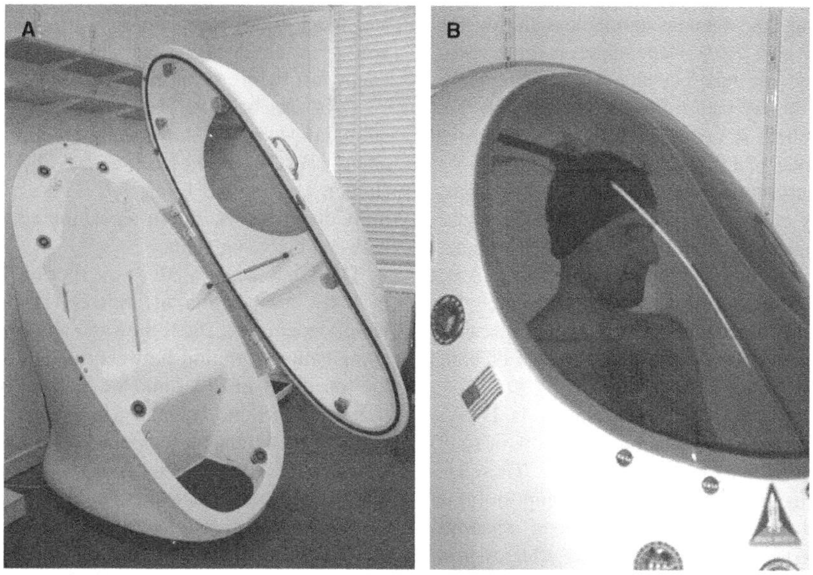

Figure 3.6.4 (a) The BOD POD chamber. (b) A volunteer being measured

concluded ADP to be reliable and valid for measuring female athletes.

Although densitometry has traditionally been used with athletes more frequently than ADP, ADP is certainly becoming more common, with its widespread use in obesity and paediatric research and its appeal for professional sports teams who wish to establish dedicated facilities. To date, studies using ADP with athletes have shown differences from densitometry (both under- and over-estimating predicted percentage fat), suggesting that, despite their similar underlying rationale, the methods are not interchangeable. Although cheaper than weighing-tank facilities, ADP is sensitive to air currents and building vibrations, which both invalidate its calibration, and the location of ADP facilities requires careful consideration.

Dual-energy X-ray absorptiometry (DXA)

Dual-energy X-ray absorptiometry (DXA) was originally developed as a scanning system for investigating bone, adapted from earlier photon absorptiometry using gamma-emitting radioisotopes. When X-rays pass through an absorbing substance, they attenuate according to the atomic number of the molecules in their path; using two energies, there is a steeper attenuation at the lower energy than the higher. As a result, the ratio of the attenuation at the two energies (the R-value) varies between tissues and can be used to measure tissue composition. Lipids predominantly comprise hydrogen, carbon, and oxygen, while lean soft tissue contains several other elements of higher atomic mass, such as sodium, chlorine, potassium, and phosphorous, and bone mineral has a predominance of calcium and phosphorous. Lipids have R-values of 1.20–1.22; other soft tissues 1.29–1.39; and bone mineral 2.86.

A DXA scanner comprises an X-ray source, scanning table, detection arm, and associated computer hardware and software. For measurement, the volunteer lies supine wearing a hospital gown or sports clothing, and remains motionless during the scan, as illustrated in Figure 3.6.5. Depending on the scanner type, a whole-body scan can take between 4 and 15 minutes, depending on whether the ray path follows an array (fan) or a pencil beam (see Pietrobelli *et al.*, 1996 for a comprehensive summary on the principles of DXA).

DXA effectively maps the R-values of each pixel in a scan area, and automatically calibrates these against known tissue equivalents. The dedicated system software then differentiates bone from non-bone, and fat tissue (FT) from fat-free soft tissue (FFST). A fat distribution model is applied which predicts the soft tissue in the pixels that contain bone; the resulting scan is a regionalized map of the body detailing the masses of bone mineral, FFST, and FT for each body segment.

Perhaps the biggest methodological weakness is DXA's inability to measure soft tissue composition in the 40% or so of pixels in a healthy adult's scan which contain bone. These are predicted according to the gradient of differing composition in 'soft-tissue shells' and projected behind the bone 'shadow' (Nord and Payne, 1995). Lean individuals and athletes may have relatively fewer non-bone pixels from which this prediction can be made.

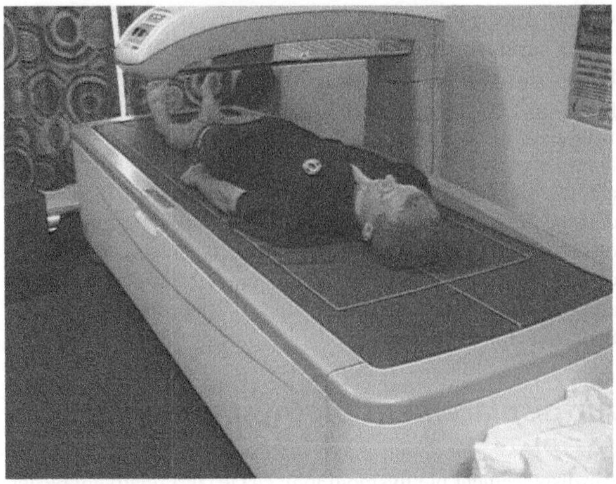

Figure 3.6.5 A volunteer being assessed by the Lunar Prodigy Scanner

Also, an individual's size can be a limitation in two ways. First, in larger or strength-trained individuals, the tissue depth is a factor attenuating the X-ray beam, and for those with chest depth greater than 25 cm, composition predictions can be unreliable (Jebb, Goldberg and Elia, 1993). Second, scanning tables are generally about 185–190 cm long and several athletes exceed this even with the legs abducted. In practice, the head and neck can be scanned separately from the torso and the data combined retrospectively (Prior *et al.*, 1997).

DXA introduces a low-dose radiation equivalent to about one transatlantic flight, and it is important to understand that compliance with regulations by government health authorities and local ethical procedures is essential. In some situations, women of childbearing age are required to undertake a pregnancy test before being scanned.

DXA has been validated for accuracy using phantom and animal models, and shows good precision in humans, typically about 3% CV for FM, 0.8% for FFST, and 0.9% for bone (Fuller, Laskey and Elia, 1992). Poorer precision is found in individual regions, due to repositioning affecting the boundary between body segments.

DXA has been used as a criterion for predictive models of body composition in athletes (De Lorenzo *et al.*, 2000; Houtkooper *et al.*, 2001; Stewart and Hannan, 2000a). Regional composition variation between athletic groups has confirmed the specificity of FFST and the 'generality' of their fat distributions (Stewart and Hannan, 2000b), and has confirmed a hierarchical paradigm of regional fat loss as a result of exercise (Nindl *et al.*, 1996). DXA is also used to predict muscle mass by summing the appendicular FFST and multiplying by an assumed constant to reflect torso muscle. The limb : torso ratio was assumed to be 3 : 1 for early predictions, leading to the constant 1.33 (Hansen *et al.*, 1993), based on previous cadaver dissection studies. However, values of 1.412 and 1.241 were derived using DXA in an athletic sample of male and females, respectively (Stewart, 2003). Later muscle-mass predictions based on CT (Visser *et al.*, 1999) or MRI (Kim *et al.*, 2002) used much larger samples, the latter producing evidence of a gender effect on muscle partitioning, despite those on training programmes being excluded.

Seasonal changes in athlete body composition have also been investigated using DXA (Egan *et al.*, 2006). These have been interpreted as alterations in FM and FFM, reflecting different emphases in conditioning at different stages of the competitive season. However, researchers attempting to replicate this should consider carefully whether changes could be attributed to fluid and glycogen fluctuations (detected as FFST by DXA), although with professional soccer players standardizing the conditions of prior exercise is problematic.

In summary, DXA offers some advantages beyond densitometry or ADP, enabling total and regional composition estimation. However, despite efforts at cross-calibration between equipment manufacturers, results currently remain specific to each scanner and software version.

3D body scanning

In the late 1990s, 3D body scanners became accurate, reliable, and affordable (Olds and Honey, 2006). They were developed largely for clothing manufacture applications and represented a dramatic step forward in enabling dimensional data to be collected easily. This has considerably enhanced the capability for measuring large groups in survey situations, and avoids transcription errors of conventional anthropometry. Their output includes linear and curved distances, cross-sectional areas, surface areas and volumes, and they have already made an impact in obesity research (Wells, Ruto and Treleaven, 2008). 3D scanners work by measuring the participant in form-fitting clothing and take approximately 10 seconds to capture body shape. There are three basic ways this can be done: using light, laser, or radio-wave technology. Each generates a 3D shape from which dimensional data can be extracted by system software.

One approach uses photogrammetry from the distortion of projected light on the body, for example TC^2 (Cary, NC, USA) or Wicks and Wilson Ltd. (Basingstoke, UK). An alternative uses low-power millimetre waves in a linear array with a rotating wand, which can profile body shape through clothing, for example Intellifit Corporation (Plymouth Meeting, PA,USA). Finally, lasers can be set in columns, with cameras triangulating the position of the reflected beam, for example the body line scanner (Hamamatsu Photonics K.K., Hamamatsu City, Japan), the Vitus smart (Vitronic, Wiesbaden, Germany), and the WBX (Cyberware Inc, Monterey, CA, USA).

Data from the different cameras are registered together, and a whole-body scan typically has 6–10 × 10^5 data points (referred to as a 'point cloud'). Primary landmarks are applied to key points such as the vertex, crotch, and axilla, which divides the scan into regions. Each data point is joined to its neighbouring points to produce a wireframe mesh, and data gaps can be filled by some software applications to produce a 'watertight' 3D object file that can be printed on a rapid prototype machine.

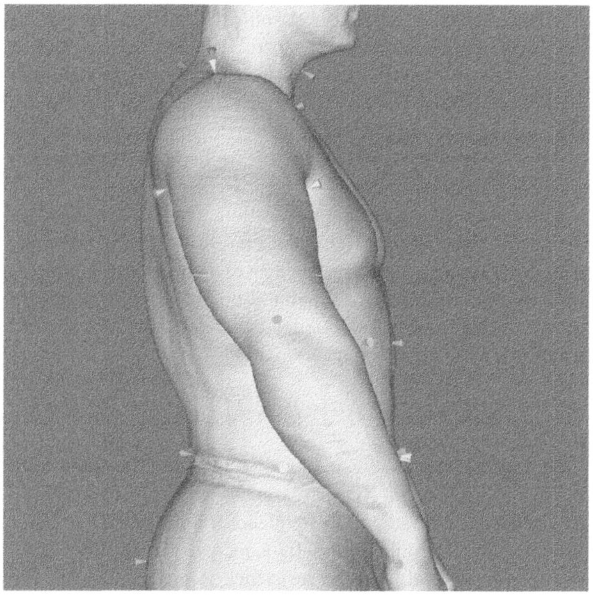

Figure 3.6.6 Automated landmarking system using the Hamamatsu body line scanner system

Secondary automated landmarks are located via proportional distance or shape-change algorithms; these are illustrated in Figure 3.6.6. They enable a range of measures to be made virtually instantaneously. Landmarks can also be placed manually, by using of reflective discs, or digitally, by software tools, to enable measurement of any region.

The capability of 3D scanners to assess body volumes enables the possibility of direct body composition using the two-compartment method. Validation of the Hamamatsu laser scanner was shown to be accurate in 92 individuals weighing 23–182 kg (Wang *et al.*, 2006), but all volunteers had to maintain a full exhalation throughout the 10-second scan. The utility of shape analysis has enabled large clothing survey data, highlighting gender differences in shape variation by BMI category (Wells, Treleaven and Cole, 2007), inter-ethnic comparison (Wells *et al.*, 2008), and age (Wells, Cole and Treleaven, 2008), and has allowed monitoring of shape change with weight loss (Stewart, Nevill and Johnstone, 2009). The utility of 3D scanners in assessing the effect of strength and conditioning training on shape is clear. Future research will no doubt exploit this technology in a range of sporting groups, as well as in clinical applications.

3.6.5.2 Field methods

Bioelectric impedance analysis

An electric current will be conducted or impeded depending on the type of body tissue it passes through. Bioelectrical impedance analysis introduces a source of alternating current (too small for the individual to perceive) and detects it in a different body location. Because water (and the salts it contains) conducts electricity well, whereas other molecules – principally lipid – are poor conductors, bioelectric impedance analysis (BIA) predicts the water content of the body and 'translates' this into composition.

Impedance (Z), the total opposition to electrical flow, is dependent on resistance (R) – the opposition within the conducting substance – and reactance (Xc) – the circuit's opposition to alternating current, due to build-up of electric or magnetic fields. The relationship is frequency-dependent, and is expressed in the following equation:

$$Z(\Omega) = \left[R^2 + Xc^2\right]^{0.5}$$

At low frequencies, Z = R as Xc approaches zero. With increasing frequency, Xc increases as a result of multiple current pathways, but at high frequencies it falls again to zero. The impedance profile across frequencies is specific to each conducting material, and the resistance is proportional to the material's cross-sectional area. Muscle has a relatively low impedance, whereas adipose tissue's impedance is high. The volume of the conductor is in proportion to the square of its length divided by the resistance. However, in the living tissues of the body, reactance is caused by cell membranes acting as capacitors to store charge, retarding conductivity. At frequencies below 50 kHz, the current flows via extracellular water, but with increasing frequency, current penetrates cells, and multi-frequency devices are theoretically able to differentiate between intra- and extracellular water; measurement bias depends on the subject's water distribution and physique (Deurenberg, Tagliabue and Schouten, 1995). Although others have questioned whether multi-frequency devices improve body composition predictions beyond those of single-frequency devices (Simpson *et al.*, 2001), higher frequencies up to 300 kHz have proved better than 50 kHz at estimating segmental muscle mass (Pietrobelli *et al.*, 1998), which could be a consideration for athletes.

In practice, BIA devices have traditionally had a tetra-polar arrangement, where the source and detection electrodes are placed on the dorsal surfaces of the right hand and foot approximately 5 cm apart, to avoid interference. Some methods use electrodes on both sides of both arms and legs (Malavolti *et al.*, 2003), which could be useful in addressing body asymmetry. The distance between the source and detection electrodes is the conductor length. This assumes that a body segment comprises an isotropic material and that the body can be represented by limbs and torso cylinders of similar impedance (the head is discounted). In practice, body tissues are considerably more complex, especially in the torso. Newer devices involving hand-to-hand or foot-to-foot arrangements (which incorporate measurements of body mass) sample only a portion of the body, and extrapolate to the whole.

The BIA measurement approach makes assumptions concerning the hydration of tissues and their proportions. However, strength training and conditioning programmes introduce factors which make it easy for participants to violate these assumptions. First, exercise elevates temperature, affecting specific resistivity. Second, intensive exercise training triggers wide fluctuations in hydration status. Third, approximately 97% of the total impedance comes from the limbs, and only 3% from the torso – where perhaps 50% of body fat might be located. Fourth, prediction models use gender, age, stature, and mass to collectively account for up to 85% of the variation in body-fat prediction. Finally, some devices only measure resistance and ignore reactance.

The problematic nature of estimating total body composition in exercising individuals using BIA has triggered work using segmental impedance in a sample of trained and untrained individuals, validated against ADP with standard errors of 5.2–5.5% (Ishiguro *et al.*, 2005), and validated by MRI to predict torso muscle volume with a standard error of 8.5% (Ishiguro *et al.*, 2006).

As with anthropometric equations, BIA predictions need to be validated against a reference method such as densitometry or DXA. The sample used for this might have specific inclusion criteria such as gender, age, and race, which are unique and not applicable to other groups. Although generalized equations have been proposed, strength athletes with high total body mass and muscle mass, but low body fat, will be misrepresented. In addition, a large number of other factors confound valid measurements, with ambient temperature, hydration status, and intake of diuretic substances all affecting the measured impedance (Heyward, 2004). In practice, these can make measurement of individuals on conditioning programmes problematic for two reasons: first, the difficulty in compliance with the conditions for valid measurement, and second, the lack of evidence that such individuals have complied. For these reasons, many practitioners avoid using BIA on individuals who are undertaking conditioning programmes.

Anthropometry

Anthropometry involves surface dimensional measures of the body. These include skinfolds, girths, and skeletal lengths and breadths. Despite their underlying simplicity of concept, these serve as useful surrogate measures for adiposity, muscularity, or frame size. Each set of measurements will be considered independently.

Skinfold measurements have been performed for about a century to indicate relative fatness, and to date in excess of 100 prediction equations relate the thickness of skinfolds to a fat mass or percentage fat, but the assumptions made in order to treat the data in this way are of questionable validity. Accurate measurement of skinfolds relies on the correct location of measurement sites and skilled use of skinfold callipers to provide a valid and reliable measurement. The skinfold score corresponds to a compressed double layer of adipose tissue plus skin, as shown in Figure 3.6.7.

Skinfold sites are marked on the skin after accurate landmarking, which usually relates to the underlying skeletal structure. The site is marked to describe the orientation of the skinfold (long axis) and the positioning of the measurer's left index finger and thumb. Landmarking is the cornerstone of reliable measuring, because much of the measurement error is attributable to skinfold location, since the depth of adipose tissue changes over a short distance on the skin surface (Hume and Marfell-Jones, 2008). A full description of these landmarks is beyond the scope of this chapter; the reader is referred to Olds and Tomkinson (2009) for further information.

The other main source of error lies in the technique employed, and it is recommended that a standardized protocol be used, such as the International Society for the Advancement of Kinanthropometry (ISAK) method, used in over 40 countries worldwide (Marfell-Jones *et al.*, 2006).

A further source of error relates to the calibration of the callipers, which alters as a result of extended use, humidity, and accidental impact damage. Calibration requires both the analogue dial (reflecting the actual aperture of the calliper blades) and the compression force to be checked. Callipers can be returned to manufacturers for calibration, or checked using material of known thickness, hardness and compressibility.

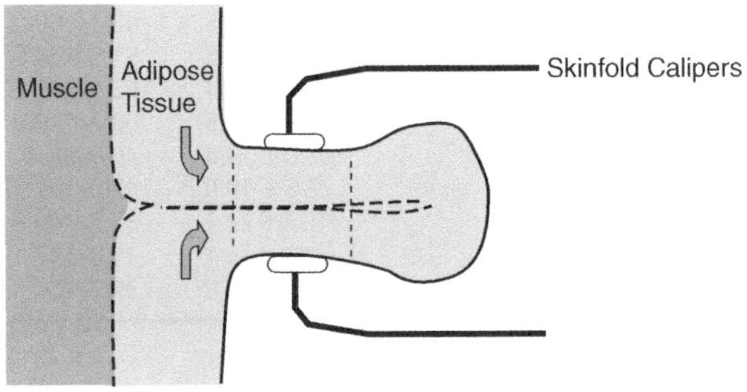

Figure 3.6.7 Cross-section of a skinfold

Skinfold measurements

For each of the following skinfolds, the calliper is held in the right hand and the fold is lifted and held by the thumb and index finger of the left hand. This skinfold, which does not include muscle, encompasses a double layer of skin, as well as the underlying adipose tissue, and must be held throughout the measurement. The calliper must be applied at right angles to the fold so that the pressure plate is 1.0 cm from the left thumb and index finger. A reading is taken two seconds after the application of the calliper and the score is recorded to the nearest 0.1 or 0.5 mm, depending on the calliper type. The ISAK protocol recommends taking six to eight skinfolds (all on the right side of the body) so as to provide a representative sample of subcutaneous fat deposition and thereby account for inter-individual differences in adipose tissue patterning. One measurement from each site should be made in the following order; this should be repeated to record either the mean of two or the median of three scores.

Triceps skinfold

The triceps site is located at the mid-position between the acromiale and radiale landmarks on the most posterior aspect of the arm when the forearm is in the mid-prone position. A vertical fold is lifted by the left thumb and index finger so that the landmark is midway between the lower edges of the thumb and index finger. The calliper jaws are applied 1 cm below the marked site, to a similar depth on the fold as the left thumb and index finger.

Subscapular skinfold

The inferior angle of the scapula (subscapulare; see Figure 3.6.8a) is marked with the subject standing erect and the arms by the side. Using a tape, the skinfold site is marked 2 cm obliquely downward and laterally from the subscapulare mark.

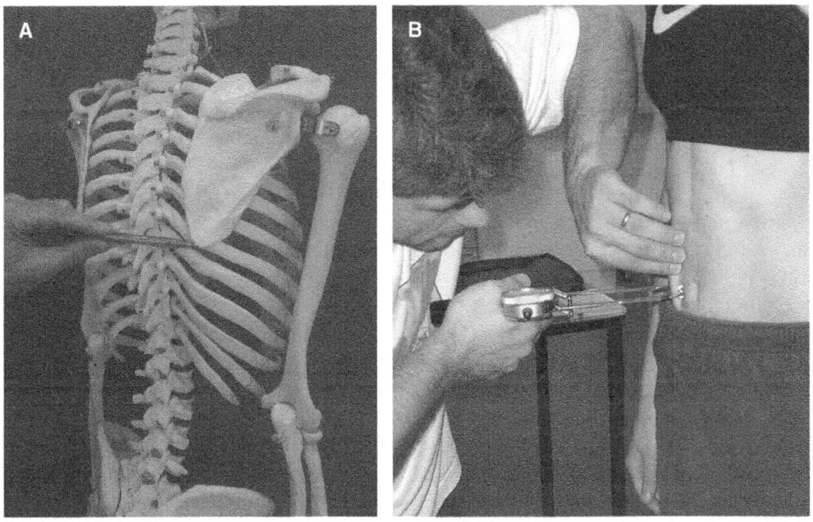

Figure 3.6.8 (a) Identifying the subscapulare landmark. (b) Measuring the abdominal skinfold

Table 3.6.1 Normative data for international- and national-level female athletes. Reprinted, with permission, from D.A. Kerr and A.D. Stewart, 2009, Body composition in sport. In Applied Anatomy and Biomechanics in Sport, 2nd ed., edited by T.R. Ackland, B.C. Elliott, and J. Bloomfield (Champaign, IL: Human Kinetics), p. 82

Sport	Level	Position/event	Skinfold sum (mm)[a]		
			Number of subjects	Mean	Range
Athletics[b]	National	SASI jumps	4	61.1 ± 12.7	41.7–72.8
		SASI throws	9	95.3 ± 49.4	53.0–203.7
		SASI sprint	7	60.3 ± 11.9	45.1–83.9
		SASI middle distance	20	59.2 ± 19.6	37.4–110.6
		SASI long distance	6	51.3 ± 8.8	40.4–68.3
		Scotland distance runners[f]	10	57.9 ± 14.9	32.2–72.6
Basketball[c]	International	Guard	64	76.6 ± 22.2	36.4–143.5
		Forward	65	76.0 ± 20.1	40.9–131.7
		Centre	47	88.0 ± 21.1	45.7–146.8
Cricket[b]	National		27	90.8 ± 19.7	55.9–141.1
Cycling, Road[b]	National		32	61.9 ± 12.0	33.8–89.5
Diving[d]	International		39	65.6 ± 17.0	32.1–114.3
Gymnastics[b]		SASI elite	68	37.9 ± 6.1	27.4–57.6
Hockey[b]		SASI senior	57	87.4 ± 18.5	48.1–140.3
Netball[b]		SA senior	33	83.4 ± 17.3	51.5–124.0
Rowing[b]		SASI lightweight	24	73.4 ± 13.4	55.5–105.2
		SASI heavyweight	30	87.5 ± 17.8	60.7–119.4
Triathlon[e]	International		19	62.8 ± 13.4	40.3–98.4
Swimming[d]	International		170	72.6 ± 19.6	37.9–147.1
Synchronized swimming[d]	International		137	81.7 ± 22.1	37.5–145.8
Volleyball[b]		SASI senior	29	90.5 ± 25.1	35.8–147.1
Waterpolo[d]	International		109	89.8 ± 23.8	39.7–151.6

[a]Sum of seven skinfolds (unless otherwise indicated) = triceps, subscapular, biceps, supraspinale, abdominal, front thigh, medial calf.
[b]Adapted with permission from the South Australian Sports Institute and published previously by Woolford *et al.*, 1993.
[c]Ackland *et al.*, 1997b.
[d]Note: sum of six skinfolds from Carter and Ackland (1994) = triceps, subscapular, supraspinale, abdominal, front thigh, medial calf.
[e]Note: sum of eight skinfolds from Ackland *et al.*(1998) = triceps, subscapular, biceps, iliac crest, supraspinale, abdominal, front thigh, medial calf.
[f]Sum of eight (all ISAK sites). Source: unpublished PhD thesis, Stewart (1999) University of Edinburgh; data collected 1996–1998.

An oblique fold is lifted and the calliper is applied 1 cm lateral to the left index finger and thumb, perpendicular to the line of the fold.

Biceps skinfold

This site is also located at the mid-position between the acromiale and radiale landmarks, but at the most anterior aspect of the arm when the forearm is in the mid-prone position. A vertical fold is lifted by the left thumb and index finger so that the landmark is midway between the lower edges of the digits. The calliper jaws are applied 1 cm below the marked site, to a similar depth on the fold as the left thumb and index finger.

Iliac crest skinfold

The subject is directed to place the right arm across the chest. A mark is then made on the iliac crest in the mid-axillary line (iliocristale). The left thumb is placed on the mark and the left index finger a sufficient distance superior to this point so as to lift an appropriate fold. A natural fold of skin (slightly downward anteriorly) is lifted with the left hand and the calliper is applied 1 cm anterior to the left index finger and thumb.

Supraspinale skinfold

To locate this site, an imaginary line is projected from the iliospinale landmark to the right anterior axillary border (armpit). A point is marked on this imaginary line at a level that is horizontal to the iliac crest mark (iliocristale). An oblique fold (downwards medially at about 45°) is lifted at this site and the calliper is applied 1 cm medial to the left index finger and thumb.

Abdominal skinfold

The site for this skinfold is 5 cm to the right side of the midpoint of the umbilicus (omphalion). A vertical fold is lifted with the left index finger and thumb at the site and the calliper is applied 1 cm inferiorly (Figure 3.6.8b).

Table 3.6.2 Normative data for international- and national-level male athletes. Reprinted, with permission, from D.A. Kerr and A.D. Stewart, 2009, Body composition in sport. In Applied Anatomy and Biomechanics in Sport, 2nd ed., edited by T.R. Ackland, B.C. Elliott, and J. Bloomfield (Champaign, IL: Human Kinetics), p. 83

Sport	Level	Position/event	Skinfold sum (mm)[a]		
			Number of subjects	Mean	Range
Athletics[b]	State	SASI pole	3	46.8 ± 0.3	46.4–47.1
		SASI sprint	4	56.1 ± 2.2	53.9–58.3
		SASI middle distance	9	38.6 ± 12.0	25.8–68.2
		SASI long distance	4	49.8 ± 6.4	41.3–56.4
Australian Rules Football[b]	National	Under 17 yrs	20	67.2 ± 6.9	44.7–104.1
Boxing[b]	State		13	57.5 ± 17.7	34.2–95.2
Cricket[b]	National		22	77.8 ± 23.0	52.3–135.2
Cycling[b]	State	Road	24	58.1 ± 11.9	42.9–85.0
	Scotland National	Road and time trial	16	69.0 ± 17.2	44.2–101.6
	National	Track	83	53.9 ± 12.7	26.4–85.3
Diving[c]	International		43	45.9 ± 11.4	28.0–79.7
Gymnastics[b]	State	SASI elite	41	41.6 ± 7.2	27.5—59.1
Hockey[b]	State	Under 21 squad	22	59.4 ± 17.0	38.7–107.2
Kayaking[b]	State	SASI senior	64	58.0 ± 14.0	37.4–96.7
Orienteering (and fell running)[d]	National	Scotland	12	60.6 ± 16.7	42.9–96.3
Racket sports	Area and National	Scotland	10	76.1 ± 27.6	42.7–121.9
Rowing[b]	State	SASI lightweight	27	45.2 ± 6.5	35.8–65.1
	State	SASI heavyweight	18	66.9 ± 18.0	46.1–111.8
	Scotland Area and National	Mixed light & heavyweight[d]	15	70.2 ± 12.5	50.7–91.1
Rugby Union[b]	State	SASI senior	58	92.2 ± 32.9	50.6–223.2
	International	Scotland team[d]	11	93.3 ± 30.3	62.0–147.8
Triathlon[e]	International		19	48.3 ± 10.2	36.8–85.9
	Area and National	Scotland[d]	10	63.7 ± 12.6	34.4–77.3
Swimming[c]	International		231	45.8 ± 9.5	26.6–99.9
Volleyball[b]	State	SASI senior	17	56.8 ± 13.2	36.9–79.6
Weightlifting[b]	State	SASI squad	47	74.9 ± 34.4	33.9–190.2
Waterpolo[c]	International		190	62.5 ± 17.7	27.9–112.1

[a]Sum of seven skinfolds (unless otherwise indicated) = triceps, subscapular, biceps, supraspinale, abdominal, front thigh, medial calf.
[b]Adapted from with permission from the South Australian Sports Institute and published previously by Woolford *et al.*, 1993.
[c]Note: Sum of six skinfolds from Carter and Ackland (1994) = triceps, subscapular, supraspinale, abdominal, front thigh, medial calf.
[d]Note: Sum of eight skinfolds from Ackland *et al.*(1998) = triceps, subscapular, biceps, iliac crest, supraspinale, abdominal, front thigh, medial calf.
[e]Note: Sum of eight (all ISAK sites). Source: unpublished PhD thesis, Stewart (1999) University of Edinburgh; data collected 1996–1998.

Front thigh skinfold

With the subject seated, the site for the front thigh skinfold is located on the anterior thigh, at the mid-position between the inguinal crease and the most anterior aspect of the patella. A fold is lifted parallel to the shaft of the femur at the site and the calliper is applied 1 cm distal to the left index finger and thumb. Where this fold is difficult to obtain, the subject should support the hamstring musculature to relieve tension from the skin.

Medial calf skinfold

This site is located on the medial aspect of the leg at the level of the greatest girth whilst the subject is standing. The subject may then raise the right leg to rest on a measurement box to assist the measurer. A vertical fold is lifted at the site and the calliper is applied 1 cm distal to the left index finger and thumb.

The most robust method of reporting skinfold data is to simply sum the scores for six or eight sites. Normative data are presented for comparison purposes in Tables 3.6.1 and 3.6.2. Alternatively, six skinfold values (triceps, subscapular, supraspinale, abdominal, front thigh, and medial calf), together with stature and mass, can be used to create O-scale adiposity and proportional weight scales for comparison over time or against population norms (Ward *et al.*, 1989).

Girth measurements

Segmental girths are often used as a surrogate measure of adiposity, particularly in obese individuals for whom skinfold

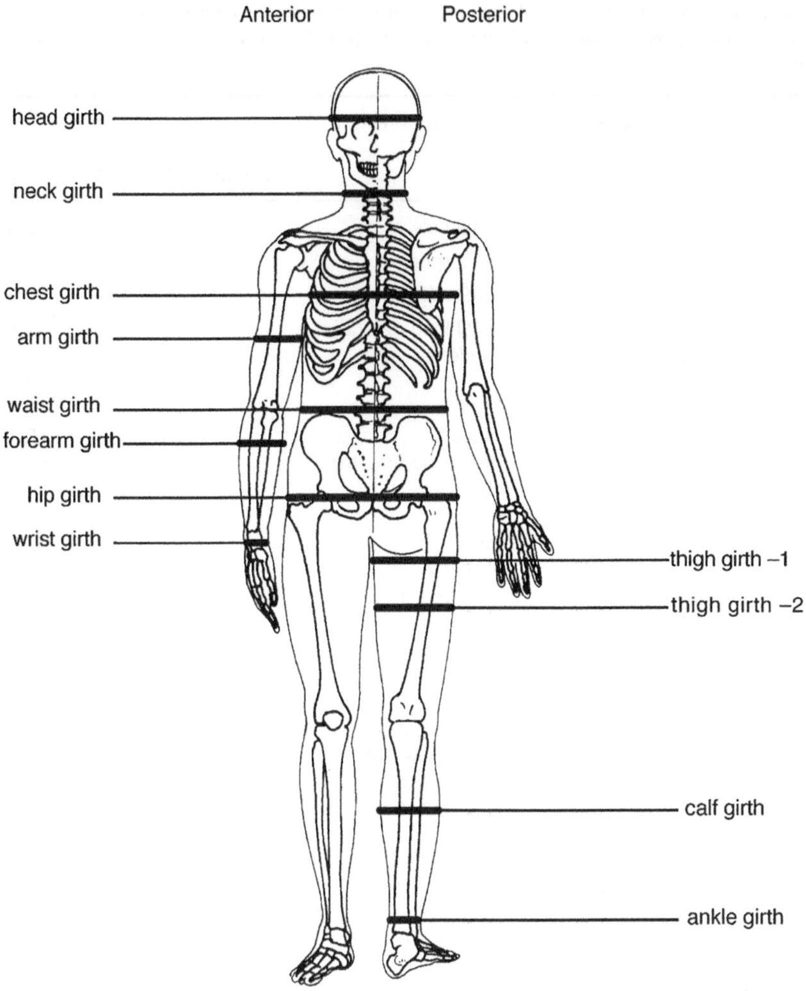

Figure 3.6.9 ISAK standard sites for girth measurement

testing is not possible. In contrast, for very muscular athletes, girth measures allow monitoring or comparison of muscularity or the robustness of physique (mesomorphy). For each of the following girths, the tape (of metal construction, preferably) must lie against the skin (spanning any concavities), but not drawn so tightly as to compress it. Using a light touch (except where indicated below), the measurer can ensure the tape follows the skin contour and is manipulated into position. Not all ISAK girth measurements are listed here, only those sites of particular relevance to athlete monitoring. The reader is referred to Marfell-Jones *et al.* (2006) for full descriptions of all anthropometric measurements. The locations of girth measurement sites are shown in Figure 3.6.9.

Head girth

With the subject seated and the head in the Frankfort plane, the tape is placed horizontally around the head immediately superior to the glabella (portion of the frontal bone between the supraorbital ridges). Firm pressure is applied to compress the hair.

Neck girth

The subject remains seated with the head in the Frankfort plane while the tape is placed around the neck, perpendicular to the long axis and at a level immediately above the larynx. Measurement is made without constricting the skin.

Relaxed arm girth

The subject is directed to stand with the right arm hanging naturally by the side. The mid-distance between acromiale and radiale landmarks is marked. The tape is placed around the arm at this level (perpendicular to the shaft of the humerus) and a measurement is made.

Flexed arm girth
The subject is directed to raise the arm forward to the horizontal, supinate, and the forearm is flexed to 90° or more. The tape is placed around the arm and the subject is encouraged to isometrically contract the forearm flexor and extensor muscles. The tape is moved to the peak of the biceps brachii muscle and measurement is made without compressing the skin.

Forearm girth
With the subject's arm supinated and placed by the side, the tape is moved to the maximum forearm girth (distal to the humeral epicondyles) and drawn together with sufficient tension that the skin is not compressed.

Wrist girth
The tape is mved down to the minimum girth at the wrist (distal to the styloid processes) and measurement is made without constricting the skin.

Chest girth
The tape is passed horizontally around the subject's chest at the level of the mesosternale. The subject may then lower the arms to a comfortable position by the sides. Chest girth is measured at the end of a normal expiration (end tidal).

Waist girth
The tape is placed horizontally around the waist and aligned at the level of the minimum girth (between the lower costal border and the iliac crest). Measurement is made without indenting the skin at the end of a normal expiration (end tidal).

Hip girth (gluteal girth)
The subject should wear only light clothing and stand erect with feet together. With the measurer standing to the side, the tape is placed horizontally around the hips at the level of the greatest posterior protuberance of the buttocks.

Thigh girth 1
The subject is directed to stand erect with the feet slightly apart and weight equally distributed. The tape is passed around the thigh at a horizontal position 1 cm below the gluteal fold. It is then drawn together with sufficient tension that the skin is not constricted.

Thigh girth 2 (mid-thigh girth)
The mid-position between trochanterion and tibiale laterale landmarks is measured and marked. With the measurer standing to the side, the tape is placed horizontally around the thigh at this level and measurement is made without constricting the skin.

Calf girth
With the subject standing, the tape is placed horizontally around the leg at the maximum girth. The tape is then drawn together with sufficient tension that the skin is not constricted.

Ankle girth
The tape is moved down to the minimum girth, which is just superior to the medial malleolus. Measurement is made without constricting the skin. Note that the tape is not horizontal for this site.

Breadth measurements
Skeletal breadths are also used as surrogate measures of frame size and robustness of physique. For each of the following breadths, the calliper (large sliding calliper or wide-blade bone calliper) must make firm contact with the skin that overlies each skeletal landmark. Not all ISAK breadth measurements (Marfell-Jones *et al.*, 2006) are listed here; only those sites of particular relevance to athlete monitoring are included.

Biacromial breadth
The subject stands with arms relaxed by the side. From behind the subject and with the pointers of the calliper angled upward at 45°, the distance between left and right acromion processes is measured. Firm pressure is applied to the acromion process during the measure, without pushing the scapulae together.

Transverse chest breadth
The subject is seated with the hands resting on the knees. The calliper is angled downward at about 30° as the breadth of the chest is measured at the mesosternale level, with care being taken to avoid the pectoral and latissimus dorsi muscle contours. Moderate pressure is required on the pointers and the measure is taken at the end of a normal expiration (end tidal).

Biiliocristal breadth
The subject stands with arms crossed over the chest. While the measurer stands in front of the subject and with the pointers angled upward at 45°, the distance between left and right iliocristale is measured. Firm pressure is applied during the measurement.

Biepicondylar humerus breadth
The subject is directed to raise the arm forward to the horizontal position and flex the forearm to 90°. The arms of the bone calliper are pointed upward at about 45° and the greatest distance between lateral and medial epicondyles of the humerus is measured. Firm pressure on the arms of the calliper is required for this measurement.

Biepicondylar femur breadth
With the subject seated so that knee is bent to 90°, the femoral epicondyles are palpated. The bone callipers are angled downward at about 45° and, with firm pressure applied, measurement is made across the femoral epicondyles.

Anterior–posterior chest depth
With the subject seated with arms by the side, the sliding calliper branch is positioned horizontally at the level of the mesosternale. The posterior arm of the calliper is positioned on the nearest vertebral spinous process. Using only light pressure, this distance is measured at the end of a normal expiration (end tidal).

The generalized equations for predicting percentage fat, using densitometery as a reference method, have been shown to be invalid for athletes (Sinning *et al.*, 1985). Even with an athletic sample, the violations of the assumptions made for the two-compartment model used in densitometry and ADP crucially undermine the accuracy of derived skinfold equations. While some skinfold equations use DXA as the reference, with standard errors of the estimate as low as 2.2% (Stewart and Hannan, 2000a), scientists are increasingly retaining skinfold totals of individuals as a surrogate for fatness, without converting them to percentage fat.

3.6.6 PROFILING

Formal athlete profiling has been carried out by coaches and sport scientists for the past two or three decades (Gulbin and Ackland, 2009). General profiling is carried out in a detrained state and administered at the commencement of a season or conditioning programme. In order to ascertain an individual's status within the group, the results of a series of tests are generally evaluated against normative data from other high-level athletes in the same sport or event. When the coach, sport scientist, and athlete have evaluated the test scores, the season's training schedule can be planned with each individual's strengths and weaknesses in mind. Body-composition testing is usually a critical component of general profiling. It should be kept in mind that general profiling is often more useful for potential elite athletes in the developmental stages; for the senior international-level athlete it will not be as valuable.

Some athletes are only profiled once every season at this level in order to identify weaknesses. Others have follow-up tests at varying intervals to monitor the progress of any intervention program. Agility athletes such as gymnasts and divers are regularly monitored for body composition, because power-to-weight ratios are crucial factors in their performances.

Specific profiling is usually carried out with elite senior athletes in events that are won by very small margins or times, as in aerobic sports like swimming, rowing, kayaking, running, and cycling, where it is important to accurately evaluate the individual's adaptation to the stress of heavy training at regular intervals. In some programmes, tests are carried out as regularly as every two weeks, but more often the tests are a month apart. When a major championship is approaching, some coaches request that the 'key' stress adaptation tests be carried out at more regular intervals, which can be as often as every week. Body-composition testing might be included in specific profiling, especially among athletes endeavouring to make a weight category for competition whilst minimizing the loss of FFM.

The process of creating a profile from test scores and comparing these values to normative data is beyond the scope of this chapter; Gulbin and Ackland (2009) cover this topic in detail. Until recently it has been difficult to perform meaningful international profile comparison, due to a lack of published data, or because the test protocols varied from country to country. It is fortuitous that both these problems are now being overcome and more reliable data are available for comparative purposes.

3.6.7 CONCLUSION

Many methods in the range of lab- and field-based techniques can be used effectively with a sample population undergoing conditioning. However, a full appreciation of the limitations of available methods is essential, especially with athletes or with those whose exercise alters key parameters affecting measurement. The utility of such measurement lies in its precision and ability to detect a meaningful change over time. This has implications not only for the techniques used, but for the training of technicians, selection of tools, and presentation of participants to be measured. Whereas future research for nutrition and medical studies is likely to continue using multi-component models, profiling of athletes will demand more pragmatic and cost-effective approaches which continue to use anthropometry and other lab methods including ADP, DXA, and 3D body scanning.

References

Ackland T.R., Blanksby B.A. Landers G. and Smith D. (1998) Anthropometric profiles of elite triathletes. *J Sci Med Sport*, **1** (1), 51–56

Ackland T.R., Schreiner A.B. and Kerr D.A. (1997) Absolute size and proportionality characteristics of world championship female basketball players. *J Sports Sci*, **15** (5), 485–490.

Adams J., Mottola M., Bagnell K.M. and McFadden K.D. (1982) Total body fat content in a group of professional football players. *Can J Appl Sport Sci*, **7**, 36–40.

Ballard T.P., Fafara L. and Vukovich M.D. (2004) Comparison of bod pod and DXA in female collegiate athletes. *Med Sci Sports Exerc*, **36**, 731–735.

Brozek J., Grande F., Anderson J.T. and Keys A. (1963) Densitometric analysis of body composition: revision of some quantitative assumptions. *Ann N Y Acad Sci*, **110**, 113–140.

Carter J.E.L. (1985) Morphological factors limiting human performance, in *Limits of Human Performance, American Academy of Physical Education Papers No 18* (eds D.H. Clark and H.M. Eckert), Human Kinetics, Champaign, IL, pp. 106–117.

Carter J.E.L. and Ackland T.R. (1994) *Kinanthropometry in Aquatic Sports*. Champaign IL: Human Kinetics Publishers.

Collins M.A., Millard-Stafford M.L., Sparling P.B. *et al.* (1999) Evaluation of the BOD POD for assessing body fat in collegiate football players. *Med Sci Sports Exerc*, **31**, 1350–1356.

Cote K.D and Adams W.C. (1993) Effect of bone density on body composition estimates in young adult black and white women. *Med Sci Sports Exerc*, **25**, 290–296.

De Lorenzo A., Bertini I., Iacopino L. *et al.* (2000) Body composition measurement in highly trained male athletes. A comparison of three methods. *J Sports Med Phys Fitness*, **40**, 178–183.

Dempster P. and Aitkens S. (1995) A new air displacement method for the determination of human body composition. *Med Sci Sports Exerc*, **27**, 1692–1697.

Deurenberg P., Tagliabue A. and Schouten F.J.M. (1995) Multi-frequency impedance for the prediction of extracellular water and total body water. *Br J Nutr*, **73**, 349–358.

Drinkwater D.T., Martin A.D., Ross W.D. and Clarys J.P. (1986) Validation by cadaver dissection of Matiegka's equations for the anthropometric estimation of anatomical body composition in adult humans, in *The 1984 Olympic Scientific Congress Proceedings: Perspectives in Kinanthropometry* (ed. J.A.P. Day), Human Kinetics, Champaign, IL, pp. 221–227.

Durnin J.V.G. and Womersley J. (1974) Body fat assessed from total body density and its estimation from skinfold thickness: measurements on 481 men and women aged from 16 to 72 years. *Br J Nutr*, **32**, 77–97.

Egan E., Wallace J., Reilly T. *et al.* (2006) Body composition and bone mineral density changes during a premier league season as measured by dual-energy X-ray absorptiometry. *Int J Body Compost Res*, **4**, 61–66.

Evans E.M., Prior B.M., Arngrimsson S.A. *et al.* (2001) Relation of bone mineral density and content to mineral content and density of the fat-free mass. *J Appl Physiol*, **91**, 2166–2172.

Fuller N.J., Laskey M.A. and Elia M. (1992) Assessment of the composition of major body regions by dual-energy X-ray absorptiometry (DEXA), with special reference to limb muscle mass. *Clin Physiol*, **12**, 253–266.

Gilbert G. and Lees A. (2003) Maximum grip width regulations in powerlifting discriminate against larger athletes, in *Kinanthropometry VIII* (eds T. Reilly and M. Marfell-Jones), Routeledge, London, UK, pp. pp. 175–180.

Gulbin J.P. and Ackland T.R. (2009) Talent identification and profiling, *Applied Anatomy and Biomechanics in Sport*, 2nd edn (eds T.R. Ackland, B.C. Elliott and J. Bloomfield), Human Kinetics, Champaign, IL, pp. 11–26.

Hall K.D. (2007) Body fat and fat-free mass inter-relationships: Forbes's theory revisited. *Br J Nutr*, **97**, 1059–1063.

Hansen N.J., Lohman T.G., Going S.B. *et al.* (1993) Prediction of body composition in premenopausal females from dual-energy x-ray absorptiometry. *J Appl Physiol*, **75**, 1637–1641.

Hawes M.R. and Sovak D. (1994) Morphological prototypes, assessment and change in elite athletes. *J Sports Sci*, **12**, 235–242.

Heymsfield S.B., Lohman T.G., Wang Z. and Going S.B. (2005) *Human Body Composition*, Human Kinetics, Champaign, IL.

Heyward V.H. (2004) *Applied Body Composition Assessment*, 2nd edn, Human Kinetics, Champaign, IL.

Houtkooper L.B., Mullins V.A., Going S.B. *et al.* (2001) Body composition profiles of elite American heptathletes. *Int J Sports Exerc Metab*, **11**, 162–173.

Hume, P. and Marfell-Jones, M. (2008) The importance of accurate site location for skinfold measurement. *J Sports Sci*, **26**: 1333–40.

Ishiguro N., Hiroaki K., Masae M. *et al.* (2005) A comparison of three bioelectrical impedance analyses for predicting lean body mass in a population with a large difference in muscularity. *Eur J Appl Physiol*, **94**, 25–35.

Ishiguro N., Hiroaki K., Masae M. *et al.* (2006) Applicability of segmental bioelectrical impedance analysis for predicting trunk skeletal muscle volume. *J Appl Physiol*, **100**, 572–578.

Jackson A.S. and Pollock M.L. (1978) Generalized equations for predicting body density of men. *Br J Nutr*, **40**, 497–504.

Jebb S.A., Goldberg G.R. and Elia M. (1993) DXA measurements of fat and bone mineral density in relation to depth and adiposity, in *Human Body Composition* (eds K.J. Ellis and J.D. Eastman), Plenum Press, New York, NY, pp. 115–119.

Kerr D.A. and Stewart A.D. (2009) Body composition in sport, *Applied Anatomy and Biomechanics in Sport*, 2nd edn (eds T.R. Ackland, B.C. Elliott and J. Bloomfield), Human Kinetics, Champaign, IL, p. 82.

Kershaw E.E. and Flier J.S. (2004) Adipose tissue as an endocrine organ. *J Clin Endocrinol Metab*, **89**, 2548–2556.

Kim J., Wang Z., Heymsfield S.B. *et al.* (2002) Total body skeletal muscle mass: estimation by a new dual-energy X-ray absorptiometry method. *Am J Clin Nutr*, **76**, 378–383.

McCrory M.A., Gomez T.D., Bernauer E.M. and Mole P.A. (1995) Evaluation of a new air displacement plethysmograph for measuring human body composition. *Med Sci Sports Exerc*, **27**, 1686–1691.

Malavolti M., Mussi C., Ploi M. *et al.* (2003) Cross-calibration of eight polar bioelectrical impedance analysis versus dual-energy X-ray absorptiometry for the assessment of total and appendicular body composition in healthy adults aged 21–82 years. *Ann Hum Biol*, **30**, 380–391.

Marfell-Jones M., Olds T., Stewart A.D. and Carter J.E.L. (2006) International Standards for Anthropometric Assessment, International Society for the Advancement of Kinanthropometry, Potchesfstroom, South Africa.

Matiegka J. (1921) The testing of physical efficiency. *Am J Phys Anthropol*, **4**, 223–230.

Nindl B.C., Friedl K.E., Marchitelli L.J. *et al.* (1996) Regional fat placement in physically fit males and changes with weight loss. *Med Sci Sports Exerc*, **28**, 786–793.

Nord R. and Payne R. (1995) Body composition by dual energy X-ray absorptiometry: a review of the technology. *Asia Pac J Clin Nutr*, **4**, 167–171.

Olds T. and Honey F. (2006) The use of 3D whole-body scanners in anthropometry, in *Kinanthropometry IX, Proceedings of the 9th International Conference of the International Society for the Advancement of Kinanthropometry* (eds M. Marfell-Jones, T. Olds, T. and A.D. Stewart), Routledge, London, UK, pp. 1–14.

Olds T.S. and Tomkinson G.R. (2009) Absolute body size, in *Applied Anatomy and Biomechanics in Sport*, 2nd edn (eds T.R. Ackland, B.C. Elliott and J. Bloomfield), Human Kinetics, Champaign, IL, pp. 29–46.

Pietrobelli A., Formica C., Wang Z. and Heymsfield S. (1996) Dual-energy X-ray absorptiometry body composition model: review of physical concepts. *Am J Physiol*, **271**(Endocrinol. Metab. 34), E941–E951.

Pietrobelli A., Morini P., Battistini N. *et al.* (1998) Appendicular skeletal muscle mass: prediction from multiple frequency segmental bioimpedance analysis. *Eur J Clin Nutr*, **52**, 507–511.

Prior B.M., Cureton K.J., Modelsky C.M. *et al.* (1997) In vivo validation of whole body composition estimates from dual-energy X-ray absorptiometry. *J Appl Physiol*, **83**, 623–630.

Prior B.M., Modlesky C.M., Evans E.M. *et al.* (2001) Muscularity and the density of fat-free mass in athletes. *J Appl Physiol*, **90**, 1523–1531.

Simpson J., Lobo A.D., Anderson D.N. *et al.* (2001) Body water compartment measurements: a comparison of bioelectrical impedance analysis with tritium and sodium bromide dilution techniques. *Clin Nutr*, **20**, 339–344.

Sinning W.E., Dolny D.G., Little K.D. *et al.* (1985) Validity of 'generalised' equations for body composition analysis in male athletes. *Med Sci Sports Exerc*, **17**, 124–130.

Siri W.E. (1956) The gross composition of the body, in *Advances in Biological and Medical Physics* (eds J.H. Lawrence and C.A. Tobias), Academic Press Inc, London and New York, pp. 239–280.

Sloan A.W. (1967) Estimation of body fat in young men. *J Appl Physiol*, **23**, 311–315.

Stewart A.D. (1999) Unpublished PhD thesis entitled 'Body Composition of Athletes, assessed by Dual X-ray Absorptiometry and other methods', University of Edinburgh. (Data collection 1996–1998).

Stewart A.D. (2001) Assessing body composition in athletes. *Nutrition*, **17**, 694–695.

Stewart A.D. (2003) Mass fractionation in male and female athletes, in *Kinanthropometry VIII, Proceedings of the 8th International Society for the Advancement of Kinanthropometry Conference* (eds T. Reilly and M. Marfell-Jones), Routledge, London, UK, pp. 203–209.

Stewart A.D. and Hannan W.J. (2000a) Body composition prediction in male athletes using dual X-ray absorptiometry as the reference method. *J Sports Sci*, **18**, 263–274.

Stewart A.D. and Hannan J. (2000b) Sub-regional tissue morphometry in male athletes and controls using dual X-Ray absorptiometry (DXA). *Int J Sport Nutr Exerc Metab*, **10**, 157–169.

Stewart A.D., Nevill A.M. and Johnstone A. (2009) Shape change assessed by 3D laser scanning following weight loss in obese men, in *Kinanthropometry XI, 2008 Pre-Olympic Congress Anthropometry Research* (eds P.A. Hume, A.D. Stewart and H. de Ridder), ISAK, Potchefstroom, pp. 20–24.

Visser M., Fuerst T., Lang T. *et al.* (1999) Validity of fan-beam dual-energy absorptiometry for measuring fat-free mass and leg muscle mass. *J Appl Physiol*, **87**, 1513–1520.

Wagner D.R. and Heyward V.H. (2000) Measures of body composition in blacks and whites: a comparative review. *Am J Clin Nutr*, **71**, 1392–1402.

Wang J., Gallagher D., Thornton J.C. *et al.* (2006) Validation of a 3-dimensional photonic scanner for the measurement of body volumes, dimensions and percentage body fat. *Am J Clin Nutr*, **83**, 809–816.

Wang Z.M., Pierson R.N., Jr. and Heymsfield S.B. (1992) The five-level model: a new approach to organizing body composition research. *Am J Clin Nutr*, **56**, 19–28.

Ward R., Ross W., Layland A. and Selbie S. (1989) *The Advanced O-Scale Physique Assessment System*, Kinemetrix Ltd, Burnaby, BC, Canada.

Wells J.C.K., Treleaven P. and Cole T.J. (2007) BMI compared with 3-dimensional shape: the UK National Sizing Survey. *Am J Clin Nutr*, **85**, 419–425.

Wells J.C.K., Ruto A. and Treleaven P. (2008) Whole-body three dimensional photonic scanning: a new technique for obesity research and clinical practice. *Int J Obes*, **32**, 232–238.

Wells J.C.K., Cole T.J., Bruner D. and Treleaven P. (2008) Body shape in American and British adults: between-country and inter-ethnic comparisons. *Int J Obes*, **32**, 152–159.

Wells J.C.K., Cole T.J. and Treleaven P. (2008) Age-variability in body shape associated with excess weight: the UK National Sizing Survey. *Obesity*, **16**, 435–441.

Wilmore J.H. and Behnke A.R. (1969) An anthropometric estimation of body density and lean body weight in young men. *J Appl Physiol*, **27**, 25–31.

Wilmore J.H., Vodak P.A., Parr R.B. *et al.* (1980) Further simplification of a method for determination of residual volume. *Med Sci Sports Exerc*, **12**, 216–218.

Woolford S., Bourdon P, Craig N. and Stanef T. (1993) Body composition and its effects on athletic performance. *Sports Coach*, **16**, 24–30.

3.7 Total Athlete Management (TAM) and Performance Diagnosis

Robert U. Newton[1] and Marco Cardinale[2,3], [1]School of Exercise, Biomedical and Health Sciences, Edith Cowan University, Joondalup, WA, Australia, [2]British Olympic Medical Institute, Institute of Sport Exercise and Health, University College London, London (UK), [3]School of Medical Sciences, University of Aberdeen, Aberdeen, Scotland (UK)

3.7.1 TOTAL ATHLETE MANAGEMENT

The modern athlete faces enormous demands on their time, body, and mind. As professional sport becomes more sophisticated and the income potential escalates; training, preparation, and competition dominate almost every waking hour. The days of being successful at the elite level based purely on natural ability are gone. Genetically determined capacity is still essential for elite performance but optimal strength and conditioning, nutrition, rest and recovery, travel management, mental preparation, skill practice, and strategy must all come together in a planned and implemented total athlete management (TAM) programme.

TAM is the ongoing process of 'plan, do, review, improve' common to management practice but applied equally effectively to athlete performance. For the elite, professional athlete the process is total in scope, addressing all aspects of the athlete's life: sporting, business, and social. While the strength and conditioning specialist is concerned less with business and social aspects, all components impact on all others, all impact on the athlete's ability to train and perform, and thus all impact on the specialist's planning and processes. Having said this, we will be focusing in this chapter on assessment of athlete physical performance and subsequent training programme design, while being cognizant that so many other factors (relationships, sponsorship commitments, family, etc.) impact on an athlete's ability to train and compete. Of the key areas for TAM we will not be discussing mental preparation or skill practice and strategy as these domains are outside the usual strength and conditioning specialist role.

3.7.1.1 Strength and conditioning

For obvious reasons, in this chapter we will spend most of our discussion on the strength and conditioning component of athlete management. The previous sections of this book have explored the most important physical tests for assessing an athlete's performance qualities and how these are performed and interpreted. Now we must determine the meaningfulness of the data and how they are to be used to inform programme design and implement adaptation and change.

Keep in mind 'plan, do, review, improve':

- Plan. The strength and conditioning programme for the year or years is written based on previous plans and experience combined with knowledge of the athlete and their future commitments in terms of competition, travel, and even social and sponsorship events.
- Do. Implement the plan as written but be prepared to adapt to changing conditions such as injury, overtraining, and unpredicted events.
- Review. Constantly assess the athlete's performance qualities to determine if the plan is producing the expected outcomes. Assessment is critical to gauging the athlete's adaptation status. Are they fatigued? Are they ill? Are social commitments or family life impacting on their abilities? How are they coping with the travel schedule? Within periodized macro- and micro-cycles decisions can then be made to modify training in order to maximize adaptation when the athlete is most responsive or to reduce training load when they appear stale or fatigued. The review process should also extend beyond assessment of the athlete to internal and external review of the strength and conditioning programme. Many professional teams bring experts from the same or different sports into clubs to review the processes and this can only be of benefit to improvement.
- Improve. TAM is of no benefit unless the strength and conditioning specialist constantly uses the review process to improve their training programmes, athlete management, and administrative processes.

3.7.1.2 Nutrition

Nutrition and strength and conditioning go hand in glove for every aspect from appropriate hydration to optimizing the anabolic state of the athlete. In some instances the strength and conditioning specialist is responsible for nutrition support but usually there is a dietician in the team to manage this aspect. However, within TAM the two roles must be complementary. For example, there is solid research indicating carbohydrate feeding prior to a resistance-training session with carbohydrate and protein immediately after facilitates the anabolic benefit. Likewise, if a training session is particularly gruelling and the environment is of high heat stress then it makes for more effective and safe management of hydration strategies if the strength and conditioning specialist relays this to the dietician. With the advent of widespread use of cold-water immersion this aspect enters the mix for TAM. Nutrition also plays a role in travel management when flights are involved as dehydration is an issue. Further, when time zones are crossed, diet can be manipulated to reduce the effects of jet lag.

3.7.1.3 Rest and recovery

Sometimes we forget that athletes get stronger, bigger, and faster while they rest and recover, not while they are actually training, when the reverse occurs. This is a critical aspect of TAM that must be planned as carefully as any training session. Sleep quality and quantity is an issue for many athletes and close cooperation between the medical team (with possible use of hypnotics where permitted), strength and conditioning specialist (testing for overtraining, rescheduling training across the day to reduce interference with sleep), psychologist (stress management and relaxation techniques), and dietician (planning meals to aid sleep) can help to manage the athlete more effectively.

Recovery is a very large topic, beyond the scope of this chapter. Suffice to state that a very important role for the strength and conditioning specialist is planning the training with adequate recovery, both within and post the session, as well as between sessions. Ongoing monitoring is critical here as performance tests can be accurate barometers of neural, physiological, and psychological fatigue across a week (Cormack *et al.*, 2008a) or a season (Cormack *et al.*, 2008b). Adjustments to short- and long-term training programme design can then be made to allow sufficient recovery and maximize training benefit. For example, with weekly testing of a couple of key performance qualities, such as those derived from a vertical jump test, it has been shown that declines in performance are associated with alterations in cortisol (Cormack *et al.*, 2008b), and can help to predict when the athlete is getting into difficulty. For many team sports the weekend competition requires careful management over the ensuing week to try and recover athletes while maintaining fitness levels. It has been reported that games impact differently on different athletes and that often the characteristics of a game in terms of time in play and intensity of work do not correlate with how well athletes recover over the week. Simple performance tests have however demonstrated sensitivity to player readiness to train (Cormack *et al.*, 2008a) and this can be a powerful tool to assist decisions on adjusting the volume, intensity, and mode of training in the week following a game.

3.7.1.4 Travel

Most athletes experience extensive travel schedules which impact all other aspects of TAM. Whether it be long bus journeys or international flights through several times zones, any and all travel will affect athletes' health, physical preparedness, sleep, hydration, nutrition, and recovery. Flying is particularly troublesome because cabin pressure and humidity are kept low, which can result in fluid in the tissues (oedema) and dehydration. Jet lag must be managed and there are several research and professional publications examining this (Reilly *et al.*, 2007; Meir, 2002). If an athlete is travelling from low to high altitude then this adds another layer of issues to be dealt with (Bartsch and Saltin, 2008). The strength and conditioning specialist has a role here in that training may have to be adjusted to cope with preparation for travel and with the condition of the athlete when they arrive at their destination. Some strength and conditioning specialists will perform basic tests such as vertical jump to gauge how the athlete has travelled, and if necessary will reduce training workload, relocate to a lower altitude, or programme recovery strategies such as light aerobic work, pool work, or even complete rest.

3.7.2 PERFORMANCE DIAGNOSIS

Certain measures represent specific or independent qualities of the athlete that contribute to their sporting ability; these qualities can be assessed and trained independently (Newton and Dugan, 2002a). Performance diagnosis is the process of determining an athlete's level of development in each of these distinct qualities. When specific performance qualities are targeted with prescribed training, greater efficiency of training effort can be achieved, resulting in enhanced athlete performance. Because elite athletes tend to be genetically predisposed to their sport and train to enhance their abilities, specificity of these qualities is inherent to a particular sport or athletic event. In other words, each sport or event requires a certain level of these performance qualities to underpin a competitive advantage.

Performance diagnosis is a critical component of TAM because the adage 'If you can't measure it, you can't manage it' applies to sport just as it does to business. There are other important outcomes of performance diagnosis. Providing feedback on physical performance is critical to motivation and learning for both the athlete and coach. The more the athlete comes to understand how their body responds to training, the more motivated they become to even the most unpleasant methods. The testing, feedback, and improvement process creates team trust across the other disciplines (nutrition, medicine, therapy, psychology) because it increases understanding of the strength and conditioning role and what it has to offer the support team and athletes. Finally, perhaps most importantly, performance diagnosis increases job security for the strength and conditioning specialist. Extensive data on athletes' performance qualities past and current provide solid evidence of how well the strength and conditioning specialist is performing their job. If the team comes bottom of the competition but the data demonstrate that the athletes are stronger, bigger, faster, less fatigable, and more agile than in previous seasons then it is harder to blame their physical preparation and sack the strength and conditioning specialist!

3.7.2.1 Optimizing training-programme design and the window of adaptation

As discussed above, almost all sports require a range of physical performance qualities. These include components of strength, power, speed, agility, endurance, cardiorespiratory fitness, flexibility, and body composition. While there are other qualities such as skill and psychological state, our discussion in this chapter will be confined to the physical capacities listed above. For any given sport, and in some cases for particular positions in team sports, an athlete will have an enhanced chance of success if their body composition and neuromuscular and cardiorespiratory systems are specifically tuned for the tasks required. It is when the underlying 'machinery' is correctly built and tuned that the skill, strategy, and psychological abilities of the athlete can best be brought to bear and the greatest chance of success can be realized.

This highlights the multifaceted nature of sports performance, with a mixed training methods approach being most effective as it develops more components of performance (Newton and Kraemer, 1994). A key principle of training methodology is that of diminishing returns, whereby the more developed a particular component, the smaller the window for adaptation. This relates to the initial level of fitness or development of a given component. Each can be developed only to a maximum level, dictated by genetic endowment, and as the athlete's development moves along this continuum, the same training effort produces ever decreasing percentage improvement. The practical application of this is that when an athlete develops one component to a high level (e.g. strength) the potential for that component to contribute to further increases in sport performance diminishes. Thus, each component can be thought of as a 'window of adaptation' to the larger window of adaptation in overall performance. This concept is summarized in Figure 3.7.1. For example, if an athlete undertakes a programme of training to develop cardiorespiratory fitness they will exhibit a shrinking window of adaptation to this form of stimulus. As this window shrinks, training time will be more efficiently spent on other training methods, such as speed development or muscle strengthening. Further, training must be targeted to increase performance in those components in which the athlete is weakest, because these have the largest window for adaptation and thus lead to the greatest increase in overall sports performance. The athlete will be limited in potential by their weakest link.

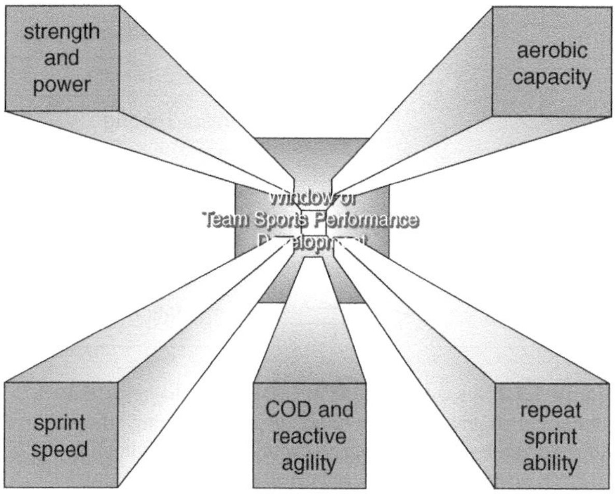

Figure 3.7.1 Window of adaptation for team sports performance. In this model five key qualities have been highlighted as critical to the overall performance

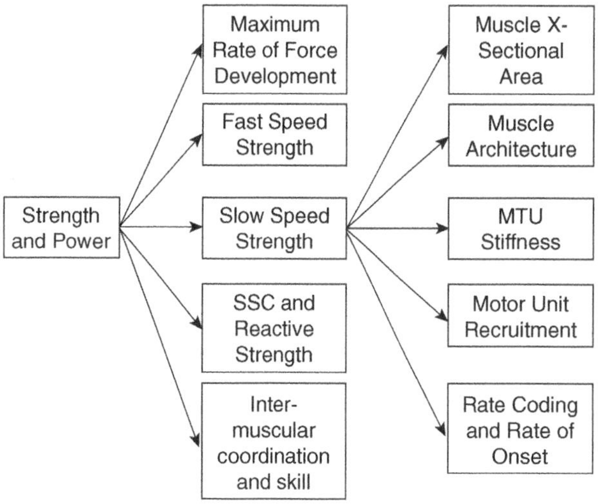

Figure 3.7.2 A performance quality such as strength and power can be modelled as a window for adaptation with several contributing windows, Such as slow speed strength and maximum rate of force development. Slow speed strength in turn has several factors that contribute to its window of adaption

Most performance qualities can be drilled down into their constituent qualities until the most basic elements of human movement are revealed. For example, if we take the quality of strength and power, this can be defined by more specific qualities, as shown in Figure 3.7.2. Each of these qualities at whatever level can be developed independently to a certain extent by specific training techniques. There are also considerable interactions between strength qualities. For example, muscle architecture contributes to both slow and fast speed strength but

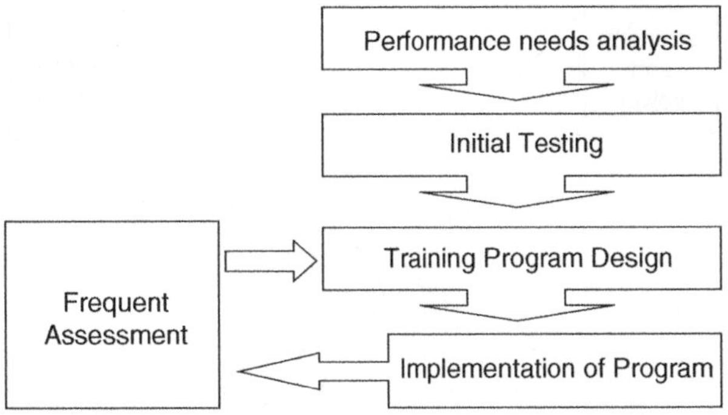

Figure 3.7.3 A test–retest cycle of performance diagnosis and subsequent programme design results in greater training efficiency. An important aspect is ongoing assessment to quantify training progress, gauge effectiveness, and detect staleness, overtraining, or injury (Newton and Dugan, 2002b)

in entirely different directions. Slow heavy-resistance training will cause pennate muscles to increase the angle at which fascicles insert on the tendon – an adaptation favouring high force output at low speeds. However, fast velocity training with lighter loads will produce a decrease in pennation angle, an adaptation favouring fast speed strength. As can be seen from Chapters 3.1–3.6, there are very specific tests for the vast range of performance qualities. By assessing these qualities, relative weaknesses and strengths can be determined and more effective training strategies developed.

The implementation of performance diagnosis and prescription to enhance maximal power performance should flow according to a logical sequence (Figure 3.7.3). The initial step requires a determination of the important qualities in the target activity (i.e. a performance needs analysis). A test battery is then established to assess these qualities in an efficient, valid, and reliable manner. A training programme is developed based on the performance diagnosis, which will improve performance in the target sport. The final aspect is perhaps the most important, as isolated testing has little utility. It is only with frequent, ongoing assessment that a complete profile of the athlete is compiled and manipulation of training variables can be coordinated to advance the athlete toward performance goals. The test–retest cycle (Figure 3.7.3), with frequent adjustments to the training programme, is a key feature of the performance diagnosis and prescription process. We address various aspects of this process in the next section.

3.7.2.2 Determination of key performance characteristics

The first step toward achieving the desired performance goal is to determine the key performance characteristics of the target activity. For example, if the task is to maximize takeoff velocity in the high jump, then those strength and power qualities that influence takeoff velocity need to be determined. This can be achieved by several processes, such as biomechanical evaluation, physiological needs analysis, analysis of high-level athletes, and regular monitoring. The best approach may be to combine all four to gain the greatest understanding of the target performance.

1. Biomechanical evaluation leads to an understanding of the forces exerted by the athlete against their environment and the implements (balls, etc.) that they must manipulate. Aspects to be considered may include: peak and average forces, impulse, power output, peak and minimum velocities, contact times, minimum and maximum joint angles and range of motion, type of muscle contraction. For example, previous research has indicated a significant difference in eccentric strength between elite and sub-elite downhill skiers (Abe *et al.*, 1992) and biomechanical analysis of the sport demonstrates very high eccentric loads during the event. Such evidence changed training methodologies for downhill skiers, with a far greater emphasis on heavy eccentric training.
2. Physiological evaluation can be conducted to determine the aerobic and anaerobic demands of the activity. This can be achieved via simulated games measuring oxygen consumption and heart rate during simulated or actual events to determine energy demands and heart rate ranges. Game analysis is also very informative for subsequent training-programme design. Video systems that track events and player movement provide a wealth of information as to the physiological demands of a sport. The advent of inexpensive and accurate GPS makes analysis of field sports in particular quite precise in terms of distance covered and time spent at various intensities of running. Combined with heart-rate data the strength and conditioning specialist can determine the performance qualities required for the sport and even differences in playing position, for example backs versus forwards in rugby. Performance qualities that might be considered include:
   a) duration, intensity, work : rest ratios
   b) power performance under fatigue, for example in rugby, football, hockey
   c) combative sports: impact of repeated physical contact
   d) environmental issues: heat, cold, altitude
   e) body composition.
3. Analysis of high-level athletes in a given sport can provide information on performance qualities. It can be assumed, though with caution, that an athlete who is performing well in the sport possesses the necessary levels of these qualities. By testing such lighthouse athletes in a range of performance qualities deemed important to the sport a profile of target capacities can be constructed. Good record keeping is critical and all strength and conditioning specialists should know or be able to look up the personal bests (PBs) of the star athletes they have worked with over the years. The cautionary note is that elite athletes may be achieving for reasons other than physical capacity; psychological tough-

ness, strategic ability, skill, and a range of other factors contribute to the overall ability of an athlete and these components may compensate for some deficiencies in physical performance.

4. Regular monitoring tests athletes before and after phases involving certain training emphases. If they respond with large improvements in the targeted strength or power quality, it could be that they are deficient in that quality and this may require further attention. Certainly, if a component is improving rapidly, it may be prudent to maintain the emphasis until some plateau occurs. There are caveats to this approach. First, it is wasteful to continue to seek improvement if a quality is not of significance to the task or if the athlete has an adequate level of the quality such that other qualities may be more limiting. Second, it may be better, for the sake of training variety, to take note of the large response and return to the given quality at a later phase of training.
5. Comparison across current and previous squads provides important knowledge of key performance qualities. Reiterating point 3 above, it is crucial that strength and conditioning specialists maintain extensive records of physical performance scores over their entire career. Each season of new results grows the database, adding to the capability of the strength and conditioning specialist to make intelligent and informed decisions on training programme design. Data on current and past squads can be used to identify an individual ranked low on a particular quality, indicating that more work is required on this aspect. Coach appraisal is very useful and can be related to performance data. Comparison of rookies versus seniors gives an indication of how the performance qualities must be manipulated to move the newer players through their careers. Comparison to past squad profiles, particularly winning squads, provides insights into where the current squad might be lacking. Finally, tracking an individual athlete's ranking in the squad over time and making changes to programme design and emphasis can help the strength and conditioning specialist develop a more rounded athlete better matched to the demands of their sport.

3.7.2.3 Testing for specific performance qualities

Tests, like training, must be specific for the performance quality in question. The previous sections have outlined issues of validity and reliability of testing, as well as the various tests that can be implemented. Avoid the scattergun approach. Choose your tests carefully and only select tests that will give you valid information on a key performance quality for that athlete.

Performance testing can often be viewed as a distraction from training and other activities. Try to minimize the time required by only choosing key tests, scheduling the athletes for maximum throughput, and minimizing the time athletes have to wait. Make sure all your equipment is working correctly and is calibrated well ahead of time. Don't drag the timing lights out 30 minutes before your first athlete is due to arrive and expect to have a fruitful and stress-free testing session. Record the information on pre-prepared data sheets, or preferably straight into a spreadsheet or database.

3.7.2.4 What to look for

Apart from assessing an athlete's relative strengths and weaknesses, performance diagnosis can also detect problems with the athlete, their training programme, or both. For example, if there is a lack of progression of a target quality then either the athlete is approaching their genetic potential for that attribute, the training is inappropriate, or the athlete has adapted to the current programme and requires variation to stimulate further gains.

If a decline in a strength/power quality is detected, this may simply be a result of changing emphasis in the training. However, if it was deemed important to develop that attribute in an earlier phase, a significant decline should be avoided. For example, if a certain level of reactive strength was attained pre-season and a 15% decline in-season was seen, some modification to the training regimen might stem the loss. Furthermore, decline of a strength and power quality may indicate tiredness, overtraining, staleness, or injury, so these causes should also be investigated.

3.7.2.5 Assessing imbalances

In the interests of both injury reduction and performance enhancement, it is instructive to investigate imbalances between agonist and antagonist muscle groups as well as between the left and right sides of the body. There is considerable literature (Aagaard *et al.*, 1995, 1998) on the former, so we will confine our discussion to assessing imbalances between the left and right leg extensors (Newton *et al.*, 2006). Most people exhibit some dominance, resulting in differences in performance between, for example, hops performed on the left versus the right leg. Differences of more than 15% may indicate existing pain and injury, inadequate recovery from previous injury, or an undesirable imbalance in muscle strength/power qualities. It is easy to assess such differences by having the athlete perform unilateral movements such as single-leg hops and comparing flight and contact times. Cutting and sidestepping tests using timing lights or contact mats to measure speed in each direction are also useful.

One should take the specific sporting movements into consideration when assessing left- to right-side imbalances. In sports that require a one-legged takeoff, for example the high jump or long jump, it may be quite normal for the dominant or takeoff leg to be stronger than the contralateral leg. This would be expected due to the nature of the event, so time spent trying to eliminate imbalances in leg power might be counterproductive.

Table 3.7.1 Descriptors for rating of perceived exertion

Rating	Descriptor	Plain English
0	Rest	Rest
1	Very, very easy	Really easy
2	Easy	Easy
3	Moderate	Moderate
4	Somewhat hard	Sort of hard
5	Hard	Hard
6	"	Hard
7	Very hard	Very hard
8	"	The coach tried to kill us
9	"	I feel like death warmed up
10	Maximal	Oh s—!

3.7.2.6 Session rating of perceived exertion

Session rating of perceived exertion (RPE) is the rating of difficulty or exertion that an athlete reports after a given session of training. The scale is provided in Table 3.7.1. Research (Foster, 1998) has indicated that a high percentage of illnesses can be accounted for when individual athletes exceed individually identifiable training thresholds, mostly related to the strain of training, which can be determined using RPE. Session RPE is an attractive tool for the strength and conditioning specialist as it is quick and easy to administer and provides very useful information for TAM.

3.7.2.7 Presenting the results

Perhaps the most important aspect of athlete assessment is the provision of simple, concise, informative, and educational reports. Detailed athlete assessment is only of use if the results are used to inform programme design for greater effectiveness and efficiency, reduce injury risk, and educate the athlete and coach about responses and adaptations to training and competition. An entire chapter could be devoted to the topic of reporting, particularly when one considers the latest advances in Internet and database technologies. However, a few key points are pertinent. Reports:

- are one of the most important aspects of TAM
- yet are often done poorly or not at all!
- must be complete, valid, and reliable
- must be easily accessed: Web-based if possible
- must be secure and confidential
- provide research, feedback, education
- identify relative weaknesses and strengths
- should be used to adjust training programmes.

Results should be reported graphically whenever possible and include benchmarks for comparison. Benchmarking might include published results for national- or international-level athletes in the same or a similar sport. If an entire team or squad has been tested then benchmarking to the mean and standard deviation is useful. Z-scores are useful in this regard as they indicate how many standard deviations the athlete's score is from the group mean. The 'radar plots' in Microsoft Excel are very effective for presenting this type of data; an example is provided in Figure 3.7.4. To produce such a graph, the key components of the athlete-assessment battery are listed in the spreadsheet. The mean and standard deviation for each parameter are calculated and then the Z-score ((athlete score − mean) / standard deviation) for each athlete's results are determined. These can then be plotted for each athlete, and comparisons between athletes or repeated test sessions can be included.

The zero ring represents the average for the team, so scores greater than zero indicate better-than-average performance, while negative scores are below the team average. Such a representation graphically displays a large amount of information in a single figure and in a form that athletes and coaches can readily understand. Relative strengths and weaknesses can immediately be highlighted. Some points to note in Figure 3.7.4 are that the senior player has a very rounded profile, indicating no qualities that are less developed. The senior player is well above the average for all measures. In contrast, the rookie has a very unbalanced profile, with weaknesses in the strength domain indicated by very low scores in squat and bench-press strength, a reflection of limited resistance training history. Apart from aerobic fitness (beep test), which is around the team average, all their other qualities are well below the average, indicating that this athlete really lacks the machinery required for the sport at this stage of their career and needs extensive training, with upper- and lower-body strength a priority.

The radar plot can also be used to display an athlete's profile over time, which indicates how they are progressing relative to the rest of the team. In the example above, one would expect that with targeted training the squat and bench-press scores could be lifted considerably to provide a more rounded profile.

The report should be short but informative as busy athletes and coaches will not take the time to read and digest long documents. If possible, one page is ideal, hence the value of the radar plots, which contain much information in a single figure. The report can be provided electronically via email or a secure Web site, perhaps with links to drill down further into the data if required.

3.7.2.8 Sophisticated is not necessarily expensive

The number of laboratories and training facilities that have sophisticated equipment for performance diagnosis is increasing as the value of this evaluation and training becomes more recognized. However, one can implement valid and reliable performance diagnoses with a minimum of equipment and expense. All of the jump tests can be performed with either

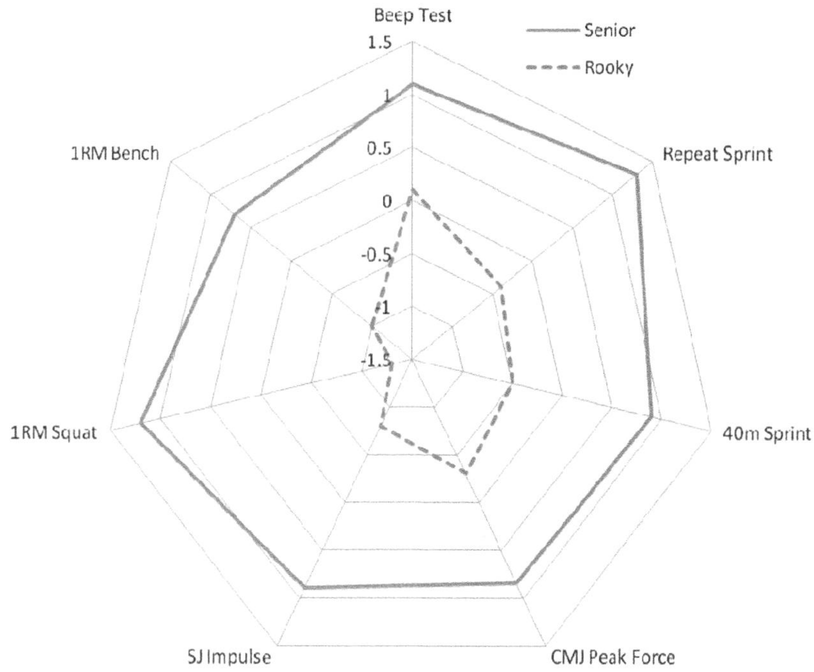

Figure 3.7.4 Radar plot of performance qualities for two football athletes, one a senior player with extensive training history and the other a rookie just recruited to the team. CMJ = countermovement jump, SJ = squat jump

jump-and-reach equipment or a contact mat system, and while not providing the same detailed measures of force and power, they provide adequate information for basic performance diagnosis. Other tests such as the standing broad jump or medicine ball throw can provide good information using only a measuring tape. Strength testing using a barbell and power rack provides as valid data as any system. Many teams use the beep test to assess aerobic capacity.

3.7.2.9 Recent advances and the future

Ubiquitous performance diagnosis

There is a strong rationale for large-scale testing of athletes on an annual, biannual, or even quarterly basis, but more commonly strength and conditioning specialists are integrating assessment into almost every workout. In the weight room this is easy, with each lift recorded, but there has been good utility in performing a basic and rapid test of a key parameter at each training session. The vertical jump is often chosen, as it is a good measure of overall body strength in a rapid movement. Strength and conditioning specialists can use data from such a test as a daily barometer of how their athletes are progressing, and research has demonstrated relationships with measures of stress such as cortisol (Cormack *et al.*, 2008b). New instrumentation is available to measure a wide array of performance qualities continuously while an athlete trains. Heart-rate monitors are routinely worn for later download or monitoring in real time back to a base station. GPS systems can track and measure speed and distance covered, with recent systems including physiological measures such as body core temperature, heart rate, blood pressure, and so on. The athlete is essentially unaware that this data is being streamed from their body so it does not interfere with their training or movement, but the data that the strength and conditioning specialist can access is very powerful for making decisions on training strategy.

Online data collection: wireless

Advances in computer technology and communications have resulted in new methods of measurement and management of athlete data. Wireless technology such as Bluetooth permits very small sensors to be placed on the athlete and motion and forces to be recorded without impeding movement. Computers collect and process these data, giving new insights into performance qualities. Using wireless networking, data collected can be uploaded to a central database across the stadium or across the world so that they can be accessed at any time and from anywhere.

Real-time reporting

The quality of performance feedback is greater the faster it is returned to the athlete. The latest systems can analyse performance and inform the athlete and coach immediately. More detailed reports can be generated immediately and printed for the athlete to receive straight after the test session. This makes the data relevant and timely, with greatest value to the athlete and coach.

Web-based report and support

The results of performance diagnosis should be available through secure Web sites so that athletes and coaches can view and analyse them from anywhere. Some sporting organizations already have these systems well developed and the data are extensive, both historical and current. When more testing is completed anywhere in the world it can be uploaded and become part of the Web site immediately. Such Web sites should have extensive information on the tests, the results, and what they mean. Training programmes should be integrated into the system to ensure assessment and training are always closely linked.

Field monitoring: training and testing

Many athletes spend much of their time travelling. They must still be assessed on a regular basis, so systems are becoming available that are portable and connected to the Internet so that results can be uploaded to the central database immediately.

Education

As has already been discussed, performance diagnosis can have a very important role in educating the athlete and coach. The extension of this builds on the online test results with a 'college' designed to teach the athlete more about themselves and their sport. This knowledge will help them to understand when things are right and when their physical or psychological status is not so good.

The Internet is a wonderful resource, but it is also filled with misinformation and inaccuracies. A dedicated Web site belonging to the squad, team, or sport, which is online, extensive, pervasive, accurate, based in science, and relevant, can be of great benefit to the athlete and coaching staff.

3.7.3 CONCLUSION

TAM is the management of a range of physical, psychological, and social aspects of an athlete's life with the goal of maximizing performance, safety, and competitive longevity. How many components are managed depends on the sport, the level of competition, athlete characteristics, and resources. For most strength and conditioning specialists, TAM will be confined to assessment and physical training of their athletes.

Performance diagnosis involves testing key qualities critical to a sport and then adjusting programme design accordingly, a process which can greatly increase training efficiency. By determining those qualities in which an athlete is most weak, the largest window of adaptation can be targeted, resulting in faster and larger improvement in overall sport performance. The most important aspect of performance diagnosis is using the information to make changes to training programmes. The second most important is providing concise, valid, and informative reports to the athlete and coach in a timely fashion – preferably at the end of the session.

New developments in technology are improving the ease and power of performance diagnosis and making TAM much more effective by being based in quality information at the strength and conditioning specialist's fingertips.

References

Aagaard P. *et al.* (1995) Isokinetic hamstring/quadriceps strength ratio: influence from joint angular velocity, gravity correction and contraction mode. *Acta Physiol Scand*, **154** (4), 421–427.

Aagaard P. *et al.* (1998) A new concept for isokinetic hamstring: quadriceps muscle strength ratio. *Am J Sports Med*, **26** (2), 231–237.

Abe T. *et al.* (1992) Isometric and isokinetic knee joint performance in Japanese alpine ski racers. *J Sports Med Phys Fitness*, **32** (4), 353–357.

Bartsch P. and Saltin B. (2008) General introduction to altitude adaptation and mountain sickness. *Scand J Med Sci Sports*, **18**, 1–10.

Cormack S.J., Newton R.U. and McGuigan M.R. (2008a) Neuromuscular and endocrine responses of elite players to an australian rules football match. *Int J Sports Physiol Perform*, **3** (3), 359–374.

Cormack S.J. *et al.* (2008b) Neuromuscular and endocrine responses of elite players during an australian rules football season. *Int J Sports Physiol Perform*, **3**, 439–453.

Foster C. (1998) Monitoring training in athletes with reference to overtraining syndrome. *Med Sci Sports Exerc*, **30** (7), 1164–1168.

Meir R. (2002) Managing transmeridian travel: guidelines for minimizing the negative impact of international travel on performance. *Strength Cond J*, **24** (4), 28–34.

Newton R.U. *et al.* (2006) Determination of functional strength imbalance of the lower extremities. *J Strength Cond Res*, **20** (4), 971–977.

Newton R.U. and Dugan E. (2002a) Application of strength diagnosis. *Strength Cond J*, **24** (5), 50–59.

Newton R.U. and Dugan E.L. (2002b) Application of strength diagnosis. *Strength Cond J*, **24** (5), 50–59.

Newton R.U. and Kraemer W.J. (1994) Developing explosive muscular power: implications for a mixed methods training strategy. *Strength Cond J*, **16** (5), 20–31.

Reilly T. *et al.* (2007) Coping with jet-lag: a position statement for the European College of Sport Science. *Eur J Sport Sci*, **7** (1), 1–7.

Section 4
Practical applications

4.1 Resistance Training Modes: A Practical Perspective

Michael H. Stone[1,2,3] and Margaret E. Stone[1], [1]Center of Excellence for Sport Science and Coach Education, East Tennessee State University, Johnson City, TN, USA, [2]Department of Physical Education, Sport and Leisure Studies, University of Edinburgh, Edinburgh, UK, [3]Edith Cowan University, Perth, Australia

4.1.1 INTRODUCTION

Resistance training is being used by large portions of the population for a variety of reasons, including bodybuilding/cosmetics, improving aspects of health, reducing or rehabiliting injuries, and as a means of improving daily or athletic performance. Resistance training-associated health benefits can include improved cardiovascular parameters, beneficial endocrine adaptations including increased insulin sensitivity and serum lipid adaptations, positive alterations in body composition, increased tissue tensile strength including bone, and decreased physiological stress (Blessing *et al.*, 1987; Brill *et al.*, 1998; Conroy *et al.*, 1993; Johnson *et al.*, 1982, 1983; McMillan *et al.*, 1993; Meka *et al.*, 2008; Poehlman *et al.*, 1992; Siegrist, 2008; Stone *et al.*, 1983, 1991). However, when considering resistance training, most people associate its effects with improved physical performance. Indeed, physical-training programmes that incorporate strength training as an integral part have consistently shown improved performance in a variety of tasks, ranging from health-related physical tasks to athletic performance. These have included ergonomics, for example lifting various loads to different heights (Asfour, Ayoub and Mital, 1984; Genaidy *et al.*, 1994), and improved gait and daily functioning among the elderly, stroke patients, and cerebral palsy patients (Alexander *et al.*, 2001; Eek *et al.*, 2008; Lexell and Flansbjer 2008). Today, resistance exercise has become an integral part of training for improved performance among athletes and sports teams.

Improvements in performance variables as a result of resistance training can include increased muscular strength, power, and both low- and high-intensity exercise endurance (McGee *et al.*, 1992; Paavolainen *et al.*, 1999; Robinson *et al.*, 1995; Stone *et al.*, 2006; Stone, Sands and Stone, 2007). More importantly, changes in these characteristics underlie alterations in measures of athletic performance such as the vertical jump, sprint times, distance-running times, and measures of agility (Harris *et al.*, 1999; Paavolainen *et al.*, 1999; Stone *et al.*, 2006; Stone, Sands and Stone, 2007; Wilson and Murphy, 1996). Observations such as these indicate that there can be a 'transfer of training effect' that results in a change in specific functional abilities and capacities. Of paramount importance is the method of training; choosing a training method (reps and sets, velocity of movement, periodization, etc.) can make a considerable difference in the outcome of a resistance-training programme (Garhammer, 1981b; Harris *et al.*, 1999; Stone and O'Bryant, 1987; Stone *et al.*, 1999a, 1999b). For example, high-volume resistance-training programmes likely have a greater influence on body composition, metabolism, and endurance factors than low-volume programmes (McCaulley *et al.*, 2009; McGee *et al.*, 1992; Stone, Sands and Stone, 2007), and properly periodized programmes produce superior performance alterations to non-periodized programmes, particularly over relatively long periods of time (Plisk and Stone, 2003, Stone, Sands and Stone, 2007). Furthermore, the type of training mode (type of equipment) can markedly influence training adaptations.

The following operational definitions will be used (Stone, Plisk and Collins, 2002; Stone, Sands and Stone, 2007):

- Free-weight: a freely moving body is used to apply resistance. This includes barbells, dumbbells, associated benches and racks, chains, medicine balls, throwing implements, body mass, and augmented body mass (i.e. weighted vests, limb weights). This allows for resistance to force-production accommodation and challenges the lifter to control, stabilize, and direct the movement.
- Machine: applies resistance in a guided or restricted manner. This includes plate-loaded and selectorizer devices, electronically braked devices, springs, and elastic bands. Generally this results in a smaller challenge for control, stabilization, and direction of movement.

The following discussion will examine the relative usefulness of various types of machines and free-weight to sports

Strength and Conditioning – Biological Principles and Practical Applications Marco Cardinale, Rob Newton, and Kazunori Nosaka.

performance enhancement in relation to (Stone, Plisk and Collins, 2002; Stone, Sands and Stone, 2007): (1) basic training principles, (2) comparison research, and (3) practical aspects based on 1 and 2.

4.1.2 BASIC TRAINING PRINCIPLES

There are three basic training principles: overload, variation, and specificity.

4.1.2.1 Overload

Overload is a stimulus of sufficient strength, duration, and frequency that it forces an organism to adapt (e.g. alterations in physiology, psychology, etc.). The adaptation will be specific to the type of overload. In the case of physical activity, overload consists of exercise and training which goes beyond normal levels of physical performance.

All stimuli (overload) have a level of intensity, relative intensity (% of maximum), frequency, and duration of application. The intensity of training is associated with the rate at which energy (ATP) is used and is typically estimated by the rate of performing work (power output) or magnitude of force production. The volume of training is an estimate of how much total work is accomplished and therefore a crude estimate of the total amount of energy expended. For resistance training, intensity (and relative intensity) is provided by the amount of weight lifted (load). The volume of training is related to the number of repetitions and sets per exercise, the number and types of exercises used (large versus small muscle mass), and the number of times per day, week, month, etc. these exercises are repeated. Volume load (repetitions × mass lifted) is the best practical estimate of the amount of work accomplished during training (Stone and O'Bryant, 1987; Stone *et al.* 1998, 1999a, 1999b, 2007). Training intensity and volume can be considered in terms of the workout (i.e. all exercises performed during a specified training session, including the warm-up sets) or in terms of a specific exercise(s).

An understanding of overload factors can aid in the selection of exercises and equipment. Whilst simple aspects of programming (i.e. sets and reps according to a specified plan) for a specific exercise are largely independent of exercise mode, the resulting overall total work (accomplished per session, week, month, etc.) is not independent. For example, in general, changes in body composition, particularly decreases in body fat, are related to the total energy expenditure (during and post-exercise) and therefore the volume of training (Stone, Sands and Stone, 2007; Votruba, Horvitz and Schoeller, 2000). Whilst there may be a few exceptions, such as leg-press devices, many, likely most, machines are devised as single-joint or small-muscle-mass exercise devices. Furthermore, many training programmes using machines are centred around a series of small-muscle-mass exercises that will result in a smaller total energy expenditure compared to using primarily large, multi-joint exercises. Thus, we make the argument that large-muscle-mass exercises and therefore larger energy expenditures are much more readily accomplished using free-weights (see Section 4.1.3.4, number 3).

4.1.2.2 Variation

Variation deals with the manipulation of training variables including volume and intensity, speed of movement, rest periods, and exercise selection. Thus, variation is a method that can be used to alter the overload stimulus. In the context of overload-stimulus variation, the prolongation of adaptations over continuous training programmes depends upon this variation (Kramer *et al.*, 1997; Stone *et al.*, 2000a, 2000b). Appropriate sequencing of training variables including exercise selection in a periodized manner can lead to superior enhancement of a variety of performance abilities (Harris *et al.*, 1999; Stone, Sands and Stone, 2007). Even though changes in volume and load are easily accomplished using machines, proper application, sequencing, and variation of movement patterns, speed–strength, and speed-orientated exercises are limited in scope. So our argument is that training limitations can result from restrictions in the movement pattern and movement characteristics imposed by the mode of training.

4.1.2.3 Specificity

Specificity of exercise and training is the most important consideration when selecting appropriate equipment for resistance training, especially if performance enhancement is a primary goal. Specificity includes both bioenergetics and mechanics of training (Stone, Sands and Stone, 2007). As this chapter deals with modes of training, this discussion will be concerned with mechanical specificity.

Mechanical specificity refers to the degree of similarity between a training exercise and a performance. So, mechanical specificity deals with kinetic and kinematic associations such as movement patterns, peak force, rate of force development (RFD), acceleration, and velocity parameters. Transfer of training effect is the degree to which training exercises transfer to performance. The more similar a training exercise is to the actual physical performance, the greater the probability of transfer (Behm, 1995; Sale, 1992; Schmidt, 1991; Stone, Sands and Stone, 2007).

Siff and Verkhoshanski (1998) argue that the basic mechanics, not necessarily the outward appearance, of training movements must be similar to those of performance in order to maximize transfer. They present a number of considerations and performance criteria that can be used to maximize transfer of training effect (Table 4.1.1).

4.1.3 STRENGTH, EXPLOSIVE STRENGTH, AND POWER

In terms of strength training, mechanical specificity has been extensively studied. Strength can be defined as the ability to produce force (Stone, 1993; Stone, Sands and Stone, 2007).

Table 4.1.1 Mechanical criteria for specificity and transfer of training effect

- Accentuated regions of force production during the movement.
- Amplitude and direction of movement.
- Dynamics of effort (i.e. static versus dynamic aspects of the movement: appropriate acceleration, velocity, and power output).
- Rate and time to peak force production.
- Contraction type(s) (e.g. eccentric vs concentric vs stretch–shortening muscle actions).

The fourth criterion, dealing with rate of force production, is especially important in selecting exercises for the training of explosive athletic movements.

Modified from Siff and Verkhoshanski (1998); Stone *et al.* 2007.

Explosive strength is associated with the ability to produce high RFDs, which is related to acceleration capabilities (Schmidtbleicher, 1992; Stone, 1993). Explosive strength can be produced dynamically or isometrically (Stone, 1993). Dynamic explosive strength exercises which result in high power outputs and high RFDs are crucial for the training of athletes in a variety of sports (McBride *et al.*, 2002; Schmidtbleicher, 1992; Stone, 1993; Stone, Sands and Stone, 2007). For selection of appropriate training modes an understanding of the components of dynamic explosive exercises is an important consideration.

Work is the product of force and the distance that the object moves in the direction of the force (force × distance). Power is the rate of doing work (P = force × (distance / time)) and can be expressed as the product of force and velocity (P = force × velocity). Power can be calculated as an average over a range of motion or as an instantaneous value occurring at a particular instant during the displacement of an object. Peak power (PP) is the highest instantaneous power value found over a range of motion under a given set of conditions (e.g. fatigue, state of training, etc.). Maximum power (MP) is the highest PP output one is capable of generating under ideal conditions. Typically the highest concentric power outputs occur at approximately 0 (e.g. jumps) to 50% (e.g. snatch, clean) of maximum isometric force capabilities, depending upon the exercise (Cormie *et al.*, 2007; Stone *et al.*, 2003; Wilson *et al.*, 1993).

Power output is likely the most important factor separating sport performances (losers from winners). Average power output is associated with performance in endurance events. For explosive activities such as jumping, sprinting, and weightlifting movements, PP is typically strongly related to success (Garhammer, 1993; Kauhanen, Garhammer and Hakkinen, 2000; McBride *et al.*, 1999; Thomas *et al.*, 1996).

Heavy weight-training can produce a rightward shift and beneficial effects across the force–velocity curve in lesser-trained subjects (Hakkinen, 1994; Stone and O'Bryant, 1987; Stone, Sands and Stone, 2007). However, evidence indicates that among well-trained athletes, high-velocity training is necessary to make further alterations in the high-velocity end of the force–velocity curve (Hakkinen, 1994; Harris *et al.*, 1999; Stone, Sands and Stone, 2007).

Isometric training with high RFDs can result in an increased RFD and movement velocity, especially in untrained subjects (Behm, 1995). However, isometric training effects on dynamic explosive force production are relatively minor (Hakkinen, 1994). Support for the use of fast ballistic movements can be found in the comparison of the peak RFDs of isometric actions with fast ballistic movements (Haff *et al.*, 1997). Studies and reviews of the scientific literature indicate that ballistic training has a greater effect on RFD and velocity of movement, whilst traditional heavy weight-training (or isometric training) primarily increases maximum strength (Hakkinen, 1994; Harris *et al.*, 1999; McBride *et al.*, 1999, 2002; Sale, 1988). Furthermore, evidence indicates that high-power training can beneficially alter a wider range of athletic performance variables than traditional heavy weight-training especially in subjects with a reasonable initial level of maximum strength (Cormie *et al.*, 2010; Stone *et al.*, 2003; Wilson and Murphy, 1996). However, improvements in strength, power, and measures of athletic performance resulting from combination and sequenced training (strength → power → speed) may be superior to either heavy resistance training or high-speed resistance training alone (Hakkinen, 1994; Harris *et al.*, 1999; Medvedev *et al.*, 1981; Plisk and Stone, 2003; Stone, 1993; Stone, Sands and Stone, 2007). For example, a longitudinal study using American collegiate football players (Harris *et al.*, 1999) indicated that a sequenced combination (heavy training followed by combination of heavy plus high-power training) resulted in improvement in a greater number of variables encompassing a wider range of performance measures than either heavy or high-power training.

Based on these arguments, in order to increase power and speed of movement, some combination of power and speed training (as well as variation) during training is a necessity. Evidence indicates that when using most machines, high-speed and high-power training may be limited relative to that available through free-weights, due to limitations on acceleration patterns (particularly in variable-resistance and semi-isokinetic devices), friction, inappropriate movement patterns, and limited ranges of motion (Cabell and Zebas, 1999; Chow, Darling and Hay, 1997; Harman, 1983). Thus, we argue that dynamic explosive exercises are more effectively performed using free-weights and body-weight exercises.

4.1.3.1 Joint-angle specificity

Although isometric training is not used often in training it can be useful in some situations, such as in injury or disease that precludes dynamic training, in overcoming sticking regions, or in 'functional isometrics' in which dynamic and isometric exercise are combined (Giorgi *et al.*, 1998). In very sedentary subjects isometric training can improve strength at a variety of angles (Marks, 1994). However, isometric training typically produces gains that are greatest at the joint angle trained (i.e. joint-angle specificity). Gains in isometric strength are progressively smaller as measurement moves away from the training angle (Atha, 1983; Folland *et al.*, 2000; Logan, 1960).

Variable-resistance devices attempt to match the encountered resistance to human strength curves by the use of various cams and lever systems (McMaster, Cronin and Mcguigan, 2009). Theoretically, provided a maximum effort is made, a maximum contraction can occur throughout the range of motion by maintaining optimum length–tension/leverage characteristics. However, there is little evidence that variable-resistance devices have been able to completely match resistance to human strength curves (Cabell and Zebas, 1999; Folland and Morris, 2008; Harman, 1983, 1994; Johnson, Colodny and Jackson, 1990). There are two possible reasons as to why a mismatch occurs between variable-resistance devices and human strength curves. First, because of differences in physical characteristics such as trunk length, limb lengths, and moment arms, there is a relatively high degree of variability among humans; thus machines in which resistance only matches average strength curves do not appropriately match an individual's strength curve (Cabell and Zebas, 1999; Harman, 1983, 1994). Even if the resistance provided by the machine could match the strength curve of an individual, a confounding factor is the force–velocity relationship. In order for an individual to maximally load the involved muscles throughout the range of motion, the speed of movement would have to be constant. This would result in an exercise pattern in which the neural control would be nonspecific to most real-life movements. Second, there is no evidence that the variable-resistance devices actually match the average human strength curves (Cabell and Zebas, 1999; Harman, 1983, 1994; Johnson, Colodny and Jackson, 1990). Furthermore, it should be noted that for most movements, particularly multi-joint movements, groups of muscles rather than a single muscle are involved. Since each of these muscles can have a different architecture and a different moment arm there may be no common force–velocity or length–tension relationship.

Because of these different physical and performance characteristics, a constant velocity of movement will not necessarily fit every muscle involved. As a result of these problems, some variable-resistance devices may apply resistance inappropriately and performance adaptation may be reduced. Although not all studies agree (Manning *et al.*, 1990), it has been noted that training with variable-resistance machines can result in strength gains that are greatest at the joint angle at which the greatest resistance is applied, and gains may be reduced at other angles (Atha, 1983; Logan, 1960) However, angle specificity as a training adaptation has not been apparent when using free-swinging or freely moving devices (Kovaleski *et al.*, 1995; Nosse and Hunter, 1985).

For a number of years it has been noted that in many free movements, particularly multi-joint movements, more external force can be exerted toward the end of a range of motion (ascending strength curve). For example, in a squat, if a maximal effort is made throughout, the bar speed will diminish through the sticking region (around 90° knee angle) and accelerate as the knee angle increases. Whilst evidence indicates that currently available variable-resistance machines likely do not adequately train these types of movement, it is theoretically possible that devices that would add resistance at the appropriate regions of movement could be useful in maintaining a force overload.

In recent years the addition of elastic bands and chains to a freely moving bar has gained considerable popularity (McMaster, Cronin and Mcguigan, 2009). As a bar is lifted vertically, elastic bands add additional resistance in a curvilinear manner and chains in a linear manner, both modes attempting to add additional resistance toward the end of the range of motion. Although the small amount of research that has been carried out so far as to the efficacy of these modes is equivocal at best, these devices may eventually be shown to be a useful tool in training, particularly the training of athletes (McMaster, Cronin and Mcguigan, 2009).

4.1.3.2 Movement-pattern specificity

Several studies and reviews have consistently noted that the magnitude of measured performance gains depends on the similarity between the strength test and the actual training exercise (Abernethy and Jurimae, 1996; Augustsson *et al.*, 1998; Behm, 1995; Fry, Powell and Kraemer, 1992; Rasch and Morehouse, 1957; Rutherford and Jones, 1986; Sale, 1988, 1992; Stone *et al.*, 2000a, 2007). Additionally, specific movement patterns appear to result in specific patterns of hypertrophy (Abe *et al.*, 2003; Antonio, 2000). These observations strongly indicate that adaptations resulting from training have a high degree of task specificity. Therefore, selecting training exercises that will have a large potential transfer of training effect can be one of the most important considerations in creating a training programme. For example, investigations indicate that exercises using free-weights can have strong mechanical relationships to a number of specific activities, such as the vertical jump (Canavan *et al.*, 1996; Garhammer, 1981a). As a result of these relationships there is a strong probability that training with specific exercises using free-weights may have a greater transfer of training to athletic (and ergonomic) tasks compared to machines (Nosse and Hunter, 1985; Stone, 1982; Stone and Garhammer, 1981). This primarily results from the ability to perform movements with free-weights that meet the necessary criteria for exercise selection (Table 4.1.1) and related athletic (and daily) tasks more effectively than machines. However, there are few studies available that actually compare training with various devices and their effects on performance or physiology.

Closed vs open kinetic chains

In the past 15 years, open (OKCE) and closed (CKCE) kinetic-chain exercises have received considerable attention in the scientific literature, particularly in terms of injury rehabilitation (Beynnon and Johnson, 1996; Grodski and Marks, 2008; Palmitier *et al.*, 1991). Although the exact definitions for various movement types have been debated and grey areas exist (Blackard, Jensen and Ebben, 1999; Dillman, Murray and Hintermeister, 1994), movements have generally been divided into exercises in which the peripheral segment moves freely (OKCE) and those in which the peripheral segment is fixed (CKCE). For this discussion, CKCE is a movement in which the foot or hand is fixed and force (in a weight-bearing manner) is transmitted directly through the foot or hand, such as a squat or bench press. An OKCE is a movement in which the foot or hand is not fixed and the peripheral segment can move freely, such as a leg extension (Palmitier *et al.*, 1991; Steindler, 1973). Typically, CKCEs produce substantially different muscle recruitment patterns and joint motions than OKCEs. For example, the isolated knee articulation of a leg extension versus the multiple articulations of the ankle, knee, and hip of a squatting movement produce different quadriceps activation patterns (Stensdotter *et al.*, 2003). These differences likely further complicate the learning and neural effects previously discussed. Although some human movements, such as walking and running, can contain a combination of open- and closed-chain aspects, it is the closed-chain aspects of movement that are crucial to performance and especially to improving performance (Palmitier *et al.*, 1991; Steindler, 1973). As many machines are OKCE devices they likely do not provide the same level of specificity for training (or testing) as CKCE training (Abernethy and Jurimae, 1996; Augustsson *et al.*, 1998; Blackburn and Morrissey, 1998; Grodski and Marks, 2008; Manske *et al.*, 2003; Palmitier *et al.*, 1991).

As a result of the activation-pattern differences (i.e. OKCE vs CKCE), rather than differences in muscle contraction type, studies comparing different modes of training may produce outcome differences (Augustsson *et al.*, 1998; Wilk *et al.*, 1996). It is also probable that mono-articular (single-joint) or small-muscle-mass training programmes (or testing) will not provide adequate movement-pattern specificity. Indeed, muscle action has been shown to be task-dependent and muscle functions during isolated movements are not the same as during multi-joint movements (Zajac and Gordon, 1989). For example, it is possible that the differences noted between semi-isokinetic devices and free-weights (Abernethy and Jurimae, 1996) may result from differences in the pattern of activation and movement (OKCE vs CKCE or mono-articular vs multi-joint), rather than actual differences in the way a muscle contracts. Perhaps if movement patterns could be made more similar, results might be more readily comparable.

4.1.3.3 Machines vs free-weights

Transfer of training effects: maximum strength gains

Semi-isokinetic devices

'Isokinetic' refers to exercise using constant angular velocity of a machine lever arm on which a body segment applies force. Theoretically, an isokinetic device will accommodate force production and maintain a constant velocity; thus a maximum-force effort can be made through the complete range of motion. Although technology is improving, there are no commercially available devices which can produce isokinetic movement throughout a complete range of motion, especially at the faster available speeds (Chow, Darling and Hay, 1997). This lack of complete isokinetic range of motion is due to acceleration at the beginning and deceleration at the end of the range of motion (Chow, Darling and Hay, 1997; Murray and Harrison, 1986). Thus, these devices are more properly termed 'semi-isokinetic'.

Many clinicians and some exercise scientists believe that semi-isokinetic training and particularly testing can offer advantages. Proponents point out that the reliability of these devices can be very good (Abernethy and Jurimae, 1996) and that movement is less technique-dependent compared to many free-weight exercises and some resistance machines (Augustsson *et al.*, 1998). Furthermore, the reliability and validity of semi-isokinetic devices may be enhanced by considering positional specificity (Wilson, Walshe and Fisher, 1997).

However, the external validity of semi-isokinetic devices can be questioned (Abernethy and Jurimae, 1996; Augustsson *et al.*, 1998; Fry, Powell and Kraemer, 1992; Kovaleski *et al.*, 1995; Ostenberg *et al.*, 1998; Tunstall, Mullineaux and Vernon, 2005). Indeed, there is considerable scientific evidence which demonstrates that isokinetic training and testing likely do not offer advantages over other forms of testing and resistance training, and in many instances the results may be inferior to other modes and methods (Augustsson *et al.*, 1998; Hakkinen, 1994; Kovaleski *et al.*, 1995; Ostenberg *et al.*, 1998; Petsching, Baron and Albrecht, 1998; Tunstall, Mullineaux and Vernon, 2005).

Although isokinetic devices can have good correlations with other forms of strength testing (Verdijk *et al.*, 2009), studies and reviews comparing semi-isokinetic and other resistance-training modes indicate that isokinetic testing may lack a degree of necessary specificity to examine some aspects of training adaptation (Hakkinen, 1994; Morrissey *et al.*, 1995; Tunstall, Mullineaux and Vernon, 2005; Wilson, Walshe and Fisher, 1997). For example, moments (forces) produced during isokinetic contractions of the same muscles at the same velocities can be different compared to the forces produced as a result of free movements. During a vertical jump, the moments produced during plantar flexion as well as the timing of muscle activation are substantially different to a semi-isokinetic movement (Bobbert and van Ingen Schenau, 1990). Furthermore, isokinetic peak angular velocity, at target values, can be substantially lower than videographically measured kinematic data because the isokinetic data are 'oversmoothed', masking important data (Tunstall, Mullineaux and Vernon, 2005).

Movement is rarely performed at a constant velocity through a full range of motion. It may be argued that a freely moving object or device will allow muscle contractions to occur which are more similar to natural motions (Stone and O'Bryant, 1987; Stone, Sands and Stone, 2007). A comparison of isotonic (freely moving leg-extension device) and semi-isokinetic leg-extension training indicates that dynamic non-isokinetic training is superior in producing both strength and power gains (Kovaleski *et al.*, 1995).

According to proponents, semi-isokinetic devices offer a degree of velocity specificity not found with other training modes. The assumption is that this results from the ability of such devices to overload at fast speeds (Watkins and Harris, 1983). However, the maximum usable testing/training speeds on commercially available semi-isokinetic devices are 400–500 °/s or less. These angular velocities can be considerably less than the either single or summated multiple-joint peak velocities that occur in many athletic activities (Colman *et al.*, 1993). Both unloaded and loaded single- and multiple-joint movements with light loads can produce angular velocities faster than semi-isokinetic devices can accurately measure. Commonly used free-weight movements, particularly those with multiple-joints actions, such as jumping, weightlifting (snatch, clean, and jerk), and throwing/striking, can result in much higher angular velocities than are possible when using the semi-isokinetic devices that are currently available (Tunstall, Mullineaux and Vernon, 2005). Maximal attempts produce angular velocities at the hip and knee during the snatch that can exceed 500 °/s (Baumann *et al.*, 1988). Lifts with sub-maximal weights that are used as part of a warm-up or during training can produce even higher values. Furthermore, these peak angular velocities occur at joint angles at which force is still being exerted on the bar (Baumann *et al.*, 1988), clearly illustrating the potential for a force-overload stimulus at angular velocities that exceeds the limits of semi-isokinetic devices.

These semi-isokinetic-testing shortcomings (position and velocity specificity) may explain why strength and power gains resulting from free-weight-training and variable-resistance training are not always demonstrable when measured on semi-isokinetic devices (Abernethy and Jurimae, 1996; Augustsson *et al.*, 1998; Fry, Powell and Kraemer, 1992). Considering the relatively poor external validity provided by semi-isokinetic devices, the potential for producing a position/velocity/force-specific stimulus that is comparable to free movements, particularly as it pertains to multi-joint movements, is very low.

Non-isokinetic devices

Surprisingly few studies have made direct comparisons of training modes, even though a variety of training devices can be found in most YMCAs, health clubs, and collegiate training facilities. Machines cause different patterns and less muscle activation than free-weights, likely as a result of position, tracking (i.e. limited body movement eliminating or reducing synergist involvement), and general body stabilization (McCaw and Friday, 1994; Stensdotter *et al.*, 2003; Wilk *et al.*, 1996). Based on different activation patterns and movement characteristics, one would assume that a high degree of specificity might be observed in strength gains, with machines producing greater gains on machines, and free-weights greater gains on free-weights (Boyer, 1990; Willoughby and Gillespie, 1990). However, short-term studies using male subjects (Jesse *et al.*, 1988; Stone *et al.*, 1979; Wathen and Shutes, 1982) and specific strength tests have indicated that free-weights can produce superior strength gains independent of testing mode. Interestingly, these studies (Jesse *et al.*, 1988; Stone *et al.*, 1979; Wathen and Shutes, 1982) indicate that, when measuring 1 RMs, free-weight-training transfers to machine testing somewhat better than machine training transfers to free-weight testing, perhaps as a result of greater muscle activation when using free-weights (McCaw and Friday, 1994; Stensdotter *et al.*, 2003). Unpublished results from our laboratory (Brindell, 1999) indicate that this effect also occurs among females. Even in studies that appear to indicate no statistical difference between groups (e.g. Boyer, 1990) actually suggest a relatively greater effect (based on larger calculated effect sizes in the free-weight groups compared to the machine groups when tested on the comparison device) when using free-weights or freely moving devices rather than machines.

Some studies have used a third, 'neutral' device to assess strength gains when comparing different modes of training, assuming that neither of the comparison devices will be favoured. These studies, in which training and testing devices have not been specific, have not indicated strength-gain differences (Messier and Dill, 1985; Saunders, 1980; Silvester *et al.*, 1982; Willoughby and Gillespie, 1990). For example, studies by Saunders (1980) and Silvester *et al.* (1982) used dynamic exercise for training but used isometric testing to evaluate gains, which can reduce any specific maximum strength gains or differences (Wilson and Murphy, 1996). Furthermore, dynamic tests of strength, in which the testing device is supposedly neutral, can favour either free-weight or machine training. This is because the dynamic-testing device has to be either free-weights or a machine.

In the study by Messier and Dill (1985) comparing Nautilus and free-weight-training, tests of leg strength were performed on a Cybex II semi-isokinetic leg-extension device, an OKCE. The Nautilus group used leg extensions as one of the training exercises. Free-weight-training was carried out using the squat, a CKCE, and no leg extensions were performed. Thus, the Nautilus group potentially had an advantage, in part because their training was mechanically more similar to the testing device, particularly in terms of position. Furthermore, in the study by Willoughby and Gillespie (1990), as pointed out by its authors, the free-weight group, whilst not statistically different, produced much greater gains on a universal machine compared to the hydra-fitness group in terms of both percentage gains and treatment effects, as calculated by the authors of this chapter (22%, ES = 1.52 vs 9%, ES = 0.41) (see 'Problems with Studies' – kinetic chain discussion below). These above studies do point out an important consideration: whilst training adaptations may be masked or muted by using a neutral or 'less' specific device to measure gains in strength (or other variables such as power), they do demonstrate a degree of transfer of training effect for strength gains.

Care must also be taken in properly describing training devices, so that they can be categorized and their functional characteristics made apparent. For example, in the study by Boyer (1990) the 'free-weight' lower-body training programme was actually carried out using a leg sled. A leg sled is not a true free-weight device as its motion is limited to a single fixed plane, which results in guided and restricted movements. Additionally, the guiding apparatus can produce considerable friction not encountered in a freely moving object.

Transfer of training effects

Very few studies (Medline and Sports Discus, 1978–2009) dealing with modes of training have investigated transfer to aspects of performance other than strength, such as sprinting or jumping, and none have investigated effects on ergonomic tasks. Only a very few studies have compared free-weights and resistance machines (Augustsson *et al.*, 1998; Bauer, Thayer and Baras, 1990; Jesse *et al.*, 1988; Silvester *et al.*, 1982; Stone *et al.*, 1979; Wathen, 1980; Wathen and Shutes, 1982) as to their effects on performance (other than strength). Interestingly, all seven studies made comparisons using the vertical jump and vertical-jump power indices. The vertical jump is often chosen as an indicator of explosive athletic performance because: (1) it is easy to measure, (2) it is a primary component of performance in many sports (e.g. basketball, volleyball), (3) it and its components, including velocity and power output, have been associated with performance ability in numerous specific sports (Anderson, Montogomery and Turcotte, 1990; Barker *et al.*, 1993; Stone *et al.*, 1980; Thissen-Milder and Mayhew, 1991), and (4) it and its components can differentiate between athletes' performance capabilities; for example, sprinters jump higher and sprint faster than distance runners (Hollings and Robson, 1991).

Five studies (Augustsson *et al.*, 1998; Bauer, Thayer and Baras, 1990; Silvester *et al.*, 1982; Stone *et al.*, 1979; Wathen, 1980) found that free-weights produce superior vertical-jump results, while two (Jesse *et al.*, 1988; Wathen and Shutes, 1982) found statistically equal results, although percentage gains and calculated effect sizes favoured the free-weight groups. No studies could be located which indicate that machine training produces superior results compared to free-weights in gains in vertical jump (or any other performance variable).

Whilst these studies generally indicate the superiority of free-weights in producing a transfer of training effect, they are not definitive. Indeed, some evidence indicates that performance adaptation, including jump performance, may be specific to a device (Caruso *et al.*, 2008). More investigation is needed to fully understand the training adaptations associated with both machines and free-weights.

As previously pointed out, specificity dictates that a number of kinetic and kinematic parameters must be appropriately overloaded to stimulate gains in performance. The use of weightlifting movements (snatch, clean, and jerk, and derivatives) and weightlifting training practices has been an area of intense study and conjecture. Although adaptations to training are always multi-factorial, one likely contributing factor is the degree of associated mechanical specificity that has been observed between weightlifting movements and the vertical jump (Canavan, Garret and Armstrong, 1996; Garhammer, 1981b; Garhammer and Gregor, 1992; Hori *et al.*, 2008). Improved weightlifting performance as a result of training has been associated with increased vertical-jump height and associated power output among novices and beginners (Stone *et al.*, 1980; Channell and Barfield, 2008), and the vertical jump is stratified by achievement level among weightlifters (i.e. better weightlifters jump higher) (Stone and Kirksey, 2000); furthermore, weightlifters have been shown to have superior weighted and un-weighted vertical-jump heights and power outputs than other athletes (McBride *et al.*, 1999; Stone *et al.*, 1991). Indeed, weightlifting training has been shown to increase jumping performance as well as or better than jump training (Tricoli *et al.*, 2005). Part of the reason for these superior performance characteristics is likely related to the mode and methods used by weightlifters in training. These factors include a combination of high power output, high rates of force development, and movement patterns that cannot easily be duplicated solely by machine use.

Some common criticisms of the use of free-weights cannot be supported when examined closely. For example, it is often assumed that throwing motions requiring twisting (trunk rotation) cannot be made and trained appropriately with free-weights and that 'rotational' machines are necessary. This idea may be more related to lack of experience with free-weight-training than the actual mechanics of free-weights. It is important to note that most throwing movements are made in a standing or upright position. For many years throwers have simulated these upright positions using weighted balls and implements; additionally, walking twists and weighted hammer-throwing exercises have been successfully used to overload upright trunk rotation and throwing motions. Furthermore, with the use of benches or pommel horses a variety of positional exercises using both weights and balls can stress trunk rotation from a variety of angles that cannot be attained using most machines.

Some additional resistance during rotational movements may be attained with the use of elastic bands (McMaster, Cronin and Mcguigan, 2009). However, care must be taken in placing the elastic bands so that the angle of pull is directed appropriately.

Whilst most physical activities can be simulated and appropriately trained solely using free-weights, there are potential exceptions: for example, some aspects of swimming, in which motion is generated in a supine or prone position largely through propulsion by the upper body, may require specialized dry-land training. In this case it may be advantageous to use a 'swim bench' or elastic tubing, in conjunction with traditional modes, in order to simulate and overload stroke mechanics (Girold *et al.*, 2006).

Problems associated with free-weight vs machine comparisons

Comparisons of training adaptations resulting from various modes of resistance exercise can be quite difficult. Several confounding factors become evident.

Study length and trained state

The subject numbers in many of the comparative studies are relatively small. This small subject number is especially apparent when the subjects were previously trained or were athletes. For example, in the study by Wathen and Shutes (1982) comparing groups training with free-weights and a mini-gym jumping device, the authors concluded that there was no difference in vertical-jump gains. However, the authors also indicated that significance favouring free-weights would likely have been reached had the subject number been higher ($n = 8$ per group).

Regardless of the purpose, a significant problem with the majority of training studies is the training length. No study making comparisons of training modes has lasted longer than 12 weeks. Study length (i.e. training time) is an important consideration as it impacts on the training status of the subjects. It is recognized that initially untrained subjects can markedly increase maximum strength using almost any reasonable training programme or device. Both neural and muscle adaptations impact maximum strength. However, initial strength improvements are more likely to involve neural alterations associated with learning to more fully activate motor units, learning a technique, or enhancing skill, rather than muscle adaptations. It is well established that in the initial stages of learning a new movement, early gains are typically rapid, with subsequent improvements becoming asymptotic (Crossman, 1964; Schmidt, 1991; Stone, Sands and Stone, 2007). For any training programme, it is highly probable that early gains in performance would primarily be due to the improved neural activation and central representation of the skill, rather than to muscle adaptations. Indeed, significant strength gains have been shown to occur through the use of mental practice alone (i.e. no physical training, only imagining an execution of the movement). Therefore, it is likely that both central representation (learning how to make a maximum effort and learning the skill) and peripheral nerve transmission/efficiency of fibre recruitment account for greater variance in the early 'strength' gains which result from training (Smith *et al.*, 1998; Yue and Cole, 1992). Although some specificity can be demonstrated over a relatively short term (Abernethy and Jurimae, 1996; Rasch and Morehouse, 1957), comparisons between devices lasting only a few weeks are likely measuring only initial changes in neural activation (e.g. learning), many of which are more general in nature, particularly intramuscular adaptations. Additionally, the magnitude of the initial changes in performance, particularly strength gains, can be quite large compared to adaptations occurring later in the training programme. Whilst these initial changes can lead to an increased ability to exert maximum force, many training-specific effects may require longer periods to become evident or may be masked by the large initial changes in the nervous system. Indeed, it is possible that the inability of beginners and novices to exert themselves maximally could limit gains in hypertrophy due to the inability to damage muscle sufficiently. Consequently, longer observation periods (>0.5 years) are likely necessary to completely determine any long-term intra- and inter-muscular task specificity or specific alterations in hypertrophy/muscle physiology as a result of training with different modes.

Three studies in the scientific literature used previously trained subjects (Stone *et al.*, 1979; Wathen, 1980; Wathen and Shutes, 1982), and only two of these used competitive athletes (Wathen, 1980; Wathen and Shutes, 1982). However, even though the subjects were already trained, it is still quite possible that learning and skill acquisition effect and may play a confounding role in these investigations. When individuals skilled in one set of movements (e.g. free-weight bench press) change to another skill movement (even though it may be related, e.g. universal bench press), the neural adaptations and skill acquisition gains previously described suggest that those switching to the novel skill will improve more rapidly than those still using the old training mode. From a learning and motor-control perspective, differences in the complexity of the exercises used can also serve to confound effects in short-term studies. Furthermore, prior skill level may also confound results through the test exercise itself. Depending on their previous practice/training levels, some subjects may have an inherent advantage over others. Consequently, only the results of investigations which have utilized repeated-measures designs or used the pre-intervention performance on the test exercise as a covariate, and which have allowed for the differential impact of learning effects, can be really trusted. Finally, few studies have used women as subjects. Obviously, the only really clear conclusion is that much more comprehensive study needs to be carried out over longer periods of time.

Work equalization

Equalizing work is very difficult to achieve even when set and repetition combinations are the same. This difficulty is due partly to the variety of methods employ by machines to produce resistance (semi-isokinetic, variable resistance, friction, springs, etc.) and partly to the problem of accurately calculating the amount of work accomplished (Augustsson *et al.*, 1998; Cabell and Zebas, 1999). Even when using free-weights, actual measurement of work requires a force plate and measurement is not practical in most situations (McCaulley *et al.*, 2009). So work is usually estimated by volume load, which is associated with actual work measures (McCaulley *et al.*, 2009; Stone, Sands and Stone, 2007). Compounding this problem is the fact that, in practice, training protocols with equal workloads are rarely chosen.

Training protocols are typically selected because they are believed to produce desired results. It is possible that equalizing work may in fact bias the results of a study in favour of one group or another. This results from removing one group (sets and reps) from its optimum set-and-rep protocol. For example, ten sets of three (a STRENGTH–endurance protocol) and three sets of ten (a strength–ENDURANCE protocol) could be performed with the same load and produce equal-volume loads and very similar work-loads, but the resulting physiological and strength gains would not be the same as the load would be optimal for 3×10 but sub-optimal for 10×3. Furthermore, the manufacturers or retailers dealing with machines often recommend training protocols that are quite different from those commonly used, particularly protocols used by serious athletes interested in improving performance (e.g. one set to failure verus multiple sets, non-ballistic versus ballistic movements). This observation is quite important as many studies have actually compared one mode plus protocol with a different mode and protocol. For example, in studying the practical application of methods/modes of training and comparing a manufacture's recommendations, Stone *et al.* (1979) used one set to failure with the Nautilus group and multiple sets with the free-weight group. Obviously, this could preclude definitive conclusions as to the effectiveness of the mode of training independent of the training programme.

Mixed protocols

At least one study has dealt with a comparison of a combination of free-weights and machines with machine-only training (Meadors, Crews and Adeyonju, 1983). Although this is a very 'real-world' approach, it makes it difficult to separate out individual effects for different devices. However, it should be noted that most of the general public as well as most athletes usually train with a combination of devices and free-weights. Thus comparisons of training programmes may lead to valuable insights into optimum training modes and methods. Few of these types of observation have been carried out, and studies using real-world protocols and mixed training modes should be pursued in the future.

4.1.3.4 Practical considerations: advantages and disadvantages associated with different modes of training

Based on the available scientific evidence, careful observation of training and training practices, and logical arguments, it is apparent that different modes of training are associated with different advantages and disadvantages (Nosse and Hunter, 1985; Stone *et al.*, 1991, 2000a; Stone, Sands and Stone, 2007).

1. A major advantage of free-weights is that short-term (months) or long-term (years) training protocols can be developed with appropriate training variation and a high degree of mechanical specificity. This results from the patterns of intra- and especially inter-muscular activation that have a higher degree of task specificity than is usually obtained through machine exercises. Free-weight exercises allow proprioceptive and kinaesthetic feedback to occur in a manner more similar to that of most athletic and daily performance movements. This is possible because free-weight movement can take place in all three planes and is not guided or otherwise restricted by the device. Machines can limit movement or exercise selection in various ways. (1) Typically only one or two exercises can be performed on a machine, thus many machines may be necessary for a complete training session. Free-weights can allow many different exercises to be performed with minimum equipment. (2) Machines typically allow relatively little mechanical exercise variation (e.g. changes in hand or foot spacing, OKCE versus CKCE movements, etc.), while free-weights allow nearly unlimited variation. (3) Many machines are limited to a single plane of movement, while free-weights require balance and therefore permit exercise in multiple planes, such as typically occur in athletic and ergonomic movements. (4) Variable-resistance and semi-isokinetic devices can alter normal proprioception and kinaesthetic feedback as a result of restricting normal acceleration and velocity patterns. For example, attempts to match human strength curves with the resistance supplied by the machine have not been entirely successful.

   From a very practical standpoint it can be argued that a primary rationale for the use of multi-joint exercises such as squats and weightlifting movements and their derivatives is that muscles act as functional task groups rather than in an isolated manner and therefore must be targeted as such. It can be argued that the greater the effort (force/power/RFD), the greater the subsequent training effect on the ability to activate the neuromuscular system, and thus the greater the training effect on force/impulse/power-output development.

   Evidence indicates that power-output development is the most important aspect for athletic and likely daily movements. Furthermore, transmission of power from the ground up through the kinetic chain is a prerequisite for the development of neuromuscular synergy, stabilization, kinesthesis, and proprioception, in turn carrying over to both daily and athletic movements.

2. Free-weight exercises typically involve more joints, used in greater complexity, and have greater degrees of freedom. As a result, free-weights can automatically confer neurogenic and skill-acquisition benefits that do not result from most machines. While the degree of specificity likely dictates the degree of training-effect transfer, some transfer is likely to occur even if the exercises are not identically matched with the target movements (Table 4.1.1). Because free-weight movements can be designed to more closely approximate sport skills and daily tasks than do machines, the potential for greater transfer and consequently better motor performance is enhanced.

3. When training both athletes and sedentary individuals, metabolic factors are not inconsequential and must be considered in the overall training process. Large-muscle-mass, multi-joint exercises can result in energy expenditures and neuromuscular, immune system, autocrine paracrine, and endocrine responses which likely influence training adaptations to a greater degree than small-muscle-mass exercises (Stone, Sands and Stone, 2007). For example, large-muscle-mass exercises require substantially more energy than small-muscle-mass exercises (Scala *et al.*, 1987; Stone *et al.*, 1983). Because metabolism, body mass, and body composition are strongly influenced by energy expenditure, large-muscle-mass exercises are likely to be more effective in causing body composition (and metabolic) changes (Stone *et al.*, 1991; Stone, Sands and Stone, 2007). A variety of large-muscle-mass exercises, such as squats and weightlifting movements and their derivatives, can be performed with free-weights. We would argue that these large-muscle-mass exercises are more easily accomplished with free-weights than with typical machines.

   The use of large-muscle-mass, multi-joint exercises can result in increased time efficiency during a training session. A single large-muscle-mass, multi-joint exercise can activate as many muscle groups as four to eight small-muscle-mass exercises. The relative advantages of a large-muscle-mass exercise compared to smaller-muscle-mass exercises and single-joint exercises are indicated by metabolic considerations (Scala *et al.*, 1987), as well as EMG findings (Stuart *et al.*, 1996; Wilk *et al.*, 1996). For example, a power snatch or squat press involves both upper- and lower-body musculature; in order to activate the same muscle mass, several upper- and lower-body single-joint isolated exercises would be required. Similar arguments can be made when comparing movements which seemingly exercise some or all of the same muscles, for example squats versus a leg press or leg extension; Wilk *et al.* (1996) have shown that in fact the amount of muscle activated steadily decreases from squats to the leg press to leg extensions. Thus, employing a few large-muscle-mass exercises rather than many small- or isolated-muscle-mass exercises can be time-efficient.

4. Some free-weight exercises, such as heavy back squats, and occasionally some machine exercises require the use of spotters. Spotters are necessary to catch the weight if a repetition is missed, and should also provide technique feedback and encouragement.

5. Time can be a major factor in some training situations. For example, time factors often become a problem at the collegiate sport level when facilities and space are limited and practice is scheduled at specific times for specific sports. Time may also be a problem for business people who have limited opportunities during the day to complete a training session (often at lunch). However, the idea that machines can always save time is a common misconception. If the rest period between sets is very short (<30 s) then the ease of moving a pin into a weight stack, for example, may be an advantage. However, in most training situations, particularly priority training, the rest time between sets is a function of the volume load per set and typically lasts about 2–3.5 minutes. Because of the relatively long rest periods, changing weights is not a problem.

6. Moving a weight-stack pin is usually easier and somewhat faster than changing weights on a bar; typical weight-stack machines offer increments of 7.5–12.5 kg. Whilst some machine manufacturers offer lighter additional weights that can be added to the weight stack, many do not. Furthermore, most gyms, health clubs, and so on do not have these smaller add-on weights available. Additionally, devices which use springs and elastic bands to produce resistance do not typically provide bands offering small increments (usually the increments would be approximately 5–10 kg). With typical free-weights and some machines, the range of incremental jumps can be from approximately 0.5 kg up to 50 kg. This wider range allows easier progression and more accurate resistance loading, especially if percentages of maximum are used in planning training programmes.

7. Learning the technique of *some* multi-joint free-weight movements, such as weightlifting movements or squats, requires an expenditure of time and effort, and in some cases specialized coaching. However, most trainees can learn these exercises adequately within two weeks. Most importantly, the cost-to-benefit ratio of learning a new skill can make it worth the effort.

8. Small-muscle-group isolation exercises and single-joint exercises can be accomplished quite easily using machines. Under some specific conditions machines may isolate small muscle masses or stress-specific parts of small muscle masses more efficiently or more easily than free-weights as less effort will be placed on balance. Training isolated muscle groups or single joints can be important in initial rehabilitation, as part of injury-prevention programmes, and in certain aspects of body-building programmes.

9. Although it is commonly believed that resistance training has a high injury rate, in actuality resistance training is a remarkably safe method of training and typically few injuries result, particularly in supervised training situations (Hamill, 1994; Stone *et al.*, 1994, 2007). It is also commonly believed that machines are safer than free-weights, but there is little evidence to support this belief (Requa, DeAvilla and Garrick, 1993), particularly in supervised settings (Hamill, 1994). The authors of this chapter have a combined weight-room/strength-training experience of well over 65 years. During this experience we have observed injuries involving both machines and free-weights, and while the number of injuries during this period has been few, there have been no more injuries among free-weight users than among those using machines. Indeed, weight-training and weightlifting produce few injuries compared to other sports and physical activities (Hamill, 1994; Pierce *et al.*, 1999, 2008; Zemper, 1990), and in fact when appropriately conducted can reduce injury potential (Peterson and Bryant, 1995; Stone *et al.*, 1994).

10. Space for training equipment is usually not a major problem in most public or private training areas such as YMCAs or health clubs, nor is it typically a problem for dedicated gyms such as the sports weight rooms at major American universities and many Sports Institutes (e.g. USOC, Colorado Springs, Australian Institute of Sport, Canberra). However, it can be a problem in some cases. For example, storage space in many private homes is limited, while in the military, space is often at a premium, such as aboard ships. Transportation and storage of equipment occasionally dictates the type of equipment which can be used. In many cases machines, especially those using springs and elastic bands, take up less space.

11. Quite often, cost is the determining factor in the selection of equipment.

12. Machines, especially semi-isokinetic devices, are usually more expensive than free-weights. Considering the cost of multi-station and single-exercise machines, free-weight equipment can be used to train the same number of

people for less money. When equipping a typical training facility, free-weight equipment can also allow more people to be trained at the same time for the same monetary cost.

4.1.4 CONCLUSION

The mode of training can make a relatively large difference to the outcome of a training process. Current information and empirical evidence indicates that transfer of training from the weight room to sport or athletic endeavours necessitates a reasonable level of mechanical specificity. Because free-weight exercises can have a greater degree of specificity than machines, free-weights should produce a more effective training transfer. As a result, training with complex, multi-joint exercises using free-weights can produce superior results to training with machines for most activities.

Therefore, the majority of resistance exercises making up a training programme should comprise free-weight exercises, particularly multi-joint exercises, with emphasis on mechanical specificity. Machines can be used as an adjunct to training, for injury rehabilitation, and to a greater or lesser extent to produce variation during specific phases of the training process (preparation, pre-competition, competition) or if there is a need to isolate specific muscle groups.

It is obvious that additional research is necessary to establish the exact effects of different modes of training on athletic, daily, and ergonomic performance. Future research should attempt to obviate several methodological problems associated with past comparison studies, particularly those associated with training state and length of study.

References

Abe T. Kojima, K., Kearns C.F., Yohena, H. and Fukuda J. (2003) Whole body muscle hypertrophy from resistance traiing: distribution and totoal mass. *Br J Sports Med*, **37**, 543–545.

Abernethy P.J. and Jurimae J. (1996) Cross-sectional and longitudinal uses of isoinertial, isometric and isokinetic dynamometry. *Med Sci Sports Exerc*, **28**, 1180–1187.

Alexander N.B., Galecki A.T., Grenier M.L. *et al.* (2001) Task-specific resistance training to improve the ability of activities of daily living-impaired older adults to rise from a bed and from a chair. *J Am Gerontol Soc*, **49**, 1418–1427.

Anderson R.E., Montogomery D.L. and Turcotte R.A. (1990) An on-site battery to evaluate giant slalom skiing performance. *J Sport Med Phys Fitness*, **30**, 276–282.

Antonio J. (2000) Nonuniform response of skeletal muscle to heavy resistance training: can bodybuilders induce regional muscle hypertrophy? *J Strength Cond Res*, **14**, 102–113.

Asfour S.S., Ayoub M.M. and Mital A. (1984) Effects of an endurance and strength training programme on lifting capability of males. *Ergonomics*, **27**, 435–442.

Atha J. (1983) Strengthening muscle, in *Exercise and Sport Sciences Reviews* (ed. A.I. Miller), The Franklin Institute Press, Plymouth, MN, pp. 1–73.

Augustsson J., Esko A., Thomes R. *et al.* (1998) Weight-training the thigh muscles using closed vs. open kinetic chain exercises: a comparison of performance enhancement. *J Orthop Sports Med Phys Ther*, **27**, 3–8.

Barker M., Wyatt T., Johnson R.L. *et al.* (1993) Performance factors, psychological factors, physical characteristics and football playing ability. *J Strength Cond Res*, **7**, 224–233.

Bauer T., Thayer R.E. and Baras G. (1990) Comparison of training modalities for power development in the lower extremity. *J Appl Sport Sci Res*, **4**, 115–121.

Baumann W., Gross V., Quade K. *et al.* (1988) The snatch technique of world class weightlifters at the 1985 world championships. *Int J Sport Biomech*, **4**, 68–89.

Behm D.G. (1995) Neuromuscular implications and applications of resistance training. *J Strength Cond Res*, **9**, 264–274.

Beynnon D. and Johnson R.J. (1996) Anterior cruciate ligament injury rehabilitation in athletics. *Sports Med*, **22**, 54–64.

Blackard, D.O., Jensen, R.L. and Ebben, W.P. (1999) Use of EMG in analysis in challenging kinetic terminology. *Med Sci Sports Exerc*, **31**: 443–8.

Blackburn J.R. and Morrissey M.C. (1998) The relationship between open and closed kinetic chain strength of the lower limb and jumping performance. *J Orthop Sports Phys Ther*, **27**, 430–435.

Blessing D., Stone M.H., Byrd R. *et al.* (1987) Blood lipid and hormonal changes from Jogging and weight-training of middle-aged men. *J Appl Sports Sci Res*, **1**, 25–29.

Bobbert M.F. and Schenau G.J. (1990) Mechanical output about the ankle joint in isokinetic flexion and jumping. *Med Sci Sports Exerc*, **22**, 660–668.

Boyer B.T. (1990) A comparison of three strength training programs on women. *J Appl Sports Sci Res*, **4**, 88–94.

Brill P.A., Probst J.C., Greenhouse D.L. *et al.* (1998) Clinical feasibility of a free-weight strength-training program for older adults. *J Am Board Fam Pract*, **11**, 445–451.

Brindell G. (1999) Efficacy of Three Different Resistance Training Modes on Performance and Physical Characteristics in Young Women, Appalachian State University, Master's Thesis.

Cabell L. and Zebas C.J. (1999) Resistive torque validation of the nautilus multi-biceps machine. *J Strength Cond Res*, **13**, 20–23.

Canavan P.K., Garret G.E. and Armstrong L.E. (1996) Kinematic and kinetic relationships between an Olympic-style lift and the vertical jump. *J Strength Cond Res*, **10**, 127–130.

Caruso J.F., Coday M.A., Ramsey C.A. *et al.* (2008) Leg and calf press training modes and their impact on jump performance. *J Strength Cond Res*, **22**, 766–772.

Channell B.T. and Barfield J.P. (2008) Effect of Olympic and traditional resistance training on vertical jump improvement in high school boys. *J Strength Cond Res*, **22**, 1522–1527.

Chow J.W., Darling W.G. and Hay J.G. (1997) Mechanical characteristics of knee extension exercises performed on an isokinetic dynamometer. *Med Sci Sports Exerc*, **29**, 794–803.

Colman S.G.S., Benham A.S. and Northcutt S.R. (1993) A three-dimensional cinematographical analysis of the volleyball spike. *J Sports Sci*, **11**, 295–302.

Conroy B.P., Kraemer W.J., Maresh C.M. *et al.* (1993) Bone mineral density in weightlifters. *Med Sci Sport Exerc*, **25**, 1103–1109.

Cormie P., McCaulley G.O., Triplett N.T. *et al.* (2007) Optimal loading for maximal power during lower-body resistance exercises. *Med Sci Sport Exerc*, **39**, 340–349.

Cormie P., McGuigan M.R. and Newton R.U. (2010) Influence of strength on magnitude and mechanisms of adaptation to power training. *Med Sci Sports Exerc*, **42**, 1566–1581.

Crossman E.R.F.W. (1964) Information processes in human skill. *Br Med Bull*, **20**, 32–37.

Dillman C.J., Murray T.A. and Hintermeister R.A. (1994) Biomechanical differences of open and closed chain exercises of the shoulder. *J Sport Rehabil*, **3**, 228–238.

Eek M.N., Tranber R., Zugner R. *et al.* (2008) Muscle strength training to improve gait function in children with cerebral palsy. *Developmental Medicine and Child Neurology*, **50**, 759–764.

Folland J. and Morris B. (2008) Variable cam resistance training machines: do they match the angle-torque relationship in humans? *J Sport Sci*, **26**, 163–169.

Folland J., Hawker K., Leach B. *et al.* (2000) Strength: training: isometric training at a range of joint angles versus dynamic training. *J Sport Sci*, **23**, 817–824.

Fry A.C., Powell D.R. and Kraemer W.J. (1992) Validity of isometric testing modalities for assessing short-term resistance exercise strength gains. *J Sport Rehabil*, **1**, 275–283.

Garhammer J. (1981a) Equipment for the development of athletic strength and power. *Natl Strength Cond Assoc J*, **3**, 24–26.

Garhammer J. (1981b) *Sports Illustrated Strength Training*, Time Inc., New York, NY.

Garhammer J.J. (1993) A review of the power output studies of Olympic and powerlifting: methodology, performance prediction and evaluation tests. *J Strength Cond Res*, **7**, 76–89.

Garhammer J. and Gregor R. (1992) Propulsion as a function of intensity for weightlifting and vertical jumping. *J Strength Cond Res*, **6**, 129–134.

Genaidy A., Davis N., Delgado E. *et al.* (1994) Effects of a job-simulated exercise programme on employees performing manual handling operations. *Ergonomics*, **37**, 95–106.

Giorgi A., Wilson G.J., Weatherby R.P. *et al.* (1998) Functional isometric training: its effects on the development of muscular function and the endocrine system over an 8-week training period. *J Strength Cond Res*, **12**, 18–25.

Girold S., Calmels P., Marurin D. *et al.* (2006) Assisted and resisted sprint training in swimming. *J Strength Cond Res*, **20**, 547–554.

Grodski M. and Marks R. (2008) Exercises following anterior cruciate ligament constructive surgery: biomechanical considerations and efficacy of current approaches. *Res Sports Med*, **16**, 75–96.

Haff G.G., Stone M.H., O'Bryant H.S. *et al.* (1997) Force-Time dependent characteristics of dynamic and isometric muscle actions. *J Strength Cond Res*, **11**, 269–272.

Hakkinen K. (1994) Neuromuscular adaptation during strength training, aging, detraining and immobilization. *Crit Rev Phys Rehabil Med*, **6**, 161–198.

Hamill B.P. (1994) Relative safety of weightlifting and weight-training. *J Strength Cond Res*, **8**, 53–57.

Harman E. (1983) Resistive torque of 5 nautilus exercise machines. *Med Sci Sports Exerc*, **15**, S113.

Harman E. (1994) Resistance training modes: a biomechanical perspective. *Strength Cond*, **16**, 59–65.

Harris G.R., Stone M.H., O'Bryant H. *et al.* (1999) Short term performance effects of high speed, high force and combined weight-training. *J Strength Cond Res*, **14**, 14–20.

Hollings S.C. and Robson G.J. (1991) Body build and performance characteristics of male adolescent track and field athletes. *J Sports Med Phys Fitness*, **31**, 178–182.

Hori N., Newton R.U., Andrews W.A. *et al.* (2008) Does performance of hang power clean differentiate performance of jumping, sprinting, and changing of direction? *Journal of Strength and Conditioning Research*, **22**, 412–418.

Jesse C., McGee D., Gibson J. *et al.* (1988) A comparison of Nautilus and free-weight-training (abstract). *J Appl Sport Sci Res*, **3** (2), 59.

Johnson C.C., Stone M.H., Byrd R.J. *et al.* (1983) The response of serum lipids and plasma androgens to weight-training exercise in sedentary males. *J Sports Med Phys Fitness*, **23**, 39–41.

Johnson C.C., Stone M.H., Lopez S.A. *et al.* (1982) Diet and exercise in middle-aged men. *J Am Diet Assoc*, **81**, 695–701.

Johnson J.H., Colodny S. and Jackson D. (1990) Human torque capability versus machine resistive torque for four eagle resistance machines. *J Appl Sport Sci Res*, **4**, 83–87.

Kauhanen H., Garhammer J. and Hakkinen K. (2000) Relationships between power output, body size and snatch performance in elite weightlifters, in *Proceedings of the 5th Annual Congress of the European College of Sports Science* (eds J. Avela, P.V. Komi and J. Komulainen), Jyvaskala, Finland, p. 383.

Kovaleski J.E., Heitman R.H., Trundle T.L. *et al.* (1995) Isotonic preload versus isokinetic knee extension resistance training. *Med Sci Sports Exerc*, **27**, 895–899.

Kramer J.B., Stone M.H., O'Bryant H.S. *et al.* (1997) Effects of single versus multiple sets of weight-training: impact of volume, intensity and variation. *J Strength Cond Res*, **11**, 143–144.

Lexell J. and Flansbjer U.B. (2008) Muscle strength training, gait performance and physiotherapy after stroke. *Minerva Med*, **99**, 353–368.

Logan G.A. (1960) Differential applications of resistance and resulting strength measured at varying degrees of knee flexion. Doctoral Dissertation, USC.

McBride J., Triplett-McBride T., Davie A. *et al.* (1999) A comparison of strength and power characteristics between power lifters, Olympic lifters and sprinters. *J Strength Cond Res*, **13**, 58–66.

McBride J., Triplett-McBride T., Davie, A., *et al.* (2002) The effect of heavy- vs light-load jump squats on the development of strength, power and speed. *J Strength Cond Res*, **16**, 75–82.

McCaulley G.O., McBride, J.I.M., Cormie, P. *et al.* (2009) Acute hormonal and neuromuscular responses to hypertrophy, strength and power type resistance exercise. *Eur J Appl Physiol*, **105**, 695–704.

McCaw S.T. and Friday J.J. (1994) A comparison of muscle activity between a free-weight and machine bench press. *J Strength Cond Res*, **8**, 259–264.

McGee D., Jesse T.C., Stone M.H. *et al.* (1992) Leg and hip endurance adaptations to three different weight-training programs. *J Appl Sports Sci Res*, **6**, 92–95.

McMaster D.T., Cronin J. and Mcguigan M. (2009) Forms of variable resistance training. *Strength Cond J*, **31**, 50–64.

McMillan J., Stone M.H., Sartain J. *et al.* (1993) The 20-hr hormonal response to a single session of weight-training. *J Strength Cond Res*, **7**, 51–54.

Manning R.J., Graves J.E., Carpenter D.M. *et al.* (1990) Constant versus variable resistance knee extension training. *Med Sci Sport Exerc*, **22**, 397–401.

Manske R.C., Smith B.S., Rogers M.E. *et al.* (2003) Closed kinetic chain (linear) isokinetic testing: relationships to functional testing. *Isokinet Exerc Sci*, **11**, 171–179.

Marks R. (1994) The effects of 16 months of angle specific isometric strengthening exercises in midrange on torque of the knee extensor muscles in osteoarthritis of the knee: a case history. *J Orthop Sport Phys Ther*, **20**, 103–109.

Meadors W.J., Crews T.R. and Adeyonju K. (1983) A comparison of three conditioning protocols and muscular strength and endurance of sedentary college women. *Athl Train*, Fall, 240–242.

Medvedev A.S., Rodionov V.F., Rogozkin V. and Gulyants A.E. (1981) [Training content of weightlifters in preparation period] (trans. M. Yessis). *Teoriya I praktika Fizcheskoi Kultury*, **12**, 5–7.

Meka N., Katragadda S., Cherian B. *et al.* (2008) Endurance exercise and resistance training in cardiovascular disease. *Ther Adv Cardiovasc Dis*, **2**, 115–121.

Messier S.P. and Dill M.E. (1985) Alterations in strength and maximal oxygen uptake consequent to Nautilus circuit weight-training. *Res Q*, **56**, 345–351.

Morrissey M.C., Harman E.A. and Johnson M.J. (1995) Resistance training modes: specificity and effectiveness. *Med Sci Sports Exerc*, **27**, 648–660.

Murray D.A. and Harrison E. (1986) Constant velocity dynamometer: an appraisal using mechanical loading. *Med Sci Sports Exerc*, **18**, 612–624.

Nauhiro H., Newton R.U., Andrews W.A. *et al.* (2008) Does performance of hang power clean differentiate performance of jumping, sprinting, and changing of direction?. *J Strength Cond Res*, **22**, 412–418.

Nosse L.J. and Hunter G.R. (1985) Free-weights: a review supporting their use in rehabilitation. *Athl Train*, Fall, 206–209.

Ostenberg A., Roos E., Ekdahl C. *et al.* (1998) Isokinetic knee extensor strength and functional performance in healthy female soccer players. *Scand J Med Sci Sports*, **8**, 257–264.

Paavolainen L., Hakkinen K., Hamalainen I. *et al.* (1999) Explosive strength-training improves 5-km running time by improving running economy and muscle power. *J Appl Physiol*, **86**, 1527–1533.

Palmitier R.A., Kai-Nan A., Scott S.G. *et al.* (1991) Kinetic chain exercise in knee rehabilitation. *Sports Med*, **11**, 402–413.

Peterson J.A. and Bryant C.X. (1995) Understanding closed chained exercise. *Fitness Handbook*, **2**, 125–130.

Petsching R., Baron R. and Albrecht M.T. (1998) The relationship between isokinetic quadriceps strength test and hop test for distance and one-legged vertical jump test following anterior cruciate ligament reconstruction. *J Orthop Sports Phys Ther*, **28**, 23–31.

Pierce K., Byrd R. and Stone M. (1999) Injuries in youth weightlifting. *J Strength Cond Res*, **13**, 430.

Pierce K.C., Brewer C., Ramsey M.W. *et al.* (2008) Youth resistance training. *UK Strength Cond Assoc J*, **10**, 9–23.

Plisk S. and Stone M.H. (2003) Periodization strategies. *Strength Cond*, **25**, 19–37.

Poehlman E.T., Gardner A.W., Ades P.A. *et al.* (1992) Resting energy metabolism and cardiovascular disease risk in resistance-trained and aerobically trained males. *Metabolism*, **41**, 1351–1360.

Rasch P.J. and Morehouse L.E. (1957) Effect of static and dynamic exercises on muscular strength and hypertrophy. *J Appl Physiol*, **11**, 29–34.

Requa R.K., DeAvilla L.N. and Garrick J.G. (1993) Injuries in recreational adult fitness activities. *Am J Sports Med*, **21**, 461–467.

Robinson J.M., Stone M.H., Johnson R.L. *et al.* (1995) Effects of different weight-training intervals on strength, power and high intensity exercise endurance. *J Strength Cond Res*, **9**, 216–221.

Rutherford O.M. and Jones D.A. (1986) The role of learning and coordination in strength training. *Eur J Appl Physiol*, **55**, 100–105.

Sale D.G. (1988) Neural adaptations to resistance training. *Med Sci Sport Exerc*, **20**, S135–S245.

Sale D.G. (1992) Neural adaptation to strength training, in *Strength and Power in Sport* (ed. P.V. Komi), Blackwell, London. pp. 249–265.

Saunders M.T. (1980) A comparison of two methods of training on the development of muscular strength and endurance. *J Orthop Sports Phys Ther*, Spring, 210–213.

Scala D., McMillan J., Blessing D. *et al.* (1987) Metabolic cost of apreparatory phase of training in weightlifting: a practical observation. *J Appl Sports Sci Res*, **1**, 48–52.

Schmidt R.A. (1991) *Motor Learning and Performance*, Human Kinetics, Champaign, IL.

Schmidtbleicher D. (1992) Training for power events, in *Strength and Power in Sports* (ed. P.V. Komi), Blackwell Scientific Publications, London, UK, pp. 381–395.

Siegrist M. (2008) Role of physical activity in the prevention of osteoporosis. *Medizinische Monatsschrift Pharmazeuten*, **31**, 259–264.

Siff M.C. and Verkhoshanski Y. (1998) *Supertraining: Strength Training for Sporting Excellence*, 3rd edn, University of the Witwatersrand, Johannesburg, South Africa.

Silvester L.J., Stiggins C., McGowen C. *et al.* (1982) The effect of variable resistance and free-weight-training programs on strength and vertical jump. *Natl Strength Cond Assoc J*, **3**, 30–33.

Smith D., Holmes P., Collins D. *et al.* (1998) The effect of mental practice on muscle strength and EMG activity. Proceedings of the 1998 Annual Conference of the British Psychological Society, 22 BPS, Leicester, UK.

Steindler A. (1973) *Kinesiology of the Human under Normal and Pathological Conditions*, Charles C. Thomas, Springfield, IL.

Stensdotter A.K., Hodges P.W., Mellor R. *et al.* (2003) Quadriceps activation in closed and in open kinetic chain exercises. *Med Sci Sport Exerc*, **35**, 2043–2047.

Stone M. (1982) Considerations in gaining a strength-power training effect. *Natl Strength Cond Assoc J*, **4**, 22–24.

Stone M.H. (1991) Physiological aspects of safety and conditioning, in *United States Weightlifting Federation*

Safety and Conditioning Manual (eds J. Chandler and M.H. Stone), USWF, Colorado Springs, CO.

Stone M.H. (1993) NSCA position stance literature review 'explosive exercise'. *Natl Strength Cond Assoc J*, **15**, 7–15.

Stone M.H. and Garhammer J. (1981) Some thoughts on strength and power: the Nautilus controversy. *Natl Strength Cond Assoc J*, **3**, 24–40.

Stone M.H. and Kirksey K.B. (2000) Weightlifting, in *Exercise and Sport Science* (eds W.E. Garret and D.T. Kirkendall), Lipincott Williams and Wilkins, Media, PA, pp. 955–964.

Stone M.H. and O'Bryant H.S. (1987) *Weight-Training: A Scientific Approach*, Burgess International, Minneapolis, MN.

Stone M.H., Byrd R., Tew J. *et al.* (1980) Relationship of anaerobic power and Olympic weightlifting performance. *J Sports Med Phys Fitness*, **20**, 99–102.

Stone M.H., Johnson R.L. and Carter D.R. (1979) A short term comparison of two different methods of resistance training on leg strength and power. *Athl Train*, **14**, 158–160.

Stone M.H., Wilson G.D., Blessing D. *et al.* (1983) Cardiovascular responses to short-term Olympic style weight-training in young men. *Can J Appl Sports Sci Res*, **8**, 134–139.

Stone M.H., Fleck S.J., Kraemer W.J. *et al.* (1991) Health and performance related adaptations to resistive training. *Sports Med*, **11**, 210–231.

Stone M.H., Fry A.C., Ritchie M. *et al.* (1994) Injury potential and safety aspects of weightlifting movements. *Strength Cond*, **16**, 15–24.

Stone M.H., Plisk S., Stone M.E. *et al.* (1998) Athletic performance development: volume load–1 set vs multiple sets, training velocity and training variation. *Strength Cond*, **2**, 22–31.

Stone M.H., O'Bryant H.S., Pierce K.C. *et al.* (1999a) Periodization: effects of manipulating volume and intensity–Part 1. *Strength Cond*, **21**, 56–62.

Stone M.H., O'Bryant H.S., Pierce K.C. *et al.* (1999b) Periodization: effects of manipulating volume and intensity–Part 2. *Strength Cond*, **21**, 54–60.

Stone M.H., Collins D., Plisk S. *et al.* (2000a) Training principles: evaluation of modes and methods of resistance training. *Strength Cond*, **22**, 65–76.

Stone M.H., Potteiger J., Proulx C.M. *et al.* (2000b) Comparison of the effects of three different weight-training programs on the 1 RM squat. *J Strength Cond Res*, **14**, 332–337.

Stone M.H., Plisk S. and Collins D. (2002) Training principles: evaluation of modes and methods of resistance training–a coaching perspective. *Sport Biomech*, **1**, 79–104.

Stone M.H., O'Bryant H.S., McCoy L. *et al.* (2003) Power and maximum strength relationships during performance of dynamic and static weighted jumps. *J Strength Cond Res*, **17**, 140–147.

Stone M.H., Stone M.E., Sands W.A. *et al.* (2006) Maximum strength and strength training: a relationship to endurance? *Strength Cond*, **28**, 44–53.

Stone M.H., Sands W.A. and Stone M.E. (2007) *Principles and Practice of Strength-Power Training*, Human Kinetics, Champaign, IL.

Stuart M.J., Meglan D.A., Lutz G.E. *et al.* (1996) Comparison of intersegmental tibiofemoral joint forces and muscle activity during various closed kinetic chain exercises. *Am J Sports Med*, **24**, 792–799.

Thissen-Milder M. and Mayhew J.L. (1991) Selection and classification of high school volleyball players from performance tests. *J Sports Med Phys Fitness*, **31**, 380–384.

Thomas M., Fiataron A. and Fielding R.A. (1996) Leg power in young women: relationship to body composition, strength and function. *Med Sci Sports Exerc*, **28**, 1321–1326.

Tricoli V., Lamas L., Carnevale R. *et al.* (2005) Short-term effects on lower-body functional power development: weightlifting vs vertical jump training programs. *J Strength Cond Res*, **19**, 433–437.

Tunstall H., Mullineaux D.R. and Vernon T. (2005) Criterion validity of an isokinetic dynamometer to assess shoulder function in tennis players. *Sport Biomech*, **4**, 101–111.

Vertijik L.B., van Loon L., Meijer K. *et al.* (2009) One-repetition maximum strength test represents a valid means to assess leg strength in vivo in humans. *J Sports Sci*, **27**, 59–68.

Votruba S.B., Horvitz M.A. and Schoeller D.A. (2000) The role of exercise in the treatment of obesity. *Nutrition*, **16**, 179–188.

Wathen D. (1980) A comparison of the effects of selected isotonic and isokinetic exercises, modalities and programs on the vertical jump in college football players. *Natl Strength Cond Assoc J*, **2**, 47–48.

Wathen D. and Shutes M. (1982) A comparison of the effects of selected isotonic and isokinetic exercises, modalities and programs on the acquisition of strength and power in collegiate football players. *Natl Strength Cond Assoc J*, **4**, 40–42.

Watkins M. and Harris B. (1983) Evaluation of isokinetic muscle performance. *Clin Sports Med*, **2**, 37–53.

Wilk K.E., Escamilla R.F., Felig G.A. *et al.* (1996) A comparison of tibiofemoral joint forces and electromyographic activity during open and closed kinetic chain exercises. *Am J Sports Med*, **24**, 518–527.

Willoughby D.S. and Gillespie J.W. (1990) A comparison of isotonic free-weights and omnikinetic exercise machines on strength. *J Hum Mov Stud*, **19**, 93–100.

Wilson G.J. and Murphy A.J. (1996) The use of isometric test of muscular function in athletic assessment. *Sports Med*, **22**, 19–37.

Wilson G.J., Newton R.U., Murphy A.J. *et al.* (1993) The optimal training load for the development of dynamic athletic performance. *Med Sci Sport Exerc*, **25**, 1279–1286.

Wilson G.J., Walshe A.D. and Fisher M.R. (1997) The development of an isokinetic squat device: reliability and relationship to functional performance. *Eur J Appl Physiol Occup Physiol*, **75**, 455–461.

Yue G. and Cole K.J. (1992) Strength increases from the motor program: comparison of training with maximal voluntary and imagined muscle contractions. *J Neurophysiol*, **67**, 1114–1123.

Zajac F.E. and Gordon M.E. (1989) Determining muscle's force and action in multi-atrticular movement, in *Exercise and Sport Sciences Reviews* (ed. K. Pandolph), Williams and Wilkins, Baltimore, MD, pp. 187–230.

Zemper E.D. (1990) Four-year study of weight-room injuries in a national sample of college football teams. *Natl Strength Cond Assoc J*, **12**, 32–34.

4.2 Training Agility and Change-of-direction Speed (CODS)

Jeremy Sheppard[1,2] and Warren Young[3], [1]Queensland Academy of Sport, Australia, [2]School of Exercise, Biomedical and Health Sciences, Edith Cowan University, Joondalup, WA, Australia, [3]University of Ballarat, Victoria, Australia

4.2.1 FACTORS AFFECTING AGILITY

An agility task may be best described as a rapid, whole-body change of direction or speed in response to a stimulus (Sheppard and Young, 2006). Many other approaches to defining agility have focused on the physical requirement only, generally a whole-body direction change (Fulton, 1992; Rigg and Reilly, 1987; Tsitskarsis, Theoharopoulus and Garefis, 2003). The unique distinction between the definition used by Sheppard and Young (2006) and other previous definitions is the inclusion of reaction to a stimulus, rather than just change-of-direction speed (CODS). Put simply, agility is an open skill, and CODS is a closed skill. Closed skills can be precisely pre-planned, whilst open skills involve movements that are composed in response to circumstances in the environment (stimuli), such as the movement of a competitor or the bounce of a ball.

Recent research has demonstrated the importance of agility, inclusive of reaction to a sport-relevant stimulus, and its distinctness from physical qualities such as sprint and CODS ability, in field running sports (Gabbett, Kelly and Sheppard, 2008; Sheppard *et al.*, 2006). Both Sheppard *et al.* (2006) and Gabbett, Kelly and Sheppard (2008) have found that players at a higher level of their sport were superior in an agility test that required reaction to a stimulus, but not superior in a task requiring a planned change of direction. Sheppard *et al.* (2006) found that performance on the agility test was not highly correlated with performance on the sprint test or the planned CODS test, whilst Gabbett, Kelly and Sheppard (2008) found a significant ($p < 0.05$) but small–moderate ($r = 0.40$–0.51) relationship between agility and speed and CODS tasks. These results suggest that agility that is inclusive of perceptual and decision-making factors (i.e. response to a stimulus) is a distinct quality from sprint abilities and CODS.

However, this is not to say that the physical factors of sprinting and CODS are not influential on agility performance. Rather, it should be noted that in a well-trained population of athletes, as was the case with the studies by both Sheppard *et al.* (2006) and Gabbett, Kelly and Sheppard (2008), it is likely that the performance of the physical qualities of sprinting and CODS are not very different between athletes (i.e. they are generally all well developed), and that it is the integration of these physical qualities with effective perceptual and decision-making abilities that will contribute to higher overall agility performance in elite athletes. Put simply, at the elite level, most athletes are fast sprinters and fast at changing direction (Baker, 1999; Gabbett, 2002; Gabbett, Georgieff and Domrow, 2007), but integrating this with effective perceptual and decision-making skills is what separates the higher and lower athletes within elite populations (Abernethy and Russell, 1987; Gabbett, Kelly and Sheppard, 2008; Sheppard *et al.*, 2006). For overall agility performance, effective development of sprint speed, CODS, and perceptual and decision-making factors, and the underpinning qualities related to these, are all relatively important to improving performance.

4.2.2 ORGANIZATION OF TRAINING

As noted by Young, James and Montgomery (2002), agility can be broken down into sub-components comprising physical qualities and perceptual and decision-making (cognitive) abilities. As such, agility performance is affected by a diverse collection of qualities. It is a complex sporting skill, and involves many relationships with trainable physical qualities such as leg strength (Cronin, McNair and Marshall, 2003; Young, James and Montgomery, 2002), speed, and CODS (Baker, 1999), as well as cognitive skills (Abernethy and Russell, 1987; Abernethy, Wood and Parks, 1999; Farrow and Abernethy, 2002).

Using the multi-component model of agility outlined by Young, James and Montgomery (2002) it could be viewed that there are two main sub-components of agility: (1) CODS and

Strength and Conditioning – Biological Principles and Practical Applications Marco Cardinale, Rob Newton, and Kazunori Nosaka.

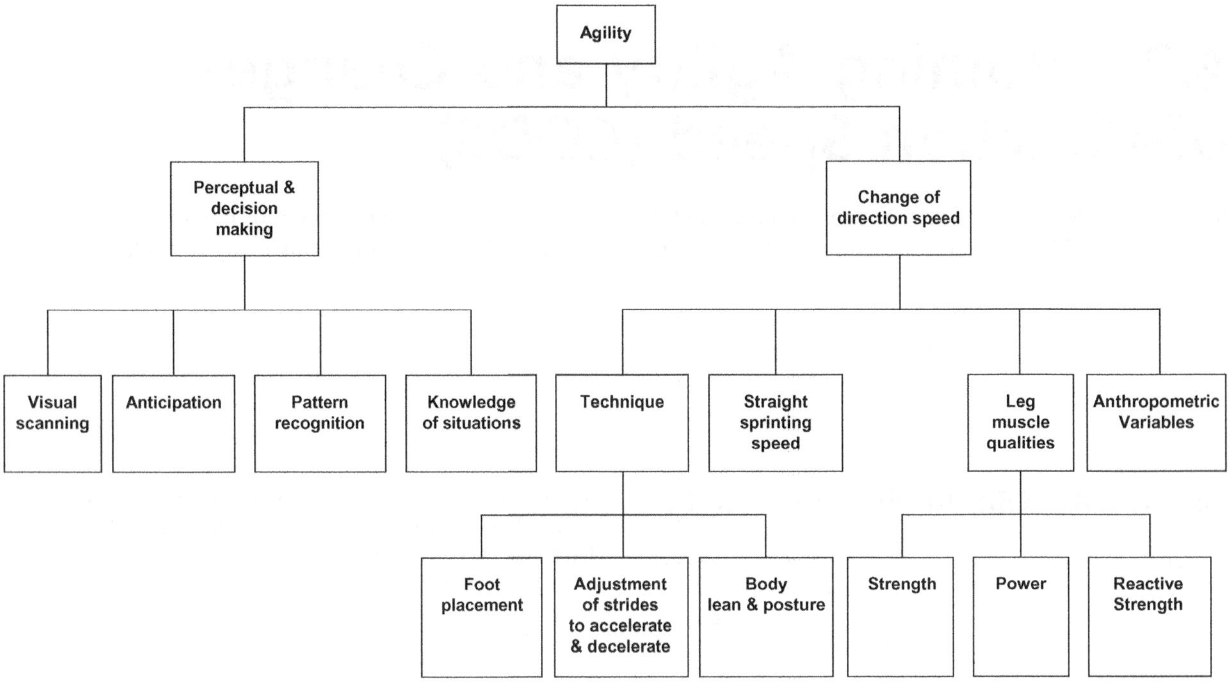

Figure 4.2.1 Agility components. Adapted from Young, James and Montgomery (2002, p. 284)

(2) perceptual and decision-making factors. Within these two main components, sub-components exist, as outlined in Figure 4.2.1. With this in mind, practitioners can target improvements in agility through training its sub-components, such as speed, change of direction speed, and perceptual factors, as well as training sport-relevant agility as a whole. The approach the practitioner adopts in improving agility performance should be based on an understanding of the numerous underpinning components of agility and the individual athlete's performance in these specific areas, in order to create an individual needs analysis for the athlete. Subsequently, the practitioner can then aim to exploit the athlete's areas of weakness and improve overall agility performance. This would appear to be a fundamentally sound approach, as agility performance appears to be trainable (Sheppard, Barker and Gabbett, 2008), and this trainability can likely be accomplished through improvements in physical factors (Young, McDowell and Scarlett, 2001; Young, James and Montgomery, 2002), as well as perceptual and decision-making training (Abernethy, Wann and Parks, 1998; Abernethy, Wood and Parks, 1999; Farrow and Abernethy, 2002; Farrow *et al.*, 1998).

As outlined in Chapter 3.2, one of the purposes of speed, change of direction, and agility assessment is to profile an athlete's individual strengths and weaknesses. From this, the practitioner can highlight which specific areas of weakness to target, as these areas likely present the largest window of adaptation to improve overall performance. For example, an athlete who is skilled at perceptual skills such as anticipation and is effective at decision-making accuracy, but who is of average speed and ability at changing direction, is likely best engaging in a sprint and CODS-orientated programme. On the other hand, a faster athlete, who performs well in tests of sprint and CODS, is likely better off engaging in perceptual and decision-making training and reactive agility training (Gabbett, Kelly and Sheppard, 2008; Sheppard, Barker and Gabbett, 2008). As outlined in Table 4.2.1, interpretation of the appropriate testing data allows the practitioner to prioritize the agility training of an individual based on their specific needs.

4.2.3 CHANGE-OF-DIRECTION SPEED

The primary physical action in most agility tasks is deceleration and changing direction. Speed in changing direction is a clear determinant of sport performance in field sports (Keogh, Weber and Dalton, 2003), and this ability to accelerate, decelerate, and change direction is a foundation component of overall agility performance. In sports such as football codes, athletes perform sprints with rapid deceleration and changes of direction throughout a game (Dawson *et al.*, 1998; Dawson *et al.*, 2004; Keogh, Weber and Dalton, 2003; Reilly *et al.*, 2000).

Although speed is a fundamental component of CODS, CODS is a unique quality that requires specific training. Athletes who are fast sprinters are not necessarily fast in changing direction, as several studies have observed (Draper and

Table 4.2.1 Interpretation and training prescription for four players with different results on the reactive agility test. Gabbett, Kelly, and Sheppard, 2008.

Player	Decision Time (ms)	Movement Time (s)	Interpretation	Prescription
"Fast Mover-Fast Thinker"	58.75	2.31	Speed and fast decision time contributes to above average anticipation skills.	Continue to develop change of direction speed and decision-making skills.
"Fast Mover-Slow Thinker"	148.75	2.33	Has speed but slow decision time contributes to below average anticipation skills.	Needs more decision-making training on (e.g. reactive agility training) and off (e.g. video-based perceptual training) the field.
"Slow Mover-Fast Thinker"	28.75	2.85	Perceptually skilled, but lacks change of direction speed.	Needs more speed/change of direction speed training to improve physical attributes.
"Slow Mover-Slow Thinker"	112.50	2.86	Poor speed and slow decision time contributes to below average anticipation skills.	Needs more decision-making and speed/change of direction speed training to improve physical attributes and perceptual skill.

"Fast Movers/Fast Thinkers" = *good change of direction speed and good perceptual skill;*
"Fast Movers/Slow Thinkers" = *good change of direction speed and below average perceptual skill;*
"Slow Movers/Fast Thinkers" = *below average change of direction speed and good perceptual skill;*
"Slow Movers/Slow Thinkers" = *below average change of direction speed and below average perceptual skill.*

Lancaster, 1985; Pauole *et al.*, 2000; Sheppard *et al.*, 2006; Tsitskarsis, Theoharopoulus and Garefis, 2003). This is not to suggest that sprint speed and acceleration are not underpinning qualities of CODS, as any change of direction involves these components. However, improving an athlete's speed in changing direction requires specific training beyond that of just sprint training. With athletes who are already generally trained in sprint running, additional sprint training may not yield noticeable improvements in highly specific CODS tasks (Young, McDowell and Scarlett, 2001). This highlights the role of performing specific change of direction speed training that is relevant to the athletes sport and even position. For example, a volleyball player might undertake training to establishing or improve basic movement patterns in their movement and change of direction at the net as this relates to blocking duties (Figure 4.2.2). Although during a match this task is performed in response to cues from team mates and the opposition, movement-efficiency and basic CODS training are fundamental and underpinning components of this reactive agility task.

Figure 4.2.2 A volleyball player performs closed-skill CODS training relevant to blocking duties

Acceleration qualities are associated with CODS performance (Baker, 1999; Gabbett, Kelly and Sheppard, 2008; Sheppard *et al.*, 2006), and are trainable qualities, even in well-trained field and court-sport athletes (Harrison and Gillian, 2009; Sheppard, Barker and Gabbett, 2008; Sheppard, Gabbett and Borgeaud, 2008; Young, McDowell and Scarlett, 2001). As such, acceleration and deceleration training are typical components of CODS and agility training programmes (Baechle, 1994; Gambetta, 2007).

Acceleration–deceleration sequences are trainable qualities (Draper and Lancaster, 1985; Young, McDowell and Scarlett, 2001) that can influence CODS and agility performance (Cronin, McNair and Marshall, 2003; Gabbett, Kelly and Sheppard, 2008). Therefore, training methods for effective acceleration and deceleration are worthwhile for the practitioner to understand.

CODS performance can generally be viewed as dependent on fast movement (i.e. acceleration) with effective and rapid deceleration and re-acceleration (towards a different direction). These abilities are likely influenced by strength and power qualities, balance and coordination, and technique. As such, it

Figure 4.2.3 The catch position in weightlifting movements, providing not only the development of strength and power qualities, but also relevant eccentric strength in arresting the load

is important to recognize the potential positive impact of improving these qualities that underpin CODS performance, and to exploit these qualities to improve overall agility performance.

4.2.3.1 Leg-strength qualities and CODS

Leg strength and power is an important contributor to speed in changing direction. The deceleration and push-off action of a change of direction involves high forces applied very rapidly (Besier *et al.*, 2001b), and as such involves high levels of strength from the athlete. As such, practitioners should aim to develop not only general strength and power qualities of the lower body through general strength training, but also weightlifting (Figure 4.2.3), as this method of training allows for the relatively safe development of relevant eccentric-strength qualities through arresting heavy loads in the 'catch' movement.

Young, James and Montgomery (2002) demonstrated that lower-extremity muscle imbalances between the left and right leg were related to CODS. Subjects were found to be significantly slower in changing direction off the weaker leg, when comparisons were made using a unilateral-drop-jump test for reactive strength. This might be due to the similar push-off action when comparing a unilateral drop jump (reactive strength movement) and that of the dynamic single-leg push-off in changing direction while running. This conclusion stands to reason in that the more similar the assessment of the strength qualities, the more influential to the task of interest. As such, practitioners should aim to include unilateral (single-leg) strength training? (Figure 4.2.4) and single-leg reactive strength training? through relevant plyometric tasks such as lateral bounds and hops, and overload exercises such as declining changes of direction (Figure 4.2.5) and lateral single-leg drop jumps (Figure 4.2.6).

4.2.3.2 Technique

Running technique has been suggested to play a key role in performance of sprints with directional changes (Bompa, 1983; Sayers, 2000). In particular, utilizing a forward lean and low centre of gravity (CG) would appear to be essential in optimizing acceleration and deceleration, as well as increasing stability. The stability afforded by a low CG, as opposed to the upright stance and high CG of track-and-field sprinters (Francis, 1997; Mann, 1981), would potentially allow for more rapid changes of direction. In order to change direction at higher speeds, athletes must first decelerate and lower their CG (Sayers, 2000). In other words, Sayers suggested that because sprinting with a high CG (as seen in track-and-field technique) would require postural adjustments (lowering of the CG and shortening stride lengths) and deceleration prior to changing direction, athletes in sports that require frequent changes of direction should run with a lower CG, greater forward lean, and perhaps shorter stride lengths compared to competitive sprinters.

Figure 4.2.4 Single-leg squat for unilateral strength development

Figure 4.2.5 Decline change of direction: on a decline, the athlete steps laterally (downhill) and then employs a dynamic leg drive to change direction and sprint uphill

When reviewing the opinions proposed by Sayers (2000), it becomes apparent that there is a greater need for specificity between training for sprinting and training for speed and agility with sports that require changes of direction, due to the apparent biomechanical differences. However, it should be considered that the taller running posture (as in track and field) can be relevant to field sport players when the opportunity arises for them to get close to top speed. Therefore there may be scope to train these general principles (i.e. use some track-and-field methodology) in team-sport players rather than just accepting that team-sport players should be taught to run low. In fact, there may be a particular need for field-sport athletes to learn to transition from running 'in traffic' to more open space and practise these changes of posture, as well as training the postural changes in 'transitions' between acceleration, high-speed running, deceleration, and changing direction.

It should also be noted that the posture and biomechanics of straight sprinting during the acceleration phase for track-and-field sprinters certainly shares some common themes with the suggestions proposed by Sayers (2000) in regards to field-sport running postures. Specifically, pronounced forward lean and low CG are an integral part of acceleration in sprinting for track and field (Francis, 1997; Mann, 1981), and these biomechanical considerations are quite similar to field-sports. The obvious exception would be that in track and field, sprinters are taught to keep their visual focus low (looking downward) for a portion of the acceleration phase, while in field sports, visual scanning of the playing field is continuous.

Technique in changing direction from a run is a critical factor of success in performing rapid direction changes. Amongst physical-preparation practitioners, there is a degree of mystique attributed to technique in direction change, possibly with the perception that there exists an *optimal technique*. However, there is limited research outlining the technique characteristics of higher- and lower-performing athletes in changing direction. We propose that there is no single optimal model that is most effective in changing direction, due to differences in individual athlete characteristics, such as strength and power, anthropometry, and the demands of the sport and even the position. However, a fast change of direction, regardless of the situation, involves rapid deceleration and arresting of force, and acceleration towards the intended direction of movement. When we view CODS in this way, some accepted principles of biomechanics and physiology can be applied to ensure effective technique.

To accelerate, we must initially overcome our body's inertia in order to get moving, a performance that is highly dependent

Figure 4.2.6 Lateral single-leg drop jumps: the athlete drops off of a box laterally and employs a dynamic single-leg drive to explosively step back on to the box

Figure 4.2.7 Forward lean in acceleration: the lean is a whole-body lean from the feet to the head

on maximal relative strength levels (Baker and Nance, 1999; Sheppard, 2003; Sleivert and Taingahue, 2004; Young, McLean and Ardagna, 1995). The importance of maximal relative strength and power to the start and acceleration phase of the sprint can be understood more clearly by examining their primary technical characteristics, namely longer time on the ground to develop higher forces (to overcome greater inertia) with larger amounts of movement through the hip, knee, and ankle (angular displacements) while the leg is on the ground through an extension action. Stronger athletes are also better able to utilize greater forward lean during acceleration, as they are able to apply the extensive forces required to complete the pushing action in the acute forward-lean position.

During acceleration, the first step is quite short, with a relatively long ground-contact time (GCT). Each subsequent step involves a shorter GCT, and in each subsequent stride the stride length becomes longer. Fundamental to this technique is having an acute, *positive* shin angle when the driving leg contacts the ground. The emphasis of acceleration with the legs is on back-side mechanics and the pushing action. To achieve this, the arms must provide propulsion and aid in lift and the athlete must utilize the aforementioned forward lean. Because the force applied to the ground is through the foot, the arm action can sometimes be under-emphasized. However, the arm drive is fundamental to effective application of force through the legs, particularly during acceleration, and poor arm action will result in inefficient acceleration technique.

In front of the body, the arms provide lift by driving forward and upwards from the shoulder, in synch with the forward drive of the opposite knee. The rearward drive of the arm downward and backward assists in the application of force against the ground. Fundamentally, the arms should remain nearly in-line with the shoulders, but variations in technique will of course result when the athlete is carrying a ball (e.g. football) or an implement (e.g. hockey stick). Again, fundamentally, during initial acceleration, there is very little shoulder and hip rotation, yet carrying a stick or a ball, or visual scanning demands (i.e. the athlete is looking in a different direction than that of their movement), may influence technique in this regard.

When an athlete decelerates from a forward run, the CG is lowered slightly and the athlete's foot-strike must occur ahead of the body. During initial deceleration, the athlete is in a rearward-lean position. It is important to note that regardless of whether we are discussing the forward lean in accelerating, or rearward lean in decelerating, the lean must be *from the ground*, not *from the waist*, to be effective (Figure 4.2.7). In other words, the athlete should not be encouraged to 'bend over' to accelerate or 'lean back' to decelerate, as this will do little to assist them. It is the angle of the body in these tasks which allows the athlete to produce (acceleration) or arrest/absorb (deceleration) force more efficiently.

During deceleration, the athlete absorbs force, primarily through flexion of the ankle, knee, and hip. This action is aided by the rearward body lean, opposite to that of acceleration. The muscles in this action are essentially decelerating the movement of the body's mass under a high eccentric (lengthening action) load, controlling the rate of deceleration to a speed where a change of direction or skill can effectively be executed. The arms continue to oppose the movements of the lower body, aiding in the absorption of force and providing balance to help control the athlete's movement, coordination, and centre of mass.

In discussing deceleration in the context of performing a change of direction, there is an immediate need to initiate a propulsive force soon after (i.e. decelerate and then push-off to change direction). Simply put, the athlete must reduce force (decelerate) and produce force (accelerate) in some manner, such as changing direction, jumping, tackling, etc. Performing this task effectively is vital for agility performance.

The underpinning component of effective reduction and then production of force, such as in decelerating from a sprint prior to changing direction, is the utilization of the stretch load inherent to the eccentric action. If utilized effectively, the stretch load provided by the eccentric action can contribute greatly to the production of force in the following concentric (shortening) action of the muscle (stretch–shortening cycle, SSC). The eccentric–concentric action sequence can provide superior performance output through increased force (Doan *et al.*, 2002; Sheppard, Newton and McGuigan, 2007; Sheppard *et al.*, 2008) and velocity (Bobbert, 1990; Bobbert *et al.*, 1987; Sheppard, McGuigan and Newton, 2008). Considering that SSC function is influenced by the rate, magnitude, and load of the stretch, and is dependent on a short delay between the eccentric action and initiation of the concentric action (Enoka, 2000), well-developed technique in the deceleration–acceleration sequences of CODS tasks will allow the athlete to change direction faster.

When decelerating and absorbing force through the lower body, the athlete must utilize a range of motion that allows enough lengthening of the muscle to reduce force and stimulate the SSC (as the SSC is influenced by the magnitude of stretch). However, too much flexion (coaching cue is to 'sit down low') places the athlete in a poor position biomechanically to exert concentric force, thereby negating the positive effects of the large magnitude of stretch and likely increasing the delay between the eccentric and concentric actions.

A simple method to reinforce this concept is to attempt counter-movement vertical jumps from several different depths. Attempt a vertical jump with the instructions to perform a very shallow dip motion, an extremely deep motion, and then finally from a depth that you feel will illicit the greatest jump height. In most athletes, the depth used to achieve their greatest jump height will be an intermediate dip distance somewhere between very shallow (minimizing muscle-length changes and maximizing speed) and extremely deep (maximizing muscle-length changes but reducing speed), as this depth optimizes the effective contribution of the SSC whilst initiating the concentric action from a position where the muscles can effectively produce force. The same principle is applicable to the magnitude of flexion of the ankles, knees, and hips in decelerating from running speeds; the amount of flexion needs to optimize the relationship between dissipating the force whilst optimizing the joint-angle-specific strength of the muscles.

The physical quality involving the absorption (arrest) of force and subsequent acceleration in deceleration–change of direction sequences is often referred to as reactive strength (Young, 1995; Young, James and Montgomery, 2002), and training this quality is related to executing effective technique in decelerating and changing direction (Cronin, McNair and Marshall, 2003; Djevalikian, 1993), indicating the utility of plyometic exercises in developing CODS. The safe execution of this skill requires well-developed strength in the legs, but also body control and awareness (Gambetta, 2007). This is particularly important when we consider the importance of open-skill agility in many sports (Gabbett, Kelly and Sheppard, 2008; Sheppard and Young, 2006; Sheppard *et al.*, 2006), and that during unplanned tasks (i.e. changes of direction in response to a stimulus) the forces that need to be absorbed through the body are much greater than in controlled, planned deceleration and change-of-direction tasks (Besier *et al.*, 2001a, 2001b).

In an investigation into knee-joint loads, comparing unplanned and planned change of direction, specific approach-speed technique differences were reported (Besier *et al.*, 2001a, 2001b). When subjects were required to react to a light stimulus to change direction, the loads on the knee were increased, which was thought to be related to a posture imposed by the time-stress in reacting to the stimulus and changing direction. This implies that when comparing planned and unplanned changes of direction, such as those found in many sports, postural techniques differ. This finding indicates that practitioners should carefully consider the possible subtle but important technical differences that exist between planned and reactive changes of direction. The management of training load between these circumstances could have profound effects on the performance, as well as the injury potential, of the athlete. For example, if an athlete only performs CODS training under planned circumstances (i.e. only uses closed-skill CODS training), yet competes in a sport that involves reactive changes of direction (i.e. open-skill agility tasks), they may not be optimally prepared physically or technically for the open-skills movements. They may also be exposed to increased injury risk as their training history will not have involved the higher loads presented by reactive training.

4.2.3.3 Anthropometrics and CODS

Very little research has attempted to correlate anthropometric measures and CODS performance. Theoretically, factors such as body fat and body segment lengths may contribute to agility performance. When comparing two athletes of equal total body mass, a fatter athlete will have less lean mass to contribute to the speed requirements of agility performance. In addition, the fatter athlete will have a greater inertia, requiring greater force production to produce a change in velocity or direction.

Test batteries have revealed that athletes in sports such as rugby and football who perform better on CODS tests tend to also have lower body fat (Reilly *et al.*, 2000; Rigg and Reilly, 1987). However, this certainly does not indicate a causal relationship. In fact, direct correlations between CODS and body fat were not performed in the aforementioned studies. A rare study that did involve correlations of body fat and CODS in rugby players found that the two variables were not significantly or strongly related (r = 0.21) (Webb and Lander, 1983), but this study used a relatively homogenous population, which may reduce the apparent significance of the relationship.

Other factors that could potentially be related to CODS performance are height, relative limb lengths, and the height of the athlete's CG. Some research (Cronin, McNair and Marshall, 2003) has suggested that limb length has a relationship with certain sporting tasks, such as lunges typical of directional changes in court-based racquet sports. A person with a lower CG may conceivably be able to apply force more horizontally, which might produce a faster change of direction.

4.2.4 PERCEPTUAL AND DECISION-MAKING FACTORS

Based on cognitive research, there are distinct differences between experts and non-experts in visual search strategies (Abernethy and Russell, 1987; Berry, 1999; Muir, 1996; Ritchie, 1999; Starkes, 1987; Tenenbaum *et al.*, 1996). Further, not only do experts utilize different cues by which to anticipate and respond to stimuli, but they also may perform tasks with a higher search rate (Abernethy and Russell, 1987; Williams *et al.*, 1993). These differences further strengthen the need for a highly specific stimulus.

This is an important consideration in the specificity of the stimulus used to develop any skill that involves anticipation and reaction. Experts utilize different cues than novices, and therefore a generic stimulus is unlikely to be very useful in improving performance in specific situations. In other words, a general reaction to a cue that is not relevant to the sporting task is limiting; although it will develop general response skills and technique without the aid of anticipation, it does not develop visual search strategies, anticipation, or decision-making relevant to the actual sport setting (Abernethy and Russell, 1987; Muir, 1996; Williams *et al.*, 1993).

The specificity of the stimulus presentation is of extremely high importance, as anticipatory expertise appears to be dependent on the specific stimulus used to test this quality (Abernethy and Russell, 1987; Muir, 1996; Tenenbaum *et al.*, 1996). When considering the human information-processing model, a stimulus produces specific mental operations that are based on the subject's retrieval of stored memory information, prior to initiating a response. The accuracy and speed of this response will clearly be dependent on the previously stored information specific to that situation (Cox, 2002). In other words, if the training stimulus is not adequately specific to the sport setting, then the response training is not be optimal in developing sporting expertise, because the participants are not developing anticipation and pattern recognition of a situation that is specific to their sport.

Higher-performance athletes have demonstrated unique visual search rates (Williams *et al.*, 1993) and unique visual cues (Abernethy and Russell, 1987; Muir, 1996; Tenenbaum *et al.*, 1996) that are specific to the domain of their sport, when compared to lower-performers. In other words, research has demonstrated that the higher-performing athlete can be distinguished by cognitive abilities (anticipation, decision-making time, decision-making accuracy) when these are tested with sport-specific cues. Therefore, practitioners can not only confidently utilize physical-based training to develop agility, but supplement this with audio-visual training, such as video-occlusion techniques, to develop cognitive skills such as anticipation, pattern recognition, and decision-making.

Research into developing anticipation and decision-making has highlighted possible methods to advance agility performance (Abernethy, Wood and Parks, 1999; Farrow and Abernethy, 2002; Farrow *et al.*, 1998; Maxwell, Masters and Eves, 2000; Muir, 1996; Ritchie, 1999). Most investigations, such as those conducted by Farrow *et al.* (1998) and Farrow and Abernethy (2002), have assessed the efficacy of various learning techniques intended to accelerate anticipation performance. The studies by Farrow *et al.* (1998) and Farrow and Abernethy (2002), as well as research by Abernethy, Wood and Parks (1999), Borgeaud and Abernethy (1987), Abernethy and Russell (1987), and Tenenbaum *et al.* (1996), investigated the efficacy of cognitive testing and training (using video clips of sporting plays) in an attempt to identify differences in expertise and increase performance in several factors that are dependent on anticipation. Temporal-occlusion techniques are generally used in these studies, wherein a video of a particular skill (e.g. tennis strokes/serves) is played, and is stopped at certain points prior to the actual execution of the skill. The subject is expected to respond to the stimulus, either generically (by pushing a button) or specifically (e.g. by mimicking the appropriate returning stroke to counter the move) in order to indicate what an appropriate reaction would be in the sport setting. The responses can be measured for accuracy (did the subject choose the right response?) and to a certain extent speed (how quickly did the subject initiate or carry out the response?).

The results of these studies suggest that anticipation and reaction in a specific sport context are indeed trainable though attention methods such as video occlusion. It is apparent that there has been a logical progression in the advancement of the methods used. Initially, studies involved anticipation and reaction time based around generic stimuli, and generic but athletic responses (Chelladurai, Yuhasz and Sipura, 1977; Hertel *et al.*, 1999). Abernethy, Wood and Parks (1999) stressed that based on their findings and those of others, training of anticipation and response time with athletes should involve a sport-specific presentation, in order to truly assess and develop the visual skills and recognition required of that specific sporting context. The authors stated that experimental evidence has demonstrated that generic visual training approaches to motor learning are ineffective, likely because they train perceptual factors that do not limit performance in the specific sporting environment.

An emerging technique that might be useful in developing cognitive expertise is the utilization of extreme temporal stress. To increase temporal stress, or 'over-speed training', anticipation and decision-making skills are trained though greater-than-normal speed of displayed cues. This is accomplished by taking normal video footage of a sport-relevant stimulus and speeding it up. The athlete can be asked to provide a physical response to the displayed cue as it is played through on video, or a traditional verbal or written response can be made once the video is occluded at a specific point. This concept of training can be likened to a downhill mountain-bike athlete training on a motocross bike, as the stimulus of riding the bike is similar enough to be relevant, but the speed is much faster. Extensive research on the efficacy of this technique has not been conducted, but anecdotal support by coaches and skill-acquisition specialists suggests that it may prove to be a useful training technique.

4.2.5 TRAINING AGILITY

Training agility as a whole is fundamental to the preparation of athletes in sports that require high levels of agility. Although, as noted previously in this chapter, there is likely a great utility in breaking agility skills into their specific parts to develop deficient sub-qualities of overall agility performance, inherent to preparation for an agility-based sport is skill work. This skill work, as for example in a team sport, will naturally involve elements of small-group work and small-sided games, as well as team tactics. The individual and small-group skills will tend to involve a relatively simple cognitive demand (i.e. simple anticipation and basic decision-making) (Figure 4.2.8). Larger-team training or game play may require very broad cognitive demands that include greater elements of pattern recognition and complex decision-making (Figure 4.2.9).

The amount of training time dedicated to small-group and larger-team or tactical training should logically depend on where the athletes are in their preparation for competition, but equally importantly, where they are in their athletic development. It is sensible to attend to and attain the necessary individual skills early in the preparation period to underpin subsequent

Figure 4.2.8 Basic agility tasks tend to require anticipation of a relatively simple stimulus such as the movement of an opposing player

Figure 4.2.9 Team training inherently involves a greater range of stimuli, increasing the cognitive demands in pattern recognition and visual search strategies

group skills and tactics. So too, it is sensible to develop individual skills and abilities early in an athlete's career, before developing complex tactical capabilities and complex pattern recognition. Put simply, using football as an example, it would not make sense to teach a group of novice footballers the subtle points of a tactical structure, if these athletes annot not defend possession in a one-on-one scenario. The individual skills and abilities are a fundamental component of the larger and more complex sporting skills. An important consideration for the practitioner is to evaluate the short- and long-term needs of their athletes, to determine the structure and relative emphasis of individual and/or team training.

Simple agility work can involve relatively general training, such as various tagging and evasion games and using agility belts (Figure 4.2.10), but can also involve more sport-specific scenarios, such as one-on-one and two-on-two drills. Different cues and scenarios can be utilized in these scenarios. For example, specific skills being executed such as one-on-one kick-and-chase drills for rugby (Figure 4.2.11) or one-on-one serve–receives for volleyball (Figure 4.2.12).

This type of agility work involves a limited range of cues, as well as distractions and irrelevant cues. This offers the benefit of allowing a very targeted development of cue recognition. For example, in a tagging game each athlete is likely only focused on the movement exhibited by their opponent, and not influenced by the movement of other players, the movement of the ball, or considerations of more complex tactics. These simpler agility-training exercises may involve greater physical demands, with more acceleration, deceleration, and changes of direction in response to relevant stimuli.

More complex agility training, such as team games and competition-like environments, generally presents greater demand on pattern recognition and complex decision-making. The physical demands may be reduced, as the individual may not have as many physical skills to execute, but this will depend on the sport context.

These Competition-like training scenarios involve greater variety of stimuli, challenging the athlete to attend to the most relevant cues in order to achieve their performance outcome. As such, this type of training represents the application of their general agility qualities to the whole sporting complex (Figure 4.2.13).

Agility-training drills can be manipulated to create additional temporal and/or spatial stress, depending on the practitioner's preference. Additional temporal stress involves drills where the athlete is required to execute an agility task with

Figure 4.2.10 Agility belts: athletes are attached to each other at the waist by a Velcro belt. The object of the game is for one designated athlete to evade their opponent and create enough space to rip the Velcro attachment

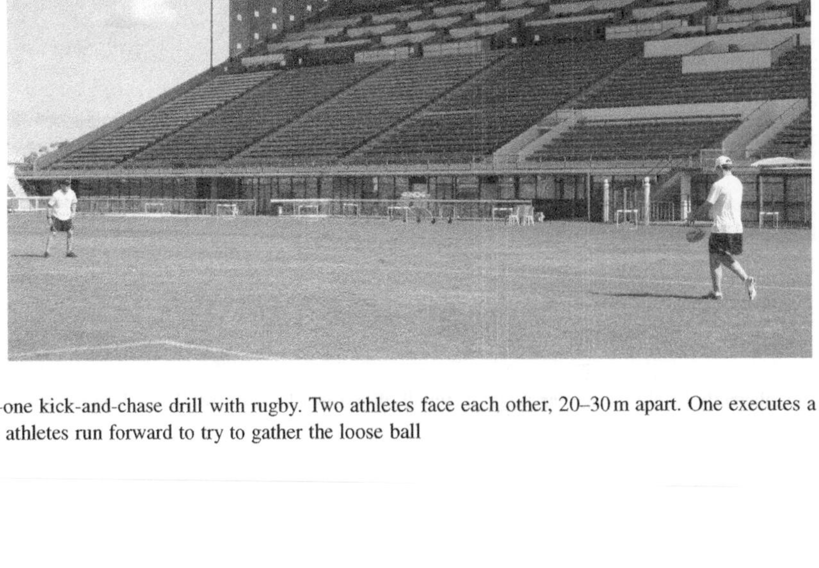

Figure 4.2.11 One-on-one kick-and-chase drill with rugby. Two athletes face each other, 20–30 m apart. One executes a chip-kick in the space between them, and both athletes run forward to try to gather the loose ball

Figure 4.2.12 One-on-one serve–receive practice in volleyball

reduced time (i.e. an increased time stress). For example, the athlete faces the opposite direction to an opponent running towards him or her. On a coach's voice command, the athlete turns around to face the oncoming opponent, and must react and move towards the opponent. Another manifestation of this drill is to have two athletes running up-field, with one athlete carrying the ball (in rugby, for example). When the athlete is instructed to turn around, they must identify the ball carrier and move to make a tackle (Figure 4.2.14). Although the coach's voice command in this drill is non-specific (i.e. is not domain-specific for this sport), the purpose of the drill is to target time stress, and in any case the athlete must still react to the movements of an opponent and make a decision on how to play the attacker, which is a relevant visual cue.

Agility drills can emphasize spatial stress, by reducing the normal space within which an athlete has to recognize relevant cues, and of course reduced space in which to move and execute skills. As such, with closer proximity, an increased spatial stress also involves an element of increased temporal stress. Common examples of this training type are to modify small-sided games such that the athletes must execute tasks in reduced space, such as playing four-on-four football 'keep-away' (or two-on-two with two 'neutrals') in a 10 m by 10 m space. Spatial stress can also be used to stress an athlete's agility skills simply by increasing the number of opponents in a given area. For example, a rugby player might perform a ball-carry drill where they must evade four or five opponents in a given area on their own, as opposed to attacking a line of two or three opponents with the benefit of passing options (Figure 4.2.15).

Figure 4.2.13 Team games and competition-like drills generally involve greater pattern recognition and more complex decision-making

Figure 4.2.14 Temporal-stress drill: reduced time to recognize the ball carrier and move to make a tackle

Figure 4.2.15 Spatial-stress drill: an increased number of opponents effectively increases the demand on a player, in a reduced space, during a ball-carry drill

4.2.6 CONCLUSION

Agility is a rapid, whole-body change of direction or speed in response to a stimulus (Sheppard and Young, 2006). Athletes at a higher level of sport tend to be superior in agility tasks that require reaction to a stimulus (Gabbett, Kelly and Sheppard, 2008; Sheppard *et al.*, 2006). Although agility is a distinct quality, it is underpinned by physical qualities such as sprint and CODS ability, and cognitive qualities such as anticipation, pattern recognition, and decision-making. These sub-qualities are trainable on their own, and agility as a whole is a trainable quality. However, designing a training philosophy that emphasizes only a few sub-qualities which influence agility will be an inadequate approach. For example, training only closed skills, or change of direction with non-relevant cues, such as with popular speed–quickness programmes, may result in poorer performances and quite possibly greater risk of injuries. The challenge for the practitioner is to determine the training needs of his or her athletes, and thereby implement training that achieves an appropriate balance to improve the physical components, cognitive components, and agility as a whole.

References

Abernethy B. and Russell D.G. (1987) Expert-novice difference in an applied selective attention task. *Journal of Sport Psychology*, **9**, 326–345.

Abernethy B., Wann J. and Parks S. (1998) In *Training in Sport: Applying Sport Science* (Ed, Elliot B.C.) John Wiley & Sons, West Sussex, pp. 426.

Abernethy B., Wood M.J. and Parks S. (1999) Can the anticipatory skills of experts be learned by novices? *Research Quarterly for Exercise and Sport*, **70**, 331–318.

Baechle T.R. (1994) *Essentials of strength and conditioning*, Human Kinetics, Champaign, IL.

Baker D. (1999) A comparison of running speed and quickness between elite professional and young rugby league players. *Strength and Conditioning Coach*, **7**, 3–7.

Baker D. and Nance S. (1999) The relationship between running speed and measures of strength and power in professional rugby league players. *Journal of Strength and Conditioning Research*, **13**, 230–235.

Berry J.T. (1999) *Pattern recognition and expertise in Australian football*, unpublished Honours, University of Ballarat.

Besier T.F., Lloyd D.G., Ackland T.R. and Cochrane J.L. (2001a) Anticipatory effects of knee joint loading during running and cutting manoeuvres. *Medicine and Science in Sports and Exercise*, **33**, 1176–1181.

Besier T.F., Lloyd D.G., Cochrane J.L. and Ackland T.R. (2001b) External loading of the knee joint during running and cutting manoeuvres. *Medicine and Science in Sports and Exercise*, **33**, 1168–1175.

Bobbert M.F. (1990) Drop jumping as a training method for jumping ability. *Sports Medicine*, **9**, 7–22.

Bobbert M.F., Huijing P.A. and Jan Van Ingen Schenau G. (1987) Drop jumping. II. The influence of jumping technique on the biomechanics of jumping. *Medicine and Science in Sports and Exercise*, **19**, 339–346.

Bompa T. (1983) *Theory and Methodology of Training*, Kendall-Hunt, Dubuque, Iowa.

Borgeaud P. and Abernethy B. (1987) Skilled perception in volleyball defense. *Journal of Sport Psychology*, **9**, 400–406.

Chelladurai P., Yuhasz M. and Sipura R. (1977) The reactive agility test. *Perceptual and Motor Skills*, **44**, 1319–1324.

Clark S., Martin D., Lee H., Fornasiero D. and Quinn A. (1998) In Australian Conference of Science and Medicine in Sport Adelaide, pp. 1–2.

Cox R.H. (2002) *Sport Psychology: concepts and applications*, McGraw-Hill, New York.

Cronin J., McNair P.J. and Marshall R.N. (2003) Lunge performance and its determinants. *Journal of Sports Sciences*, **21**, 49–57.

Dawson B., Hopkinson R., Appleby B., Stewart G. and Roberts C. (2004) Player movement patterns and game activities in the Australian Football League. *Journal of Science and Medicine in Sport*, **7**, 278–291.

Djevalikian R. (1993) The relationship between asymetrical leg power and change of running direction, unpublished Master's, University of North Carolina.

Doan B.K., Newton R.U., Marsit J.L., Triplett-McBride T.N., Koziris P.L., Fry A.C. and Kraemer W.J. (2002) Effects of Increased Eccentric Loading on Bench Press 1RM. *Journal of Strength and Conditioning Research*, **16**, 9–13.

Docherty D., Wenger H.A. and Neary P. (1998) Time-motion analysis related to the physiological demands of rugby. *Journal of Human Movement Studies*, **14**, 269–277.

Draper J.A. and Lancaster M.G. (1985) The 505 test: A test for agility in the horizontal plane. *Australian Journal for Science and Medicine in Sport*, **1**, 15–18.

Enoka R. (2000) *Neuromechanics of Human Movement*, Human Kinetics, Champaign, Ill.

Farrow D. and Abernethy B. (2002) Can anticipatory skills be learned through implicit video-based perceptual training? *Journal of Sports Sciences*, **20**, 471–485.

Farrow D., Chivers P., Hardingham C. and Sachse S. (1998) The effect of video-based perceptual training on the tennis return of serve. *International Journal of Sport Psychology*, **29**, 231–242.

Francis C. (1997) *Training for Speed*, Faccioni, Canberra, ACT.

Fulton K.T. (1992) Off-season strength training for basketball. *National Strength and Conditioning Association Journal*, **14**, 31–33.

Gabbett T., Georgieff B. and Domrow N. (2007) The use of physiological, anthropometric, and skill data to predict selection in a talent-identified junior volleyball squad. *Journal of Sports Sciences*, **25**, 1337–1344.

Gabbett T.J. (2002) Influence of physiological characteristics on selection in a semi-professional first grade rugby league team: a case study. *Journal of Sports Sciences*, **20**, 399–405.

Gabbett T.J., Kelly J.N. and Sheppard J.M. (2008) Speed, change of direction speed, and reactive agility of rugby league players. *Journal of Strength and Conditioning Research*, **22**, 174–181.

Gambetta V. (2007) *Athletic Development: the art and science of functional sports conditioning*, Human Kinetics, Champaign, IL.

Harrison A.J. and Gillian B. (2009) The effect of resisted sprint training on speed and strength performance in male rugby players. *Journal of Strength and Conditioning Research*, **23**, 275–283.

Hertel J., Denegar C.R., Johnson P.D., Hale S.A. and Buckely W.E. (1999) Reliability of the cybex reactor in the assessment of an agility task. *Journal of Sport Rehabilitation*, **8**, 24–31.

Keogh J., Weber C.L. and Dalton C.T. (2003) Evaluation of anthropometric, physiological, and skill-related tests for talent identification in female field hockey. *Canadian Journal of Applied Physiology*, **28**, 397–350.

Mann R.V. (1981) A kinetic analysis of sprinting. *Medicine and Science in Sports and Exercise*, **13**, 325–328.

Maxwell J.P., Masters R.S.W. and Eves F.F. (2000) From novice to no know-how: A longitudinal study of implicit learning. *Journal of Sport Sciences*, **18**, 111–120.

Muir P.A. (1996) *Expertise in surfing: nature of the perceptual advantage*, Honours, University of Ballarat.

Pauole K., Madole K., Garhammer J., Lacourse M. and Rozenek R. (2000) Reliability and validity of the t-test as a measure of agility, leg power, and leg speed in college-aged men and women. *Journal of Strength and Conditioning Research*, **14**, 443–450.

Reilly T., Williams A.M., Nevill A. and Franks A. (2000) A multidisciplinary approach to talent identification in soccer. *Journal of Sports Sciences*, **18**, 695–702.

Rigg P. and Reilly T. (1987) A fitness profile and anthropometric analysis of first and second class rugby union players. In Proceedings of the First World Congress on Science and Football (Ed, Rigg P.) E & F.N. Spon, Liverpool, England, pp. 194–200.

Ritchie N. (1999) An investigation of pattern recognition anticipation within Australian rules football, unpublished Honours, University of Ballarat.

Sayers M. (2000) Running techniques for field sport players. Sports Coach, Autumn, pp. 26–27.

Sheppard J. (2003) Strength and conditioning exercise selection in speed development. *Strength and Conditioning Journal*, **25**, 26–30.

Sheppard J.M., Barker M. and Gabbett T. (2008a) Training agility in elite rugby players: a case study. *Journal of Australian Strength and Conditioning*, **16**, 15–19.

Sheppard J.M., Gabbett T. and Borgeaud R. (2008b) Training repeated effort ability in national team male volleyball players. *International Journal of Sports Physiology and Performance*, **3**, 397–400.

Sheppard J.M., Hobson S., Chapman D., Taylor K.L., McGuigan M. and Newton R.U. (2008c) The effect of training with accentuated eccentric load counter-movement jumps on strength and power characteristics of high-performance volleyball players. *International Journal of Sports Science and Coaching*, **3**, 355–363.

Sheppard J.M., McGuigan M.R. and Newton R.U. (2008d) The effects of depth-jumping on vertical jump performance of elite volleyball players: an examination of the transfer of increased stretch-load tolerance to spike jump performance. *Journal of Australian Strength and Conditioning*, **16**, 3–10.

Sheppard J.M. and Young W. (2006) Agility Literature Review: classifications, training and testing. *Journal of Sport Sciences*, **24**, 919–932.

Sheppard J.M., Young W.B., Doyle T.L.A., Sheppard T.A. and Newton R.U. (2006) An evaluation of a new test of reactive agility and its relationship to sprint speed and change of direction speed. *Journal of Science and Medicine in Sport*, **9**, 342–349.

Sleivert G. and Taingahue M. (2004) The relationship between maximal jump-squat power and sprint acceleration in athletes. *European Journal of Applied Physiology*, **91**, 46–52.

Starkes J. (1987) Skill in field hockey: The nature of the cognitive advantage. *Journal of Sport Psychology*, **9**, 146–160.

Tenenbaum G., Levy-Kolker N., Sade S., Lieberman D. and Lidor R. (1996) Anticipation and confidence of decisions related to skilled performance. *International Journal of Sport Psychology*, **27**, 293–307.

Tsitskarsis G., Theoharopoulus A. and Garefis A. (2003) Speed, speed dribble and agility of male basketball players playing in different positions. *Journal of Human Movement Studies*, **45**, 21–30.

Webb P. and Lander J. (1983) An economical fitness testing battery for high school and college rugby teams. *Sports Coach*, **7**, 44–46.

Williams A.M., Davids K., Burwitz L. and Williams J.G. (1993) Visual search and sports performance. *The Australian Journal for Science and Medicine in Sport*, **25**, 55–65.

Young W. (1995) Laboratory strength assessment of athletes. *New Studies in Athletics*, **10**, 89–96.

Young W.B., James R. and Montgomery I. (2002) Is muscle power related to running speed with changes of direction? *Journal of Sports Medicine and Physical Fitness*, **43**, 282–288.

Young W.B., McDowell M.H. and Scarlett B.J. (2001) Specificity of sprint and agility training methods. *Journal of Strength and Conditioning Research*, **15**, 315–319.

Young W.B., McLean B. and Ardagna J. (1995) Relationship between strength qualities and sprinting performance. *Journal of Sports Medicine and Physical Fitness*, **35**, 13–19.

4.3 Nutrition for Strength Training

Christopher S. Shaw[1] and Kevin D. Tipton[2], [1]School of Sport and Exercise Sciences, The University of Birmingham, Edgbaston, Birmingham, UK, [2]Department of Sports Studies, University of Stirling, Stirling, Scotland (UK)

4.3.1 INTRODUCTION

Muscle strength is related primarily to the size of the muscle mass (Maughan, Watson and Weir, 1983). Approximately 20% of muscle tissue is composed of protein and the vast majority (~65%) of the body's protein is found in skeletal muscle. Thus muscle protein plays a huge role in whole-body protein metabolism. The aim of training to improve strength therefore is to stimulate the synthesis of new muscle proteins to expand the size of the musculature (hypertrophy). The synthesis of new proteins following strength training involves the incorporation of amino acids into proteins primarily involved in muscle contraction. These proteins (actin and myosin) are referred to as the myofibrillar proteins and are the key components of the contractile machinery responsible for producing force during muscle contractions. Myofibrillar proteins make up ~80% of all muscle protein. The main stimulus for synthesizing myofibrillar proteins and increasing muscle size is the volume and intensity of resistance-type training, but that stimulus can be modulated by nutrition. The combination of resistance training and nutrition can therefore be used in an attempt to optimize this adaptation process. In this chapter we describe the role of nutrition in muscle hypertrophy from strength training and the scientific evidence supporting nutritional strategies to sustain strength training.

4.3.2 THE METABOLIC BASIS OF MUSCLE HYPERTROPHY

Amino acids are the building blocks of proteins. They can be categorized as non-essential amino acids (NEAAs), which can be synthesized within the body, and essential amino acids (EAAs), which must be supplied in the diet. The synthesis of new proteins occurs when signals within the muscle stimulate the assembly of a specific sequence of amino acids into a chain, which eventually forms the new protein. Muscle protein mass is determined by the balance between the rate of synthesis of new proteins and the breakdown of existing proteins (Phillips, 2004; Phillips, Hartman and Wilkinson, 2005; Tipton, Jeukendrup and Hespel, 2007). The difference between synthesis and breakdown is termed net muscle protein balance (NMPB). The processes of synthesis and breakdown of proteins are constant and concurrent, but NMPB will fluctuate throughout the day depending on the exercise and feeding situation. In the postprandial (after feeding) period, a rise in anabolic signals will result in rates of protein synthesis exceeding rates of protein breakdown, resulting in a positive NMPB where the muscle will be in a state of anabolism. During periods of fasting, on the other hand, rates of protein breakdown will exceed rates of protein synthesis; NMPB will become negative and the muscle will be in a catabolic state. Thus, over time, the mass of the musculature will reflect the duration and magnitude of the respective periods of positive and negative NMPB.

NMPB is influenced by exercise and nutritional intake. Resistance exercise, at least of sufficient intensity (Kumar *et al.*, 2009), stimulates muscle protein synthesis and breakdown (Biolo *et al.*, 1995b). However, the increase in synthesis is greater than the increase in breakdown, resulting in increased NMPB. The increase in muscle protein synthesis may persist for as much as 48 hours after the exercise. In the fasting condition, the increase in NMPB does not result in positive balance. Thus, without food intake, in particular a source of EAAs, NMPB remains negative. When a source of amino acids, whether intact proteins, protein hydrolysates, or free amino acids, is ingested in conjunction with resistance exercise, muscle protein synthesis is further increased, resulting in positive NMPB (Biolo *et al.*, 1997). Thus, with each bout of resistance exercise and protein intake, small amounts of protein are deposited, and if continued, will result in muscle growth (Tipton and Witard, 2007).

4.3.3 OPTIMAL PROTEIN INTAKE

For centuries it has been almost universally accepted that the protein requirements to support increased rates of protein synthesis from strength training become elevated above the recommended daily allowance (RDA) (Wolfe, 2000). This assumption

Strength and Conditioning – Biological Principles and Practical Applications Marco Cardinale, Rob Newton, and Kazunori Nosaka.

is somewhat unsurprising given that muscle mass gains rely on increased rates of protein synthesis after resistance exercise, which is dependent on increased amino acid provision from the diet. In addition, it has been suggested that protein requirements may theoretically be increased to support elevated amino acid oxidation during exercise and muscle repair following elevations in post-exercise protein breakdown (Phillips, 2006). This belief has spawned the growth of a multi-million dollar industry of nutritional supplements to support greater protein ingestion.

Protein intake may be expressed in several ways. Probably the most common is in terms of grams protein ingested per kilogram body weight per day (g/kg/day). The RDA is 0.8 g/kg/day, but many athletes and coaches support protein consumption even above 3.0 g/kg/day (>210 g per day for a 70 kg person) (Alway *et al.*, 1992; Phillips, 2004). In fact, even the American College of Sports Medicine (ACSM) guidelines advise that a protein intake more than double the RDA (>1.6 g/kg/day) be consumed by athletes involved in strength training.

Despite this long-held belief, the scientific evidence supporting the requirement for consumption of large amounts of protein during strength training is at best equivocal and certainly controversial (Phillips, 2004; Tipton and Witard, 2007; Wolfe, 2000). Many scientists believe that intake of protein at the RDA level is more than sufficient to support optimal athletic performance (Phillips, 2006; Rennie and Tipton, 2000; Tipton and Witard, 2007). A major point supporting the argument that protein needs are not increased with training is that the efficiencies of utilization of the supplied amino acids and of reutilization of intracellular amino acids derived from protein breakdown are improved with training (Phillips, 2006). For example, 12 weeks of resistance training that increased muscle mass and strength was associated with nitrogen retention, indicating that the conservation of existing protein was improved (Hartman, Moore and Phillips, 2006; Moore *et al.*, 2007). Thus, the protein requirements for trained individuals may actually be less than for untrained. The excess amino acids from ingested protein that cannot be incorporated into new proteins are simply oxidized, resulting in increased urea excretion (Tarnopolsky, 2004); it may therefore be futile to consume large quantities of protein. Nevertheless, whereas habitual protein intakes >2 g/kg/day seem unnecessary for muscle hypertrophy and gains in strength (Tipton, Jeukendrup and Hespel, 2007), they are unlikely to be detrimental to the gains in muscle mass unless total energy intake or carbohydrate consumption is compromised.

Regardless of the controversy over the requirement for protein for athletes, it is clear that the optimal amount of dietary protein required for muscle hypertrophy has yet to be established. In any case, given the enormous variety of athletes performing a strength-based programme as part of training, it appears somewhat irrational to suggest a certain protein intake to cover all athletes. Furthermore, as we will discuss later, it could be argued that the total protein content ingested is irrelevant. Instead, other factors that affect the relatively short and transient rises in protein synthesis that occur in response to acute exercise and nutritional interventions may have a greater bearing on muscle adaptation to strength training (Tipton and Witard, 2007).

There are some athletes who may need more careful consideration of the amount of protein in their diet. Whereas high protein intakes – even those several-fold greater than the RDA – are unlikely to be detrimental to healthy individuals, they may be dangerous to those with existing complications, such as renal disease (Phillips, 2004; Tipton, Jeukendrup and Hespel, 2007). On the other hand, some athletes may be at risk of insufficient protein intake. For example, since it is possible that animal protein sources are superior to plant sources for utilization of amino acids by peripheral tissues (i.e. muscle) (Fouillet *et al.*, 2002), vegetarians/vegans may need to increase their protein intake to account for the lower efficiency of plant proteins. Of course, these sorts of decisions need to be made on an individual basis and should not be made based on group characteristics. Finally, as will be discussed below, energy intake during training may profoundly influence the necessity for protein intake.

The issue of energy imbalance could be particularly relevant for athletes who restrict energy intake in order to meet weight limits (boxing, martial arts, rowing, etc.). It is likely that some if not many of these athletes will wish to increase their protein intake in an attempt to minimize reductions in lean body mass resulting from dietary energy restriction (Mettler, Mitchell and Tipton, 2010). This issue will be presented in more detail below.

In reality, the argument over increased protein requirements for athletes is likely a moot point for most. The increased energy requirements with high training loads mean that most athletes eat ample protein to support even the higher estimates. Studies investigating the eating habits of resistance-trained athletes demonstrate that the vast majority naturally ingest much more than 1.2–1.5 g/kg/day of protein in their habitual diet without the incorporation of protein supplements (Phillips, 2006; Tarnopolsky, 2004). Even such moderate (at least moderate in the eyes of many athletes) protein intakes are clearly sufficient to allow gains in lean body mass during a resistance-training program (Moore *et al.*, 2007). Thus, whereas there is an interesting academic argument raging over the amount of protein necessary for optimal gains in mass and strength, it is unlikely to be important, or even relevant, for most athletes.

4.3.3.1 The importance of energy balance

For most athletes and coaches, the nutritional focus for increasing mass and strength is protein. However, a more important nutritional factor is likely the total energy balance. Certainly, negative energy balance should be avoided if maximal muscle hypertrophy is a goal. Inducing a state of negative energy balance through dietary restriction and exercise induces a state of negative nitrogen balance which is indicative of muscle loss (Todd, Butterfield and Calloway, 1984). On the other hand, it seems clear that when energy balance is maintained, muscle mass may be gained on a wide range of protein intakes. For example, over 100 years ago it was demonstrated that muscle mass was maintained and strength was gained during training when energy balance was maintained with even modest rates of protein intake (1 g/kg/day) (Chittenden, 1907). Finally, the notion that energy intake is often more limiting than protein intake for maximal muscle gains is supported by results of a recent study. Weightlifters ingested 2000 calories more than their normal dietary intake during a training programme. The energy for one group came entirely from carbohydrate, while the other consumed extra calories in the form of both protein and carbohydrate. Surprisingly, both groups gained equal amounts of muscle mass and strength (Rozenek *et al.*, 2002). These results clearly demonstrate that – given a sufficient amount of protein intake – muscle mass and strength may be limited more by energy than by protein intake.

The importance of energy expenditure to the maintenance of NMPB has important implications for athletes who are competing in weight-category sports. The prevailing opinion is that gains in muscle mass are not possible in a state of negative energy balance. However, protein intake may be more important for maintenance of muscle mass during reduced energy intake combined with maintenance of training intensity and volume. A recent study from our laboratory investigated the impact of high protein intakes on muscle mass and strength during severe calorie restriction in trained weightlifters (Mettler, Mitchell and Tipton, 2010). The results suggest that a high-protein diet may be beneficial for the maintenance of muscle mass during training when energy is restricted. During 60% calorie restriction, those athletes consuming a high-protein diet (2.3 g/kg/day or 35% of total energy intake) maintained muscle mass to a greater degree than those on a modest protein intake (1.0 g/kg/day), though fat loss was similar. Thus loss of total mass was greater in the group that had moderate protein intakes. Similar results have been reported in obese individuals (Layman *et al.*, 2005). Hence, with sufficient protein intake, energy balance appears to be crucial for increasing muscle mass. On the other hand, during energy restriction, a high protein intake may help to preserve the muscle mass during periods of weight loss.

Clearly, energy intake is an important consideration for those concerned with gaining muscle. It is a consideration that is often neglected, especially given the emphasis on protein in many magazines, Web sites, and other popular sources. However, there is no question that sufficient, if not ample, energy intake is critical for maximal gains in muscle mass. Muscle mass may be gained on a wide range of protein intakes when sufficient energy is consumed.

4.3.4 ACUTE EFFECTS OF AMINO ACID/PROTEIN INGESTION

In addition to energy intake, there are other aspects of the diet that influence mass gains other than the total amount of protein ingested. The acute protein synthetic response to exercise can be affected by the type and amount of protein ingested, the timing of protein consumption in relation to exercise, and the co-ingestion of other nutrients (Tipton, 2008; Tipton and Ferrando, 2008; Wolfe, 2000), in addition to the mode, intensity, and duration of the exercise bout itself (Kumar *et al.*, 2009). These factors will be examined in more detail.

Much of the information available on the impact of various nutritional factors stems from acute metabolic studies, rather than longitudinal training studies. It is now clear that the acute response of NMPB is a good predictor of the changes in muscle mass in response to a nutritional intervention during training (Hartman, Moore and Phillips, 2006; Tipton and Witard, 2007; Tipton and Wolfe, 2004; Wilkinson *et al.*, 2007). Therefore, analysis of the acute response to exercise and nutrient intake can be used to determine optimal feeding strategies over prolonged training periods (Phillips, Hartman and Wilkinson, 2005; Tipton and Witard, 2007).

Numerous studies have shown that increasing the concentration of circulating amino acids will increase the delivery of amino acids to the muscle and result in elevations in muscle protein synthesis after exercise (Rennie and Tipton, 2000). This increase in protein synthesis occurs in the absence of any other exogenous substrate and without large increases in the anabolic hormone insulin (Koopman *et al.*, 2007). Interestingly, this response occurs in the absence of NEAAs (Tipton *et al.*, 1999). Thus, it is the provision of EAAs that is the important factor.

EAAs can be supplied in different forms, from the whole protein down to protein hydrolysates or even single or multiple free amino acids. Further, the amino acid source may have an important effect upon the anabolic response. Timing of ingestion, concurrent ingestion of other nutrients, and still other factors are important determinants of the utilization of ingested amino acids (Tipton and Witard, 2007). Thus, it is important to examine the various factors that influence the utilization of amino acids from an ingested amino acid source.

4.3.4.1 Amino acid source

Whole proteins

The type of protein ingested may be an important component delineating the response of skeletal muscle protein synthesis in the period after exercise. Characteristics unique to each protein, such as digestive properties or amino acid composition, will influence the utilization of the amino acids from that protein. The digestion of proteins is not simple and relies on the breakdown of proteins by enzymes and subsequent absorption from the digestive system into the blood before amino acids from that protein are available for use by muscle and other tissues. Therefore there is a delay between the ingestion of the amino acid source and amino acid delivery to the muscle. This delay will depend upon the digestive properties of the protein source (Boirie *et al.*, 1997).

Each protein may exhibit different digestive properties. For example, whey and soy proteins are rapidly emptied from the stomach and result in higher peak amino acid concentrations which return to basal levels relatively quickly. These proteins have been termed ‘fast’ proteins. On the other hand, blood amino acid levels following casein ingestion (a 'slow' protein) are of lower magnitude but take longer to return to basal levels due to slower emptying from the stomach (Tipton *et al.*, 2004). Resting studies suggest that amino acid retention is greater following ingestion of slow proteins than fast proteins (Dangin *et al.*, 2001). However, in the period after resistance exercise there appears to be little difference in the response of NMPB to the ingestion of casein and whey proteins (Tipton *et al.*, 2004).

Interestingly, recent evidence suggests that milk proteins, such as whey and casein protein (in a ~1 : 4 ratio in milk), may be superior to soy as a protein source for maximal gains of muscle. While soy proteins are digested more rapidly, it seems that at rest the spanchnic tissues (primarily the liver) take up a higher proportion of the amino acids. These amino acids likely contribute to protein synthesis in the liver or are directed to deamination pathways (Fouillet *et al.*, 2002). On the other hand, there is evidence that the ingestion of milk proteins, rather than soy protein, leads to increased muscle protein synthesis and NMPB (Wilkinson *et al.*, 2007), and chronically increased muscle mass (Hartman *et al.*, 2007) following exercise. This apparently superior response of NMPB with animal proteins over plant proteins could have particular implications for vegetarians, especially vegans. However, this relationship needs to be explored in more detail before specific recommendations can be made.

Essential amino acids

In addition to digestive properties, the amino acid composition of a protein will influence the utilization of those amino acids for muscle protein synthesis. The EAAs are the nine amino acids that cannot be synthesized within the body and can only be obtained from dietary sources. They are all found within skeletal muscle and are responsible for the increases in muscle protein synthesis following protein/amino acid consumption. Ingestion of only EAAs (without NEAAs) stimulates muscle protein synthesis, resulting in positive NMPB following resistance exercise (Tipton *et al.*, 1999). Additional studies have shown that small amounts of ingested EAA (10 g) are enough to activate the anabolic signalling pathways responsible for the stimulation of protein synthesis in skeletal muscle (Cuthbertson *et al.*, 2005). The importance of the EAAs to protein synthesis may explain why the ingestion of animal proteins in the form of milk, which elicit greater uptake of EAAs, appears to bring about greater protein accretion in muscle after exercise in comparison to soy proteins (Wilkinson *et al.*, 2007).

As the elevation in protein synthesis following feeding, both at rest and following resistance exercise, occurs in response to EAA ingestion only, it could be argued that the consumption of EAAs is preferable to the consumption of whole proteins containing a mixture of EAAs and NEAAs. Certainly at rest the ingestion of 15 g EAAs stimulates protein synthesis to a greater extent than an isocaloric amount of whey protein (Paddon-Jones *et al.*, 2006). However, normalized for EAA content, whey protein stimulates a greater anabolic response than an amount of free EAAs equivalent to that found in the whey protein (Katsanos *et al.*, 2008). Thus, if energy intake must be minimized, EAA ingestion might be preferable to whole proteins, but these data suggest that whey proteins provide an anabolic stimulus above and beyond their EAA content; what that is, exactly, is unknown. So if energy is not critical, then perhaps whey proteins may be a better choice, especially given the pragmatic issues of increased cost and bad taste associated with free amino acids. Of course, more research must be conducted to determine how other proteins may influence this response. Perhaps more importantly, these studies were conducted in resting subjects; thus, at this point, it is uncertain how exercise influences this relationship.

Branched-chain amino acids

The EAAs that tend to be given the most attention are the branched-chain amino acids (BCAA): leucine, isoleucine, and valine. BCAA supplements are marketed for a wide range of ergogenic properties. In regard to muscle mass, BCAAs are associated with increased muscle protein synthesis and decreased protein breakdown. BCAAs, in particular leucine, stimulate the anabolic signalling pathways in the muscle that lead to muscle protein synthesis (Anthony *et al.*, 2000). Many *in vitro* and animal studies demonstrate that the BCAAs are capable of stimulating protein synthesis at rest (Matthews, 2005). In addition, following resistance exercise, ingestion of BCAAs increases anabolic signalling over the effects of exercise alone (Karlsson *et al.*, 2004). Furthermore, following treadmill exercise which depressed muscle protein synthesis in rats, leucine ingestion restored protein synthesis back to normal resting levels (Anthony *et al.*, 2000). These data are often used to support the claims that BCAA supplementation is necessary for increased muscle mass in humans.

However, direct information from studies in exercising humans is inconsistent at best. Many human studies suggest that BCAA (or leucine) supplementation at rest actually attenuates rates of protein breakdown, but has little effect upon protein synthesis despite appearing to activate some of the anabolic signalling processes (Matthews, 2005). Leucine is a known insulin secretagogue, and the hyperinsulinemia elicited by leucine ingestion may explain the observed reductions in protein breakdown (discussed in more detail later). Furthermore, the addition of leucine to a beverage containing whey protein and carbohydrate ingested after resistance exercise has been associated with higher insulin concentrations and increased rates of protein synthesis compared to carbohydrate alone (Koopman *et al.*, 2005). However, when compared to whey protein plus carbohydrate, the addition of leucine did not further stimulate muscle protein synthesis (Koopman *et al.*, 2005). Nevertheless, these data are often used to support the efficacy of BCAA/leucine supplementation for muscle growth. On the other hand, recent data clearly demonstrate that adding leucine to whey protein does not result in increased NMPB following resistance exercise (Tipton *et al.*, 2009). Thus, as yet there remains little evidence that the ingestion of leucine alone is any more effective at increasing muscle anabolism than the ingestion of whole proteins following exercise.

This lack of a clear impact on human muscle following resistance exercise, despite the clear impact in exercising rats, may have a reasonable metabolic explanation. When the muscle is catabolic, as in the rat exercise model (Anthony *et al.*, 2000), leucine ingestion is quite effective for restoring muscle protein synthesis back to normal levels. However, in an anabolic situation – such as following resistance exercise in humans, especially with protein ingestion – leucine does not result in further stimulation of muscle protein anabolism (Koopman *et al.*, 2005; Tipton *et al.*, 2009). Despite increased signalling in the muscle (Karlsson *et al.*, 2004), muscle protein synthesis is likely already maximally stimulated by the exercise and protein; there is a ceiling effect. Furthermore, leucine ingestion decreases the availability of other EAAs in the blood, limiting muscle protein synthesis (Nair *et al.*, 1992).

This interpretation of the leucine data suggests that supplementation of leucine or BCAAs may not be particularly effective in healthy humans desiring increased muscle mass. However, there may be situations in which BCAA supplementation could be important. For instance, during immobilization due to exercise-induced injury, muscle atrophies, at least partially due to a resistance to the anabolic stimulation of amino acids (Glover *et al.*, 2008). Additional leucine seems to be effective for overcoming the anabolic resistance of muscle, at least in the elderly (Katsanos *et al.*, 2006). Thus, it is possible that leucine supplementation during immobilization may help reduce muscle loss. At this point, this notion is purely speculative, and studies are needed to determine the efficacy of leucine supplementation for muscle hypertrophy during immobilization and during resistance-exercise training.

4.3.4.2 Timing

Another aspect of nutrient ingestion that must be considered in the context of muscle hypertrophy is the timing of nutrient ingestion. It is clear that combining resistance exercise with amino acid ingestion results in a synergistic effect upon protein synthesis (Biolo *et al.*, 1997). Recently, the timing of amino acid and protein intake has been examined and determined to markedly influence the response of NMPB (Levenhagen *et al.*, 2001; Tipton *et al.*, 2001, 2007). Whereas it is clear that the timing of nutrient ingestion will influence the response of muscle anabolism, the optimal timing of ingestion is not clear and likely depends on the protein type, EAA content, and other nutrients ingested concurrently (Tipton *et al.*, 2001, 2007).

A commonly held belief has developed that a 'metabolic window' of opportunity exists where proteins need to be ingested within a short period after exercise in order to maximize elevations in protein synthesis following exercise (Ivy and Portman, 2004). There are suggestions that this window may last for as little as 45 minutes (Ivy and Portman, 2004). In the context of muscle growth, this notion is often supported by recent studies (Esmarck *et al.*, 2001; Levenhagen *et al.*, 2001). In one, elderly volunteers ingested a protein containing supplement either immediately or two hours post-exercise during a training programme. The subjects ingesting the supplement two hours after exercise showed no gains in muscle mass, whereas muscle growth was substantial in those ingesting the supplement immediately post-exercise (Esmarck *et al.*, 2001).

While these data seem to support the notion that protein should be ingested immediately post-exercise, questions certainly should be raised. It should be pointed out that there was no muscle growth in the subjects who ingested the protein two hours after exercise (Esmarck *et al.*, 2001), despite many other examples of muscle growth in elderly subjects with this type of training programme (Fiatarone *et al.*, 1994). It is clear that muscle NMPB responds to EAA ingestion as much as three hours following resistance exercise (Rasmussen *et al.*, 2000). Thus, the ingestion of protein two hours after exercise can be said to actually inhibit muscle growth. This result is very puzzling, but may be related to the satiety effect of protein (Veldhorst *et al.*, 2008). It is likely that ingestion of the supplement so late after exercise caused the subjects to consume less in subsequent meals (Fiatarone *et al.*, 1994), thus blunting the overall response. Therefore, the lack of muscle growth has less to do with the metabolic window in relation to the exercise than with nutrient ingestion in relation to subsequent meals. Hence, the evidence for this metabolic window is not as clear as is often claimed.

Moreover, several lines of evidence have raised question marks over the presence of the metabolic window for protein/amino acid ingestion. First, there is no evidence that rates of protein synthesis are diminished when feeding occurs either one or three hours after the completion of exercise (Rasmussen *et al.*, 2000). Second, the elevations in protein synthesis may not actually depend upon post-exercise feeding at all, as several studies have shown that protein synthesis is at least equal, if not elevated, when feeding takes place prior to exercise (Tipton *et al.*, 2001, 2007). In fact, there is evidence that when feeding takes place prior to exercise, protein synthesis rates increase during the exercise session itself (Beelen *et al.*, 2008; Tipton *et al.*, 1999) and are higher in the period following exercise compared to ingestion immediately after exercise (Tipton *et al.*, 2001). A recent study demonstrated that the exercise-induced rise in muscle protein synthesis when resistance exercise is performed two hours following a meal (Witard *et al.*, 2009) is similar to that observed when amino acids are provided following exercise (Phillips *et al.*, 2002). Finally, since the impact of resistance exercise on muscle protein synthesis lasts for 36–72 hours, there is no reason to believe that the additive response of exercise and protein/amino acid intake will not occur beyond the first hour or so following exercise. That is, exercise results in a prolonged sensitivity of the muscle to the anabolic stimulus of protein ingestion. These results support the notion that the muscle is responsive to nutrients well past the purported metabolic window.

The lack of support for the strict necessity of ingesting protein within the first few minutes following exercise should not be interpreted as an argument that this practice is to be avoided. On the contrary, ingestion of some sort of protein soon after exercise is often convenient for many athletes and is certainly not likely to be harmful. Thus, from a practical standpoint, nutrient ingestion, including protein, soon after resistance exercise is likely to be an important aspect of a programme designed to increase muscle mass.

4.3.4.3 Dose

Another aspect of protein ingestion that may impact muscle mass is the optimal dose. As mentioned above, ingestion of as little as 3 g of EAAs is required to stimulate protein synthesis after resistance exercise (Miller *et al.*, 2003). At rest, EAA ingestion above ~10 g did not further stimulate muscle protein synthesis (Cuthbertson *et al.*, 2005). Following exercise, a dose response was first addressed from the results of two separate studies. At least in small quantities, there appeared to be a positive relationship between the dose of EAAs and protein synthesis; that is, muscle protein synthesis rates doubled when EAAs were increased from 3 to 6 g (Miller *et al.*, 2003; Rasmussen *et al.*, 2000). Recently, Moore *et al.* (2009) directly addressed this issue, showing that protein synthesis rates after resistance exercise are dose-dependent when whole-egg protein ingestion after exercise is increased from 0 to 20 g (Moore *et al.*, 2009). They also showed that maximal rates of protein synthesis after exercise were achieved with the ingestion of 20 g of protein and that there was no further increase in protein synthesis when protein intake was increased up to 40 g. Interestingly, this amount of protein provided approximately the same amount of EAAs as in the previously mentioned resting study (Cuthbertson *et al.*, 2005). Thus, it seems the amount of EAAs provided may be the key.

This dose-dependency begs the question: what happens when the protein is consumed above the optimal dose? According to the evidence, amino acid oxidation increased only when whole protein intake was above 20 g (Moore *et al.*, 2009), indicating that when protein intake is increased above the upper limit, any excess amino acids are simply oxidized. The optimal post-exercise dose may be influenced by other factors, such as the type of whole protein or amino acids, the type or intensity of exercise, and other nutrients co-ingested with the protein. The influence of these factors on the dose response has yet to be determined.

4.3.4.4 Co-ingestion of other nutrients

Protein is not often consumed in isolation. The acute metabolic response to protein/amino acid ingestion with resistance exercise is clearly influenced by concurrent ingestion of other nutrients, such as carbohydrate and maybe even fat (Tipton and Witard, 2007). Utilization of amino acids from ingested proteins may be increased by ingestion of carbohydrates or fats at the same time.

Carbohydrate

The influence of carbohydrate on NMPB is almost certainly due to the resulting insulin response. The influence of insulin on muscle protein synthesis is somewhat controversial. At rest, insulin stimulates anabolic signalling processes and can stimulate increases in protein synthesis provided amino acid availability is maintained (Biolo *et al.*, 1999). Therefore, the ingestion of carbohydrate, which elicits increases in plasma insulin, may be beneficial to the anabolic response (Rennie *et al.*, 2006). Additionally, in the resting state insulin increases total blood flow (Baron *et al.*, 1995), redistributes blood to the muscle bed (Vincent *et al.*, 2004), and therefore also increases amino acid delivery and uptake into skeletal muscle (Biolo *et al.*, 1995a). However, large increases in muscle protein synthesis occur with amino acid ingestion in the presence of low insulin concentrations (Rennie *et al.*, 2006). Therefore, carbohydrate ingestion coupled to large elevations in circulating insulin does not appear to be imperative for large increases in muscle protein synthesis.

The influence of carbohydrate following resistance exercise is not the same as that at rest. Following resistance exercise, it seems that insulin provides little further impact on muscle protein synthesis (Biolo *et al.*, 1999; Borsheim *et al.*, 2004; Miller *et al.*, 2003), despite increased anabolic signalling in the muscle (Dreyer *et al.*, 2008). In addition, leg blood flow remains elevated three hours after exercise, whereas the infusion of insulin fails to elicit any further increases in leg blood flow (Biolo *et al.*, 1999). However, moderate increases in insulin increase NMPB in the period after exercise by reducing protein breakdown (Biolo *et al.*, 1999). An attenuation of muscle protein breakdown will further improve NMPB, leading to muscle protein accretion. Therefore, although the addition of a small/moderate amount of carbohydrate to the ingestion of an amino acid source after exercise may have little effect upon rates of muscle protein synthesis per se, improvements in NMPB may occur through the inhibition of protein breakdown.

Carbohydrate ingestion after resistance exercise may also be beneficial for other reasons than its direct effect upon NMPB. The ATP needed for muscle contraction during resistance exercise comes primarily from the carbohydrates stored within the muscle as glycogen for fuel (Koopman *et al.*, 2006). If these carbohydrate stores become depleted during exercise then the athlete will become prematurely fatigued, performance capacity will decline, and the quality of training will be compromised. Likewise, if the carbohydrate stores are not fully replenished before the next training session, the athlete's performance capacity will deteriorate, reducing the quality of the training session. In addition, there are suggestions that the anabolic signals derived from training cannot be maximal when the muscle is in a glycogen-depleted state (Churchley *et al.*, 2007; Creer *et al.*, 2005), which may further result in suboptimal muscular adaptations.

Therefore, co-ingestion of carbohydrate and an amino acid source may result in the optimal state of muscle protein accretion in the period after exercise. However, what may be more important is that sufficient carbohydrate is consumed within the diet to replenish glycogen stores depleted by exercise, in order to prevent premature fatigue and optimize the anabolic status of the muscle protein balance.

Fat

The notion that fat intake might influence the anabolic response of muscle has not received much attention. Research into the effect of dietary fat on protein synthesis has so far been overlooked. The idea that fat ingestion will benefit the resistance-trained athlete who desires large volumes of lean mass with minimal fat mass may be somewhat counterintuitive. In a similar manner to muscle glycogen, fat stored within the muscle is utilized during high-intensity resistance exercise (Koopman *et al.*, 2006), although it does appear that the replenishment of intramuscular fat stores does not rely on post-exercise feeding and is probably derived from endogenous fat stores.

At rest, fat does not seem to have any impact on NMPB (Svanberg *et al.*, 1999). However, there is now preliminary evidence that fat may play a role in utilization of amino acids following resistance exercise. Whole milk ingestion results in greater utilization of available amino acids than an isonitrogenous (containing the same amount of protein) or isoenergetic (containing the same amount of calories) amount of fat-free milk (Elliot *et al.*, 2006). Since there is no evidence that fat alone is able to stimulate protein synthesis (Svanberg *et al.*, 1999), the mechanism for this effect remains unknown. Nonetheless, it is clear that the inclusion of fat does not impair any anabolic response and may be a way to ensure sufficient energy provision to maintain a positive energy balance. More work is required to fully understand the effect of dietary fat upon muscle anabolism following exercise.

4.3.4.5 Protein supplements

The form in which protein is ingested should merit some attention. It seems clear that protein ingested as part of food is just as effective as that ingested in supplemental form.

Recent evidence supports the efficacy of protein in the form of food: Symons *et al.* (2007) demonstrated that ingestion of beef stimulates muscle protein synthesis, while the fact that milk ingestion stimulates positive NMPB following exercise (Elliot *et al.*, 2006), in addition to the data mentioned earlier (Hartman *et al.*, 2007; Wilkinson *et al.*, 2007), should illustrate that the effectiveness of protein in supplement form is no greater than that of protein from food. In fact, since carbohydrate and fat seem to improve amino acid utilization, it could be argued that food may be a better form in which to ingest protein.

4.3.4.6 Other supplements

Over the years, in addition to amino acid and protein supplementation, a large number of other substances have been alleged to benefit muscle hypertrophy when combined with strength training. Boron, vanadium, chromium, beta-hydroxylmethylbutyrate (HMB), conjugated linoleic acid (CLA), lipoic acid, creatine, and other hormones and pro-hormones have all been marketed as muscle-building supplements. There is very little evidence to support such claims for the majority of these supplements (with the exception of creatine) and it would be unwise to recommend such products when giving nutritional advice to athletes involved in strength training.

Creatine does appear to aid muscle hypertrophy during periods of strength training, as increases in lean body mass and strength have been reported with creatine supplementation (Becque, Lochmann and Melrose, 2000). The mechanisms for the gain in muscle mass associated with creatine supplementation are unclear. Creatine supplementation has no effect upon rates of protein synthesis or protein breakdown either at rest of after resistance exercise (Louis *et al.*, 2003a, 2003b; Parise *et al.*, 2001). These results suggest that the increase in energy stores associated with creatine supplementation (from increased muscle content of creatine phosphate) enables a greater amount of contractile work to be performed during training bouts, which results in a greater anabolic stimulus. Therefore, creatine supplementation may be considered to maximize muscle hypertrophy with strength training. The details of the use of creatine supplementation to increase muscle mass have been widely publicized, and we will thus refer the reader to other sources for further information (Tipton, Jeukendrup and Hespel, 2007; Volek and Rawson, 2004).

4.3.5 CONCLUSION

It seems nonsensical to provide specific recommendations about the amount of protein required to support strength training. There are too many factors which influence the response more than the amount of protein ingested in the diet. Clearly, the broad range of training programmes is a factor. Furthermore, the total amount of protein may not be particularly relevant as the need to maintain a positive energy balance appears to be more important, so long as a sufficient amount of protein is ingested. Finally, the type, dose, and timing of protein ingestion, along with other nutrients, may be more relevant and have greater implications for gains in muscle mass during strength training than the total amount. Clearly, the anabolic nature of resistance training increases the utilization of ingested proteins, suggesting that training may decrease requirements. The bottom line is that muscle mass may be gained from a wide range of protein intakes and there is little evidence to support very large (>2.0 g/kg/day) protein intakes.

From the available evidence we can recommend some general nutritional guidelines to support strength training:

- The vast majority of strength-trained athletes consume enough protein to support muscle hypertrophy without the requirement for protein/amino acid supplements. However, protein consumption beyond that required is unlikely to be detrimental to the adaptation process or to health.
- Total calorie intake should be sufficient to maintain a positive energy balance that will allow gains in muscle mass and strength. Energy intake will depend on the individual and the training volume.
- High protein ingestion rates may be beneficial to athletes undergoing calorie restriction as they will prevent losses in muscle mass.
- The athlete should receive sufficient carbohydrate intake to prevent compromised training capacity and to optimize training adaptations.

- While immediate feeding post-exercise may not be as important as once believed, it certainly will not harm muscle hypertrophy. It may be beneficial to consume amino acids prior to exercise.
- The resistance-trained athlete should receive sufficient fat in the diet to provide essential fatty acids and to make the diet palatable.

In order to optimize the increases in protein synthesis associated with resistance exercise, an acute feeding strategy may be more important than the total mass of protein ingested in a day. However, at this moment in time we don't know the optimal composition of the supplement, the optimal time at which it should be ingested, or the amount of EAAs to be ingested. From the available scientific evidence a few guidelines can be given:

- EAA provision is a requirement for a positive NMPB.
- The ingestion of animal proteins appears to be more beneficial than that of plant proteins.
- Recent reports suggest that 20 g of whole protein is enough to elicit the maximal protein synthesis response.
- The addition of carbohydrate is important in optimizing net protein balance and replenishing glycogen stores.
- Ingestion of an amino acid source before and/or after exercise is beneficial to gains in muscle protein.
- Fat may play a role in the adaptation process.

It may be that the consumption of protein from a food source containing other nutrients (CHO and fat), such as whole milk, is optimal for supporting strength training.

References

Alway S.E., Grumbt W.H., Stray-Gundersen J. and Gonyea W.J. (1992) Effects of resistance training on elbow flexors of highly competitive bodybuilders. *J Appl Physiol*, **72**, 1512–1521.

Anthony J.C., Anthony T.G., Kimball S.R. *et al.* (2000) Orally administered leucine stimulates protein synthesis in skeletal muscle of postabsorptive rats in association with increased eIF4F formation. *J Nutr*, **130**, 139–145.

Baron A.D., Steinberg H.O., Chaker H. *et al.* (1995) Insulin-mediated skeletal muscle vasodilation contributes to both insulin sensitivity and responsiveness in lean humans. *J Clin Invest*, **96**, 786–792.

Becque M.D., Lochmann J.D. and Melrose D.R. (2000) Effects of oral creatine supplementation on muscular strength and body composition. *Med Sci Sports Exerc*, **32**, 654–658.

Beelen M., Koopman R., Gijsen A.P. *et al.* (2008) Protein coingestion stimulates muscle protein synthesis during resistance-type exercise. *Am J Physiol Endocrinol Metab*, **295**, E70–E77.

Biolo G., Fleming R.Y., Maggi S.P. and Wolfe R.R. (1995a) Transmembrane transport and intracellular kinetics of amino acids in human skeletal muscle. *Am J Physiol*, **268**, E75–E84.

Biolo G., Maggi S.P., Williams B.D. *et al.* (1995b) Increased rates of muscle protein turnover and amino acid transport after resistance exercise in humans. *Am J Physiol*, **268**, E514–E520.

Biolo G., Tipton K.D., Klein S. and Wolfe R.R. (1997) An abundant supply of amino acids enhances the metabolic effect of exercise on muscle protein. *Am J Physiol*, **273**, E122–E129.

Biolo G., Williams B.D., Fleming R.Y. and Wolfe R.R. (1999) Insulin action on muscle protein kinetics and amino acid transport during recovery after resistance exercise. *Diabetes*, **48**, 949–957.

Boirie Y., Dangin M., Gachon P. *et al.* (1997) Slow and fast dietary proteins differently modulate postprandial proteinaccretion. *Proc Natl Acad Sci*, **94**, 14930–14935.

Borsheim E., Cree M.G., Tipton K.D. *et al.* (2004) Effect of carbohydrate intake on net muscle protein synthesis during recovery from resistance exercise. *J Appl Physiol*, **96**, 674–678.

Chittenden R.H. (1907) *The Nutrition of Man*, Heinemann, London, UK.

Churchley E.G., Coffey V.G., Pedersen D.J. *et al.* (2007) Influence of preexercise muscle glycogen content on transcriptional activity of metabolic and myogenic genes in well-trained humans. *J Appl Physiol*, **102**, 1604–1611.

Creer A., Gallagher P., Slivka D. *et al.* (2005) Influence of muscle glycogen availability on ERK1/2 and Akt signaling after resistance exercise in human skeletal muscle. *J Appl Physiol*, **99**, 950–956.

Cuthbertson D., Smith K., Babraj J. *et al.* (2005) Anabolic signaling deficits underlie amino acid resistance of wasting, aging muscle. *FASEB J*, **19**, 422–424.

Dangin M., Boirie Y., Garcia-Rodenas C. *et al.* (2001) The digestion rate of protein is an independent regulating factor of postprandial protein retention. *Am J Physiol Endocrinol Metab*, **280**, E340–E348.

Dreyer H.C., Drummond M.J., Pennings B. *et al.* (2008) Leucine-enriched essential amino acid and carbohydrate ingestion following resistance exercise enhances mTOR signaling and protein synthesis in human muscle. *Am J Physiol Endocrinol Metab*, **294**, E392–E400.

Elliot T.A., Cree M.G., Sanford A.P. *et al.* (2006) Milk ingestion stimulates net muscle protein synthesis following resistance exercise. *Med Sci Sports Exerc*, **38**, 667–674.

Esmarck B., Andersen J.L., Olsen S. *et al.* (2001) Timing of postexercise protein intake is important for muscle hypertrophy with resistance training in elderly humans. *J Physiol*, **535**, 301–311.

Fiatarone M.A., O'Neill E.F., Ryan N.D. *et al.* (1994) Exercise training and nutritional supplementation for physical frailty in very elderly people. *N Engl J Med*, **330**, 1769–1775.

Fouillet H., Mariotti F., Gaudichon C. *et al.* (2002) Peripheral and splanchnic metabolism of dietary nitrogen are differently affected by the protein source in humans as assessed by compartmental modeling. *J Nutr*, **132**, 125–133.

Glover E.I., Phillips S.M., Oates B.R. *et al.* (2008) Immobilization induces anabolic resistance in human myofibrillar protein synthesis with low and high dose amino acid infusion. *J Physiol*, **586**, 6049–6061.

Hartman J.W., Moore D.R. and Phillips S.M. (2006) Resistance training reduces whole-body protein turnover and improves net protein retention in untrained young males. *Appl Physiol Nutr Metab*, **31**, 557–564.

Hartman J.W., Tang J.E., Wilkinson S.B. *et al.* (2007) Consumption of fat-free fluid milk after resistance exercise promotes greater lean mass accretion than does consumption of soy or carbohydrate in young, novice, male weightlifters. *Am J Clin Nutr*, **86**, 373–381.

Ivy J.L. and Portman R. (2004) *Nutrient Timing: The Future of Sports Nutrition*, Basic Health Publications, North Bergen, NJ.

Karlsson H.K., Nilsson P.A., Nilsson J. *et al.* (2004) Branched-chain amino acids increase p70S6k phosphorylation in human skeletal muscle after resistance exercise. *Am J Physiol Endocrinol Metab*, **287**, E1–E7.

Katsanos C.S., Kobayashi H., Sheffield-Moore M. *et al.* (2006) A high proportion of leucine is required for optimal stimulation of the rate of muscle protein synthesis by essential amino acids in the elderly. *Am J Physiol Endocrinol Metab*, **291**, E381–E387.

Katsanos C.S., Chinkes D.L., Paddon-Jones D. *et al.* (2008) Whey protein ingestion in elderly persons results in greater muscle protein accrual than ingestion of its constituent essential amino acid content. *Nutr Res*, **28**, 651–658.

Koopman R., Wagenmakers A.J., Manders R.J. *et al.* (2005) Combined ingestion of protein and free leucine with carbohydrate increases postexercise muscle protein synthesis in vivo in male subjects. *Am J Physiol Endocrinol Metab*, **288**, E645–E653.

Koopman R., Manders R.J., Jonkers R.A. *et al.* (2006) Intramyocellular lipid and glycogen content are reduced following resistance exercise in untrained healthy males. *Eur J Appl Physiol*, **96**, 525–534.

Koopman R., Beelen M., Stellingwerff T. *et al.* (2007) Coingestion of carbohydrate with protein does not further augment postexercise muscle protein synthesis. *Am J Physiol Endocrinol Metab*, **293**, E833–E842.

Kumar V., Selby A., Rankin D. *et al.* (2009) Age-related differences in the dose-response relationship of muscle protein synthesis to resistance exercise in young and old men. *J Physiol*, **587**, 211–217.

Layman D.K., Evans E., Baum J.I. *et al.* (2005) Dietary protein and exercise have additive effects on body composition during weight loss in adult women. *J Nutr*, **135**, 1903–1910.

Levenhagen D.K., Gresham J.D., Carlson M.G. *et al.* (2001) Postexercise nutrient intake timing in humans is critical to recovery of leg glucose and protein homeostasis. *Am J Physiol Endocrinol Metab*, **280**, E982–E993.

Louis M., Poortmans J.R., Francaux M. *et al.* (2003a) No effect of creatine supplementation on human myofibrillar and sarcoplasmic protein synthesis after resistance exercise. *Am J Physiol Endocrinol Metab*, **285**, E1089–E1094.

Louis M., Poortmans J.R., Francaux M. *et al.* (2003b) Creatine supplementation has no effect on human muscle protein turnover at rest in the postabsorptive or fed states. *Am J Physiol Endocrinol Metab*, **284**, E764–E770.

Matthews D.E. (2005) Observations of branched-chain amino acid administration in humans. *J Nutr*, **135**, S1580–S1584.

Maughan R.J., Watson J.S. and Weir J. (1983) Relationships between muscle strength and muscle cross-sectional area in male sprinters and endurance runners. *Eur J Appl Physiol Occup Physiol*, **50**, 309–318.

Mettler S., Mitchell N. and Tipton K.D. (2010) Increased protein intake reduces lean body mass loss during weight loss in athletes, in review.

Miller S.L., Tipton K.D., Chinkes D.L. *et al.* (2003) Independent and combined effects of amino acids and glucose after resistance exercise. *Med Sci Sports Exerc*, **35**, 449–455.

Moore D.R., Del Bel N.C., Nizi K.I. *et al.* (2007) Resistance training reduces fasted- and fed-state leucine turnover and increases dietary nitrogen retention in previously untrained young men. *J Nutr*, **137**, 985–991.

Moore D.R., Robinson M.J., Fry J.L. *et al.* (2009) Ingested protein dose response of muscle and albumin protein synthesis after resistance exercise in young men. *Am J Clin Nutr*, **89**, 161–168.

Nair K.S., Matthews D.E., Welle S.L. and Braiman T. (1992) Effect of leucine on amino-acid and glucose-metabolism in humans. *Metabolism*, **41**, 643–648.

Paddon-Jones D., Sheffield-Moore M., Katsanos C.S. *et al.* (2006) Differential stimulation of muscle protein synthesis in elderly humans following isocaloric ingestion of amino acids or whey protein. *Exp Gerontol*, **41**, 215–219.

Parise G., Mihic S., MacLennan D. *et al.* (2001) Effects of acute creatine monohydrate supplementation on leucine kinetics and mixed-muscle protein synthesis. *J Appl Physiol*, **91**, 1041–1047.

Phillips S.M. (2004) Protein requirements and supplementation in strength sports. *Nutrition*, **20**, 689–695.

Phillips S.M. (2006) Dietary protein for athletes: from requirements to metabolic advantage. *Appl Physiol Nutr Metab*, **31**, 647–654.

Phillips S.M., Parise G., Roy B.D. *et al.* (2002) Resistance-training-induced adaptations in skeletal muscle protein turnover in the fed state. *Can J Physiol Pharmacol*, **80**, 1045–1053.

Phillips S.M., Hartman J.W. and Wilkinson S.B. (2005) Dietary protein to support anabolism with resistance exercise in young men. *J Am Coll Nutr*, **24**, S134–S139.

Rasmussen B.B., Tipton K.D., Miller S.L. *et al.* (2000) An oral essential amino acid-carbohydrate supplement enhances muscle protein anabolism after resistance exercise. *J Appl Physiol*, **88**, 386–392.

Rennie M.J. and Tipton K.D. (2000) Protein and amino acid metabolism during and after exercise and the effects of nutrition. *Annu Rev Nutr*, **20**, 457–483.

Rennie M.J., Bohe J., Smith K. *et al.* (2006) Branched-chain amino acids as fuels and anabolic signals in human muscle. *J Nutr*, **136**, S264–S268.

Rozenek R., Ward P., Long S. and Garhammer J. (2002) Effects of high-calorie supplements on body composition and muscular strength following resistance training. *J Sports Med Phys Fitness*, **42**, 340–347.

Svanberg E., Moller-Loswick A.C., Matthews D.E. *et al.* (1999) The role of glucose, long-chain triglycerides and amino acids for promotion of amino acid balance across peripheral tissues in man. *Clin Physiol*, **19**, 311–320.

Symons T.B., Schutzler S.E., Cocke T.L. *et al.* (2007) Aging does not impair the anabolic response to a protein-rich meal. *Am J Clin Nutr*, **86**, 451–456.

Tarnopolsky M. (2004) Protein requirements for endurance athletes. *Nutrition*, **20**, 662–668.

Tipton K.D. (2008) Protein for adaptations to exercise training. *Eur J Sport Sci*, **8**, 107–118.

Tipton K.D. and Ferrando A.A. (2008) Improving muscle mass: response of muscle metabolism to exercise, nutrition and anabolic agents. *Essays Biochem*, **44**, 85–98.

Tipton K.D. and Witard O.C. (2007) Protein requirements and recommendations for athletes: relevance of ivory tower arguments for practical recommendations. *Clin Sports Med*, **26**, 17–36.

Tipton K.D. and Wolfe R.R. (2004) Protein and amino acids for athletes. *J Sports Sci*, **22**, 65–79.

Tipton K.D., Ferrando A.A., Phillips S.M. *et al.* (1999) Postexercise net protein synthesis in human muscle from orally administered amino acids. *Am J Physiol*, **276**, E628–E634.

Tipton K.D., Rasmussen B.B., Miller S.L. *et al.* (2001) Timing of amino acid-carbohydrate ingestion alters anabolic response of muscle to resistance exercise. *Am J Physiol Endocrinol Metab*, **281**, E197–E206.

Tipton K.D., Elliott T.A., Cree M.G. *et al.* (2004) Ingestion of casein and whey proteins result in muscle anabolism after resistance exercise. *Med Sci Sports Exerc*, **36**, 2073–2081.

Tipton K.D., Elliott T.A., Cree M.G. *et al.* (2007) Stimulation of net muscle protein synthesis by whey protein ingestion before and after exercise. *Am J Physiol Endocrinol Metab*, **292**, E71–E76.

Tipton K.D., Jeukendrup A.E. and Hespel P. (2007) Nutrition for the sprinter. *J Sports Sci*, **25** (Suppl. 1), S5–15.

Tipton K.D., Elliot T.A., Ferrando A.A., Aarsland A.A. and Wolfe R.R. (2009) Stimulation of net muscle protein synthesis by resistance exercise and ingestion of leucine plus protein. *Appl Physiol Nutr Met*, **34** (2), 151–161.

Todd K.S., Butterfield G.E. and Calloway D.H. (1984) Nitrogen balance in men with adequate and deficient energy intake at three levels of work. *J Nutr*, **114**, 2107–2118.

Veldhorst M., Smeets A., Soenen S. *et al.* (2008) Protein-induced satiety: effects and mechanisms of different proteins. *Physiol Behav*, **94**, 300–307.

Vincent M.A., Clerk L.H., Lindner J.R. *et al.* (2004) Microvascular recruitment is an early insulin effect that regulates skeletal muscle glucose uptake in vivo. *Diabetes*, **53**, 1418–1423.

Volek J.S. and Rawson E.S. (2004) Scientific basis and practical aspects of creatine supplementation for athletes. *Nutrition*, **20**, 609–614.

Wilkinson S.B., Tarnopolsky M.A., Macdonald M.J. *et al.* (2007) Consumption of fluid skim milk promotes greater muscle protein accretion after resistance exercise than does consumption of an isonitrogenous and isoenergetic soy-protein beverage. *Am J Clin Nutr*, **85**, 1031–1040.

Witard O.C., Tieland M., Beelen M. *et al.* (2009) Resistance exercise increases postprandial muscle protein synthesis in humans. *Med Sci Sports Exerc*, **41**, 144–154.

Wolfe R.R. (2000) Protein supplements and exercise. *Am J Clin Nutr*, **72**, S551–S557.

4.4 Flexibility

William A. Sands, Monfort Family Human Performance Research Laboratory, Mesa State College – Kinesiology, Grand Junction, CO, USA

4.4.1 DEFINITIONS

Flexibility is considered one of the pillars of physical fitness, and stretching has been included or emphasized in many aspects of health and performance. Unfortunately, the terms 'flexibility' and 'stretching' have often been used interchangeably. This has caused some confusion, similar to the problem of differentiating the stretch–shortening cycle (SSC) from plyometrics; they are the mechanism and the method, respectively. Flexibility is the outcome or property, the increase in range of motion (ROM) of a joint or a related series of joints (Alter, 1996; Bloomfield and Wilson, 1998; Sands, 1985). Stretching is the means by which an increased ROM is obtained (Alter, 1988; Anderson, 1980; Bloomfield and Wilson, 1998; Stone *et al.*, 2006). However, beliefs about flexibility and the optimal means of stretching have often proceeded from assumptions that have never been tested and from an almost religious zeal regarding the perceived benefits of stretching by a few (Holt, Holt and Pelham, 1995a, 1995b, 1995c; Holt, Pelham and Holt, 2008; Shrier, 1999, 2004). A more encompassing, and perhaps more practical, definition was provided decades ago by Heyward (1984) and Metheny (1952) in separate works: flexibility is the freedom to move, the capacity of a joint to move fluidly through its full ROM. This definition helps include aspects of motion that are not simply determined by a range from beginning to end. The addition of the phrases 'freedom to move' and 'fluidly' helps distinguish movement that could be measured in an unconscious or dead organism from the elegant movements of a highly trained athlete or the grace of a simple well-coordinated running stride.

Practical and scientific experiences have shown that the typical definition of flexibility (the ROM in a joint or related series of joints) is inadequate. ROM is almost always measured statically (e.g. sit-and-reach test) (Hubley-Kozey, 1991); the ability of the athlete to use the ROM is seldom considered (Siff, 1998), the agility of the athlete with hypo- and hyperflexibility is often neglected (Van Gelder and Bartz, 2009), and little if any consideration is given to the fact that tissues undergo constant dynamic changes while moving (e.g. temperature, metabolic and/or excitatory state, relative stiffness, and so forth) (Siff, 1998). Moreover, it defies common sense to use a single static measurement as a snapshot of the characteristics of a dynamic system (Siff, 1993). It is not uncommon to observe athletes, particularly in aesthetic sports, who have more passive ROM available than they can use actively; even when they swing a limb, for example, they can use inertia to move it to an extreme position. It is common to see female gymnasts who can place their feet on a raised object in a split position while the ischial tuberosities of their raised legs sit on the mat or floor, indicating that they are extremely flexible in this passive position (Figure 4.4.1), but who cannot perform a split leap with legs split to 180 degrees (which requires other qualities, particularly strength and skill) (Sands and McNeal, 2000).

In non-aesthetic sports (McNeal and Sands, 2006) such as ball games, some combat sports, running, and cycling, and aesthetic sports (Sands, 1988, 1994, 2002) such as gymnastics, diving, figure skating, and some martial arts, the most flexible athletes are not necessarily the most successful; flexibility for athletes may be an optimization rather than a maximization problem (McNeal and Sands, 2006). Moreover, the commonly accepted idea that increased ROM and stretching prior to activity prevents injuries has been challenged and found to be on the shakiest of scientific foundations, or to come from such a paucity of data that no reasonable conclusions can be drawn. For example, reviews by Shrier (1999, 2004) and others have shown that short-term stretching and stretching for warm-up do not appear to prevent injuries or enhance immediately subsequent performance (Hubley-Kozey and Stanish, 1990; Kirby *et al.*, 1981; Knapik *et al.*, 1992; Shrier, 2004; Stone *et al.*, 2006; Thacker *et al.*, 2004; Witvrouw *et al.*, 2004). Interestingly, both strength and flexibility appear to be enhanced by long-term rather than short-term stretching (Armiger and Martyn, 2010; Bloomfield and Wilson, 1998; Cureton, 1941; Harvey and Mansfield, 2000; Krivickas and Feinberg, 1996; McCann, 1979). Paradoxically, long-term stretching results in cumulative positive effects, ranging from increased ROM to increased strength (Ford, 2007; Kokkonen *et al.*, 2007; LaRoche, Lussier and Roy, 2008; Reid and McNair, 2004; Stone *et al.*, 2006).

Strength and Conditioning – Biological Principles and Practical Applications Marco Cardinale, Rob Newton, and Kazunori Nosaka.

Figure 4.4.1 Gymnasts performing an 'oversplit'

Confusing the picture of flexibility and stretching even more are a few publications that proclaim the efficacy of stretching for all kinds of ailments while attributing the mechanisms and effects of stretching to fanciful ideas. These sources name underlying anatomical structures that may not exist as referenced and certainly do not possess the properties attributed to them; for example: 'Unfortunately, since fascia as a system is still a relatively new concept, there has not been a human dissection or even a good drawing that shows it in its entirety. But if we had a complete computerized model depicting the full extent of our fascia, you would still see the entire nervous and circulatory systems, since both are formed by distinct specialized connective tissues. While we are most occupied here with the fascia of muscle, called myofascia, be aware that when we talk about flexibility we are talking about all of the connective tissue that makes up the fascial network, or simply, the fascial net. If muscles get tight, so do the nerves and blood vessels that supply them. When you stretch the fascial net, you need to know what you are stretching as well as how, why, and when to do it, since what you stretch and how you stretch it dictates the location and type of effect generated by the stretch' (Frederick and Frederick, 2006, p. 23). If it is true that there is a 'fascial net' and that one must understand it before stretching, how can this be done when there is little evidence from which to make clinical/practical exercise selections and judgments? While it is true that connective tissue supports and helps hold the form of nearly every structure in human anatomy, the idea that stretching the skin of the foot results in measurable changes elsewhere remains to be demonstrated. Or: 'Again, think of the fascia as the body's internet, allowing for body-wide communication and *embodying intelligence of movement*. Recent biological discoveries have promoted the concept of *sophisticated intelligence systems residing outside the brain* in the body. For instance, science has recently classified a third nervous system (the central and autonomic systems being the other two) called the enteric nervous system, which appears to govern many functions of the digestive tract independent of the brain. Similarly, there is mounting evidence for independent roles played by the connective tissue system. One of our teachers and a colleague, James Oschman (2003), calls this system "the living matrix" because of *its reach beyond the connective tissues and into "every organ, tissue, cell, molecule, atom, and subatomic particle* within the body as well as the *energy fields*" that are within and around us. Because the connective tissue system has such extensive reach and influence, it offers great potential for improving more than flexibility. We think that understanding the form and function of your fascia will help you realize that flexibility training is much more than just stretching. You will benefit on many levels besides flexibility with you stick with this system' (emphasis mine) (Frederick and Frederick, 2006, p. 26). Or finally: '... all the cells get this mechanical message of movement as it undulates and reverberates through the fascial *network at the speed of light* ...' (emphasis mine) (Frederick and Frederick, 2006, p. 28). There is little to be gained by these types of statements. The enteric nervous system involves the digestive system and is classified as a subsystem of the autonomic nervous system; while it performs some interesting and potentially far-reaching functions in terms of a brain–gut axis, this does not mean that it possesses cognition (Goyal and Hirano, 1996; Thompson, 2002).

Of course, there is a great financial incentive to present pseudoscientific information as fact in order to gain a competitive advantage in the marketplace and to constantly use 'name dropping' rather than data and peer-reviewed studies to establish a market niche. This approach was called 'cargo cult science' by the late Richard Feynman, a Nobel prize winner in physics (Feynman, 1985, 1988). The basic idea is that pseudoscientific treatments often include scientific terms from other fields used incorrectly, along with some of the trappings of science without the evidence or substance. There are plenty of unanswered questions concerning flexibility and stretching that demand investigation without resorting to fringe areas and fanciful ideas.

4.4.2 WHAT IS STRETCHING?

Stretching is the means by which flexibility is enhanced. In general, stretching is simply the act of placing a joint in an extreme position, placing a tensile stress on the muscles that oppose that position. Stretching has been categorized in several ways:

1. Active versus passive, depending on whether the subject is using their own muscle tension to move a limb to an extreme position or is using gravity, inertia, a machine, or a partner (Guissard and Duchateau, 2004; Sands, 2002, 2007; Winters *et al.*, 2004).
2. Acute versus chronic, depending on whether the outcomes are studied immediately post-stretching or over a longer

period (Dent *et al.*, 2009; Guissard and Duchateau, 2006; Kokkonen, Nelson and Cornwell, 1998; Kokkonen *et al.*, 2007; Nelson, Cornwell and Heise, 1996; Schilling and Stone, 2000; Stone *et al.*, 2006).

3. Static versus dynamic, distinguished by whether the extreme position is held or is changed, with the subject moving to and from the extreme position in slow or rapid movements (Bacurau *et al.*, 2009; Bandy, Irion and Briggler, 1998; Bradley, Olsen and Portas, 2007; Costa *et al.*, 2009; Cottrell, 2009; Cross and Worrell, 1999; Davis *et al.*, 2005; O'Sullivan, Murray and Sainsbury, 2009; Young, 2007).

The primary mechanisms involved in these three categories of stretching involve enhanced 'stretch tolerance' (ability to tolerate the discomfort of stretching) (LaRoche and Connolly, 2006; Magnusson, 1998; Magnusson *et al.*, 1996, 1997, 1998) and changes in tissue elastic (i.e. stress/strain) and length characteristics (Alter, 1996, 2004; Hutton, 1992; Kurz, 1994; Siff, 1993). Applied heat to the stretched muscles and connective tissue may enhance the effects of stretching by increasing the creep of molecular bonds within the connective-tissue molecular structure or fibrils and fibres (Draper *et al.*, 2004; Funk *et al.*, 2001b; Lentell *et al.*, 1992; Peres *et al.*, 2002). However, there is conflicting evidence that applied external heat does not improve flexibility (Brucker *et al.*, 2005; Burke *et al.*, 2001; Draper *et al.*, 2002; Funk *et al.*, 2001a).

Sport-specific movement has been postulated to increase flexibility simply by repetition, and separate 'generic' stretching positions and exercises, while common, have been questioned (Chandler *et al.*, 1990; Mann and Jones, 1999; Marshall *et al.*, 1980; Toskovic, Blessing and Williford, 2004; Volver, Viru and Viru, 2000). It has been argued that the sport itself should provide all the stretching stimuli needed to attain and maintain an appropriate ROM in the sport-specific movements. The late Mel Siff wrote: 'It is almost heretical to question this stretching doctrine, yet it is important to disclose that there is no research which proves categorically that there is any need for separate stretching sessions, phases or exercises to be conducted to improve performance and safety. To appreciate this fact, it is useful to return to one of the clinical definitions of flexibility, namely that flexibility refers to the range of movement of a specific joint or group of anatomical tissues. Moreover, flexibility cannot be considered separate from other fitness factors such as strength and stamina' (Siff, 1998, p. 123). He also wrote: 'Furthermore, there is no real need to prescribe separate stretching exercises or sessions, since logically structured training should take every joint progressively through its full range of static and dynamic movement. In other words, every movement should be performed to enhance flexibility, strength, speed, local muscular endurance and skill, so that separate stretching sessions then become largely redundant' (Siff, 1998, p. 123). This contrasts with the idea, largely from aesthetic sports, that one should first increase the ROM by whatever means, and then allow the acquisition of new flexibility by training the strength to control the newly acquired range (McNeal and Sands, 2006).

The purpose of stretching depends on whether the flexibility attained is going to be used for aesthetic reasons, such as a split in gymnastics or dance, or to increase the ROM for enhanced skill performance via increased impulse development by increasing the ROM through which force is applied (Hubley-Kozey, 1991; McNeal and Sands, 2006). Bloomfield and Wilson support this division with the following: 'Being able to increase the range of various movements throughout the body has enabled athletes to place themselves into more aesthetic positions in almost all sports' (Bloomfield and Wilson, 1998, p. 240), and 'When an athlete is able to increase the range of motion in any skill in a ballistic sport, the potential to produce more force or velocity becomes possible, because a greater range of movement increases the distance and time over which a force can be developed' (Bloomfield and Wilson, 1998, p. 240).

A separate category of stretching is proprioceptive neuromuscular facilitation (PNF). PNF is more than a stretching programme and involves strengthening, position awareness, and other factors (Etnyre and Lee, 1987; McAtee, 1993; Siff, 2000). PNF conditioning methods have been developed over many years to embrace the idea of both stretching and strengthening limb motion. While PNF is often offered as a panacea for stretching and strengthening, and is shown to be effective in many studies (Cornelius *et al.*, 1992; Ford and McChesney, 2007; Funk *et al.*, 2003; Haigh, Jones and Bampouras, 2005; Higgs and Winter, 2009; Kokkonen and Lauritzen, 1995; Rowlands, Marginson and Lee, 2003; Sady, Wortman and Blanke, 1982), there are examples from the literature that do not support its superiority; moreover, some cautions are applied to PNF methods to prevent overstrain of muscles (Bonnar, Deivert and Gould, 2004; Bradley, Olsen and Portas, 2007; Church *et al.*, 2001; Cornelius and Hands, 1992; Davis *et al.*, 2005; Feland and Marin, 2004; Knappstein, Stanley and Whatman, 2004; Osternig *et al.*, 1990). Certainly, achieving the positions in effective PNF training requires a clever instructor and participant, and great care by a stretching partner to ensure that the participant does not overstretch a particular muscle. Experience with PNF approaches and athletes has demonstrated that strict attention to detail, feedback from the person being stretched, and a level of maturity must be present in order to ensure safety (Sands, 1984).

Vibration is a relatively new and promising addition to the world of flexibility and stretching. Bierman (1960) may have been the first to study vibration and flexibility via trunk flexion, although the vibrator was not described and application to the backs of the subjects led to the conclusion that the mechanism of increased ROM was muscle relaxation from the vibration. Atha and Wheatley (1976) used vibration as an adjunct to enhanced stretching and flexibility in forward trunk flexion, with the subjects seated and using vibration on their buttocks and low back. Issurin, Liebermann and Tenenbaum (1993, 1994) used a vibrating ring suspended by a cable, in which the foot of the subject was placed while they stretched forward over the raised leg, targeting the hamstrings. The resulting increase in ROM was astonishing. These researchers demonstrated that vibration could enhance flexibility. Bierman and

Atha did not include stretching during vibration but Issurin and colleagues did.

Since then, a number of studies have confirmed and expanded on these earlier results. Several studies have looked specifically at limb vibration and enhanced flexibility, with acute effects improving ROM as much as 400% in one four-minute treatment (Kinser *et al.*, 2008; McNeal and Sands, 2006; Sands, 2007; Sands *et al.*, 2005, 2006, 2008a, 2008b; Sands, McNeal and Stone, 2009). Other studies have found that whole-body vibration, not specifically stretching, also enhanced flexibility (Abercromby *et al.*, 2007a, 2007b; Burns, Beekhuizen and Jacobs, 2005; Cochrane and Stannard, 2005; Edwards, Skidmore and Signorile, 2009; Fagnani *et al.*, 2006; Feland, Hopkins and Hunter, 2005; van den Tillaar, 2006). The mechanism(s) postulated for the effects of vibration include: increased muscle temperature, increased muscle relaxation, decreased myotatic reflex sensitivity, and decreased pain perception resulting in increased stretch tolerance (Issurin, Liebermann and Tenenbaum, 1993, 1994; Sands *et al.*, 2006, 2008a, 2008b). Anecdotally, in our studies it is quite common to hear athletes say that the stretch position does not hurt in the way they're accustomed to feeling with the addition of vibration (Figures 4.4.2 and 4.4.3). Also anecdotally, a study of US National Team pairs figure skaters (Sands, 2007) showed that 12 of 18 male figure skaters were able to achieve a flat split position following one treatment of vibration and stretching in a split position; none of the 12 had ever achieved this position in prior training. The remaining six skaters were either already able to perform splits or else dramatically improved their split position without quite getting flat to the floor.

Strength training and strength training coupled with stretching have been shown to enhance flexibility. High-performance gymnasts have usually been stretching by various methods for years, sometimes for more than two decades. Increasing flexibility in this population is particularly difficult. In a study of female gymnasts who had been stretching for years, and who showed little or no improvement in the most recent years, subjects were divided into experimental and control groups (Sands and McNeal, 2000). Each group did their normal training. The experimental group ended each workout with standing straight-leg kicks with each leg forward, sideward, and rearward, and performed side split jumps and straddle split jumps using Theraband strips tied to each ankle so that each kick or split

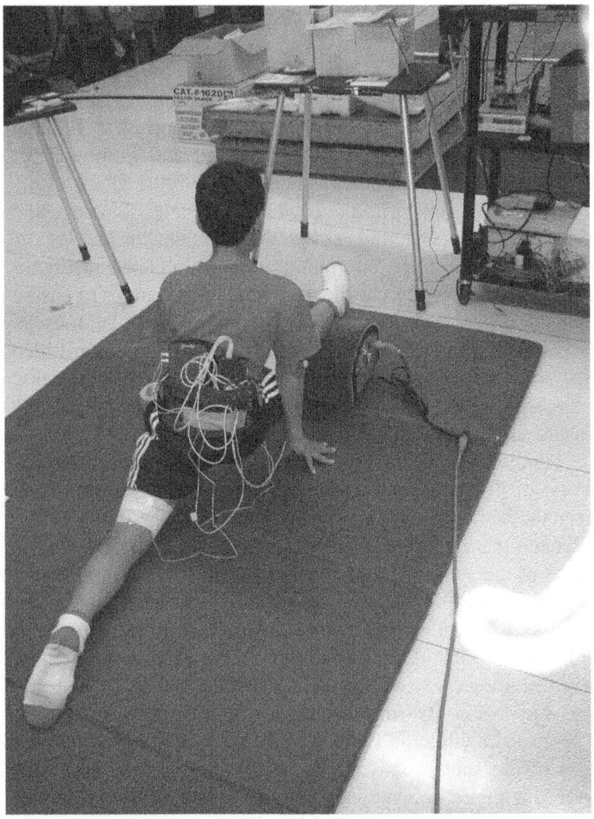

Figure 4.4.2 Split stretching with the forward leg placed on top of a vibration device

Figure 4.4.3 (a) ROM before application of vibration for four minutes to the forward and rearward legs in a split position. (b) ROM after application of vibration on the four muscle groups listed above

movement was resisted by the elastic band. The control group performed their normal stretching and conditioning without the addition of the elastic bands (Figure 4.4.4). The study was interesting for two reasons. First, the experimental group improved their flexibility quickly, as shown in a 6° average increase in their split leaps as determined dynamically via kinematic analysis of split-leap ROM. Second, although the study was planned for four weeks, it had to be terminated after two weeks because the control group would not comply with the control condition. The control-group athletes saw how rapidly and effectively the experimental group was improving and started to 'sneak' training with the elastic bands after the first two weeks of the study so that they would not be outperformed by the experimental group (Sands and McNeal, 2000). At this point in the study, the investigators determined that it was unethical to withhold the treatment from the control group. While 6° may seem like a small change, the difference in a split-leap performance of 6° is the penalty of a two to three tenth deduction or zero deduction. The author has obtained unpublished data of young male gymnasts undertaking four weeks of training using only Theraband elastic bands to perform the same kicks and jumps as the study cited above and no other focused static stretching during the study; the pre-test–post-test differences were statistically significant, showing that their flexibility in a split position improved. Unfortunately, no control group was available. This study demonstrates that strength enhancements may simultaneously enhance passive flexibility.

Figure 4.4.4 Dynamic stretching with resistance provided by an elastic band

4.4.3 A MODEL OF EFFECTIVE MOVEMENT: THE INTEGRATION OF FLEXIBILITY AND STRENGTH

Most discussions of flexibility begin with obvious anatomical constraints. The reader is referred to a host of texts covering the ultra-structure of muscle and connective tissue to gain insight into the influence of anatomy on flexibility (Alter, 1996, 2004; Norkin and Levangie, 1992; Ylinen, 2008). Following anatomy, neural control, muscle spindles, Golgi tendon organs, mechanoreceptors, and reflexes are covered. Typically these topics are followed by various limitations on flexibility, such as gender, age, anthropometry, circadian rhythms, and so forth. Most texts then finish with dozens of exercises showing how to stretch various muscles and muscle groups. While these typical approaches are extraordinarily helpful in understanding flexibility by itself, they rarely discuss the integration of flexibility with strength, speed, stamina, body size and shape, genetic predispositions, current tissue status, or other concepts.

The model shown in Figure 4.4.5 provides a framework of integration for flexibility with strength. Flexibility *requires*

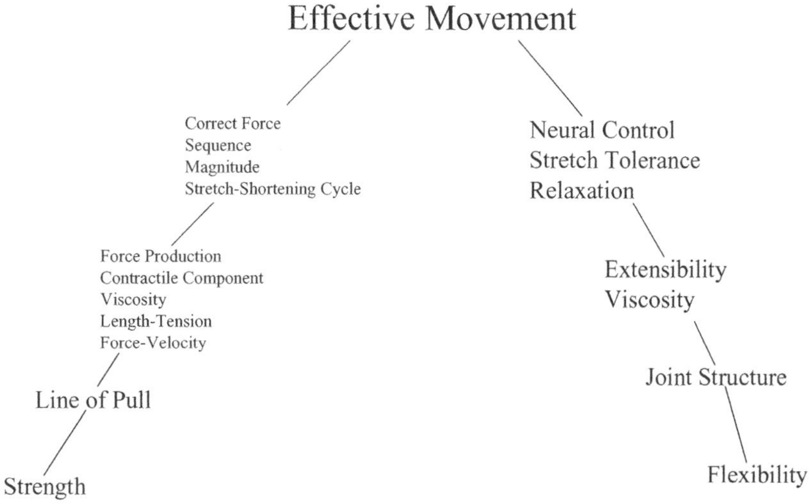

Figure 4.4.5 Model of effective movement integrating strength and flexibility

integration with other movement components to produce skilled movement. The model is an oversimplification, with a litany of factors underlying each labeled component, but it also amplifies the idea that flexibility is useless without strength, and vice versa. For example, training muscles to be extraordinarily strong with the intent of improving performance and preventing injury does little good if those muscles cannot move through an appropriate ROM, cannot be summoned by the nervous system at the proper time and with the proper tension, and are not inhibited by antagonistic muscles. In other words, a strong muscle is useless if it doesn't 'fire' at the right time and in the right amount. The same can be said for flexibility. Increasing a person's ROM is useless if that increased range is not under exquisitely precise nervous control and the muscles controlling the movement are not strong enough to control the position.

The information presented above should support the premise that flexibility is not as simple as ROM about a joint. Flexibility, or motion around a joint, is under the control of numerous factors; changing one without consideration of all the others is folly and often results in dysfunction. Increasing flexibility without simultaneously increasing strength in the new range, nervous control, stamina, and other factors invites puzzling failure. The challenge of the future is to discover how the integration of these factors produces skilled movements. This will require a breakdown of academic turf barriers, and new and broader training of exercise and sport scientists so that integration of systems becomes the focus.

References

Abercromby A.F.J., Amonette W.E., Layne C.S. *et al.* (2007a) Variation in neuromuscular responses during acute whole-body vibration exercise. *Med Sci Sports Exerc*, **39** (9), 1642–1650.

Abercromby A.F.J., Amonette W.E., Layne C.S. *et al.* (2007b) Vibration exposure and biodynamic responses during whole-body vibration training. *Med Sci Sports Exerc*, **39** (10), 1794–1800.

Alter M.J. (1988) *Science of Stretching*, Human Kinetics, Champaign, IL.

Alter M.J. (1996) *Science of Flexibility*, Human Kinetics, Champaign, IL.

Alter M.J. (2004) *Science of Flexibility*, Human Kinetics, Champaign, IL.

Anderson B. (1980) *Stretching*, Shelter Pub, Bolinas, CA.

Armiger P. and Martyn M.A. (2010) *Stretching for Functional Flexibility*, Lippincott Williams and Wilkins, Philadelphia, PA.

Atha J. and Wheatley D.W. (1976) Joint mobility changes due to low frequency vibration and stretching exercises. *Br J Sports Med*, **10** (1), 26–34.

Bacurau R.F., Monteiro G.A., Ugrinowitsch C. *et al.* (2009) Acute effect of a ballistic and a static stretching exercise bout on flexibility and maximal strength. *J Strength Cond Res*, **23** (1), 304–308.

Bandy W.D., Irion J.M. and Briggler M. (1998) The effect of static stretch and dynamic range of motion training on the flexibility of the hamstring muscles. *J Orthop Sports Phys Ther*, **27** (4), 295–300.

Bierman W. (1960) Influence of cycloid vibration massage on trunk flexion. *Am J Phys Med*, **39**, 219–224.

Bloomfield J. and Wilson G. (1998) Flexibility in sport, in *Training in Sport* (ed. B. Elliott), John Wiley & Sons, Inc., New York, NY, pp. 239–285.

Bonnar B.P., Deivert R.G. and Gould T.E. (2004) The relationship between isometric contraction durations during hold-relax stretching and improvement of hamstring flexibility. *J Sports Med Phys Fitness*, **44** (3), 258–261.

Bradley P.S., Olsen P.D. and Portas M.D. (2007) The effect of static, ballistic and proprioceptive neuromuscular facilitation stretching on vertical jump performance. *J Strength Cond Res*, **21** (1), 223–226.

Brucker J.B., Knight K.L., Rubley M.D. and Draper D.O. (2005) An 18-day stretching regimen, with or without pulsed, shortwave diathermy and ankle dorsiflexion after 3 weeks. *J Athl Train*, **40** (4), 276–280.

Burke D.G., Holt L.E., Rasmussen R. *et al.* (2001) Effects of hot or cold water immersion and modified proprioceptive neuromuscular facilitation flexibility exercise on hamstring length. *J Athl Train*, **36** (1), 16–19.

Burns P.A., Beekhuizen K.S. and Jacobs P.L. (2005) Acute effects of whole-body vibration and bicycle ergometry on muscular strength and flexibility. *Med Sci Sports Exerc*, **37** (Suppl. 5), S262–S263.

Chandler T.J., Kibler W.B., Uhl T.L. *et al.* (1990) Flexibility comparisons of junior elite tennis players to other athletes. *Am J Sports Med*, **18** (2), 134–136.

Church J.B., Wiggins M.S., Moode F.M. and Crist R. (2001) Effect of warm-up and flexibility treatments on vertical jump performance. *J Strength Cond Res*, **15** (3), 332–336.

Cochrane D.J. and Stannard S.R. (2005) Acute whole body vibration training increases vertical jump and flexibility performance in elite female field hockey players. *Br J Sports Med*, **39** (11), 860–865.

Cornelius W.L. and Hands M.R. (1992) The effects of a warm-up on acute hip joint flexibility using a modified pnf stretching technique. *J Athl Train*, **27** (2), 112–114.

Cornelius W.L., Ebrahim K., Watson J. and Hill D.W. (1992) The effects of cold application and modified PNF stretching techniques on hip joint flexibility in college males. *Res Q Exerc Sport*, **63** (3), 311–314.

Costa P.B., Graves B.S., Whitehurst M. and Jacobs P.L. (2009) The acute effects of different durations of static stretching on dynamic balance performance. *J Strength Cond Res*, **23** (1), 141–147.

Cottrell T. (2009) The effects of static and dynamic stretching protocols on knee torque at varying angular velocities. *Med Sci Sports Exerc*, **41** (Suppl. 5), S64.

Cross K.M. and Worrell T.W. (1999) Effects of a static stretching program on the incidence of lower extremity musculotendinous strains. *J Athl Train*, **34** (1), 11–14.

Cureton T.K. (1941) Flexibility as an aspect of physical fitness. *Res Q*, **12**, 381–390.

Davis D.S., Ashby P.E., McCale K.L. *et al.* (2005) The effectiveness of 3 stretching techniques on hamstring flexibility using consistent stretching parameters. *J Strength Cond Res*, **19** (1), 27–32.

Dent J., O'Brien J., Bushman T. *et al.* (2009) Acute and prolonged effects of static stretching and dynamic warm-up on muscular power and strength. *Med Sci Sports Exerc*, **41** (Suppl. 5), S64.

Draper D.O., Miner L., Knight K.L. and Ricard M.D. (2002) The carry-over effects of diathermy and stretching in developing hamstring flexibility. *J Athl Train*, **37** (1), 37–42.

Draper D.O., Castro J.L., Feland B. *et al.* (2004) Shortwave diathermy and prolonged stretching increase hamstring flexibility. *J Orthop Sports Phys Ther*, **34** (1), 13–20.

Edwards D.A., Skidmore E.C. and Signorile J.F. (2009) The acute effects of whole body vibration on hamstring flexibility. *Med Sci Sports Exerc*, **41** (5), S255.

Etnyre B.R. and Lee E.J. (1987) Comments on proprioceptive neuromuscular facilitation stretching techniques. *Res Q Exerc Sport*, **58** (2), 184–188.

Fagnani F., Giombini A., Di Cesare A. *et al.* (2006) The effects of a whole-body vibration program on muscle performance and flexibility in female athletes. *Am J Phys Med Rehabil*, **85** (12), 956–962.

Feland J.B. and Marin H.N. (2004) Effect of submaximal contraction intensity in contract-relax proprioceptive neuromuscular facilitation stretching. *Br J Sports Med*, **38** (4), E18.

Feland J.B., Hopkins T. and Hunter I. (2005) Acute changes in hamstring flexibility using a whole-body-vibration platform with static stretch. *Med Sci Sports Exerc*, **37** (Suppl. 5), S410.

Feynman R.P. (1985) *Surely You're Joking Mr Feynman!*, Bantam Books, New York, NY.

Feynman R.P. (1988) *What Do You Care What Other People Think?*, Bantam Books, New York, NY.

Ford P.M.J. (2007) Duration of maintained hamstring ROM following termination of three stretching protocols. *J Sport Rehabil*, **16** (1), 18–27.

Ford P. and McChesney J.M. (2007) Duration of maintained hamstring ROM following termination of three stretching protocols. *J Sport Rehabil*, **16**, 18–27.

Frederick A. and Frederick C. (2006) *Stretch to Win*, Human Kinetics, Champaign, IL.

Funk D., Swank A.M., Adams K.J. and Treolo D. (2001a) Efficacy of moist heat pack application over static stretching on hamstring flexibility. *J Strength Cond Res*, **15** (1), 123–126.

Funk D., Swank A.M., Adams K.J. and Treolo D. (2001b) Efficacy of moist heat pack application over static stretching on hamstring flexibility. *J Strength Cond Res*, **15** (1), 123–126.

Funk D.C., Swank A.M., Mikla B.M. *et al.* (2003) Impact of prior exercise on hamstring flexibility: a comparison of proprioceptive neuromuscular facilitation and static stretching. *J Strength Cond Res*, **17** (3), 489–492.

Goyal R.K. and Hirano I. (1996) The enteric. *N Engl J Med*, **334**, 1106.

Guissard N. and Duchateau J. (2004) Effect of static stretch training on neural and mechanical properties of the human plantar-flexor muscles. *Muscle Nerve*, **29** (2), 248–255.

Guissard N. and Duchateau J. (2006) Neural aspects of muscle stretching. *Med Sci Sports Exerc*, **34** (4), 154–158.

Haigh A., Jones M.A. and Bampouras T. (2005) The influence of three different warm-up stretching routines on muscle performance. *J Sports Sci*, **23**, 195–196.

Harvey D. and Mansfield C. (2000) Measuring flexibility for performance and injury prevention, in *Physiological Tests for Elite Athletes* (ed. C.J. Gore), Human Kinetics, Champaign, IL, pp. 98–113.

Heyward V.H. (1984) *Designs for Fitness: A Guide to Physical Fitness Appraisal and Exercise Prescription*, Burgess, Minneapolis, MN.

Higgs F. and Winter S.L. (2009) The effect of a four-week proprioceptive neuromuscular facilitation stretching program on isokinetic torque production. *J Strength Cond Res* **23** (5), 1442–1447.

Holt J., Holt L.E. and Pelham T.W. (1995a) Flexibility redefined, in *Proceedings XIII International Symposium on Biomechanics in Sports, International Society of Biomechanics in Sports* (ed. T. Bauer), Dalhousie University, Nova Scotia, pp. 170–174.

Holt J., Holt L.E. and Pelham T.W. (1995b) What research tells us about flexibility, in *Proceedings XIII International Symposium on Biomechanics in Sports, International Society of Biomechanics in Sports* (ed. T. Bauer), Dalhousie University, Nova Scotia, pp. 175–179.

Holt J., Holt L.E. and Pelham T.W. (1995c) What research tells us about flexibility, in *Proceedings XIII International Symposium on Biomechanics in Sports, International Society of Biomechanics in Sports* (ed. T. Bauer), Dalhousie University, Nova Scotia, pp. 180–183.

Holt L.E., Pelham T.W. and Holt J. (2008) *Flexibility*, Humana Press.

Hubley-Kozey C.L. (1991) Testing flexibility, in *Physiological Testing of the High-Performance Athlete* (eds J. Duncan MacDougall, H.A. Wenger and H.J. Green), Human Kinetics, Champaign, IL, pp. 309–359.

Hubley-Kozey C.L. and Stanish W.D. (1990) Can stretching prevent athletic injuries? *J Musculoskelet Med*, **7** (3), 21–31.

Hutton R.S. (1992) Neuromuscular basis of stretching exercises, in *Strength and Power in Sport* (ed. P.V. Komi), Blackwell Scientific Publications, Oxford, UK, pp. 29–38.

Issurin V.B., Liebermann D.G. and Tenenbaum G. (1993) Vibratory stimulation training: a new approach for developing strength and flexibility in athletes. 2nd Maccabiah-Wingate International Congress on Sport and Coaching Sciences, The Wingate Institute for Physical Education and Sport.

Issurin V.B., Liebermann D.G. and Tenenbaum G. (1994) Effect of vibratory stimulation on maximal force and flexibility. *J Sports Sci*, **12**, 561–566.

Kinser A.M., Ramsey M.W., O'Bryant H.S. *et al.* (2008) Vibration and stretching effects on flexibility and explosive strength in young gymnasts. *Med Sci Sports Exerc*, **40** (1), 133–140.

Kirby R.L., Simms F.C., Symington V.J. and Garner J.B. (1981) Flexibility and musculoskeletal symptomatology in female gymnasts and age-matched controls. *Am J Sports Med*, **9** (3), 160–164.

Knapik J.J., Jones B.H., Bauman C.L. and Harris J.M. (1992) Strength, flexibility and athletic injuries. *Sports Med*, **14** (5), 277–288.

Knappstein A., Stanley S. and Whatman C. (2004) Range of motion immediately post and seven minutes post, PNF stretching. *N Z J Sports Med*, **32** (2), 42–46.

Kokkonen J. and Lauritzen S. (1995) Isotonic strength and endurance gains through PNF stretching. *Med Sci Sports Exerc*, **27**, S22.

Kokkonen J., Nelson A.G. and Cornwell A. (1998) Acute muscle stretching inhibits maximal strength performance. *Res Q Exerc Sport*, **69** (4), 411–415.

Kokkonen J., Nelson A.G., Eldredge C. and Winchester J.B. (2007) Chronic static stretching improves exercise performance. *Med Sci Sports Exerc*, **39** (10), 1825–1831.

Krivickas L.S. and Feinberg J.H. (1996) Lower extremity injuries in college athletes: Relation between ligamentous laxity and lower extremity muscle tightness. *Arch Phys Med Rehabil*, **77**, 1139–1143.

Kurz T. (1994) *Stretching Scientifically: A Guide to Flexibility Training*, Stadion, Island Pond, VT.

LaRoche D.P. and Connolly D.A. (2006) Effects of stretching on passive muscle tension and response to eccentric exercise. *Am J Sports Med*, **34** (6), 1000–1007.

LaRoche D.P., Lussier M.V. and Roy S.J. (2008) Chronic stretching and voluntary muscle force. *J Strength Cond Res*, **22** (2), 589–596.

Lentell G., Hetherington T., Eagan J. and Morgan M. (1992) The use of thermal agents to influence the effectiveness of a low-load prolonged stretch. *J Orthop Sports Phys Ther*, **16** (5), 200–207.

Magnusson S.P. (1998) Passive properties of human skeletal muscle during stretch maneuvers. A review. *Scand J Med Sci Sports*, **8** (2), 65–77.

Magnusson S.P., Aagard P., Simonsen E. and Bojsen-Møller F. (1998) A biomechanical evaluation of cyclic and static stretch in human skeletal muscle. *Int J Sports Med*, **19** (5), 310–316.

Magnusson S.P., Simonsen E.B., Aagaard P. *et al.* (1997) Determinants of musculoskeletal flexibility: viscoelastic properties, cross-sectional area, EMG and stretch tolerance. *Scand J Med Sci Sports*, **7**, 195–202.

Magnusson S.P., Simonsen E.B., Aagaard P. *et al.* (1996) A mechanism for altered flexibility in human skeletal muscle. *J Physiol*, **497** (Pt 1), 291–298.

Mann D.P. and Jones M.T. (1999) Guidelines to the implementation of a dynamic stretching program. *Strength Cond J*, **21** (6), 53–55.

Marshall J.L., Johanson N., Wickiewicz T.L. *et al.* (1980) Joint looseness: a function of the person and the joint. *Med Sci Sports Exerc*, **12** (3), 189–194.

McAtee R.E. (1993) *Facilitated Stretching*, Human Kinetics, Champaign, IL.

McCann C. (1979) Young gymnasts: injury prone, less flexible. *Physician Sports Med*, **7** (1), 23–24.

McNeal J.R. and Sands W.A. (2006) Stretching for performance enhancement. *Curr Sports Med Rep*, **5**, 141–146.

Metheny E. (1952) *Body Dynamics*, McGraw-Hill, New York, NY.

Nelson A.G., Cornwell A. and Heise G.D. (1996) Acute stretching exercises and vertical jump stored elastic energy. *Med Sci Sports Exerc*, **28**, S156.

Norkin C.C. and Levangie P.K. (1992) *Joint Structure and Function*, F.A. Davis Co, Philadelphia, PA.

O'Sullivan K., Murray E. and Sainsbury D. (2009) The effect of warm-up, static stretching and dynamic stretching on hamstring flexibility in previously injured subjects. *BMC Musculoskeletal Disorders*, **16**, 10–37.

Osternig L.R., Robertson R.N., Troxel R.K. and Hansen P. (1990) Differential responses to proprioceptive neuromuscular facilitation (PNF) stretch techniques. *Med Sci Sports Exerc*, **22** (1), 106–111.

Peres S.E., Draper D.O., Knight K.L. and Ricard M.D. (2002) Pulsed shortwave diathermy and prolonged long-duration stretching increase dorsiflexion range of motion more than identical stretching without diathermy. *J Athl Train*, **37** (1), 43–50.

Reid D.A. and McNair P.J. (2004) Passive force, angle and stiffness changes after stretching of hamstring muscles. *Med Sci Sports Exercs*, **36** (11), 1944–1948.

Rowlands A.V., Marginson V.F. and Lee J. (2003) Chronic flexibility gains: effect of isometric contraction duration during proprioceptive neuromuscular facilitation stretching techniques. *Res Q Exerc Sport*, **74** (1), 47–51.

Sady S.P., Wortman M. and Blanke D. (1982) Flexibility training: ballistic, static or proprioceptive neuromuscular facilitation? *Arch Phys Med Rehabil*, **63** (6), 261–263.

Sands B. (1984) *Coaching Women's Gymnastics*, Human Kinetics, Champaign, IL.

Sands W.A. (1985) A Comparison of Leg to Trunk Flexibility Tests. PhD thesis, University of Utah.

Sands W. (1988) US Gymnastics Federation physical abilities testing for women. *Technique*, **8** (3–4), 27–32.

Sands W.A. (1994) Physical abilities profiles – 1993 national TOPs testing. *Technique*, **14** (8), 15–20.

Sands W.A. (2002) Physiology, in *Scientific Aspects of Women's Gymnastics* (eds W.A. Sands, D.J. Caine and J. Borms), Karger, Basel, Switzerland, pp. 128–161.

Sands W.A. (2007) Shake it up, tests confirm that vibration can improve flexibility. *Skating*, Aug/Sep: 46–47.

Sands W.A. and McNeal J.R. (2000) Enhancing flexibility in gymnastics. *Technique*, **20**, 6–9.

Sands W.A., McNeal J.R. and Stone M.H. (2009) Vibration, split stretching and static vertical jump performance in young male gymnasts. *Med Sci Sports Exerc*, **41** (5), S255.

Sands W.A., McNeal J.R., Stone M.H. *et al.* (2008a) Effect of vibration on forward split flexibility and pain perception in young male gymnasts. *Int J Sports Physiol Perform*, **3** (4), 469–481.

Sands W.A., McNeal J.R., Stone M.H. *et al.* (2008b) The effect of vibration on active and passive range of motion in elite female synchronized swimmers. *Eur J Sport Sci*, **8** (4), 217–233.

Sands W.A., McNeal J.R., Stone M.H. *et al.* (2006) Flexibility enhancement with vibration: acute and long-term. *Med Sci Sports Exerc*, **38** (4), 720–725.

Sands W.A., Stone M.H., Smith S.L. and McNeal J.R. (2005) Enhancing forward split flexibility: USA synchronized swimming. *Synchro Swimming USA*, **13** (2), 15–16.

Schilling B.K. and Stone M.H. (2000) Stretching: acute effects on strength and power performance. *Strength Cond*, **22** (1), 44–47.

Shrier I. (1999) Stretching before exercise does not reduce the risk of local muscle injury: a critical review of the clinical and basic science literature. *Clin J Sport Med*, **9** (4), 221–227.

Shrier I. (2004) Does stretching improve performance? A systematic and critical review of the literature. *Clin J Sport Med*, **14** (5), 267–273.

Siff M.C. (1993) Exercise and the soft tissues. *Fitness Sports Rev Int*, **28** (1), 32.

Siff M.C. (1998) *Facts and Fallacies of Fitness*, Mel Siff, University of Witwatersrand, Johannesburg, South Africa.

Siff M.C. (2000) *Supertraining*, Supertraining Institute, Denver, CO.

Stone M., Ramsey M.W., Kinser A.M. *et al.* (2006) Stretching: acute and chronic? The potential consequences. *Strength Cond J*, **26** (6), 66–74.

Thacker S.B., Gilchrist J., Stroup D.F. and Kimsey C.D., Jr. (2004) The impact of stretching on sports injury risk: a systematic review of the literature. *Med Sci Sports Exerc*, **36** (3), 371–378.

Thompson W.G. (2002) Review article: the treatment of irritable bowel syndrome. *Aliment Pharmacol Ther*, **16** (8), 1395–1406.

Toskovic N.N., Blessing D. and Williford H.N. (2004) Physiologic profile of recreational male and female novice and experienced Tae Kwon Do practitioners. *J Sports Med Phys Fitness*, **44**, 164–172.

van den Tillaar R. (2006) Will whole-body vibration training help increase the range of motion of the hamstrings? *J Strength Cond Res*, **20** (1), 192–196.

Van Gelder L.H. and Bartz S. (2009) The effects of stretching on agility performance. *Med Sci Sports Exerc*, **41** (Suppl. 5), S64.

Volver A., Viru A. and Viru M. (2000) Improvement of motor abilities in pubertal girls. *J Sports Med Phys Fitness*, **40**, 17–25.

Winters M.V., Blake C.G., Trost J.S. *et al.* (2004) Passive versus active stretching of hip flexor muscles in subjects with limited hip extension: a randomized clinical trial. *Phys Ther*, **84** (9), 800–807.

Witvrouw E., Mahieu N., Danneels L. and McNair P. (2004) Stretching and injury prevention. *Sports Med*, **34** (7), 443–449.

Ylinen J. (2008) *Stretching Therapy*, Churchill Livingstone Elsevier, Edinburgh, UK.

Young W.B. (2007) The use of static stretching in warm-up for training and competition. *Int J Sports Physiol Perform*, **2** (2), 212–216.

4.5 Sensorimotor Training

Urs Granacher,[1] Thomas Muehlbauer,[2] Wolfgang Taube,[3] Albert Gollhofer[4] and Markus Gruber[5], [1]Institute of Sport Science, Friedrich Schiller University, Jena, Germany, [2]Institute of Exercise and Health Sciences, University of Basel, Basel, Switzerland, [3]Unit of Sports Science, University of Fribourg, Fribourg, Switzerland, [4]Institute of Sport and Sport Science, University of Freiburg, Freiburg, Germany, [5]Department of Training and Movement Sciences, University of Potsdam, Potsdam, Germany

4.5.1 INTRODUCTION

Over the last two decades, a large number of studies have investigated the impact of training protocols containing sensorimotor tasks on various neuromuscular capacities (e.g. joint position sense, postural control, maximal and explosive force production, etc.) in healthy young, middle-aged, and elderly subjects (Granacher, Zahner and Gollhofer, 2008; Taube, Gruber and Gollhofer, 2008), as well as in sports injury (Hupperets, Verhagen and van Mechelen, 2009) and fall-prevention (Gillespie *et al.*, 2009). A thorough literature search indicates that different terms were synonymously used for these training protocols (Banaschewski *et al.*, 2001; Bernier and Perrin, 1998; Chong *et al.*, 2001; Freeman, Dean and Hanham, 1965; Gruber and Gollhofer, 2004; Heitkamp *et al.*, 2001a; Wulker and Rudert, 1999). In an early study by Freeman, Dean and Hanham (1965) the expression 'coordination exercise' was used to describe a balance task on a see-saw board for rehabilitation of functional ankle instability. In other studies, 'balance training' referred to balance exercises conducted largely in bipedal and monopedal stances on stable and unstable surfaces with the primary intention of improving balance (Bernier and Perrin, 1998; Heitkamp *et al.*, 2001a). Later on the term 'proprioceptive training' was applied, because it seemed to be more specific in terms of denoting the actual cause of injury-related coordination deficits (impairment of peripheral sensation originating from mechanoreceptors) (Chong *et al.*, 2001; Wulker and Rudert, 1999). However, 'proprioceptive training' is limited to the perception of afferent input and does not take into account adaptive processes that occur on the motor side (Ashton-Miller *et al.*, 2001), and so the term 'proprioceptive' might be misleading. Other authors have therefore used 'sensorimotor training' (Banaschewski *et al.*, 2001; Gruber and Gollhofer, 2004).

'Sensorimotor' explicitly denotes the potential system in which training-induced adaptive processes may occur. The sensorimotor system describes mechanisms involved in the perception of a sensory stimulus through the proprioceptive, visual, and vestibular systems. It also comprises the subsequent conversion of the stimulus to a neural signal and the transmission of the signal via afferent pathways to the central nervous system (CNS), the processing and integration of the signal by the various sites within the CNS, and the motor responses resulting in muscle activation and thus the production of the forces that are necessary to stand, walk, or run (Lephart, Riemann and Fu, 2000) (Figure 4.5.1). Therefore, sensorimotor training is not restricted to afferent sensory contributions, since it contains both the afferent and the efferent side as possible adaptive mechanisms. However, the wide-ranging character of this term leaves room for the inclusion of other exercise protocols such as resistance training. The term 'sensorimotor training' thus needs a distinct definition to distinguish it from other training protocols. Given these circumstances, sensorimotor training can be defined as a training regimen that primarily aims at an improved perception and integration of sensory signals on a spinal and supraspinal level, as well as an optimized conversion of the integrative processes in an adequate neuromuscular response/motor action. Note that in this particular context, training-induced motor-performance enhancement (e.g. improved force production) is considered to be a secondary outcome. The term 'sensorimotor training' seems to be superior to all others because it (1) denotes the potential training-induced adaptive mechanisms and (2) includes core stability exercises (Marshall and Murphy, 2005) as well as exercises for the lower (Gruber and Gollhofer, 2004) and the upper (Borsa *et al.*, 1994) extremities.

Given this wide range of application in research and prevention/rehabilitation, the purpose of this chapter is to describe and summarize the effects of sensorimotor training on variables of postural control and strength in healthy young, middle-aged, and elderly subjects, and to provide information for practitioners, therapists, and scientists on the equipment, methodological design, and volume of sensorimotor training.

Strength and Conditioning – Biological Principles and Practical Applications Marco Cardinale, Rob Newton, and Kazunori Nosaka.

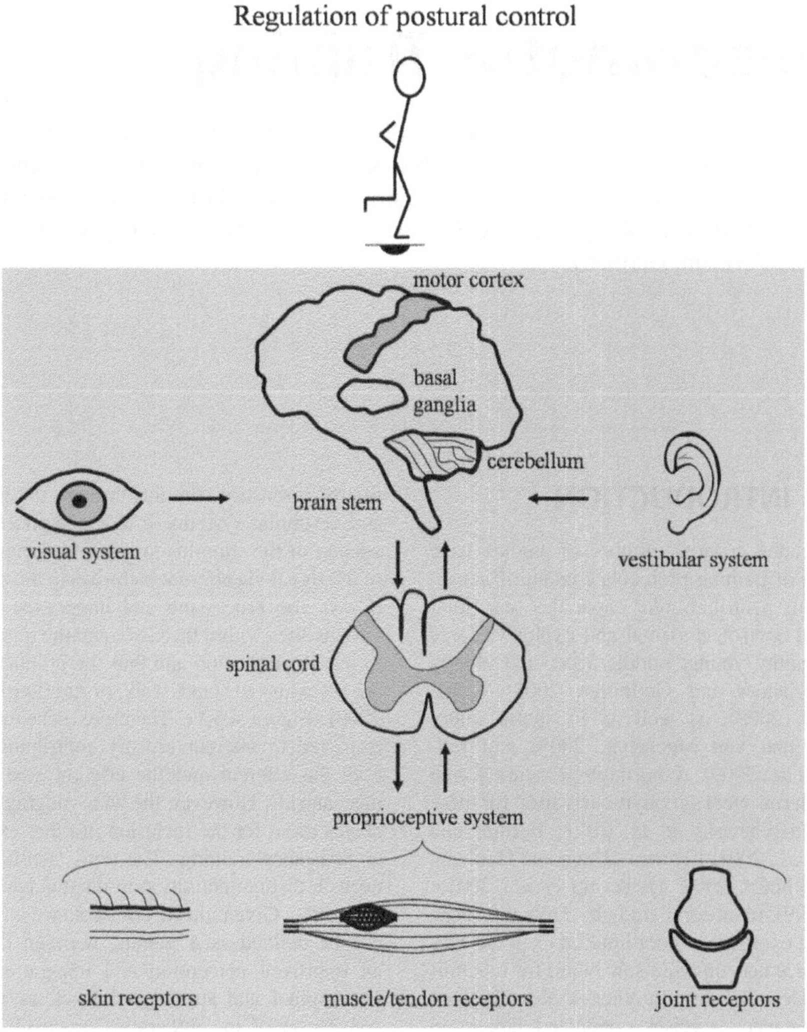

Figure 4.5.1 Simplified illustration of the regulation of postural control within the sensorimotor system. Neural connections/pathways between the different sites are illustrated by slim arrows

4.5.2 THE IMPORTANCE OF SENSORIMOTOR TRAINING TO THE PROMOTION OF POSTURAL CONTROL AND STRENGTH

Postural control and strength are two fundamental motor skills in the successful performance of sports-related activities and activities of daily living. Whereas postural control has been defined as the control of the body's position in space for the purpose of balance and orientation (Shumway-Cook and Woollacott, 2001), muscle strength is the maximal force that a muscle or muscle group can generate at a specified velocity (Knuttgen and Kraemer, 1987).

The ability to control posture and to produce force can be described as a dynamic process across the life span. These two capacities develop during childhood and adolescence, reach their peak in young adulthood, remain relatively stable in middle age, and decline in old age (Bosco and Komi, 1980; Era *et al.*, 2006; Hytonen *et al.*, 1993; Lephart, Riemann and Fu, 2000; Oberg, Karsznia and Oberg, 1993). Therefore, a U-shaped dependency between measures of static balance and age (Hytonen *et al.*, 1993), and an inverted U-shaped dependency between variables of strength and age (Larsson, Grimby and Karlsson, 1979), can be postulated. It has been recommended that during the early years of life, these two capacities should be promoted to avoid maturational deficits (Cumberworth *et al.*, 2007; Kanehisa *et al.*, 1995) which make children prone to sustaining falls and injuries (Ellsässer and Diepgen, 2002).

Since the occurrence of sports-related injuries is particularly high in mid-life (Steinbruck, 1999), insufficient postural control and strength – two major internal injury-related risk factors – should be specifically exercised. The prevalence of sustaining a fall is exceptional in old age due to deficits in postural control and muscle strength (Granacher, Zahner and Gollhofer, 2008). As a consequence, these two capacities should be exercised for fall-preventive purposes.

4.5.3 THE EFFECTS OF SENSORIMOTOR TRAINING ON POSTURAL CONTROL AND STRENGTH

Traditionally, sensorimotor training has been used to rehabilitate ankle and knee injuries. In fact, prospective studies have shown preventive effects in the incidence rate of ankle and knee-joint injuries (Caraffa *et al.*, 1996; Emery *et al.*, 2005; Myklebust *et al.*, 2003; Wedderkopp *et al.*, 1999). It has also been found that sensorimotor training is effective in reducing fall incidence rate in old age (Madureira *et al.*, 2007; Sihvonen *et al.*, 2004; Steadman, Donaldson and Kalra, 2003). Recent studies have extended the base of knowledge on the preventive effects of sensorimotor training by gaining new insights into the impact of sensorimotor training on measures of postural control, strength, and jumping performance in healthy young, middle-aged, and old subjects (Beck *et al.*, 2007; Bruhn, Kullmann and Gollhofer, 2004, 2006; Granacher, Gollhofer and Strass, 2006; Granacher, Bergmann and Gollhofer, 2007; Granacher, Gollhofer and Kriemler, 2010; Gruber, Bruhn and Gollhofer, 2006; Gruber *et al.*, 2007a, 2007b; Heitkamp *et al.*, 2001a; Schubert *et al.*, 2008; Taube *et al.*, 2007b).

4.5.3.1 Sensorimotor training in healthy children and adolescents

Hadders-Algra, Brogren and Forssberg (1996) scrutinized the impact of two months of sensorimotor training on postural responses during sitting on a moveable platform in healthy infants aged 5–10 months. The authors reported that training facilitated the most complete direction-specific postural response pattern and that it accelerated the development of response modulation. These results indicate that sensorimotor training has the potential to promote motor development in infants.

In a recent study, we obtained preliminary data on the effects of sensorimotor training and detraining on measures of postural control and strength in prepubertal children (age 7 ± 1 years; Tanner stages I and II). Four weeks of sensorimotor training resulted in tendencies in terms of small-to-medium interaction effects but no statistically significant improvements in postural sway, force production of the plantar flexors, or jumping height (countermovement jump, CMJ). Immaturity of the postural control system and/or deficits in attentional focus during practice of balance exercises could account for the observed findings.

In another study (Granacher *et al.* 2010), the effects of sensorimotor training on postural sway, leg extensor strength, and jumping height were investigated in a cohort of adolescent high-school students (age 18 ± 1 years). Four weeks of sensorimotor training induced significant decreases in postural sway, increases in jumping height (CMJ and SJ (squat jump)), and rate of force development of the leg extensors.

In addition, Taube *et al.* (2007b) observed a modified reflex activation during stance perturbation as well as a significant increase in jumping height in SJ and CMJ following six weeks of sensorimotor training in adolescent elite athletes (age 15 ± 1 years). These results indicate that sensorimotor training is not only beneficial for prevention and rehabilitation but also for the improvement of athletic performance.

4.5.3.2 Sensorimotor training in healthy adults

Heitkamp *et al.* (2001a, 2001b) investigated the impact of six weeks of sensorimotor training in healthy active adults (age 32 ± 6 years) on static postural control and strength of the knee extensors and flexors. They observed that sensorimotor training significantly improved one-legged stance balance and isokinetic torque of the knee extensors and flexors.

Following four weeks of sensorimotor training, Gruber and Gollhofer (2004) found a significant increase in RFD but not in maximum voluntary contraction of the leg extensors in healthy active adults (age 28 ± 7 years). The gain in RFD was accompanied by an increased EMG of the M. vastus medialis. Kean, Behm and Young (2006) reported a significantly enhanced vertical jump height in CMJ after six weeks of sensorimotor training in young, recreationally active women (age 24 ± 4 years). Yaggie and Campbell (2006) were able to show that four weeks of combined sensorimotor and coordination training significantly decreased postural sway and time to complete a shuttle-run course consisting of sprinting, backpedaling, sidestepping, starting, and stopping movements. The authors concluded that sensorimotor training improved performance of selected sport-related activities and postural control measures in young, recreationally active, healthy adults (age 23 ± 2 years).

4.5.3.3 Sensorimotor training in healthy seniors

Steadman, Donaldson and Kalra (2003) reported that six weeks of sensorimotor training improved performance in clinical balance and mobility tests in seniors aged 83 ± 6 years. Hu and Woollacott (1994) investigated the impact of two weeks of sensorimotor training on the temporal and spatial organization of postural responses in subjects aged 65–90 years. The authors observed a training-induced shortened onset latency of postural muscles. In a more recent study, Granacher, Gollhofer and

Strass (2006) examined the effects of 13 weeks of sensorimotor or heavy-resistance strength training on the ability to compensate for gait perturbations in elderly men. Sensorimotor training resulted in a decrease in onset latency and an enhanced reflex activity in the prime mover, compensating for the decelerating perturbation impulse. No statistically significant changes were observed in the heavy-resistance strength group or the control group. In another study, Granacher *et al.* (2007) were able to show that 13 weeks of sensorimotor training improved maximal- and explosive-force production capacity of the leg extensors in a cohort of 40 healthy, elderly males between the ages of 60 and 80 years.

Silsupadol *et al.* (2009) investigated the effects of single-task training (sensorimotor exercises only) versus dual training (sensorimotor exercises while concurrently performing cognitive or motor-interference tasks) on gait speed under single (walking only) and dual (walking while concurrently performing an arithmetic task) task conditions in elderly subjects (age 75 ± 6 years). After the four-week intervention programme, participants in all training groups significantly improved performance on single-task gait speed. However, dual-task training was superior to single-task training in improving walking under dual-task conditions. In fact, only participants who received dual-task training walked significantly faster after the training when simultaneously performing a cognitive task. This finding suggests that older adults are able to improve their walking performance under dual-task conditions only after specific types of training and that training balance under single-task conditions may not generalize to balance control in dual-task contexts (Silsupadol *et al.*, 2009).

This is in accordance with another recent approach in which more specifically designed sensorimotor training programmes, so-called 'perturbation-based training regimes', have begun to receive attention (Maki and McIlroy, 2005). This approach is based on the assumption that neural control of volitional limb movements differs in some fundamental ways to reactions that are evoked by postural perturbation (Maki and McIlroy, 1997). Thus, Maki *et al.* (2008) argue that the most effective training programmes involve the use of perturbations. Recently, Sakai *et al.* (2008) were able to show that a short-term perturbation-based sensorimotor training programme on a treadmill significantly decreased postural sway in a cohort of 45 community-dwelling elderly subjects (age 71 ± 4 years). Therefore, specificity of training should receive attention not only in competitive sports but also in fall-preventive approaches for seniors.

4.5.4 ADAPTIVE PROCESSES FOLLOWING SENSORIMOTOR TRAINING

The underlying mechanisms responsible for improvements in postural control and strength following sensorimotor training have specifically been investigated in young populations (Beck *et al.*, 2007; Gruber and Gollhofer, 2004; Schubert *et al.*, 2008; Taube *et al.*, 2007a, 2007b). For a recent review see Taube, Gruber and Gollhofer (2008).

In one of the earlier publications, Gollhofer (2003) assumed that adaptive processes following sensorimotor training mainly take place at a spinal level due to the high intermuscular activation frequencies observed during stabilization tasks on unstable platforms. Recently, Taube *et al.* (2007a) investigated cortical and spinal adaptations in young subjects (mean age 25 years) following sensorimotor training by means of H-reflex stimulation, transcranial magnetic stimulation (TMS), and conditioning of the H-reflex by TMS. After four weeks of sensorimotor training, the authors observed an improved postural stability accompanied by a decrease in motor evoked potentials during stance perturbation on a treadmill. At the same time, H-reflexes were decreased despite an unchanged background EMG during a balance task. This could imply that sensorimotor training induced changes in the regulation of human erect posture in terms of a shift from cortical to subcortical areas. Thus, supraspinal rather than spinal mechanisms seem to be responsible for the training-induced postural improvement (Taube *et al.*, 2007a). In fact, brain-imaging as well as electrophysiological studies have provided evidence for primary motor cortex plasticity during the early phase of motor-skill acquisition (Lotze *et al.*, 2003; Muellbacher *et al.*, 2001; Puttemans, Wenderoth and Swinnen, 2005). Furthermore, it was reported that skilled compared to non-skilled subjects show reduced neural activation in primary and secondary motor areas when performing the same motor action (Haslinger *et al.*, 2004; Jancke *et al.*, 2000). This finding has been interpreted as reflecting diminished neural effort required for a particular motor performance following intensive motor training (Schubert *et al.*, 2008).

Further research is necessary to clarify whether the training-induced adaptive processes responsible for improvements in postural control and strength observed in young adults can be transferred to different age (e.g. children, adolescents, seniors) or even patient (e.g. Parkinsonians, patients with chronic ankle instability) groups.

4.5.5 CHARACTERISTICS OF SENSORIMOTOR TRAINING

Unlike resistance training, there are no scientific guidelines concerning the optimal duration and intensity of sensorimotor training. Thus, there is large variation in these parameters between studies. In order to provide practitioners with guidelines regarding the planning of their exercise programmes, we have reviewed the training protocols of studies that were successful in improving balance performance in healthy young, middle-aged, and elderly subjects.

4.5.5.1 Activities

In general, conditions inducing training-related changes in postural control include the type of training activity and the

Table 4.5.1 List of exercises, materials, and additional conditions used during sensorimotor training

Training activities	
Exercises	
Standing	Walking
Two/one-legged stance	Walk forward/backward
Semi-tandem/tandem stance	Walk sideways
Skipping	Tandem walk
Perturbed standing	Walk with stop and go
Complete turns	Walk over/around obstacles
Additional conditions	Materials and tools
For/back/sideward sways	Soft mat
Eyes open/closed	Ankle disk
Dominant/non-dominant leg	Balance/wobble/tilt board
Left/right head turns	Air cushion
Name/spell words	Sissle
Count for/backward	Spinning top
Throw/catch/kick a ball	Rope
Bounce/juggle a ball	Obstacle, ball

treatment dose (Kraemer and Ratamess, 2004). With respect to the type of activity, static exercises (e.g. sitting, standing) are distinguished from dynamic ones (e.g. walking). Both can be performed in terms of steady-state, reactive, and/or proactive balance (Shumway-Cook and Woollacott, 2007). The applied-balance exercises may be modified by incorporating additional conditions or by implementing different materials/tools (Table 4.5.1); for example, maintenance of balance during two-legged stance can be modified by bouncing or juggling a ball, while balance during walking can be modified by crossing/circulating obstacles.

4.5.5.2 Load dimensions

Figure 4.5.2 illustrates an examples of a standardized sensorimotor training protocol that has been used in a number of recent studies (Beck *et al.*, 2007; Bruhn, Kullmann and Gollhofer, 2004, 2006; Granacher, Gollhofer and Strass, 2006; Gruber and Gollhofer, 2004; Gruber *et al.*, 2007a, 2007b; Schubert *et al.*, 2008; Taube *et al.*, 2007a, 2007b). Typically, participants performed one-legged exercises on unstable support surfaces

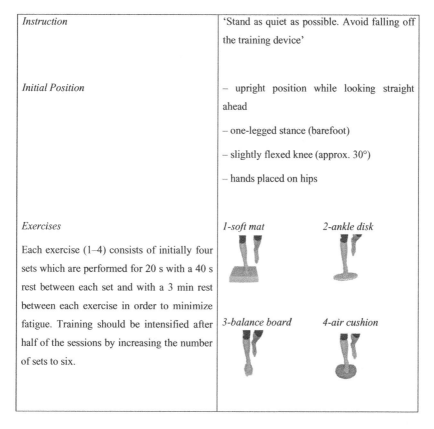

Instruction	'Stand as quiet as possible. Avoid falling off the training device'
Initial Position	– upright position while looking straight ahead – one-legged stance (barefoot) – slightly flexed knee (approx. 30°) – hands placed on hips
Exercises Each exercise (1–4) consists of initially four sets which are performed for 20 s with a 40 s rest between each set and with a 3 min rest between each exercise in order to minimize fatigue. Training should be intensified after half of the sessions by increasing the number of sets to six.	*1-soft mat*  *2-ankle disk*  *3-balance board*  *4-air cushion* 

Figure 4.5.2 Example of exercise conditions used during sensorimotor training. Modified from Gruber *et al.*, (2007). Differential effects of ballistic versus sensorimotor training on rate of force development and neural activation in humans. J Strength Cond Res. 21(1):274–82 with permission from the National Strength and Conditioning Association, Colorado Springs, CO, USA

Table 4.5.2 Characteristics of treatment load/training volume for sensorimotor training used in a number of recent studies

Characteristics of training	General specification
Warm-up	10 min
Cool-down	10 min
Duration of training	4–6 wk
Frequency of training	2–3/wk
Duration of session	25–45 min
Duration of exercise	20–40 s
Number of sets	3–5 (early training phase) 6–8 (late training phase)
Duration of rests	20–40 s between sets 2–5 min between exercises
Progression	Challenging (e.g. (1) two-legged stance, (2) semi-tandem stance, (3) tandem stance, (4) one-legged stance)

(e.g. soft mat, ankle disc, balance board, air cushion, etc.) and were instructed to stand as quietly as possible and avoid falling off the training device. All exercises were performed in an upright position while looking straight ahead. The participants stood barefoot, the knee slightly bent (approximately 30°), and with hands placed on hips.

Table 4.5.2 includes characteristics of treatment load typically applied in studies investigating sensorimotor training effects. Significant improvements in variables of postural control were shown after four to six weeks of training in healthy young and middle-aged subjects (Eils and Rosenbaum, 2001; Gruber, Bruhn and Gollhofer, 2006; Verhagen *et al.*, 2005). Training frequency amounted to two to three times a week on average (Bruhn, Kullmann and Gollhofer, 2004, 2006; Heitkamp *et al.*, 2001b; Taube *et al.*, 2007b). The total duration of each training session was approximately 60 minutes, including a 10-minute warm-up and 10-minute cool-down. During session, each exercise consisted of four sets performed for 20 seconds each, with a 40 second rest between sets and a 3 minute break in between exercises. In terms of exercise duration, it is of interest to note that most studies applied a time interval of 20–40 seconds (Bruhn, Kullmann and Gollhofer, 2004; Granacher, Gollhofer and Strass, 2006; Gruber and Gollhofer, 2004; Gruber *et al.*, 2007a).

To the best of our knowledge, there is currently no scientific evidence of an optimal exercise duration in sensorimotor training. Therefore, we analysed muscle activity of M. tibialis anterior and M. peroneus longus in relation to joint movement (flexion–extension and inversion–eversion movements at the ankle joint) while subjects (six men and six women, 24 ± 2 years) performed sensorimotor training on different training devices (wobbling board, spinning top, soft mat, and cushion) for 120 seconds each. It was found that irrespective of the training device, the ratio of the joint movements to muscle activities was at its minimum during the time interval 20–40 seconds. This was mainly due to the drastic decrease in ankle-joint movement (approximately 20%) from the first analysed time interval (0–20 seconds) to the second interval (20–40 seconds). During the interval of 40–60 seconds, ankle-joint movement increased again, whereas muscle activity declined slightly from the beginning onwards until the interval of 60–80 seconds. These results indicate that specific time-related processes might take place during the execution of a sensorimotor training task such as learning or adapting postural strategies in response to the unstable environment in the initial phase (0–20 seconds) and fatigue during later phases (starting from 40–60 seconds onwards). At the very beginning of sensorimotor training, durations of 20–40 seconds might therefore be most suitable for taking advantage of the neuromuscular learning process and avoiding fatigue.

Furthermore, it can be assumed that the neuromuscular system adapts specifically and progressively to the initial training volume and the applied training intensity. To appropriately challenge the sensorimotor system in a longer-lasting training intervention and to induce more profound/intense neuromuscular adaptations, it seems advisable to alter the training volume and intensity. This was actually done in some studies by increasing the duration of the exercises from 20 to 40 seconds or the number of sets from four to six after subjects had attended half of the training sessions (Beck *et al.*, 2007; Gruber *et al.*, 2007a, 2007b; Taube *et al.*, 2007a).

Regarding progression of training, almost all reviewed studies reported that balance exercises should be challenging to the participants. For example, it is possible to create balance tasks that gradually reduce the base of support by starting at two-legged stance, progressing to semi-tandem and tandem stance, and finishing at one-legged stance. In accordance with the American College of Sports Medicine (Chodzko-Zajko *et al.*, 2009), additional options for a progressive increase in intensity of sensorimotor training include:

1. reducing stability of the ground (e.g. standing on foam surfaces)
2. dynamic movements that perturb the centre of gravity (e.g. tandem walk, circle turns)
3. stressing postural muscle groups (e.g. heel stands, toe stands)
4. reducing sensory input (e.g. standing with eyes closed)
5. applying unexpected situations (e.g. light nudge)
6. adding tools (e.g. balls, obstacles)
7. varying cognitive demands (e.g. counting numbers, naming animals).

Scientific evidence regarding a potential progression model in sensorimotor training has been offered by Nashner (1992). He developed the Sensory Organization Test, which is targeted at the progressive increase in manipulation of visual and kinaesthetic stimuli during different conditions. This protocol was applied by Cohen *et al.* (1996), who investigated four different age groups: young (18–44 years), middle-aged (45–69 years), old (70–79 years), and elderly (80–89 years) adults. They found significant effects of test condition and age, as well as a significant test condition–age interaction, where young adults performed better than old or elderly adults. From their findings, the authors suggested that younger and older adults use different strategies to maintain their balance, so that training paradigms to improve postural control should vary with age group. Lajoie *et al.* (1996) reported similar results when comparing the attentional requirements for maintaining postural control between young (22–34 years) and elderly (66–79 years) people with increasing postural demand (from seated position, through broad/narrow-support upright standing position, to walking). Results showed a significant group effect and task effect, and a significant group–task interaction. For both young and older adults, reaction times for giving a verbal response to an unpredictable auditory stimulus were slower in the standing and walking tasks than in the sitting task, suggesting that attentional demands increased with increasing balance requirements.

On the other hand, the effects of increased attentional demand with respect to balancing were investigated using a dual-task paradigm: the execution of a concurrent cognitive and/or motor task while performing a static or dynamic postural control task (Fraizer and Mitra, 2008). When combining upright standing with an additional cognitive task (e.g. digit reversal/classification, counting backward), Pellecchia (2003) was able to show that in 20 healthy adults (10 men and 10 women, aged 18–30 years) the attentional demand of the concurrent task had an impact on postural sway, with the most difficult cognitive task (counting backward) having the greatest effect. Furthermore, in elderly people (age 70 ± 3 years) an increase in centre-of-pressure (COP) displacements has been observed when performing more demanding cognitive tasks (Huxhold *et al.*, 2006). When combining upright one-legged standing with an additional motor task (rhythmically throwing and catching a ball), Wilke, Froböse and Schulz (2003) reported a significant increase in activity of M. tibialis anterior, M. gastrocnemius medialis and lateralis, M. semimembranosus and semitendinosus, M. vastus medialis and lateralis, and M. tensor fasciae latae compared to a control condition without the ball in 20 healthy young adults (aged 21–33 years). In a recent study, preliminary data from our own laboratory were obtained on the combinatory effects of a cognitive and a motor task on postural control; 36 healthy adults (18 young, age 22 ± 3 years; 18 elderly, age 74 ± 6 years) conducted a two-legged stance for 30 seconds on a balance platform and a 10-minute walk using an instrumented walkway. In the elderly but not in the young subjects, COP displacements increased with increasing task complexity (i.e. from single, through dual, to triple tasking). In both age groups, stride-to-stride variability during triple-task conditions was significantly larger than during single-task conditions. These results agree with another study investigating stride-to-stride variability under triple-task conditions (Laessoe *et al.*, 2008).

Furthermore, a number of studies have examined the effects of different balance exercises or exercise devices on postural control. For example, Marshall and Murphy (2005) found significantly increased activation levels of the M. rectus abdominis, M. transversus abdominis, and M. internal obliques abdominis when performing different core-stability exercises on a Swiss ball (or gym ball) compared to a stable surface in eight healthy young subjects. In another study, Kavcic, Grenier and McGill (2004) examined spine stability by means of muscle activation patterns measured when performing eight stabilization exercises in 10 male subjects (age 21 ± 3 years). The aim of the study was to find out the most appropriate exercise for specific patients and specific objectives. Exercises were ranked according to the magnitude of stability versus compression, as well as which exercises focussed on training the abdominals versus the extensors. In addition, Anders, Wenzel and Scholle (2008) investigated the effects of an oscillating pole on levels of trunk-muscle activation in 30 healthy subjects (15 men, age 25 ± 6 years; 15 women, age 23 ± 2 years). Independent of subjects' sex, the authors reported that the electromyographic amplitude levels of M. rectus abdominis and M. external obliques abdominis were proportionally elevated while oscillation frequency was increased from 3.0 through 3.5 to 4.5 Hz. Irrespective of the oscillation plane (horizontal or vertical), all abdominal muscles exhibited continuous activation patterns. However, back muscles changed from a continuous activation in horizontal plane to phasic pattern in vertical plane.

4.5.5.3 Supervision

Although the feasibility of sensorimotor training has been demonstrated for several age groups, care is needed when applying such a programme, particularly for the elderly. The performance of balance exercises can pose a threat to postural stability and thus a potential risk of falling; therefore, any training programme needs a detailed schedule and has to be professionally supervised. Changes in the ability to maintain balance have to be documented in order to make a reasonable decision on progress regarding level of difficulty/complexity, number of sets, and exercise duration. Participants should also be instructed on how to perform balance exercises (with feedback when executing the task and with manual support during early phases of training).

4.5.5.4 Efficiency

Sensorimotor training meets several criteria of efficiency. For example, many exercises can be conducted in pairs or by using circuit-training (Eils and Rosenbaum, 2001; Olsen *et al.*, 2005; Verhagen *et al.*, 2004, 2005), and balance skills, strength, and coordination abilities can thus be exercised efficiently in a short period of time. Moreover, an instructor-to-participant ratio of approximately 1 : 10 is reported to be sufficient (Granacher, Bergmann and Gollhofer, 2007). Only a small amount of equipment is needed during circuit-training, and most devices are low-cost and widely available (Eils and Rosenbaum, 2001). Furthermore, the compliance rates of approximately 90%

reported in some studies are fairly high (Eils and Rosenbaum, 2001; Granacher, 2006; Granacher, Gruber and Gollhofer, 2009).

4.5.6 CONCLUSION

Over the last two decades, sensorimotor training has been successfully applied as a tool for (1) the prevention of lower-limb sports injuries, (2) fall-prevention in seniors, (3) rehabilitation of ankle and knee-joint injuries, and (4) enhancement of athletic performance. Various studies have proved the effectiveness of sensorimotor training in these settings. Recently, scientific evidence of the underlying mechanisms responsible for the training-induced adaptive processes has been found. However, guidelines concerning the optimal duration, frequency, and intensity of sensorimotor training are rare and lack (satisfactory) scientific validation.

References

Anders C., Wenzel B. and Scholle H.C. (2008) Activation characteristics of trunk muscles during cyclic upper-body perturbations caused by an oscillating pole. *Arch Phys Med Rehabil*, **89**, 1314–1322.

Ashton-Miller J.A., Wojtys E.M., Huston L.J. and Fry-Welch D. (2001) Can proprioception really be improved by exercises? *Knee Surg Sports Traumatol Arthrosc*, **9**, 128–136.

Banaschewski T., Besmens F., Zieger H. and Rothenberger A. (2001) Evaluation of sensorimotor training in children with ADHD. *Percept Mot Skills*, **92**, 137–149.

Beck S., Taube W., Gruber M. *et al.* (2007) Task-specific changes in motor evoked potentials of lower limb muscles after different training interventions. *Brain Res*, **1179**, 51–60.

Bernier J.N. and Perrin D.H. (1998) Effect of coordination training on proprioception of the functionally unstable ankle. *J Orthop Sports Phys Ther*, **27**, 264–275.

Borsa P.A., Lephart S.M., Kocher M.S. and Lephart S.P. (1994) Functional assessment and rehabilitation of shoulder proprioception for glenohumeral instability. *J Sport Rehabil*, **3**, 84–104.

Bosco C. and Komi P.V. (1980) Influence of aging on the mechanical behavior of leg extensor muscles. *Eur J Appl Physiol*, **45**, 209–219.

Bruhn S., Kullmann N. and Gollhofer A. (2004) The effects of a sensorimotor training and a strength training on postural stabilisation, maximum isometric contraction and jump performance. *Int J Sports Med*, **25**, 56–60.

Bruhn S., Kullmann N. and Gollhofer A. (2006) Combinatory effects of high-intensity-strength training and sensorimotor training on muscle strength. *Int J Sports Med*, **27**, 401–406.

Caraffa A., Cerulli G., Projetti M. *et al.* (1996) Prevention of anterior cruciate ligament injuries in soccer. A prospective controlled study of proprioceptive training. *Knee Surg Sports Traumatol Arthrosc*, **4**, 19–21.

Chodzko-Zajko W.J., Proctor D.N., Fiatarone Singh M.A. *et al.* (2009) American College of Sports Medicine position stand. Exercise and physical activity for older adults. *Med Sci Sports Exerc*, **41**, 1510–1530.

Chong R.K., Ambrose A., Carzoli J. *et al.* (2001) Source of improvement in balance control after a training program for ankle proprioception. *Percept Mot Skills*, **92**, 265–272.

Cohen H., Heaton L.G., Congdon S.L. and Jenkins H.A. (1996) Changes in sensory organization test scores with age. *Age Ageing*, **25**, 39–44.

Cumberworth V.L., Patel N.N., Rogers W. and Kenyon G.S. (2007) The maturation of balance in children. *J Laryngol Otol*, **121**, 449–454.

Eils E. and Rosenbaum D. (2001) A multi-station proprioceptive exercise program in patients with ankle instability. *Med Sci Sports Exerc*, **33**, 1991–1998.

Ellsässer G. and Diepgen T.L. (2002) Epidemiologische Analyse von Sturzunfällen im Kindesalter (<15 Jahre): Konsequenzen für die Prävention. *Bundesgesundheitsblatt Gesundheitsforschung Gesundheitsschutz*, **45**, 267–276.

Emery C.A., Cassidy J.D., Klassen T.P. *et al.* (2005) Effectiveness of a home-based balance-training program in reducing sports-related injuries among healthy adolescents: a cluster randomized controlled trial. *Can Med Ass J*, **172**, 749–754.

Era P., Sainio P., Koskinen S. *et al.* (2006) Postural balance in a random sample of 7979 subjects aged 30 years and over. *Gerontology*, **52**, 204–213.

Fraizer E.V. and Mitra S. (2008) Methodological and interpretive issues in posture-cognition dual-tasking in upright stance. *Gait Posture*, **27**, 271–279.

Freeman M.A.R., Dean M.R.E. and Hanham I.W.F. (1965) The etiology and prevention of functional instability of the foot. *J Bone Joint Surg*, **47** (B), 678–685.

Gillespie L.D., Robertson M.C., Gillespie W.J. *et al.* (2009) Interventions for preventing falls in older people living in the community. *Cochrane Database Syst Rev*, CD007146.

Gollhofer A. (2003) Proprioceptive training: considerations for strength and power production, in *Strength and Power in Sport* (ed. P.V. Komi), Blackwell Publishing, Oxford, UK., pp. 331–342.

Granacher U. (2006) *Neuromuskuläre Leistungsfähigkeit Im Alter (>60 Jahre): Auswirkungen Von Kraft- Und Sensomotorischem Training*, Maurer Verlag, Geislingen/Steige.

Granacher U., Gollhofer A. and Strass D. (2006) Training induced adaptations in characteristics of postural reflexes in elderly men. *Gait Posture*, **24**, 459–466.

Granacher U., Bergmann S. and Gollhofer A. (2007) Allgemeine Richtlinien für den Einsatz von sensomotorischem Training im Schulsport. *Sportunterricht*, **9**, 259–265.

Granacher U., Gruber M., Strass D. and Gollhofer A. (2007) Die Auswirkungen von sensomotorischem Training im Alter auf die Maximal- und Explosivkraft. *Deut Z Sportmed*, **12**, 446–451.

Granacher U., Zahner L. and Gollhofer A. (2008) Strength, power, and postural control in seniors: considerations for functional adaptations and for fall prevention. *Eur J Sport Sci*, **8**, 325–340.

Granacher U., Gruber M. and Gollhofer A. (2009) The impact of sensorimotor training on postural control in elderly men. *Deut Z Sportmed*, **60** (12), 16–22.

Granacher U., Gollhofer A. and Kriemler S. (2010) Effects of balance training on postural sway, leg extensor strength and jumping height in adolescents. *Res Q Exerc Sport*, **81** (3) 245–251.

Gruber M. and Gollhofer A. (2004) Impact of sensorimotor training on the rate of force development and neural activation. *Eur J Appl Physiol*, **92**, 98–105.

Gruber M., Bruhn S. and Gollhofer A. (2006) Specific adaptations of neuromuscular control and knee joint stiffness following sensorimotor training. *Int J Sports Med*, **27**, 636–641.

Gruber M., Gruber S.B., Taube W. *et al.* (2007a) Differential effects of ballistic versus sensorimotor training on rate of force development and neural activation in humans. *J Strength Cond Res*, **21**, 274–282.

Gruber M., Taube W., Gollhofer A. *et al.* (2007b) Training-specific adaptations of H- and stretch reflexes in human soleus muscle. *J Mot Behav*, **39**, 68–78.

Hadders-Algra M., Brogren E. and Forssberg H. (1996) Training affects the development of postural adjustments in sitting infants. *J Physiol*, **493** (Pt 1), 289–298.

Haslinger B., Erhard P., Altenmuller E. *et al.* (2004) Reduced recruitment of motor association areas during bimanual coordination in concert pianists. *Hum Brain Mapp*, **22**, 206–215.

Heitkamp H.C., Horstmann T., Mayer F. *et al.* (2001a) Balance training in men and women: effect on knee extensors and flexors. *Isokin Exerc Sci*, **9**, 41–44.

Heitkamp H.C., Horstmann T., Mayer F. *et al.* (2001b) Gain in strength and muscular balance after balance training. *Int J Sports Med*, **22**, 285–290.

Hu M.H. and Woollacott M.H. (1994) Multisensory training of standing balance in older adults: II. Kinematic and electromyographic postural responses. *J Gerontol*, **49**, M62–M71.

Hupperets M.D., Verhagen E.A. and van Mechelen W. (2009) Effect of sensorimotor training on morphological, neurophysiological and functional characteristics of the ankle: a critical review. *Sports Med*, **39**, 591–605.

Huxhold O., Li S.C., Schmiedek F. and Lindenberger U. (2006) Dual-tasking postural control: aging and the effects of cognitive demand in conjunction with focus of attention. *Brain Res Bull*, **69**, 294–305.

Hytonen M., Pyykko I., Aalto H. and Starck J. (1993) Postural control and age. *Acta Otolaryngol*, **113**, 119–122.

Jancke L., Himmelbach M., Shah N.J. and Zilles K. (2000) The effect of switching between sequential and repetitive movements on cortical activation. *Neuroimage*, **12**, 528–537.

Kanehisa H., Yata H., Ikegawa S. and Fukunaga T. (1995) A cross-sectional study of the size and strength of the lower leg muscles during growth. *Eur J Appl Physiol*, **72**, 150–156.

Kavcic N., Grenier S. and McGill S.M. (2004) Quantifying tissue loads and spine stability while performing commonly prescribed low back stabilization exercises. *Spine*, **29**, 2319–2329.

Kean C.O., Behm D.G. and Young W.B. (2006) Fixed foot balance training increases rectus femoris activation during landing and jump height in recreationally active women. *J Sports Sci Med*, **5**, 138–148.

Knuttgen H.G. and Kraemer W.J. (1987) Terminology and measurement in exercise performance. *J Strength Cond Res*, **1**, 1–10.

Kraemer W.J. and Ratamess N.A. (2004) Fundamentals of resistance training: progression and exercise prescription. *Med Sci Sports Exerc*, **36**, 674–688.

Laessoe U., Hoeck H.C., Simonsen O. and Voigt M. (2008) Residual attentional capacity amongst young and elderly during dual and triple task walking. *Hum Mov Sci*, **27**, 496–512.

Lajoie Y., Teasdale N., Bard C. and Fleury M. (1996) Upright standing and gait: are there changes in attentional requirements related to normal aging? *Exp Aging Res*, **22**, 185–198.

Larsson L., Grimby G. and Karlsson J. (1979) Muscle strength and speed of movement in relation to age and muscle morphology. *J Appl Physiol*, **46**, 451–456.

Lephart S.M., Riemann B.L. and Fu F.H. (2000) Introduction to the sensorimotor system, in *Proprioception and Neuromuscular Control in Joint Stability* (eds S.M. Lephart and F.H. Fu), Human Kinetics, Champaign, IL., pp. 17–24.

Lotze M., Braun C., Birbaumer N. *et al.* (2003) Motor learning elicited by voluntary drive. *Brain*, **126**, 866–872.

Madureira M.M., Takayama L., Gallinaro A.L. *et al.* (2007) Balance training program is highly effective in improving functional status and reducing the risk of falls in elderly women with osteoporosis: a randomized controlled trial. *Osteoporos Int*, **18**, 419–425.

Maki B.E. and McIlroy W.E. (1997) The role of limb movements in maintaining upright stance: the 'change-in-support' strategy. *Phys Ther*, **77**, 488–507.

Maki B.E. and McIlroy W.E. (2005) Change-in-support balance reactions in older persons: an emerging research area of clinical importance. *Neurol Clin*, **23**, 751–vii.

Maki B.E., Cheng K.C., Mansfield A. *et al.* (2008) Preventing falls in older adults: new interventions to promote more effective change-in-support balance reactions. *J Electromyogr Kinesiol*, **18**, 243–254.

Marshall P.W. and Murphy B.A. (2005) Core stability exercises on and off a Swiss ball. *Arch Phys Med Rehabil*, **86**, 242–249.

Muellbacher W., Ziemann U., Boroojerdi B. *et al.* (2001) Role of the human motor cortex in rapid motor learning. *Exp Brain Res*, **136**, 431–438.

Myklebust G., Engebretsen L., Braekken I.H. *et al.* (2003) Prevention of anterior cruciate ligament injuries in female team handball players: a prospective intervention study over three seasons. *Clin J Sport Med*, **13**, 71–78.

Nashner L.M. (1992) *EquiTest System Operator's Manual, Version 4.04*, Clackamas OR NeuroCom International, Inc.

Oberg T., Karsznia A. and Oberg K. (1993) Basic gait parameters: reference data for normal subjects, 10–79 years of age. *J Rehabil Res Dev*, **30**, 210–223.

Olsen O.E., Myklebust G., Engebretsen L. *et al.* (2005) Exercises to prevent lower limb injuries in youth sports: cluster randomised controlled trial. *Br Med J*, **330**, 449.

Pellecchia G.L. (2003) Postural sway increases with attentional demands of concurrent cognitive task. *Gait Posture*, **18**, 29–34.

Puttemans V., Wenderoth N. and Swinnen S.P. (2005) Changes in brain activation during the acquisition of a multifrequency bimanual coordination task: from the cognitive stage to advanced levels of automaticity. *J Neurosci*, **25**, 4270–4278.

Sakai M., Shiba Y., Sato H. and Takahira N. (2008) Motor adaptations during slip-perturbed gait in older adults. *J Phys Ther Sci*, **20**, 109–115.

Schubert M., Beck S., Taube W. *et al.* (2008) Balance training and ballistic strength training are associated with task-specific corticospinal adaptations. *Eur J Neurosci*, **27**, 2007–2018.

Shumway-Cook A. and Woollacott M. (2001) *Motor Control: Theory and Practical Applications*, Lippincott Williams & Wilkins, Philadelphia, PA.

Shumway-Cook A. and Woollacott M.H. (2007) *Motor Control: Translating Research into Clinical Practice*, 3rd edn, Lippincott Williams & Wilkins, Philadelphia, PA.

Sihvonen S., Sipila S., Taskinen S. and Era P. (2004) Fall incidence in frail older women after individualized visual feedback-based balance training. *Gerontology*, **50**, 411–416.

Silsupadol P., Shumway-Cook A., Lugade V. *et al.* (2009) Effects of single-task versus dual-task training on balance performance in older adults: a double-blind, randomized controlled trial. *Arch Phys Med Rehabil*, **90**, 381–387.

Steadman J., Donaldson N. and Kalra L. (2003) A randomized controlled trial of an enhanced balance training program to improve mobility and reduce falls in elderly patients. *J Am Geriatr Soc*, **51**, 847–852.

Steinbruck K. (1999) Epidemiology of sports injuries – 25-year-analysis of sports orthopedic-traumatologic ambulatory care. *Sportverletzung Sportschaden*, **13**, 38–52.

Taube W., Gruber M., Beck S. *et al.* (2007a) Cortical and spinal adaptations induced by balance training: correlation between stance stability and corticospinal activation. *Acta Physiol*, **189**, 347–358.

Taube W., Kullmann N., Leukel C. *et al.* (2007b) Differential reflex adaptations following sensorimotor and strength training in young elite athletes. *Int J Sports Med*, **28**, 999–1005.

Taube W., Gruber M. and Gollhofer A. (2008) Spinal and supraspinal adaptations associated with balance training and their functional relevance. *Acta Physiol*, **193**, 101–116.

Verhagen E., van der B.A., Twisk J. *et al.* (2004) The effect of a proprioceptive balance board training program for the prevention of ankle sprains: a prospective controlled trial. *Am J Sports Med*, **32**, 1385–1393.

Verhagen, E., Bobbert, M., Inklaar, M. *et al.* (2005) The effect of a balance training programme on centre of pressure excursion in one-leg stance. *Clin Biomech*, **20**, 1094–1100.

Wedderkopp N., Kaltoft M., Lundgaard B. *et al.* (1999) Prevention of injuries in young female players in European team handball. A prospective intervention study. *Scand J Med Sci Sports*, **9**, 41–47.

Wilke C., Froböse I. and Schulz A. (2003) Einsatzmöglichkeiten des Posturomed im Rahmen des sensomotorischen Trainings für die untere Extremität. *Gesundheitssport Sporttherapie*, **19**, 9–14.

Wulker N. and Rudert M. (1999) Lateral ankle ligament rupture. When is surgical management indicated and when conservative therapy preferred? *Orthopade*, **28**, 476–482.

Yaggie J.A. and Campbell B.M. (2006) Effects of balance training on selected skills. *J Strength Cond Res*, **20**, 422–428.

Section 5
Strength and Conditioning special cases

5.1 Strength and Conditioning as a Rehabilitation Tool

Andreas Schlumberger, EDEN Reha, Rehabilitation Clinic, Donaustauf, Germany

5.1.1 INTRODUCTION

Strength and conditioning has become an important aspect of the complex rehabilitation strategies developed in the last decades. The increasing popularity of strength and conditioning in the rehabilitation setting may be due to three main factors:

1. The need for retraining because of a specific effect of injury on neuromuscular performance.
2. The need for detailed and differentiated approaches to retraining due to the well-known difficulties of recovering athletic performance to pre-injury levels.
3. The fact that the speed and safety with which an athlete returns to athletic activity may be highly dependent upon the quality and the characteristics of the rehabilitation programme.

Traditionally, the classical strength and conditioning methods, like training for muscle hypertrophy or training for increasing neural drive to the muscle, are employed to restore the normal functioning of the neuromuscular system. However, based on the current knowledge of the effects of injury and reduced muscle use on the neuromuscular system, it seems reasonable to suggest the use of additional training techniques to address specific neuromuscular deficits (training to improve local stabilizer activation). In this context, recent research indicates that the correct identification of the type of injury has a great influence on the choice of training exercise and method (Schlumberger *et al.*, 2006).

In muscle injuries, it is not clear whether the loss of neuromuscular performance is the cause or the effect of the injury. One possible way of integrating this basic problem into the planning of rehabilitation strategies is through the analysis of the risk factors of specific injuries. For example, it is accepted that misalignment of the lower limb may contribute to the occurrence of several knee injuries or disorders (such as anterior cruciate ligament (ACL) rupture or patellofemoral pain). Consequently, to allow optimal recovery and reduce the risk of re-injury such neuromuscular deficits should be targeted in rehabilitation, independent of whether they were the cause or the effect of the injury.

Another important aspect assisting in the planning of training in rehabilitation is the knowledge of injury mechanisms. The analysis of injury mechanisms helps to outline which critical movements have to be improved and which basic and complex aspects of muscle function are necessary to cope with injury-critical situations. This knowledge helps in delineating targeted exercise strategies. Such considerations seem to be especially important since the risk of injury is higher in athletes previously who have previously been affected by injuries (see for example Sherry and Best, 2004).

Finally, the integration of sports-specific requirements is an important part of planning an appropriate neuromuscular training programme. Since the final goal of rehabilitation is full participation in training and competition, the retraining of sports-specific movements should be considered an important part of targeted strength and conditioning (Ellenbecker, De Carlo and DeRosa, 2009). This seems to hold true from a physiological as well as a psychological perspective.

The main goal of this chapter is to examine actual trends and principles in strength and conditioning after athletic injury and to consider the injury-specific neuromuscular characteristics determining different injuries. In particular, a differentiated approach integrating recovery from local to complex muscle function is presented. In this regard, one big challenge for the therapeutic staff is to integrate neuromuscular recovery programmes with injury-specific and sports-specific performance requirements in rehabilitation. The main focus is on injuries of the lower limb, but the same principles seem to be applicable to other common athletic injuries such as low-back disorders (McGill, 2007) and shoulder injuries (Ellenbecker, De Carlo and DeRosa, 2009).

5.1.2 NEUROMUSCULAR EFFECTS OF INJURY AS A BASIS FOR REHABILITATION STRATEGIES

Injury can induce pronounced performance decreases in the neuromuscular system. According to scientific results and practical experience, full neuromuscular recovery is often difficult to reach (e.g. recovery of knee extensor strength after ACL reconstruction, Lorentzon *et al.*, 1989). For this reason, it is important to define which factors influence and limit the trainability of the neuromuscular system, and a detailed analysis of the trainability of the neuromuscular system after injury seems to be warranted. One main factor inducing performance decreases seems to be the injury- or surgery-related reduced use of muscles during immobilization or other types of unloading. Therefore, analysis of the isolated effects of immobilization on the neuromuscular system may give important insights into the necessities for retraining.

In this context, a study by Hortobagyi *et al.* (2000) made interesting observations regarding the necessary efforts for recovery of knee-extensor strength after knee immobilization in healthy subjects. Three weeks of knee immobilization caused a reduction in knee-extensor strength of 47% and in cross-sectional area (CSA) of slow- and fast-twitch fibres of 13% and 10% respectively. Interestingly, in a subgroup of subjects, two weeks of spontaneous recovery (normal daily activity without any training effort) reduced knee-extensor strength to an average strength deficit of only 11%. In addition, CSA of slow- and fast-twitch fibres was 5% smaller than baseline values after spontaneous recovery. These findings indicate that after knee immobilization in healthy subjects a significant amount of strength and muscle-size recovery is possible within a relatively short time frame (two weeks), without specific training efforts. Furthermore, the authors observed that with active retraining three to five weeks are necessary for full recovery of knee-extensor strength. The most effective training mode seemed to be resistance training, including eccentric actions (pure eccentric or mixed eccentric–concentric training).

Additional insights regarding the necessary effort for recovery of muscle size after immobilization were obtained by Hespel *et al.* (2001). They found that two weeks of knee immobilization in healthy subjects resulted in a 10% decrease in the size of the quadriceps muscle. While the main focus of their study was on the effects of creatine supplementation on regaining of muscle size, one group which performed active rehabilitation (strength training for the quadriceps muscle) without supplementation needed only three weeks to reach similar levels of muscle size compared to pre-immobilization level.

Two conclusions may be reached from such observations. First, in subjects without knee injury, regaining of muscle size may be relatively rapid, especially when assisted by active training. Consequently, muscle hypertrophy immediately after a phase of short-term muscle atrophy (two to three weeks) in healthy subjects seems to occur much faster than muscle hypertrophy in athletes trying to reach a new level of muscle size (generally accepted as at least six to seven weeks, Philipps, 2000). In other words, regaining of muscle size after short-term atrophy may be viewed as a re-normotrophy, not as a real hypertrophy. Thus, the time frame for reaching re-normotrophy seems to be shorter than that for hypertrophy. Second, since regaining of muscle size of the quadriceps after knee injury is very difficult (see above), other factors must be responsible for the well-known limited trainability of quadriceps muscle after knee injury/surgery.

Factors which could limit the trainability of muscles after injury include a loss of sensorimotor interaction (e.g. induced by a loss of mechanoreceptors, Hartigan, Axe and Snyder-Mackler, 2009) and a reduction in quadriceps neuromuscular activity as a consequence of knee effusion. Regarding the latter, saline injection into healthy knee joints has been demonstrated to induce a reduction in quadriceps and an increase in hamstring activity in the gait cycle, a reduction of muscle activity and force output of the quadriceps muscle in knee-extensor exercise, and a distinct reduction of neuromuscular activity of the vastus medialis muscle as measured by electromyography (Torry *et al.*, 2000). These alterations in muscle activity may be caused by the knee-joint effusion and the related joint swelling.

Such observations seem to indicate that joint-induced inhibitory effects frequently seen in post-surgery rehabilitation limit the trainability of important target muscles (such as the quadriceps in the knee patient). Consequently, regaining of quadriceps strength cannot be optimized only by direct training influences but probably also by strategies which reduce joint effusion-induced inhibition on important target muscles (e.g. physical therapy for reduction of effusion, giving enough time for joint healing) and which assist in regaining optimal sensorimotor interaction in terms of integrating afferent input and efferent output.

Further evidence for a limited trainability and/or recovery of muscles in some post-surgery phases comes from a recent study which investigated the effects of creatine supplementation in ACL rehabilitation from weeks 6 to 12 (Tyler *et al.*, 2004). In contrast to findings in strength training with non-injured athletes, the authors of this study observed that creatine supplementation had no additional effect on strength increases in this phase. One possible cause is the lack of sufficient overload for the knee-extensor muscles. These data indicate a limited trainability of the knee-extensor muscles in this phase, which may be caused by inhibitory effects that allow optimal structural healing.

The influence of structural healing on neuromuscular recovery has been shown in a study with ACL-reconstructed patients by Aune *et al.* (2001). These authors were able to demonstrate different strength adaptations in the knee extensors and flexors post-operatively depending on the graft choice. Patients with semitendinosus–gracilis graft exhibited better knee-extensor strength six months post-surgery than patients undergoing patella–tendon bone grafts (PTBGs). However, 12 and 24 months after surgery no differences were found. On the other hand, the semitendinosus–gracilis group showed significantly weaker knee-flexor strength 12 months post-operation than its PTBG counterpart.

In conclusion, there is evidence that trainability of important target muscles after injury is influenced by the time course of structural healing and the amount of inhibitory effect on these muscles. Further studies are needed to investigate and define the exact time course in which the reduction of inhibition is the main goal (including all targeted complementary training strategies which support this goal) and in which full trainability of muscle is given, thereby allowing the use of strength-training methods which are able to induce gains in muscle size.

5.1.3 STRENGTH AND CONDITIONING IN RETRAINING OF THE NEUROMUSCULAR SYSTEM

It is well known that muscle size is a strong predictor of maximum strength. In addition, maximum strength is a necessary prerequisite for power production in skeletal muscle. Due to the resulting interdependency of strength and power with muscle size, the normalization of muscle size is a predominant goal in rehabilitation.

In general, the training methods for increasing and regaining muscle strength, muscle power, and muscle size are similar in patients and in healthy athletes. For instance, resistance training which aims at muscle hypertrophy is performed using the same methods after injury as in a healthy athlete. However, after injury several injury- and healing-specific circumstances determine the possibilities of loading the neuromuscular system (see above). As a consequence, while the basic training methods are the same, the specific applications differ depending on the injury-specific conditions. It is the main goal of this section to consider and describe the essential factors that determine the loading in strength and conditioning in rehabilitation.

5.1.3.1 Targeted muscle overloading: criteria for exercise choice

Similarly to strength training in healthy (non-injured) athletes, the main goal of muscle strengthening in patients is to stimulate the neuromuscular system with an overload stimulus.

The main question when selecting the appropriate exercises for retraining of the neuromuscular system is whether the target muscles are activated in the way that is intended. Choosing exercises to optimize the muscle activity of injury-specific important muscles for retraining seems to be much more complicated than the same in healthy subjects. The main reason may be the above-mentioned changes in muscle activation due to post-surgery disuse, pain avoidance, and activity changes in synergistic muscles to unload healing structures.

A good example of the basic problem is the difficulty of retraining the quadriceps muscles after injuries of the lower extremity, and in particular after knee injuries (Lorentzon *et al.*, 1989). In non-injured athletes exercises like the squat or the leg press are widely accepted for use in improving strength and sports-related functional activities. However, after knee injury it has been shown that the probability of successful quadriceps strength recovery after ACL reconstruction is increased if the single-joint knee extension is used in addition to the so-called functional multi-joint exercises like the leg press or squat (Mikkelsen, Werner and Eriksson, 2000). Such observations indicate that isolated activation of target muscles, such as the quadriceps, may be a critical factor in retraining strategies after injury.

Muscle targeting of the muscle vastus medialis (especially its distal fibres, sometimes named M. vastus medialis obliquus (VMO)) seems to be an integral part of successful quadriceps strengthening. As a consequence, exercises which activate the VMO in terms of an adequate overload stimulus seem to be essential.

Since it has been hypothesized that adductor–quadriceps coactivation may fulfil this requirement, we investigated the effects of a non-weight-bearing exercise with simultaneous quadriceps and adductor activity in a sitting position and with an extended knee (see Figure 5.1.1) (Sander and Schlumberger, unpublished). In 16 post-surgery knee patients (on average, 14 weeks after ACL reconstruction and/or internal meniscal repair) an addition of low-intensity static adductor activity in the static quadriceps contraction task increased the electromyography (EMG) activity of the VMO in the injured leg by 32.1% and in the non-injured leg by 44.2%. These observations lead to the assumption that an additional adductor activity can increase the activity of the VMO. As a consequence, exercises with quadriceps–adductor coactivation seem to be a promising approach in retraining strategies for the quadriceps muscles in general and the VMO in particular.

The normalization of maximal- and explosive-force generation capacity of the knee extensors is a major goal in rehabilitation after knee injury. However, additional deficits in knee-extensor function occur that are directed towards altered

Figure 5.1.1 Isometric quadriceps contraction with low-intensity isometric adductor contraction

activity patterns of the knee extensors in terms of a 'hyperactivity' of this muscle group; for example, Williams *et al.* (2004) found in ACL-deficient patients the inability to put the quadriceps off when performing knee-flexion tasks immediately after a knee-extension task. When performing such a knee flexion, the knee extensors are usually silent. This altered muscle activity pattern may be interpreted as 'hyperactivity' of this muscle group in relation to its normal physiological function in the non-injured state. These observations demonstrate that altered muscle activity patterns are another feature of neuromuscular performance deficits after injury.

Recent findings in knee patients help in extending the views on adequate training strategies for the quadriceps and the lower-leg muscles. There is increasing evidence that the misalignment of segments (like the lower-leg chain) is a critical factor which has to be considered. Mascal, Landel and Powers (2003) used such an approach in a single-case-based study design. Based on movement analysis which showed misalignment of the leg axis in a step-down task (excessive femoral internal rotation and adduction in combination with knee valgus), they used a three-phase programme of neuromuscular retraining. This programme consisted of a continuum of exercises from single-joint hip external and abduction training in non-weight-bearing conditions to complex static and dynamic multi-joint exercise in weight-bearing conditions, with simultaneous accentuated activation of the hip external rotators and abductors. The results of this programme were significant gains in strength of hip external, internal, and adductor muscles, as well as in the knee and hip extensor muscles. In addition, alignment in the step-down tasks was normalized after the programme.

Besides the decreases in function of muscles directly related to an injured joint, the performance capacity of muscles located more proximal or distal to the site of injury may also be considered. We observed in high-level athletes significant side differences in maximum isokinetic strength in the plantar flexor (−8.3%) and invertor muscles (−24.6%) on average nine weeks after knee injury (ACL reconstruction and/or internal meniscal repair) (Schlumberger, 2002). Furthermore, Friel *et al.* (2006) found negative effects of injury on strength of proximal muscles. They observed in subjects with chronic ankle sprains significant reductions in strength of the hip abductor muscles. The reason for such strength reductions of distal and proximal muscles is probably the reduced use or the pain-avoiding compensatory muscle use in the whole extremity as a secondary consequence of the injury.

A typical strategy in intermediate and late rehabilitation phases is the integration of so-called functional exercises. In this context, the bilateral squat exercise is frequently used in rehabilitation of lower-leg injuries, especially when trying to regain muscle size and strength in the leg extensors after knee injury. Salem, Salinas and Harding (2003) found in ACL-reconstructed patients 30 weeks post-surgery differences in the relative force contribution of the hip and knee extensors in the bilateral squat between the injured and the non-injured side. In the non-involved limb, muscular effort between hip and knee extensors was equally distributed. However, in the involved limb patients used a strategy that increased the muscular effort of the hip extensors and reduced the effort of the knee extensors. Therefore, the bilateral squat may be not an appropriate exercise when trying to stimulate the knee extensors in the injured limb. Unilateral exercises may in most phases of rehabilitation be the more targeted approach.

In summary, it seems that exercise choice for patients has to be designed in a more differentiated manner than for healthy people. In addition, the main focus in patients has to be on the normalization of activity of injury-specific target muscles (stabilizers and prime movers) in terms of normalizing relative force contribution within several exercises or functional tasks. Normalization of local muscle activation as well as intermuscular coordination in complex tasks seems to be an important goal in retraining strategies of the neuromuscular system. The generalized use of exercises which are proven to be effective in non-injured subjects may sometimes be misleading. In addition, besides the training of muscles directly related to an injured area (such as the knee extensors or flexors for the knee joint), proximal and distal muscle training may also be addressed to regain optimal function in the whole limb.

Training intensity and training volume

Besides the exercise choice in terms of stimulating important target muscles, training intensity is the second most important factor in regulating a sufficient overload stimulus to retrain the neuromuscular system. In general, intensity in resistance training depends on experience in strength training (quality of exercise technique) as well as on the goals. Additional factors which have an influence on muscle activation have to be considered when regulating the intensity of resistance exercises after injury.

Post-operatively, resistance training with excessive intensity can induce pain and swelling. As a consequence, minor adaptations may occur when intensity is too high. Horstmann *et al.* (1994) demonstrated in a pilot study the importance of the adequate choice of exercise intensity in ACL-reconstructed patients. They compared single-joint isokinetic strength training for the knee extensors and flexors with 60 or 80% of the individual maximum strength on average 16 weeks after ACL reconstruction (4 weeks of training, 20 training sessions). Interestingly, the 60% group showed better gains in maximum strength than the 80% group. At 80% training intensity, pain and swelling were reported more frequently and training had to be interrupted, resulting in overall lower strengthening effects. Consequently, regulation of training load in post-operative strength training should be done with caution.

Another important aspect of regulation of training intensity was demonstrated by Ludwig (1997). He observed that ACL-reconstructed patients are able to increase training intensity in a single training session. Within a five-set training regimen, isokinetic torque production as well as EMG activity of the knee extensors increased from set to set. In addition, simultaneous increases in frequency content of the EMG signal indicated additional recruitment of fast motor units in the course of the five-set training. Probably both phenomena are the result of a reduction of inhibition within a single training session. Several consequences may be drawn from such observations. First, it seems that in some time phases in rehabilitation, training load can be increased within a training session in a multiple-set training regimen. Second, isokinetic systems, with their accommodating resistance, may offer an optimal technical solution for using this phenomenon due to the possibility of individual regulation of acute increases in force generation within a training session. Third, training volume may be dependent on the amount of reduction of inhibition in the course of a single training session. Further studies are needed to reveal a differentiated application of such strategies in rehabilitation training.

When considering training intensity in strength training in rehabilitation, the importance of submaximal training intensity seems to be further substantiated by findings from Hortobagyi *et al.* (2004). These authors observed in patients with knee osteoarthritis significant deficits in submaximal force matching of the quadriceps muscle. Deficits in submaximal force matching may represent an inappropriate basis for the optimal control of muscle forces, especially for submaximal efforts in activities of daily living. As a consequence, training in rehabilitation should be orientated not only towards optimizing maximum-force-generation capacity but also to improving the control of submaximal force generation.

Due to the temporarily limited joint or structural loading capacity in patients, muscle loading seems to be limited, at least in the early and intermediate phases of rehabilitation after injury. With regard to training practice, this means that in early and intermediate rehabilitation stages low- to medium-intensity muscle-training exercises specific to the improvement of muscle function at the site of injury have to be employed. The corresponding method for improving muscle function may be termed 'training for improving intermuscular coordination' in terms of optimizing synergistic and antagonistic force contribution within a given exercise task. The suggested training variables can be found in Table 5.1.1.

Table 5.1.1 Training for improvement of intermuscular coordination

Intensity	20–60%
Repetitions	10–15
Series	3–6
Rest	1–2 min

Pre-surgery interventions/strategies

Frequently, injury requires repair of the injured structures to restore their function (e.g. ACL reconstruction). Besides optimized rehabilitation programmes, another strategy for assisting in optimization of neuromuscular recovery is the utilization of pre-surgery strength and conditioning programmes. In this context, Eitzen, Holm and Risberg (2009) found that functional recovery was improved if pre-surgery asymmetries in knee-extensor strength were reduced below a borderline of 20%. In another recently published study, Hartigan, Axe and Snyder-Mackler (2009) investigated whether a pre-operative therapy programme including perturbation training and progressive quadriceps strengthening would be helpful in regaining quadriceps strength and knee function compared to pure quadriceps strengthening in non-copers. These authors observed that recovery of quadriceps strength and gait were more symmetrical in the combined group (perturbation training and quadriceps strengthening) than in the pure strength-training group. Perturbation training increases knee motion and decreases co-activation in the involved limb in the weight-acceptance phase of the gait. In addition, the better quadriceps recovery may decrease the risk of knee osteoarthritis. Consequently, pre-operative training interventions may be a promising strategy in optimizing functional recovery in ACL patients.

In addition, since patients with pre-operative deficits in knee extension are more likely to develop motion complications, methods to restore full extension pre-operatively seem to be an important part of pre-surgery strategy (see Cascio, Culp and Cosgarea, 2004). Besides classical techniques of physiotherapy, stretching methods may be used to optimize range of motion (ROM) before surgery.

In summary, it seems that restoration of full extension, normalization of knee-extensor strength, and improvement of intermuscular coordination in functional tasks are important and adequate pre-operative strategies in strength and conditioning for the optimization of functional outcome in ACL rehabilitation.

5.1.3.2 Active lengthening/ eccentric training

Muscle strengthening with an emphasis on eccentric force generation or eccentric muscle function is an important aspect of strength and conditioning in athletes. In recent times, eccentric training has also become an important tool for normalizing strength and muscle function in rehabilitation.

An actual trend in rehabilitation of overuse tendon injuries (especially tendinopathies) is the use of eccentric training as the main strengthening tool. Due to their frequent chronic nature, tendinopathies are a common and major problem in rehabilitation of athletes.

Several studies have shown that a pure eccentric strengthening programme for the calf muscles is able to remove pain in most patients with mid-portion Achilles tendinopathy (for review see Langberg and Kongsgaard, 2008). In addition to pain relief, eccentric calf training has been demonstrated to induce recovery of tendon structure (see for example Öhberg and Alfredson, 2004). These last observations in particular support the assumption that the specific mechanical stimulus of an eccentric contraction may be a critical factor in normalizing tendon structure.

Regarding the exercise choice and the related training methods, in Achilles tendon patients the unilateral eccentric calf raise is the main recommended exercise (see Figure 5.1.2). Based on the results of several studies, it seems that training this eccentric calf raise one or two times per day with three sets of 10–15 repetitions is an adequate training stimulus.

However, eccentric training has been shown not to be successful in all tendon injuries. While there is some positive trend,

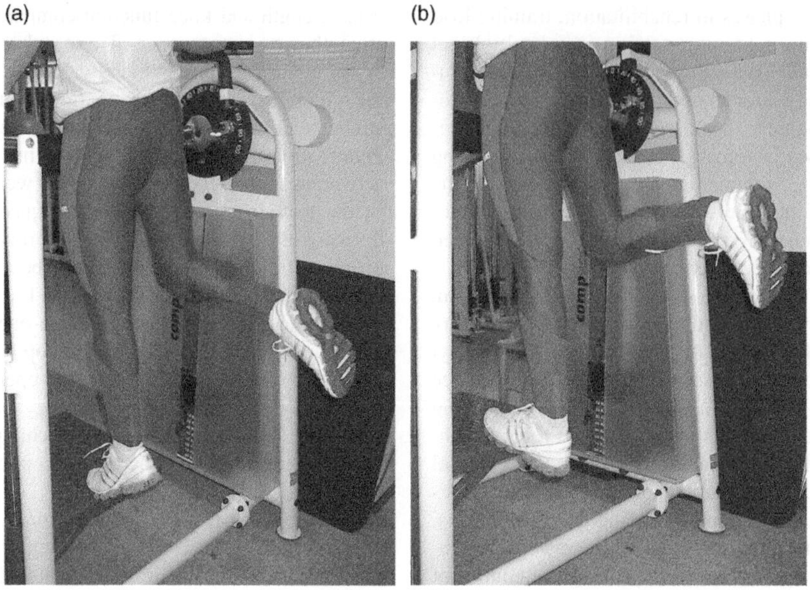

Figure 5.1.2 Eccentric calf raise

eccentric training seems not to be as successful in pain reduction in patellar, supraspinatus and insertional Achilles tendinopathies, and lateral epicondylite (Langberg and Kongsgaard, 2008).

Another promising aspect of eccentric training is its application in the rehabilitation of muscle injuries. Croisier *et al.* (2002) found poor eccentric strength to be a discriminating factor in the explanation of hamstring muscle re-injury. Consequently, eccentric strengthening of the hamstring muscles may be an important strengthening method for the optimization of rehabilitation strategies in order to reach full recovery and thereby reduce the risk of re-injury. Regarding the exercise choice, it seems plausible to restore eccentric strength and function of the hamstring muscles by incorporating both single-joint and multi-joint eccentric training. Single-joint eccentric training can be performed with isotonic devices and with an isokinetic system; the advantage of the latter is the possibility of regulating training intensity with the principle of accommodation on an individual basis; this is very safe because of the ability to fix torque limits. Another simple multi-joint eccentric exercise is the Nordic hamstring (see Figure 5.1.3), where both active lengthening of the hamstrings and synergistic activation of distal calf and proximal abdominal muscles occur. In injury-prevention research it has been shown that the Nordic hamstring is an effective exercise in the reduction of hamstring-injury risk.

It is a common finding that knee injuries such as ACL rupture occur in landing or deceleration movements. Consequently, regarding the type of muscle action, these movements are dominated by eccentric work of the muscles of the leg-extensor chain. Taken together with actual ACL-related injury-prevention research, it seems that appropriate eccentric movement control in landing and deceleration actions is a main goal in reducing the risk of re-injury (Schlumberger *et al.*, 2006). Therefore, exercise strategies to emphasize eccentric force generation and eccentric movement control may be important in normalizing muscle function and strength as well

Figure 5.1.3 Nordic hamstring

as injury-related movement behaviour. Regarding the exercise choice, a functional progression from single-joint (knee-extension) or multi-joint (leg-press) eccentric strengthening exercises to training of uni- and bilateral landing technique after jumps seems to be recommendable.

Since eccentric training in terms of landing training is associated with relatively high quadriceps-muscle and joint-structure loading, strategies to allow a stepwise integration of loading seem to be necessary after knee injury. In this context, Blackburn and Padua (2009) found that landing in a more flexed position of the trunk results in less quadriceps activity, possibly inducing a lower impact on the ACL. Consequently, training of landing technique with a continuum of trunk positions from more flexed to more upright might be an adequate strategy in a step-by-step increase of the amount of quadriceps force generation and thus ACL loading.

In conclusion, targeting and normalizing of eccentric strength and function as well as movement control under eccentric conditions seems to play a significant role in normalizing neuromuscular performance after several types of injury.

5.1.3.3 Passive lengthening/stretching

Stretching or passive muscle lengthening is commonly used in rehabilitation training. Recent developments in research on the effectiveness of stretching help to further use stretching stimuli in a targeted manner. While passive stretching seems to have some disadvantages in warm-up before speed and power events due to its inhibitory effects on the stretched muscles, several advantages of the passive lengthening stimulus can be identified.

Loss of normal joint ROM is a typical deficit after injury. Consequently, stretching is frequently used in rehabilitation as the main training method to improve or regain normal joint ROM. From a methodological standpoint, all effective stretching techniques can be used to reach ROM improvements (static stretching with or without antagonistic contraction, dynamic stretching, post-isometric stretching). However, while improving ROM and bringing it to pre-injury levels is a major goal in nearly all injuries, the amount of ROM normalization seems to depend on the phase of rehabilitation; for example, regaining full knee extension is an important goal in early phases after ACL reconstruction (Cascio, Culp and Cosgarea, 2004). In contrast, normalization of knee flexion after ACL reconstruction is a major goal in the intermediate phases of rehabilitation (typically between weeks 10 and 16 after ACL reconstruction). In addition, in the post-surgical state after a disc prolapse in the lumbar spine, regaining of full ROM has to be retarded in order to guarantee optimal structural healing without too much shear loading on the disc induced by too-high amplitudes of movement.

Besides the regaining of full ROM, another major goal in many injuries is the reduction of tightness in some muscles. Tightness may be a consequence of long-term work by muscles in reduced joint amplitudes, or it may be caused by muscle hyperactivity (an increase in the amount of muscle activity or of activity in phases of movement where these muscles are typically inactive).

Since stretching can be regarded as an adequate method for relaxation of tight muscles, it may be used to relax specific muscles after injury or pain syndromes (Frederick and Frederick, 2006). Tight muscles are observed frequently in several injuries and diagnoses; for example, the M. rectus femoris frequently shows tightness after knee injury/surgery (ACL reconstruction, meniscal repair) and knee overuse injuries (such as femoropatellar pain or patellar tendionopathy). Further, the muscle tensor fascia latae typically shows tightness after all knee injuries, particularly in the iliotibal band syndrome. In back-pain patients, the muscles quadratus lumborum and psoas major are typically tight. Such examples demonstrate that in each injury some muscles react to some type of unusual overloading by tightening. Consequently, stretching to reduce tightness by lengthening and relaxing muscles seems to be an important aspect of recovery of muscle function after injury.

Stretching of single muscles which show tightness requires exercises that integrate necessities of positioning to optimally stretch the target muscle; for example, stretching of hip flexors is typically performed by bringing the hip joint to an extended position. This strategy is adequate when optimizing normal hip-extension ROM. However, in order to optimally lengthen and thereby relax the muscle psoas major, a combination of hip extension, hip internal rotation, and spinal lateral flexion to the opposite side is needed (see Figure 5.1.4). In contrast, stretching the hip flexor muscle tensor fascia latae requires a combination of hip extension, hip external rotation, and hip adduction (see Figure 5.1.5). As opposed to the one-dimensional stretching techniques used for increases in joint ROM, muscle-specific stretching is characterized by accurate three-dimensional positioning for optimal lengthening of the target muscle.

The rectus femoris, acting as hip flexor and knee extensor, also needs specific positioning for optimal lengthening since in normal one-dimensional stretching in maximum knee flexion the lengthening stimulus for the rectus femoris is not sufficiently specific. Two examples are shown in Figures 5.1.6 and 5.1.7.

The importance of stretching to regain normal muscle function in rehabilitation can also be shown by observations made with trigger points. Trigger points are local areas within a muscle that show a local energy deficit combined with contracture of sarcomeres (building a so-called taut band) (Mense and Simons, 2001). A possible cause of the development of trigger points after injury/surgery is (more or less) excessive muscle overloading (e.g. induced by overloading due to chronic muscle hyperactivity in daily living or by resistance training with a high eccentric component). Stretching seems to be an effective method of reducing trigger point-induced pain, especially when trigger points are newly activated simply by moderate muscle overloading. According to Mense and Simons (2001), discomfort release can be reached with static stretching, post-isometric relaxation techniques, and the technique of reciprocal inhibition.

Figure 5.1.4 Stretching of the M. psoas major

Figure 5.1.5 Stretching of the M. tensor fascia latae

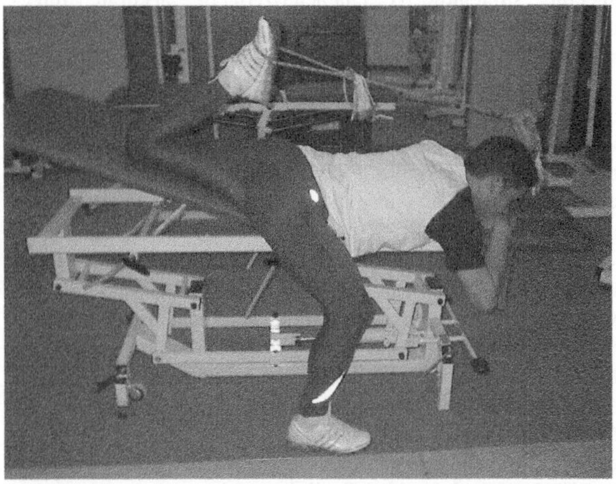

Figure 5.1.6 Stretching of the M. rectus femoris in supine position

Figure 5.1.7 Stretching of the M. rectus femoris in kneeling position

Stretching has also been shown to play an important role in normalizing functional activities after muscle strain of the hamstrings. In general, static stretching is adequately suited to a safe restoration of ROM after muscle injury. In this context, Malliaropoulos *et al.* (2004) demonstrated that athletes with hamstring muscle injuries returned faster to full training when using an intensive stretching programme in rehabilitation. A training volume of 4 × 30 seconds three or four times per day has been shown to be more effective than the same once per day. Stretching may assist in regaining normal tensile strength of the muscle after muscle injury by reducing scar tissue and normal collagen fibre alignment. It may be speculated that the amount of stretching after injury has to be higher than in healthy people due to the reduced viscoelasticity of the muscle.

Another aspect of the value of stretching in general as well as in rehabilitation was considered by Witvrouw *et al.* (2007), regarding its effects on tendon function. The basis of their approach is consideration of the type of sports activity. For stretch–shortening-type muscle actions, sufficient compliance of the tendon is an important prerequisite. According to Witvrouw *et al.* (2007), promoting tendon compliance by stretching techniques may be an effective training strategy in tendon rehabilitation.

In summary, stretching is an effective means of relaxing muscles with some type of tightness or related neuromuscular deficits. Consequently, stretching seems to be an important method of regaining normal muscle function. For targeted use in rehabilitation, the specific positioning of the particular muscle is an important feature. In addition, stretching may assist in normalization of tendon characteristics.

5.1.3.4 Training of the muscles of the lumbopelvic hip complex

The importance of the function of the muscles of the lumbopelvic hip complex (LPHC) (also named 'core stability') to force generation in functional activities (like running, sprinting, or cutting) is increasingly being recognized. In fact, strengthening the LPHC may be of great importance in recovering neuromuscular function after injury, particularly after injuries of the lower leg. Recently, differential experimental approaches have highlighted the positive impact of LPHC strengthening for regaining neuromuscular performance after lower-leg injury (Myer *et al.*, 2009).

The general effect of strengthening the LPHC on lower-leg function has been observed in a recent pilot study (Hajduk and Schlumberger, unpublished). We compared the effects of traditional leg-strength training with a training programme combining leg and trunk strengthening in healthy female subjects. It was observed that the combined leg- and trunk-strengthening programme was significantly more effective in increasing maximum strength of the leg extensors (1 RM in the squat) than an isolated leg-strengthening programme (Table 5.1.2). These results seem to indicate the positive influence of strengthening the lumbopelvic-hip muscles on force generation of the leg extensors working in the kinetic chain. Future studies are needed to discover whether trunk-muscle strengthening is also able to assist in regaining maximum strength of the leg-extensor chain after injury (e.g. after ACL reconstruction).

Within the LPHC, the function and strength of the gluteus medius and maximus muscle seem to have an important influence on the function of the whole lower limb; for example, it is commonly observed that gluteal muscle weakness is associated with several lower-extremity injuries, including patellofemoral pain syndrome, iliotibial band friction syndrome, ACL sprains, and chronic ankle instability. Weakness in the gluteal muscles contributes to poor lower-limb control (e.g. excessive hip internal rotation and adduction). Consequently, strengthening programmes following such injuries have to integrate exercises which improve function and strength of these muscles. Regarding the most effective exercises, Di Stefano *et al.* (2009) found in a recent study that the best exercise for the gluteus medius is the side-lying hip abduction, while the single-leg squat and the single-leg deadlift seem to be appropriate for strengthening the gluteus maximus muscle.

Furthermore, strengthening of LPHC muscles has been shown to be an effective training method in rehabilitation after muscle injury. Sherry and Best (2004) observed that agility and trunk-strengthening exercises assist in a safe and fast recovery after hamstring muscle injury. The approach using agility and trunk-strengthening was significantly more effective in this study than a rehabilitation programme based mainly on hamstring stretching. From these data it may be speculated that an improved neuromuscular control of the lumbopelvic region is an important prerequisite for optimal recovery of hamstring function in terms of optimizing length–tension or force–velocity relationships.

It has been shown that ROM limitations and muscle tightness frequently occur after injury (see above). There is some preliminary evidence that LPHC training may assist in optimizing ROM. Kuszewski, Gnat and Saulicz (2009) investigated the effects of a four-week LPHC-strengthening programme challenging local and global stabilizer activity on hamstring flexibility. In this study, the LPHC training was conducted with the sling-exercise device. Two exercise examples of LPHC strengthening using this device are given in Figure 5.1.8. This LPHC-strengthening programme showed a tendency towards

Table 5.1.2 1 RM (in kg) in the bilateral squat before and after six-week training with leg strengthening (group leg, $n = 12$), leg and trunk strengthening (group leg + trunk, $n = 12$), or no training (control, $n = 12$) (Hajduk and Schlumberger, unpublished)

	Pre-training	Post-training	Change (%)
Group leg	100.8 ± 20.9	108.0 ±	+7.5[a]
Group leg + trunk	104.8 ± 21.3	120.6 ± 19.1	+15.9[a,b]
Control	104.8 ± 17.0	101.9 ± 17.3	−3.3

[a] Significant change.
[b] Significantly higher increase than in group leg.

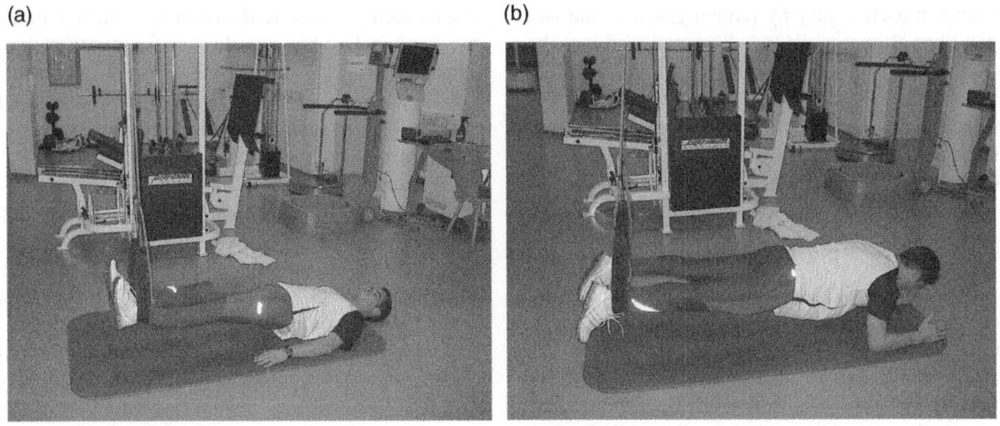

Figure 5.1.8 LPHC strengthening with sling exercise. (a) Static activation of lumbar extensors. (b) Static coactivation of abdominal muscles and lumbar and thoracic extensors

increases in flexibility of the hamstrings. Such evidence indicates that LPHC function may have an influence on lengthening characteristics of lower-limb muscles like the hamstrings. However, future research is needed to clarify the influence of mechanics and LPHC stability on lower-extremity flexibility in injured and non-injured subjects.

5.1.3.5 Training of sport-specific movements

Restoration of athletic performance requires the regaining of local and complex muscular function, strength, and basic movement patterns (like walking, running, or stepping). In addition, the retraining of sports-specific movement patterns seems to play a major role in strength and conditioning in rehabilitation (Schlumberger, 2009). This appears to be especially true in athletes suffering from injury with long-term absence from sports-specific movement training (e.g. in the ACL-reconstructed patient). Due to the high loading on muscles and passive structures in sports-specific movements, the related training methods have to be used in the final phase of rehabilitation.

Direct evidence for the effectiveness of various methodological approaches to sports-specific movement training is lacking. However, important conclusions can be drawn from injury-prevention research. In order to identify possible mechanisms contributing to injury occurrence in sports-specific situations, Besier, Lloyd and Ackland (2003) investigated the muscle-activation strategies of the knee during running and cutting movements. These authors compared the effects of changes of direction after linear running under pre-planned conditions (subjects knew the direction of the change of direction) and unanticipated conditions (direction was indicated by a light signal during running). It was found that under pre-planned conditions there was selective activation of knee medio-lateral and internal/external rotation stabilizers in combination with a co-contraction of the knee flexors and extensors. Under unanticipated conditions, subjects reacted with a generalized co-contraction without selective stabilizer activation. While these results were obtained in healthy subjects, it might be speculated that training of specific movements under variable external stimuli in team sports (such as soccer, basketball, handball, or ice hockey) could be an important aspect in regaining appropriate and economic specific movement patterns, thereby decreasing the risk of re-injury.

This optimization of sports-specific movement patterns (in terms of normalizing the intermuscular coordination between prime movers and stabilizers) could help improve rehabilitation training strategies. Since movement patterns are specific to a given sports discipline, an important step in planning the final phases of rehabilitation training of an athlete is careful evaluation of the sport-specific movement patterns. In Table 5.1.3 an example is given of the characteristic movement patterns in soccer. Their optimization should be the final goal of rehabilitation.

Table 5.1.3 Specific movement patterns in soccer (movement coordination is influenced by opponent behaviour)

Linear run
Linear-sprint acceleration (5–15 m)
Quasi-maximum linear-sprint velocity (20–40 m)
Sprinting from standing and out of movement (variable positions, bilateral symmetrical standing, one leg front, one leg behind)
Rapid changes of direction (multidirectional moving)
Rapid deceleration
Jumping (powerful actions in all three main directions of impulse: vertical, horizontal, lateral; one- and two-legged actions)
Running/moving with ball
Kicking actions (coordination of standing and kicking leg)
Movement coordination is influenced by opponent behavior

The last step after restoration of basic and sports-specific movement patterns is to increase the endurance capacity in these movements in terms of building fatigue resistance. The classical methods of endurance training and energy-system development are used.

5.1.4 CONCLUSION

Rehabilitation of the neuromuscular system is an outstanding challenge for therapeutic staff. As presented in this chapter, strength and conditioning in rehabilitation seems to have an important impact on quality, quantity, and speed of neuromuscular recovery. Several main features should be considered when planning a rehabilitation programme:

1. The normalization of local muscle function (improvement of local activation, reduction of tightness, reduction of muscle hyperactivity).
2. The regaining of strength and conditioning levels necessary for basic functional activities in early and intermediate phases of rehabilitation (normalization of intermuscular coordination in terms of physiological relative force contribution of stabilizer and prime-mover muscles for typical daily activities, such as the gait cycle).
3. The utilization of strength and conditioning programmes aimed at full structural and functional recovery (e.g. optimization of muscle size, leg alignment in landing) as well as adequate sports-specific movement patterns.
4. The maintenance of adaptations and further improvements in neuromuscular function while returning to sport-specific training and competition.

A major goal in rehabilitation is the safe return of an athlete to training and competition. In this sense, training strategies in rehabilitation may follow a progressive reintroduction of activities depending on the severity of injury and the constraints of the healing process. An essential feature of targeted rehabilitation training is appropriate exercise choice, based on injury-specific deficits of neuromuscular function, combined with adequate use of the main training variables (intensity and volume of training). Finally, complete rehabilitation should incorporate and allow sufficient and appropriate time for several stages of healing and treatment/training.

References

Aune A.K., Holm I., Risberg M.A. *et al.* (2001) Four-strand hamstring tendon autograft compared with patellar-tendon-bone autograft for anterior cruciate ligament reconstruction. a randomized study with two-year follow-up. *Am J Sports Med*, **29**, 722–728.

Besier T.F., Lloyd D.G. and Ackland T.R. (2003) Muscle activation strategies at the knee during running and cutting maneuvres. *Med Sci Sports Exerc*, **35**, 119–127.

Blackburn J.T. and Padua D.A. (2009) Sagittal-plane trunk position, landing forces, and quadriceps electromyographic activity. *J Athletic Train*, **44**, 174–179.

Cascio B.M., Culp L. and Cosgarea A.J. (2004) Return to play after anterior cruciate ligament reconstruction. *Clin Sports Med*, **23**, 395–408.

Croisier J.L., Forthomme B., Namurois M.H. *et al.* (2002) Hamstring muscle strain recurrence and strength performance disorders. *Am J Sports Med*, **30**, 199–203.

Di Stefano L.J., Blackburn J.T., Marshall S.W. and Padua D.A. (2009) Gluteal muscle activation during common therapeutic exercises. *J Orthop Sports Phys Ther*, **39**, 532–540.

Eitzen I., Holm I. and Risberg M.A. (2009) Preoperative quadriceps strength is a significant predictor of knee function two years after anterior ligament reconstruction. *Br J Sports Med*, **43**, 371–376.

Ellenbecker T., De Carlo M. and DeRosa C. (2009) *Effective Functional Progressions in Sport Rehabilitation*, Human Kinetics, Champaign, IL.

Frederick A. and Frederick C. (2006) *Stretch to Win*, Human Kinetics, Champaign, IL.

Friel K., McLean N., Myers C. and Caceres M. (2006) Ipsilateral hip abductor weakness after inversion ankle sprains. *J Athl Train*, **41** (1), 74–78.

Hartigan E., Axe M.J. and Snyder-Mackler L. (2009) Perturbation training prior to ACL reconstruction improves gait asyymetries in non-copers. *J Orthop Res*, **27**, 724–729.

Hespel P., Op't Eijnde B., Van Leemputte M. *et al.* (2001) Oral creatine supplementation facilitates the rehabilitation of disuse atrophy and alters the expression of muscle myogenic factors in humans. *J Physiol*, **536**, 625–633.

Horstmann T., Mayer F., Greiner T. *et al.* (1994) Isokinetic muscle hypertrophy training after ACL-rupture dependent on training intensity, in *Mechanisms of Regulation and Repair* (eds H. Liesen, M. Weiß and M. Baum), Deutscher Aerzteverlag, pp. 887–890.

Hortobagyi T., Dempsey L., Fraser D. *et al.* (2000) Changes in muscle strength, muscle fibre size and myofibrillar gene expression after immobilization and retraining in humans. *J Physiol*, **524**, 293–304.

Hortobagyi T., Garry J., Holbert D. and Devita P. (2004) Aberrations in the control of quadriceps muscle force in patients with knee osteoarthritis. *Arthritis Rheum*, **51**, 562–569.

Kuszewski M., Gnat R. and Saulicz E. (2009) Stability training of the lumbo-pelvic-hip complex influence stiffness of the hamstrings. *Scand J Med Sci Sports*, **19**, 260–266.

Langberg H. and Kongsgaard M. (2008) Eccentric training in tendinopathy – more questions than answers (editorial). *Scand J Med Sci Sports*, **18**, 541–542.

Lorentzon R., Elmqvist L.G., Sjöström M. *et al.* (1989) Thigh musculature in relation to chronic anterior cruciate ligament tear: muscle size, morphology, and mechanical output before reconstruction. *Am J Sports Med*, **17**, 423–429.

Ludwig M. (1997) [Analysis of rehabilitative strength training after ACL-reconstruction]. *Deutsche Zeitschrift Sportmedizin*, **48**, 193–200.

Malliaropoulos N., Papalexandris S., Papalada A. and Papacostas E. (2004) The role of stretching in rehabilitation of hamstring injuries: 80 athletes follow-up. *Med Sci Sports Exerc*, **36** (5), 756–759.

Mascal C.L., Landel R. and Powers C. (2003) Management of patellofemoral pain targeting hip, pelvis, and trunk muscles function: 2 case reports. *J Orthop Sports Phys Ther*, **33**, 642–660.

McGill S. (2007) *Low Back Disorders*, Human Kinetics, Champaign, IL.

Mense S. and Simons D.G. (2001) *Muscle Pain: Understanding Its Nature, Diagnosis, and Treatment*, Lippincott, Philadelphia, PA.

Mikkelsen C., Werner S. and Eriksson E. (2000) Closed kinetic chain alone compared to combined open and closed kinetic chain exercises for quadriceps strengthening after anterior cruciate ligament reconstruction with respect to return to sports: a prospective matched follow-up study. *Knee Surg Sports Traumatol Arthrosc*, **8** (6), 337–342.

Myer G.D., Paterno M.V., Ford K.R. and Hewett T.E. (2009) Neuromuscular training techniques to target deficits before return to sport after anterior cruciate ligament reconstruction. *J Strength Cond Res*, **22**, 987–1014.

Öhberg L. and Alfredson H. (2004) Effects of neovascularisation behind the good results with eccentric training in chronic mid-portion Achilles tendinosis? *Knee Surg Sports Traumatol Arthrosc*, **12**, 465–470.

Philipps S.M. (2000) Short-term training: when do repeated bouts of resistance exercise become training? *Can J Appl Physiol*, **25**, 185–193.

Salem G.J., Salinas R. and Harding F.V. (2003) Bilateral kinematic and kinetic analysis of the squat exercise after anterior ligament reconstruction. *Arch Phys Med Rehabil*, **84**, 1211–1216.

Schlumberger A. (2002) Strength of the ankle muscles in high-level athletes after knee surgery. 3rd International Conference on Strength Training (Abstract Book), Budapest.

Schlumberger A. (2009) Training for sensorimotor performance after injury in soccer, in *The Sensorimotor*

System (ed. W. Laube), Thieme, Stuttgart, Germany, pp. 600–617.

Schlumberger A., Laube W., Bruhn S. *et al.* (2006) Muscle imbalances – fact or fiction? *Isokin Exerc Sci*, **14**, 3–11.

Sherry M.A. and Best T.M. (2004) A comparison of 2 rehabilitation programs in the treatment of acute hamstring strains. *J Orthop Sports Phys Ther*, **34**, 116–125.

Torry M.R., Decker M.J., Viola R.W. *et al.* (2000) Intra-articular knee joint effusion induces quadriceps avoidance gait patterns. *Clin Biomech*, **15**, 147–159.

Tyler T.F., Nicholas S.J., Hershman E.B. *et al.* (2004) The effect of creatine supplementation on strength recovery after anterior cruciate ligament (ACL) reconstruction. *Am J Sports Med*, **32**, 383–388.

Williams G.N., Barrance P.J., Snyder-Mackler L. and Buchanan T.S. (2004) Altered quadriceps control in people with anterior cruciate ligament deficiency. *Med Sci Sports Exerc*, **36** (7), 1089–1097.

Witvrouw E., Mahieu N., Roosen P. and McNair P. (2007) The role of stretching in tendon injuries. *Br J Sports Med*, **41**, 224–226.

5.2 Strength Training for Children and Adolescents

Avery D. Faigenbaum, Department of Health and Exercise Science, The College of New Jersey, Ewing, NJ, USA

5.2.1 INTRODUCTION

Physical activity is essential for normal growth and development during childhood and adolescence. It is generally agreed that school-age youth should participate regularly in at least 60 minutes or more of moderate to vigorous physical activity that is developmentally appropriate and enjoyable (Department of Health, 2004; Department of Health and Human Services, 2008). While a variety of activities should be recommended, research increasingly indicates that strength training can offer unique benefits for children and adolescents when appropriately prescribed and supervised (Faigenbaum and Myer, 2010; Ortega *et al.*, 2008; Strong *et al.*, 2005; Vaughn and Micheli, 2008). Despite outdated concerns regarding the safety and effectiveness of strength training for children and adolescents, the qualified acceptance of youth strength training by medical and fitness organizations is becoming universal (American College of Sports Medicine, 2010; Australian Strength and Conditioning Association, 2009; Behm *et al.*, 2008; British Association of Sport and Exercise Sciences, 2004; Faigenbaum *et al.*, 2010; Mountjoy *et al.*, 2008).

Nowadays, public health guidelines aim to increase the number of boys and girls who regularly engage in muscle-strengthening physical activities (Department of Health, 2004; Department of Health and Human Services, 2008). Many school-based physical education programmes are specifically designed to enhance health-related components of physical fitness, including muscular strength, and a growing number of young athletes now strength-train to enhance their sports performance (Lee *et al.*, 2007; National Association for Sport and Physical Education, 2005; Vaughn and Micheli, 2008). As more children and adolescents strength-train in schools, fitness centres, and sport-training facilities, it is important to understand safe and effective practices by which strength training can improve the health, fitness, and sports performance of younger populations. Therefore, the main focus of this chapter is to review the risks and concerns associated with youth strength training, examine the trainability of muscular strength in younger populations, and highlight programme design considerations for healthy children and adolescents.

For the purpose of this chapter, the term 'children' refers to boys and girls who have not yet developed secondary sex characteristics such as changes in voice pitch, facial hair, and body configuration. This period of development is referred to as 'pre-adolescence' and generally includes girls and boys up to the age of roughly 12 years. The term 'adolescence' refers to a period of time between childhood and adulthood (typically ages 13–18 years). The terms 'youths' and 'young athletes' are broadly defined in this chapter to include both children and adolescents. By definition, the term 'strength training' (also called 'resistance training') refers to a specialized method of conditioning which involves the progressive use of a wide range of resistive loads (including body mass) and a variety of training modalities designed to enhance health, fitness, and sports performance. Strength training should be distinguished from weightlifting, which is a competitive sport in which athletes attempt to lift maximal amounts of weight in the clean-and-jerk and snatch exercises.

5.2.2 RISKS AND CONCERNS ASSOCIATED WITH YOUTH STRENGTH TRAINING

The traditional concerns associated with youth strength training stemmed from three general misconceptions about strength exercise. First, that any type of exercise that involved moderate to heavy lifting would be unsafe and inappropriate for children and adolescents. Second, that strength training would damage developing growth plates and possibly stunt the linear growth of children and adolescents. Third, that children would not benefit from strength training because youths lacked adequate amounts of circulating androgens. Categorically, all of these misperceptions have been disproved by research evidence, which clearly indicates that regular participation in a well-designed and competently supervised strength-training programme can be safe, effective and worthwhile for healthy children and adolescents (Faigenbaum and Myer, 2010; Falk and Eliakim, 2003; Malina, 2006; Pierce *et al.*, 2008).

Strength and Conditioning – Biological Principles and Practical Applications Marco Cardinale, Rob Newton, and Kazunori Nosaka.

A careful evaluation of research findings indicates a relatively low risk of injury in children and adolescents who follow age-appropriate strength-training guidelines (Faigenbaum *et al.*, 2010; Malina, 2006; Pierce *et al.*, 2008). Based on an analysis of strength training-related injuries that resulted in visits to US emergency rooms, Myer *et al.* (2010) noted that children had a lower risk of strength training-related joint sprains and muscle strains than adults. In support of these observations, others reported no evidence of either musculoskeletal injury (measured by biphasic scintigraphy) or muscle necrosis (determined by serum creatine phosphokinase levels) in children following 14 weeks of strength training (Rians, Weltman and Cahill, 1987).

Only three published studies have reported strength training-related injuries in children (a shoulder strain which resolved within one week of rest (Rians, Weltman and Cahill, 1987); a shoulder strain which resulted in one missed training session (Lillegard *et al.*, 1997); and nonspecific anterior thigh pain which resolved with five minutes of rest (Sadres *et al.*, 2001)). In the vast majority of prospective published reports, no serious injuries are reported in young lifters who participated in supervised strength-training programmes that were appropriately prescribed to ensure they were matched to each participant's initial capabilities. Although strength training, like most physical activities, does have an inherent risk of musculoskeletal injury, the available data suggest that this risk is no greater than that in other sports and recreational activities in which youths regularly participate.

Despite these noteworthy findings, a recurring concern among some youth coaches and health-care providers centres around the safety and appropriateness of weightlifting and plyometric exercises for youths. Unlike traditional strength-building exercises such as the chest press or biceps curl, which are relatively easy to learn and perform, weightlifting movements and plyometrics are explosive but highly controlled movements that require a relatively high degree of technical skill. For example, to accomplish the clean and jerk, the barbell must be lifted from the platform to the shoulders and then to the overhead position to complete the two-part lift. While this movement involves more complex neural activation patterns than most strength exercises, the belief that weightlifting movements are riskier than other sports and activities is not supported by research finding (Byrd *et al.*, 2003; Hamill, 1994; Pierce, Byrd and Stone, 1999).

In one retrospective evaluation of injury rates in adolescents, it was revealed that strength training and weightlifting were markedly safer than many other sports and activities in which youths regularly participate (Hamill, 1994). In this report the overall injury rate per 100 participant hours was 0.8000 for rugby and 0.0120 and 0.0013 for strength training and weightlifting, respectively. This latter finding may be explained, at least in part, by the observation that weightlifting is typically characterized by well-informed coaches and a gradual progression from basic exercises (e.g. front squat) to skill-transfer exercises (e.g. overhead squat) and finally to competitive lifts (snatch and clean and jerk). Others have reported significant gains in muscular strength without any report of injury when weightlifting movements such as the snatch, clean and jerk, and modified cleans, pulls, and presses are incorporated into youth strength-training programmes (Faigenbaum *et al.*, 2007a; Gonzales-Badillo *et al.*, 2005; Sadres *et al.*, 2001).

A related concern associated with youth strength training regards the safety of plyometric exercises for children and adolescents. Although plyometric training typically includes hops and jumps that exploit the muscles' cycle of lengthening and shortening to increase muscle power, watching children on a playground supports the premise that the movement patterns of boys and girls as they skip and jump can be considered plyometric (Chu, Faigenbaum and Falkel, 2006). The belief that age-appropriate plyometric training is unsafe for youths, or that a pre-determined baseline level of strength (e.g. one-repetition maximum (1 RM) squat should be 1.5 times body weight) should be a prerequisite for lower-body plyometric training, is not supported by current research and clinical observations. Indeed, well-designed strength-training programmes that include plyometric exercises have been found to enhance movement biomechanics, improve functional abilities, and decrease the number of sports-related injuries in young athletes (Hewett *et al.*, 1999; Mandelbaum *et al.*, 2005; Myer *et al.*, 2005; Thomas, French and Hayes, 2009).

Perhaps the most enduring concern related to youth strength training regards the potential for training-induced damage to the growth cartilage. Since growth cartilage is 'pre-bone', it is weaker than adjacent connective tissue and therefore more easily damaged by repetitive microtrauma (Micheli, 2006). A few retrospective case reports published in the 1970s and 1980s noted injury to the growth cartilage in young lifters (Gumbs *et al.*, 1982; Jenkins and Mintowt-Czyz, 1986; Rowe, 1979; Ryan and Salciccioli, 1976). However, most of these injuries were due to improper lifting techniques, maximal lifts, or lack of qualified adult supervision. To date, injury to the growth cartilage has not been reported in any prospective youth strength-training research study. Furthermore, there is no evidence to suggest that strength training will negatively impact growth and maturation during childhood and adolescence (Falk and Eliakim, 2003; Malina, 2006). If age-specific training guidelines are followed and if nutritional recommendations (e.g. adequate calcium) are adhered to, weight-bearing physical activity (including strength training) will likely have a favourable influence on growth during childhood and adolescence, but will not affect the genotypic maximum.

It is worth noting that there is an increased risk of injury to children and adolescents who use exercise equipment at home without supervision (Gould and DeJong, 1994; Jones, Christensen and Young, 2000). However, the risk of injury while strength training can be minimized by qualified supervision, appropriate programme design, careful selection of training equipment, and a safe training environment. In addition, the risk of injury can be reduced by systematically varying the training programme, limiting the number of heavy lifts during a workout, and allowing for adequate recovery between training sessions.

5.2.3 THE EFFECTIVENESS OF YOUTH RESISTANCE TRAINING

A compelling body of scientific evidence indicates that children and adolescents can significantly increase their muscular strength given a training programme of sufficient intensity, volume, and duration (Behm *et al.*, 2008; Blimkie and Bar-Or, 2008; Faigenbaum and Myer, 2010; Myer and Wall, 2006; Pierce *et al.*, 2008; Vaughn and Micheli, 2008). In addition, two meta-analyses on youth strength training (Falk and Tenenbaum, 1996; Payne *et al.*, 1997), along with clinical observations, indicate that well-designed strength-training programmes can enhance the muscular strength of children and adolescents beyond that produced by normal growth and development.

A majority of youth strength-training studies lasted 8–20 weeks and most subjects were between 7 and 15 years of age. A wide variety of strength-training programmes, from single-set sessions on weight machines to progressive, multi-set training protocols on different types of equipment, have proven to be effective (Annesi *et al.*, 2005; Faigenbaum *et al.*, 2007a; Gonzales-Badillo *et al.*, 2005; Ramsay *et al.*, 1990; Sadres *et al.*, 2001; Westcott, 1992). Training modalities have included weight machines (both adult- and child-size), free weights (i.e. barbells and dumbbells), medicine balls, elastic bands, and body-weight exercises.

Strength gains of roughly 30% are common following short-term (8–20 weeks) youth strength-training programmes. Figure 5.2.1 illustrates training-induced lower-body-strength gains in children following an eight-week strength-training programme. While it is evident that all children responded favourably to the training stimulus (1–2 sets of 10–15 repetitions at 60–70% 1 RM), the individual response was variable. Subject 1 demonstrated relatively small gains in muscle strength, while subject 20 experienced the largest gains. Although the group mean strength gain was significant, the variation in the individual response to the training programme suggests that other factors (e.g. genetics, training experience, motivation) need to be considered when evaluating such data. From a practical standpoint, coaches and teachers should be aware of the individual responses to strength exercise and may need to identify participants who might warrant more attention and/or a modification of their strength-training programme.

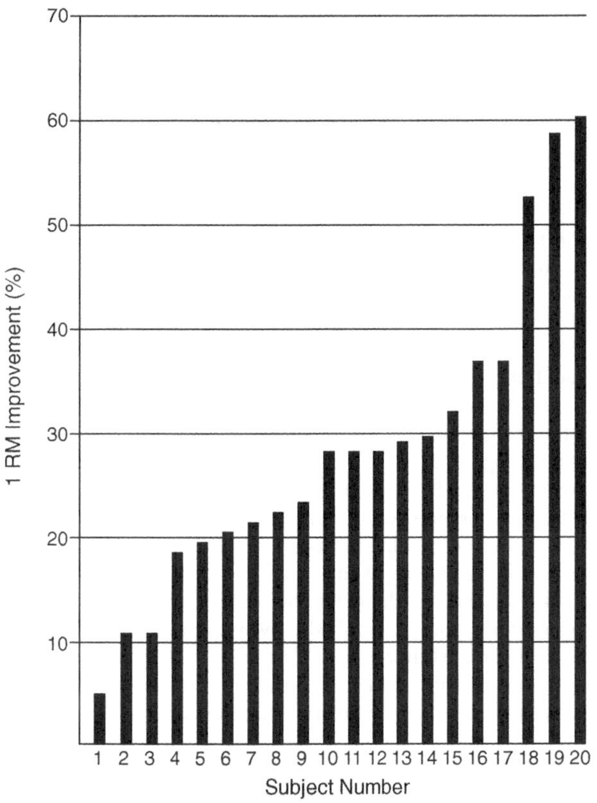

Figure 5.2.1 Individual changes in muscle strength in 20 children in response to eight weeks of strength training. Unpublished data from Avery Faigenbaum, The College of New Jersey, USA

5.2.3.1 Persistence of training-induced strength gains

The temporary or permanent reduction or withdrawal of a training stimulus is referred to as 'de-training'. The evaluation of strength changes in youths following a de-training period is complicated by the concomitant growth-related strength increases in the same time period. The available data suggest that training-induced gains in strength in youths are impermanent and tend to regress towards untrained control-group values during the de-training period (Faigenbaum *et al.*, 1996; Ingle *et al.*, 2006; Tsolakis, Vagenas and Dessypris, 2004). Although the precise nature of the de-training response and the physiological adaptations that occur during this period remain uncertain, it seems that changes in neuromuscular functioning and the hormonal responses to de-training should be considered.

The effects of training frequency on the maintenance of training-induced strength gains in children and adolescents are also worthy of further study. Following 20 weeks of strength training, Blimkie *et al.* (1989) found that a once-weekly maintenance training programme was not adequate to maintain the training-induced strength gains in pre-adolescent males. Conversely, a once-weekly maintenance programme was just as sufficient as a twice-weekly maintenance programme in retaining the strength gains made after 12 weeks of strength training in a group of adolescent male athletes (DeRenne *et al.*, 1996).

5.2.3.2 Programme evaluation and testing

The degree of measured strength change following a training programme can be influenced by many factors, including training experience, programme design, and specificity of testing and training. In addition, the methods for evaluating training-induced changes in muscle strength need to be considered. In some studies subjects were trained and tested using different modalities (Pfeiffer and Francis, 1986; Sewall and Micheli, 1986; Weltman *et al.*, 1986), and in other published reports strength changes were evaluated by relatively high RM values (e.g. 10RM) (Faigenbaum *et al.*, 1993; Lillegard *et al.*, 1997).

Strength changes have also been evaluated by maximal load lifting (e.g. 1RM) on the equipment used in training (DeRenne *et al.*, 1996; Faigenbaum *et al.*, 2002; Pikosky *et al.*, 2002; Ramsay *et al.*, 1990; Volek *et al.*, 2003). However, some practitioners and researchers have not used 1RM testing to evaluate training-induced changes in muscular strength because of the presumption that high-intensity loading may cause structural damage in children. Yet no injuries have been reported in prospective studies that utilized adequate warm-up periods, appropriate progression of loads, close and qualified supervision, and critically chosen maximal-strength tests to evaluate training-induced changes in young lifters.

In one report, 96 children performed a 1RM strength test on upper-body and lower-body weight-machine exercises (Faigenbaum, Milliken and Westcott, 2003). No abnormal responses or injuries occurred during the study period and the testing protocol was reportedly well-tolerated by the subjects. In other reports, children and adolescents safely performed 1RM strength tests using free-weight exercises (Baker, 2002; Hetzler *et al.*, 1997; Sadres *et al.*, 2001; Volek *et al.*, 2003). These observations suggest that the maximal force-producing capabilities of healthy children and adolescents can be safely evaluated by 1RM testing procedures provided that youths participate in a habituation period prior to testing and that qualified professionals closely supervise and administer each test. Since most of the forces that youths are exposed to in sports and recreational activities are likely to be greater in both duration and magnitude than carefully performed 1RM testing, the careful evaluation of maximal muscle strength in children and adolescents should be supported by qualified professionals.

However, when properly administered, 1RM tests are time-consuming and labour-intensive; in some instances, such as physical-education classes, field-based measures may be more appropriate and time-efficient. Milliken *et al.* (2008) and Holm *et al.* (2008) have documented significant correlations between 1RM strength and common field measures such as handgrip strength and long jump in children. In any case, unsupervised and improper strength testing characterized by inadequate progression of loading and poor exercise technique should not be performed by children or adolescents under any circumstances due to the real risk of injury (Risser, 1991).

5.2.4 PHYSIOLOGICAL MECHANISMS FOR STRENGTH DEVELOPMENT

Training-induced strength gains in children are more related to neurological mechanisms than to morphological changes in muscle size (Malina, 2006; Sale, 1989). Without adequate levels of circulating testosterone to stimulate increases in muscle size, children experience greater difficulty increasing their muscle mass consequent to a strength-training programme as compared to older populations (Ozmun, Mikesky and Surburg, 1994; Ramsay *et al.*, 1990). However, since some findings are at variance with this suggestion (Fukunga, Funato and Ikegawa, 1992; Mersch and Stoboy, 1989), it is possible that more intensive training programmes, longer training durations, and more sensitive measuring techniques that are ethically appropriate for this population may be needed to partition the effects of training on fat-free mass from expected gains due to growth and maturation.

Without corresponding increases in fat-free mass, neuromuscular adaptations (i.e. a trend towards increased motor-unit activation and changes in motor-unit coordination, recruitment, and firing) and possibly intrinsic muscle adaptations appear to be primarily responsible for training-induced strength gains during pre-adolescence (Ozmun, Mikesky and Surburg, 1994; Ramsay *et al.*, 1990). Using the interpolated twitch technique, Ramsay *et al.* (1990) found an increase of 12 and 14% in motor-unit activation of the elbow flexors and knee extensors, respectively, in pre-adolescent boys following 20 weeks of strength training. Likewise, Ozmun, Mikesky and Surburg (1994) used integrated electromyography amplitude to demonstrate an increase in neuromuscular activation in agonist muscles following eight weeks of strength training in children.

In both of the aforementioned studies (Ozmun, Mikesky and Surburg, 1994; Ramsay *et al.*, 1990), measured increases in training-induced strength were greater than changes in neuromuscular activation. Thus, it is likely that improvements in motor-skill performance and the coordination of the involved muscle groups also play a significant role. In support of these observations, several training studies have reported significant improvements in strength during pre-adolescence without corresponding increases in gross limb morphology, as compared to a similar control group (Faigenbaum *et al.*, 1993; Lillegard *et al.*, 1997; Ramsay *et al.*, 1990). Since most children have limited experience of strength training, it is reasonable to suggest that the first few weeks of training involve neuromuscular learning or optimization of intermuscular coordination (agonists, synergists, stabilizers) (Behm *et al.*, 2008). During and after puberty, training-induced gains in muscle strength may be associated with changes in hypertrophic factors in males, since testosterone and other hormonal influences on muscle hypertrophy will be operant (Kraemer *et al.*, 1989).

5.2.5 POTENTIAL HEALTH AND FITNESS BENEFITS

While a majority of the paediatric research has focused on activities that enhance cardiorespiratory fitness (Rowland, 2005), recent findings indicate that strength training can offer unique benefits to children and adolescents. In addition to

Table 5.2.1 Potential benefits of youth strength training

- Increased muscle strength.
- Increased muscle power.
- Increased local muscular endurance.
- Improved bone health.
- Improved body composition.
- Improved motor performance skills.
- Enhanced sports performance.
- Increased resistance to sports-related injuries.
- A more positive attitude towards lifetime physical activity.

enhancing musculoskeletal strength, regular participation in youth strength training can improve cardiovascular risk profile, facilitate weight control, improve motor performance skills, and increase resistance to sports-related injuries. Moreover, since good health habits established during childhood may carry over into adulthood, the potential positive influence on the adult lifestyle should be recognized (Telama *et al.*, 2005; Trudeau, Laurencelle and Shephard, 2004). A summary of the potential benefits of regular participation in a youth strength-training programme is given in Table 5.2.1.

5.2.5.1 Cardiovascular risk profile

The potential influence of strength training on body composition (the percentage of total body weight that is fat verus fat-free) has become an important topic of investigation given that the prevalence of obesity among children and adolescents continues to increase worldwide (Wang and Lobstein, 2006). Although regular physical activity is the cornerstone of treatment, obese youths often lack the motor skills and confidence to be physically active, and they may actually perceive prolonged periods of aerobic exercise to be boring or discomforting. Excess body weight also hinders the performance of weight-bearing physical activities such as jogging and increases the risk of musculoskeletal injuries.

Recently, it has been suggested that strength training may offer observable health value to obese children and adolescents (Benson, Torade and Fiatarone Singh, 2008a; Faigenbaum and Westcott, 2007). Several studies have reported favourable changes in body composition following participation in a strength-training programme or a circuit weight-training (i.e. combined strength and aerobic training) programme in children and adolescents who were obese or at risk for obesity (Benson, Torade and Fiatarone Singh, 2008b; McGuigan *et al.*, 2009; Shaibi *et al.*, 2006; Sothern *et al.*, 2000). Of note, Shaibi *et al.* (2006) found that participation in a 16-week strength-training programme significantly decreased body fat and significantly increased insulin sensitivity in adolescent males who were at risk for obesity. Since the increase in insulin sensitively remained significant after adjustment for changes in total fat mass and total lean mass, it appeared that regular strength training may have resulted in qualitative changes in skeletal muscle that contributed to enhanced insulin action. In support of these observations, Benson, Torade and Singh (2006) found that muscular strength was an independent and powerful predictor of better insulin sensitivity in youth.

There is no clear association between strength training and reductions in blood pressure or improvements in the blood lipid profile in healthy youths. Limited data suggest that strength training may be an effective nonpharmacologic intervention in hypertensive adolescents (Hagberg *et al.*, 1984), and others have suggested that strength training characterized by moderate loads and a high number of repetitions can have a positive influence on the blood lipid profile of children and adolescents (Fripp and Hodgson, 1987; Sung *et al.*, 2002; Weltman *et al.*, 1987). Although further research is warranted, a comprehensive health-enhancing programme that includes regular physical activity (both aerobic and strength exercise), behavioural counselling, and nutrition education may be most effective for improving the blood pressure in hypertensive youths and the blood lipid profile in children and adolescents with dyslipidemia.

5.2.5.2 Bone health

Current observations suggest that childhood and adolescence may be the most opportune time for the bone-modelling and remodelling process to respond to the tensile and compressive forces associated with weight-bearing activities (Bass, 2000; Hind and Borrows, 2007). Since 50% of adult peak bone mass is acquired before puberty (Magarey *et al.*, 1999; Sabatier *et al.*, 1996), it is critical to maximize bone formation during this developmental period. If age-specific strength-training guidelines are followed and if nutritional recommendations are adhered to, regular participation in a strength-training programme can be a potent osteogenic stimulus during childhood and adolescence.

Results from several research studies indicate that regular participation in sports and specialized fitness activities that include strength training can enhance bone health in youth (MacKelvie *et al.*, 2004; Morris *et al.*, 1997; Ward *et al.*, 2005). Moreover, it has been observed that adolescent weightlifters displayed levels of bone-mineral density (Conroy *et al.*, 1993) and bone-mineral content (Virvidakis *et al.*, 1990) well above the values of age-matched controls. Others reported that pre-adolescent gymnasts whose training involved high-impact loading had significantly thicker cortical bone at the tibia and radius than the control group (Ward *et al.*, 2005). McKay *et al.* (2005) found that a school-based physical activity intervention which included body-weight jump training enhanced bone mass at the weight-bearing proximal femur in children.

Strength training at a young age has also been associated with a decreased risk of osteoporotic fractures later in life (Bass *et al.*, 1998; Heinonen *et al.*, 2000). However, the importance of maintaining participation in weight-bearing physical activities as an ongoing lifestyle choice must not be overlooked as training-induced improvements in bone health may be lost over time if the programme is not continued (Gustavsson, Olsson and Nordstrom, 2003).

5.2.5.3 Motor performance skills and sports performance

Improvements in selected motor performance skills (e.g. long jump, vertical jump, sprint speed, and medicine-ball toss) have been observed in children and adolescents following strength training (Faigenbaum and Mediate, 2006; Falk and Mor, 1996; Flanagan *et al.*, 2002; Hetzler *et al.*, 1997). As previously observed in adults, researchers have reported that the combination of strength training and plyometric training may offer the most benefit for children and adolescents (Faigenbaum *et al.*, 2007b; Lephart *et al.*, 2005; Myer *et al.*, 2005). The available data indicate that the effects of strength training and plyometric training may actually be synergistic, with their combined effect being greater than that of each programme performed alone.

Although the potential for strength training to enhance the sports performance of young athletes seems reasonable, scientific evaluations of this observation are difficult. Two studies (Blanksby and Gregor, 1981; Bulgakova, Vorontsov and Fomichenko, 1990) reported favourable changes in swim performance in age-group swimmers, although one study found no significant difference in freestyle turning performance in adolescent swimmers who performed 15 minutes of plyometric training for 20 weeks (Cossor *et al.*, 1999). Other researchers who studied young basketball, rugby, and soccer players noted the importance of incorporating strength training into sports practice sessions in order to maximize gains in muscular strength and power (Christou *et al.*, 2006; Gabbett, Johns and Riemann, 2008; Vamvakoudis *et al.*, 2007). Although most published reports and anecdotal comments from youth coaches suggest that regular participation in a well-designed strength-training programme will enhance athletic performance, further research is still required in this important field of study.

5.2.5.4 Sports-related injuries

Appropriately designed and sensibly progressed conditioning programmes that include strength training may help to reduce the likelihood of sports-related injuries in young athletes (Abernethy and Bleakley, 2007; Hewett, Myer and Ford, 2005; Renstrom *et al.*, 2008). By addressing the risk factors associated with youth sport injuries (e.g. low fitness level, muscle imbalances, errors in training), it has been suggested that both acute and overuse injuries could be reduced by 15% to 50% (Micheli, 2006). While there are many mechanisms to potentially reduce sports-related injuries in young athletes (e.g. coaching education, safe equipment, proper nutrition), enhancing physical fitness as a preventative health measure should be considered a cornerstone of multi-component treatment programmes.

Comprehensive conditioning programmes that include strength training have proven to be an effective strategy for reducing sports-related injuries in adolescent athletes (Heidt *et al.*, 2000; Hewett *et al.*, 1999; Mandelbaum *et al.*, 2005) and it is possible that similar effects would be observed in children, although additional research is needed to support this contention. Pre-season conditioning programmes that included strength training decreased the number and severity of injuries in adolescent American football players (Cahill and Griffith, 1978) and, similarly, decreased the incidence of injury in adolescent soccer players (Heidt *et al.*, 2000). Others observed that balance training and strengthening exercises were effective in reducing sports-related injuries in adolescent athletes (Wedderkopp *et al.*, 1999, 2003).

In addition, pre-season conditioning programmes that included strength training and education on jumping mechanics significantly reduced the number of serious knee injuries in adolescent female athletes (Hewett *et al.*, 1999; Mandelbaum *et al.*, 2005). Due to the sedentary lifestyle of a growing number of children and adolescents (Hill, King and Armstrong, 2007), there is a distinct need to ensure that all aspiring young athletes participate in some type of pre-season conditioning programme prior to sports practice and competition.

5.2.6 YOUTH STRENGTH-TRAINING GUIDELINES

Although there is no minimum age at which children can begin strength training, all participants must be mentally and physically ready to comply with coaching instructions and undergo the stress of a training programme. In general, if a child is ready for participation in sports activities (generally age seven or eight), then they may be ready for some type of strength training. A medical examination prior to participation in a youth strength-training programme is not mandatory for apparently healthy children, but a medical examination is recommended for youths with known medical conditions, including diabetes, obesity and orthopedic ailments (Behm *et al.*, 2008; Faigenbaum *et al.*, 2009).

Instruction and supervision should be provided by qualified adults who have an understanding of youth strength-training guidelines and knowledge of the physical and psychosocial uniqueness of children and adolescents. Qualified and enthusiastic instruction not only enhances participant safety and enjoyment, but can improve programme adherence and optimize strength gains (Coutts, Murphy and Dascombe, 2004). Instructors should provide basic education on weight-room etiquette, spotting procedures, and exercise technique. Since visual feedback can help young lifters learn proper form and become cognizant of poor lifting biomechanics, exercise demonstrations, mirrors, or video equipment can be used to make youths aware of training errors.

It is important that youth coaches teaching advanced training programmes have the appropriate practical experience and training (e.g. Certified Strength and Conditioning Specialist or Accredited Strength and Conditioning Coach). While less experienced coaches and volunteers can assist in the implementation and supervision of an advanced strength-training workout, it is unlikely that they will be able to provide the level of technical expertise and instruction that is needed to safely and effectively learn advanced training procedures. If qualified supervision and a safe training environment are not available, youths should not perform strength exercise, due to the increased risk of injury.

Prior to every strength-training session, youths should participate in warm-up activities. Since long-held beliefs regarding the routine practise of warm-up static stretching have recently been questioned (Shrier, 2004; Thacker *et al.*, 2004), there has been rising interest in dynamic warm-up procedures. Dynamic warm-up involves the performance of various hops, skips, jumps, and movement-based exercises for the upper and lower body, designed to elevate core body temperature, enhance motor-unit excitability, improve kinaesthetic awareness, and maximize active ranges of motion (Faigenbaum and McFarland, 2007; Robbins, 2005). A dynamic warm-up that includes moderate- and high-intensity movements has been shown to enhance power performance in youths (Faigenbaum *et al.*, 2005, 2006a, 2006b; Siatras *et al.*, 2003). Without evidence to endorse pre-event static stretching, a reasonable suggestion is to perform five to ten minutes of dynamic activities during the warm-up period, and less-intense callisthenics and static stretching at the end of the workout.

Other programme variables that should be considered when designing a youth strength-training programme include: (1) choice and order of exercise, (2) training intensity and volume, (3) rest intervals between sets and exercises, (4) repetition velocity, (5) training frequency, and (6) programme variation. Table 5.2.2 summarizes youth strength-training guidelines.

5.2.6.1 Choice and order of exercise

Although a limitless number of exercises can be used to enhance muscular strength, it is important to select exercises that are appropriate for a child's body size, fitness level, and exercise technique experience. The choice of exercises should promote muscle balance across joints and between opposing muscle groups (e.g. quadriceps and hamstrings). Weight machines (both child-sized and adult-sized), as well as free weights (barbells and dumbbells), elastic bands, medicine balls, and body-weight exercises, have been used by children and adolescents in clinical and school-based fitness programmes. While weight-machine and body-weight exercises help to facilitate a safe environment when supervision is limited, training with free weights and medicine balls may offer the best opportunity to enhance motor performance skills and athletic performance.

Regardless of the mode of training, it is reasonable to start with relatively simple exercises and gradually progress to more advanced multi-joint movements as confidence and competence improve. With qualified supervision and instruction, youths can learn how to perform plyometric exercises as well as weightlifting movements such as the snatch and clean and jerk. As shown in Figure 5.2.2, young weightlifters should learn how to perform advanced exercises with a light load.

An important issue concerning the choice of exercise is the inclusion of exercises for the *core* of a young lifter's body (i.e. abdomen, gluteals, and lower back) (Hibbs *et al.*, 2008). In several reports, lower-back pain was the most frequent injury in adolescent athletes who participated in a strength-training programme (Brady, Cahill and Bodnar, 1982; Brown and Kimball, 1983). Although many factors need to be considered when evaluating these data (e.g. exercise technique

Table 5.2.2 General youth strength-training guidelines

- Provide qualified instruction and supervision.
- Ensure the exercise environment is safe and free of hazards.
- Start each training session with a 5–10 minute dynamic warm-up.
- Begin with relatively light loads and focus on learning the correct exercise technique.
- Perform one to three sets of 6–15 repetitions on a variety of strength exercises.
- Include specific exercises that strengthen the core muscles.
- Gradually progress the intensity and volume of training, depending on goals and abilities.
- Increase the resistance gradually (5–10%) as strength improves.
- Cool-down with less-intense callisthenics and static stretching.
- Strength-train two to three times per week on non-consecutive days.
- Systematically vary the training programme over time.

Figure 5.2.2 A 12-year-old child completing the clean-and-jerk exercise

and progression of training loads), the importance of general physical fitness and lower-back health should not be overlooked (Andersen, Wedderkopp and Leboeuf-Yde, 2006). Because of the potential for lower-back injuries, there is a need for pre-habilitation interventions for the core musculature in order to attempt to reduce the prevalence and/or severity of lower-back pain in youth. That is, exercises that can be prescribed for the rehabilitation of an injury should be prescribed beforehand as part of a preventative health measure. Since there is no one single exercise that activates all of the core muscles, a combination of different exercises will likely offer the most benefit.

In regards to the order of exercises, most youths will perform total-body workouts involving multiple exercises several times per week, stressing all major muscle groups each session. In this type of workout, large-muscle-group exercises should be performed before smaller-muscle-group exercises, and multiple-joint exercises should be performed before single-joint exercises. It is also helpful to perform more challenging exercises earlier in the workout, when the neuromuscular system is less fatigued. Thus, if weightlifting or plyometric exercises are part of a child's workout programme, these should be performed early in the training session so that the child can perform them properly without undue fatigue.

5.2.6.2 Training intensity and volume

'Training intensity' typically refers to the amount of resistance used for a specific exercise, whereas 'training volume' generally refers to the total amount of work performed in a training session. While both of these programme variables are significant, training intensity is one of the more important factors in the design of a strength-training programme. Nonetheless, in order to maximize gains in muscular fitness and minimize the risk of injury, youths must first learn how to perform each exercise correctly with a light load (e.g. unloaded barbell) and then gradually progress the training intensity and/or volume to the desired level without compromising exercise technique.

A simple approach is to first establish the repetition range, and then by trial and error determine the maximum load that can be handled for the prescribed range. For example, a child might begin strength training with one set of 10–15 repetitions with a relatively light load in order to develop proper exercise technique (Faigenbaum *et al.*, 1999). Depending on individuals needs, goals, and abilities, over time the programme can be progressed to two to three sets with heavier loads (e.g. 6–10RM) to maximize gains in muscular strength and power (Faigenbaum *et al.*, 2010). While all exercises do not need to be performed for the same number of sets, multiple-set training protocols have proven to be more effective than single-set protocols in adults, and it appears that similar findings occur in children and adolescents (Ratamess *et al.*, 2009). Note that due to the relatively intense nature of weightlifting and plyometric movements, fewer than six to eight repetitions per set are typically recommended in order to maintain movement speed and efficiency for all repetitions within a set.

5.2.6.3 Rest intervals between sets and exercises

The length of the rest interval between sets and exercises is of primary importance. While a rest interval of at least two to three minutes for primary exercises is typically recommended during adult strength-training programmes (Ratamess *et al.*, 2009), this may not be consistent with the needs and abilities of children and adolescents due to growth- and maturation-related differences in response to physical exertion (Falk and Dotan, 2006). For example, it has been reported that children have a higher oxidative capacity than adults and a tendency towards faster phosphocreatine resynthesis following high-intensity exercise (Kuno *et al.*, 1995; Taylor *et al.*, 1997).

The available data suggest that strength-training recommendations for rest interval length may need to be age-specific (Faigenbaum *et al.*, 2008; Zafeiridis *et al.*, 2005). For example, Faigenbaum *et al.* (2008) reported significant differences in lifting performance between boys, teenagers, and men in response to various rest-interval lengths on the bench-press exercise. In this study, pre-adolescent boys (age 11.3 ± 0.8 years), adolescent boys (age 13.6 ± 0.6 years), and men (age 21.4 ± 2.1 years) performed three sets with a 10RM load and a one-, two-, and three-minute rest interval between sets. As shown in Figure 5.2.3, boys and adolescents performed significantly more total repetitions than adults following protocols with a one-, two-, and three-minute rest interval. While adults

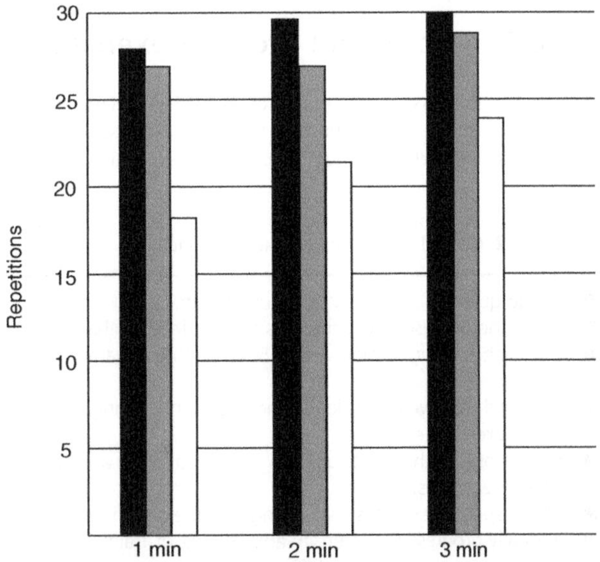

Figure 5.2.3 Effect of rest-interval length on bench-press lifting performance in boys (black bar), teenagers (hatched bar), and men (white bar). Subjects performed three sets with a 10 RM load and a one-, two-, and three-minute rest interval between sets. Total repetitions completed for three sets at each rest interval are shown. Based on data from Faigenbaum *et al.* (2008)

may require up to three minutes of recovery between sets if strength training is the primary goal, these findings suggest that a rest interval of only one to two minutes is needed to minimize loading reductions while maintaining a high lifting volume in youths.

5.2.6.4 Repetition velocity

Since youths need to learn how to perform each exercise correctly with a relatively light load, it is generally recommended that they strength-train in a controlled manner at a moderate velocity. However, different training velocities may be used depending on the choice of exercise. For example, plyometric and weightlifting movements are explosive but highly controlled and should be performed at a high velocity. Although additional research is needed, it is likely that the performance of different training velocities within a training programme may provide the most effective strength-training stimulus.

5.2.6.5 Training frequency

A strength-training frequency of two to three times per week on non-consecutive days will allow for adequate recovery between sessions (48–72 hours) and will be effective for enhancing muscular fitness in children and adolescents. Although once-per-week training may be effective in retaining the strength gains made after strength training (DeRenne *et al.*, 1996), it may be suboptimal for enhancing muscular strength in youth (Blimkie *et al.*, 1989; Faigenbaum *et al.*, 2002). While some young athletes may participate in strength and conditioning activities more than three days per week, factors such as the training volume, training intensity, exercise selection, nutritional intake, and sleep habits need to be considered as they may influence the athlete's ability to recover from and adapt to the training programme.

5.2.6.6 Programme variation

Systematic variance of a training programme over time is known as periodization. In the long term, periodized training programmes (with adequate recovery between training sessions) will reduce the risk of overtraining and allow participants to make even greater gains as the body will be challenged to adapt to even greater demands (Ratamess *et al.*, 2009). While additional research involving younger populations is needed, it is reasonable to suggest that children and adolescents who participate in periodized strength-training programmes and continue to improve their health and fitness may be more likely to adhere to their exercise programmes. Furthermore, planned changes in the programme variables can help to prevent training plateaus, which are not uncommon after the first weeks of strength training.

In order to maximize long-term gains in physical fitness, youth conditioning-training programmes should also include educational sessions on lifestyle factors and behaviours that are conducive to high performance. For example the importance of proper nutrition, sufficient hydration, and adequate sleep should not be overlooked. Detailed information on designing youth strength-training programmes is beyond the scope of this chapter; see Faigenbaum and Westcott (2009), Jeffreys (2008), Kraemer and Fleck (2005), or Mediate and Faigenbaum (2007) for further information.

5.2.7 CONCLUSION

Despite outdated concerns regarding the safety and effectiveness of youth strength training, a compelling body of scientific evidence now indicates that strength training has the potential to offer observable health and fitness value to children and adolescents provided that appropriate training guidelines are followed and qualified instruction is available. In addition to fitness-related benefits, the effects of strength training on selected health-related measures, including bone health, body composition, and sports-injury reduction, should be recognized by teachers, coaches, and health-care providers. If youth strength-training programmes are well designed and sensibly progressed over time, children and adolescents can gain the knowledge, skills, and self-motivation to regularly strength-train as a lifestyle choice. An important future research goal should be to establish the combination of programme variables that enhance long-term training adaptations in young athletes and youths with various medical conditions.

ACKNOWLEDGEMENTS

The author thanks Krissi Pennisi for creating Figures 5.2.1 and 5.2.3.

References

Abernethy L. and Bleakley C. (2007) Strategies to prevent injury in adolescent sport: a systematic review. *Br J Sports Med*, **41**, 627–638.

American College of Sports Medicine. (2010) *ACSM's Guidelines for Exercise Testing and Prescription*. Philadelphia, PA: Lippincott, Williams & Wilkins.

Andersen L., Wedderkopp N. and Leboeuf-Yde C. (2006) Association between back pain and physical fitness in adolescents. *Spine*, **31**, 1740–1744.

Annesi J., Westcott W., Faigenbaum A. and Unruh J. (2005) Effects of a 12 week physical activity program delivered by YMCA after-school counselors (Youth Fit for Life) on fitness and self-efficacy changes in 5–12 year old boys and girls. *Res Q Exerc Sport*, **76**, 468–476.

Australian Strength and Conditioning Association. (2007). Resistance training for children and youth: A position stand from the Australian Strength and Conditioning Association. www.strength and conditioning.org. Accessed February 9, 2009.

Baker D. (2002) Differences in strength and power among junior-high, senior-high, college-aged and elite professional rugby league players. *J Strength Cond Res*, **16**, 581–585.

Bass S. (2000) The prepubertal years. A uniquely opportune stage of growth when the skeleton is most responsive to exercise? *Sports Med*, **39**, 73–78.

Bass S., Pearce G., Bradney M. *et al.* (1998) Exercise before puberty may confer Residual benefits in bone density in adulthood: studies in active prepubertal and retired female gymnasts. *J Bone Miner Res*, **13**, 500–507.

Behm D., Faigenbaum A., Falk B. and Klentrou P. (2008) Canadian Society for Exercise Physiology position paper: resistance training in children and adolescents. *J Appl Physiol Nutr Metab*, **33**, 547–561.

Benson A., Torade M. and Singh M. (2006) Muscular strength and cardiorespiratory fitness is associated with higher insulin sensitivity in children and adolescents. *Int J Pediatr Obes*, **1**, 222–231.

Benson A., Torade M. and Fiatarone Singh M. (2008a) The effect of high-intensity progressive resistance training on adiposity in children: a randomized controlled trial. *Int J Obes*, **32**, 1016–1027.

Benson A., Torade M. and Fiatarone Singh M. (2008b) Effects of resistance training on metabolic fitness in children and adolescents. *Obes Rev*, **9**, 43–66.

Blanksby B. and Gregor J. (1981) Anthropometric, strength and physiological changes in male and female swimmers with progressive resistance training. *Aust J Sport Sci*, **1**, 3–6.

Blimkie C. and Bar-Or O. (2008) Muscle strength, endurance and power: trainability during childhood, in *The Young Athlete* (eds H. Hebestreit and O. Bar-Or), Blackwell Publishing, MA, pp. 65–83.

Blimkie C., Martin J., Ramsay D. *et al.* (1989) The effects of detraining and maintenance weight training on strength development in prepubertal boys. *Can J Sport Sci*, **14**, 104P.

Brady T., Cahill B. and Bodnar L. (1982) Weight training related injuries in the high school athlete. *Am J Sports Med*, **10**, 1–5.

British Association of Sport and Exercise Sciences (2004) BASES position statement on guidelines for resistance exercise in young people. *J Sports Sci*, **22**, 383–390.

Brown E. and Kimball R. (1983) Medical history associated with adolescent power lifting. *Pediatrics*, **72**, 636–644.

Bulgakova N., Vorontsov A. and Fomichenko T. (1990) Improving the technical preparedness of young swimmers by using strength training. *Sov Sports Rev*, **25**, 102–104.

Byrd R., Pierce K., Rielly L. and Brady J. (2003) Young weightlifters' performance across time. *Sports Biomech*, **2**, 133–140.

Cahill B. and Griffith E. (1978) Effect of preseason conditioning on the incidence and severity of high school football knee injuries. *Am J Sports Med*, **6**, 180–184.

Christou M., Smilios I., Sptiropoulos K. *et al.* (2006) Effects of resistance training on the physical capacities of adolescent soccer players. *J Strength Cond Res*, **20**, 783–791.

Chu D., Faigenbaum A. and Falkel J. (2006) *Progressive Plyometrics for Kids*, Healthy Learning, Monterey, CA.

Conroy B., Kraemer W., Maresh C. *et al.* (1993) Bone mineral density in elite junior Olympic weightlifters. *Med Sci Sports Exerc*, **25**, 1103–1109.

Cossor J., Blanksby B. and Elliot B. (1999) The influence of plyometric training on the freestyle tumble turn. *J Sci Med Sport*, **2**, 106–116.

Coutts A., Murphy A. and Dascombe B. (2004) Effect of direct supervision of a strength coach on measures of muscular strength and power in young rugby league players. *J Str Cond Res*, **18**, 316–323.

Department of Health (2004) At Least Five a Week. Evidence on the Impact of Physical Activity and its Relationship to Health. A Report from the Chief Medical Officer, Department of Health, London.

Department of Health and Human Services (2008) *Physical Activity Guidelines for Americans*, Department of Health and Human Services, Washington, DC.

DeRenne C., Hetzler R., Buxton B. and Ho K. (1996) Effects of training frequency on strength maintenance in pubescent baseball players. *J Strength Cond Res*, **10**, 8–14.

Faigenbaum A. and Myer G. (2010) Resistance training among young athletes: safety, efficacy and injury prevention effects. *Brit J Sports Med*, **44**, 56–63.

Faigenbaum A. and McFarland J. (2007) Guidelines for implementing a dynamic warm-up for physical education. *J Phys Ed Rec Dance*, **78**, 25–28.

Faigenbaum A. and Mediate P. (2006) The effects of medicine ball training on physical fitness in high school physical education students. *Phys Educ*, **63**, 160–167.

Faigenbaum A. and Westcott W. (2007) Resistance training for obese children and adolescents. *President's Council Physical Fitness Sport Res Digest*, **8**, 1–8.

Faigenbaum A. and Westcott W. (2009) *Youth Strength Training: Programs for Health, Fitness and Sport*, Human Kinetics, Champaign, IL.

Faigenbaum A., Zaichkowsky L., Westcott W. *et al.* (1993) The effects of a twice per week strength training program on children. *Pediatr Exerc Sci*, **5**, 339–346.

Faigenbaum A., Westcott W., Micheli L. *et al.* (1996) The effects of strength training and detraining on children. *J Strength Cond Res*, **10**, 109–114.

Faigenbaum A., Westcott W., Loud R. and Long C. (1999) The effects of different resistance training protocols on muscular strength and endurance development in children. *Pediatrics*, **104**, E5.

Faigenbaum A., Milliken L., LaRosa Loud R. *et al.* (2002) Comparison of 1 day and 2 days per week of strength training in children. *Res Q Exerc Sport*, **73**, 416–424.

Faigenbaum A., Milliken L. and Westcott W. (2003) Maximal strength testing in children. *J Strength Cond Res*, **17**, 162–166.

Faigenbaum A., Bellucci M., Bernieri A., Bakker B. and Hoorens K. (2005). Acute effects of different warm-up protocols on fitness performance in children. *J Str Cond Res*, **19**, 376–381.

Faigenbaum A., Kang J., McFarland J. *et al.* (2006a) Acute Effects of Different Warm-up Protocols on Anaerobic Performance in Teenage Athletes. *Pediatr Exerc Sci*, **17**, 64–75.

Faigenbaum A., McFarland J., Schwerdtman J. *et al.* (2006b) Dynamic warm-up protocols, with and without a weighted vest and fitness performance in high school female athletes. *J Athl Train*, **41**, 357–363.

Faigenbaum A., McFarland J., Johnson L. *et al.* (2007a) Preliminary evaluation of an after-school resistance training program. *Percept Mot Skills*, **104**, 407–415.

Faigenbaum A., McFarland J., Keiper F. *et al.* (2007b) Effects of a short term plyometric and resistance training program on fitness performance in boys age 12 to 15 years. *J Sports Sci Med*, **6**, 519–525.

Faigenbaum A., Ratamess N., McFarland J. *et al.* (2008) Effect of rest interval length on bench press performance in boys, teens and men. *Pediatr Exerc Sci*, **20**, 457–469.

Faigenbaum A., Kraemer W., Blimkie C. *et al.* (2009) Youth resistance training: Updated position statement paper from the National Strength and Conditioning Association. *J Strength Cond Res*, **23**, S60–S79.

Falk B. and Dotan R. (2006) Child-adult differences in the recovery from high intensity exercise. *Exerc Sport Sci Rev*, **34**, 107–112.

Falk B. and Eliakim A. (2003) Resistance training, skeletal muscle and growth. *Pediatr Endocrinol Rev*, **1**, 120–127.

Falk B. and Mor G. (1996) The effects of resistance and martial arts training in 6- to 8-year-old boys. *Pediatr Exerc Sci*, **8**, 48–56.

Falk B. and Tenenbaum G. (1996) The effectiveness of resistance training in children. A meta-analysis. *Sports Med*, **22**, 176–186.

Flanagan S., Laubach L., DeMarco G. *et al.* (2002) Effects of two different strength training modes on motor performance in children. *Res Q Exerc Sport*, **73**, 340–344.

Fripp R. and Hodgson J. (1987) Effect of resistive training on plasma lipid and lipoprotein levels in male adolescents. *J Pediatr*, **111**, 926–931.

Fukunga T., Funato K. and Ikegawa S. (1992) The effects of resistance training on muscle area and strength in prepubescent age. *Ann Physiol Anthropol*, **11**, 357–364.

Gabbett T., Johns J. and Riemann M. (2008) Performance changes following training in junior rugby league players. *J Strength Cond Res*, **22**, 910–917.

Gonzales-Badillo J., Gorostiaga E., Arellano R. and Izquierdo M. (2005) Moderate resistance training volume produces more favorable strength gains than high or low volumes during a short-term training cycle. *J Strength Cond Res*, **19**, 689–697.

Gould J. and DeJong A. (1994) Injuries to children involving home exercise equipment. *Arch Pediatr Adol Med*, **148**, 1107–1109.

Gumbs V., Segal D., Halligan J. and Lower G. (1982) Bilateral distal radius and ulnar fractures in adolescent weight lifters. *Am J Sports Med*, **10**, 375–379.

Gustavsson A., Olsson T. and Nordstrom P. (2003) Rapid loss of bone mineral density of the femoral neck after cessation of ice hockey training: a 6 year longitudinal study in males. *J Bone Miner Res*, **18**, 1964–1969.

Hagberg J., Ehsani A., Goldring D. *et al.* (1984) Effect of weight training on blood pressure and hemodynamics in hypertensive adolescents. *J Pediatr*, **104**, 147–151.

Hamill B. (1994) Relative safety of weight lifting and weight training. *J Strength Cond Res*, **8**, 53–57.

Heidt R., Swetterman L., Carlonas R. *et al.* (2000) Avoidance of soccer injuries with preseason conditioning. *Am J Sports Med*, **28**, 659–662.

Heinonen A., Sievanen H., Kannus P. *et al.* (2000) High impact exercise and bones of growing girls: a 9-month control trial. *Osteoporos Int*, **11**, 1010–1017.

Hetzler R., DeRenne C., Buxton B. *et al.* (1997) Effects of 12 weeks of strength training on anaerobic power in prepubescent male athletes. *J Strength Cond Res*, **11**, 174–181.

Hewett T., Riccobene J., Lindenfeld T. and Noyes F. (1999) The effects of neuromuscular training on the incidence of knee injury in female athletes: a prospective study. *Am J Sports Med*, **27**, 699–706.

Hewett T., Myer G. and Ford K. (2005) Reducing knee and anterior cruciate ligament injuries among female athletes. *J Knee Surg*, **18**, 82–88.

Hibbs A., Thompson K., French D. *et al.* (2008) Optimizing performance by improving core stability and core strength. *Sports Med*, **38**, 995–1008.

Hill A., King N. and Armstrong T. (2007) The contribution of physical activity and sedentary behaviours to the growth and development of children and adolescents. *Sports Med*, **37**, 533–545.

Hind K. and Borrows M. (2007) Weight-bearing exercise and bone mineral accrual in children and adolescents: a review of controlled trials. *Bone*, **51**, 81–101.

Holm I., Fredriksen P., Fosdahl M. and Vollestad N. (2008) A normative sample of isotonic and isokinetic muscle strength measurements in children 7 to 12 years of age. *Acta Paediatr*, **97**, 602–607.

Ingle L., Sleap M. and Tolfrey K. (2006) The effect of a complex training and detraining programme on selected strength and power variables in early prepubertal boys. *J Sports Sci*, **24**, 987–997.

Jeffreys I. (2008) *Coaches Guide to Enhancing Recovery in Athletes: A Multidimensional Approach to Developing the Performance Lifestyle*, Healthy Learning, California.

Jenkins N. and Mintowt-Czyz W. (1986) Bilateral fracture separations of the distal radial epiphyses during weight-lifting. *Br J Sports Med*, **20**, 72–73.

Jones C., Christensen C. and Young M. (2000) Weight training injury trends. *Phys Sports Med*, **28**, 61–72.

Kraemer W. and Fleck S. (2005) *Strength Training for Young Athletes*, 2nd edn, Human Kinetics, Champaign, IL.

Kraemer W., Fry A., Frykman P. *et al.* (1989) Resistance training and youth. *Pediatr Exerc Sci*, **1**, 336–350.

Kuno S.H., Takahashi K., Fujimoto H. *et al.* (1995) Muscle metabolism during exercise using phosphorus-31 nuclear magnetic resonance spectroscopy in adolescents. *Eur J Occup Physiol*, **70**, 301–304.

Lee S., Burgeson C., Fulton J. and Spain C. (2007) Physical education and physical activity: results from the school health policies and programs study 2006. *J School Health*, **77**, 435–463.

Lephart S., Abt J., Ferris C. *et al.* (2005) Neuromuscular and biomechanical characteristic changes in high school athletes: a plyometric versus basic resistance program. *Br J Sports Med*, **39**, 932–938.

Lillegard W., Brown E., Wilson D. *et al.* (1997) Efficacy of strength training in prepubescent to early postpubescent males and females: effects of gender and maturity. *Pediatr Rehabil*, **1**, 147–157.

McGuigan M., Tatasciore M., Newtown R. and Pettigrew S. (2009) Eight weeks of resistance training can significantly alter body composition in children who are overweight or obese. *J Strength Cond Res*, **23**, 80–85.

McKay H., Maclean L., Petit M. *et al.* (2005) 'Bounce at the Bell': a novel program of short bouts of exercise improves proximal femur bone mass in early pubertal children. *Br J Sports Med*, **39**, 521–526.

MacKelvie K., Petit M., Khan K. *et al.* (2004) Bone mass and structure and enhanced following a 2-year randomized controlled trial of exercise in prepubertal boys. *Bone*, **34**, 755–764.

Magarey A., Boulton T., Chatterton B. *et al.* (1999) Bone growth from 11 to 17 years: relationship to growth, gender and changes with pubertal status including timing of menarche. *Acta Paediatr*, **88**, 139–146.

Malina R. (2006) Weight training in youth-growth, maturation and safety: an evidence based review. *Clin J Sports Med*, **16**, 478–487.

Mandelbaum B., Silvers H., Watanabe D. *et al.* (2005) Effectiveness of a neuromuscular and proprioceptive training program in preventing anterior cruciate ligament injuries in female athletes. *Am J Sports Med*, **33**, 1003–1010.

Mediate P. and Faigenbaum A. (2007) *Medicine Ball for All Kids*, Healthy Learning, California.

Mersch F. and Stoboy H. (1989) Strength training and muscle hypertrophy in children, in *Children and Exercise XIII* (eds S. Oseid and K. Carlsen), Human Kinetics, Champaign, IL, pp. 165–182.

Micheli L. (2006) Preventing injuries in sports: what the team physician needs to know, *F.I.M.S. Team Physician Manual*, 2nd edn (eds K. Chan, L. Micheli, A. Smith *et al.*), CD Concept, Hong Kong, pp. 555–572.

Milliken L., Faigenbaum A., LaRosa Loud R. and Westcott W. (2008) Correlates of upper and lower body muscular strength in children. *J Strength Cond Res*, **22**, 1339–1346.

Morris F., Naughton G., Gibbs J. *et al.* (1997) Prospective ten-month exercise intervention in premenarcheal girls: positive effects on bone and lean mass. *J Bone Min Res*, **12**, 1453–1462.

Mountjoy M., Armstrong N., Bizzini L. *et al.* (2008) IOC consensus statement: 'training the elite young athlete. *Clin J Sport Med*, **18**, 122–123.

Myer G. and Wall E. (2006) Resistance training in the young athlete. *Oper Tech Sports Med*, **14**, 218–230.

Myer G., Ford K., Palumbo J. and Hewitt T. (2005) Neuromuscular training improves performance and lower extremity biomechanics in female athletes. *J Strength Cond Res*, **19**, 51–60.

Myer G., Quatman C., Khoury J. *et al.* (2009) Youth vs. adult 'weightlifting' injuries presenting to United States emergency rooms. *J Str Cond Res*, **23**, 2054–2060.

National Association for Sport and Physical Education (2005) *Physical Education for Lifelong Fitness*, 2nd edn, Human Kinetics, Champaign, IL.

Ortega F., Ruiz J., Castillo M. and Sjostrom M. (2008) Physical fitness in children and adolescence: a powerful marker of health. *Int J Obes*, **32**, 1–11.

Ozmun J., Mikesky A. and Surburg P. (1994) Neuromuscular adaptations following prepubescent strength training. *Med Sci Sports Exerc*, **26**, 510–514.

Payne V., Morrow J., Johnson L. and Dalton S. (1997) Resistance training in children and youth: a meta-analysis. *Res Q Exerc Sport*, **68**, 80–88.

Pfeiffer R. and Francis R. (1986) Effects of strength training on muscle development in prepubescent, pubescent and postpubescent males. *Physician Sports Med*, **14**, 134–143.

Pierce K., Byrd R. and Stone M. (1999) Youth weightlifting – is it safe? *Weightlifting USA*, **17**, 5.

Pierce K., Brewer C., Ramsey M. *et al.* (2008) Youth resistance training. *Prof Strength Cond*, **10**, 9–23.

Pikosky M., Faigenbaum A., Westcott W. and Rodriquez N. (2002) Effects of resistance training on protein utilization in healthy children. *Med Sci Sports Exerc*, **34**, 820–827.

Ramsay J., Blimkie C., Smith K. *et al.* (1990) Strength training effects in prepubescent boys. *Med Sci Sports Exerc*, **22**, 605–614.

Ratamess N., Alvar B., Evetoch T. *et al.* (2009) Progression models in resistance training for healthy adults. *Med Sci Sports Exerc*, **41**, 687–708.

Renstrom P., Ljungqvist A., Arendt E. *et al.* (2008) Non-contact ACL injuries in female athletes: an International Olympic Committee current concepts statement. *Br J Sports Med*, **42**, 394–424.

Rians C., Weltman A. and Cahill B. (1987) Strength training for prepubescent males: is it safe? *Am J Sports Med*, **15**, 483–489.

Risser W. (1991) Weight-training injuries in children and adolescents. *Am Fam Phys*, **44**, 2104–2110.

Robbins D. (2005) Postactivation potentiation and its practical application: a brief review. *J Strength Cond Res*, **19**, 453–458.

Rowe P. (1979) Cartilage fracture due to weight lifting. *Br J Sports Med*, **13**, 130–131.

Rowland T. (2005) *Children's Exercise Physiology*, 2nd edn, Human Kinetics, Champaign, IL.

Ryan J. and Salciccioli G. (1976) Fractures of the distal radial epiphysis in adolescent weight lifters. *Am J Sports Med*, **4**, 26–27.

Sabatier J., Guaydier-Souquieres D., Laroche D. *et al.* (1996) Bone mineral acquisition during adolescence and early adulthood: a study of 574 healthy females 10–24 years of age. *Osteoporos Int*, **6**, 141–148.

Sadres E., Eliakim A., Constantini N. *et al.* (2001) The effect of long-term resistance training on anthropometric measures, muscle strength and self-concept in pre-pubertal boys. *Ped Exerc Sci*, **13**, 357–372.

Sale D. (1989) Strength training in children, in *Perspectives in Exercise Science and Sports Medicine* (eds G. Gisolfi and D. Lamb), Benchmark Press, Indiana, pp. 165–216.

Sewall L. and Micheli L. (1986) Strength training for children. *J Pediatr Orthop*, **6**, 143–146.

Shaibi G., Cruz M., Ball G. *et al.* (2006) Effects of resistance training on insulin sensitivity in overweight Latino Adolescent males. *Med Sci Sports Exerc*, **38**, 1208–1215.

Shrier I. (2004) Does stretching improve performance? A systematic and critical review of the literature. *Clin J Sports Med*, **14**, 267–273.

Siatras T., Papadopoulos G., Mameletzi D. *et al.* (2003) Static and dynamic acute stretching effect on gymnasts' speed in vaulting. *Ped Exerc Sci*, **15**, 383–391.

Sothern M., Loftin J., Udall J. *et al.* (2000) Safety, feasibility and efficacy of a resistance training program in preadolescent obese youth. *Am J Med Sci*, **319**, 370–375.

Sung R., Yu C., Chang S. *et al.* (2002) Effects of dietary intervention and strength training on blood lipid level in obese children. *Arch Disord Child*, **86**, 407–410.

Strong W., Malina R., Blimkie C. *et al.* (2005) Evidence-based physical activity for school age youth. *J Pediatrics*, **146**, 732–737.

Taylor D., Kemp G., Thompson C. and Radda G. (1997) Ageing: effects on oxidative function of skeletal muscle in vivo. *Mol Cell Biochem*, **174**, 321–324.

Telama R., Yang X., Viikari J. *et al.* (2005) Physical activity from childhood to adulthood: a 21 year tracking study. *Am J Prev Med*, **28**, 267–273.

Thacker S., Gilchrist J., Stroup D. and Kimsey C. (2004) The impact of stretching on sports injury risk: a systematic review of the literature. *Med Sci Sports Exerc*, **36**, 371–378.

Thomas K., French D. and Hayes P. (2009) The effect of two plyometric training techniques on muscular power and agility in youth soccer players. *J Strength Cond Res*, **23**, 332–335.

Trudeau F., Laurencelle L. and Shephard R. (2004) Tracking of physical activity from childhood to adulthood. *Med Sci Sports Exerc*, **36**, 1937–1943.

Tsolakis C., Vagenas G. and Dessypris A. (2004) Strength adaptations and hormonal responses to resistance training and detraining in preadolescent males. *J Strength Cond Res*, **18**, 625–629.

Vamvakoudis E., Vrabas I., Galazoulas C. *et al.* (2007) Effects of basketball training on maximal oxygen uptake, muscle strength and joint mobility in young basketball players. *J Strength Cond Res*, **21**, 930–936.

Vaughn J. and Micheli L. (2008) Strength training recommendations for the young athlete. *Phys Med Rehabil Clin N Am*, **19**, 235–245.

Virvidakis K., Georgiu E., Korkotsidis A. *et al.* (1990) Bone mineral content of junior competitive weightlifters. *Int J Sports Med*, **11**, 244–246.

Volek J., Gomez A., Scheett T. *et al.* (2003) Increasing fluid milk intake favorably affects bone mineral density responses to resistance training in adolescent boys. *J Am Diet Assoc*, **103**, 1353–1356.

Wang Y. and Lobstein T. (2006) Worldwide trends in childhood overweight and obesity. *Int J Pediatr Obes*, **1**, 11–25.

Ward K., Robets S., Adams J. and Mughal M. (2005) Bone geometry and density in the skeleton of prepubertal gymnasts and school children. *Bone*, **26**, 1012–1018.

Wedderkopp N., Kaltoft B., Lundgaard M. *et al.* (1999) Prevention of injuries in young female players in European team handball: a prospective intervention study. *Scand J Med Sci Sports*, **9**, 41–47.

Wedderkopp N., Kaltoft B., Holm R. and Froberg K. (2003) Comparison of two intervention programmes in young female players in European handball: with and without ankle disc. *Scand J Med Sci Sports*, **13**, 371–375.

Weltman A., Janney C., Rians C. *et al.* (1986) The effects of hydraulic resistance strength training in pre-pubertal males. *Med Sci Sports Med*, **18**, 629–638.

Weltman A., Janney C., Rians C. *et al.* (1987) Effects of hydraulic-resistance strength training on serum lipid levels in prepubertal boys. *Am J Dis Child*, **141**, 777–780.

Westcott W. (1992) A new look at youth fitness. *Am Fitness Q*, **11**, 16–19.

Zafeiridis A., Dalamitros A., Dipla K. *et al.* (2005) Recovery during high-intensity intermittent anaerobic exercise in boys, teens and men. *Med Sci Sports Exerc*, **37**, 505–512.

5.3 Strength and Conditioning Considerations for the Paralympic Athlete

Mark Jarvis[1], Matthew Cook[2] and Paul Davies[3], [1]English Institute of Sport, Alexander Stadium, WM & EM, GB Wheelchair Basketball, Perry Barr, Birmingham, UK, [2]English Institute of Sport, Sportcity, Manchester, UK, [3]English Institute of Sport, Bisham Abbey Performance Centre, Marlow, Bucks, UK

5.3.1 INTRODUCTION

Exercise participation within disabled populations, whilst still lagging behind that in able-bodied groups, is becoming more prevalent.[1] Many things contribute towards this; not least the understanding that exercise can be of considerable benefit in increasing quality of life, through its ability to preserve function, which in turn results in greater independence in many cases. The positive move towards an increased acceptance of disability within most cultures has made people with a disability more visible within society, and many more people now have a basic understanding of disability and at least a basic familiarity with some of the interventions that can be put in place to support the integration of individuals with a disability into the day-to-day activities that many of us take for granted.

Paralympic sport has been a very visible force in the breaking down of perceptions of disability, focusing upon what can be achieved in spite of physical or learning limitations. As the Paralympic movement has grown, both the size and the strength of competition have increased; medals are won or lost by the same small margins that are seen in Olympic competition. Athletes will therefore seek to maximize their performances by engaging the same support networks that have become commonplace within elite able-bodied sport.

As the birthplace of organized sport for the disabled, the UK has for many years led the way in the support services that it delivers to athletes who represent Great Britain and Northern Ireland at major competitions, including the Paralympics. The following paragraphs are a distillation of what has been learnt through supporting Paralympic and aspirant-Paralympic athletes in their preparations over a number of Paralympic cycles dating as far back as Atlanta 1996.

This chapter will focus specifically upon the disability groups eligible to compete at the Paralympic Games, which are spinal-cord injury, visual impairment, cerebral palsy, amputee, learning-disabled, and *les autres* (from the French for 'the others'; a broad category encompassing many varied disabilities, including those such as dwarfism and multiple sclerosis).

There are some common and notable disability and impairment groups which are not included in the Paralympic programme, the reason for which is largely historical. The earliest multi-sport, multi-disability 'Paralympic' competitions were coordinated by the International Stoke Mandeville Wheelchair Sports Federation, which later became amalgamated into the International Coordination Committee of World Sports Organizations for the Disabled: a coming together of the four disability-specific international sports federations (IOSDs). Between them, these groups represent the six disability groups that currently compete at the Paralympic Games (see Table 5.3.1).

5.3.2 PROGRAMMING CONSIDERATIONS

Before beginning the programming of a training regime for a Paralympic athlete, the fundamental philosophy behind its design must be considered. Any elite strength and conditioning programme should reflect both the demands of the event and the athlete's own unique physical characteristics; Paralympic strength and conditioning programmes should be no different. However, the need to individualize the exercise selection will often take on additional significance due to the functional limitation experienced by the athlete. It should be noted that it is

[1]For example, Sport England (2002) Adults with a Disability and Sport National survey – 2000/2001; Sport England (2009) Active People Survey 2 – 2007/2008.

Table 5.3.1 Disability groups at the Paralympic Games

Acronym	Federation	Disability group(s)
IBSA	International Blind Sports Federation	Visual impairment
INAS-FID	International Sports Federations for Persons with an Intellectual Disability	Intellectual disability (often referred to as learning-disabled)
IWAS	International Wheelchair and Amputee Sports Federation	SCI Amputee *Les autres*
CPISRA	Cerebral Palsy International Sports and Recreation Association	CP

for this reason that many strength and conditioning coaches believe that constructing Paralympic programmes drives them towards a greater level of lateral thinking and innovation within their programme design.

Many coaches believe in an approach which primarily treats the athlete as if able-bodied. Under such a philosophy, the athlete's disability is only taken into consideration when it prevents them from completing an element of the programme. The most common differences occur in exercise selection, when the Paralympic athlete is unable to perform one of the exercises as it would be performed by an able-bodied athlete. Provided that effective solutions are found to adapt the exercises without changing the emphasis of the programme, this strategy will generally lead to a positive outcome: achieving the key goal of enhancing sports performance. However, such a strategy used in isolation may neglect the opportunity to make further performance gains by addressing disability-specific issues. For example, when working with an individual with ataxic CP, additional gains may be made by including balance training within the athlete's conditioning programme.

Based on the above considerations, the schematic in Figure 5.3.1 proposes a methodology which places primary focus on the demands of the sport, while at the same time allowing the opportunity for individualized and disability-specific adaptations.

A common supplementary benefit to the Paralympic athlete who carries out a strength and conditioning regime is an improved quality of daily life. This may come in many forms, but will be related to an increased ease in performing activities of daily living, such as transferring in and out of a wheelchair (through strength adaptations) or reduced breathlessness (through enhanced cardiovascular fitness). It is easy for these benefits to be overlooked by the strength and conditioning coach who focuses solely on sports performance. It should be borne in mind that a positive cycle of improvement can be started, whereby improvements in the ease with which activities of daily living are performed will positively impact upon the energies that can be directed towards performance-based activities.

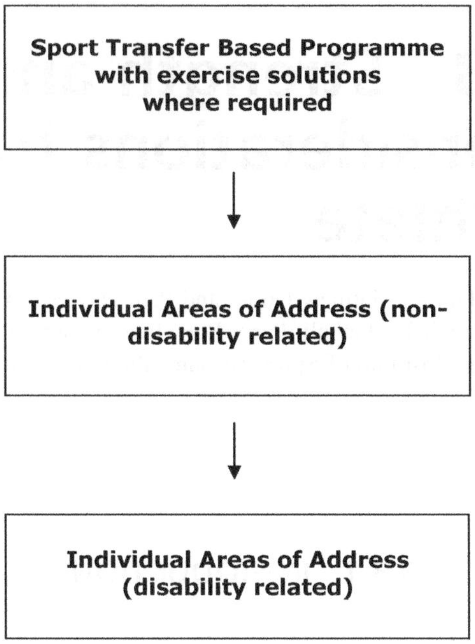

Figure 5.3.1 Schematic for Paralympic programme design

A final fundamental principle which should be considered with regard to exercise selection is that of experimentation. Such variety exists in the precise nature of disabilities, even within a specific disability group or class, that programming must be truly individual. It is often the case that traditional exercises require minor adaptations, but in some instances such a large adaptation is needed that the end result is the development of a completely new exercise. Such a process of experimentation and adaptation is to be applauded, as long as it is based on scientifically robust principles and does not alter the nature of the adaptations that are achieved. Also, within this process the practitioner should be encouraged to share their successes and failures in these endeavours within the strength and conditioning community, in order that the knowledge within the discipline is grown. Finally, it is likely that the best results will be achieved through practical experimentation and by taking a problem-solving approach which involves the athlete themselves. The obvious benefits of this are that an open dialogue about the athlete's disability is started and the athlete gains a greater sense of ownership and control of the programme.

5.3.3 CURRENT CONTROVERSIES IN PARALYMPIC STRENGTH AND CONDITIONING

The first of the slightly more controversial topics which should be considered when coaching disabled athletes is their training age. Disabled athletes often have lower training ages than their able-bodied counterparts of the same chronological age. This is certainly not true of all disabled athletes, especially those with minor disabilities able to compete with able-bodied athletes at a young age and those who have pursued a career as an able-bodied athlete before becoming eligible for a disability classification at a later stage (e.g. after a traumatic, disabling accident). However, the relative lack of provision and opportunity for those athletes who have been disabled from birth has often resulted in mature disability athletes with a lower training status than their able-bodied peers. Training age is an important consideration when planning any training programme, but is particularly important when basing your approach on treating the athlete as if they were able-bodied. Furthermore, both our observations and the growing volume of anecdotal evidence indicate that some disabilities significantly restrict the rate and extent of adaptation that can be achieved. Some forms of CP, for example, can cause athletes to experience an increase in neuromuscular tone in one or more limbs, or systemically. This, at its most severe, can completely prevent the use of the affected limb(s), or at least significantly limit the range of movement possible and the potential for adaptation to flexibility training (active and dynamic). Increased neuromuscular tone also often increases the rate at which athletes become fatigued, thus limiting the total workload and volume, especially of high-intensity work. The underlying mechanisms of this will become apparent as we discuss CP in more detail later in this chapter. As a final point on training age, whilst coaches working with disabled athletes must be aware of the potential issues, the nature of every disability is highly individual and therefore each training programme must be tailored to accommodate individual differences.

Potentially the most controversial topic for all coaches working with disabled athletes is that of classification, particularly when it comes to considering whether it is possible for an athlete to make such improvements as a result of training that they no longer qualify for their classification group. The controversy exists because disability classification is a complex process that involves several stages and differs between sports. It usually involves the assessment of both basic functional movements and sports-specific movements, and often includes assessment within competition. When an athlete is first assigned to a classification it is usually not permanent, especially for young athletes who are still maturing, as the extent of their disability is not yet considered to be stable. Classifications are open to challenge from other competitors and can be reviewed by the sports authorities at any point.

The potential complications for the coach come not only from the effects of maturation-related changes to athletes' functional abilities, but also from training-related changes. The explicit purpose of all training is to elicit physical and physiological changes that enhance sports performance. We know that the human neuromuscular system is capable of change and learning throughout life, through a recognized process known as 'plasticity' (Doidge, 2007). New motor skills are acquired through the formation of neural circuits or maps which are developed and enhanced through practice. The majority of motor learning occurs during childhood and adolescence, but it continues into adult life. This is of course excellent news for all coaches and athletes, as it allows them to refine sport skills; however, some disabilities prevent or at least reduce an athlete's ability to perform some functional movements and to acquire certain skills. In the areas where neuromuscular plasticity is possible, and where training is designed to take advantage of this, it is possible to develop new functional skills that may have a significant impact not only upon their sporting performance and quality of daily living, but on the classification to which they are assigned.

5.3.4 SPECIALIST EQUIPMENT

The increased profile of Paralympic sports, and of disability sports in general, has encouraged an increased rate of sports participation amongst adults with a disability. In order to meet the needs of this increased interest in physical activity, a number of companies have started to produce exercise equipment specifically adapted for use by the disabled performer.

Bodies such as the Inclusive Fitness Initiative (IFI) (a UK initiative) have been set up to promote the design, development, and manufacture of adapted fitness equipment, as well as to train strength and conditioning coaches and to accredit training facilities that have the requisite access, equipment, and expertise to support the training needs of an athlete with a disability.

Initiatives such as this have generated far more opportunities than ever seen before for those with Paralympic aspirations to engage in strength and fitness training in public facilities.

Perhaps the two most significant aids that a strength and conditioning coach will now find readily available to them are arm-crank ergometers, which are currently being mass-produced by numerous manufacturers and have become commonplace in many gyms, and a wide variety of gripping aids, which can increase the range of strength-training options for those with an amputation or other loss of full grip.

5.3.5 CONSIDERATIONS FOR SPECIFIC DISABILITY GROUPS

As previously discussed, whilst numerous disabilities exist, the eligibility and classification criteria laid down by Paralympic sports mean that only a limited range will be seen by the strength and conditioning coach working specifically with Paralympic athletes. The following paragraphs offer some guidance based upon the authors' experience of delivering strength and conditioning programmes to elite Paralympic performers. We will discuss the six Paralympic disability groups in turn, first giving a brief overview of the disability and its manifestation, then moving on to discuss implications for training and testing, and finally highlighting any other key areas for consideration.

5.3.5.1 Spinal-cord injuries

Spinal-cord-injured (SCI) athletes have at some stage had some form of insult to the spinal cord that affects the proper functioning of the neural pathway. This insult can originate from a number of causes, which are either congenital, such as spina bifida, or acquired, such as traumatic injury, tumour, or viral infection (e.g. polio).

The extent of functional limitation that a SCI athlete experiences will be influenced by the level (distance from the brain) at which the spinal cord is injured and the completeness of the loss of nerve transmission. The impact of a SCI is therefore often described using a letter and a number (e.g. T11), which indicates the most distal uninvolved segment of the cord, followed by the comment 'complete' or 'incomplete', indicating the extent of interruption of nerve transmission. 'Complete' indicates an absence of sensory or motor function below the lowest sacral segment, and 'incomplete' indicates that 'there is partial preservation of sensory and/or motor function below the neurological level and includes the lowest sacral segment'.[2]

To the uninitiated, this type of terminology can at first seem daunting. However, the important point to note is that, due to the variability in response and adaptation, functional ability between two athletes with the same classification can vary considerably, and working assumptions should be based upon investigation and communication with the athlete, and not on injury level or pathology alone.

Most athletes will be able to provide detailed information on what level of spinal injury they have sustained, and whether they are complete or incomplete. They will also have a very good understanding of their resulting level of function. As with all athletes, the best method of gaining insight into the particular circumstance is to ask the athlete themselves.

Special considerations

Besides paralysis, several other complicating factors must be considered during training and testing. Spasticity is a common consequence of SCI, with stimuli producing uncontrollable reflex spasms below the level of the injury. These may manifest as extensor spasms, where the legs straighten and become rigid, or as flexor spasms, where the legs are pulled toward the chest. A third manifestation, clonus, is the repetitive twitching of the muscle, often seen at the ankle; such spasms are often sufficiently strong to cause the foot to visibly and repeatedly bounce up and down.

In addition to trying to identify and remove or limit the cause of the reflex action, our experience shows that spastic muscle groups benefit from being stretched up to seven times a day, as this appears to help in reducing motor-neuron excitability. Wheelchair athletes should also attempt to extend hips and knees when at home, to avoid unwanted adaptations to prolonged periods of sitting.

Athletes with a lesion at T6 or above are at risk of a condition known as autonomic dysreflexia. This is a result of an irritation below the level of injury and leads to concomitant lower-body vasoconstriction and upper-body vasodilation, resulting in large and often dangerous increases in systolic blood pressure. Athletes and coaches should make themselves aware of the warning signs of this condition, which include headache, sweating, and reddening of flesh above the level of the injury, along with cold clammy flesh and goosebumps below the injury. Many good resources exist that explain autonomic dysreflexia in greater detail than is possible here, and the reader is referred to the University of Alabama at Birmingham's very clear guidance notes and educational resources for the identification and treatment of autonomic dysreflexia.[3]

Coaches should also be aware that some athletes have historically deliberately induced autonomic dysreflexia in order to seek a physiological advantage. This technique, known as 'boosting', has been shown to increase performance, peak heart rate, and peak oxygen consumption as a result of an enhanced catecholomine response to exercise (Schmid *et al.*, 2001). This practice carries very serious health risks, including death, and is regarded as a prohibited manipulation by the International Paralympic Committee.

Training and testing

The capacity for strength training in SCI athletes will be significantly influenced by the level of trunk function. Strapping may be used to assist athletes with a higher-level injury and poor trunk function in performing strength exercises within their wheelchair, as can manual assistance. Even if strapping is used, postural stability may still be a limiting factor to the performance of exercises. If the athlete has some level of trunk function, this may provide a postural training effect that will contribute to sports performance. However, in athletes with little or no trunk stability this is unlikely to occur and training loads are likely to be limited by balance. It may therefore be difficult to maximally overload an exercise within a wheelchair. Under such circumstances, exercises where a bench provides stability (such as bench press or pull) may be more suitable for stressing the athlete's ability to produce force. Again, adaptive equipment, such as a wider bench, may provide the extra stability needed to perform exercises safely and with good form.

Athletes with lesions at T1 and above may suffer from a weakened grip. This can generally be overcome through various aids allowing weights, handles, and so forth to be strapped to the hand. However, care must be taken to ensure that blood flow is not compromised, especially if the athlete has reduced sensation.

The repeated pushing movement involved in wheelchair propulsion, coupled with extended periods in a sitting position, results in SCI athletes commonly presenting with postural issues, which may be addressed by a strength-training programme. Such individuals are at a high risk of shoulder injury due to a combination of overuse and movement dysfunction. The difficulty in resting an injured shoulder for a wheelchair user means that avoidance of such an injury is paramount. Typical postural issues include protracted shoulders, internally rotated humerus, and thoracic kyphosis. It is recommended that the use of strengthening exercises, soft-tissue treatments, and

[2]Definition taken from International Standards Classifications of SCI, Maynard *et al.* (1997).

[3]http://www.spinalcord.uab.edu/show.asp?durki=21479.

stretching in combination represents the most effective strategy for addressing these issues.

Thermoregulation

The loss of autonomic-nervous-system function associated with SCI leads to reduced or absent sweat response to exercise. Again, the manifestation of this will be influenced by the level and completeness of the lesion. In warm conditions (25–40 °C), affected athletes will be at an increased risk of heat injury (Price, 2006). Increased thermoregulatory stress will occur when heat production is no longer matched by heat loss through sweat capacity and/or convective cooling; this situation is often encountered during the kind of high-intensity, intermittent activities characteristic of indoor team sports. Foot and hand cooling have been found to be effective in reducing both core temperature and heart rate during breaks in exercise (personal observation and Hagobian *et al.*, 2004), as have pre-exercise cooling strategies (e.g. Webbon *et al.*, 2008). It should be noted that whilst aggressive pre-cooling of the core in SCI athletes can prolong performance in the heat that would otherwise be limited by thermoregulatory factors, there is some evidence that this is at the expense of short-term power output and sprint capacity in some athletes (Webborn *et al.*, 2005, personal observation, and athlete report).

During metabolic conditioning it should be noted that in individuals with lesions around T6 or higher, the sympathetic-nervous-system impairment is sufficient to influence heart function, and the maximum achievable heart rate may be limited to 110–130 bpm. As a result, programming based on HR targets may prove unsuitable; in this instance Rating of Perceived Exertion (RPE) is recommended as the best measure of exercise intensity (Tolfrey, Goosey-Tolfrey and Campbell, 2001).

5.3.5.2 Amputees

Like SCI, amputation can arise from a number of causes, both congenital and acquired. In the general population there is clear evidence of a weighting in the distribution of cause of amputation towards vascular and circulatory disorders. In the Paralympic population there is a greater prevalence of amputation as a result of trauma or congenital deformities.

Athletes with an amputation therefore tend to suffer relatively few secondary complications in comparison with other Paralympic groups. Those who have lost limbs as a result of road traffic accidents and other traumatic incidents will generally only be restricted by mode of conditioning-exercise and resistance-exercise selection. Athletes who have suffered a lower-extremity amputation as a result of peripheral vascular disease or diabetes may need to take appropriate precautions during demanding metabolic training sessions. In either case the strength and conditioning coach must be mindful of the debilitating consequences of poor stump care, which can include the overloading of a weight-bearing stump, reduced blood circulation through a too-tight-fitting prosthetic socket, and poor aeration and drying of the skin, which causes the skin to stay moist. From experience, minor skin disorders that are not appropriately identified and treated can rapidly evolve into more serious conditions that impact on health, performance, and the ability to carry out activities of daily living.

Training and testing

During cross-training conditioning work the mode of exercise will be limited by the athlete's physical capabilities. If the athlete is involved in a wheelchair-based sport it may be advantageous to perform some conditioning in a mode which primarily works the posterior musculature of the trunk (to counter the repetitive pushing motion). It has been demonstrated that sprint kinematics in elite amputee sprinters are similar in both injured and non-injured limbs (Buckley, 1999). However, the reduced capacity for eccentric work and the absence of a stretch–shortening cycle at the Achilles tendon is in part compensated for by a larger reliance on concentric power production at the hip and knee (Buckley, 2000). These alterations in biomechanics appear to be highly individual but should be considered key factors in the training approach of both track athletes and those who may seek to use running as a training tool.

When designing a strength-training programme, asymmetry is often an issue. Lower-extremity amputees may still have the ability to exercise various squat patterns using the non-injured leg. Whilst this is encouraged, it does also pose the potential problem of causing imbalances elsewhere in the body. For example, an athlete performing a single-leg squat on their uninjured leg will not only affect the target musculature of that limb (prime movers) but also aspects of the hip, trunk, and even shoulder girdle though the influence of myofascial slings (Myers, 2001). Consequently, further consideration must be given as to how the injured side of the body may be trained through alternative exercise selection. The coach must make a judgement as to the significance of any potential asymmetry versus the potential benefits of the exercise.

Upper-extremity amputees are likely to encounter difficulty in balance during exercises, some of which it may only be possible to perform in a unilateral fashion. If a prosthesis which ensures near-equal limb length is not available then exercises may have to be adapted accordingly. The use of assistance may also be required in order to make exercises which rely on body weight for load feasible. This might include using bands for assistance during pull-ups or providing manual assistance.

Contraindications for specific exercise-programme components will vary with the nature and cause of the disability. For example, the risk factors associated with training a young upper-limb amputee athlete with an extensive pre-trauma training history are significantly different from those with an athlete who has lost a limb through vascular disease. Indeed, the motivation for training each of these performers may be quite different, with one perhaps seeking to optimize sports performance, and the other hoping to reduce risk factors for the development of a secondary disability. It is the responsibility of the strength and conditioning coach to seek out as much relevant information relating to the athlete's disability and underlying pathology as possible, in order to support them in programming decisions.

5.3.5.3 Cerebral palsy

Cerebral palsy (CP) is a lesion of the brain that occurs either *in utero*, during birth, or soon after, which disrupts normal brain development. In Paralympic sport, other conditions that result in the exhibition of CP-like symptoms are often included in this category. The most common of these is acquired brain injury, often as the result of a traumatic event; while whilst not 'true' CP, it has many of the same symptoms, and importantly the adaptations that need to be considered by the coach translate well between the two conditions.

CP is characterized by the impairment of voluntary movement or motor control resulting from lesions to the upper motor neurones in the brain, which control muscle tone and spinal reflexes. Crucially, CP is not a progressive disorder, but one that is characterized by restrictions in the ability to perform some movements and/or control the quality of movement.

The motor involvement and difficulties experienced vary hugely between CP athletes. A minimally impaired athlete may have very few observable movement dysfunctions, whereas a severely impaired athlete may not have enough motor control to perform activities of daily living such as feeding themselves. Consequently, CP athletes present across many classification groups, and across a range of sports. As a result, the training programmes and improvements that can be expected will vary significantly from sport to sport, and from athlete to athlete. Generally the effects of CP are categorized by the area of the body affected (see Table 5.3.2). However, this categorization system gives little insight into the nature of the disorder or associated conditions. It is therefore difficult to extrapolate from category alone the wider implications for training and testing. Our experience shows us that these are best understood by considering the region of the brain that is injured and the pattern of motor involvement, which can be considered as either spastic, athetoid, ataxic, or mixed CP.

Training and testing

Spastic CP is the most common type, and is characterized by an increased neuromuscular tone, resulting in uncontrolled grouped muscular contractions (synergies) with a lack of isolated control. For example, in a common dysfunctional movement pattern an athlete is not able to flex the knee when the hip is extended but is able to flex both together. This spasticity usually involves the flexor muscles of the upper extremity and extensor muscles of the lower body, resulting in common chronic faulty movement patterns including glehohumeral adduction and internal rotation; elbow, wrist and finger flexion; hip flexion, internal rotation and adduction; and ankle plantar flexion and inversion (Lockette and Keynes, 1994).

Table 5.3.2 Categorization of the effects of cerebral palsy

Type	Effect
Monoplegia	One extremity affected
Hemiplegia	Two extremities affected on the same side of the body
Triplegia	Primary involvement of three extremities
Quadriplegia	All four extremities and trunk affected. Upper body is usually more affected than lower body
Diplegia	All four extremities and trunk affected. Upper body only mildly affected

Athetoid CP is the second most common type, and is characterized by involuntary movement in the extremities (and sometimes the trunk), and often speech difficulties (Miller, 1995). These involuntary movements may be slow and smooth or abrupt and jerky in nature and can usually be expected to increase as intensity of effort, emotional stress, and/or fatigue levels rise.

Ataxic CP is characterized by unsteadiness and involuntary movements, impaired balance, poor trunk control, and difficulties with fine movement skills.

Pure ataxic CP is not common, and is often combined with athetoid conditions. In fact, the most common type of CP is a mixture of spastic and athetoid conditions, with athletes displaying some but not all of the symptoms of each.

Clearly all of the conditions have significant implications for exercise selection, with some limbs unable to perform particular movements, or only able to do so at a significantly reduced range of movement or workload. The level of coordination and control in the trunk and extremities will dictate the movements that can be trained and whether control is sufficient to use free weights or machines. Free-weight exercises help athletes to improve control of movement and thus have greater transfer to sporting movements; however, for those athletes with involuntary or very jerky movements, safety considerations dictate that they are better using machine weights or partner resistance during what would otherwise be dangerous (e.g. overhead pressing) movements. This assists control in both the direction and the range of movement, with the ultimate goal of developing the ability to actively control the movement unaided. When deciding whether to train a heavily affected limb or movement, the coach must consider two aspects: first, whether it is required for sporting performance, for example whether a swimmer actually uses an affected arm for propulsion when swimming or just holds it to the side of the body; and second, whether attempting to improve the use of the limb will assist activities of everyday living and thus reduce the workload on the contra-lateral limb. For example, improving a wheelchair basketball player's ability to weight-bear through the affected limb may not affect their performance, but will allow them to move or transfer more easily, reducing the daily workload for their arms and thus improving performance.

Trunk strength and stability will also have a significant impact on the exercises that can be performed, but use of assistive equipment such as elastic binders or belts around the trunk may provide sufficient support for free-weight exercises to be performed. As previously mentioned, difficulties with balance caused by ataxic conditions may be improved through additional work; difficulties with trunk control may require that more training time be devoted to improving the strength and control of this area, whereas severe difficulties with fine movement control may dictate that only simple gross motor exercises are included in the programme and that the time devoted to learning new skills should be extended. Upper-body resistance-training exercises are often limited by grip strength in CP athletes, and the use of straps and/or the previously mentioned commercially available 'hook gloves' may be required to help to overcome this issue. However, athletes should be encouraged to use their own grip in order to develop strength in this area.

Special considerations

If unmanaged, spastic conditions can increase neuromuscular tone and can cause a shortening of the muscular and connective tissue, the result of which is poor movement patterns such as scissor gait and toe walking. The extent of an athlete's spasticity can fluctuate as a result of factors such as the external temperature, fatigue, and emotional stress. Specifically, cold environments, high levels of fatigue, and stressful situations tend to increase spasticity (Miller, 1995), and with it the potential for injury. From experience, key strategies for countering these problems, preventing injury, and maximizing gains in performance have included:

- slow passive stretches
- regular soft tissue therapy (massage)
- strengthening of the muscles opposing those affected by spasticity, thus decreasing tone through reciprocal inhibition
- ensuring a suitable training environment
- employing an extensive warm-up (especially in cold climates)
- making gradual progressive increases in training load and intensity with regular unloading phases.

Some coaches have expressed concern that strength training may lead to an increase in muscle stiffness, and thus compound the problematic effects of spastic CP. However, our observations, along with those of research colleagues, have indicated that the athlete who is given an appropriate training prescription shows no evidence that strengthening spastic muscles has a negative effect on range of movement or the level of spasticity (Damiano and Abel, 1998; Dodd *et al.*, 2002; personal observation).

Additional physical conditions that are commonly associated with CP include convulsive disorders, perceptual and motor disorders, and primitive reflexes. Convulsive disorders occur in approximately 25% of the CP population and are most often seen in those with hemiplegia (Lockette and Keyes, 1994). They are more likely to occur at rest than during sporting activities and should be managed by removing hazardous objects from the area.

Perceptual-motor disorders are also relatively common among CP athletes and can affect the ability to perceive an object's position relative to oneself; this means that affected athletes will have difficulty in judging direction and distance. This can affect activities such as maintaining a straight line when running, swimming, or cycling, and cause issues in catching or dodging objects, and must be taken into account when planning the training programme.

Primitive reflexes such as the tonic neck reflex, which causes the upper extremities to flex when the neck is moved towards the chest and extend when moved backwards, are usually integrated in the early stages of physical development but can persist in CP athletes. These can be dangerous during the extreme muscle effort involved in exercises such as the bench press, as the neck flexion which occurs may cause complete elbow flexion, rather than the forceful elbow extension required. It is therefore important that the coach checks whether the athlete's condition causes them to experience primitive reflexes, and adapts the training programme accordingly.

When considering the adaptations that can be expected as a result of training in athletes with CP, Ito *et al.* (1996) found a significant predominance of type I and deficiency of type IIB gastrocnemius muscle fibres when compared to non-disabled participants. This research did not specifically involve *athletes* with CP and did not distinguish whether the participants experienced spasticity in the gastrocnemius muscle, but if this is a consistent affect of the condition it would suggest that the potential for developing strength and power among this group of athletes is likely to be limited.

The additional energy demands that result from neuromuscular stiffness and involuntary or unnecessary movements significantly reduce mechanical efficiency and result in an increased energy expenditure for comparable tasks (Lundburg, 1984). MacPhail and Kramer (1995) showed a direct relationship between knee-extensor strength and walking efficiency, and postulated that improvements in coordination and strength resulting from strength training have the potential to significantly improve performance among athletes with CP.

The tendency for athletes with CP to fatigue quickly under highly intense training loads raises concerns about the reliability of testing. Despite these concerns, high levels of reliability have been demonstrated for sub-maximal graded exercise tests (Mossberg and Green, 2005), VO_{2max} tests (Holland, 1994), and the anaerobic Wingate test (Tirosh *et al.*, 1990) among individuals with CP. However, it remains good practice to evaluate the athlete's physical status, training load, sleep, nutrition, and fatigue prior to testing.

5.3.5.4 Visual impairment

Visual impairment can include the loss of central visual acuity, peripheral visual acuity, or both. Visual acuity is a measure of the sharpness or clearness of vision, and athletes eligible for Paralympic sports vary in the extent of their impairment; many sports attempt to ensure fairness by requiring athletes to use blacked-out goggles (e.g. goalball and football 5-a-side). Athletes with very little sight, particularly those visually impaired from birth, may show low proprioceptive awareness and postural control as a result of difficulties learning in early years due to the lack of visual cues. However, athletes without any coexisting conditions can be expected to achieve normal training adaptation, such as improvements in strength and metabolic fitness. In fact, from our experience of working with this population, many visually-impaired athletes can be seen to have developed a heightened awareness of other sensory skills such as touch, hearing, and movement, and with them the ability to learn well from verbal cues and kinaesthetic feedback.

Training and testing

The key considerations for the coach are twofold: first, how to ensure the athlete's safety, and second, how best to tap into the athlete's sensory skills and learning style in order to coach them effectively. Some basic safety guidelines include:

- Take time to introduce the athlete to the training facility and establish points of reference. This may only need to be done once for those with partial sight or very well-developed directional skills, but it is more likely to take several times, and it is often necessary to repeat this exercise if long breaks between visits occur.
- Ensure that the training facility is tidy (e.g. dumbbells or other floor-level equipment are not left lying around), and that the athlete can access the facility at times that are less busy.
- Encourage the athlete to train with a non-visually-impaired partner where possible.
- Programme conditioning activities such as running that can be done with a guide wire or partner and are performed on relatively smooth surfaces.
- Check that the athlete can see and read any written material (e.g. training programmes) or signs around the facility. This may necessitate the use of large print with high-contrast colours.
- When coaching new exercises, place more emphasis on concise verbal information and kinaesthetic feedback, rather than visual demonstrations. This may involve physically touching the athlete more than you would a sighted athlete, to assist them in achieving good positions. Prior consent should be obtained and it is best practice to give a verbal warning immediately before any contact.

5.3.5.5 Intellectual disabilities

The nature and severity of intellectual disabilities cover a wide spectrum (Horvat, 1990). The key challenges for coaches working with intellectually-disabled athletes are similar to those with the visually impaired, in that it is not the nature and extent of the physical adaptations that can achieved, but rather how best to communicate effectively and structure training sessions that is the key to maximizing performance gains. As with all disabilities, intellectual disabilities are highly specific to the individual and it is not uncommon for the intellectual disability to be multi-factorial.

While not mutually exclusive, intellectual disabilities can generally be categorized as those that primarily affect the organization and control of movement (e.g. dyspraxia) and those that primarily affect the ability to interact socially (e.g. autism and Asperger syndrome).

It can be difficult to recognize the symptoms of those with mild forms of socially orientated intellectual disabilities, as these athletes are generally without any clear visible symptoms. However, it is important to know and be able to recognize the signs, as this will have significant implications for the best methods of communicating with these athletes. For example, those with the 'autistic spectrum' of disorders will have difficulty understanding subtlety in both verbal and non-verbal language. They are often unable to recognize body language and will take a very literal interpretation of the spoken word. Consequently, the use of facial expressions, hand gestures, tone of voice, jokes, and sarcasm, all of which are an important part of a coach's skill set, is irrelevant and likely to be misinterpreted by the autistic athlete. Some may not speak, or have fairly limited speech, but will usually understand what other people say to them, though they may prefer to use alternative means of communication themselves, such as visual symbols (Farrell, 2006). Others will have good language skills, but may still find it hard to understand the give-and-take nature of conversations, perhaps repeating what the other person has just said (echolalia), which can make it difficult to develop effective two-way communication (Poinet, 1993). Moderate amounts of choice, free time, and distraction can also easily confuse and worry athletes with these conditions.

In these circumstances the coach should consider the following guidelines:

- Instructions and feedback must be clear, concise and consistent.
- The athlete must be given time to process what has been said to them and the coach must take time to fully understand the athlete's response.
- Experiment with a variety of visual, auditory, and kinaesthetic learning styles, as you would with able-bodied athletes.
- Be patient; it will take a long time for the coach–athlete relationship to develop.
- Set highly structured training programmes; introduce athlete choice gradually and where the decision caries little or no risk.
- Train in conditions with limited distractions, only introducing distraction gradually. This trains the athlete to learn to deal with distractions appropriately, training them to be able to concentrate on performance, whether this be in training or in competition.

Coaching athletes with the types of intellectual disability that effect the organization and control of movement requires a thorough understanding of the learning process. Schmidt (1975) proposed that the acquisition of skill requires four elements:

1. the initial conditions of the movements
2. the characteristics of the generalized motor programme
3. knowledge of results
4. sensory consequences of the movement.

For example, a wheelchair basketball player must constantly survey the match situation, so that when they receive a pass they can make a decision (initial conditions) as to what to do (pass, dribble, shoot, etc.). Having selected the course of action, they initiate a motor programme to complete the necessary skill, and having completed the skill, they evaluate the outcome (knowledge of results) and the sensory feedback, and modify the decision-making process and/or motor programme for use the next time they encounter a similar situation.

Athletes who have acquired the ability to perform isolated skills, but who often make incorrect decisions, may benefit from conditioning tasks which emphasize the ability to converge on important cues, track slow-moving objects (e.g. sponge balls or balloons), and recognize or discriminate between different cues (e.g. using colour-coded cones). Athletes for whom sensory perception and integration into movement is inhibited may require a different style of coaching, one which emphasizes recognition of the sensory experiences and directs the athlete's attention to the different sensory systems. Including challenges for an athlete's balance will heighten the sensitivity of the vestibular system and may enhance sensory feedback. The corollary is that in some athletes this may induce feelings of motion sickness, which may impede skill acquisition.

Focusing on the sound of particular exercises (e.g. foot contact) can stimulate the auditory system and improve timing during plyometric activities. Altering the surface may stimulate the tactile senses and improve sensory feedback in activities such as running and skipping.

5.3.5.6 Les autres

This somewhat ambiguous category includes multiple sclerosis and dwarfism. Due to the broadness of the category, brief descriptions of the potential training issues within the groups most likely to be encountered are given below.

Dwarfism is one of the most commonly seen *les autres* categories in Paralympic sports. There are a wide variety of potential causes of the condition. Areas which strength and conditioning coaches should be aware of include hypotonia (lack of muscle tone) and increased spinal curvature. Both of these have clear potential implications for resistance exercise. It is also noteworthy that pituitary dwarfs may demand particular attention due to their growth-hormone deficiency. If untreated, the athlete may not respond as expected to some resistance-training protocols, due to an attenuated hormonal response. Conversely, those who are undergoing growth-hormone treatment may achieve much better than average gains. Both the ethics and the legality of this may be subject to review.

Multiple sclerosis is a disease affecting an individual's central nervous system, with common symptoms including fatigue, muscular weakness, and loss of coordination and balance. The condition may be described as being either relapse-remitting or progressive. In a progressive condition, training emphasis may shift over time from being primarily focused on the athletic event to attempting to counter the debilitating effects of the condition. Large differences in the effects of the condition are seen on both an inter- and intra-individual basis. There is likely to be a need for the strength and conditioning coach to adapt training on an almost day-to-day basis, depending on the condition of the athlete. Exercise has been shown to be an effective tool in alleviating symptoms and reducing, or even regressing, the development of the disease (de Souza-Teixeria *et al.*, 2009). Whilst there is a very strong case for multiple sclerosis athletes benefiting from strength and conditioning activities, it may be wise to avoid pushing the athlete to their physical limits due to the risk of inducing excessive fatigue and leaving them unfit to cope with daily activities (Dalgas, Stenager and Ingemann-Hansen, 2008). Multiple sclerosis certainly presents a strong case for the strength and conditioning coach working as part of a multidisciplinary team, which may include physiotherapy and medical and nutritional input.

5.3.6 TIPS FOR MORE EFFECTIVE PROGRAMMING

Before seeing an athlete for the first time, seek to understand their sport. Don't assume that the same rules and format apply as in the able-bodied game. Make use of the resources available online, which include the Web sites of the International Paralympic Committee, the British Paralympic Association, and the four disability-specific international sports federations (CPISRA, IBSA, INAS-FID, and IWAS). Recent innovations like WebTV allow viewing of and familiarization with Paralympic sports through dedicated channels such as ParalympicSport.tv.

Involve the athlete in programme discussion and decisions; breaking down barriers is key to forming a positive working relationship that will allow delivery to be optimized. From experience, most athletes will be happy to discuss what is and isn't possible with their condition once they have built a good relationship with their coach.

A thorough and realistic evaluation of the athlete's functional ability should be performed at the start of any programme; from this, achievable goals and objectives can be set. Again, the athlete's knowledge of themselves and their disability should be considered a key resource in this process.

The age of Paralympic athletes varies tremendously, with the youngest competitor at the Beijing Paralympics being 13 and the eldest being 69 years old. Within this age range there will also be significant variability in both the training age of athletes and the amount of time that athletes have trained with their disability. Collating and considering as much of this information as you can will be helpful in designing and monitoring the progress of the conditioning programme.

Be aware of the complications and contraindications of exercise for each of the disability groups you work with. Conditions such as autonomic dysreflexia are potentially life-threatening and working with susceptible athletes carries with it the responsibility of ensuring that both health and well-being are maintained.

References

Buckley J.G. (1999) Sprint kinematics of athletes with lower-limb amputations. *Arch Phys Med Rehabil*, **80** (5), 501–508.

Buckley J.G. (2000) Biomechanical adaptations of transtibial amputee sprinting in athletes using dedicated prostheses. *Clin Biomech Jun*, **15** (5), 352–358.

Dalgas U., Stenager E. and Ingemann-Hansen T. (2008). Multiple sclerosis and physical exercise: recommendations for the application of resistance, endurance and combined training. *Multiple sclerosis*, **14** (1), 35–53.

Damiano D.L. and Abel M.F. (1998) Functional outcomes of strength training in spastic cerebral palsy. *Archives of Physical and Medical Rehabilitation*, **79** (2), 119–125.

Dodd K.J., Taylor N.F. and Damiano D.L. (2002) A systematic review of the effectiveness of strength-training programs for people with cerebral palsy. *Archives of Physical and Medical Rehabilitation*, **83** (8), 1157–1164.

Doidge N. (2007) The Brain that Changes Itself. Penduin Group, New York.

Farrell M. (2006) *The effective teacher's guide to behavioural, emotional and social difficulties: practical strategies*. London: Routledge.

Hagobian T.A., Jacobs K.A., Kiratli B.J. and Friedlander A.J. (2004) Foot cooling reduces exercise-induced hyperthermia in men with spinal cord injury. *Med Sci Sports Exerc*, **36** (3), 411–417.

Holland L.J., Bhambhani Y.N., Ferrara M.S. and Steadward R.D. (1994) Reliability of the maximal aerobic power and ventilatory threshold in adults with cerebral palsy. *Archives of Physical and Medical Rehabilitation*, **75** (6), 687–691.

Horvat M. (1990) Physical Education and Sport for Exceptional Students. Dubuque, Iowa.

Ito J., Araki A., Tanaka H., Tasaki T., Cho K. and Yamazaki R. (1996) Muscle histopathology in spastic cerebral palsy. *Brain Development*, **18** (4), 299–303.

Lockete K.F. and Keyes A.M. (1994) *Conditioning with Physical Disabilities*. Human Kinetics, Champaign IL.

Lundberg A. (1984) Longitudinal study of physical working capacity of young people with spastic cerebral palsy. *Developmental Medicine and Child Neurology*, **26** (3), 328–334.

MacPhail H.E. and Kramer J.F. (1995) Effect of isokinetic strength-training on functional ability and walking efficiency in adolescents with cerebral palsy. *Developmental Medicine and Child Neurology*, **37** (9), 763–775.

Maynard F.M., Jr. Bracken M.B. and Creasey G. (1997). International Standards for Functional Classification of Spinal Cord Injury. *Spinal Cord*, **35**, 266–274.

Miller P.D. (1995) *Fitness Programming and Physical Disability*. Human Kinetics, Champaign, IL.

Mossberg K.A. and Greene B.P. (2005) Reliability of graded exercise testing after traumatic brain injury: submaximal and peak responses. *American Journal of Physical Medicine & Rehabilitation*, **84** (7), 492–500.

Myers T.W. (2001) *Anatomy Trains*, Elsevier Ltd.

Poinet B. (1993) Movement Activities for Children with Learning Difficulties Jessica Kingsley Publishers: London and Philadelphia.

Price M.J. (2006) Thermoregulation during exercise in individuals with spinal cord injuries. *Sports Med*, **36** (10), 863–879.

Schmidt R.A. (1975) A schema theory of discrete motor skill learning. *Psychological Review*, **82**, 225–260.

Schmid A., Schmidt-Truckass A., Huonker M. *et al.* (2001) Catecholamines response of high performance wheelchair athletes at rest and during exercise with autonomic dysreflexia. *Int J Sports Med*, **22** (1), 2–7.

de Souza-Teixeria F., Costilla S., Ayan C. *et al.* (2009) Effects of resistance training in multiple sclerosis. *Int J Sports Med*, **30** (4), 245–250.

Tirosh E., Bar-Or O. and Rosenbaum P. (1990) New muscle power test in neuromuscular disease. Feasibility and reliability. *American Journal of Disceases of Children*, **144** (10), 1083–1087.

Tolfrey K., Goosey-Tolfrey V.L. and Campbell I.G. (2001) The oxygen uptake–heart rate relationship in elite wheelchair racers. *Eur J Appl Physiol*, **86** (2), 174–178.

Webborn N., Price M.J., Castle P.C. and Goosey-Tolfrey V.L. (2005) Effects of two cooling strategies on thermoregulatory responses of tetraplegic athletes during repeated intermittent exercise in the heat. *J Appl Physiol*, **98**, 2101–2107.

Webbon N., Price M.J., Castle P.C. and Goosey-Tolfrey V.L. (2008) Cooling strategies improve intermittent sprint performance in the heat of athletes with tetraplegia. *Br J Sports Med.*

Index

Printed and bound by CPI Group (UK) Ltd, Croydon, CR0 4YY
27/10/2024
14580203-0005